Springer Texts in Statistics

Series Editors

R. DeVeaux
S.E. Fienberg
I. Olkin

For further volumes:
http://www.springer.com/series/417

Matthew A. Carlton • Jay L. Devore

Probability with Applications in Engineering, Science, and Technology

 Springer

Matthew A. Carlton
Department of Statistics
California Polytechnic State University
San Luis Obispo, CA 93407
USA

Jay L. Devore
Department of Statistics
California Polytechnic State University
San Luis Obispo, CA 93407
USA

ISSN 1431-875X   ISSN 2197-4136 (electronic)
ISBN 978-1-4939-0394-8   ISBN 978-1-4939-0395-5 (eBook)
DOI 10.1007/978-1-4939-0395-5
Springer New York Heidelberg Dordrecht London

Library of Congress Control Number: 2014942253

Printed on acid-free paper

Springer is part of Springer Science+Business Media (www.springer.com)

Preface

Purpose

Our objective is to provide a post-calculus introduction to the subject of probability that

- Has mathematical integrity and contains some underlying theory
- Shows students a broad range of applications involving real problem scenarios
- Is current in its selection of topics
- Is accessible to a wide audience, including mathematics and statistics majors (yes, there are a few of the latter, and their numbers are growing), prospective engineers and scientists, and business and social science majors interested in the quantitative aspects of their disciplines
- Illustrates the importance of software for carrying out simulations when answers to questions cannot be obtained analytically

A number of currently available probability texts are heavily oriented toward a rigorous mathematical development of probability, with much emphasis on theorems, proofs, and derivations. Even when applied material is included, the scenarios are often contrived (many examples and exercises involving dice, coins, cards, and widgets). So in our exposition we have tried to achieve a balance between mathematical foundations and the application of probability to real-world problems. It is our belief that the theory of probability by itself is often not enough of a "hook" to get students interested in further work in the subject. We think that the best way to persuade students to continue their probabilistic education beyond a first course is to show them how the methodology is used in practice. Let's first seduce them (figuratively speaking, of course) with intriguing problem scenarios and applications. Opportunities for exposure to mathematical rigor will follow in due course.

Content

The book begins with an Introduction, which contains our attempt to address the following question: "Why study probability?" Here we are trying to tantalize students with a number of intriguing problem scenarios—coupon collection, birth and death processes, reliability engineering, finance, queuing models, and various conundrums involving the misinterpretation of probabilistic information (e.g., Benford's Law and the detection of fraudulent data, birthday problems, and the likelihood of having a rare disease when a diagnostic test result is positive). Most of the exposition contains references to recently published results. It is not necessary or even desirable to cover very much of this motivational material in the classroom. Instead, we suggest that instructors ask their students to read selectively outside class (a bit of pleasure reading at the very beginning of the term should not be an undue burden!). Subsequent chapters make little reference to the examples herein, and separating out our "pep talk" should make it easier to cover as little or much as an instructor deems appropriate.

Chapter 1 covers sample spaces and events, the axioms of probability and derived properties, counting, conditional probability, and independence. Discrete random variables and distributions are the subject of Chap. 2, and Chap. 3 introduces continuous random variables and their distributions. Joint probability distributions are the focus of Chap. 4, including marginal and conditional distributions, expectation of a function of several variables, correlation, modes of convergence, the Central Limit Theorem, reliability of systems of components, the distribution of a linear combination, and some results on order statistics. These four chapters constitute the core of the book.

The remaining chapters build on the core in various ways. Chapter 5 introduces methods of statistical inference—point estimation, the use of statistical intervals, and hypothesis testing. In Chap. 6 we cover basic properties of discrete-time Markov chains. Various other random processes and their properties, including stationarity and its consequences, Poisson processes, Brownian motion, and continuous-time Markov chains, are discussed in Chap. 7. The final chapter presents some elementary concepts and methods in the area of signal processing.

One feature of our book that distinguishes it from the competition is a section at the end of almost every chapter that considers simulation methods for getting approximate answers when exact results are difficult or impossible to obtain. Both the R software and Matlab are employed for this purpose.

Another noteworthy aspect of the book is the inclusion of roughly 1100 exercises; the first four core chapters together have about 700 exercises. There are numerous exercises at the end of each section and also supplementary exercises at the end of every chapter. Probability at its heart is concerned with problem solving. A student cannot hope to really learn the material simply by sitting passively in the classroom and listening to the instructor. He/she must get actively involved in working problems. To this end, we have provided a wide spectrum of exercises, ranging from straightforward to reasonably challenging. It should be easy for an instructor to find enough problems at various levels of difficulty to keep students gainfully occupied.

Mathematical Level

The challenge for students at this level should be to master the concepts and methods to a sufficient degree that problems encountered in the real world can be solved. Most of our exercises are of this type, and relatively few ask for proofs or derivations. Consequently, the mathematical prerequisites and demands are reasonably modest. Mathematical sophistication and quantitative reasoning ability are, of course, crucial to the enterprise. Univariate calculus is employed in the continuous distribution calculations of Chap. 3 as well as in obtaining maximum likelihood estimators in the inference chapter. But even here the functions we ask students to work with are straightforward—generally polynomials, exponentials, and logs. A stronger background is required for the signal processing material at the end of the book (we have included a brief mathematical appendix as a refresher for relevant properties). Multivariate calculus is used in the section on joint distributions in Chap. 4 and thereafter appears rather rarely. Exposure to matrix algebra is needed for the Markov chain material.

Recommended Coverage

Our book contains enough material for a year-long course. An instructor must be selective when using it in a course of shorter duration. To give a sense of what might be reasonable, we now describe two different courses at our home institution, California Polytechnic State University (in San Luis Obispo, CA), for which this book is appropriate. The university is on a quarter rather than a semester calendar. Depending on the quarter, there are between 38 and 41 class meetings (not including the final exam), each one lasting for 50 min. Courses are more commonly taught on the semester calendar, with three 50-min meetings or two 75-min meetings per week. This would allow for either more intensive work in the core chapters or a bit more time spent on selected material from Chaps. 5 to 8 than what we indicate below.

The first course is *Introduction to Probability Models*. It has calculus and linear algebra prerequisites, but no prior exposure to probability or statistics is assumed. The course is taken by mathematics and statistics majors plus a handful of engineering and quantitative finance students who are either in their sophomore or junior year. Coverage of material is roughly as follows:

Topic	Number of class meetings
Introduction to Probability (Ch. 1)	7
Discrete Random Variables and Distributions (Ch. 2)	7
Continuous Random Variables and Distributions (Ch. 3)	7
Joint Probability Distributions (Ch. 4)	7
Markov Chains (Ch. 6)	5
Selected Material from Reliability/Random Processes (Chs. 4, 7)	3
	36

Virtually all of the probability material in Chap. 1 is covered as well as much of what appears in Chap. 2. General properties of continuous distributions, the normal distribution, and the exponential distribution are featured, whereas other continuous distributions and transformations receive little attention. In Chap. 4, joint continuous distributions are covered very lightly and quickly; correlation, the Central Limit Theorem, and linear combinations are emphasized, but properties of conditional expectation and transformations are just briefly mentioned and order statistics are off limits. The Markov chain chapter contains more material than can be covered in the course, and no inference or signal processing is included.

The second course for which the book is intended is *Probability and Random Processes for Engineers*. The audience consists primarily of computer and electrical engineering majors in their third year of study. Their mathematical background is more extensive than that of at least some of the students in the probability models course, with exposure to differential equations and Fourier analysis. The following table describes coverage of material:

Topic	Number of class meetings
Introduction to Probability (Ch. 1)	6
Discrete Random Variables and Distributions (Ch. 2)	6
Continuous Random Variables and Distributions (Ch. 3)	6
Joint Probability Distributions (Ch. 4)	5
Selected Material from Random Processes (Ch. 7)	6
Continuous-Time Signal Processing (Ch. 8)	7
	36

Once again instructors must be judicious in covering material from the core chapters in order to leave sufficient time for the more advanced material from Chaps. 7 and 8 (this syllabus was developed jointly with the relevant engineering departments).

We are able to cover as much material as indicated on the foregoing syllabi with the aid of a not-so-secret weapon: we prepare and require that students bring to class a course booklet. The booklet contains most of the examples we present as well as some surrounding material. A typical example begins with a problem statement and then poses several questions (as in the exercises in this book). After each posed question there is some blank space so that the student can either take notes as the solution is developed in class or else work the problem on his/her own if asked to do so. Because students have a booklet, the instructor does not have to write as much on the board as would otherwise be necessary and the student does not have to do as much writing to take notes. Both the instructor and the students benefit.

We also like to think that students can be asked to read an occasional subsection or even section on their own and then work exercises to demonstrate understanding, so that not everything needs to be presented in class. For example, we have found that assigning a take-home exam problem that requires reading about the Weibull and/or lognormal distributions is a good way to acquaint students with them. But instructors should always keep in mind that there is never enough time in a course of any duration to teach students all that we'd like them to know. Hopefully students will like the book enough to keep it after the course is over and use it as a basis for extending their knowledge of probability!

Acknowledgments

We gratefully acknowledge the plentiful feedback provided by the following reviewers: Allan Gut, Murad Taqqu, Mark Schilling and Robert Heiny.

We very much appreciate the production services provided by the folks at SPi Technologies. Our production editors, Ahmad Ejaz and Sashi Shivakumar did a first-rate job of moving the book through the production process and were always prompt and considerate in communications with us. Thanks to our copyeditors at SPi for employing a light touch and not taking us too much to task for our occasional grammatical and stylistic lapses. The staff at Springer U.S. has been especially supportive during both the developmental and production stages; special kudos go to Marc Strauss, Jon Gurstelle, Michael Koy, and Hannah Bracken.

A Final Thought

It is our hope that students completing a course taught from this book will feel as passionately about the subject of probability as we still do after so many years of living with it. Only teachers can really appreciate how gratifying it is to hear from a student after he/she has completed a course that the experience had a positive impact and maybe even affected a career choice.

San Luis Obispo, CA   Matthew A. Carlton
San Luis Obispo, CA   Jay L. Devore

Contents

Introduction: Why Study Probability?

Some of you may enjoy mathematics for its own sake—it is a beautiful subject which provides many wonderful intellectual challenges. Of course students of philosophy would say the same thing about their discipline, ditto for students of linguistics, and so on. However, many of us are not satisfied just with aesthetics and mental gymnastics. We want what we're studying to have some utility, some applicability to real-world problems. Fortunately, mathematics in general and probability in particular provide a plethora of tools for answering important professional and societal questions. In this section, we'll attempt to provide some preliminary motivation before forging ahead.

The initial development of probability as a branch of mathematics goes back over 300 years, where it had its genesis in connection with questions involving games of chance. One of the earliest recorded instances of probability calculation appeared in correspondence between the two very famous mathematicians, Blaise Pascal and Pierre de Fermat. The issue was which of the following two outcomes of die-tossing was more favorable to a bettor: (1) getting at least one 6 in four rolls of a fair die ("fair" here means that each of the six outcomes 1, 2, 3, 4, 5, and 6 is equally likely to occur) or (2) getting at least one pair of 6s when two fair dice are rolled 24 times in succession. By the end of this chapter, you shouldn't have any difficulty showing that there is a slightly better than 50-50 chance of (1) occurring, whereas the odds are slightly against (2) occurring.

Games of chance have continued to be a fruitful area for the application of probability methodology. Savvy poker players certainly need to know the odds of being dealt various hands, such as a full house or straight (such knowledge is necessary but not at all sufficient for achieving success in card games, as such endeavors also involve much psychology). The same holds true for the game of blackjack. In fact, in 1962 the mathematics professor Edward O. Thorp published the book *Beat the Dealer*; in it he employed probability arguments to show that as cards were dealt sequentially from a deck, there were situations in which the likelihood of success favored the player rather than the dealer. Because of this work, casinos changed the way cards were dealt in order to prevent card-counting strategies from bankrupting them. A recent variant of this is described in the paper "Card Counting in Continuous Time" (*Journal of Applied Probability*, 2012: 184-198), in which the number of decks utilized is large enough to justify the use of a continuous approximation to find an optimal betting strategy.

In the last few decades, *game theory* has developed as a significant branch of mathematics devoted to the modeling of competition, cooperation, and conflict. Much of this work involves the use of probability properties, with applications in such diverse fields as economics, political science, and biology. However, especially over the course of the last 60 years, the scope of probability applications has expanded way beyond gambling and games. In this section, we present some contemporary examples of how probability is being used to solve important problems.

Software Use in Probability

Modern probability applications often require the use of a calculator or software. Of course, we rely on machines to perform every conceivable computation from adding numbers to evaluating definite integrals. Many calculators and most computer software packages even have built-in functions that make a number of specific probability calculations more convenient; we will highlight these throughout the text. But the real utility of modern software comes from its ability to *simulate* random phenomena, which proves invaluable in the analysis of very complicated probability models. We will introduce the key elements of probability simulation in Sect. 1.7 and then revisit simulation in a variety of settings throughout the book.

Numerous software packages can be used to implement a simulation. We will focus on two: Matlab and R. Matlab is a powerful engineering software package published by MathWorks; many universities and technology companies have a license for Matlab. A freeware package called Octave has been designed to implement the majority of Matlab functions using identical syntax; consult http://www.gnu.org/software/octave/. (Readers using Mac OS or Windows rather than GNU/Linux will find links to compatible versions of Octave on this same website.) R is a freeware statistical software package maintained by a core user group. The R base package and numerous add-ons are available at http://cran.r-project.org/.

Throughout this textbook, we will provide side by side Matlab and R code for both probability computations and simulation. It is not the goal, however, to serve as a primer in either language (certainly, some prior knowledge of elementary programming is required). Both software packages have extensive help menus and active online user support groups. Readers interested in a more thorough treatment of these software packages should consult *Matlab Primer* by Timothy A. Davis or *The R Book* by Michael J. Crawley.

Modern Application of Classic Probability Problems

The *coupon collector problem* has been well known for decades in the probability community. As an example, suppose each box of a certain type of cereal contains a small toy. The manufacturer of this cereal has included a total of ten toys in its cereal boxes, with each box being equally likely to yield one of the ten toys.

Suppose you want to obtain a complete set of these toys for a young relative or friend. Clearly you will have to purchase at least ten boxes, and intuitively it would seem as though you might have to purchase many more than that. How many boxes would you expect to have to purchase in order to achieve your goal? Methods from Chap. 4 can be used to show that the average number of boxes required is $10(1 + 1/2 + 1/3 + \cdots + 1/10)$. If instead there are n toys, then n replaces 10 in this expression. And when n is large, more sophisticated mathematical arguments yield the approximation $n(\ln(n) + .58)$.

The article "A Generalized Coupon Collector Problem" (*Journal of Applied Probability*, 2011: 1081-1094) mentions applications of the classic problem to dynamic resource allocation, hashing in computer science, and the analysis of delays in certain wireless communication channels (in this latter application, there are n users, each receiving packets of data from a transmitter). The generalization considered in the article involves each cereal box containing d different toys with the purchaser then selecting the least collected toy thus far. The expected number of purchases to obtain a complete collection is again investigated, with special attention to the case of n being quite large. An application to the wireless communication scenario is mentioned.

Applications to Business

The article "Newsvendor-Type Models with Decision-Dependent Uncertainty" (*Mathematical Methods of Operations Research*, 2012, published online) begins with an overview of a class of decision problems involving uncertainty. In the classical *newsvendor problem*, a seller has to choose the amount of inventory to obtain at the beginning of a selling season. This ordering decision is made only once, with no opportunity to replenish inventory during the season. The amount of demand D is uncertain (what we will call in Chap. 2 a *random variable*). The cost of obtaining inventory is c per unit ordered, the sale price is r per unit, and any unsold inventory at the end of the season has a salvage value of v per unit. The optimal policy, that which maximizes expected profit, is easily characterized in terms of the *probability distribution* of D (this distribution specifies how likely it is that various values of D will occur).

In the *revenue management problem*, there are S units of inventory to sell. Each unit is sold for a price of either r_1 or r_2 ($r_1 > r_2$). During the first phase of the selling season, customers arrive who will buy at the price r_2 but not at r_1. In the second phase, customers arrive who will pay the higher price. The seller wishes to know how much of the initial inventory should be held in reserve for the second phase. Again the general form of the optimal policy that maximizes expected profit is easily determined in terms of the distributions for demands in the two periods. The article cited in the previous paragraph goes on to consider situations in which the distribution(s) of demand(s) must be estimated from data and how such estimation affects decision making.

A cornerstone of probabilistic inventory modeling is a general result established more than 50 years ago: Suppose that the amount of inventory of a commodity is reviewed every T time periods to decide whether more should be ordered. Under rather general conditions, it was shown that the optimal policy—the policy that minimizes the long-run expected cost—is to order nothing if the current level of inventory is at least an amount s but to order enough to bring the inventory level up to an amount S if the current level is below s. The values of s and S are determined by various costs, the price of the commodity, and the nature of demand for the commodity (how customer orders and order amounts occur over time).

The article "A Periodic-Review Base-Stock Inventory System with Sales Rejection" (*Operations Research*, 2011: 742-753) considers a policy appropriate when backorders are possible and lost sales may occur. In particular, an order is placed every T time periods to bring inventory up to some level S. Demand for the commodity is filled until the inventory level reaches a sales rejection threshold M for some $M < S$. Various properties of the optimal values of M and S are investigated.

Applications to the Life Sciences

Examples of the use of probability and probabilistic modeling can be found in many subdisciplines of the life sciences. For example, *Pseudomonas syringae* is a bacterium which lives in leaf surfaces. The article "Stochastic Modeling of *Pseudomonas Syringae* Growth in the Phyllosphere" (*Mathematical Biosciences*, 2012: 106-116) proposed a probabilistic (synonymous with "stochastic") model called a *birth and death process with migration* to describe the aggregate distribution of such bacteria and determine the mechanisms which generated experimental data. The topic of birth and death processes is considered briefly in Chap. 7 of our book.

Another example of such modeling appears in the article "Means and Variances in Stochastic Multistage Cancer Models" (*Journal of Applied Probability*, 2012: 590-594). The authors discuss a widely used model of carcinogenesis in which division of a healthy cell may give rise to a healthy cell and a mutant cell, whereas division of a mutant cell may result in two mutant cells of the same type or possibly one of the same types and one with a further mutation. The objective is to obtain an expression for the expected number of cells at each stage and also a quantitative assessment of how much the actual number might deviate from what is expected (that is what "variance" does).

Epidemiology is the branch of medicine and public health that studies the causes and spread of various diseases. Of particular interest to epidemiologists is how epidemics are propagated in one or more populations. The general stochastic epidemic model assumes that a newly infected individual is infectious for a random amount of time having an *exponential distribution* (this distribution is discussed in Chap. 3) and during this infectious period encounters other individuals at times determined by a *Poisson process* (one of the topics in Chap. 7). The article "The Basic Reproduction Number and the Probability of Extinction for a Dynamic

Epidemic Model" (*Mathematical Biosciences*, 2012: 31-35) considers an extension in which the population of interest consists of a fixed number of subpopulations. Individuals move between these subpopulations according to a *Markov transition matrix* (the subject of Chap. 6) and infectives can only make infectious contact with members of their current subpopulation. The effect of variation in the infectious period on the probability that the epidemic ultimately dies out is investigated.

Another approach to the spread of epidemics is based on *branching processes*. In the simplest such process, a single individual gives birth to a random number of individuals; each of these in turn gives birth to a random number of progeny, and so on. The article "The Probability of Containment for Multitype Branching Process Models for Emerging Epidemics" (*Journal of Applied Probability*, 2011: 173-188) uses a model in which each individual "born" to an existing individual can have one of a finite number of severity levels of the disease. The resulting theory is applied to construct a simulation model of how influenza spread in rural Thailand.

Applications to Engineering and Operations Research

We want products that we purchase and systems that we rely on (e.g., communication networks, electric power grids) to be highly reliable—have long lifetimes and work properly during those lifetimes. Product manufacturers and system designers therefore need to have testing methods that will assess various aspects of reliability. In the best of all possible worlds, data bearing on reliability could be obtained under normal operating conditions. However, this may be very time consuming when investigating components and products that have very long lifetimes. For this reason, there has been much research on "accelerated" testing methods which induce failure or degradation in a much shorter time frame. For products that are used only a fraction of the time in a typical day, such as home appliances and automobile tires, acceleration might entail operating continuously in time but under otherwise normal conditions. Alternatively, a sample of units could be subjected to stresses (e.g., temperature, vibration, voltage) substantially more severe than what is usually experienced. Acceleration can also be applied to entities in which degradation occurs over time—stiffness of springs, corrosion of metals, and wearing of mechanical components. In all these cases, probability models must then be developed to relate lifetime behavior under such acceleration to behavior in more customary situations. The article "Overview of Reliability Testing" (*IEEE Transactions on Reliability*, 2012: 282-291) gives a survey of various testing methodologies and models. The article "A Methodology for Accelerated Testing by Mechanical Actuation of MEMS Devices" (*Microelectronics Reliability*, 2012: 1382-1388) applies some of these ideas in the context of predicting lifetimes for micro-electro-mechanical systems.

An important part of modern reliability engineering deals with building redundancy into various systems in order to decrease substantially the likelihood of failure. A *k-out-of-n:G system* works or is good only if at least k amongst the n constituent components work or are good, whereas a *k-out-of-n:F system* fails if and only if at least k of the n components fail. The article "Redundancy Issues in Software and Hardware Systems: An Overview" (*Intl. Journal of Reliability, Quality, and Safety Engineering*, 2011: 61-98) surveys these and various other systems that can improve the performance of computer software and hardware. The so-called triple modular redundant systems, with 2-out-of-3:G configuration, are now commonplace (e.g., Hewlett-Packard's original NonStop server, and a variety of aero, auto, and rail systems). The article "Reliability of Various 2-Out-of-4:G Redundant Systems with Minimal Repair" (*IEEE Transactions on Reliability*, 2012: 170-179) considers using a Poisson process with time-varying rate function to model how component failures occur over time so that the rate of failure increases as a component ages; in addition, a component that fails undergoes repair so that it can be placed back in service. Several failure modes for combined k-out-of-n systems are studied in the article "Reliability of Combined m-Consecutive-k-out-of-n:F and Consecutive-k_c-out-of-n:F Systems" (*IEEE Transactions on Reliability*, 2012: 215-219); these have applications in the areas of infrared detecting and signal processing.

A compelling reason for manufacturers to be interested in reliability information about their products is that they can establish warranty policies and periods that help control costs. Many warranties are "one dimensional," typically characterized by an interval of age (time). However, some warranties are "two dimensional" in that warranty conditions depend on both age and cumulative usage; these are common in the automotive industry. The article "Effect of Use-Rate on System Lifetime and Failure Models for 2D Warranty" (*Intl. Journal of Quality and Reliability Management*, 2011: 464-482) describes how certain *bivariate* probability models for jointly describing the behavior of time and usage can be used to investigate the reliability of various system configurations.

The word *queue* is used chiefly by the British to mean "waiting line," i.e., a line of customers or other entities waiting to be served or brought into service. The mathematical development of models for how a waiting line expands and contracts as customers arrive at a service facility, enter service, and then finish began in earnest in the middle part of the 1900s and continues unabated today as new application scenarios are encountered.

For example, the arrival and service of patients at some type of medical unit are often described by the notation $M/M/s$, where the first M signifies that arrivals occur according to a Poisson process, the second M indicates that the service time of each patient is governed by an exponential probability distribution, and there are s servers available for the patients. The article "Nurse Staffing in Medical Units: A Queueing Perspective" (*Operations Research*, 2011: 1320-1331) proposes an alternative closed queueing model in which there are s nurses within a single medical unit servicing n patients, where each patient alternates between requiring assistance and not needing assistance. The performance of the unit is characterized by the likelihood that delay in serving a patient needing assistance will exceed some critical threshold. A staffing rule based on the model and assumptions is developed; the resulting rule differs significantly from the fixed nurse-to-patient staffing ratios mandated by the state of California.

A variation on the medical unit situation just described occurs in the context of call centers, where effective management entails a trade-off between operational costs and the quality of service offered to customers. The article "Staffing Call Centers with Impatient Customers" (*Operations Research*, 2012: 461-474) considers an $M/M/s$ queue in which customers who have to wait for service may become frustrated and abandon the facility (don't you sometimes feel like doing that in a doctor's office?). The behavior of such a system when n is large is investigated, with particular attention to the staffing principle that relates the number of servers to the square root of the workload offered to the call center.

The methodology of queueing can also be applied to find optimal settings for traffic signals. The article "Delays at Signalized Intersections with Exhaustive Traffic Control" (*Probability in Engineering and Informational Sciences*, 2012: 337-373) utilizes a "polling model," which entails multiple queues of customers (corresponding to different traffic flows) served by a single server in cyclic order. The proposed vehicle-actuated rule is that traffic lights stay green until all lanes within a group are emptied. The mean traffic delay is studied for a variety of vehicle interarrival-time distributions in both light-traffic and heavy-traffic situations.

Suppose two different types of customers, primary and secondary, arrive for service at a facility where the servers have different service rates. How should customers be assigned to the servers? The article "Managing Queues with Hetero-geneous Servers" (*Journal of Applied Probability*, 2011: 435-452) shows that the optimal policy for minimizing mean wait time has a "threshold structure": for each server, there is a different threshold such that a primary customer will be assigned to that server if and only if the queue length of primary customers meets or exceeds the threshold.

Applications to Finance

The most explosive growth in the use of probability theory and methodology over the course of the last several decades has undoubtedly been in the area of finance. This has provided wonderful career opportunities for people with advanced degrees in statistics, mathematics, engineering, and physics (the son-in-law of one of the authors earned a Ph.D. in mechanical engineering and taught for several years, but then switched to finance). Edward O. Thorp, whom we previously met as the man who figured out how to beat blackjack, subsequently went on to success in finance, where he earned much more money managing hedge funds and giving advice than he could ever have hoped to earn in academia (those of us in academia love it for the intangible rewards we get—psychic income, if you will).

One of the central results in mathematical finance is the *Black-Scholes theorem*, named after the two Nobel-prize-winning economists who discovered it. To get the flavor of what is involved here, a bit of background is needed. Suppose the present price of a stock is $20 per share, and it is known that at the end of 1 year, the price will either double to $40 or decrease to $10 per share (where those prices are expressed in current dollars, i.e., taking account of inflation over the 1-year period).

You can enter into an agreement, called an option contract, that allows you to purchase y shares of this stock (for any value y) 1 year from now for the amount cy (again in current dollars). In addition, right now you can buy x shares of the stock for $20x$ with the objective of possibly selling those shares 1 year from now. The values x and y are both allowed to be negative; if, for example, x were negative, then you would actually be selling shares of the stock now that you would have to purchase at either a cost of \$40 per share or \$10 per share 1 year from now. It can then be shown that there is only one value of c, specifically 50/3, for which the gain from this investment activity is 0 regardless of the choices of x and y and the value of the stock 1 year from now. If c is anything other than 50/3, then there is an *arbitrage*, an investment strategy involving choices of x and y that is guaranteed to result in a positive gain.

A general result called the *Arbitrage Theorem* specifies conditions under which a collection of investments (or bets) has expected return 0 as opposed to there being an arbitrage strategy. The basis for the Black-Sholes theorem is that the variation in the price of an asset over time is described by a stochastic process called *geometric Brownian motion* (see Sect. 7.6). The theorem then specifies a fair price for an option contract on that asset so that no arbitrage is possible.

Modern quantitative finance is very complex, and many of the basic ideas are unfamiliar to most novices (like the authors of this text!). It is therefore difficult to summarize the content of recently published articles as we have done for some other application areas. But a sampling of recently published titles emphasizes the role of probability modeling. Articles that appeared in the 2012 *Annals of Finance* included "Option Pricing Under a Stressed Beta Model" and "Stochastic Volatility and Stochastic Leverage"; in the 2012 *Applied Mathematical Finance*, we found "Determination of Probability Distribution Measures from Market Prices Using the Method of Maximum Entropy in the Mean" and "On Cross-Currency Models with Stochastic Volatility and Correlated Interest Rates"; the 2012 *Quantitative Finance* yielded "Probability Unbiased Value-at-Risk Estimators" and "A Generalized Birth-Death Stochastic Model for High-Frequency Order Book Dynamics."

If the application of mathematics to problems in finance is of interest to you, there are now many excellent masters-level graduate programs in quantitative finance. Entrance to these programs typically requires a very solid background in undergraduate mathematics and statistics (including especially the course for which you are using this book). Be forewarned, though, that not all financially savvy individuals are impressed with the direction in which finance has recently moved. Former Federal Reserve Chairman Paul Volcker was quoted not long ago as saying that the ATM cash machine was the most significant financial innovation of the last 20 years; he has been a very vocal critic of the razzle-dazzle of modern finance.

Probability in Everyday Life

In the hopefully unlikely event that you do not end up using probability concepts and methods in your professional life, you still need to face the fact that ideas surrounding uncertainty are pervasive in our world. We now present some amusing and intriguing examples to illustrate this.

The behavioral psychologists Amos Tversky and Daniel Kahneman spent much of their academic careers carrying out studies to demonstrate that human beings frequently make logical errors when processing information about uncertainty (Kahneman won a Nobel prize in economics for his work, and Tversky would surely have also done so had the awards been given posthumously). Consider the following variant of one Tversky-Kahneman scenario. Which of the following two statements is more likely?

(A) Dr. D is a former professor.
(B) Dr. D is a former professor who was accused of inappropriate relations with some students, investigation substantiated the charges, and he was stripped of tenure.

T-K's research indicated that many people would regard statement B as being more likely, since it gives a more detailed explanation of why Dr. D is no longer a professor. However, this is incorrect. Statement B implies statement A. One of our basic probability rules will be that if one event B is contained in another event A (i.e., if B implies A), then the smaller event B is less likely to occur or have occurred than the larger event A. After all, other possible explanations for A are that Dr. D is deceased or that he is retired or that he deserted academia for investment banking—all of those plus B would figure in to the likelihood of A.

The survey article "Judgment under Uncertainty; Heuristics and Biases" (*Science*, 1974: 1124-1131) by T-K described a certain town served by two hospitals. In the larger hospital about 45 babies are born each day, whereas about 15 are born each day in the smaller one. About 50% of births are boys, but of course the percentage fluctuates from day to day. For a 1-year period, each hospital recorded days on which more than 60% of babies born were boys. Each of a number of individuals was then asked which of the following statements he/she thought was correct: (1) the larger hospital recorded more such days, (2) the smaller hospital recorded more such days, or (3) the number of such days was about the same for the two hospitals. Of the 95 participants, 21 chose (1), 21 chose (2), and 53 chose (3). In Chap. 5 we present a general result which implies that the correct answer is in fact (2), because the sample percentage is less likely to stray from the true percentage (in this case about 50%) when the sample size is larger rather than small.

In case you think that mistakes of this sort are made only by those who are unsophisticated or uneducated, here is yet another T-K scenario. Each of a sample of 80 physicians was presented with the following information on treatment for a particular disease:

> With surgery, 10% will die during treatment, 32% will die within a year, 66% will die within 5 years. With radiation, 0% will die during treatment, 23% will die within a year, 78% will die within 5 years.

Each of the 87 physicians in a second sample was presented with the following information:

> With surgery, 90% will survive the treatment, 68% will survive at least 1 year, and 34% will survive at least 5 years. With radiation, 100% will survive the treatment, 77% will survive at least 1 year, and 22% will survive at least 5 years.

When each physician was asked to indicate whether he/she would recommend surgery or radiation based on the supplied information, 50% of those in the first group said surgery whereas 84% of those in the second group said surgery.

The distressing thing about this conclusion is that the information provided to the first group of physicians is identical to that provided to the second group, but described in a slightly different way. If the physicians were really processing information rationally, there should be no significant difference between the two percentages.

It would be hard to find a book containing even a brief exposition of probability that did not contain examples or exercises involving coin tossing. Many such scenarios involve tossing a "fair" coin, one that is equally likely to result in H (head side up) or T (tail side up) on any particular toss. Are real coins actually fair, or is there a bias of some sort? Various analyses have shown that the result of a coin toss is predicable at least to some degree if initial conditions (position, velocity, angular momentum) are known. In practice, most people who toss coins (e.g., referees in a football game trying to determine which team will kick off and which will receive) are not conversant in the physics of coin tossing. The mathematician and statistician Persi Diaconis, who was a professional magician for 10 years prior to earning his Ph.D. and mastered many coin and card tricks, has engaged in ongoing collaboration with other researchers to study coin tossing. One result of these investigations was the conclusion based on physics that for a caught coin, there is a slight bias toward heads—about .51 versus .49. It is not, however, clear under which real-world circumstances this or some other bias will occur.

Simulation of fair-coin tossing can be done using a random number generator available in many software packages (about which we'll say more shortly). If the resulting random number is between 0 and .5, we say that the outcome of the toss was H, and if the number is between .5 and 1, then a T occurred (there is an obvious modification of this to incorporate bias). Now consider the following sequence of 200 Hs and Ts:

```
THTHTTTHTTTTTTHTHTTTHTTTHHHTHHTHTHTHTTTTTHHTTTHHTTHHHT
HHHTTHHHTTTHHHTHHHHTTTHTHTHHHHTHTTTHHHTHHTHTTTTHHTH
HHTHHHHTTHTHHTHHHTTTHTHHHTHHTTTHHHTTTTHHHTHTHHHHTH
TTHHTTTTHTHTHTHTHTHHTHTTTHTTTTHHHHTHTHHHTHHHHHTHH
```

Did this sequence result from actually tossing a fair coin (equivalently, using computer simulation as described), or did it come from someone who was asked to write down a sequence of 200 Hs and Ts that he/she thought would come from tossing a fair coin? One way to address this question is to focus on the longest run of Hs in the sequence of tosses. This run is of length 4 for the foregoing sequence.

Probability theory tells us that the expected length of the longest run in a sequence of n fair-coin tosses is approximately $\log_2(n) - 2/3$. For $n = 200$, this formula gives an expected longest run of length about 7. It can also be shown that there is less than a 10% chance that the longest run will have a length of 4 or less. This suggests that the given sequence is fictitious rather than real, as in fact was the case; see the very nice expository article "The Longest Run of Heads" (*Mathematics Magazine*, 1990, 196-207).

As another example, consider giving a fair coin to each of the two authors of this textbook. Carlton tosses his coin repeatedly until obtaining the sequence HTT. Devore tosses his coin repeatedly until the sequence HTH is observed. Is Carlton's expected number of tosses to obtain his desired sequence the same as Devore's, or is one expected number of tosses smaller than the other? Most students to whom we have asked these questions initially answered that the two expected numbers should be the same. But this is not true. Some rather tricky probability arguments can be used to show that Carlton's expected number of tosses is eight, whereas Devore expects to have to make ten tosses. Very surprising, no? A bit of intuition makes this more plausible. Suppose Carlton merrily tosses away until at some point he has just gotten HT. So he is very excited, thinking that just one more toss will enable him to stop tossing the coin and move on to some more interesting pursuit. Unfortunately his hopes are dashed because the next toss is an H. However, all is not lost, as even though he must continue tossing, at this point he is partway toward reaching his goal of HTT. If Devore sees HT at some point and gets excited by light at the end of the tunnel but then is crushed by the appearance of a T rather than an H, he essentially has to start over again from scratch. The charming nontechnical book *Probabilities: The Little Numbers That Rule Our Lives* by Peter Olofsson has more detail on this and other probability conundrums.

One of the all-time classic probability puzzles that stump most people is called the *Birthday Problem*. Consider a group of individuals, all of whom were born in the same year (one that did not have a February 29). If the group size is 400, how likely is it that at least two members of the group share the same birthday? Hopefully a moment's reflection will bring you to the realization that a shared birthday here is a sure thing (100% chance), since there are only 365 possible birthdays for the 400 people. On the other hand, it is intuitively quite unlikely that there will be a shared birthday if the group size is only five; in this case we would expect that all five individuals would have different birthdays.

Clearly as the group size increases, it becomes more likely that two or more individuals will have the same birthday. So how large does the group size have to be in order for it to be more likely than not that at least two people share a birthday (i.e., that the likelihood of a shared birthday is more than 50%)? Which one of the following four group-size categories do you believe contains the correct answer to this question?

(1)	At least 100	(2)	At least 50 but less than 100
(3)	At least 25 but less than 50	(4)	Fewer than 25

When we have asked this of students in our classes, a substantial majority opted for the first two categories. Very surprisingly, the correct answer is category (4). In Chapter 1 we will show that with as few as 23 people in the group, it is a bit more likely than not that at least two group members will have the same birthday.

Two people having the same birthday implies that they were born within 24 h of one another, but the converse is not true; e.g., one person might be born just before midnight on a particular day and another person just after midnight on the next day. This implies that it is more likely that two people will have been born within 24 h of one another than it is that they have the same birthday. It follows that a smaller group size than 23 is needed to make it more likely than not that at least two people will have been born within 24 h of one another. In Sect. 4.9 we show how this group size can be determined.

Two people in a group having the same birthday is an example of a coincidence, an accidental and seemingly surprising occurrence of events. The fact that even for a relatively small group size it is more likely than not that this coincidence will occur should suggest that coincidences are often not as surprising as they might seem. This is because even if a particular coincidence (e.g., "graduated from the same high school" or "visited the same small town in Croatia") is quite unlikely, there are so many opportunities for coincidences that quite a few are sure to occur.

Back to the follies of misunderstanding medical information: Suppose the incidence rate of a particular disease in a certain population is 1 in 1000. The presence of the disease cannot be detected visually, but a diagnostic test is available. The diagnostic test correctly detects 98% of all diseased individuals (this is the *sensitivity* of the test, its ability to detect the presence of the disease), and 93% of non-diseased individuals test negative for the disease (this is the *specificity* of the test, an indicator of how specific the test is to the disease under consideration). Suppose a single individual randomly selected from the population is given the test and the test result is positive. In light of this information, how likely is it that the individual will have the disease?

First note that if the sensitivity and the specificity were both 100%, then it would be a sure thing that the selected individual has the disease. The reason this is not a sure thing is that the test sometimes makes mistakes. Which one of the following five categories contains the actual likelihood of having the disease under the described conditions?

1. At least a 75% chance (quite likely)
2. At least 50% but less than 75% (moderately likely)
3. At least 25% but less than 50% (somewhat likely)
4. At least 10% but less than 25% (rather unlikely)
5. Less than 10% (quite unlikely)

Student responses to this question have overwhelmingly been in categories (1) or (2)—another case of intuition going awry. The correct answer turns out to be category (5). In fact, even in light of the positive test result, there is still only a bit more than a 1% chance that the individual is diseased!

What is the explanation for this counterintuitive result? Suppose we start with 100,000 individuals from the population. Then we'd expect 100 of those, or 100, to be diseased (from the 1 in 1000 incidence rate) and 99,900 to be disease free. From the 100 we expect to be diseased, we'd expect 98 positive test results (98% sensitivity). And from the 99,900 we expect to be disease free, we'd expect 7% of those or 6993 to yield positive test results. Thus we expect many more false positives than true positives. This is because the disease is quite rare and the diagnostic test is rather good but not stunningly so. (In case you think our sensitivity and specificity are low, consider a certain D-dimer test for the presence of a coronary embolism; its sensitivity and specificity are 88% and 75%, respectively.)

Later in Chapter 1 (Example 1.31) we develop probability rules which can be used to show that the *posterior* probability of having the disease conditional on a positive test result is .0138—a bit over 1%. This should make you very cautious about interpreting the results of diagnostic tests. Before you panic in light of a positive test result, you need to know the incidence rate for the condition under consideration and both the sensitivity and specificity of the test. There are also implications for situations involving detection of something other than a disease. Consider airport procedures that are used to detect the presence of a terrorist. What do you think is the incidence rate of terrorists at a given airport, and how sensitive and specific do you think detection procedures are? The overwhelming number of positive test results will be false, greatly inconveniencing those who test positive!

Here's one final example of probability applied in everyday life: One of the following columns contains the value of the closing stock index as of August 8, 2012, for each of a number of countries, and the other column contains fake data obtained with a random number generator. Just by looking at the numbers, without considering context, can you tell which column is fake and which is real?

China	2264	3058
Japan	8881	9546
Britain	5846	7140
Canada	11,781	6519
Euro area	797	511
Austria	2053	4995
France	3438	2097
Germany	6966	4628
Italy	14,665	8461
Spain	722	598
Norway	480	1133
Russia	1445	4100
Sweden	1080	2594
Turkey	64,699	35,027
Hong Kong	20,066	42,182
India	17,601	3388
Pakistan	14,744	10,076
Singapore	3052	5227
Thailand	1214	7460
Argentina	2459	2159
⋮	⋮	⋮

The key to answering this question is a result called *Benford's Law*. Suppose you start reading through a particular issue of a publication like the *New York Times* or *The Economist*, and each time you encounter any number (the amount of donations to a particular political candidate, the age of an actor, the number of members of a union, and so on), you record the first digit of that number. Possible first digits are 1, 2, 3, ..., or 9. In the long run, how frequently do you think each of these nine possible first digits will be encountered? Your first thought might be that each one should have the same long-run frequency, 1/9 (roughly 11%). But for many sets of numbers this turns out not to be the case. Instead the long-run frequency is given by the formula $\log_{10}[(x+1)/x]$, which gives .301, .176, .125, ..., .051, .046 for $x=1$, 2, 3, ..., 8, 9. Thus a leading digit is much more likely to be 1, 2, or 3 than 7, 8, or 9.

Examination of the foregoing lists of numbers shows that the first column conforms much more closely to Benford's Law than does the second column. In fact, the first column is real, whereas the second one is fake. For Benford's Law to be valid, it is generally required that the set of numbers under consideration span several orders of magnitude. It does not work, for example, with batting averages of major league baseball players, most of which are between .200 and .299, or with fuel efficiency ratings (miles per gallon) for automobiles, most of which are currently between 15 and 30. Benford's Law has been employed to detect fraud in accounting reports, and in particular to detect fraudulent tax returns. So beware when you file your taxes next year!

This list of amusing probability appetizers could be continued for quite a while. Hopefully what we have shown thus far has sparked your interest in knowing more about the discipline. So without further ado ....

Probability

1

Probability is the subdiscipline of mathematics that focuses on a systematic study of randomness and uncertainty. In any situation in which one of a number of possible outcomes may occur, the theory of probability provides methods for quantifying the chances, or likelihoods, associated with the various outcomes. The language of probability is constantly used in an informal manner in both written and spoken contexts. Examples include such statements as "It is likely that the Dow Jones Industrial Average will increase by the end of the year," "There is a 50–50 chance that the incumbent will seek reelection," "There will probably be at least one section of that course offered next year," "The odds favor a quick settlement of the strike," and "It is expected that at least 20,000 concert tickets will be sold." In this chapter, we introduce some elementary probability concepts, indicate how probabilities can be interpreted, and show how the rules of probability can be applied to compute the chances of many interesting events. The methodology of probability will then permit us to express in precise language such informal statements as those given above.

1.1   Sample Spaces and Events

In probability, an **experiment** refers to any action or activity whose outcome is subject to uncertainty. Although the word *experiment* generally suggests a planned or carefully controlled laboratory testing situation, we use it here in a much wider sense. Thus experiments that may be of interest include tossing a coin once or several times, selecting a card or cards from a deck, weighing a loaf of bread, measuring the commute time from home to work on a particular morning, determining blood types from a group of individuals, or calling people to conduct a survey.

M.A. Carlton and J.L. Devore, *Probability with Applications in Engineering, Science, and Technology*, Springer Texts in Statistics, DOI 10.1007/978-1-4939-0395-5_1, © Springer Science+Business Media New York 2014

1.1.1 The Sample Space of an Experiment

> **DEFINITION**
> The **sample space** of an experiment, denoted by $\mathcal{S}$, is the set of all possible outcomes of that experiment.

Example 1.1 The simplest experiment to which probability applies is one with two possible outcomes. One such experiment consists of examining a single fuse to see whether it is defective. The sample space for this experiment can be abbreviated as $\mathcal{S} = \{N, D\}$, where N represents not defective, D represents defective, and the braces are used to enclose the elements of a set. Another such experiment would involve tossing a thumbtack and noting whether it landed point up or point down, with sample space $\mathcal{S} = \{U, D\}$, and yet another would consist of observing the gender of the next child born at the local hospital, with $\mathcal{S} = \{M, F\}$. ∎

Example 1.2 If we examine three fuses in sequence and note the result of each examination, then an outcome for the entire experiment is any sequence of Ns and Ds of length 3, so

$$\mathcal{S} = \{NNN, NND, NDN, NDD, DNN, DND, DDN, DDD\}$$

If we had tossed a thumbtack three times, the sample space would be obtained by replacing N by U in $\mathcal{S}$ above. A similar notational change would yield the sample space for the experiment in which the genders of three newborn children are observed. ∎

Example 1.3 Two gas stations are located at a certain intersection. Each one has six gas pumps. Consider the experiment in which the number of pumps in use at a particular time of day is observed for each of the stations. An experimental outcome specifies how many pumps are in use at the first station and how many are in use at the second one. One possible outcome is (2, 2), another is (4, 1), and yet another is (1, 4). The 49 outcomes in $\mathcal{S}$ are displayed in the accompanying table.

First station	Second station						
	0	1	2	3	4	5	6
0	(0, 0)	(0, 1)	(0, 2)	(0, 3)	(0, 4)	(0, 5)	(0, 6)
1	(1, 0)	(1, 1)	(1, 2)	(1, 3)	(1, 4)	(1, 5)	(1, 6)
2	(2, 0)	(2, 1)	(2, 2)	(2, 3)	(2, 4)	(2, 5)	(2, 6)
3	(3, 0)	(3, 1)	(3, 2)	(3, 3)	(3, 4)	(3, 5)	(3, 6)
4	(4, 0)	(4, 1)	(4, 2)	(4, 3)	(4, 4)	(4, 5)	(4, 6)
5	(5, 0)	(5, 1)	(5, 2)	(5, 3)	(5, 4)	(5, 5)	(5, 6)
6	(6, 0)	(6, 1)	(6, 2)	(6, 3)	(6, 4)	(6, 5)	(6, 6)

The sample space for the experiment in which a six-sided die is thrown twice results from deleting the 0 row and 0 column from the table, giving 36 outcomes. ∎

Example 1.4 A reasonably large percentage of C++ programs written at a particular company compile on the first run, but some do not. Suppose an experiment consists of selecting and compiling C++ programs at this location until encountering a program that compiles on the first run. Denote a program that compiles on the first run by S (for success) and one that doesn't do so by F (for failure). Although it may not be very likely, a possible outcome of this experiment is that the first 5 (or 10 or 20 or . . .) are Fs and the next one is an S. That is, for any positive integer n we may have to examine n programs before seeing the first S. The sample space is $\mathscr{S} = \{S, FS, FFS, FFFS, \ldots\}$, which contains an infinite number of possible outcomes. The same abbreviated form of the sample space is appropriate for an experiment in which, starting at a specified time, the gender of each newborn infant is recorded until the birth of a female is observed. ∎

1.1.2 Events

In our study of probability, we will be interested not only in the individual outcomes of $\mathscr{S}$ but also in any collection of outcomes from $\mathscr{S}$.

DEFINITION

An **event** is any collection (subset) of outcomes contained in the sample space $\mathscr{S}$. An event is said to be **simple** if it consists of exactly one outcome and **compound** if it consists of more than one outcome.

When an experiment is performed, a particular event A is said to occur if the resulting experimental outcome is contained in A. In general, exactly one simple event will occur, but many compound events will occur simultaneously.

Example 1.5 Consider an experiment in which each of three vehicles taking a particular freeway exit turns left (L) or right (R) at the end of the off-ramp. The eight possible outcomes that comprise the sample space are LLL, RLL, LRL, LLR, LRR, RLR, RRL, and RRR. Thus there are eight simple events, among which are $E_1 = \{LLL\}$ and $E_5 = \{LRR\}$. Some compound events include

$A = \{RLL, LRL, LLR\} = $ the event that exactly one of the three vehicles turns right
$B = \{LLL, RLL, LRL, LLR\} = $ the event that at most one of the vehicles turns right
$C = \{LLL, RRR\} = $ the event that all three vehicles turn in the same direction

Suppose that when the experiment is performed, the outcome is *LLL*. Then the simple event E_1 has occurred and so also have the events B and C (but not A). ∎

Example 1.6 (Example 1.3 continued) When the number of pumps in use at each of two six-pump gas stations is observed, there are 49 possible outcomes, so there are 49 simple events: $E_1 = \{(0, 0)\}, E_2 = \{(0, 1)\}, \ldots, E_{49} = \{(6, 6)\}$. Examples of compound events are

$A = \{(0, 0), (1, 1), (2, 2), (3, 3), (4, 4), (5, 5), (6, 6)\}$ = the event that the number of pumps in use is the same for both stations
$B = \{(0, 4), (1, 3), (2, 2), (3, 1), (4, 0)\}$ = the event that the total number of pumps in use is four
$C = \{(0, 0), (0, 1), (1, 0), (1, 1)\}$ = the event that at most one pump is in use at each station ∎

Example 1.7 (Example 1.4 continued) The sample space for the program compilation experiment contains an infinite number of outcomes, so there are an infinite number of simple events. Compound events include

$A = \{S, FS, FFS\}$ = the event that at most three programs are examined
$B = \{S, FFS, FFFFS\}$ = the event that exactly one, three, or five programs are examined
$C = \{FS, FFFS, FFFFFS, \ldots\}$ = the event that an even number of programs are examined ∎

1.1.3 Some Relations from Set Theory

An event is nothing but a set, so relationships and results from elementary set theory can be used to study events. The following operations will be used to construct new events from given events.

DEFINITION

1. The **complement** of an event A, denoted by A', is the set of all outcomes in $\mathcal{S}$ that are *not* contained in A.
2. The **intersection** of two events A and B, denoted by $A \cap B$ and read "A and B," is the event consisting of all outcomes that are in *both A and B*.
3. The **union** of two events A and B, denoted by $A \cup B$ and read "A or B," is the event consisting of all outcomes that are *either in A or in B or in both events* (so that the union includes outcomes for which both A and B occur as well as outcomes for which exactly one occurs)—that is, all outcomes in at least one of the events.

Example 1.8 (Example 1.3 continued) For the experiment in which the number of pumps in use at a single six-pump gas station is observed, let $A = \{0, 1, 2, 3, 4\}$, $B = \{3, 4, 5, 6\}$, and $C = \{1, 3, 5\}$. Then

$$A \cup B = \{0, 1, 2, 3, 4, 5, 6\} = \mathscr{S} \quad A \cup C = \{0, 1, 2, 3, 4, 5\}$$

$$A \cap B = \{3, 4\} \quad A \cap C = \{1, 3\} \quad A' = \{5, 6\} \quad (A \cup C)' = \{6\} \qquad \blacksquare$$

Example 1.9 (Example 1.4 continued) In the program compilation experiment, define A, B, and C as in Example 1.7. Then

$A \cup B = \{S, FS, FFS, FFFFS\}$
$A \cap B = \{S, FFS\}$
$A' = \{FFFS, FFFFS, FFFFFS, \ldots\}$
 and
$C' = \{S, FFS, FFFFS, \ldots\} = $ the event that an odd number of programs are examined $\qquad \blacksquare$

The complement, intersection, and union operators from set theory correspond to the *not*, *and*, and *or* operators from computer science. Readers with prior programming experience may be aware of an important relationship between these three operators, first discovered by the nineteenth-century British mathematician Augustus De Morgan.

DE MORGAN'S LAWS
Let A and B be two events in the sample space of some experiment. Then
1. $(A \cup B)' = A' \cap B'$
2. $(A \cap B)' = A' \cup B'$

De Morgan's laws state that the complement of a union is an intersection of complements, and the complement of an intersection is a union of complements.

Sometimes A and B have no outcomes in common, so that the intersection of A and B contains no outcomes (see Exercise 11).

DEFINITION
When A and B have no outcomes in common, they are said to be **disjoint** or **mutually exclusive** events. Mathematicians write this compactly as $A \cap B = \varnothing$, where $\varnothing$ denotes the event consisting of no outcomes whatsoever (the "null" or "empty" event).

Example 1.10 A small city has three automobile dealerships: a GM dealer selling Chevrolets and Buicks; a Ford dealer selling Fords and Lincolns; and a Chrysler dealer selling Jeeps and Chryslers. If an experiment consists of observing the brand of the next car sold, then the events $A = \{$Chevrolet, Buick$\}$ and $B = \{$Ford, Lincoln$\}$ are mutually exclusive because the next car sold cannot be both a GM product and a Ford product. ∎

A pictorial representation of events and manipulations with events is obtained by using **Venn diagrams**. To construct a Venn diagram, draw a rectangle whose interior will represent the sample space $\mathscr{S}$. Then any event A is represented as the interior of a closed curve (often a circle) contained in $\mathscr{S}$. Figure 1.1 shows examples of Venn diagrams.

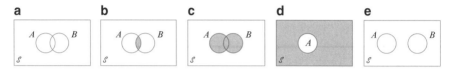

Fig. 1.1 Venn diagrams. (**a**) Venn diagram of events A and B (**b**) Shaded region is $A \cap B$ (**c**) Shaded region is $A \cup B$ (**d**) Shaded region is A' (**e**) Mutually exclusive events

The operations of union and intersection can be extended to more than two events. For any three events A, B, and C, the event $A \cap B \cap C$ is the set of outcomes contained in *all* three events, whereas $A \cup B \cup C$ is the set of outcomes contained in *at least one* of the three events. A collection of several events is said to be mutually exclusive (or pairwise disjoint) if no two events have any outcomes in common.

1.1.4 Exercises: Section 1.1 (1–12)

1. Ann and Bev have each applied for several jobs at a local university. Let A be the event that Ann is hired and let B be the event that Bev is hired. Express in terms of A and B the events
   (a) Ann is hired but not Bev.
   (b) At least one of them is hired.
   (c) Exactly one of them is hired.
2. Two voters, Al and Bill, are each choosing between one of three candidates—1, 2, and 3—who are running for city council. An experimental outcome specifies both Al's choice and Bill's choice, e.g., the pair (3,2).
   (a) List all elements of $\mathscr{S}$.
   (b) List all outcomes in the event A that Al and Bill make the same choice.
   (c) List all outcomes in the event B that neither of them votes for candidate 2.
3. Four universities—1, 2, 3, and 4—are participating in a holiday basketball tournament. In the first round, 1 will play 2 and 3 will play 4. Then the two winners will play for the championship, and the two losers will also play. One

possible outcome can be denoted by 1324: 1 beats 2 and 3 beats 4 in first-round games, and then 1 beats 3 and 2 beats 4.

(a) List all outcomes in $\mathscr{S}$.

(b) Let A denote the event that 1 wins the tournament. List outcomes in A.

(c) Let B denote the event that 2 gets into the championship game. List outcomes in B.

(d) What are the outcomes in $A \cup B$ and in $A \cap B$? What are the outcomes in A'?

4. Suppose that vehicles taking a particular freeway exit can turn right (R), turn left (L), or go straight (S). Consider observing the direction for each of three successive vehicles.

(a) List all outcomes in the event A that all three vehicles go in the same direction.

(b) List all outcomes in the event B that all three vehicles take different directions.

(c) List all outcomes in the event C that exactly two of the three vehicles turn right.

(d) List all outcomes in the event D that exactly two vehicles go in the same direction.

(e) List the outcomes in D', $C \cup D$, and $C \cap D$.

5. Three components are connected to form a system as shown in the accompanying diagram. Because the components in the 2–3 subsystem are connected in parallel, that subsystem will function if at least one of the two individual components functions. For the entire system to function, component 1 must function and so must the 2–3 subsystem.

The experiment consists of determining the condition of each component: S (success) for a functioning component and F (failure) for a nonfunctioning component.

(a) What outcomes are contained in the event A that exactly two out of the three components function?

(b) What outcomes are contained in the event B that at least two of the components function?

(c) What outcomes are contained in the event C that the system functions?

(d) List outcomes in C', $A \cup C$, $A \cap C$, $B \cup C$, and $B \cap C$.

6. Each of a sample of four home mortgages is classified as fixed rate (F) or variable rate (V).

(a) What are the 16 outcomes in $\mathscr{S}$?

(b) Which outcomes are in the event that exactly three of the selected mortgages are fixed rate?

   (c) Which outcomes are in the event that all four mortgages are of the same
       type?
   (d) Which outcomes are in the event that at most one of the four is a variable-
       rate mortgage?
   (e) What is the union of the events in parts (c) and (d), and what is the
       intersection of these two events?
   (f) What are the union and intersection of the two events in parts (b) and (c)?
7. A family consisting of three persons—A, B, and C—belongs to a medical clinic
   that always has a doctor at each of stations 1, 2, and 3. During a certain week,
   each member of the family visits the clinic once and is assigned at random to a
   station. The experiment consists of recording the station number for each
   member. One outcome is (1, 2, 1) for A to station 1, B to station 2, and C to
   station 1.
   (a) List the 27 outcomes in the sample space.
   (b) List all outcomes in the event that all three members go to the same station.
   (c) List all outcomes in the event that all members go to different stations.
   (d) List all outcomes in the event that no one goes to station 2.
8. A college library has five copies of a certain text on reserve. Two copies (1 and
   2) are first printings, and the other three (3, 4, and 5) are second printings. A
   student examines these books in random order, stopping only when a second
   printing has been selected. One possible outcome is 5, and another is 213.
   (a) List the outcomes in $\mathscr{S}$.
   (b) Let A denote the event that exactly one book must be examined. What
       outcomes are in A?
   (c) Let B be the event that book 5 is the one selected. What outcomes are in B?
   (d) Let C be the event that book 1 is not examined. What outcomes are in C?
9. An academic department has just completed voting by secret ballot for a
   department head. The ballot box contains four slips with votes for candidate
   A and three slips with votes for candidate B. Suppose these slips are removed
   from the box one by one.
   (a) List all possible outcomes.
   (b) Suppose a running tally is kept as slips are removed. For what outcomes
       does A remain ahead of B throughout the tally?
10. A construction firm is currently working on three different buildings. Let A_i
    denote the event that the ith building is completed by the contract date. Use the
    operations of union, intersection, and complementation to describe each of the
    following events in terms of A_1, A_2, and A_3, draw a Venn diagram, and shade
    the region corresponding to each one.
    (a) At least one building is completed by the contract date.
    (b) All buildings are completed by the contract date.
    (c) Only the first building is completed by the contract date.
    (d) Exactly one building is completed by the contract date.
    (e) Either the first building or both of the other two buildings are completed by
        the contract date.

11. Use Venn diagrams to verify De Morgan's laws:
    (a) $(A \cup B)' = A' \cap B'$
    (b) $(A \cap B)' = A' \cup B'$
12. (a) In Example 1.10, identify three events that are mutually exclusive.
    (b) Suppose there is no outcome common to all three of the events A, B, and C. Are these three events necessarily mutually exclusive? If your answer is yes, explain why; if your answer is no, give a counterexample using the experiment of Example 1.10.

1.2   Axioms, Interpretations, and Properties of Probability

Given an experiment and its sample space $\mathscr{S}$, the objective of probability is to assign to each event A a number $P(A)$, called **the probability of the event A**, which will give a precise measure of the chance that A will occur. To ensure that the probability assignments will be consistent with our intuitive notions of probability, all assignments should satisfy the following axioms (basic properties) of probability.

AXIOM 1

For any event A, $P(A) \geq 0$.

AXIOM 2

$P(\mathscr{S}) = 1$.

AXIOM 3

If $A_1, A_2, A_3, \ldots$ is an infinite collection of disjoint events, then

$$P(A_1 \cup A_2 \cup A_3 \cup \cdots) = \sum_{i=1}^{\infty} P(A_i)$$

Axiom 1 reflects the intuitive notion that the chance of A occurring should be nonnegative. The sample space is by definition the event that must occur when the experiment is performed ($\mathscr{S}$ contains all possible outcomes), so Axiom 2 says that the maximum possible probability of 1 is assigned to $\mathscr{S}$. The third axiom formalizes the idea that if we wish the probability that at least one of a number of events will occur and no two of the events can occur simultaneously, then the chance of at least one occurring is the sum of the chances of the individual events.

You might wonder why the third axiom contains no reference to a *finite* collection of disjoint events. It is because the corresponding property for a finite collection can be derived from our three axioms. We want our axiom list to be as short as possible and not contain any property that can be derived from others on the list.

PROPOSITION

$P(\varnothing) = 0$, where $\varnothing$ is the null event. This, in turn, implies that the property contained in Axiom 3 is valid for a *finite* collection of events.

Proof First consider the infinite collection $A_1 = \varnothing$, $A_2 = \varnothing$, $A_3 = \varnothing$, ... Since $\varnothing \cap \varnothing = \varnothing$, the events in this collection are disjoint and $\cup A_i = \varnothing$. Axiom 3 then gives

$$P(\varnothing) = \sum P(\varnothing)$$

This can happen only if $P(\varnothing) = 0$.

Now suppose that A_1, A_2, ..., A_k are disjoint events, and append to these the infinite collection $A_{k+1} = \varnothing$, $A_{k+2} = \varnothing$, $A_{k+3} = \varnothing$, ... Then the events A_1, A_2, ..., A_k, A_{k+1}, ... are disjoint, since $A \cap \varnothing = \varnothing$ for all events. Again invoking Axiom 3,

$$P\left(\bigcup_{i=1}^{k} A_i\right) = P\left(\bigcup_{i=1}^{\infty} A_i\right) = \sum_{i=1}^{\infty} P(A_i) = \sum_{i=1}^{k} P(A_i) + \sum_{i=k+1}^{\infty} P(A_i)$$

$$= \sum_{i=1}^{k} P(A_i) + \sum_{i=k+1}^{\infty} 0 = \sum_{i=1}^{k} P(A_i)$$

as desired. ∎

Example 1.11 Consider tossing a thumbtack in the air. When it comes to rest on the ground, either its point will be up (the outcome U) or down (the outcome D). The sample space for this event is therefore $\mathcal{S} = \{U, D\}$. The axioms specify $P(\mathcal{S}) = 1$, so the probability assignment will be completed by determining $P(U)$ and $P(D)$. Since U and D are disjoint and their union is $\mathcal{S}$, the foregoing proposition implies that

$$1 = P(\mathcal{S}) = P(U) + P(D)$$

It follows that $P(D) = 1 - P(U)$. One possible assignment of probabilities is $P(U) = .5$, $P(D) = .5$, whereas another possible assignment is $P(U) = .75$, $P(D) = .25$. In fact, letting p represent any fixed number between 0 and 1, $P(U) = p$, $P(D) = 1 - p$ is an assignment consistent with the axioms. ∎

Example 1.12 Consider testing batteries coming off an assembly line one by one until a battery having a voltage within prescribed limits is found. The simple events are $E_1 = \{S\}, E_2 = \{FS\}, E_3 = \{FFS\}, E_4 = \{FFFS\}, \ldots$ Suppose the probability of any particular battery being satisfactory is .99. Then it can be shown that the probability assignment $P(E_1) = .99$, $P(E_2) = (.01)(.99)$, $P(E_3) = (.01)^2(.99)$, $\ldots$ satisfies the axioms. In particular, because the E_is are disjoint and $\mathcal{S} = E_1 \cup E_2 \cup E_3 \cup \ldots$, Axioms 2 and 3 require that

$$
\begin{aligned}
1 = P(\mathcal{S}) &= P(E_1) + P(E_2) + P(E_3) + \cdots \\
&= .99\left[1 + .01 + (.01)^2 + (.01)^3 + \cdots\right]
\end{aligned}
$$

This can be verified using the formula for the sum of a geometric series:

$$
a + ar + ar^2 + ar^3 + \cdots = \frac{a}{1 - r}
$$

However, another legitimate (according to the axioms) probability assignment of the same "geometric" type is obtained by replacing .99 by any other number p between 0 and 1 (and .01 by $1 - p$). ■

1.2.1 Interpreting Probability

Examples 1.11 and 1.12 show that the axioms do not completely determine an assignment of probabilities to events. The axioms serve only to rule out assignments inconsistent with our intuitive notions of probability. In the tack-tossing experiment of Example 1.11, two particular assignments were suggested. The appropriate or correct assignment depends on the nature of the thumbtack and also on one's interpretation of probability. The interpretation that is most often used and most easily understood is based on the notion of relative frequencies.

Consider an experiment that can be repeatedly performed in an identical and independent fashion, and let A be an event consisting of a fixed set of outcomes of the experiment. Simple examples of such repeatable experiments include the tack-tossing and die-tossing experiments previously discussed. If the experiment is performed n times, on some of the replications the event A will occur (the outcome will be in the set A), and on others, A will not occur. Let $n(A)$ denote the number of replications on which A does occur. Then the ratio $n(A)/n$ is called the *relative frequency* of occurrence of the event A in the sequence of n replications.

For example, let A be the event that a package sent within the state of California for 2-day delivery actually arrives within 1 day. The results from sending ten such packages (the first ten replications) are as follows.

Package #	1	2	3	4	5	6	7	8	9	10
Did A occur?	N	Y	Y	Y	N	N	Y	Y	N	N
Relative frequency of A	0	.5	.667	.75	.6	.5	.571	.625	.556	.5

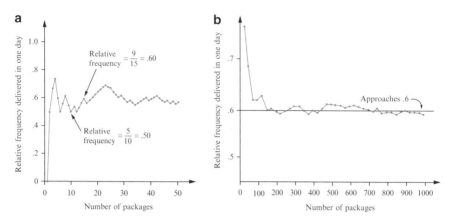

Fig. 1.2 Behavior of relative frequency: (**a**) Initial fluctuation (**b**) Long-run stabilization

Figure 1.2a shows how the relative frequency $n(A)/n$ fluctuates rather substantially over the course of the first 50 replications. But as the number of replications continues to increase, Fig. 1.2b illustrates how the relative frequency stabilizes.

More generally, empirical evidence, based on the results of many such repeatable experiments, indicates that any relative frequency of this sort will stabilize as the number of replications n increases. That is, as n gets arbitrarily large, $n(A)/n$ approaches a limiting value we refer to as the *limiting* (or *long-run*) *relative frequency* of the event A. The *objective interpretation of probability* identifies this limiting relative frequency with $P(A)$. A formal justification of this interpretation is provided by the *Law of Large Numbers*, a theorem we'll encounter in Chap. 4.

Suppose that probabilities are assigned to events in accordance with their limiting relative frequencies. Then a statement such as "the probability of a package being delivered within 1 day of mailing is .6" means that of a large number of mailed packages, roughly 60% will arrive within 1 day. Similarly, if B is the event that a certain brand of dishwasher will need service while under warranty, then $P(B) = .1$ is interpreted to mean that in the long run 10% of such dishwashers will need warranty service. This does *not* mean that exactly 1 out of 10 will need service, or exactly 20 out of 200 will need service, because 10 and 200 are not the long run. Such mis-interpretations of probability as a guarantee on short-term outcomes are at the heart of the infamous *gambler's fallacy*.

This relative frequency interpretation of probability is said to be objective because it rests on a property of the experiment rather than on any particular individual concerned with the experiment. For example, two different observers of a sequence of coin tosses should both use the same probability assignments since the observers have nothing to do with limiting relative frequency.

In practice, this interpretation is not as objective as it might seem, because the limiting relative frequency of an event will not be known. Thus we will have to assign probabilities based on our beliefs about the limiting relative frequency of events under study. Fortunately, there are many experiments for which there will be a consensus with respect to probability assignments. When we speak of a fair coin, we shall mean $P(H) = P(T) = .5$, and a fair die is one for which limiting relative frequencies of the six outcomes are all equal, suggesting probability assignments $P(\boxdot) = \cdots = P(\boxplus) = 1/6$.

Because the objective interpretation of probability is based on the notion of limiting frequency, its applicability is limited to experimental situations that are repeatable. Yet the language of probability is often used in connection with situations that are inherently unrepeatable. Examples include: "The chances are good for a peace agreement"; "It is likely that our company will be awarded the contract"; and "Because their best quarterback is injured, I expect them to score no more than 10 points against us." In such situations we would like, as before, to assign numerical probabilities to various outcomes and events (e.g., the probability is .9 that we will get the contract). We must therefore adopt an alternative interpretation of these probabilities. Because different observers may have different prior information and opinions concerning such experimental situations, probability assignments may now differ from individual to individual. Interpretations in such situations are thus referred to as *subjective*. The book by Winkler listed in the references gives a very readable survey of several subjective interpretations. Importantly, even subjective interpretations of probability must satisfy the three axioms (and all properties that follow from the axioms) in order to be valid.

1.2.2 More Probability Properties

COMPLEMENT RULE
For any event A, $P(A) = 1 - P(A')$.

Proof Since by definition of A', $A \cup A' = \mathscr{S}$ while A and A' are disjoint, $1 = P(\mathscr{S}) = P(A \cup A') = P(A) + P(A')$, from which the desired result follows. ∎

This proposition is surprisingly useful because there are many situations in which $P(A')$ is more easily obtained by direct methods than is $P(A)$.

Example 1.13 Consider a system of five identical components connected in series, as illustrated in Fig. 1.3.

Denote a component that fails by F and one that doesn't fail by S (for success). Let A be the event that the system fails. For A to occur, at least one of the individual components must fail. Outcomes in A include $SSFSS$ (1, 2, 4, and 5 all work, but 3 does not), $FFSSS$, and so on. There are, in fact, 31 different outcomes in A! However, A', the event that the system works, consists of the single outcome $SSSSS$. We will see in Sect. 1.5 that if 90% of all these components do not fail and different

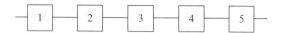

Fig. 1.3 A system of five components connected in series

components fail independently of one another, then $P(A') = .9^5 = .59$. Thus $P(A) = 1 - .59 = .41$; so among a large number of such systems, roughly 41% will fail. ∎

In general, the Complement Rule is useful when the event of interest can be expressed as "at least ...," because the complement "less than ..." may be easier to work with. (In some problems, "more than ..." is easier to deal with than "at most ...") When you are having difficulty calculating $P(A)$ directly, think of determining $P(A')$.

PROPOSITION

For any event A, $P(A) \leq 1$.

This follows from the previous proposition: $1 = P(A) + P(A') \geq P(A)$, because $P(A') \geq 0$ by Axiom 1.

When A and B are disjoint, we know that $P(A \cup B) = P(A) + P(B)$. How can this union probability be obtained when the events are not disjoint?

ADDITION RULE

For any events A and B,

$$P(A \cup B) = P(A) + P(B) - P(A \cap B).$$

Notice that the proposition is valid even if A and B are disjoint, since then $P(A \cap B) = 0$. The key idea is that, in adding $P(A)$ and $P(B)$, the probability of the intersection $A \cap B$ is actually counted twice, so $P(A \cap B)$ must be subtracted out.

Proof Note first that $A \cup B = A \cup (B \cap A')$, as illustrated in Fig. 1.4. Because A and $(B \cap A')$ are disjoint, $P(A \cup B) = P(A) + P(B \cap A')$. But $B = (B \cap A) \cup (B \cap A')$ (the union of that part of B in A and that part of B not in A). Furthermore, $(B \cap A)$ and $(B \cap A')$ are disjoint, so that $P(B) = P(B \cap A) + P(B \cap A')$. Combining these results gives

$$P(A \cup B) = P(A) + P(B \cap A') = P(A) + [P(B) - P(A \cap B)]$$
$$= P(A) + P(B) - P(A \cap B)$$

Fig. 1.4 Representing $A \cup B$ as a union of disjoint events   ∎

Example 1.14 In a certain residential suburb, 60% of all households get internet service from the local cable company, 80% get television service from that company, and 50% get both services from the company. If a household is randomly selected, what is the probability that it gets at least one of these two services from the company, and what is the probability that it gets exactly one of the services from the company?

With $A = \{$gets internet service from the cable company$\}$ and $B = \{$gets television service from the cable company$\}$, the given information implies that $P(A) = .6$, $P(B) = .8$, and $P(A \cap B) = .5$. The Addition Rule then applies to give

$$P(\text{gets at least one of these two services from the company}) =$$
$$P(A \cup B) = P(A) + P(B) - P(A \cap B) = .6 + .8 - .5 = .9$$

The event that a household gets only television service from the company can be written as $A' \cap B$, i.e., (not internet) and television. Now Fig. 1.4 implies that

$$.9 = P(A \cup B) = P(A) + P(A' \cap B) = .6 + P(A' \cap B)$$

from which $P(A' \cap B) = .3$. Similarly, $P(A \cap B') = P(A \cup B) - P(B) = .1$. This is all illustrated in Fig. 1.5, from which we see that

$$P(\text{exactly one}) = P(A \cap B') + P(A' \cap B) = .1 + .3 = .4$$

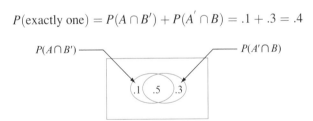

Fig. 1.5 Probabilities for Example 1.14   ∎

The probability of a union of more than two events can be computed analogously. For three events A, B, and C, the result is

$$P(A \cup B \cup C) = P(A) + P(B) + P(C) - P(A \cap B) - P(A \cap C) - P(B \cap C)$$
$$+ P(A \cap B \cap C)$$

This can be seen by examining a Venn diagram of $A \cup B \cup C$, which is shown in Fig. 1.6. When $P(A)$, $P(B)$, and $P(C)$ are added, outcomes in certain intersections

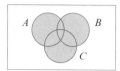

Fig. 1.6 $A \cup B \cup C$

are double counted and the corresponding probabilities must be subtracted. But this results in $P(A \cap B \cap C)$ being subtracted once too often, so it must be added back. One formal proof involves applying the Addition Rule to $P((A \cup B) \cup C)$, the probability of the union of the two events $A \cup B$ and C; see Exercise 30. More generally, a result concerning $P(A_1 \cup \cdots \cup A_k)$ can be proved by induction or by other methods. The pattern of additions and subtractions (or, equivalently, the method of deriving such union probability formulas) is often called the **inclusion–exclusion principle**.

1.2.3   Determining Probabilities Systematically

When the number of possible outcomes (simple events) is large, there will be many compound events. A simple way to determine probabilities for these events that avoids violating the axioms and derived properties is to first determine probabilities $P(E_i)$ for all simple events. These should satisfy $P(E_i) \geq 0$ and $\Sigma_{\text{all } i} P(E_i) = 1$. Then the probability of any compound event A is computed by adding together the $P(E_i)$s for all E_is in A:

$$P(A) = \sum_{\text{all } E_i s \text{ in } A} P(E_i)$$

Example 1.15 During off-peak hours a commuter train has five cars. Suppose a commuter is twice as likely to select the middle car (#3) as to select either adjacent car (#2 or #4), and is twice as likely to select either adjacent car as to select either end car (#1 or #5). Let $p_i = P(\text{car } i \text{ is selected}) = P(E_i)$. Then we have $p_3 = 2p_2 = 2p_4$ and $p_2 = 2p_1 = 2p_5 = p_4$. This gives

$$1 = \sum P(E_i) = p_1 + 2p_1 + 4p_1 + 2p_1 + p_1 = 10p_1$$

implying $p_1 = p_5 = .1, p_2 = p_4 = .2$, and $p_3 = .4$. The probability that one of the three middle cars is selected (a compound event) is then $p_2 + p_3 + p_4 = .8$.   ∎

1.2.4   Equally Likely Outcomes

In many experiments consisting of N outcomes, it is reasonable to assign equal probabilities to all N simple events. These include such obvious examples as tossing a fair coin or fair die once (or any fixed number of times), or selecting one or several cards from a well-shuffled deck of 52. With $p = P(E_i)$ for every i,

$$1 = \sum_{i=1}^{N} P(E_i) = \sum_{i=1}^{N} p = p \cdot N \quad \text{so} \quad p = \frac{1}{N}$$

That is, if there are N possible outcomes, then the probability assigned to each is $1/N$.

Now consider an event A, with $N(A)$ denoting the number of outcomes contained in A. Then

$$P(A) = \sum_{E_i \text{ in } A} P(E_i) = \sum_{E_i \text{ in } A} \frac{1}{N} = \frac{N(A)}{N}$$

Once we have counted the number N of outcomes in the sample space, to compute the probability of any event we must count the number of outcomes contained in that event and take the ratio of the two numbers. Thus when outcomes are equally likely, computing probabilities reduces to counting.

Example 1.16 When two dice are rolled separately, there are $N = 36$ outcomes (delete the first row and column from the table in Example 1.3). If both the dice are fair, all 36 outcomes are equally likely, so $P(E_i) = 1/36$. Then the event $A = \{$sum of two numbers is 8$\}$ consists of the five outcomes (⚁, ⚅), (⚂, ⚄), (⚃, ⚃), (⚄, ⚂), and (⚅, ⚁), so

$$P(A) = \frac{N(A)}{N} = \frac{5}{36}$$ ∎

The next section of this book investigates counting methods in depth.

1.2.5 Exercises: Section 1.2 (13–30)

13. A mutual fund company offers its customers several different funds: a money-market fund, three different bond funds (short, intermediate, and long-term), two stock funds (moderate and high-risk), and a balanced fund. Among customers who own shares in just one fund, the percentages of customers in the different funds are as follows:

Money-market	20%	High-risk stock	18%
Short bond	15%	Moderate-risk stock	25%
Intermediate bond	10%	Balanced	7%
Long bond	5%		

A customer who owns shares in just one fund is randomly selected.
(a) What is the probability that the selected individual owns shares in the balanced fund?
(b) What is the probability that the individual owns shares in a bond fund?
(c) What is the probability that the selected individual does not own shares in a stock fund?

14. Consider randomly selecting a student at a certain university, and let A denote the event that the selected individual has a Visa credit card and B be the analogous event for a MasterCard. Suppose that $P(A) = .5$, $P(B) = .4$, and $P(A \cap B) = .25$.
(a) Compute the probability that the selected individual has at least one of the two types of cards (i.e., the probability of the event $A \cup B$).
(b) What is the probability that the selected individual has neither type of card?
(c) Describe, in terms of A and B, the event that the selected student has a Visa card but not a MasterCard, and then calculate the probability of this event.

15. A computer consulting firm presently has bids out on three projects. Let $A_i = \{$awarded project $i\}$, for $i = 1, 2, 3$, and suppose that $P(A_1) = .22$, $P(A_2) = .25, P(A_3) = .28, P(A_1 \cap A_2) = .11, P(A_1 \cap A_3) = .05, P(A_2 \cap A_3) = .07$, $P(A_1 \cap A_2 \cap A_3) = .01$. Express in words each of the following events, and compute the probability of each event:
(a) $A_1 \cup A_2$
(b) $A_1' \cap A_2'$ [Hint: Use De Morgan's Law.]
(c) $A_1 \cup A_2 \cup A_3$
(d) $A_1' \cap A_2' \cap A_3'$
(e) $A_1' \cap A_2' \cap A_3$
(f) $(A_1' \cap A_2') \cup A_3$

16. Suppose that 55% of all adults regularly consume coffee, 45% regularly consume soda, and 70% regularly consume at least one of these two products.
(a) What is the probability that a randomly selected adult regularly consumes both coffee and soda?
(b) What is the probability that a randomly selected adult doesn't regularly consume either of these two products?

17. Consider the type of clothes dryer (gas or electric) purchased by each of five different customers at a certain store.
(a) If the probability that at most one of these customers purchases an electric dryer is .428, what is the probability that at least two purchase an electric dryer?
(b) If P(all five purchase gas) $= .116$ and P(all five purchase electric) $= .005$, what is the probability that at least one of each type is purchased?

18. An individual is presented with three different glasses of cola, labeled C, D, and P. He is asked to taste all three and then list them in order of preference. Suppose the same cola has actually been put into all three glasses.

    (a) What are the simple events in this ranking experiment, and what probability would you assign to each one?

    (b) What is the probability that C is ranked first?

    (c) What is the probability that C is ranked first and D is ranked last?

19. Let A denote the event that the next request for assistance from a statistical software consultant relates to the SPSS package, and let B be the event that the next request is for help with SAS. Suppose that $P(A) = .30$ and $P(B) = .50$.

    (a) Why is it not the case that $P(A) + P(B) = 1$?

    (b) Calculate $P(A')$.

    (c) Calculate $P(A \cup B)$.

    (d) Calculate $P(A' \cap B')$.

20. A box contains six 40-W bulbs, five 60-W bulbs, and four 75-W bulbs. If bulbs are selected one by one in random order, what is the probability that at least two bulbs must be selected to obtain one that is rated 75 W?

21. Human visual inspection of solder joints on printed circuit boards can be very subjective. Part of the problem stems from the numerous types of solder defects (e.g., pad nonwetting, knee visibility, voids) and even the degree to which a joint possesses one or more of these defects. Consequently, even highly trained inspectors can disagree on the disposition of a particular joint. In one batch of 10,000 joints, inspector A found 724 that were judged defective, inspector B found 751 such joints, and 1159 of the joints were judged defective by at least one of the inspectors. Suppose that one of the 10,000 joints is randomly selected.

    (a) What is the probability that the selected joint was judged to be defective by neither of the two inspectors?

    (b) What is the probability that the selected joint was judged to be defective by inspector B but not by inspector A?

22. A factory operates three different shifts. Over the last year, 200 accidents have occurred at the factory. Some of these can be attributed at least in part to unsafe working conditions, whereas the others are unrelated to working conditions. The accompanying table gives the percentage of accidents falling in each type of accident–shift category.

Shift	Unsafe conditions	Unrelated to conditions
Day	10%	35%
Swing	8%	20%
Night	5%	22%

    Suppose one of the 200 accident reports is randomly selected from a file of reports, and the shift and type of accident are determined.

    (a) What are the simple events?

    (b) What is the probability that the selected accident was attributed to unsafe conditions?

    (c) What is the probability that the selected accident did not occur on the day shift?

23. An insurance company offers four different deductible levels—none, low, medium, and high—for its homeowner's policyholders and three different levels—low, medium, and high—for its automobile policyholders. The accompanying table gives proportions for the various categories of policyholders who have both types of insurance. For example, the proportion of individuals with both low homeowner's deductible and low auto deductible is .06 (6% of all such individuals).

Auto	Homeowner's			
	N	L	M	H
L	.04	.06	.05	.03
M	.07	.10	.20	.10
H	.02	.03	.15	.15

Suppose an individual having both types of policies is randomly selected.
(a) What is the probability that the individual has a medium auto deductible and a high homeowner's deductible?
(b) What is the probability that the individual has a low auto deductible? A low homeowner's deductible?
(c) What is the probability that the individual is in the same category for both auto and homeowner's deductibles?
(d) Based on your answer in part (c), what is the probability that the two categories are different?
(e) What is the probability that the individual has at least one low deductible level?
(f) Using the answer in part (e), what is the probability that neither deductible level is low?

24. The route used by a driver in commuting to work contains two intersections with traffic signals. The probability that he must stop at the first signal is .4, the analogous probability for the second signal is .5, and the probability that he must stop at one or more of the two signals is .6. What is the probability that he must stop
(a) At both signals?
(b) At the first signal but not at the second one?
(c) At exactly one signal?

25. The computers of six faculty members in a certain department are to be replaced. Two of the faculty members have selected laptop machines and the other four have chosen desktop machines. Suppose that only two of the setups can be done on a particular day, and the two computers to be set up are randomly selected from the six (implying 15 equally likely outcomes; if the computers are numbered 1, 2, ... , 6, then one outcome consists of computers 1 and 2, another consists of computers 1 and 3, and so on).

   (a) What is the probability that both selected setups are for laptop computers?
   (b) What is the probability that both selected setups are desktop machines?
   (c) What is the probability that at least one selected setup is for a desktop computer?
   (d) What is the probability that at least one computer of each type is chosen for setup?

26. Show that if one event A is contained in another event B (i.e., A is a subset of B), then $P(A) \leq P(B)$. [*Hint*: For such A and B, A and $B \cap A'$ are disjoint and $B = A \cup (B \cap A')$, as can be seen from a Venn diagram.] For general A and B, what does this imply about the relationship among $P(A \cap B)$, $P(A)$, and $P(A \cup B)$?

27. The three most popular options on a certain type of new car are a built-in GPS (A), a sunroof (B), and an automatic transmission (C). If 40% of all purchasers request A, 55% request B, 70% request C, 63% request A or B, 77% request A or C, 80% request B or C, and 85% request A or B or C, compute the probabilities of the following events.
   (a) The next purchaser will request at least one of the three options.
   (b) The next purchaser will select none of the three options.
   (c) The next purchaser will request only an automatic transmission and neither of the other two options.
   (d) The next purchaser will select exactly one of these three options.
   [*Hint*: "A or B" is the event that at least one of the two options is requested; try drawing a Venn diagram and labeling all regions.]

28. A certain system can experience three different types of defects. Let A_i ($i = 1$, 2, 3) denote the event that the system has a defect of type i. Suppose that

   $$P(A_1) = .12 \qquad\qquad P(A_2) = .07 \qquad\qquad P(A_3) = .05$$
   $$P(A_1 \cup A_2) = .13 \qquad\qquad P(A_1 \cup A_3) = .14$$
   $$P(A_2 \cup A_3) = .10 \qquad\qquad P(A_1 \cap A_2 \cap A_3) = .01$$

   (a) What is the probability that the system does not have a type 1 defect?
   (b) What is the probability that the system has both type 1 and type 2 defects?
   (c) What is the probability that the system has both type 1 and type 2 defects but not a type 3 defect?
   (d) What is the probability that the system has at most two of these defects?

29. In Exercise 7, suppose that any incoming individual is equally likely to be assigned to any of the three stations irrespective of where other individuals have been assigned. What is the probability that
   (a) All three family members are assigned to the same station?
   (b) At most two family members are assigned to the same station?
   (c) Every family member is assigned to a different station?

30. Apply the proposition involving the probability of $A \cup B$ to the union of the two events $(A \cup B)$ and C in order to verify the result for $P(A \cup B \cup C)$.

1.3   Counting Methods

When the various outcomes of an experiment are equally likely (the same probability is assigned to each simple event), the task of computing probabilities reduces
to counting. Equally likely outcomes arise in many games, including the six sides of
a fair die, the two sides of a fair coin, and the 38 slots of a fair roulette wheel. As
mentioned at the end of the last section, if N is the number of outcomes in a sample
space and $N(A)$ is the number of outcomes contained in an event A, then

$$P(A) = \frac{N(A)}{N} \tag{1.1}$$

If a list of the outcomes is available or easy to construct and N is small, then the
numerator and denominator of Eq. (1.1) can be obtained without the benefit of any
general counting principles.

There are, however, many experiments for which the effort involved in
constructing such a list is prohibitive because N is quite large. By exploiting
some general counting rules, it is possible to compute probabilities of the form
(1.1) without a listing of outcomes. These rules are also useful in many problems
involving outcomes that are not equally likely. Several of the rules developed here
will be used in studying probability distributions in the next chapter.

1.3.1   The Fundamental Counting Principle

Our first counting rule applies to any situation in which an event consists of ordered
pairs of objects and we wish to count the number of such pairs. By an ordered pair,
we mean that, if O_1 and O_2 are objects, then the pair (O_1, O_2) is different from the
pair (O_2, O_1). For example, if an individual selects one airline for a trip from Los
Angeles to Chicago and a second one for continuing on to New York, one
possibility is (American, United), another is (United, American), and still another
is (United, United).

> **PROPOSITION**
> If the first element or object of an ordered pair can be selected in n_1 ways, and
> for each of these n_1 ways the second element of the pair can be selected in n_2
> ways, then the number of pairs is $n_1 n_2$.

Example 1.17 A homeowner doing some remodeling requires the services of both
a plumbing contractor and an electrical contractor. If there are 12 plumbing
contractors and 9 electrical contractors available in the area, in how many ways
can the contractors be chosen? If we denote the plumbers by $P_1, \ldots, P_{12}$ and the
electricians by $Q_1, \ldots, Q_9$, then we wish the number of pairs of the form (P_i, Q_j).

With $n_1 = 12$ and $n_2 = 9$, the proposition yields $N = (12)(9) = 108$ possible ways of choosing the two types of contractors. ∎

In Example 1.17, the choice of the second element of the pair did not depend on which first element was chosen or occurred. As long as there is the same number of choices of the second element for each first element, the proposition above is valid even when the set of possible second elements depends on the first element.

Example 1.18 A family has just moved to a new city and requires the services of both an obstetrician and a pediatrician. There are two easily accessible medical clinics, each having two obstetricians and three pediatricians. The family will obtain maximum health insurance benefits by joining a clinic and selecting both doctors from that clinic. In how many ways can this be done? Denote the obstetricians by O_1, O_2, O_3, and O_4 and the pediatricians by $P_1, \ldots, P_6$. Then we wish the number of pairs (O_i, P_j) for which O_i and P_j are associated with the same clinic. Because there are four obstetricians, $n_1 = 4$, and for each there are three choices of pediatrician, so $n_2 = 3$. Applying the proposition gives $N = n_1 n_2 = 12$ possible choices. ∎

If a six-sided die is tossed five times in succession, then each possible outcome is an ordered collection of five numbers such as (⊡, ⊡, ⊡, ⊡, ⊡) or (⊞, ⊠, ⊡, ⊡, ⊡). We will call an ordered collection of k objects a **k-tuple** (so a pair is a 2-tuple and a triple is a 3-tuple). Each outcome of the die-tossing experiment is then a 5-tuple. The following theorem, called the Fundamental Counting Principle, generalizes the previous proposition to k-tuples.

FUNDAMENTAL COUNTING PRINCIPLE
Suppose a set consists of ordered collections of k elements (k-tuples) and that there are n_1 possible choices for the first element; for each choice of the first element, there are n_2 possible choices of the second element; ...; for each possible choice of the first $k - 1$ elements, there are n_k choices of the kth element. Then there are $n_1 n_2 \cdots n_k$ possible k-tuples.

Example 1.19 (Example 1.17 continued) Suppose the home remodeling job involves first purchasing several kitchen appliances. They will all be purchased from the same dealer, and there are five dealers in the area. With the dealers denoted by $D_1, \ldots, D_5$, there are $N = n_1 n_2 n_3 = (5)(12)(9) = 540$ 3-tuples of the form (D_i, P_j, Q_k), so there are 540 ways to choose first an appliance dealer, then a plumbing contractor, and finally an electrical contractor. ∎

Example 1.20 (Example 1.18 continued) If each clinic has both three specialists in internal medicine and two general surgeons, there are $n_1 n_2 n_3 n_4 = (4)(3)(3)(2) = 72$ ways to select one doctor of each type such that all doctors practice at the same clinic. ∎

Fig. 1.7 Tree diagram for
Example 1.18

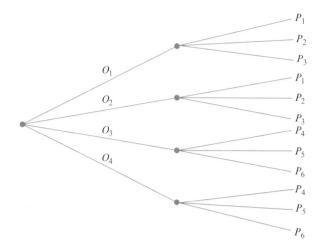

1.3.2 Tree Diagrams

In many counting and probability problems, a configuration called a *tree diagram* can be used to represent pictorially all the possibilities. The tree diagram associated with Example 1.18 appears in Fig. 1.7. Starting from a point on the left side of the diagram, for each possible first element of a pair a straight-line segment emanates rightward. Each of these lines is referred to as a first-generation branch. Now for any given first-generation branch we construct another line segment emanating from the tip of the branch for each possible choice of a second element of the pair. Each such line segment is a second-generation branch. Because there are four obstetricians, there are four first-generation branches, and three pediatricians for each obstetrician yields three second-generation branches emanating from each first-generation branch.

Generalizing, suppose there are n_1 first-generation branches, and for each first-generation branch there are n_2 second-generation branches. The total number of second-generation branches is then $n_1 n_2$. Since the end of each second-generation branch corresponds to exactly one possible pair (choosing a first element and then a second puts us at the end of exactly one second-generation branch), there are $n_1 n_2$ pairs, verifying our first proposition.

The Fundamental Counting Principle can also be illustrated by a tree diagram; simply construct a more elaborate diagram by adding third-generation branches emanating from the tip of each second generation, then fourth-generation branches, and so on, until finally kth-generation branches are added.

The construction of a tree diagram does not depend on having the same number of second-generation branches emanating from each first-generation branch. If the second clinic had four pediatricians, then there would be only three branches emanating from two of the first-generation branches and four emanating from each of the other two first-generation branches. A tree diagram can thus be used to represent experiments for which the Fundamental Counting Principle does not apply.

1.3.3 Permutations

So far the successive elements of a k-tuple were selected from entirely different sets (e.g., appliance dealers, then plumbers, and finally electricians). In several tosses of a die, the set from which successive elements are chosen is always $\{\boxdot, \boxdot, \boxdot, \boxdot, \boxdot, \boxdot\}$, but the choices are made "with replacement" so that the same element can appear more than once. If the die is rolled once, there are obviously 6 possible outcomes; for two rolls, there are $6^2 = 36$ possibilities, since we distinguish $(\boxdot, \boxdot)$ from $(\boxdot, \boxdot)$. In general, if k selections are made with replacement from a set of n distinct objects (such as the six sides of a die), then the total number of possible outcomes is n^k.

We now consider a fixed set consisting of n distinct elements and suppose that a k-tuple is formed by selecting successively from this set *without replacement*, so that an element can appear in at most one of the k positions.

> **DEFINITION**
> Any ordered sequence of k objects taken without replacement from a set of n distinct objects is called a **permutation** of size k of the objects. The number of permutations of size k that can be constructed from the n objects is denoted by $_nP_k$.

The number of permutations of size k is obtained immediately from the Fundamental Counting Principle. The first element can be chosen in n ways; for each of these n ways the second element can be chosen in $n - 1$ ways; and so on. Finally, for each way of choosing the first $k - 1$ elements, the kth element can be chosen in $n - (k - 1) = n - k + 1$ ways, so

$$_nP_k = n(n - 1)(n - 2) \cdots (n - k + 2)(n - k + 1)$$

Example 1.21 Ten teaching assistants are available for grading papers in a particular course. The first exam consists of four questions, and the professor wishes to select a different assistant to grade each question (only one assistant per question). In how many ways can assistants be chosen to grade the exam? Here $n =$ the number of assistants $= 10$ and $k =$ the number of questions $= 4$. The number of different grading assignments is then $_{10}P_4 = (10)(9)(8)(7) = 5040$. ∎

Example 1.22 *The Birthday Problem.* Disregarding the possibility of a February 29 birthday, suppose a randomly selected individual is equally likely to have been born on any one of the other 365 days. If ten people are randomly selected, what is the probability that all have different birthdays?

Imagine selecting 10 days, *with replacement*, from the calendar to represent the birthdays of the ten randomly selected people. One possible outcome of this selection would be (March 31, December 30, ..., September 27, February 12). There are 365^{10} such outcomes. The number of outcomes among them with no repeated birthdays is

$$(365)(364)\cdots(356) = {}_{365}P_{10}$$

(any of the 365 calendar days may be selected first; if March 31 is chosen, any of the other 364 days is acceptable for the second selection; and so on). Hence, the probability all ten randomly selected people have different birthdays equals ${}_{365}P_{10}/365^{10} = .883$. Equivalently, there's only a .117 chance that at least two people out of these ten will share a birthday. It's worth noting that the first probability can be rewritten as

$$\frac{{}_{365}P_{10}}{365^{10}} = \frac{365}{365} \cdot \frac{364}{365} \cdots \frac{356}{365}$$

We may think of each fraction as representing the chance the next birthday selected will be different from all previous ones. (This is an example of *conditional probability*, the topic of the next section.)

Now replace 10 with k (i.e., k randomly selected birthdays); what is the smallest k for which there is at least a 50–50 chance that two or more people will have the same birthday? Most people incorrectly guess that we need a very large group of people for this to be true; the most common guess is that 183 people are required (half the days on the calendar). But the required value of k is actually much smaller: the probability that k randomly selected people all have different birthdays equals ${}_{365}P_k/365^k$, which not surprisingly decreases as k increases. Figure 1.8 displays this probability for increasing values of k. As it turns out, the smallest k for which this probability falls below .5 is just $k = 23$. That is, there is less than a 50–50 chance (.4927, to be precise) of 23 randomly selected people all having different birthdays, and thus a probability .5073 that at least two people in a random sample of 23 will share a birthday.

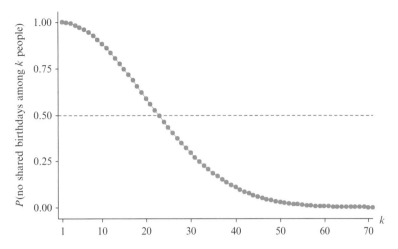

Fig. 1.8 P(no birthday match) in Example 1.22 ■

The expression for $_nP_k$ can be rewritten with the aid of factorial notation. Recall that 7! (read "7 factorial") is compact notation for the descending product of integers $(7)(6)(5)(4)(3)(2)(1)$. More generally, for any positive integer m, $m! = m(m-1)(m-2)\cdots(2)(1)$. This gives $1! = 1$, and we also define $0! = 1$.

Using factorial notation, $(10)(9)(8)(7) = (10)(9)(8)(7)(6!)/6! = 10!/6!$. More generally,

$$_nP_k = n(n-1)\cdots\cdots(n-k+1)$$
$$= \frac{n(n-1)\cdots\cdots(n-k+1)(n-k)(n-k-1)\cdots\cdots(2)(1)}{(n-k)(n-k-1)\cdots\cdots(2)(1)}$$

which becomes

$$_nP_k = \frac{n!}{(n-k)!}$$

For example, $_9P_3 = 9!/(9-3)! = 9!/6! = 9\cdot8\cdot7\cdot6!/6! = 9\cdot8\cdot7$. Note also that because $0! = 1$, $_nP_n = n!/(n-n)! = n!/0! = n!/1 = n!$, as it should.

1.3.4 Combinations

Often the objective is to count the number of *unordered* subsets of size k that can be formed from a set consisting of n distinct objects. For example, in bridge it is only the 13 cards in a hand and not the order in which they are dealt that is important; in the formation of a committee, the order in which committee members are listed is frequently unimportant.

> **DEFINITION**
> Given a set of n distinct objects, any unordered subset of size k of the objects is called a **combination**. The number of combinations of size k that can be formed from n distinct objects will be denoted by $\binom{n}{k}$ or $_nC_k$.

The number of combinations of size k from a particular set is smaller than the number of permutations because, when order is disregarded, some of the permutations correspond to the same combination. Consider, for example, the set $\{A, B, C, D, E\}$ consisting of five elements. There are $_5P_3 = 5!/(5-3)! = 60$ permutations of size 3. There are six permutations of size 3 consisting of the elements A, B, and C because these three can be ordered $3 \cdot 2 \cdot 1 = 3! = 6$ ways: (A, B, C), (A, C, B), (B, A, C), (B, C, A), (C, A, B), and (C, B, A). These six *permutations* are equivalent to the single *combination* $\{A, B, C\}$. Similarly, for any other combination of size 3, there are 3! permutations, each obtained by ordering the three objects. Thus,

$$60 = {_5P_3} = \binom{5}{3} \cdot 3! \quad \text{so} \quad \binom{5}{3} = \frac{60}{3!} = 10$$

These ten combinations are

$$\{A,B,C\}\{A,B,D\}\{A,B,E\}\{A,C,D\}\{A,C,E\}$$

$$\{A,D,E\}\{B,C,D\}\{B,C,E\}\{B,D,E\}\{C,D,E\}$$

When there are n distinct objects, any permutation of size k is obtained by ordering the k unordered objects of a combination in one of $k!$ ways, so the number of permutations is the product of $k!$ and the number of combinations. This gives

$$_nC_k = \binom{n}{k} = \frac{_nP_k}{k!} = \frac{n!}{k!(n-k)!}$$

Notice that $\binom{n}{n} = 1$ and $\binom{n}{0} = 1$ because there is only one way to choose a set of (all) n elements or of no elements, and $\binom{n}{1} = n$ since there are n subsets of size 1.

Example 1.23 A bridge hand consists of any 13 cards selected from a 52-card deck without regard to order. There are $\binom{52}{13} = 52!/(13! \cdot 39!)$ different bridge hands, which works out to approximately 635 billion. Since there are 13 cards in each suit, the number of hands consisting entirely of clubs and/or spades (no red cards) is $\binom{26}{13} = 26!/(13! \cdot 13!) = 10,400,600$. One of these $\binom{26}{13}$ hands consists entirely of spades, and one consists entirely of clubs, so there are $[\binom{26}{13} - 2]$ hands that consist entirely of clubs and spades with both suits represented in the hand. Suppose a bridge hand is dealt from a well-shuffled deck (i.e., 13 cards are randomly selected from among the 52 possibilities) and let

A = {the hand consists entirely of spades and clubs with both suits represented}
B = {the hand consists of exactly two suits}

The $N = \binom{52}{13}$ possible outcomes are equally likely, so

$$P(A) = \frac{N(A)}{N} = \frac{\binom{26}{13} - 2}{\binom{52}{13}} = .0000164$$

Since there are $\binom{4}{2} = 6$ combinations consisting of two suits, of which spades and clubs is one such combination,

$$P(B) = \frac{N(B)}{N} = \frac{6\left[\binom{26}{13} - 2\right]}{\binom{52}{13}} = .0000983$$

That is, a hand consisting entirely of cards from exactly two of the four suits will occur roughly once in every 10,000 hands. If you play bridge only once a month, it is likely that you will never be dealt such a hand. ∎

Example 1.24 A university warehouse has received a shipment of 25 printers, of which 10 are laser printers and 15 are inkjet models. If 6 of these 25 are selected at random to be checked by a particular technician, what is the probability that exactly 3 of those selected are laser printers (so that the other 3 are inkjets)?

Let D_3 = {exactly 3 of the 6 selected are inkjet printers}. Assuming that any particular set of 6 printers is as likely to be chosen as is any other set of 6, we have equally likely outcomes, so $P(D_3) = N(D_3)/N$, where N is the number of ways of choosing 6 printers from the 25 and $N(D_3)$ is the number of ways of choosing 3 laser printers and 3 inkjet models. Thus $N = \binom{25}{6}$. To obtain $N(D_3)$, think of first choosing 3 of the 15 inkjet models and then 3 of the laser printers. There are $\binom{15}{3}$ ways of choosing the 3 inkjet models, and there are $\binom{10}{3}$ ways of choosing the 3 laser printers; by the Fundamental Counting Principle, $N(D_3)$ is the product of these two numbers. So

$$P(D_3) = \frac{N(D_3)}{N} = \frac{\binom{15}{3}\binom{10}{3}}{\binom{25}{6}} = \frac{\frac{15!}{3!12!} \cdot \frac{10!}{3!7!}}{\frac{25!}{6!19!}} = .3083$$

Let $D_4 = \{$exactly 4 of the 6 printers selected are inkjet models$\}$ and define D_5 and D_6 in an analogous manner. Notice that the events D_3, D_4, D_5, and D_6 are disjoint. Thus, the probability that *at least* 3 inkjet printers are selected is

$$P(D_3 \cup D_4 \cup D_5 \cup D_6) = P(D_3) + P(D_4) + P(D_5) + P(D_6)$$

$$= \frac{\binom{15}{3}\binom{10}{3}}{\binom{25}{6}} + \frac{\binom{15}{4}\binom{10}{2}}{\binom{25}{6}} + \frac{\binom{15}{5}\binom{10}{1}}{\binom{25}{6}} + \frac{\binom{15}{6}\binom{10}{0}}{\binom{25}{6}} = .8530$$

∎

Example 1.25 The article "Does Your iPod *Really* Play Favorites?" (*The Amer. Statistician*, 2009: 263-268) investigated the randomness of the iPod's shuffling process. One professor's iPod playlist contains 100 songs, of which 10 are by the Beatles. Suppose the shuffle feature is used to play the songs in random order. What is the probability that the first Beatles song heard is the fifth song played?

In order for this event to occur, it must be the case that the first four songs played are not Beatles songs (NBs) and that the fifth song is by the Beatles (B). The total number of ways to select the first five songs is (100)(99)(98)(97)(96), while the number of ways to select these five songs so that the first four are NBs and the next is a B is (90)(89)(88)(87)(10). The random shuffle assumption implies that every sequence of 5 songs from amongst the 100 has the same chance of being selected as the first 5 played, i.e., each outcome (a list of 5 songs) is equally likely. Therefore the desired probability is

$$P\left(1^{st} \text{ B is the } 5^{th}\text{song played}\right) = \frac{90 \cdot 89 \cdot 88 \cdot 87 \cdot 10}{100 \cdot 99 \cdot 98 \cdot 97 \cdot 96} = \frac{_{90}P_4 \cdot 10}{_{100}P_5} = .0679$$

Here is an alternative line of reasoning involving combinations. Rather than focusing on selecting just the first 5 songs, think of playing all 100 songs in random order. The number of ways of choosing 10 of these songs to be the Bs (without regard to the order in which they are played) is $\binom{100}{10}$. Now if we choose 9 of the last 95 songs to be Bs, which can be done in $\binom{95}{9}$ ways, that leaves four NBs and one B for the first five songs. Finally, there is only one way for these first five songs to start with four NBs and then follow with a B (remember that we are considering *unordered* subsets). Thus

$$P\left(1^{\text{st}} \text{ B is the } 5^{\text{th}}\text{song played}\right) = \frac{\dbinom{95}{9}}{\dbinom{100}{10}}$$

It is easily verified that this latter expression is in fact identical to the previous expression for the desired probability, so the numerical result is again .0679.

By similar reasoning, the probability that one of the first five songs played is a Beatles song is

P(1st B is the 1st or 2nd or 3rd or 4th or 5th song played)

$$= \frac{\dbinom{99}{9}}{\dbinom{100}{10}} + \frac{\dbinom{98}{9}}{\dbinom{100}{10}} + \frac{\dbinom{97}{9}}{\dbinom{100}{10}} + \frac{\dbinom{96}{9}}{\dbinom{100}{10}} + \frac{\dbinom{95}{9}}{\dbinom{100}{10}} = .4162$$

It is thus rather likely that a Beatles song will be one of the first five songs played. Such a "coincidence" is not as surprising as might first appear to be the case. ∎

1.3.5 Exercises: Section 1.3 (31–49)

31. An ATM personal identification number (PIN) consists of a four-digit sequence.
    (a) How many different possible PINs are there if there are no restrictions on the possible choice of digits?
    (b) According to a representative at the authors' local branch of Chase Bank, there are in fact restrictions on the choice of digits. The following choices are prohibited: (1) all four digits identical; (2) sequences of consecutive ascending or descending digits, such as 6543; (3) any sequence starting with 19 (birth years are too easy to guess). So if one of the PINs in (a) is randomly selected, what is the probability that it will be a legitimate PIN (i.e., not be one of the prohibited sequences)?
    (c) Someone has stolen an ATM card and knows the first and last digits of the PIN are 8 and 1, respectively. He also knows about the restrictions described in (b). If he gets three chances to guess the middle two digits before the ATM "eats" the card, what is the probability the thief gains access to the account?
    (d) Recalculate the probability in (c) if the first and last digits are 1 and 1.
32. The College of Science Council has one student representative from each of the five science departments (biology, chemistry, statistics, mathematics, physics). In how many ways can
    (a) Both a council president and a vice president be selected?

(b) A president, a vice president, and a secretary be selected?

(c) Two council members be selected for the Dean's Council?

33. A friend of ours is giving a dinner party. Her current wine supply includes 8 bottles of zinfandel, 10 of merlot, and 12 of cabernet (she drinks only red wine), all from different wineries.

    (a) If she wants to serve 3 bottles of zinfandel and serving order is important, how many ways are there to do this?

    (b) If 6 bottles of wine are to be randomly selected from the 30 for serving, how many ways are there to do this?

    (c) If 6 bottles are randomly selected, how many ways are there to obtain two bottles of each variety?

    (d) If 6 bottles are randomly selected, what is the probability that this results in two bottles of each variety being chosen?

    (e) If 6 bottles are randomly selected, what is the probability that all of them are the same variety?

34. (a) Beethoven wrote 9 symphonies and Mozart wrote 27 piano concertos. If a university radio station announcer wishes to play first a Beethoven symphony and then a Mozart concerto, in how many ways can this be done?

    (b) The station manager decides that on each successive night (7 days per week), a Beethoven symphony will be played, followed by a Mozart piano concerto, followed by a Schubert string quartet (of which there are 15). For roughly how many years could this policy be continued before exactly the same program would have to be repeated?

35. A stereo store is offering a special price on a complete set of components (receiver, compact disc player, speakers, turntable). A purchaser is offered a choice of manufacturer for each component:

Receiver:	Kenwood, Onkyo, Pioneer, Sony, Yamaha
Compact disc player:	Onkyo, Pioneer, Sony, Panasonic
Speakers:	Boston, Infinity, Polk
Turntable:	Onkyo, Sony, Teac, Technics

    A switchboard display in the store allows a customer to hook together any selection of components (consisting of one of each type). Use the Fundamental Counting Principle to answer the following questions:

    (a) In how many ways can one component of each type be selected?

    (b) In how many ways can components be selected if both the receiver and the compact disc player are to be Sony?

    (c) In how many ways can components be selected if none is to be Sony?

    (d) In how many ways can a selection be made if at least one Sony component is to be included?

    (e) If someone flips switches on the selection in a completely random fashion, what is the probability that the system selected contains at least one Sony component? Exactly one Sony component?

36. In five-card poker, a straight consists of five cards in adjacent ranks (e.g., 9 of clubs, 10 of hearts, jack of hearts, queen of spades, king of clubs). Assuming that aces can be high or low, if you are dealt a five-card hand, what is the probability that it will be a straight with high card 10? What is the probability that it will be a straight? What is the probability that it will be a straight flush (all cards in the same suit)?

37. A local bar stocks 12 American beers, 8 Mexican beers, and 9 German beers. You ask the bartender to pick out a five-beer "sampler" for you. Assume the bartender makes the five selections at random and without replacement.
    (a) What is the probability you get at least four American beers?
    (b) What is the probability you get five beers from the same country?

38. Computer keyboard failures can be attributed to electrical defects or mechanical defects. A repair facility currently has 25 failed keyboards, 6 of which have electrical defects and 19 of which have mechanical defects.
    (a) How many ways are there to randomly select 5 of these keyboards for a thorough inspection (without regard to order)?
    (b) In how many ways can a sample of 5 keyboards be selected so that exactly 2 have an electrical defect?
    (c) If a sample of 5 keyboards is randomly selected, what is the probability that at least 4 of these will have a mechanical defect?

39. The statistics department at the authors' university participates in an annual volleyball tournament. Suppose that all 16 department members are willing to play.
    (a) How many different six-person volleyball rosters could be generated? (That is, how many years could the department participate in the tournament without repeating the same six-person team?)
    (b) The statistics department faculty consist of 5 women and 11 men. How many rosters comprising exactly 2 women and 4 men be generated?
    (c) The tournament's rules actually require that each team include *at least* two women. Under this rule, how many valid teams could be generated?
    (d) Suppose this year the department decides to randomly select its six players. What is the probability the randomly selected team has exactly two women? At least two women?

40. A production facility employs 20 workers on the day shift, 15 workers on the swing shift, and 10 workers on the graveyard shift. A quality control consultant is to select 6 of these workers for in-depth interviews. Suppose the selection is made in such a way that any particular group of 6 workers has the same chance of being selected as does any other group (drawing 6 slips without replacement from among 45).
    (a) How many selections result in all 6 workers coming from the day shift? What is the probability that all 6 selected workers will be from the day shift?
    (b) What is the probability that all 6 selected workers will be from the same shift?

(c) What is the probability that at least two different shifts will be represented among the selected workers?

(d) What is the probability that at least one of the shifts will be unrepresented in the sample of workers?

41. An academic department with five faculty members narrowed its choice for department head to either candidate A or candidate B. Each member then voted on a slip of paper for one of the candidates. Suppose there are actually three votes for A and two for B. If the slips are selected for tallying in random order, what is the probability that A remains ahead of B throughout the vote count (e.g., this event occurs if the selected ordering is $AABAB$, but not for $ABBAA$)?

42. An experimenter is studying the effects of temperature, pressure, and type of catalyst on yield from a chemical reaction. Three different temperatures, four different pressures, and five different catalysts are under consideration.

(a) If any particular experimental run involves the use of a single temperature, pressure, and catalyst, how many experimental runs are possible?

(b) How many experimental runs involve use of the lowest temperature and two lowest pressures?

(c) Suppose that five different experimental runs are to be made on the first day of experimentation. If the five are randomly selected from among all the possibilities, so that any group of five has the same probability of selection, what is the probability that a different catalyst is used on each run?

43. A box in a certain supply room contains four 40-W lightbulbs, five 60-W bulbs, and six 75-W bulbs. Suppose that three bulbs are randomly selected.

(a) What is the probability that exactly two of the selected bulbs are rated 75 W?

(b) What is the probability that all three of the selected bulbs have the same rating?

(c) What is the probability that one bulb of each type is selected?

(d) Suppose now that bulbs are to be selected one by one until a 75-W bulb is found. What is the probability that it is necessary to examine at least six bulbs?

44. Fifteen telephones have just been received at an authorized service center. Five of these telephones are cellular, five are cordless, and the other five are corded phones. Suppose that these components are randomly allocated the numbers 1, 2, ..., 15 to establish the order in which they will be serviced.

(a) What is the probability that all the cordless phones are among the first ten to be serviced?

(b) What is the probability that after servicing ten of these phones, phones of only two of the three types remain to be serviced?

(c) What is the probability that two phones of each type are among the first six serviced?

45. Three molecules of type A, three of type B, three of type C, and three of type D are to be linked together to form a chain molecule. One such chain molecule is $ABCDABCDABCD$, and another is $BCDDAAABDBCC$.
    (a) How many such chain molecules are there? [*Hint*: If the three A's were distinguishable from one another—A_1, A_2, A_3—and the B's, C's, and D's were also, how many molecules would there be? How is this number reduced when the subscripts are removed from the A's?]
    (b) Suppose a chain molecule of the type described is randomly selected. What is the probability that all three molecules of each type end up next to each other (such as in $BBBAAADDDCCC$)?
46. A popular Dilbert cartoon strip (popular among statisticians, anyway) shows an allegedly "random" number generator produce the sequence 999999 with the accompanying comment, "That's the problem with randomness: you can never be sure." Most people would agree that 999999 seems less "random" than, say, 703928, but in what sense is that true? Imagine we randomly generate a six-digit number, i.e., we make six draws with replacement from the digits 0 through 9.
    (a) What is the probability of generating 999999?
    (b) What is the probability of generating 703928?
    (c) What is the probability of generating a sequence of six identical digits?
    (d) What is the probability of generating a sequence with *no* identical digits? (Comparing the answers to (c) and (d) gives some sense of why some sequences feel intuitively more random than others.)
    (e) Here's a real challenge: what is the probability of generating a sequence with exactly one repeated digit?
47. Three married couples have purchased theater tickets and are seated in a row consisting of just six seats. If they take their seats in a completely random fashion (random order), what is the probability that Jim and Paula (husband and wife) sit in the two seats on the far left? What is the probability that Jim and Paula end up sitting next to one another? What is the probability that at least one of the wives ends up sitting next to her husband?
48. A starting lineup in basketball consists of two guards, two forwards, and a center.
    (a) A certain college team has on its roster five guards, four forwards, and three centers. How many different starting lineups can be created?
    (b) Their opposing team in one particular game has three centers, four guards, four forwards, and one individual (X) who can play either guard or forward. How many different starting lineups can the opposing team create? [*Hint*: Consider lineups without X, with X as a guard, and with X as a forward.]
    (c) Now suppose a team has 4 guards, 4 forwards, 2 centers, and two players (X and Y) who can play either guard or forward. If 5 of the 12 players on this team are randomly selected, what is the probability that they constitute a legitimate starting lineup?
49. Show that $\binom{n}{k} = \binom{n}{n-k}$. Give an interpretation involving subsets.

1.4   Conditional Probability

The probabilities assigned to various events depend on what is known about the experimental situation when the assignment is made. Subsequent to the initial assignment, partial information about or relevant to the outcome of the experiment may become available. Such information may cause us to revise some of our probability assignments. For a particular event A, we have used $P(A)$ to represent the probability assigned to A; we now think of $P(A)$ as the original or "unconditional" probability of the event A.

In this section, we examine how the information "an event B has occurred" affects the probability assigned to A. For example, A might refer to an individual having a particular disease in the presence of certain symptoms. If a blood test is performed on the individual and the result is negative ($B =$ negative blood test), then the probability of having the disease will change (it should decrease, but not usually to zero, since blood tests are not infallible).

Example 1.26 Complex components are assembled in a plant that uses two different assembly lines, A and A'. Line A uses older equipment than A', so it is somewhat slower and less reliable. Suppose on a given day line A has assembled 8 components, of which 2 have been identified as defective (B) and 6 as nondefective (B'), whereas A' has produced 1 defective and 9 nondefective components. This information is summarized in the accompanying table.

Line	Condition	
	B	B'
A	2	6
A'	1	9

Unaware of this information, the sales manager randomly selects 1 of these 18 components for a demonstration. Prior to the demonstration

$$P(\text{line } A \text{ component selected}) = P(A) = \frac{N(A)}{N} = \frac{8}{18} = .444$$

However, if the chosen component turns out to be defective, then the event B has occurred, so the component must have been one of the 3 in the B column of the table. Since these 3 components are equally likely among themselves, the probability the component was selected from line A, *given that event B has occurred*, is

$$P(A, \text{given } B) = \frac{2}{3} = \frac{2/18}{3/18} = \frac{P(A \cap B)}{P(B)} \qquad (1.2)$$

∎

In Eq. (1.2), the conditional probability is expressed as a ratio of unconditional probabilities. The numerator is the probability of the intersection of the two events, whereas the denominator is the probability of the conditioning event B. A Venn diagram illuminates this relationship (Fig. 1.9).

Given that B has occurred, the relevant sample space is no longer $\mathscr{S}$ but consists of just outcomes in B, and A has occurred if and only if one of the outcomes in the intersection $A \cap B$ occurred. So the conditional probability of A given B should, logically, be the ratio of the likelihoods of these two events.

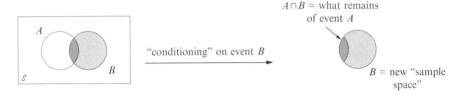

Fig. 1.9 Motivating the definition of conditional probability

1.4.1 The Definition of Conditional Probability

Example 1.26 demonstrates that when outcomes are equally likely, computation of conditional probabilities can be based on intuition. When experiments are more complicated, though intuition may fail us, we want to have a general definition of conditional probability that will yield intuitive answers in simple problems. Figure 1.9 and Eq. (1.2) suggest the appropriate definition.

> **DEFINITION**
> For any two events A and B with $P(B) > 0$, the **conditional probability of A given that B has occurred**, denoted $P(A|B)$, is defined by
>
> $$P(A \mid B) = \frac{P(A \cap B)}{P(B)}. \tag{1.3}$$

Example 1.27 Suppose that of all individuals buying a certain digital camera, 60% include an optional memory card in their purchase, 40% include an extra battery, and 30% include both a card and battery. Consider randomly selecting a buyer and let $A = \{\text{memory card purchased}\}$ and $B = \{\text{battery purchased}\}$. Then $P(A) = .60$, $P(B) = .40$, and $P(\text{both purchased}) = P(A \cap B) = .30$. Given that the selected individual purchased an extra battery, the probability that an optional card was also purchased is

$$P\left(A \mid B\right) = \frac{P(A \cap B)}{P(B)} = \frac{.30}{.40} = .75$$

That is, of all those purchasing an extra battery, 75% purchased an optional memory card. Similarly,

$$P\left(\text{battery} \mid \text{memory card}\right) = P\left(B \mid A\right) = \frac{P(A \cap B)}{P(A)} = \frac{.30}{.60} = .50$$

Notice that $P(A|B) \neq P(A)$ and $P(B|A) \neq P(B)$. Notice also that $P(A|B) \neq P(B|A)$: these represent two different probabilities computed using difference pieces of "given" information. ■

Example 1.28 A news magazine includes three columns entitled "Art" (A), "Books" (B), and "Cinema" (C). Reading habits of a randomly selected reader with respect to these columns are

Read regularly	A	B	C	$A \cap B$	$A \cap C$	$B \cap C$	$A \cap B \cap C$
Probability	.14	.23	.37	.08	.09	.13	.05

(See Fig. 1.10 on the next page.)
We thus have

$$P\left(A \mid B\right) = \frac{P(A \cap B)}{P(B)} = \frac{.08}{.23} = .348$$

$$P\left(A \mid B \cup C\right) = \frac{P(A \cap (B \cup C))}{P(B \cup C)} = \frac{.04 + .05 + .03}{.47} = \frac{.12}{.47} = .255$$

$$P\left(A \mid \text{reads at least one}\right) = P\left(A \mid A \cup B \cup C\right) = \frac{P(A \cap (A \cup B \cup C))}{P(A \cup B \cup C)}$$

$$= \frac{P(A)}{P(A \cup B \cup C)} = \frac{.14}{.49} = .286$$

and

Fig. 1.10 Venn diagram for
Example 1.28

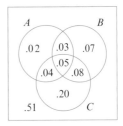

$$P\big(A\cup B \mid C\big) = \frac{P((A\cup B)\cap C)}{P(C)} = \frac{.04 + .05 + .08}{.37} = .459 \qquad\blacksquare$$

1.4.2 The Multiplication Rule for P(A ∩ B)

The definition of conditional probability yields the following result, obtained by multiplying both sides of Eq. (1.3) by $P(B)$.

> **MULTIPLICATION RULE**
>
> $$P(A\cap B) = P(A|B)\cdot P(B)$$

This rule is important because it is often the case that $P(A\cap B)$ is desired, whereas both $P(B)$ and $P(A|B)$ can be specified from the problem description. By reversing the roles of A and B, the Multiplication Rule can also be written as $P(A\cap B) = P(B|A)\cdot P(A)$.

Example 1.29 Four individuals have responded to a request by a blood bank for blood donations. None of them has donated before, so their blood types are unknown. Suppose only type O+ is desired and only one of the four actually has this type. If the potential donors are selected in random order for typing, what is the probability that at least three individuals must be typed to obtain the desired type?

Define $B = \{$first type not O+$\}$ and $A = \{$second type not O+$\}$. Since three of the four potential donors are not O+, $P(B) = 3/4$. Given that the first person typed is not O+, two of the three individuals left are not O+, and so $P(A|B) = 2/3$. The Multiplication Rule now gives

$$P(\text{at least three individuals are typed}) = P(\text{first two typed are not O+})$$
$$= P(A \cap B)$$
$$= P(A \mid B) \cdot P(B)$$
$$= \frac{2}{3} \cdot \frac{3}{4} = \frac{6}{12}$$
$$= .5 \qquad \blacksquare$$

The Multiplication Rule is most useful when the experiment consists of several stages in succession. The conditioning event B then describes the outcome of the first stage and A the outcome of the second, so that $P(A|B)$—conditioning on what occurs first—will often be known. The rule is easily extended to experiments involving more than two stages. For example,

$$P(A_1 \cap A_2 \cap A_3) = P\left(A_3 \mid A_1 \cap A_2\right) \cdot P\left(A_1 \cap A_2\right)$$
$$= P\left(A_3 \mid A_1 \cap A_2\right) \cdot P\left(A_2 \mid A_1\right) \cdot P(A_1)$$

where A_1 occurs first, followed by A_2, and finally A_3.

Example 1.30 For the blood typing experiment of Example 1.29,

$$P(\text{third type is O+}) =$$
$$P\left(\text{third is}\mid\text{first isn't} \cap \text{second isn't}\right) \cdot P\left(\text{second isn't}\mid\text{first isn't}\right) \cdot P(\text{first isn't})$$
$$= \frac{1}{2} \cdot \frac{2}{3} \cdot \frac{3}{4} = \frac{1}{4} = .25 \qquad \blacksquare$$

When the experiment of interest consists of a sequence of several stages, it is convenient to represent these with a tree diagram. Once we have an appropriate tree diagram, probabilities and conditional probabilities can be entered on the various branches; this will make repeated use of the Multiplication Rule quite straightforward.

Example 1.31 A chain of electronics stores sells three different brands of DVD players. Of its DVD player sales, 50% are brand 1 (the least expensive), 30% are brand 2, and 20% are brand 3. Each manufacturer offers a 1-year warranty on parts and labor. It is known that 25% of brand 1's DVD players require warranty repair work, whereas the corresponding percentages for brands 2 and 3 are 20% and 10%, respectively.
1. What is the probability that a randomly selected purchaser has bought a brand 1 DVD player that will need repair while under warranty?
2. What is the probability that a randomly selected purchaser has a DVD player that will need repair while under warranty?
3. If a customer returns to the store with a DVD player that needs warranty repair work, what is the probability that it is a brand 1 DVD player? A brand 2 DVD player? A brand 3 DVD player?

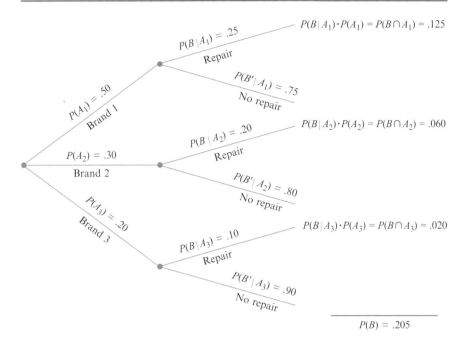

Fig. 1.11 Tree diagram for Example 1.31

The first stage of the problem involves a customer selecting one of the three brands of DVD player. Let $A_i = \{$brand i is purchased$\}$, for $i = 1$, 2, and 3. Then $P(A_1) = .50$, $P(A_2) = .30$, and $P(A_3) = .20$. Once a brand of DVD player is selected, the second stage involves observing whether the selected DVD player needs warranty repair. With $B = \{$needs repair$\}$ and $B' = \{$doesn't need repair$\}$, the given information implies that $P(B|A_1) = .25$, $P(B|A_2) = .20$, and $P(B|A_3) = .10$.

The tree diagram representing this experimental situation is shown in Fig. 1.11. The initial branches correspond to different brands of DVD players; there are two second-generation branches emanating from the tip of each initial branch, one for "needs repair" and the other for "doesn't need repair." The probability $P(A_i)$ appears on the ith initial branch, whereas the conditional probabilities $P(B|A_i)$ and $P(B'|A_i)$ appear on the second-generation branches. To the right of each second-generation branch corresponding to the occurrence of B, we display the product of probabilities on the branches leading out to that point. This is simply the Multiplication Rule in action. The answer to question 1 is thus $P(A_1 \cap B) = P(B|A_1) \cdot P(A_1) = .125$. The answer to question 2 is

$P(B) = P[(\text{brand 1 and repair}) \text{ or } (\text{brand 2 and repair}) \text{ or } (\text{brand 3 and repair})]$
$\qquad = P(A_1 \cap B) + P(A_2 \cap B) + P(A_3 \cap B)$
$\qquad = .125 + .060 + .020 = .205$

Finally,

$$P(A_1 \mid B) = \frac{P(A_1 \cap B)}{P(B)} = \frac{.125}{.205} = .61$$

$$P(A_2 \mid B) = \frac{P(A_2 \cap B)}{P(B)} = \frac{.060}{.205} = .29$$

and

$$P(A_3 \mid B) = 1 - P(A_1 \mid B) - P(A_2 \mid B) = .10$$

Notice that the initial or *prior probability* of brand 1 is .50, whereas once it is known that the selected DVD player needed repair, the *posterior probability* of brand 1 increases to .61. This is because brand 1 DVD players are more likely to need warranty repair than are the other brands. The posterior probability of brand 3 is $P(A_3|B) = .10$, which is much less than the prior probability $P(A_3) = .20$. ∎

1.4.3 The Law of Total Probability and Bayes' Theorem

The computation of a posterior probability $P(A_j|B)$ from given prior probabilities $P(A_i)$ and conditional probabilities $P(B|A_i)$ occupies a central position in elementary probability. The general rule for such computations, which is really just a simple application of the Multiplication Rule, goes back to the Reverend Thomas Bayes, who lived in the eighteenth century. To state it we first need another result. Recall that events $A_1, \ldots, A_k$ are mutually exclusive if no two have any common outcomes. The events are *exhaustive* if one A_i must occur, so that $A_1 \cup \cdots \cup A_k = \mathcal{S}$.

LAW OF TOTAL PROBABILITY
Let $A_1, \ldots, A_k$ be mutually exclusive and exhaustive events. Then for any other event B,

$$P(B) = P(B \mid A_1) \cdot P(A_1) + \cdots + P(B \mid A_k) \cdot P(A_k)$$
$$\qquad = \sum_{i=1}^{k} P(B \mid A_i) P(A_i) \tag{1.4}$$

Proof Because the A_is are mutually exclusive and exhaustive, if B occurs it must be in conjunction with exactly one of the A_is. That is, $B = (A_1 \text{ and } B)$ or ... or $(A_k \text{ and } B) = (A_1 \cap B) \cup \cdots \cup (A_k \cap B)$, where the events $(A_i \cap B)$ are mutually exclusive. This "partitioning of B" is illustrated in Fig. 1.12. Thus

$$P(B) = \sum_{i=1}^{k} P(A_i \cap B) = \sum_{i=1}^{k} P(B \mid A_i) P(A_i)$$

as desired.

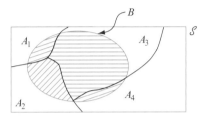

Fig. 1.12 Partition of B by mutually exclusive and exhaustive A_is   ∎

An example of the use of Eq. (1.4) appeared in answering question 2 of Example 1.31, where $A_1 = \{\text{brand } 1\}$, $A_2 = \{\text{brand } 2\}$, $A_3 = \{\text{brand } 3\}$, and $B = \{\text{repair}\}$.

Example 1.32 A student has three different e-mail accounts. Most of her messages, in fact 70%, come into account #1, whereas 20% come into account #2 and the remaining 10% into account #3. Of the messages coming into account #1, only 1% are spam, compared to 2% and 5% for account #2 and account #3, respectively. What is the student's overall spam rate, i.e., what is the probability a randomly selected e-mail message received by her is spam?

To answer this question, let's first establish some notation:

$A_i = \{\text{message is from account } \#i\}$ for $i = 1, 2, 3$; $B = \{\text{message is spam}\}$

The given percentages imply that

$$P(A_1) = .70, P(A_2) = .20, P(A_3) = .10$$
$$P(B \mid A_1) = .01, P(B \mid A_2) = .02, P(B \mid A_3) = .05$$

Now it's simply a matter of substituting into the equation for the Law of Total Probability:

$$P(B) = (.01)(.70) + (.02)(.20) + (.05)(.10) = .016$$

In the long run, 1.6% of her messages will be spam.   ∎

BAYES' THEOREM

Let $A_1, \ldots, A_k$ be a collection of mutually exclusive and exhaustive events with $P(A_i) > 0$ for $i = 1, \ldots, k$. Then for any other event B for which $P(B) > 0$,

$$P(A_j \mid B) = \frac{P(A_j \cap B)}{P(B)} = \frac{P(B \mid A_j)P(A_j)}{\sum_{i=1}^{k} P(B \mid A_i)P(A_i)} \qquad j = 1, \ldots, k \quad (1.5)$$

The transition from the second to the third expression in Eq. (1.5) rests on using the Multiplication Rule in the numerator and the Law of Total Probability in the denominator.

The proliferation of events and subscripts in Eq. (1.5) can be a bit intimidating to probability newcomers. When $k = 2$, so that the partition of $\mathcal{S}$ consists of just $A_1 = A$ and $A_2 = A'$, Bayes' Theorem becomes

$$P(A \mid B) = \frac{P(A)P(B \mid A)}{P(A)P(B \mid A) + P(A')P(B \mid A')}$$

As long as there are relatively few events in the partition, a tree diagram (as in Example 1.29) can be used as a basis for calculating posterior probabilities without ever referring explicitly to Bayes' Theorem.

Example 1.33 *Incidence of a rare disease.* In the book's Introduction, we presented the following example as a common misunderstanding of probability in everyday life. Only 1 in 1000 adults is afflicted with a rare disease for which a diagnostic test has been developed. The test is such that when an individual actually has the disease, a positive result will occur 99% of the time, whereas an individual without the disease will show a positive test result only 2% of the time. If a randomly selected individual is tested and the result is positive, what is the probability that the individual has the disease?

[*Note*: The *sensitivity* of this test is 99%, whereas the *specificity*—how specific positive results are to this disease—is 98%. As an indication of the accuracy of medical tests, an article in the October 29, 2010 *New York Times* reported that the sensitivity and specificity for a new DNA test for colon cancer were 86% and 93%, respectively. The PSA test for prostate cancer has sensitivity 85% and specificity about 30%, while the mammogram for breast cancer has sensitivity 75% and specificity 92%. All tests are less than perfect.]

To use Bayes' Theorem, let $A_1 = \{$individual has the disease$\}$, $A_2 = \{$individual does not have the disease$\}$, and $B = \{$positive test result$\}$. Then $P(A_1) = .001$, $P(A_2) = .999$, $P(B|A_1) = .99$, and $P(B|A_2) = .02$. The tree diagram for this problem is in Fig. 1.13.

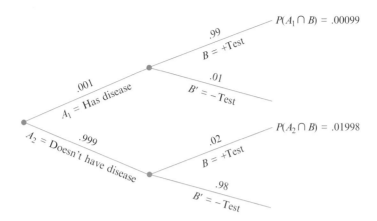

Fig. 1.13 Tree diagram for the rare-disease problem

Next to each branch corresponding to a positive test result, the Multiplication Rule yields the recorded probabilities. Therefore, $P(B) = .00099 + .01998 = .02097$, from which we have

$$P\left(A_1 \mid B\right) = \frac{P(A_1 \cap B)}{P(B)} = \frac{.00099}{.02097} = .047$$

This result seems counterintuitive; because the diagnostic test appears so accurate, we expect someone with a positive test result to be highly likely to have the disease, whereas the computed conditional probability is only .047. However, because the disease is rare and the test only moderately reliable, most positive test results arise from errors rather than from diseased individuals. The probability of having the disease has increased by a multiplicative factor of 47 (from prior .001 to posterior .047); but to get a further increase in the posterior probability, a diagnostic test with much smaller error rates is needed. If the disease were not so rare (e.g., 25% incidence in the population), then the error rates for the present test would provide good diagnoses.

This example shows why it makes sense to be tested for a rare disease only if you are in a high-risk group. For example, most of us are at low risk for HIV infection, so testing would not be indicated, but those who are in a high-risk group should be tested for HIV. For some diseases the degree of risk is strongly influenced by age. Young women are at low risk for breast cancer and should not be tested, but older women do have increased risk and need to be tested. There is some argument about where to draw the line. If we can find the incidence rate for our group and the sensitivity and specificity for the test, then we can do our own calculation to see if a positive test result would be informative.   ■

1.4.4   Exercises: Section 1.4 (50–78)

50. The population of a particular country consists of three ethnic groups. Each individual belongs to one of the four major blood groups. The accompanying *joint probability table* gives the proportions of individuals in the various ethnic group–blood group combinations.

Ethnic group	Blood group			
	O	A	B	AB
1	.082	.106	.008	.004
2	.135	.141	.018	.006
3	.215	.200	.065	.020

Suppose that an individual is randomly selected from the population, and define events by $A = \{$type A selected$\}$, $B = \{$type B selected$\}$, and $C = \{$ethnic group 3 selected$\}$.
(a) Calculate $P(A)$, $P(C)$, and $P(A \cap C)$.
(b) Calculate both $P(A|C)$ and $P(C|A)$ and explain in context what each of these probabilities represents.
(c) If the selected individual does not have type B blood, what is the probability that he or she is from ethnic group 1?

51. Suppose an individual is randomly selected from the population of all adult males living in the USA. Let A be the event that the selected individual is over 6 ft in height, and let B be the event that the selected individual is a professional basketball player. Which do you think is larger, $P(A|B)$ or $P(B|A)$? Why?

52. Return to the credit card scenario of Exercise 14, where $A = \{$Visa$\}$, $B = \{$MasterCard$\}$, $P(A) = .5$, $P(B) = .4$, and $P(A \cap B) = .25$. Calculate and interpret each of the following probabilities (a Venn diagram might help).
(a) $P(B|A)$
(b) $P(B'|A)$
(c) $P(A|B)$
(d) $P(A'|B)$
(e) Given that the selected individual has at least one card, what is the probability that he or she has a Visa card?

53. Reconsider the system defect situation described in Exercise 28.
(a) Given that the system has a type 1 defect, what is the probability that it has a type 2 defect?
(b) Given that the system has a type 1 defect, what is the probability that it has all three types of defects?
(c) Given that the system has at least one type of defect, what is the probability that it has exactly one type of defect?
(d) Given that the system has both of the first two types of defects, what is the probability that it does not have the third type of defect?

54. The accompanying table gives information on the type of coffee selected by someone purchasing a single cup at a particular airport kiosk.

	Small	Medium	Large
Regular	14%	20%	26%
Decaf	20%	10%	10%

   Consider randomly selecting such a coffee purchaser.
   (a) What is the probability that the individual purchased a small cup? A cup of decaf coffee?
   (b) If we learn that the selected individual purchased a small cup, what now is the probability that s/he chose decaf coffee, and how would you interpret this probability?
   (c) If we learn that the selected individual purchased decaf, what now is the probability that a small size was selected, and yow does this compare to the corresponding unconditional probability from (a)?

55. A department store sells sport shirts in three sizes (small, medium, and large), three patterns (plaid, print, and stripe), and two sleeve lengths (long and short). The accompanying tables give the proportions of shirts sold in the various category combinations.

   Short-sleeved

	Pattern		
Size	Plaid	Print	Stripe
S	.04	.02	.05
M	.08	.07	.12
L	.03	.07	.08

   Long-sleeved

	Pattern		
Size	Plaid	Print	Stripe
S	.03	.02	.03
M	.10	.05	.07
L	.04	.02	.08

   (a) What is the probability that the next shirt sold is a medium, long-sleeved, print shirt?
   (b) What is the probability that the next shirt sold is a medium print shirt?
   (c) What is the probability that the next shirt sold is a short-sleeved shirt? A long-sleeved shirt?
   (d) What is the probability that the size of the next shirt sold is medium? That the pattern of the next shirt sold is a print?

(e) Given that the shirt just sold was a short-sleeved plaid, what is the probability that its size was medium?

(f) Given that the shirt just sold was a medium plaid, what is the probability that it was short-sleeved? Long-sleeved?

56. One box contains six red balls and four green balls, and a second box contains seven red balls and three green balls. A ball is randomly chosen from the first box and placed in the second box. Then a ball is randomly selected from the second box and placed in the first box.

(a) What is the probability that a red ball is selected from the first box and a red ball is selected from the second box?

(b) At the conclusion of the selection process, what is the probability that the numbers of red and green balls in the first box are identical to the numbers at the beginning?

57. A system consists of two identical pumps, #1 and #2. If one pump fails, the system will still operate. However, because of the added strain, the extra remaining pump is now more likely to fail than was originally the case. That is, $r = P(\#2 \text{ fails} \mid \#1 \text{ fails}) > P(\#2 \text{ fails}) = q$. If at least one pump fails by the end of the pump design life in 7% of all systems and both pumps fail during that period in only 1%, what is the probability that pump #1 will fail during the pump design life?

58. A certain shop repairs both audio and video components. Let A denote the event that the next component brought in for repair is an audio component, and let B be the event that the next component is a compact disc player (so the event B is contained in A). Suppose that $P(A) = .6$ and $P(B) = .05$. What is $P(B|A)$?

59. In Exercise 15, $A_i = \{\text{awarded project } i\}$, for $i = 1, 2, 3$. Use the probabilities given there to compute the following probabilities, and explain in words the meaning of each one.

(a) $P(A_2|A_1)$

(b) $P(A_2 \cap A_3|A_1)$

(c) $P(A_2 \cup A_3|A_1)$

(d) $P(A_1 \cap A_2 \cap A_3|A_1 \cup A_2 \cup A_3)$

60. Three plants manufacture hard drives and ship them to a warehouse for distribution. Plant I produces 54% of the warehouse's inventory with a 4% defect rate. Plant II produces 35% of the warehouse's inventory with an 8% defect rate. Plant III produces the remainder of the warehouse's inventory with a 12% defect rate.

(a) Draw a tree diagram to represent this information.

(b) A warehouse inspector selects one hard drive at random. What is the probability that it is a defective hard drive and from Plant II?

(c) What is the probability that a randomly selected hard drive is defective?

(d) Suppose a hard drive is defective. What is the probability that it came from Plant II?

61. For any events A and B with $P(B) > 0$, show that $P(A|B) + P(A'|B) = 1$.

62. If $P(B|A) > P(B)$ show that $P(B'|A) < P(B')$. [*Hint*: Add $P(B'|A)$ to both sides of the given inequality and then use the result of the previous exercise.]

63. Show that for any three events A, B, and C with $P(C) > 0$, $P(A \cup B|C) = P(A|C) + P(B|C) - P(A \cap B|C)$.

64. At a certain gas station, 40% of the customers use regular gas (A_1), 35% use mid-grade gas (A_2), and 25% use premium gas (A_3). Of those customers using regular gas, only 30% fill their tanks (event B). Of those customers using mid-grade gas, 60% fill their tanks, whereas of those using premium, 50% fill their tanks.
    (a) What is the probability that the next customer will request mid-grade gas and fill the tank ($A_2 \cap B$)?
    (b) What is the probability that the next customer fills the tank?
    (c) If the next customer fills the tank, what is the probability that regular gas is requested? mid-grade gas? Premium gas?

65. Suppose a single gene controls the color of hamsters: black (B) is dominant and brown (b) is recessive. Hence, a hamster will be black unless its genotype is bb. Two hamsters, each with genotype Bb, mate and produce a single offspring. The laws of genetic recombination state that each parent is equally likely to donate either of its two alleles (B or b), so the offspring is equally likely to be any of BB, Bb, bB, or bb (the middle two are genetically equivalent).
    (a) What is the probability their offspring has black fur?
    (b) Given that their offspring has black fur, what is the probability its genotype is Bb?

66. Refer back to the scenario of the previous exercise. In the figure below, the genotypes of both members of Generation I are known, as is the genotype of the male member of Generation II. We know that hamster II2 must be black-colored thanks to her father, but suppose that we don't know her genotype exactly (as indicated by $B-$ in the figure).

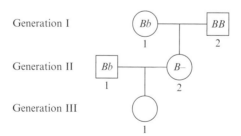

(a) What are the possible genotypes of hamster II2, and what are the corresponding probabilities?
(b) If we observe that hamster III1 has a black coat (and hence at least one B gene), what is the probability her genotype is Bb?

(c) If we later discover (through DNA testing on poor little hamster III1) that her genotype in *BB*, what is the posterior probability that her mom is also *BB*?

67. Seventy percent of the light aircraft that disappear while in flight in a certain country are subsequently discovered. Of the aircraft that are discovered, 60% have an emergency locator, whereas 90% of the aircraft not discovered do not have such a locator. Suppose a light aircraft has disappeared.

(a) If it has an emergency locator, what is the probability that it will not be discovered?

(b) If it does not have an emergency locator, what is the probability that it will be discovered?

68. Components of a certain type are shipped to a supplier in batches of ten. Suppose that 50% of all such batches contain no defective components, 30% contain one defective component, and 20% contain two defective components. Two components from a batch are randomly selected and tested. What are the probabilities associated with 0, 1, and 2 defective components being in the batch under each of the following conditions?

(a) Neither tested component is defective.

(b) One of the two tested components is defective.

[*Hint*: Draw a tree diagram with three first-generation branches for the three different types of batches.]

69. Show that $P(A \cap B|C) = P(A|B \cap C) \cdot P(B|C)$.

70. For customers purchasing a full set of tires at a particular tire store, consider the events

$A = \{$tires purchased were made in the USA$\}$
$B = \{$purchaser has tires balanced immediately$\}$
$C = \{$purchaser requests front-end alignment$\}$

along with A', B', and C'. Assume the following unconditional and conditional probabilities:

$$P(A) = .75 \qquad P(B|A) = .9 \qquad P(B|A') = .8 \qquad P(C|A \cap B) = .8$$
$$P(C|A \cap B') = .6 \qquad P(C|A' \cap B) = .7 \qquad P(C|A' \cap B') = .3$$

(a) Construct a tree diagram consisting of first-, second-, and third-generation branches and place an event label and appropriate probability next to each branch.

(b) Compute $P(A \cap B \cap C)$.

(c) Compute $P(B \cap C)$.

(d) Compute $P(C)$.

(e) Compute $P(A|B \cap C)$, the probability of a purchase of US tires given that both balancing and an alignment were requested.

71. A professional organization (for statisticians, of course) sells term life insurance and major medical insurance. Of those who have just life insurance, 70% will renew next year, and 80% of those with only a major medical policy will renew next year. However, 90% of policyholders who have both types of policy will renew at least one of them next year. Of the policy holders, 75% have term life insurance, 45% have major medical, and 20% have both.
    (a) Calculate the percentage of policyholders that will renew at least one policy next year.
    (b) If a randomly selected policy holder does in fact renew next year, what is the probability that he or she has both life and major medical insurance?

72. The Reviews editor for a certain scientific journal decides whether the review for any particular book should be short (1–2 pages), medium (3–4 pages) or long (5–6 pages). Data on recent reviews indicate that 60% of them are short, 30% are medium, and the other 10% are long. Reviews are submitted in either Word or LaTeX. For short reviews, 80% are in Word, whereas 50% of medium reviews and 30% of long reviews are in Word. Suppose a recent review is randomly selected.
    (a) What is the probability that the selected review was submitted in Word?
    (b) If the selected review was submitted in Word, what are the posterior probabilities of it being short, medium, and long?

73. A large operator of timeshare complexes requires anyone interested in making a purchase to first visit the site of interest. Historical data indicates that 20% of all potential purchasers select a day visit, 50% choose a one-night visit, and 30% opt for a two-night visit. In addition, 10% of day visitors ultimately make a purchase, 30% of night visitors buy a unit, and 20% of those visiting for two nights decide to buy. Suppose a visitor is randomly selected and found to have bought a timeshare. How likely is it that this person made a day visit? A one-night visit? A two-night visit?

74. Consider the following information about travelers (based partly on a recent Travelocity poll): 40% check work e-mail, 30% use a cell phone to stay connected to work, 25% bring a laptop with them, 23% both check work e-mail and use a cell phone to stay connected, and 51% neither check work e-mail nor use a cell phone to stay connected nor bring a laptop. Finally, 88 out of every 100 who bring a laptop check work e-mail, and 70 out of every 100 who use a cell phone to stay connected also bring a laptop.
    (a) What is the probability that a randomly selected traveler who checks work e-mail also uses a cell phone to stay connected?
    (b) What is the probability that someone who brings a laptop on vacation also uses a cell phone to stay connected?
    (c) If a randomly selected traveler checked work e-mail and brought a laptop, what is the probability that s/he uses a cell phone to stay connected?

75. There has been a great deal of controversy over the last several years regarding what types of surveillance are appropriate to prevent terrorism. Suppose a particular surveillance system has a 99% chance of correctly identifying a future terrorist and a 99.9% chance of correctly identifying someone who is not a future terrorist. Imagine there are 1000 future terrorists in a population of

300 million (roughly the US population). If one of these 300 million people is randomly selected and the system determines him/her to be a future terrorist, what is the probability the system is correct? Does your answer make you uneasy about using the surveillance system? Explain.

76. At a large university, in the never-ending quest for a satisfactory textbook, the Statistics Department has tried a different text during each of the last three quarters. During the fall quarter, 500 students used the text by Professor Mean; during the winter quarter, 300 students used the text by Professor Median; and during the spring quarter, 200 students used the text by Professor Mode. A survey at the end of each quarter showed that 200 students were satisfied with Mean's book, 150 were satisfied with Median's book, and 160 were satisfied with Mode's book. If a student who took statistics during one of these quarters is selected at random and admits to having been satisfied with the text, is the student most likely to have used the book by Mean, Median, or Mode? Who is the least likely author? [*Hint*: Draw a tree-diagram or use Bayes' theorem.]

77. A friend who lives in Los Angeles makes frequent consulting trips to Washington, D.C.; 50% of the time she travels on airline #1, 30% of the time on airline #2, and the remaining 20% of the time on airline #3. For airline #1, flights are late into D.C. 30% of the time and late into L.A. 10% of the time. For airline #2, these percentages are 25% and 20%, whereas for airline #3 the percentages are 40% and 25%. If we learn that on a particular trip she arrived late at exactly one of the two destinations, what are the posterior probabilities of having flown on airlines #1, #2, and #3? Assume that the chance of a late arrival in L.A. is unaffected by what happens on the flight to D.C. [*Hint*: From the tip of each first-generation branch on a tree diagram, draw three second-generation branches labeled, respectively, 0 late, 1 late, and 2 late.]

78. In Exercise 64, consider the following additional information on credit card usage:
   70% of all regular fill-up customers use a credit card.
   50% of all regular non-fill-up customers use a credit card.
   60% of all mid-grade fill-up customers use a credit card.
   50% of all mid-grade non-fill-up customers use a credit card.
   50% of all premium fill-up customers use a credit card.
   40% of all premium non-fill-up customers use a credit card.

   Compute the probability of each of the following events for the next customer to arrive (a tree diagram might help).
   (a) {mid-grade and fill-up and credit card}
   (b) {premium and non-fill-up and credit card}
   (c) {premium and credit card}
   (d) {fill-up and credit card}
   (e) {credit card}
   (f) If the next customer uses a credit card, what is the probability that s/he purchased premium gasoline?

1.5   Independence

The definition of conditional probability enables us to revise the probability $P(A)$ originally assigned to A when we are subsequently informed that another event B has occurred; the new probability of A is $P(A|B)$. In our examples, it was frequently the case that $P(A|B)$ was unequal to the unconditional probability $P(A)$, indicating that the information "B has occurred" resulted in a change in the chance of A occurring. There are other situations, however, in which the chance that A will occur or has occurred is not affected by knowledge that B has occurred, so that $P(A|B) = P(A)$. It is then natural to think of A and B as *independent* events, meaning that the occurrence or nonoccurrence of one event has no bearing on the chance that the other will occur.

DEFINITION

Two events A and B are **independent** if $P(A|B) = P(A)$ and are **dependent** otherwise.

The definition of independence might seem "unsymmetrical" because we do not demand that $P(B|A) = P(B)$ also. However, using the definition of conditional probability and the Multiplication Rule,

$$P(B \mid A) = \frac{P(A \cap B)}{P(A)} = \frac{P(A \mid B)P(B)}{P(A)} \qquad (1.6)$$

The right-hand side of Eq. (1.6) is $P(B)$ if and only if $P(A|B) = P(A)$ (independence), so the equality in the definition implies the other equality (and vice versa). It is also straightforward to show that if A and B are independent, then so are the following pairs of events: (1) A' and B, (2) A and B', and (3) A' and B'. See Exercise 82.

Example 1.34 Consider an ordinary deck of 52 cards comprising the four suits spades, hearts, diamonds, and clubs, with each suit consisting of the 13 ranks ace, king, queen, jack, ten, . . ., and two. Suppose someone randomly selects a card from the deck and reveals to you that it is a picture card (that is, a king, queen, or jack). What now is the probability that the card is a spade? If we let $A = \{\text{spade}\}$ and $B = \{\text{face card}\}$, then $P(A) = 13/52$, $P(B) = 12/52$ (there are three face cards in each of the four suits), and $P(A \cap B) = P(\text{spade and face card}) = 3/52$. Thus

$$P(A \mid B) = \frac{P(A \cap B)}{P(B)} = \frac{3/52}{12/52} = \frac{3}{12} = \frac{1}{4} = \frac{13}{52} = P(A)$$

Therefore, the likelihood of getting a spade is not affected by knowledge that a face card had been selected. Intuitively this is because the fraction of spades among face cards (3 out of 12) is the same as the fraction of spades in the entire deck

(13 out of 52). It is also easily verified that $P(B|A) = P(B)$, so knowledge that a spade has been selected does not affect the likelihood of the card being a jack, queen, or king. ∎

Example 1.35 Consider a gas station with six pumps numbered 1, 2, ..., 6 and let E_i denote the simple event that a randomly selected customer uses pump i. Suppose that

$$P(E_1) = P(E_6) = .10, \quad P(E_2) = P(E_5) = .15, \quad P(E_3) = P(E_4) = .25$$

Define events A, B, C by

$$A = \{2, 4, 6\}, B = \{1, 2, 3\}, C = \{2, 3, 4, 5\}$$

It is easy to determine that $P(A) = .50$, $P(A|B) = .30$, and $P(A|C) = .50$. Therefore, events A and B are dependent, whereas events A and C are independent. Intuitively, A and C are independent because the relative division of probability among even- and odd-numbered pumps is the same among pumps 2, 3, 4, 5 as it is among all six pumps. ∎

Example 1.36 Let A and B be any two mutually exclusive events with $P(A) > 0$. For example, for a randomly chosen automobile, let $A = \{$car is blue$\}$ and $B = \{$car is red$\}$. Since the events are mutually exclusive, if B occurs, then A cannot possibly have occurred, so $P(A|B) = 0 \neq P(A)$. The message here is that *if two events are mutually exclusive, they cannot be independent.* When A and B are mutually exclusive, the information that A occurred says something about the chance of B (namely, it cannot have occurred), so independence is precluded. ∎

1.5.1 $P(A \cap B)$ When Events Are Independent

Frequently the nature of an experiment suggests that two events A and B should be assumed independent. This is the case, for example, if a manufacturer receives a circuit board from each of two different suppliers, each board is tested on arrival, and $A = \{$first is defective$\}$ and $B = \{$second is defective$\}$. If $P(A) = .1$, it should also be the case that $P(A|B) = .1$; knowing the condition of the second board shouldn't provide information about the condition of the first. Our next result shows how to compute $P(A \cap B)$ when the events are independent.

PROPOSITION

A and B are independent if and only if

$$P(A \cap B) = P(A) \cdot P(B) \tag{1.7}$$

Proof By the Multiplication Rule, $P(A \cap B) = P(A|B) \cdot P(B)$, and this equals $P(A) \cdot P(B)$ if and only if $P(A|B) = P(A)$. ∎

Because of the equivalence of independence with Eq. (1.7), the latter can be used as a definition of independence.[1]

Example 1.37 It is known that 30% of a certain company's washing machines require service while under warranty, whereas only 10% of its dryers need such service. If someone purchases both a washer and a dryer made by this company, what is the probability that both machines need warranty service?

Let A denote the event that the washer needs service while under warranty, and let B be defined analogously for the dryer. Then $P(A) = .30$ and $P(B) = .10$. Assuming that the two machines function independently of each other, the desired probability is

$$P(A \cap B) = P(A) \cdot P(B) = (.30)(.10) = .03$$

The probability that neither machine needs service is

$$P(A' \cap B') = P(A') \cdot P(B') = (.70)(.90) = .63$$

Note that, although the independence assumption is reasonable here, it can be questioned. In particular, if heavy usage causes a breakdown in one machine, it could also cause trouble for the other one. ∎

Example 1.38 Each day, Monday through Friday, a batch of components sent by a first supplier arrives at a certain inspection facility. Two days a week, a batch also arrives from a second supplier. Eighty percent of all supplier 1's batches pass inspection, and 90% of supplier 2's do likewise. What is the probability that, on a randomly selected day, two batches pass inspection? We will answer this assuming that on days when two batches are tested, whether the first batch passes is independent of whether the second batch does so. Figure 1.14 displays the relevant information.

---

[1] However, the multiplication property is satisfied if $P(B) = 0$, yet $P(A|B)$ is not defined in this case. To make the multiplication property completely equivalent to the definition of independence, we should append to that definition that A and B are also independent if either $P(A) = 0$ or $P(B) = 0$.

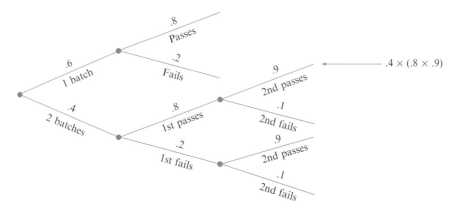

Fig. 1.14   Tree diagram for Example 1.38

$$P(\text{two pass}) = P(\text{two received} \cap \text{both pass})$$
$$= P(\text{both pass}|\text{two received}) \cdot P(\text{two received})$$
$$= [(.8)(.9)](.4) = .288 \qquad \blacksquare$$

1.5.2   Independence of More than Two Events

The notion of independence can be extended to collections of more than two events. Although it is possible to extend the definition for two independent events by working in terms of conditional and unconditional probabilities, it is more direct and less cumbersome to proceed along the lines of the last proposition.

> **DEFINITION**
>
> Events $A_1, \ldots, A_n$ are **mutually independent** if for every k ($k = 2, 3, \ldots, n$) and every subset of indices $i_1, i_2, \ldots, i_k$,
>
> $$P(A_{i_1} \cap A_{i_2} \cap \cdots \cap A_{i_k}) = P(A_{i_1}) \cdot P(A_{i_2}) \cdot \cdots \cdot P(A_{i_k})$$

To paraphrase the definition, the events are mutually independent if the probability of the intersection of any subset of the n events is equal to the product of the individual probabilities. In using this multiplication property for more than two independent events, it is legitimate to replace one or more of the A_is by their complements (e.g., if A_1, A_2, and A_3 are independent events, then so are A_1', A_2', and A_3'.) As was the case with two events, we frequently specify at the outset of a problem the independence of certain events. The definition can then be used to calculate the probability of an intersection.

Example 1.39 The article "Reliability Evaluation of Solar Photovoltaic Arrays" (*Solar Energy*, 2002: 129–141) presents various configurations of solar photovoltaic arrays consisting of crystalline silicon solar cells. Consider first the system illustrated in Fig. 1.15a. There are two subsystems connected in parallel, each one containing three cells. In order for the system to function, at least one of the two parallel subsystems must work. Within each subsystem, the three cells are connected in series, so a subsystem will work only if all cells in the subsystem work. Consider a particular lifetime value t_0, and suppose we want to determine the probability that the system lifetime exceeds t_0. Let A_i denote the event that the lifetime of cell i exceeds t_0 ($i = 1, 2, \ldots, 6$). We assume that the A_is are independent events (whether any particular cell lasts more than t_0 hours has no bearing on whether any other cell does) and that $P(A_i) = .9$ for every i since the cells are identical. Then applying the Addition Rule followed by independence,

$$
\begin{aligned}
P(\text{system lifetime exceeds } t_0) &= P\big[(A_1 \cap A_2 \cap A_3) \cup (A_4 \cap A_5 \cap A_6)\big] \\
&= P(A_1 \cap A_2 \cap A_3) + P(A_4 \cap A_5 \cap A_6) \\
&\quad - P\big[(A_1 \cap A_2 \cap A_3) \cap (A_4 \cap A_5 \cap A_6)\big] \\
&= (.9)(.9)(.9) + (.9)(.9)(.9) \\
&\quad - (.9)(.9)(.9)(.9)(.9)(.9) \\
&= .927
\end{aligned}
$$

Alternatively,

$$
\begin{aligned}
P(\text{system lifetime exceeds } t_0) &= 1 - P(\text{both subsystem lives are } \leq t_0) \\
&= 1 - [P(\text{subsystem life is } \leq t_0)]^2 \\
&= 1 - [1 - P(\text{subsystem life is } > t_0)]^2 \\
&= 1 - \left[1 - (.9)^3\right]^2 = .927
\end{aligned}
$$

Next consider the total-cross-tied system shown in Fig. 1.15b, obtained from the series–parallel array by connecting ties across each column of junctions. Now the system fails as soon as an entire column fails, and system lifetime exceeds t_0 only if the life of every column does so. For this configuration,

$$
\begin{aligned}
P(\text{system lifetime exceeds } t_0) &= [P(\text{column lifetime exceeds } t_0)]^3 \\
&= [1 - P(\text{column lifetime is } \leq t_0)]^3 \\
&= [1 - P(\text{both cells in a column have lifetime } \leq t_0)]^3 \\
&= 1 - \left[1 - (.9)^2\right]^3 = .970
\end{aligned}
$$

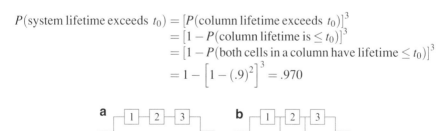

Fig. 1.15 System configurations for Example 1.39: (**a**) series–parallel; (**b**) total-cross-tied ■

Probabilities like those calculated in Example 1.39 are often referred to as the *reliability* of a system. In Sect. 4.8, we consider in more detail the analysis of system reliability.

1.5.3   Exercises: Section 1.5 (79–100)

79. Reconsider the credit card scenario of Exercise 52, and show that A and B are dependent first by using the definition of independence and then by verifying that the multiplication property does not hold.
80. An oil exploration company currently has two active projects, one in Asia and the other in Europe. Let A be the event that the Asian project is successful and B be the event that the European project is successful. Suppose that A and B are independent events with $P(A) = .4$ and $P(B) = .7$.
    (a) If the Asian project is not successful, what is the probability that the European project is also not successful? Explain your reasoning.
    (b) What is the probability that at least one of the two projects will be successful?
    (c) Given that at least one of the two projects is successful, what is the probability that only the Asian project is successful?
81. In Exercise 15, is any A_i independent of any other A_j? Answer using the multiplication property for independent events.
82. If A and B are independent events, show that A' and B are also independent. [*Hint*: First use a Venn diagram to establish a relationship among $P(A' \cap B)$, $P(B)$, and $P(A \cap B)$.]
83. Suppose that the proportions of blood phenotypes in a particular population are as follows:

A	B	AB	O
.40	.11	.04	.45

    Assuming that the phenotypes of two randomly selected individuals are independent of each other, what is the probability that both phenotypes are O? What is the probability that the phenotypes of two randomly selected individuals match?
84. The probability that a grader will make a marking error on any particular question of a multiple-choice exam is .1. If there are ten questions on the exam and questions are marked independently, what is the probability that no errors are made? That at least one error is made? If there are n questions on the exam and the probability of a marking error is p rather than .1, give expressions for these two probabilities.
85. In October, 1994, a flaw in a certain Pentium chip installed in computers was discovered that could result in a wrong answer when performing a division. The manufacturer initially claimed that the chance of any particular division being incorrect was only 1 in 9 billion, so that it would take thousands of years

before a typical user encountered a mistake. However, statisticians are not typical users; some modern statistical techniques are so computationally intensive that a billion divisions over a short time period is not unrealistic. Assuming that the 1 in 9 billion figure is correct and that results of divisions are independent from one another, what is the probability that at least one error occurs in 1 billion divisions with this chip?

86. An aircraft seam requires 25 rivets. The seam will have to be reworked if any of these rivets is defective. Suppose rivets are defective independently of one another, each with the same probability.
    (a) If 20% of all seams need reworking, what is the probability that a rivet is defective?
    (b) How small should the probability of a defective rivet be to ensure that only 10% of all seams need reworking?

87. A boiler has five identical relief valves. The probability that any particular valve will open on demand is .95. Assuming independent operation of the valves, calculate P(at least one valve opens) and P(at least one valve fails to open).

88. Two pumps connected in parallel fail independently of each other on any given day. The probability that only the older pump will fail is .10, and the probability that only the newer pump will fail is .05. What is the probability that the pumping system will fail on any given day (which happens if both pumps fail)?

89. Consider the system of components connected as in the accompanying picture. Components 1 and 2 are connected in parallel, so that subsystem works iff either 1 or 2 works; since 3 and 4 are connected in series, that subsystem works iff both 3 and 4 work. If components work independently of one another and P(component works) $= .9$, calculate P(system works).

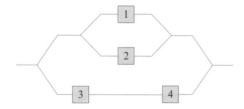

90. Refer back to the series–parallel system configuration introduced in Example 1.39, and suppose that there are only two cells rather than three in each parallel subsystem [in Fig. 1.15a, eliminate cells 3 and 6, and renumber cells 4 and 5 as 3 and 4]. Using $P(A_i) = .9$, the probability that system lifetime exceeds t_0 is easily seen to be .9639. To what value would .9 have to be changed in order to increase the system lifetime reliability from .9639 to .99? [*Hint*: Let $P(A_i) = p$, express system reliability in terms of p, and then let $x = p^2$.]

91. Consider independently rolling two fair dice, one red and the other green. Let A be the event that the red die shows 3 dots, B be the event that the green die

shows 4 dots, and C be the event that the total number of dots showing on the two dice is 7.

(a) Are these events pairwise independent (i.e., are A and B independent events, are A and C independent, and are B and C independent)?

(b) Are the three events mutually independent?

92. Components arriving at a distributor are checked for defects by two different inspectors (each component is checked by both inspectors). The first inspector detects 90% of all defectives that are present, and the second inspector does likewise. At least one inspector fails to detect a defect on 20% of all defective components. What is the probability that the following occur?

(a) A defective component will be detected only by the first inspector? By exactly one of the two inspectors?

(b) All three defective components in a batch escape detection by both inspectors (assuming inspections of different components are independent of one another)?

93. Seventy percent of all vehicles examined at a certain emissions inspection station pass the inspection. Assuming that successive vehicles pass or fail independently of one another, calculate the following probabilities:

(a) P(all of the next three vehicles inspected pass)

(b) P(at least one of the next three inspected fails)

(c) P(exactly one of the next three inspected passes)

(d) P(at most one of the next three vehicles inspected passes)

(e) Given that at least one of the next three vehicles passes inspection, what is the probability that all three pass (a conditional probability)?

94. A quality control inspector is inspecting newly produced items for faults. The inspector searches an item for faults in a series of independent fixations, each of a fixed duration. Given that a flaw is actually present, let p denote the probability that the flaw is detected during any one fixation (this model is discussed in "Human Performance in Sampling Inspection," *Human Factors*, 1979: 99–105).

(a) Assuming that an item has a flaw, what is the probability that it is detected by the end of the second fixation (once a flaw has been detected, the sequence of fixations terminates)?

(b) Give an expression for the probability that a flaw will be detected by the end of the nth fixation.

(c) If when a flaw has not been detected in three fixations, the item is passed, what is the probability that a flawed item will pass inspection?

(d) Suppose 10% of all items contain a flaw [P(randomly chosen item is flawed) $= .1$]. With the assumption of part (c), what is the probability that a randomly chosen item will pass inspection (it will automatically pass if it is not flawed, but could also pass if it is flawed)?

(e) Given that an item has passed inspection (no flaws in three fixations), what is the probability that it is actually flawed? Calculate for $p = .5$.

95. (a) A lumber company has just taken delivery on a lot of 10,000 2 × 4 boards. Suppose that 20% of these boards (2000) are actually too green to be

used in first-quality construction. Two boards are selected at random, one after the other. Let $A = \{$the first board is green$\}$ and $B = \{$the second board is green$\}$. Compute $P(A)$, $P(B)$, and $P(A \cap B)$ (a tree diagram might help). Are A and B independent?

(b) With A and B independent and $P(A) = P(B) = .2$, what is $P(A \cap B)$? How much difference is there between this answer and $P(A \cap B)$ in part (a)? For purposes of calculating $P(A \cap B)$, can we assume that A and B of part (a) are independent to obtain essentially the correct probability?

(c) Suppose the lot consists of ten boards, of which two are green. Does the assumption of independence now yield approximately the correct answer for $P(A \cap B)$? What is the critical difference between the situation here and that of part (a)? When do you think that an independence assumption would be valid in obtaining an approximately correct answer to $P(A \cap B)$?

96. Refer to the assumptions stated in Exercise 89 and answer the question posed there for the system in the accompanying picture. How would the probability change if this were a subsystem connected in parallel to the subsystem pictured in Fig. 1.15a?

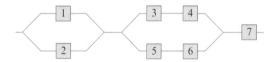

97. Professor Stander Deviation can take one of two routes on his way home from work. On the first route, there are four railroad crossings. The probability that he will be stopped by a train at any particular one of the crossings is .1, and trains operate independently at the four crossings. The other route is longer but there are only two crossings, independent of each other, with the same stoppage probability for each as on the first route. On a particular day, Professor Deviation has a meeting scheduled at home for a certain time. Whichever route he takes, he calculates that he will be late if he is stopped by trains at at least half the crossings encountered.

(a) Which route should he take to minimize the probability of being late to the meeting?

(b) If he tosses a fair coin to decide on a route and he is late, what is the probability that he took the four-crossing route?

98. For a customer who test drives three vehicles, define events $A_i = $ customer likes vehicle #i for $i = 1$, 2, 3. Suppose that $P(A_1) = .55$, $P(A_2) = .65$, $P(A_3) = .70$, $P(A_1 \cup A_2) = .80$, $P(A_2 \cap A_3) = .40$, and $P(A_1 \cup A_2 \cup A_3) = .88$.

(a) What is the probability that a customer likes both vehicle #1 and vehicle #2?

(b) Determine and interpret $P(A_2 | A_3)$.

(c) Are A_2 and A_3 independent events? Answer in two different ways.

(d) If you learn that the customer did not like vehicle #1, what now is the probability that s/he liked at least one of the other two vehicles?

99. It's a commonly held misconception that if you play the lottery n times, and the probability of winning each time is $1/N$, then your chance of winning at least once is n/N. That's true if you buy n tickets in 1 week, but not if you buy a single ticket in each of n independent weeks. Let's explore further.

   (a) Suppose you play a game n independent times, with $P(\text{win}) = 1/N$ each time. Find an expression for the probability you win at least once. [*Hint*: Consider the complement.]

   (b) How does your answer to (a) compare to n/N for the easy task of rolling a ⊡ on a fair die (so $1/N = 1/6$) in $n = 3$ tries? In $n = 6$ tries? In $n = 10$ tries?

   (c) How does your answer to (a) compare to n/N in the setting of Exercise 85: probability = 1 in 9 billion, number of tries = 1 billion?

   (d) Show that when n is much smaller than N, the fraction n/N is not a bad approximation to (a). [*Hint*: Use the binomial theorem from high school algebra.]

100. Suppose identical tags are placed on both the left ear and the right ear of a fox. The fox is then let loose for a period of time. Consider the two events $C_1 = \{\text{left ear tag is lost}\}$ and $C_2 = \{\text{right ear tag is lost}\}$. Let $p = P(C_1) = P(C_2)$, and assume C_1 and C_2 are independent events. Derive an expression (involving p) for the probability that exactly one tag is lost, given that at most one is lost ("Ear Tag Loss in Red Foxes," *J. Wildlife Manag.*, 1976: 164–167). [*Hint*: Draw a tree diagram in which the two initial branches refer to whether the left ear tag was lost.]

1.6   Simulation of Random Events

As probability models in engineering and the sciences have grown in complexity, many problems have arisen that are too difficult to attack "analytically," i.e., using mathematical tools such as those in the previous sections. Instead, computer simulation provides us an effective way to estimate probabilities of very complicated events (and, in later chapters, of other properties of random phenomena). Here we introduce the principles of probability simulation, demonstrate a few examples with Matlab and R code, and discuss the precision of simulated probabilities.

Suppose an investigator wishes to determine $P(A)$, but either the experiment on which A is defined or the A event itself is so complicated as to preclude the use of probability rules and properties. The general method for *estimating* this probability via simulation software is as follows:

– Write a program that simulates (mimics) the underlying random experiment.
– Run the program many times, with each run independent of all others.
– During each run, record whether or not the event A of interest occurs.

If the simulation is run a total of n independent times, then the estimate of $P(A)$, denoted by $\hat{P}(A)$, is

$$\hat{P}(A) = \frac{\text{number of times } A \text{ occurs}}{\text{number of runs}} = \frac{n(A)}{n}$$

For example, if we run a simulation program 10,000 times and the event of interest A occurs in 6174 of those runs, then our estimate of $P(A)$ is $\hat{P}(A) = 6174/10{,}000 = .6174$. Notice that our definition is consistent with the long-run relative frequency interpretation of probability discussed in Sect. 1.2.

1.6.1 The Backbone of Simulation: Random Number Generators

All modern software packages are equipped with a function called a **random number generator (RNG)**. A typical call to this function (such as ran or rand) will return a single, supposedly "random" number, though such functions typically permit the user to request a vector or even a matrix of "random" numbers. It is more proper to call these results pseudo-random numbers, since there is actually a deterministic (i.e., non-random) algorithm by which the software generates these values. We will not discuss the details of such algorithms here; see the book by Law listed in the references. What will matter to us are the following two characteristics:
1. Each number created by an RNG is as likely to be any particular number in the interval [0, 1) as it is to be any other number in this interval (up to computer precision, anyway).[2]
2. Successive values created by RNGs are independent, in the sense that we cannot predict the next value to be generated from the current value (unless we somehow know the exact parameters of the underlying algorithm).

A typical simulation program manipulates numbers on the interval [0, 1) in a way that mimics the experiment of interest; several examples are provided below. Arguably the most important building block for such programs is the ability to simulate a basic event that occurs with a *known* probability, p. Since RNGs produce values equally likely to be anywhere in the interval [0, 1), it follows that in the long run a proportion p of them will lie in the interval [0, p). So, suppose we need to simulate an event B with $P(B) = p$. In each run of our simulation program, we can call for a single "random" number, which we'll call u, and apply the following rules:

– If $0 \leq u < p$, then event B has occurred on this run of the program.
– If $p \leq u < 1$, then event B has *not* occurred on this run of the program.

---

[2] In the language of Chap. 3, the numbers produced by an RNG follow essentially a *uniform* distribution on the interval [0, 1).

Example 1.40 Let's begin with an example in which the exact probability can be obtained analytically, so that we may verify that our simulation method works. Suppose we have two independent devices which function with probabilities .6 and .7, respectively. What is the probability both devices function? That at least one device functions?

Let B_1 and B_2 denote the events that the first and second devices function, respectively; we know that $P(B_1) = .6$, $P(B_2) = .7$, and B_1 and B_2 are independent. Our first goal is to estimate the probability of $A = B_1 \cap B_2$, the event that both devices function. The following "pseudo-code" will allow us to find $\hat{P}(A)$.

0. Set a counter for the number of times A occurs to zero.

    Repeat n times:

1. Generate two random numbers, u_1 and u_2. (These will help us determine whether B_1 and B_2 occur, respectively.)

2. If $u_1 < .6$ AND $u_2 < .7$, then A has occurred. Add 1 to the count of occurrences of A.

    Once the n runs are complete, then $\hat{P}(A) = (\text{count of the occurrences of } A)/n$.

Figure 1.16 shows actual implementation code in both Matlab and R. We ran each program with $n = 10,000$ (as in the code); the event A occurred 4215 times in Matlab and 4181 times in R, providing estimated probabilities of $\hat{P}(A) = .4215$ and .4181, respectively. Compare this to the exact probability of A: by independence, $P(A) = P(B_1)P(B_2) = (.6)(.7) = .42$. Both of our simulation estimates were "in the ballpark" of the right answer. We'll discuss the precision of these estimates shortly.

By replacing the "and" operators && in Fig. 1.16 with "or" operators ||, we can estimate the probability at least one device functions, $P(B_1 \cup B_2)$. In one simulation (again with $n = 10,000$), the event $B_1 \cup B_2$ occurred 8,802 times, giving the estimate $\hat{P}(B_1 \cup B_2) = .8802$. This is quite close to the exact probability:

$$P(B_1 \cup B_2) = 1 - P\left(B_1' \cap B_2'\right) = 1 - (1 - .6)(1 - .7) = .88$$

a
```
A=0;
for i=1:10000
    u1=rand; u2=rand;
    if u1<.6 && u2<.7
        A=A+1;
    end
end
```

b
```
A<-0
for(i in 1:10000){
    u1<-runif(1); u2<-runif(1)
    if(u1<.6 && u2<.7){
        A<-A+1
    }
}
```

Fig. 1.16 Code for Example 1.40: (**a**) Matlab; (**b**) R   ∎

Example 1.41 Consider the following game: You'll flip a coin 25 times, winning $1 each time it lands heads (H) and losing $1 each time it lands tails (T). Unfortunately for you, the coin is biased in such a way that $P(H) = .4$ and $P(T) = .6$. What's the probability you come out ahead, i.e., you have more money at the end of the game than you had at the beginning? We'll use simulation to find out.

Now each run of the simulation requires 25 "random" objects: the results of the 25 coin tosses. What's more, we need to keep track of how much money we've won or lost at the end of the 25 tosses. Let $A = \{$we come out ahead$\}$, and use the following pseudo-code:

0. Set a counter for the number of times A occurs to zero.

Repeat n times:

1. Set your initial dollar amount to zero.
2. Generate 25 random numbers $u_1, \ldots, u_{25}$.
3. For each $u_i < .4$, heads was tossed, so add 1 to your dollar amount. For each $u_i \geq .4$, the flip was tails and 1 is deducted.
4. If the final dollar amount is positive (i.e., $1 or greater), add 1 to the count of occurrences for A.

Once the n runs are complete, then $\hat{P}(A) = (\text{count of the occurrences of } A)/n$.

Matlab and R code for Example 1.41 appear in Fig. 1.17. Our R code gave a final count of 1,567 occurrences of A, out of 10,000 runs. Thus, the estimated probability that we come out ahead in this game is $\hat{P}(A) = 1567/10{,}000 = .1567$.

a
```
A=0;
for i=1:10000
    dollar=0;
    for j=1:25
        u=rand;
        if u<.4
            dollar=dollar+1;
        else
            dollar=dollar-1;
        end
    end
    if dollar>0
        A=A+1;
    end
end
```

b
```
A <- 0
for (i in 1:10000){
    dollar<-0
    for (j in 1:25){
        u<-runif(1)
        if (u<.4){
            dollar<-dollar+1
        }
        else{dollar<-dollar-1}
    }
    if (dollar>0){
        A<-A+1
    }
}
```

Fig. 1.17 Code for Example 1.41: (**a**) Matlab; (**b**) R   ■

Throughout this textbook, we will illustrate repeated simulation through "for" loops, as in Figs. 1.16 and 1.17. Though this isn't necessarily the most efficient way to code these examples, we do so for clarity's sake. Readers familiar with basic programming may realize that such operations can be sped up by *vectorization*, i.e., by using a function call that generates all the required random numbers simultaneously, rather than one at a time. Similarly, the if/else statements used in the preceding programs to determine whether a random number lies in an interval can be rewritten in terms of true/false bits, which automatically generate a 1 if a statement is true and a 0 otherwise. For example, the Matlab code

```
if u<.5
  A=A+1;
end
```

can be replaced by the single line of code

```
A=A+(u<.5);
```

If the statement in parentheses is true, Matlab assigns a value 1 to (u<.5), and so 1 is added to the count A. Similar code works in R.

The previous two examples have both assumed independence of certain events: the functionality of neighboring devices, or the outcomes of successive coin flips. With the aid of some built-in packages within Matlab and R, we can also simulate counting experiments similar to those in Sect. 1.3, even though draws without replacement from a finite population are not independent. To illustrate, let's use simulation to estimate some of the combinatorial probabilities from Sect. 1.3.

Example 1.42 Consider again the situation presented in Example 1.24: A university warehouse has received a shipment of 25 printers, of which 10 are laser printers and 15 are inkjet models; a particular technician will check 6 of these 25 printers, selected at random. Of interest is the probability of the event $D_3 = \{$exactly 3 of the 6 selected are inkjet printers$\}$. Although the initial probability of selecting an inkjet printer is 15/25, successive selections are not independent (the conditional probability that the next printer is also an inkjet is *not* 15/25). So, the method of the preceding examples does not apply.

Instead, we use the sampling tool built into our software, as follows:

0. Set a counter for the number of times D_3 occurs to zero.
   Repeat n times:
1. Sample 6 numbers, *without replacement*, from the integers 1 through 25. (1–15 correspond to the labels for the 15 inkjet printers and 16–25 identify the 10 laser printers.)
2. Count how many of these 6 numbers fall between 1 and 15, inclusive.
3. If exactly 3 of these 6 numbers fall between 1 and 15, add 1 to the count of occurrences for D_3.
   Once the n runs are complete, then $\hat{P}(D_3) = $ (count of the occurrences of D_3)/n.

Matlab and R code for this example appear in Fig. 1.18. Vital to the execution of this simulation is the fact that both software packages have a built-in mechanism for randomly sampling without replacement from a finite set of objects (here, the integers 1–25). For more information on these functions, type help randsample in Matlab or help(sample) in R.

In both sets of code, the line sum(printers<=15) performs two actions. First, printers<=15 converts each of the 6 numbers in the vector printers into a 1 if the entry is between 1 and 15 (and into a 0 otherwise). Second, sum() adds up the 1s and 0s, which is equivalent to identifying how many 1s appear (i.e., how many of the 6 numbers fell between 1 and 15).

a
```
D=0;
for i=1:10000
    printers=randsample(25,6);
    inkjet=sum(printers<=15);
    if inkjet==3
        D=D+1;
    end
end
```

b
```
D<-0
for (i in 1:10000){
    printers<-sample(25,6)
    inkjet<-sum(printers<=15)
    if (inkjet==3){
        D<-D+1
    }
}
```

Fig. 1.18 Matlab and R code for Example 1.42

Our R code resulted in event D_3 occurring 3054 times, so $\hat{P}(D_3) = 3054/10{,}000 = .3054$, which is quite close to the "exact" answer of .3083 found in Example 1.24. The other probability of interest, the chance of randomly selecting *at least* 3 inkjet printers, can be estimated by modifying one line of code: change `inkjet==3` to `inkjet>=3`. One simulation provided a count of 8522 occurrences in 10,000 trials, for an estimated probability of .8522 (close to the combinatorial solution of .8530). ∎

1.6.2 Precision of Simulation

In Example 1.40, we gave two different estimates $\hat{P}(A)$ for a probability $P(A)$. Which is more "correct"? Without knowing $P(A)$ itself, there's no way to tell. However, thanks to the theory we will develop in subsequent chapters, we can quantify the precision of simulated probabilities. Of course, we must have written code that faithfully simulates the random experiment of interest. Further, we assume that the results of each single run of our program are independent of the results of all other runs. (This generally follows from the aforementioned independence of computer-generated random numbers.)

If this is the case, then a measure of the disparity between the true probability $P(A)$ and the estimated probability $\hat{P}(A)$ based on n runs of the simulation is given by:

$$\sqrt{\frac{\hat{P}(A)\left[1 - \hat{P}(A)\right]}{n}} \tag{1.8}$$

This measure of precision is called the **(estimated) standard error** of the estimate $\hat{P}(A)$; see Sect. 2.4 for a derivation. Expression (1.8) tells us that the amount by which $\hat{P}(A)$ typically differs from $P(A)$ depends upon two values: $\hat{P}(A)$ itself, and the number of runs n. You can make sense of the former this way: if $P(A)$ is very small, then $\hat{P}(A)$ will presumably be small as well, in which case they cannot deviate by very much since both are bounded below by zero. (Standard error quantifies the *absolute* difference between them, not the relative difference.) A similar comment applies if $P(A)$ is very large, i.e., near 1.

As for the relationship to n, Expression (1.8) indicates that the amount by which $\hat{P}(A)$ will typically differ from $P(A)$ is proportional to the reciprocal of the square root of n. So, in particular, as n increases the tendency is for $\hat{P}(A)$ to vary less and less. This speaks to the precision of $\hat{P}(A)$: our estimate becomes more precise as n increases, but not at a very fast rate.

Let's think a bit more about this relationship: suppose your simulation results thus far were too imprecise for your tastes. By how much would you have to increase the number of runs to gain one additional decimal place of precision? That's equivalent to reducing the estimated standard error by a factor of 10. Since precision is proportional to $1/\sqrt{n}$, you would need to increase n by a factor of 100 to achieve the desired improvement, e.g., if using $n = 10{,}000$ runs is insufficient for your purposes, then you'll need 1,000,000 runs to get one additional decimal place of precision. Typically, this will mean running your program 100 times longer—not a big deal if 10,000 runs only take a nanosecond but prohibitive if they require, say, an hour.

Example 1.43 (Example 1.41 continued) Based on $n = 10{,}000$ runs, we estimated the probability of coming out ahead in a certain game to be $\hat{P}(A) = .1567$. Substituting into Eq. (1.8), we get

$$\sqrt{\frac{.1567[1 - .1567]}{10{,}000}} = .0036$$

This is the (estimated) standard error of our estimate .1567. We interpret as follows: some simulation experiments with $n = 10{,}000$ will result in an estimated probability that is within .0036 of the actual probability, whereas other such experiments will give an estimated probability that deviates by more than .0036 from the actual $P(A)$; .0036 is roughly the size of a typical deviation between the estimate and what it is estimating. ∎

In Chap. 5, we will return to the notion of standard error and develop a so-called *confidence interval* estimate for $P(A)$: a range of numbers in which we can be very certain where $P(A)$ lies.

1.6.3   Exercises: Section 1.6 (101–120)

101. Refer to Example 1.40.
     (a) Modify the code in Fig. 1.16 to estimate the probability that *exactly* one of the two devices functions properly. Then find the exact probability using the techniques from earlier sections of this chapter, and compare it to your estimated probability.

(b) Calculate the estimated standard error for the estimated probability in (a).

102. Imagine you have five independently operating components, each working properly with probability .8. Use simulation to estimate the probability that

(a) All five components work properly.

(b) At least one of the five components works properly.

[Hints for (a) and (b): You can adapt the code from Example 1.40, but the and/or statements will become tedious. Consider using the max and min functions instead.]

(c) Calculate the estimated standard errors for your answers in (a) and (b).

103. Consider the system depicted in Exercise 96. Assume the seven components operate independently with the following probabilities of functioning properly: .9 for components 1 and 2; .8 for each of components 3, 4, 5, 6; and .95 for component 7. Write a program to estimate the reliability of the system (i.e., the probability the system functions properly).

104. You have an opportunity to answer six trivia questions about your favorite sports team, and you will win a pair of tickets to their next game if you can correctly answer at least three of the questions. Write a simulation program to estimate the chance you win the tickets under each of the following assumptions.

(a) You have a 50–50 chance of getting any question right, independent of all others.

(b) Being a true fan, you have a 75% chance of getting any question right, independent of all others.

(c) The first three questions are fairly easy, so you have a .75 chance of getting each of those right. However, the last three questions are much harder, and you only have a .3 probability of correctly answering each of those.

105. In the game "Now or Then" on the television show *The Price is Right*, the contestant faces a wheel with six sectors. Each sector contains a grocery item and a price, and the contestant must decide whether the price is "now" (i.e., the item's price the day of the taping) or "then" (the price at some specified past date, such as September 2003). The contestant wins a prize (bedroom furniture, a Caribbean cruise, etc.) if s/he guesses correctly on three *adjacent* sectors. That is, numbering the sectors 1–6 clockwise, correct guesses on sectors 5, 6, and 1 wins the prize but not on sectors 5, 6, and 3, since the latter are not all adjacent. (The contestant gets to guess on all six sectors, if need be.)

Write a simulation program to estimate the probability the contestant wins the prize, assuming her/his guesses are independent from item to item. Provide estimated probabilities under each of the following assumptions: (1) each guess is "wild" and thus has probability .5 of being correct, and (2) the contestant is a good shopper, with probability .8 of being correct on any item.

106. Refer to the game in Example 1.41. Under the same settings as in that example, estimate the probability the player is ahead *at any time* during

the 25 plays. [*Hint*: This occurs if the player's dollar amount is positive at any of the 25 steps in the loop. So, you will need to keep track of every value of the `dollar` variable, not just the final result.]

107. Refer again to Example 1.41. Estimate the probability that the player experiences a "swing" of at least $5 during the game. That is, estimate the chance that the difference between the largest and smallest dollar amounts during the game is at least 5. (This would happen, for instance, if the player was at one point ahead at +$2 but later fell behind to −$3.)

108. Each of this book's authors has a fair coin. Carlton tosses his coin repeatedly until obtaining the sequence HTT. Devore tosses his coin until the sequence HTH is obtained.
    (a) Write a program to simulate Carlton's coin tossing and, separately, Devore's. Your program should keep track of the number of tosses each author requires on each simulation run to achieve his target sequence.
    (b) Estimate the probability that Devore obtains his sequence with fewer tosses than Carlton requires to obtain his sequence.

109. There's a 40-question multiple-choice exam we sometimes administer in our lower-level statistics classes. The exam has a peculiar feature: 10 of the questions have two options, 13 have three options, 13 have four options, and the other 4 have five options. (FYI, this is completely real!) What is the probability that, purely by guessing, a student could get at least half of these questions correct? Write a simulation program to answer this question.

110. Major League Baseball teams play a 162-game season, during which fans are often excited by long winning streaks and frustrated by long losing streaks. But how unusual are these streaks, really? How long a streak would you expect if the team's performance were independent from game to game?

    Write a program that simulates a 162-game season, i.e., a string of 162 wins and losses, with $P(\text{win}) = p$ for each game (the value of p to be specified later). Use your program with at least 10,000 runs to answer the following questions.
    (a) Suppose you're rooting for a ".500" team—that is, $p = .5$. What is the probability of observing a streak of at least five wins in a 162-game season? Estimate this probability with your program, and include a standard error.
    (b) Suppose instead your team is quite good: a .600 team overall, so $p = .6$. Intuitively, should the probability of a winning streak of at least five games be higher or lower? Explain.
    (c) Use your program with $p = .6$ to estimate the probability alluded to in (b). Is your answer higher or lower than (a)? Is that what you anticipated?

111. A *derangement* of the numbers 1 through n is a permutation of all n those numbers such that none of them is in the "right place." For example, 34251 is a derangement of 1 through 5, but 24351 is not because 3 is in the 3rd position. We will use simulation to estimate the number of derangements of the numbers 1 through 12.

    (a) Write a program that generates random permutations of the integers 1, 2, ..., 12. Your program should determine whether or not each permutation is a derangement.

    (b) Based on your program, estimate $P(D)$, where $D = \{$a permutation of 1–12 is a derangement$\}$.

    (c) From Sect. 1.3, we know the number of permutations of n items. (How many is that for $n = 12$?) Use this information and your answer to part (b) to estimate the *number* of derangements of the numbers 1 through 12.

    [Hint for part (a): Use random sampling without replacement as in Example 1.42. Alternatively, the `randperm` command in Matlab can also be employed.]

112. The book's Introduction discussed the famous *Birthday Problem*, which was solved in Example 1.22 of Sect. 1.3. Now suppose you have 500 Facebook friends. Make the same assumptions here as in the Birthday Problem.

    (a) Write a program to estimate the probability that, on at least 1 day during the year, Facebook tells you three (or more) of your friends share that birthday. Based on your answer, should you be surprised by this occurrence?

    (b) Write a program to estimate the probability that, on at least 1 day during the year, Facebook tells you *five* (or more) of your friends share that birthday. Based on your answer, should you be surprised by this occurrence?

    [*Hint*: Generate 500 birthdays *with* replacement, then determine whether any birthday occurs three or more times (five or more for part (b)). The `table` function in R or `tabulate` in Matlab may prove useful.]

113. Consider the following game: you begin with \$20. You flip a fair coin, winning \$10 if the coin lands heads and losing \$10 if the coin lands tails. Play continues until you either go broke or have \$100 (i.e., a net profit of \$80). Write a simulation program to estimate:

    (a) The probability you win the game.

    (b) The probability the game ends within ten coin flips.

    [*Note*: This is a special case of the *Gambler's Ruin* problem, which we'll explore in much greater depth in Exercise 145 and again in Chap. 6.]

114. Consider the *Coupon Collector's Problem* described in the Introduction: 10 different coupons are distributed into cereal boxes, one per box, so that any randomly selected box is equally likely to have any of the 10 coupons inside. Write a program to simulate the process of buying cereal boxes until all 10 distinct coupons have been collected. For each run, keep track of how many cereal boxes you purchased to collect the complete set of coupons. Then use your program to answer the following questions.

    (a) What is the probability you collect all 10 coupons with just 10 cereal boxes?

    (b) Use counting techniques to determine the exact probability in (a). [*Hint*: Relate this to the Birthday Problem.]

(c) What is the probability you require more than 20 boxes to collect all 10 coupons?

(d) Using techniques from Chap. 4, it can be shown that it takes about 29.3 boxes, on the average, to collect all 10 coupons. What's the probability of collecting all 10 coupons in fewer than average boxes (i.e., less than 29.3)?

115. In the Introduction we mentioned a famous puzzle from the early days of probability, investigated by Pascal and Fermat. Which of the following events is more likely: to roll at least one ⚅ in four rolls of a fair die, or to roll at least one ⚅⚅ in 24 rolls of two fair dice?

(a) Write a program to simulate a set of four die rolls many times, and use the results to estimate P(at least one ⚅ in 4 rolls).

(b) Now adapt your program to simulate rolling a pair of dice 24 times. Repeat this simulation many times, and use your results to estimate P(at least one ⚅⚅ in 24 rolls).

116. *The Problem of the Points.* Pascal and Fermat also explored a question concerning how to divide the stakes in a game that has been interrupted. Suppose two players, Blaise and Pierre, are playing a game where the winner is the first to achieve a certain number of points. The game gets interrupted at a moment when Blaise needs n more points to win and Pierre needs m more to win. How should the game's prize money be divvied up? Fermat argued that Blaise should receive a proportion of the total stake equal to the chance he would have won if the game hadn't been interrupted (and Pierre receives the remainder).

Assume the game is played in rounds, the winner of each round gets 1 point, rounds are independent, and the two players are equally likely to win any particular round.

(a) Write a program to simulate the rounds of the game that would have happened after play was interrupted. A single simulation run should terminate as soon as Blaise has n wins or Pierre has m wins (equivalently, Blaise has m losses). Use your program to estimate P(Blaise gets 10 wins before 15 losses), which is the proportion of the total stake Blaise should receive if $n = 10$ and $m = 15$.

(b) Use your same program to estimate the relevant probability when $n = m = 10$. Logically, what should the answer be? Is your estimated probability close to that?

(c) Finally, let's assume Pierre is actually the better player: P(Blaise wins a round) $= .4$. Again with $n = 10$ and $m = 15$, what proportion of the stake should be awarded to Blaise?

117. Twenty faculty members in a certain department have just participated in a department chair election. Suppose that candidate A has received 12 of the votes and candidate B the other 8 votes. If the ballots are opened one by one in random order and the candidate selected on each ballot is recorded, use simulation to estimate the probability that candidate A remains ahead of candidate B throughout the vote count (which happens if, for example, the result is AA...AB...B but not if the result is AABABB...).

118. Show that the (estimated) standard error for $\hat{P}(A)$ is at most $1/\sqrt{4n}$.

119. Simulation can be used to estimate numerical constants, such as π. Here's one approach: consider the part of a disk of radius 1 that lies in the first quadrant (a quarter-circle). Imagine two random numbers, x and y, both between 0 and 1. The pair (x, y) lies somewhere in the first quadrant; let A denote the event that (x, y) falls inside the quarter-circle.

   (a) Write a program that simulates pairs (x, y) in order to estimate $P(A)$, the probability that a randomly selected pair of points in the square $[0, 1] \times [0, 1]$ lies in the quarter-circle of radius 1.

   (b) Using techniques from Chap. 4, it can be shown that the exact probability of A is $\pi/4$ (which makes sense, because that's the ratio of the quarter-circle's area to the square's area). Use that fact to come up with an estimate of π from your simulation. How close is your estimate to $3.14159\ldots$?

120. Consider the quadratic equation $ax^2 + bx + c = 0$. Suppose that a, b, and c are random numbers between 0 and 1 (like those produced by an RNG). Estimate the probability that the roots of this quadratic equation are real. [*Hint*: Think about the discriminant.] This probability can be computed exactly using methods from Chap. 4, but a triple integral is required.

1.7   Supplementary Exercises (121–150)

121. A small manufacturing company will start operating a night shift. There are 20 machinists employed by the company.

   (a) If a night crew consists of 3 machinists, how many different crews are possible?

   (b) If the machinists are ranked 1, 2, $\ldots$, 20 in order of competence, how many of these crews would not have the best machinist?

   (c) How many of the crews would have at least 1 of the 10 best machinists?

   (d) If one of these crews is selected at random to work on a particular night, what is the probability that the best machinist will not work that night?

122. A factory uses three production lines to manufacture cans of a certain type. The accompanying table gives percentages of nonconforming cans, categorized by type of nonconformance, for each of the three lines during a particular time period.

	Line 1	Line 2	Line 3
Blemish	15	12	20
Crack	50	44	40
Pull-Tab Problem	21	28	24
Surface Defect	10	8	15
Other	4	8	2

During this period, line 1 produced 500 nonconforming cans, line 2 produced 400 such cans, and line 3 was responsible for 600 nonconforming cans. Suppose that one of these 1,500 cans is randomly selected.
  (a) What is the probability that the can was produced by line 1? That the reason for nonconformance is a crack?
  (b) If the selected can came from line 1, what is the probability that it had a blemish?
  (c) Given that the selected can had a surface defect, what is the probability that it came from line 1?

123. An employee of the records office at a university currently has ten forms on his desk awaiting processing. Six of these are withdrawal petitions and the other four are course substitution requests.
  (a) If he randomly selects six of these forms to give to a subordinate, what is the probability that only one of the two types of forms remains on his desk?
  (b) Suppose he has time to process only four of these forms before leaving for the day. If these four are randomly selected one by one, what is the probability that each succeeding form is of a different type from its predecessor?

124. One satellite is scheduled to be launched from Cape Canaveral in Florida, and another launching is scheduled for Vandenberg Air Force Base in California. Let A denote the event that the Vandenberg launch goes off on schedule, and let B represent the event that the Cape Canaveral launch goes off on schedule. If A and B are independent events with $P(A) > P(B)$, $P(A \cup B) = .626$, and $P(A \cap B) = .144$, determine the values of $P(A)$ and $P(B)$.

125. A transmitter is sending a message by using a binary code, namely, a sequence of 0s and 1s. Each transmitted bit (0 or 1) must pass through three relays to reach the receiver. At each relay, the probability is .20 that the bit sent will be different from the bit received (a reversal). Assume that the relays operate independently of one another.

   Transmitter $\rightarrow$ Relay 1 $\rightarrow$ Relay 2 $\rightarrow$ Relay 3 $\rightarrow$ Receiver
  (a) If a 1 is sent from the transmitter, what is the probability that a 1 is sent by all three relays?
  (b) If a 1 is sent from the transmitter, what is the probability that a 1 is received by the receiver? [*Hint*: The eight experimental outcomes can be displayed on a tree diagram with three generations of branches, one generation for each relay.]
  (c) Suppose 70% of all bits sent from the transmitter are 1s. If a 1 is received by the receiver, what is the probability that a 1 was sent?

126. Individual A has a circle of five close friends (B, C, D, E, and F). A has heard a certain rumor from outside the circle and has invited the five friends to a party to circulate the rumor. To begin, A selects one of the five at random and tells the rumor to the chosen individual. That individual then selects at random one of the four remaining individuals and repeats the rumor. Continuing, a new individual is selected from those not already having

heard the rumor by the individual who has just heard it, until everyone has been told.

(a) What is the probability that the rumor is repeated in the order B, C, D, E, and F?

(b) What is the probability that F is the third person at the party to be told the rumor?

(c) What is the probability that F is the last person to hear the rumor?

127. Refer to the previous exercise. If at each stage the person who currently "has" the rumor does not know who has already heard it and selects the next recipient at random from all five possible individuals, what is the probability that F has still not heard the rumor after it has been told ten times at the party?

128. According to the article "Optimization of Distribution Parameters for Estimating Probability of Crack Detection" (*J. of Aircraft*, 2009: 2090-2097), the following "Palmberg" equation is commonly used to determine the probability $P_d(c)$ of detecting a crack of size c in an aircraft structure:

$$P_d(c) = \frac{(c/c^*)^\beta}{1 + (c/c^*)^\beta}$$

where c^* is the crack size that corresponds to a .5 detection probability (and thus is an assessment of the quality of the inspection process).

(a) Verify that $P_d(c^*) = .5$.

(b) What is $P_d(2c^*)$ when $\beta = 4$?

(c) Suppose an inspector inspects two different panels, one with a crack size of c^* and the other with a crack size of $2c^*$. Again assuming $\beta = 4$ and also that the results of the two inspections are independent of one another, what is the probability that exactly one of the two cracks will be detected?

(d) What happens to $P_d(c)$ as $\beta \to \infty$?

129. A sonnet is a 14 line poem in which certain rhyming patterns are followed. The writer Raymond Queneau published a book containing just 10 sonnets, each on a different page. However, these were such that the first line of a sonnet could come from the first line on any of the 10 pages, the second line could come from the second line on any of the ten pages, and so on (successive lines were perforated for this purpose).

(a) How many sonnets can be created from the 10 in the book?

(b) If one of the sonnets counted in (a) is selected at random, what is the probability that all 14 lines come from exactly two of the ten pages?

130. A chemical engineer is interested in determining whether a certain trace impurity is present in a product. An experiment has a probability of .80 of detecting the impurity if it is present. The probability of not detecting the impurity if it is absent is .90. The prior probabilities of the impurity being present and being absent are .40 and .60, respectively. Three separate

experiments result in only two detections. What is the posterior probability that the impurity is present?

131. Fasteners used in aircraft manufacturing are slightly crimped so that they lock enough to avoid loosening during vibration. Suppose that 95% of all fasteners pass an initial inspection. Of the 5% that fail, 20% are so seriously defective that they must be scrapped. The remaining fasteners are sent to a recrimping operation, where 40% cannot be salvaged and are discarded. The other 60% of these fasteners are corrected by the recrimping process and subsequently pass inspection.
    (a) What is the probability that a randomly selected incoming fastener will pass inspection either initially or after recrimping?
    (b) Given that a fastener passed inspection, what is the probability that it passed the initial inspection and did not need recrimping?

132. One percent of all individuals in a certain population are carriers of a particular disease. A diagnostic test for this disease has a 90% detection rate for carriers and a 5% detection rate for noncarriers. Suppose the test is applied independently to two different blood samples from the same randomly selected individual.
    (a) What is the probability that both tests yield the same result?
    (b) If both tests are positive, what is the probability that the selected individual is a carrier?

133. A system consists of two components. The probability that the second component functions in a satisfactory manner during its design life is .9, the probability that at least one of the two components does so is .96, and the probability that both components do so is .75. Given that the first component functions in a satisfactory manner throughout its design life, what is the probability that the second one does also?

134. A certain company sends 40% of its overnight mail parcels via express mail service E_1. Of these parcels, 2% arrive after the guaranteed delivery time (denote the event "late delivery" by L). If a record of an overnight mailing is randomly selected from the company's file, what is the probability that the parcel went via E_1 and was late?

135. Refer to the previous exercise. Suppose that 50% of the overnight parcels are sent via express mail service E_2 and the remaining 10% are sent via E_3. Of those sent via E_2, only 1% arrive late, whereas 5% of the parcels handled by E_3 arrive late.
    (a) What is the probability that a randomly selected parcel arrived late?
    (b) If a randomly selected parcel has arrived on time, what is the probability that it was not sent via E_1?

136. A company uses three different assembly lines—A_1, A_2, and A_3—to manufacture a particular component. Of those manufactured by line A_1, 5% need rework to remedy a defect, whereas 8% of A_2's components need rework and 10% of A_3's need rework. Suppose that 50% of all components are produced by line A_1, 30% are produced by line A_2, and 20% come from line

A_3. If a randomly selected component needs rework, what is the probability that it came from line A_1? From line A_2? From line A_3?

137. Disregarding the possibility of a February 29 birthday, suppose a randomly selected individual is equally likely to have been born on any one of the other 365 days. If ten people are randomly selected, what is the probability that either at least two have the same birthday or at least two have the same last three digits of their Social Security numbers? [*Note*: The article "Methods for Studying Coincidences" (F. Mosteller and P. Diaconis, *J. Amer. Statist. Assoc.*, 1989: 853–861) discusses problems of this type.]

138. One method used to distinguish between granitic (G) and basaltic (B) rocks is to examine a portion of the infrared spectrum of the sun's energy reflected from the rock surface. Let R_1, R_2, and R_3 denote measured spectrum intensities at three different wavelengths; typically, for granite $R_1 < R_2 < R_3$, whereas for basalt $R_3 < R_1 < R_2$. When measurements are made remotely (using aircraft), various orderings of the R_is may arise whether the rock is basalt or granite. Flights over regions of known composition have yielded the following information:

	Granite	Basalt
$R_1 < R_2 < R_3$	60%	10%
$R_1 < R_3 < R_2$	25%	20%
$R_3 < R_1 < R_2$	15%	70%

Suppose that for a randomly selected rock in a certain region, $P(\text{granite}) = .25$ and $P(\text{basalt}) = .75$.

(a) Show that $P(\text{granite}|R_1 < R_2 < R_3) > P(\text{basalt}|R_1 < R_2 < R_3)$. If measurements yielded $R_1 < R_2 < R_3$, would you classify the rock as granite or basalt?

(b) If measurements yielded $R_1 < R_3 < R_2$, how would you classify the rock? Answer the same question for $R_3 < R_1 < R_2$.

(c) Using the classification rules indicated in parts (a) and (b), when selecting a rock from this region, what is the probability of an erroneous classification? [*Hint*: Either G could be classified as B or B as G, and $P(B)$ and $P(G)$ are known.]

(d) If $P(\text{granite}) = p$ rather than .25, are there values of p (other than 1) for which a rock would always be classified as granite?

139. In a Little League baseball game, team A's pitcher throws a strike 50% of the time and a ball 50% of the time, successive pitches are independent of each other, and the pitcher never hits a batter. Knowing this, team B's manager has instructed the first batter not to swing at anything. Calculate the probability that

(a) The batter walks on the fourth pitch.

(b) The batter walks on the sixth pitch (so two of the first five must be strikes), using a counting argument or constructing a tree diagram.

(c) The batter walks.

(d) The first batter up scores while no one is out (assuming that each batter pursues a no-swing strategy).

140. Consider a woman whose brother is afflicted with hemophilia, which implies that the woman's mother has the hemophilia gene on one of her two X chromosomes (almost surely not both, since that is generally fatal). Thus there is a 50–50 chance that the woman's mother has passed on the bad gene to her. The woman has two sons, each of whom will independently inherit the gene from one of her two chromosomes. If the woman herself has a bad gene, there is a 50–50 chance she will pass this on to a son. Suppose that neither of her two sons is afflicted with hemophilia. What then is the probability that the woman is indeed the carrier of the hemophilia gene? What is this probability if she has a third son who is also not afflicted?

141. A particular airline has 10 a.m. flights from Chicago to New York, Atlanta, and Los Angeles. Let A denote the event that the New York flight is full and define events B and C analogously for the other two flights. Suppose $P(A) = .6$, $P(B) = .5$, $P(C) = .4$ and the three events are independent. What is the probability that
   (a) All three flights are full? That at least one flight is not full?
   (b) Only the New York flight is full? That exactly one of the three flights is full?

142. Consider four independent events A_1, A_2, A_3, and A_4 and let $p_i = P(A_i)$ for $i = 1, 2, 3, 4$. Express the probability that at least one of these four events occurs in terms of the p_is, and do the same for the probability that at least two of the events occur.

143. A box contains the following four slips of paper, each having exactly the same dimensions: (1) win prize 1; (2) win prize 2; (3) win prize 3; (4) win prizes 1, 2, and 3. One slip will be randomly selected. Let $A_1 = \{\text{win prize 1}\}$, $A_2 = \{\text{win prize 2}\}$, and $A_3 = \{\text{win prize 3}\}$. Show that A_1 and A_2 are independent, that A_1 and A_3 are independent, and that A_2 and A_3 are also independent (this is *pairwise* independence). However, show that $P(A_1 \cap A_2 \cap A_3) \neq P(A_1) \cdot P(A_2) \cdot P(A_3)$, so the three events are not *mutually* independent.

144. Jurors may be a priori biased for or against the prosecution in a criminal trial. Each juror is questioned by both the prosecution and the defense (the voir dire process), but this may not reveal bias. Even if bias is revealed, the judge may not excuse the juror for cause because of the narrow legal definition of bias. For a randomly selected candidate for the jury, define events B_0, B_1, and B_2 as the juror being unbiased, biased against the prosecution, and biased against the defense, respectively. Also let C be the event that bias is revealed during the questioning and D be the event that the juror is eliminated for cause. Let $b_i = P(B_i)$ $(i = 0, 1, 2)$, $c = P(C|B_1) = P(C|B_2)$, and $d = P(D|B_1 \cap C) = P(D|B_2 \cap C)$ ["Fair Number of Peremptory Challenges in Jury Trials," *J. Amer. Statist. Assoc.*, 1979: 747–753].
   (a) If a juror survives the voir dire process, what is the probability that he/she is unbiased (in terms of the b_is, c, and d)? What is the probability that he/she is biased against the prosecution? What is the probability that

he/she is biased against the defense? [*Hint*: Represent this situation using a tree diagram with three generations of branches.]

(b) What are the probabilities requested in (a) if $b_0 = .50$, $b_1 = .10$, $b_2 = .40$ (all based on data relating to the famous trial of the Florida murderer Ted Bundy), $c = .85$ (corresponding to the extensive questioning appropriate in a capital case), and $d = .7$ (a "moderate" judge)?

145. *Gambler's Ruin.* Allan and Beth currently have \$2 and \$3, respectively. A fair coin is tossed. If the result of the toss is heads, Allan wins \$1 from Beth, whereas if the coin toss results in tails, then Beth wins \$1 from Allan. This process is then repeated, with a coin toss followed by the exchange of \$1, until one of the two players goes broke (one of the two gamblers is ruined). We wish to determine $a_2 = P(\text{Allan is the winner | he starts with \$2})$. To do so, let's also consider $a_i = P(\text{Allan wins | he starts with \$}i)$ for $i = 0, 1, 3, 4,$ and 5.

   (a) What are the values of a_0 and a_5?

   (b) Use the Law of Total Probability to obtain an equation relating a_2 to a_1 and a_3. [*Hint*: Condition on the result of the first coin toss, realizing that if it is heads, then from that point Allan starts with \$3.]

   (c) Using the logic described in (b), develop a system of equations relating a_i ($i = 1, 2, 3, 4$) to a_{i-1} and a_{i+1}. Then solve these equations. [*Hint*: Write each equation so that $a_i - a_{i-1}$ is on the left hand side. Then use the result of the first equation to express each other $a_i - a_{i-1}$ as a function of a_1, and add together all four of these expressions ($i = 2, 3, 4, 5$).]

   (d) Generalize the result to the situation in which Allan's initial fortune is \$a and Beth's is \$b. [*Note*: The solution is a bit more complicated if $p = P(\text{Allan wins \$1}) \neq .5$. We'll explore Gambler's Ruin again in Chap. 6.]

146. *The Matching Problem.* Four friends—Allison, Beth, Carol, and Diane—who have identical calculators are studying for a statistics exam. They set their calculators down in a pile before taking a study break and then pick them up in random order when they return from the break.

   (a) What is the probability all four friends pick up the correct calculator?

   (b) What is the probability that at least one of the four gets her own calculator? [*Hint*: Let A be the event that Alice gets her own calculator, and define events B, C, and D analogously for the other three students. How can the event {at least one gets her own calculator} be expressed in terms of the four events $A, B, C,$ and D? Now use a general law of probability.]

   (c) Generalize the answer from part (b) to n individuals. Can you recognize the result when n is large (the approximation to the resulting series)?

147. An event A is said to *attract* event B if $P(B|A) > P(B)$ and *repel* B if $P(B|A) < P(B)$. (This refines the notion of dependent events by specifying whether A makes B more likely or less likely to occur.)

   (a) Show that if A attracts B, then A repels B'.

   (b) Show that if A attracts B, then A' repels B.

   (c) Prove the *Law of Mutual Attraction*: event A attracts event B if, and only if, B attracts A.

148. *The Secretary Problem*. A personnel manager is to interview four candidates
     for a job. These are ranked 1, 2, 3, and 4 in order of preference and will be
     interviewed in random order. However, at the conclusion of each interview,
     the manager will know only how the current candidate compares to those
     previously interviewed. For example, the interview order 3, 4, 1, 2 generates
     no information after the first interview, shows that the second candidate is
     worse than the first, and that the third is better than the first two. However, the
     order 3, 4, 2, 1 would generate the same information after each of the first
     three interviews. The manager wants to hire the best candidate but must make
     an irrevocable hire/no hire decision after each interview. Consider the follow-
     ing strategy: Automatically reject the first s candidates and then hire the first
     subsequent candidate who is best among those already interviewed (if no such
     candidate appears, the last one interviewed is hired).

     For example, with $s = 2$, the order 3, 4, 1, 2 would result in the best being
     hired, whereas the order 3, 1, 2, 4 would not. Of the four possible s values
     (0, 1, 2, and 3), which one maximizes P(best is hired)? [*Hint*: Write out the
     24 equally likely interview orderings: $s = 0$ means that the first candidate is
     automatically hired.]

149. Jay and Maurice are playing a tennis match. In one particular game, they have
     reached deuce, which means each player won three points. Now in order to
     finish the game, one of the two players must get two points ahead of the other.
     For example, Jay will win if he wins the next two points (JJ), or if Maurice
     wins the next point and Jay the three points after that ($MJJJ$), or if the result of
     the next six points is $JMMJJJ$, etc.

     (a) Suppose that the probability of Jay winning a point is .6 and outcomes of
         successive points are independent of one another. What is the probability
         that Jay wins the game? [*Hint*: In the law of total probability, let $A_1 = \{$Jay
         wins each of the next two points$\}$, $A_2 = \{$Maurice wins each of the next
         two points$\}$, and $A_3 = \{$each player wins one of the next two points$\}$. Also
         let $p = P$(Jay wins the game). How does p compare to P(Jay wins the
         game$|A_3$)?]

     (b) If Jay wins the game, what is the probability that he needed only two
         points to do so?

150. Here is a variant on one of the puzzles mentioned in the book's Introduction.
     A fair coin is tossed repeatedly until either the sequence TTH or the sequence
     THT is observed. Let B be the event that stopping occurs because TTH was
     observed (i.e., that TTH is observed before THT). Calculate $P(B)$. [*Hint*:
     Consider the following partition of the sample space: $A_1 = \{$1st toss is H$\}$,
     $A_2 = \{$1st two tosses are TT$\}$, $A_3 = \{$1st three tosses are THT$\}$, and $A_4 = \{$1st
     three tosses are THH$\}$. Also denote $P(B)$ by p. Apply the Law of Total
     Probability, and p will appear on both sides in various places. The resulting
     equation is easily solved for p.]

Discrete Random Variables and Probability Distributions

2

Suppose a city's traffic engineering department monitors a certain intersection during a one-hour period in the middle of the day. Many characteristics might be of interest to the observers, including the number of vehicles that enter the intersection, the largest number of vehicles in the left turn lane during a signal cycle, the speed of the fastest vehicle going through the intersection, the average speed of all vehicles entering the intersection. The value of each one of the foregoing variable quantities is subject to uncertainty—we don't know a priori how many vehicles will enter, what the maximum speed will be, etc. So each of these is referred to as a *random variable*—a variable quantity whose value is determined by what happens in a chance experiment.

There are two fundamentally different types of random variables, discrete and continuous. In this chapter we examine the basic properties and introduce the most important examples of discrete random variables. Chapter 3 covers the same territory for continuous random variables.

2.1 Random Variables

In any experiment, numerous characteristics can be observed or measured, but in most cases an experimenter will focus on some specific aspect or aspects of a sample. For example, in a study of commuting patterns in a metropolitan area, each individual in a sample might be asked about commuting distance and the number of people commuting in the same vehicle, but not about IQ, income, family size, and other such characteristics. Alternatively, a researcher may test a sample of components and record only the number that have failed within 1000 hours, rather than record the individual failure times.

In general, each outcome of an experiment can be associated with a number by specifying a rule of association (e.g., the number among the sample of ten

M.A. Carlton and J.L. Devore, *Probability with Applications in Engineering, Science, and Technology*, Springer Texts in Statistics, DOI 10.1007/978-1-4939-0395-5_2, © Springer Science+Business Media New York 2014

Fig. 2.1 A random variable

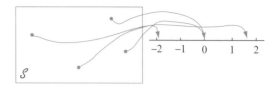

components that fail to last 1,000 h or the total weight of baggage for a sample of 25 airline passengers). Such a rule of association is called a **random variable**—a variable because different numerical values are possible and random because the observed value depends on which of the possible experimental outcomes results (Fig. 2.1).

DEFINITION

For a given sample space $\mathscr{S}$ of some experiment, a **random variable (rv)** is any rule that associates a number with each outcome in $\mathscr{S}$. In mathematical language, a random variable is a function whose domain is the sample space and whose range is some subset of real numbers.

Random variables are customarily denoted by uppercase letters, such as X and Y, near the end of our alphabet. We will use lowercase letters to represent some particular value of the corresponding random variable. The notation $X(s) = x$ means that x is the value associated with the outcome s by the rv X.

Example 2.1   When a student attempts to connect to a university computer system, either there is a failure (F) or there is a success (S). With $\mathscr{S} = \{S, F\}$, define an rv X by $X(S) = 1, X(F) = 0$. The rv X indicates whether (1) or not (0) the student can connect.   ∎

In Example 2.1, the rv X was specified by explicitly listing each element of $\mathscr{S}$ and the associated number. If $\mathscr{S}$ contains more than a few outcomes, such a listing is tedious, but it can frequently be avoided.

Example 2.2   Consider the experiment in which a telephone number in a certain area code is dialed using a random number dialer (such devices are used extensively by polling organizations), and define an rv Y by

$$Y = \begin{cases} 1 & \text{if the selected number is unlisted} \\ 0 & \text{if the selected number is listed in the directory} \end{cases}$$

For example, if 5282966 appears in the telephone directory, then $Y(5282966) = 0$, whereas $Y(7727350) = 1$ tells us that the number 7727350 is unlisted. A word description of this sort is more economical than a complete listing, so we will use such a description whenever possible.   ∎

In Examples 2.1 and 2.2, the only possible values of the random variable were 0 and 1. Such a random variable arises frequently enough to be given a special name, after the individual who first studied it.

> **DEFINITION**
> Any random variable whose only possible values are 0 and 1 is called a **Bernoulli random variable**.

We will often want to define and study several different random variables from the same sample space.

Example 2.3 Example 1.3 described an experiment in which the number of pumps in use at each of two gas stations was determined. Define rvs X, Y, and U by

X = the total number of pumps in use at the two stations

Y = the difference between the number of pumps in use at station 1 and the number in use at station 2

U = the maximum of the numbers of pumps in use at the two stations

If this experiment is performed and $s = (2, 3)$ results, then $X((2, 3)) = 2 + 3 = 5$, so we say that the observed value of X is $x = 5$. Similarly, the observed value of Y would be $y = 2 - 3 = -1$, and the observed value of U would be $u = \max(2, 3) = 3$. ■

Each of the random variables of Examples 2.1–2.3 can assume only a finite number of possible values. This need not be the case.

Example 2.4 Consider an experiment in which 9-V batteries are examined until one with an acceptable voltage (S) is obtained. The sample space is $\mathscr{S} = \{S, FS, FFS, \dots\}$. Define an rv X by

X = the number of batteries examined before the experiment terminates

Then $X(S) = 1$, $X(FS) = 2$, $X(FFS) = 3$, ..., $X(FFFFFFS) = 7$, and so on. Any positive integer is a possible value of X, so the set of possible values is infinite. ■

Example 2.5 Suppose that in some random fashion, a location (latitude and longitude) in the continental USA is selected. Define an rv Y by

Y = the height, in feet, above sea level at the selected location

For example, if the selected location were (39°50′N, 98°35′W), then we might have $Y((39°50′N, 98°35′W)) = 1748.26$ ft. The largest possible value of Y is 14,494 (Mt. Whitney), and the smallest possible value is −282 (Death Valley). The set of all possible values of Y is the set of all numbers in the interval between -282 and 14,494; that is, the range of Y is

$$\{y : -282 \leq y \leq 14{,}494\} = [-282, 14{,}494]$$

and there are an infinite number of numbers in this interval.   ∎

2.1.1   Two Types of Random Variables

Determining the values of variables such as the number of visits to a Web site during a 24-h period or the number of patients in an emergency room at a particular time requires only counting. On the other hand, determining values of variables such as fuel efficiency of a vehicle (mpg) or reaction time to a stimulus necessitates making a measurement of some sort. The following definition formalizes the distinction between these two different kinds of variables.

DEFINITION

A **discrete** random variable is an rv whose possible values constitute either a finite set or a countably infinite set (e.g., the set of all integers, or the set of all positive integers).

A random variable is **continuous** if *both* of the following apply:

1. Its set of possible values consists either of all numbers in a single interval on the number line (possibly infinite in extent, e.g., from $-\infty$ to ∞) or all numbers in a disjoint union of such intervals (e.g., [0, 10] ∪ [20, 30]).
2. No possible value of the variable has positive probability, that is, $P(X = c) = 0$ for any possible value c.

Although any interval on the number line contains an infinite number of numbers, it can be shown that there is no way to create an infinite listing of all these values—there are just too many of them. The second condition describing a continuous random variable is perhaps counterintuitive, since it would seem to imply a total probability of zero for all possible values. But we shall see in Chap. 3 that *intervals* of values have positive probability; the probability of an interval will decrease to zero as the width of the interval shrinks to zero. In practice, discrete variables virtually always involve counting the number of something, whereas continuous variables entail making measurements of some sort.

Example 2.6 All random variables in Examples 2.1–2.4 are discrete. As another example, suppose we select married couples at random and do a blood test on each person until we find a husband and wife who both have the same Rh factor.

With X = the number of blood tests to be performed, possible values of X are $\{2, 4, 6, 8, \ldots\}$. Since the possible values have been listed in sequence, X is a discrete rv. ∎

To study basic properties of discrete rvs, only the tools of discrete mathematics—summation and differences—are required. The study of continuous variables in Chap. 3 will require the continuous mathematics of the calculus—integrals and derivatives.

2.1.2   Exercises: Section 2.1 (1–10)

1. A concrete beam may fail either by shear (S) or flexure (F). Suppose that three failed beams are randomly selected and the type of failure is determined for each one. Let X = the number of beams among the three selected that failed by shear. List each outcome in the sample space along with the associated value of X.
2. Give three examples of Bernoulli rvs (other than those in the text).
3. Using the experiment in Example 2.3, define two more random variables and list the possible values of each.
4. Let X = the number of nonzero digits in a randomly selected zip code. What are the possible values of X? Give three possible outcomes and their associated X values.
5. If the sample space $\mathscr{S}$ is an infinite set, does this necessarily imply that any rv X defined from $\mathscr{S}$ will have an infinite set of possible values? If yes, say why. If no, give an example.
6. Starting at a fixed time, each car entering an intersection is observed to see whether it turns left (L), right (R), or goes straight ahead (A). The experiment terminates as soon as a car is observed to turn left. Let X = the number of cars observed. What are possible X values? List five outcomes and their associated X values.
7. For each random variable defined here, describe the set of possible values for the variable, and state whether the variable is discrete.
   (a) X = the number of unbroken eggs in a randomly chosen standard egg carton
   (b) Y = the number of students on a class list for a particular course who are absent on the first day of classes
   (c) U = the number of times a duffer has to swing at a golf ball before hitting it
   (d) X = the length of a randomly selected rattlesnake
   (e) Z = the amount of royalties earned from the sale of a first edition of 10,000 textbooks
   (f) Y = the acidity level (pH) of a randomly chosen soil sample
   (g) X = the tension (psi) at which a randomly selected tennis racket has been strung
   (h) X = the total number of coin tosses required for three individuals to obtain a match (HHH or TTT)
8. Each time a component is tested, the trial is a success (S) or failure (F). Suppose the component is tested repeatedly until a success occurs on three

consecutive trials. Let Y denote the number of trials necessary to achieve this. List all outcomes corresponding to the five smallest possible values of Y, and state which Y value is associated with each one.

9. An individual named Claudius is located at the point 0 in the accompanying diagram.

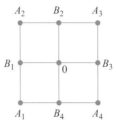

Using an appropriate randomization device (such as a tetrahedral die, one having four sides), Claudius first moves to one of the four locations B_1, B_2, B_3, B_4. Once at one of these locations, he uses another randomization device to decide whether he next returns to 0 or next visits one of the other two adjacent points. This process then continues; after each move, another move to one of the (new) adjacent points is determined by tossing an appropriate die or coin.

(a) Let $X =$ the number of moves that Claudius makes before first returning to 0. What are possible values of X? Is X discrete or continuous?

(b) If moves are allowed also along the diagonal paths connecting 0 to A_1, A_2, A_3, and A_4, respectively, answer the questions in part (a).

10. The number of pumps in use at both a six-pump station and a four-pump station will be determined. Give the possible values for each of the following random variables:

(a) $T =$ the total number of pumps in use

(b) $X =$ the difference between the numbers in use at stations 1 and 2

(c) $U =$ the maximum number of pumps in use at either station

(d) $Z =$ the number of stations having exactly two pumps in use

2.2   Probability Distributions for Discrete Random Variables

When probabilities are assigned to various outcomes in $\mathcal{S}$, these in turn determine probabilities associated with the values of any particular rv X. The *probability distribution of X* says how the total probability of 1 is distributed among (allocated to) the various possible X values.

Example 2.7 Six batches of components are ready to be shipped by a supplier. The number of defective components in each batch is as follows:

Batch	#1	#2	#3	#4	#5	#6
Number of defectives	0	2	0	1	2	0

One of these lots is to be randomly selected for shipment to a customer. Let X be the number of defectives in the selected lot. The three possible X values are 0, 1, and 2. Of the six equally likely simple events, three result in $X = 0$, one in $X = 1$, and the other two in $X = 2$. Let $p(0)$ denote the probability that $X = 0$ and $p(1)$ and $p(2)$ represent the probabilities of the other two possible values of X. Then

$$p(0) = P(X = 0) = P(\text{lot 1 or 3 or 6 is sent}) = \frac{3}{6} = .500$$

$$p(1) = P(X = 1) = P(\text{lot 4 is sent}) = \frac{1}{6} = .167$$

$$p(2) = P(X = 2) = P(\text{lot 2 or 5 is sent}) = \frac{2}{6} = .333$$

That is, a probability of .500 is distributed to the X value 0, a probability of .167 is placed on the X value 1, and the remaining probability, .333, is associated with the X value 2. The values of X along with their probabilities collectively specify the probability distribution or *probability mass function of X*. If this experiment were repeated over and over again, in the long run $X = 0$ would occur one-half of the time, $X = 1$ one-sixth of the time, and $X = 2$ one-third of the time. ∎

DEFINITION

The **probability distribution** or **probability mass function** (pmf) of a discrete rv is defined for every number x by
$$p(x) = P(X = x) = P(\text{all } s \in \mathcal{S}: X(s) = x).^{1}$$

In words, for every possible value x of the random variable, the pmf specifies the probability of observing that value when the experiment is performed. The conditions $p(x) \geq 0$ and $\Sigma p(x) = 1$, where the summation is over all possible x, are required of any pmf.

Example 2.8 Consider randomly selecting a student at a large public university, and define a Bernoulli rv by $X = 1$ if the selected student does not qualify for in-state tuition (a success from the university administration's point of view) and $X = 0$ if the student does qualify. If 20% of all students do not qualify, the pmf for X is
$$p(0) = P(X = 0) = P(\text{the selected student does qualify}) = .8$$
$$p(1) = P(X = 1) = P(\text{the selected student does not qualify}) = .2$$
$$p(x) = P(X = x) = 0 \text{ for } x \neq 0 \text{ or } 1.$$

---

[1] $P(X = x)$ is read "the probability that the rv X assumes the value x." For example, $P(X = 2)$ denotes the probability that the resulting X value is 2.

$$p(x) = \begin{cases} .8 & \text{if } x = 0 \\ .2 & \text{if } x = 1 \\ 0 & \text{if } x \neq 0 \text{ or } 1 \end{cases}$$

Figure 2.2 is a picture of this pmf, called a *line graph*.

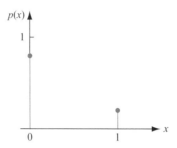

Fig. 2.2 The line graph for the pmf in Example 2.8   ■

Example 2.9 Consider a group of five potential blood donors—A, B, C, D, and E—of whom only A and B have type O+ blood. Five blood samples, one from each individual, will be typed in random order until an O+ individual is identified. Let the rv $Y =$ the number of typings necessary to identify an O+ individual. Then the pmf of Y is

$$p(1) = P(Y = 1) = P(\text{A or B typed first}) = \frac{2}{5} = .4$$

$$p(2) = P(Y = 2) = P(\text{C, D, or E first, and then A or B})$$
$$= P(\text{C, D, or E first}) \cdot P(\text{A or B next} \mid \text{C, D, or E first}) = \frac{3}{5} \cdot \frac{2}{4} = .3$$

$$p(3) = P(Y = 3) = P(\text{C, D, or E first and second, and then A or B}) = \frac{3}{5} \cdot \frac{2}{4} \cdot \frac{2}{3} = .2$$

$$p(4) = P(Y = 4) = P(\text{C, D, and E all done first}) = \frac{3}{5} \cdot \frac{2}{4} \cdot \frac{1}{3} = .1$$

$$p(y) = 0 \text{ for } y \neq 1, 2, 3, 4.$$

The pmf can be presented compactly in tabular form:

y	1	2	3	4
$p(y)$	.4	.3	.2	.1

where any y value not listed receives zero probability. Figure 2.3 shows the line graph for this pmf.

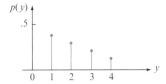

Fig. 2.3 The line graph for the pmf in Example 2.9   ∎

The name "probability mass function" is suggested by a model used in physics for a system of "point masses." In this model, masses are distributed at various locations *x* along a one-dimensional axis. Our pmf describes how the total probability mass of 1 is distributed at various points along the axis of possible values of the random variable (where and how much mass at each *x*).

Another useful pictorial representation of a pmf is called a **probability histogram**. Above each *y* with $p(y) > 0$, construct a rectangle centered at *y*. The height of each rectangle is proportional to $p(y)$, and the base is the same for all rectangles. When possible values are equally spaced, the base is frequently chosen as the distance between successive *y* values (though it could be smaller). Figure 2.4 shows two probability histograms.

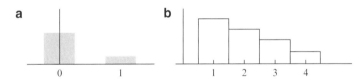

Fig. 2.4 Probability histograms: (**a**) Example 2.8; (**b**) Example 2.9

2.2.1   A Parameter of a Probability Distribution

In Example 2.8, we had $p(0) = .8$ and $p(1) = .2$. At another university, it may be the case that $p(0) = .9$ and $p(1) = .1$. More generally, the pmf of any Bernoulli rv can be expressed in the form $p(1) = \alpha$ and $p(0) = 1 - \alpha$, where $0 < \alpha < 1$. Because the pmf depends on the particular value of α, we often write $p(x; \alpha)$ rather than just $p(x)$:

$$p(x; \alpha) = \begin{cases} 1 - \alpha & \text{if } x = 0 \\ \alpha & \text{if } x = 1 \\ 0 & \text{otherwise} \end{cases} \tag{2.1}$$

Then each choice of α in Expression (2.1) yields a different pmf.

> **DEFINITION**
>
> Suppose $p(x)$ depends on a quantity that can be assigned any one of a number of possible values, with each different value determining a different probability distribution. Such a quantity is called a **parameter** of the distribution. The collection of all probability distributions for different values of the parameter is called a **family** of probability distributions.

The quantity α in Expression (2.1) is a parameter. Each different number α between 0 and 1 determines a different member of a family of distributions; two such members are

$$p(x; .6) = \begin{cases} .4 & \text{if } x = 0 \\ .6 & \text{if } x = 1 \\ 0 & \text{otherwise} \end{cases} \quad \text{and} \quad p(x; .5) = \begin{cases} .5 & \text{if } x = 0 \\ .5 & \text{if } x = 1 \\ 0 & \text{otherwise} \end{cases}$$

Every probability distribution for a Bernoulli rv has the form of Expression (2.1), so it is called the *family of Bernoulli distributions*.

Example 2.10 Starting at a fixed time, we observe the gender of each newborn child at a certain hospital until a boy (B) is born. Let $p = P(B)$, assume that successive births are independent, and define the rv X by X = number of births observed. Then

$$p(1) = P(X = 1) = P(B) = p$$
$$p(2) = P(X = 2) = P(GB) = P(G) \cdot P(B) = (1 - p)p$$

and

$$p(3) = P(X = 3) = P(GGB) = P(G) \cdot P(G) \cdot P(B) = (1 - p)^2 p$$

Continuing in this way, a general formula emerges:

$$p(x) = \begin{cases} (1 - p)^{x-1} p & x = 1, 2, 3, \ldots \\ 0 & \text{otherwise} \end{cases} \tag{2.2}$$

The quantity p in Expression (2.2) represents a number between 0 and 1 and is a parameter of the probability distribution. In the gender example, $p = .51$ might be appropriate, but if we were looking for the first child with Rh-positive blood, then we might have $p = .85$. The random variable X has what is known as a *geometric distribution*, which we will discuss in Sect. 2.6. ∎

2.2.2 The Cumulative Distribution Function

For some fixed value x, we often wish to compute the probability that the observed value of X will be *at most* x. For example, let X be the number of beds occupied in a hospital's emergency room at a certain time of day, and suppose the pmf of X is given by

x	0	1	2	3	4
$p(x)$	.20	.25	.30	.15	.10

Then the probability that at most two beds are occupied is $P(X \leq 2) = p(0) + p(1) + p(2) = .75$. Furthermore, since $X \leq 2.7$ iff $X \leq 2$, we also have $P(X \leq 2.7) = .75$, and similarly $P(X \leq 2.999) = .75$. Since 0 is the smallest possible X value, $P(X \leq -1.5) = 0$, $P(X \leq -10) = 0$, and in fact for any negative number x, $P(X \leq x) = 0$. And because 4 is the largest possible value of X, $P(X \leq 4) = 1$, $P(X \leq 9.8) = 1$, and so on.

Very importantly, $P(X < 2) = p(0) + p(1) = .45 < .75 = P(X \leq 2)$, because the latter probability includes the probability mass at the x value 2 whereas the former probability does not. More generally, $P(X < x) < P(X \leq x)$ whenever x is a possible value of X. Furthermore, $P(X \leq x)$ is a well-defined and computable probability for *any* number x.

DEFINITION

The **cumulative distribution function** (cdf) $F(x)$ of a discrete rv X with pmf $p(x)$ is defined for every number x by

$$F(x) = P(X \leq x) = \sum_{y:y \leq x} p(y) \qquad (2.3)$$

For any number x, $F(x)$ is the probability that the observed value of X will be at most x.

Example 2.11 A store carries flash drives with 1, 2, 4, 8, or 16 GB of memory. The accompanying table gives the distribution of $X =$ the amount of memory in a purchased drive:

x	1	2	4	8	16
$p(x)$	.05	.10	.35	.40	.10

Let's first determine $F(x)$ for each of the five possible values of X:

$$F(1) = P(X \leq 1) = P(X = 1) = p(1) = .05$$
$$F(2) = P(X \leq 2) = P(X = 1 \text{ or } 2) = p(1) + p(2) = .15$$
$$F(4) = P(X \leq 4) = P(X = 1 \text{ or } 2 \text{ or } 4) = p(1) + p(2) + p(4) = .50$$
$$F(8) = P(X \leq 8) = p(1) + p(2) + p(4) + p(8) = .90$$
$$F(16) = P(X \leq 16) = 1$$

Now for any other number x, $F(x)$ will equal the value of F at the closest possible value of X to the left of x. For example,

$$F(2.7) = P(X \leq 2.7) = P(X \leq 2) = F(2) = .15$$
$$F(7.999) = P(X \leq 7.999) = P(X \leq 4) = F(4) = .50$$

If x is less than 1, $F(x) = 0$ [e.g., $F(.58) = 0$], and if x is at least 16, $F(x) = 1$ [e.g., $F(25) = 1$]. The cdf is thus

$$F(x) = \begin{cases} 0 & x < 1 \\ .05 & 1 \leq x < 2 \\ .15 & 2 \leq x < 4 \\ .50 & 4 \leq x < 8 \\ .90 & 8 \leq x < 16 \\ 1 & 16 \leq x \end{cases}$$

A graph of this cdf is shown in Fig. 2.5.

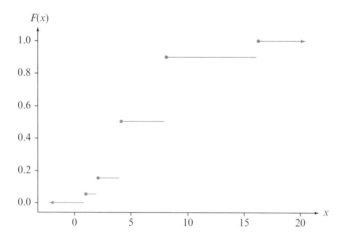

Fig. 2.5 A graph of the cdf of Example 2.11   ∎

For X a discrete rv, the graph of $F(x)$ will have a jump at every possible value of X and will be flat between possible values. Such a graph is called a **step function**.

Example 2.12 In Example 2.10, any positive integer was a possible X value, and the pmf was

$$p(x) = \begin{cases} (1-p)^{x-1}p & x = 1, 2, 3, \ldots \\ 0 & \text{otherwise} \end{cases}$$

For any positive integer x,

$$F(x) = \sum_{y \leq x} p(y) = \sum_{y=1}^{x} (1-p)^{y-1}p = p\sum_{y=0}^{x-1}(1-p)^{y} \qquad (2.4)$$

To evaluate this sum, we use the fact that the partial sum of a geometric series is

$$\sum_{y=0}^{k} a^{y} = \frac{1-a^{k+1}}{1-a}$$

Using this in Eq. (2.4), with $a = 1 - p$ and $k = x - 1$, gives

$$F(x) = p \cdot \frac{1 - (1-p)^{x}}{1 - (1-p)} = 1 - (1-p)^{x} \qquad x \text{ a positive integer}$$

Since F is constant in between positive integers,

$$F(x) = \begin{cases} 0 & x < 1 \\ 1 - (1-p)^{[x]} & x \geq 1 \end{cases} \qquad (2.5)$$

where $[x]$ is the largest integer $\leq x$ (e.g., $[2.7] = 2$). Thus if $p = .51$ as in the birth example, then the probability of having to examine at most five births to see the first boy is $F(5) = 1 - (.49)^{5} = 1 - .0282 = .9718$, whereas $F(10) \approx 1.0000$. This cdf is graphed in Fig. 2.6.

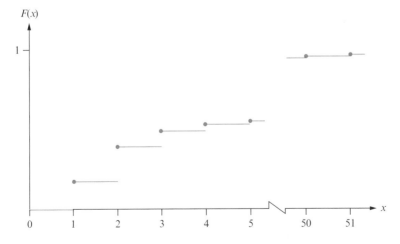

Fig. 2.6 A graph of $F(x)$ for Example 2.12   ∎

In our examples thus far, the cdf has been derived from the pmf. This process can be reversed to obtain the pmf from the cdf whenever the latter function is available. Suppose, for example, that X represents the number of defective components in a shipment consisting of six components, so that possible X values are 0, 1, ..., 6. Then

$$
\begin{aligned}
p(3) &= P(X = 3) \\
&= [p(0) + p(1) + p(2) + p(3)] - [p(0) + p(1) + p(2)] \\
&= P(X \le 3) - P(X \le 2) \\
&= F(3) - F(2)
\end{aligned}
$$

More generally, the probability that X falls in a specified interval is easily obtained from the cdf. For example,

$$
\begin{aligned}
P(2 \le X \le 4) &= p(2) + p(3) + p(4) \\
&= [p(0) + \cdots + p(4)] - [p(0) + p(1)] \\
&= P(X \le 4) - P(X \le 1) \\
&= F(4) - F(1)
\end{aligned}
$$

Notice that $P(2 \le X \le 4) \ne F(4) - F(2)$. This is because the X value 2 is included in $2 \le X \le 4$, so we do not want to subtract out its probability. However, $P(2 < X \le 4) = F(4) - F(2)$ because $X = 2$ is not included in the interval $2 < X \le 4$.

PROPOSITION

For any two numbers a and b with $a \le b$,

$$
P(a \le X \le b) = F(b) - F(a-)
$$

where "$a-$" represents the largest possible X value that is strictly less than a. In particular, if the only possible values are integers and if a and b are integers, then

$$
\begin{aligned}
P(a \le X \le b) &= P(X = a \text{ or } a + 1 \text{ or} \ldots \text{or } b) \\
&= F(b) - F(a - 1)
\end{aligned}
$$

Taking $a = b$ yields $P(X = a) = F(a) - F(a - 1)$ in this case.

The reason for subtracting $F(a-)$ rather than $F(a)$ is that we want to include $P(X = a)$; $F(b) - F(a)$ gives $P(a < X \le b)$. This proposition will be used extensively when computing binomial and Poisson probabilities in Sects. 2.4 and 2.5.

Example 2.13 Let $X =$ the number of days of sick leave taken by a randomly selected employee of a large company during a particular year. If the maximum number of allowable sick days per year is 14, possible values of X are 0, 1, ..., 14. With $F(0) = .58$, $F(1) = .72$, $F(2) = .76$, $F(3) = .81$, $F(4) = .88$, and $F(5) = .94$,

$$P(2 \leq X \leq 5) = P(X = 2, 3, 4, \text{ or } 5) = F(5) - F(1) = .22$$

and

$$P(X = 3) = F(3) - F(2) = .05 \qquad \blacksquare$$

2.2.3 Another View of Probability Mass Functions

It is often helpful to think of a pmf as specifying a mathematical model for a discrete population.

Example 2.14 Consider selecting at random a household in a certain region, and Let $X =$ the number of individuals in the selected household. Suppose the pmf of X is as follows:

x	1	2	3	4	5	6	7	8	9	10
$p(x)$	.140	.175	.220	.260	.155	.025	.015	.005	.004	.001

This is very close to the household size distribution for rural Thailand given in the article "The Probability of Containment for Multitype Branching Process Models for Emerging Epidemics" (*J. of Applied Probability*, 2011: 173–188), which modeled influenza transmission.

Suppose this is based on one million households. One way to view this situation is to think of the population as consisting of 1,000,000 households, each one having its own X value; the proportion with each X value is given by $p(x)$ in the above table. An alternative viewpoint is to forget about the households and think of the population itself as consisting of X values—14% of these values are 1, 17.5% are 2, and so on. The pmf then describes the distribution of the possible population values 1, 2, ..., 10. $\blacksquare$

Once we have such a population model, we will use it to compute values of various population characteristics such as the *mean*, which describes the center of the population distribution, and the *standard deviation*, which describes the extent of spread about the center. Both of these are developed in the next section.

2.2.4   Exercises: Section 2.2 (11–28)

11. Let X be the number of students who show up at a professor's office hours on a particular day. Suppose that the only possible values of X are 0, 1, 2, 3, and 4, and that $p(0) = .30$, $p(1) = .25$, $p(2) = .20$, and $p(3) = .15$.
    (a) What is $p(4)$?
    (b) Draw both a line graph and a probability histogram for the pmf of X.
    (c) What is the probability that at least two students come to the office hour? What is the probability that more than two students come to the office hour?
    (d) What is the probability that the professor shows up for his office hour?

12. Airlines sometimes overbook flights. Suppose that for a plane with 50 seats, 55 passengers have tickets. Define the random variable Y as the number of ticketed passengers who actually show up for the flight. The probability mass function of Y appears in the accompanying table.

y	45	46	47	48	49	50	51	52	53	54	55
$p(y)$	.05	.10	.12	.14	.25	.17	.06	.05	.03	.02	.01

   (a) What is the probability that the flight will accommodate all ticketed passengers who show up?
   (b) What is the probability that not all ticketed passengers who show up can be accommodated?
   (c) If you are the first person on the standby list (which means you will be the first one to get on the plane if there are any seats available after all ticketed passengers have been accommodated), what is the probability that you will be able to take the flight? What is this probability if you are the third person on the standby list?

13. A mail-order computer business has six telephone lines. Let X denote the number of lines in use at a specified time. Suppose the pmf of X is as given in the accompanying table.

x	0	1	2	3	4	5	6
$p(x)$	.10	.15	.20	.25	.20	.06	.04

   Calculate the probability of each of the following events.
   (a) {at most three lines are in use}
   (b) {fewer than three lines are in use}
   (c) {at least three lines are in use}
   (d) {between two and five lines, inclusive, are in use}
   (e) {between two and four lines, inclusive, are not in use}
   (f) {at least four lines are not in use}

14. A contractor is required by a county planning department to submit one, two, three, four, or five forms (depending on the nature of the project) in applying

for a building permit. Let $Y =$ the number of forms required of the next applicant. The probability that y forms are required is known to be proportional to y—that is, $p(y) = ky$ for $y = 1, \ldots, 5$.

(a) What is the value of k? [*Hint:* $\sum_{y=1}^{5} p(y) = 1$.]

(b) What is the probability that at most three forms are required?

(c) What is the probability that between two and four forms (inclusive) are required?

(d) Could $p(y) = y^2/50$ for $y = 1, \ldots, 5$ be the pmf of Y?

15. Many manufacturers have quality control programs that include inspection of incoming materials for defects. Suppose a computer manufacturer receives computer boards in lots of five. Two boards are selected from each lot for inspection. We can represent possible outcomes of the selection process by pairs. For example, the pair $(1, 2)$ represents the selection of boards 1 and 2 for inspection.

(a) List the ten different possible outcomes.

(b) Suppose that boards 1 and 2 are the only defective boards in a lot of five. Two boards are to be chosen at random. Define X to be the number of defective boards observed among those inspected. Find the probability distribution of X.

(c) Let $F(x)$ denote the cdf of X. First determine $F(0) = P(X \le 0)$, $F(1)$, and $F(2)$, and then obtain $F(x)$ for all other x.

16. Some parts of California are particularly earthquake-prone. Suppose that in one such area, 25% of all homeowners are insured against earthquake damage. Four homeowners are to be selected at random; let X denote the number among the four who have earthquake insurance.

(a) Find the probability distribution of X. [*Hint:* Let S denote a homeowner who has insurance and F one who does not. Then one possible outcome is $SFSS$, with probability $(.25)(.75)(.25)(.25)$ and associated X value 3. There are 15 other outcomes.]

(b) Draw the corresponding probability histogram.

(c) What is the most likely value for X?

(d) What is the probability that at least two of the four selected have earthquake insurance?

17. A new battery's voltage may be acceptable (A) or unacceptable (U). A certain flashlight requires two batteries, so batteries will be independently selected and tested until two acceptable ones have been found. Suppose that 90% of all batteries have acceptable voltages. Let Y denote the number of batteries that must be tested.

(a) What is $p(2)$, that is, $P(Y = 2)$?

(b) What is $p(3)$? [*Hint:* There are two different outcomes that result in $Y = 3$.]

(c) To have $Y = 5$, what must be true of the fifth battery selected? List the four outcomes for which $Y = 5$ and then determine $p(5)$.

(d) Use the pattern in your answers for parts (a)–(c) to obtain a general formula for $p(y)$.

18. Two fair six-sided dice are tossed independently. Let $M =$ the maximum of the two tosses, so $M(1, 5) = 5$, $M(3, 3) = 3$, etc.
    (a) What is the pmf of M? [*Hint:* First determine $p(1)$, then $p(2)$, and so on.]
    (b) Determine the cdf of M and graph it.

19. A library subscribes to two different weekly news magazines, each of which is supposed to arrive in Wednesday's mail. In actuality, each one may arrive on Wednesday, Thursday, Friday, or Saturday. Suppose the two arrive independently of one another, and for each one $P(W) = .3, P(Th) = .4, P(F) = .2$, and $P(S) = .1$. Let $Y =$ the number of days beyond Wednesday that it takes for both magazines to arrive (so possible Y values are 0, 1, 2, or 3). Compute the pmf of Y. [*Hint:* There are 16 possible outcomes; $Y(W, W) = 0$, $Y(F, Th) = 2$, and so on.]

20. Three couples and two single individuals have been invited to an investment seminar and have agreed to attend. Suppose the probability that any particular couple or individual arrives late is .4 (a couple will travel together in the same vehicle, so either both people will be on time or else both will arrive late). Assume that different couples and individuals are on time or late independently of one another. Let $X =$ the number of people who arrive late for the seminar.
    (a) Determine the probability mass function of X. [*Hint:* label the three couples #1, #2, and #3 and the two individuals #4 and #5.]
    (b) Obtain the cumulative distribution function of X, and use it to calculate $P(2 \leq X \leq 6)$.

21. As described in the book's Introduction, *Benford's Law* arises in a variety of situations as a model for the first digit of a number:

    $$p(x) = P(\text{1st digit is } x) = \log_{10}\left(\frac{x+1}{x}\right), \quad x = 1, 2, \ldots, 9$$

    (a) Without computing individual probabilities from this formula, show that it specifies a legitimate pmf.
    (b) Now compute the individual probabilities and compare to the distribution where 1, 2, ... , 9 are equally likely.
    (c) Obtain the cdf of X, a rv following Benford's law.
    (d) Using the cdf, what is the probability that the leading digit is at most 3? At least 5?

22. Refer to Exercise 13, and calculate and graph the cdf $F(x)$. Then use it to calculate the probabilities of the events given in parts (a)–(d) of that problem.

23. Let X denote the number of vehicles queued up at a bank's drive-up window at a particular time of day. The cdf of X is as follows:

$$F(x) = \begin{cases} 0 & x < 0 \\ .06 & 0 \le x < 1 \\ .19 & 1 \le x < 2 \\ .39 & 2 \le x < 3 \\ .67 & 3 \le x < 4 \\ .92 & 4 \le x < 5 \\ .97 & 5 \le x < 6 \\ 1 & 6 \le x \end{cases}$$

Calculate the following probabilities directly from the cdf:

(a) $p(2)$, that is, $P(X = 2)$
(b) $P(X > 3)$
(c) $P(2 \le X \le 5)$
(d) $P(2 < X < 5)$

24. An insurance company offers its policyholders a number of different premium payment options. For a randomly selected policyholder, let $X =$ the number of months between successive payments. The cdf of X is as follows:

$$F(x) = \begin{cases} 0 & x < 1 \\ .30 & 1 \le x < 3 \\ .40 & 3 \le x < 4 \\ .45 & 4 \le x < 6 \\ .60 & 6 \le x < 12 \\ 1 & 12 \le x \end{cases}$$

(a) What is the pmf of X?
(b) Using just the cdf, compute $P(3 \le X \le 6)$ and $P(4 \le X)$.

25. In Example 2.10, let $Y =$ the number of girls born before the experiment terminates. With $p = P(B)$ and $1 - p = P(G)$, what is the pmf of Y? [*Hint:* First list the possible values of Y, starting with the smallest, and proceed until you see a general formula.]

26. Alvie Singer lives at 0 in the accompanying diagram and has four friends who live at $A, B, C,$ and D. One day Alvie decides to go visiting, so he tosses a fair coin twice to decide which of the four to visit. Once at a friend's house, he will either return home or else proceed to one of the two adjacent houses (such as $0, A,$ or C when at B), with each of the three possibilities having probability 1/3. In this way, Alvie continues to visit friends until he returns home.

(a) Let $X =$ the number of times that Alvie visits a friend. Derive the pmf of X.

(b) Let $Y =$ the number of straight-line segments that Alvie traverses (including those leading to and from 0). What is the pmf of Y?

(c) Suppose that female friends live at A and C and male friends at B and D. If $Z =$ the number of visits to female friends, what is the pmf of Z?

27. After all students have left the classroom, a statistics professor notices that four copies of the text were left under desks. At the beginning of the next lecture, the professor distributes the four books in a completely random fashion to each of the four students (1, 2, 3, and 4) who claim to have left books. One possible outcome is that 1 receives 2's book, 2 receives 4's book, 3 receives his or her own book, and 4 receives 1's book. This outcome can be abbreviated as (2, 4, 3, 1).

(a) List the other 23 possible outcomes.

(b) Let X denote the number of students who receive their own book. Determine the pmf of X.

28. Show that the cdf $F(x)$ is a nondecreasing function; that is, $x_1 < x_2$ implies that $F(x_1) \le F(x_2)$. Under what condition will $F(x_1) = F(x_2)$?

2.3   Expected Value and Standard Deviation

Consider a university with 15,000 students and let $X =$ the number of courses for which a randomly selected student is registered. The pmf of X follows. Since $p(1) = .01$, we know that $(.01) \cdot (15,000) = 150$ of the students are registered for one course, and similarly for the other x values.

x	1	2	3	4	5	6	7	
$p(x)$	.01	.03	.13	.25	.39	.17	.02	(2.6)
Number registered	150	450	1950	3750	5850	2550	300	

To compute the average number of courses per student, i.e., the average value of X in the population, we should calculate the total number of courses and divide by the total number of students. Since each of 150 students is taking one course, these 150 contribute 150 courses to the total. Similarly, 450 students contribute 2(450) courses, and so on. The population average value of X is then

$$\frac{1(150) + 2(450) + 3(1950) + \cdots + 7(300)}{15{,}000} = 4.57 \qquad (2.7)$$

Since $150/15{,}000 = .01 = p(1)$, $450/15{,}000 = .03 = p(2)$, and so on, an alternative expression for Eq. (2.7) is

$$1 \cdot p(1) + 2 \cdot p(2) + \cdots + 7 \cdot p(7) \qquad (2.8)$$

Expression (2.8) shows that to compute the population average value of X, we need only the possible values of X along with their probabilities (proportions).

In particular, the population size is irrelevant as long as the pmf is given by (2.6). The average or mean value of X is then a *weighted* average of the possible values $1, \ldots, 7$, where the weights are the probabilities of those values.

2.3.1   The Expected Value of X

DEFINITION
Let X be a discrete rv with set of possible values D and pmf $p(x)$. The **expected value** or **mean value** of X, denoted by $E(X)$ or μ_X or just μ, is

$$E(X) = \mu_X = \mu = \sum_{x \in D} x \cdot p(x)$$

Example 2.15 For the pmf of $X =$ number of courses in (2.6),

$$\begin{aligned}
\mu &= 1 \cdot p(1) + 2 \cdot p(2) + \cdots + 7 \cdot p(7) \\
&= (1)(.01) + (2)(.03) + \cdots + (7)(.02) \\
&= .01 + .06 + .39 + 1.00 + 1.95 + 1.02 + .14 = 4.57
\end{aligned}$$

If we think of the population as consisting of the X values $1, 2, \ldots, 7$, then $\mu = 4.57$ is the population mean (we will often refer to μ as the *population mean* rather than the mean of X in the population). Notice that μ here is not 4, the ordinary average of $1, \ldots, 7$, because the distribution puts more weight on 4, 5, and 6 than on other X values. ∎

In Example 2.15, the expected value μ was 4.57, which is not a possible value of X. The word *expected* should be interpreted with caution because one would not expect to see an X value of 4.57 when a single student is selected.

Example 2.16 Just after birth, each newborn child is rated on a scale called the Apgar scale. The possible ratings are $0, 1, \ldots, 10$, with the child's rating determined by color, muscle tone, respiratory effort, heartbeat, and reflex irritability (the best possible score is 10). Let X be the Apgar score of a randomly selected child born at a certain hospital during the next year, and suppose that the pmf of X is

x	0	1	2	3	4	5	6	7	8	9	10
$p(x)$	.002	.001	.002	.005	.02	.04	.18	.37	.25	.12	.01

Then the mean value of X is

$$E(X) = \mu = (0)(.002) + (1)(.001) + (2)(.002) + \cdots + (8)(.25) + (9)(.12) + (10)(.01)$$
$$= 7.15$$

(Again, μ is not a possible value of the variable X.) If the stated model is correct, then the mean Apgar score for the population of all children born at this hospital next year will be 7.15.   ∎

Example 2.17 Let $X = 1$ if a randomly selected component needs warranty service and $= 0$ otherwise. If the chance a component needs warranty service is p, then X is a Bernoulli rv with pmf $p(1) = p$ and $p(0) = 1 - p$, from which

$$E(X) = 0 \cdot p(0) + 1 \cdot p(1) = 0(1 - p) + 1(p) = p$$

That is, the expected value of X is just the probability that X takes on the value 1. If we conceptualize a population consisting of 0s in proportion $1 - p$ and 1s in proportion p, then the population average is $\mu = p$.   ∎

There is another frequently used interpretation of μ. Consider observing a first value x_1 of X, then a second value x_2, a third value x_3, and so on. After doing this a large number of times, calculate the sample average of the observed x_is. This average will typically be close to μ; a more rigorous version of this statement is provided by the *Law of Large Numbers* in Chap. 4. That is, μ can be interpreted as the long-run average value of X when the experiment is performed repeatedly. This interpretation is often appropriate for games of chance, where the "population" is not a concrete set of individuals but rather the results of all hypothetical future instances of playing the game.

Example 2.18 A standard American roulette wheel has 38 spaces. Players bet on which space a marble will land in once the wheel has been spun. One of the simplest bets is based on the color of the space: 18 spaces are black, 18 are red, and 2 are green. So, if a player "bets on black," s/he has an 18/38 chance of winning. Casinos consider color bets an "even wager," meaning that a player who wages $1 on black, say, will profit $1 if the marble lands in a black space (and lose the wagered $1 otherwise).

Let $X =$ the return on a $1 wager on black. Then the pmf of X is

x	$-\$1$	$+\$1$
$p(x)$	20/38	18/38

and the expected value of X is $E(X) = (-1)(20/38) + (1)(18/38) = -2/38 = -\$.0526$. If a player makes $1 bets on black on successive spins of the roulette wheel, in the long run s/he can expect to lose about 5.26 cents per wager. Since players don't necessarily make a large number of wagers, this long-run average interpretation is perhaps more apt from the casino's perspective: in the long run, they will gain an average of 5.26 cents for every \$1 wagered on black at the roulette table. ∎

Thus far, we have assumed that the mean of any given distribution exists. If the set of possible values of X is unbounded, so that the sum for μ_X is actually an infinite series, the expected value of X might or might not exist (depending on whether the series converges or diverges).

Example 2.19 From Example 2.10, the general form for the pmf of $X =$ the number of children born up to and including the first boy is

$$p(x) = \begin{cases} (1-p)^{x-1}p & x = 1, 2, 3, \dots \\ 0 & \text{otherwise} \end{cases}$$

The expected value of X therefore entails evaluating an infinite summation:

$$E(X) = \sum_D x \cdot p(x) = \sum_{x=1}^{\infty} xp(1-p)^{x-1} = p\sum_{x=1}^{\infty} x(1-p)^{x-1}$$

$$= p\sum_{x=1}^{\infty} \left[-\frac{d}{dp}(1-p)^x \right] \tag{2.9}$$

If we interchange the order of taking the derivative and the summation in Eq. (2.9), the sum is that of a geometric series. (In particular, the infinite series converges for $0 < p < 1$.)

After the sum is computed and the derivative is taken, the final result is $E(X) = 1/p$. That is, the expected number of children born up to and including the first boy is the reciprocal of the chance of getting a boy. This is actually quite intuitive: if p is near 1, we expect to see a boy very soon, whereas if p is near 0, we expect many births before the first boy. For $p = .5$, $E(X) = 2$.

Exercise 48 at the end of this section presents an alternative method for computing the mean of this particular distribution. ∎

Example 2.20 Let X, the number of interviews a student has prior to getting a job, have pmf

$$p(x) = \begin{cases} k/x^2 & x = 1, 2, 3, \dots \\ 0 & \text{otherwise} \end{cases}$$

where k is such that $\sum_{x=1}^{\infty} (k/x^2) = 1$. (Because $\sum_{x=1}^{\infty} (1/x^2) = \pi^2/6$, the value of k is $6/\pi^2$.) The expected value of X is

$$\mu = E(X) = \sum_{x=1}^{\infty} x \frac{k}{x^2} = k \sum_{x=1}^{\infty} \frac{1}{x} \tag{2.10}$$

The sum on the right of Eq. (2.10) is the famous *harmonic series* of mathematics and can be shown to diverge. $E(X)$ is not finite here because $p(x)$ does not decrease sufficiently fast as x increases; statisticians say that the probability distribution of X has "a heavy tail." If a sequence of X values is chosen using this distribution, the sample average will not settle down to some finite number but will tend to grow without bound. ∎

2.3.2   The Expected Value of a Function

Often we will be interested in the expected value of some function $h(X)$ rather than X itself. An easy way of computing the expected value of $h(X)$ is suggested by the following example.

Example 2.21 The cost of a certain vehicle diagnostic test depends on the number of cylinders X in the vehicle's engine. Suppose the cost function is $h(X) = 20 + 3X + .5X^2$. Since X is a random variable, so is $Y = h(X)$. The pmf of X and the derived pmf of Y are as follows:

x	4	6	8		y	40	56	76
$p(x)$	.5	.3	.2	⇒	$p(y)$	.5	.3	.2

With D^* denoting possible values of Y,

$$\begin{aligned}
E(Y) = E\big[h(X)\big] &= \sum_{y \in D*} y \cdot p(y) \\
&= (40)(.5) + (56)(.3) + (76)(.2) = \$52 \\
&= h(4) \cdot (.5) + h(6) \cdot (.3) + h(8) \cdot (.2) \\
&= \sum_{D} h(x) \cdot p(x)
\end{aligned} \tag{2.11}$$

According to Eq. (2.11), it was not necessary to determine the pmf of Y to obtain $E(Y)$; instead, the desired expected value is a weighted average of the possible $h(x)$ (rather than x) values. ∎

PROPOSITION

If the rv X has a set of possible values D and pmf $p(x)$, then the expected value of any function $h(X)$, denoted by $E[h(X)]$ or $\mu_{h(X)}$, is computed by

(continued)

$$E[h(X)] = \sum_D h(x) \cdot p(x)$$

This is sometimes referred to as the *Law of the Unconscious Statistician.*

According to this proposition, $E[h(X)]$ is computed in the same way that $E(X)$ itself is, except that $h(x)$ is substituted in place of x. That is, $E[h(X)]$ is a weighted average of possible $h(X)$ values, where the weights are the probabilities of the corresponding original X values.

Example 2.22 A computer store has purchased three computers at $500 apiece. It will sell them for $1,000 apiece. The manufacturer has agreed to repurchase any computers still unsold after a specified period at $200 apiece. Let X denote the number of computers sold, and suppose that $p(0) = .1$, $p(1) = .2$, $p(2) = .3$, and $p(3) = .4$. With $h(X)$ denoting the profit associated with selling X units, the given information implies that $h(X) = $ revenue $-$ cost $= 1000X + 200(3 - X) - 1500 = 800X - 900$. The expected profit is then

$$
\begin{aligned}
E[h(X)] &= h(0) \cdot p(0) + h(1) \cdot p(1) + h(2) \cdot p(2) + h(3) \cdot p(3) \\
&= (800(0) - 900)(.1) + (800(1) - 900)(.2) + (800(2) - 900)(.3) \\
&\quad + (800(3) - 900)(.4) \\
&= (-900)(.1) + (-100)(.2) + (700)(.3) + (1500)(.4) \\
&= \$700
\end{aligned}
$$ ■

Because an expected value is a sum, it possesses the same properties as any summation; specifically, the expected value "operator" can be distributed across addition and across multiplication by constants. This important property is known as *linearity of expectation.*

LINEARITY OF EXPECTATION

For any functions $h_1(X)$ and $h_2(X)$ and any constants a_1, a_2, and b,

$$E[a_1h_1(X) + a_2h_2(X) + b] = a_1E[h_1(X)] + a_2E[h_2(X)] + b$$

In particular, for any linear function $aX + b$,

$$E(aX + b) = a \cdot E(X) + b \qquad (2.12)$$

(or, using alternative notation, $\mu_{aX+b} = a \cdot \mu_X + b$).

Proof Let $h(X) = a_1 h_1(X) + a_2 h_2(X) + b$, and apply the previous proposition:

$$E\big[a_1 h_1(X) + a_2 h_2(X) + b\big] = \sum_D \big(a_1 h_1(x) + a_2 h_2(x) + b\big) \cdot p(x)$$

$$= a_1 \sum_D h_1(x) \cdot p(x) + a_2 \sum_D h_2(x) \cdot p(x)$$

$$+ b \sum_D p(x) \qquad \text{distributive property of addition}$$

$$= a_1 E\big[h_1(X)\big] + a_2 E\big[h_2(X)\big] + b[1]$$
$$= a_1 E\big[h_1(X)\big] + a_2 E\big[h_2(X)\big] + b$$

The special case of $aX + b$ is obtained by setting $a_1 = a$, $h_1(X) = X$, and $a_2 = 0$.   ∎

By induction, linearity of expectation applies to any finite number of terms. In Example 2.21, it is easily computed that $E(X) = 4(.5) + 6(.3) + 8(.2) = 5.4$ and $E(X^2) = \sum x^2 \cdot p(x) = 4^2(.5) + 6^2(.3) + 8^2(.2) = 31.6$. Applying linearity of expectation to $Y = h(X) = 20 + 3X + .5X^2$, we obtain

$$\mu_Y = E\big[20 + 3X + .5X^2\big] = 20 + 3E(X) + .5E(X^2) = 20 + 3(5.4) + .5(31.6) = \$52,$$

which matches the result of Example 2.21.

The special case Eq. (2.12) states that the expected value of a linear function equals the linear function evaluated at the expected value $E(X)$. Since $h(X)$ in Example 2.22 is linear and $E(X) = 2$, $E[h(X)] = 800(2) - 900 = \$700$, as before. Two special cases of Eq. (2.12) yield two important rules of expected value.

1. For any constant a, $\mu_{aX} = a \cdot \mu_X$ (take $b = 0$).
2. For any constant b, $\mu_{X+b} = \mu_X + b = E(X) + b$ (take $a = 1$).

Multiplication of X by a constant a changes the unit of measurement (from dollars to cents, where $a = 100$, inches to cm, where $a = 2.54$, etc.). Rule 1 says that the expected value in the new units equals the expected value in the old units multiplied by the conversion factor a. Similarly, if the constant b is added to each possible value of X, then the expected value will be shifted by that same amount.

One commonly made error is to substitute μ_X directly into the function $h(X)$ when h is a nonlinear function, in which case Eq. (2.12) does not apply. Consider Example 2.21: the mean of X is 5.4, and it's tempting to infer that the mean of $Y = h(X)$ is simply $h(5.4)$. However, since the function $h(X) = 20 + 3X + .5X^2$ is *not* linear, this does not yield the correct answer:

$$h(5.4) = 20 + 3(5.4) + .5(5.4)^2 = 50.78 \neq 52 = \mu_Y$$

In general, $\mu_{h(X)}$ *does not equal* $h(\mu_X)$ *unless the function* $h(x)$ *is linear.*

2.3.3 The Variance and Standard Deviation of X

The expected value of X describes where the probability distribution is centered. Using the physical analogy of placing point mass $p(x)$ at the value x on a one-dimensional axis, if the axis were then supported by a fulcrum placed at μ, there would be no tendency for the axis to tilt. This is illustrated for two different distributions in Fig. 2.7.

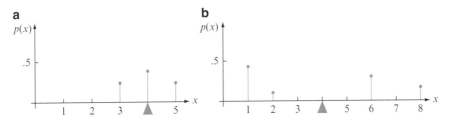

Fig. 2.7 Two different probability distributions with $\mu = 4$

Although both distributions pictured in Fig. 2.7 have the same mean/fulcrum μ, the distribution of Fig. 2.7b has greater spread or variability or dispersion than does that of Fig. 2.7a. Our goal now is to obtain a quantitative assessment of the extent to which the distribution spreads out about its mean value.

DEFINITION

Let X have pmf $p(x)$ and expected value μ. Then the **variance** of X, denoted by $\mathrm{Var}(X)$ or σ_X^2 or just σ^2, is

$$\mathrm{Var}(X) = \sum_D \left[(x - \mu)^2 \cdot p(x) \right] = E\left[(X - \mu)^2 \right]$$

The **standard deviation** (SD) of X, denoted by $\mathrm{SD}(X)$ or σ_X or just σ, is

$$\sigma_X = \sqrt{\mathrm{Var}(X)}$$

The quantity $h(X) = (X - \mu)^2$ is the squared deviation of X from its mean, and σ^2 is the expected squared deviation—i.e., a weighted average of the squared deviations from μ. Taking the square root of the variance to obtain standard deviation returns us to the original units of the variable, e.g., if X is measured in dollars, then both μ and σ also have units of dollars. If most of the probability distribution is close to μ, as in Fig. 2.7a, then σ will typically be relatively small. However, if there are x values far from μ that have large probabilities (as in Fig. 2.7b), then σ will be larger.

Example 2.23 Consider again the distribution of the Apgar score X of a randomly selected newborn described in Example 2.16. The mean value of X was calculated as $\mu = 7.15$, so

$$\text{Var}(X) = \sigma^2 = \sum_{x=0}^{10} (x - 7.15)^2 \cdot p(x)$$

$$= (0 - 7.15)^2(.002) + \ldots + (10 - 7.15)^2(.01) = 1.5815$$

The standard deviation of X is $\text{SD}(X) = \sigma = \sqrt{1.5815} = 1.26$. ∎

A rough interpretation of σ is that its value gives the size of a typical or representative distance from μ (hence, "standard deviation"). Because $\sigma = 1.26$ in the preceding example, we can say that some of the possible X values differ by more than 1.26 from the mean value 7.15 whereas other possible X values are closer than this to 7.15; roughly, 1.26 is the size of a typical deviation from the mean Apgar score.

Example 2.24 (Example 2.18 continued) The variance of X = the return on a \$1 bet on black is

$$\sigma_X^2 = (-1 - (-2/38))^2 \cdot (20/38) + (1 - (-2/38))^2 \cdot 18/38 = 0.99723$$

and the standard deviation is $\sigma_X = \sqrt{0.99723} = 0.9986 \approx \$1$. The two possible values of X are $-\$1$ and $+\$1$; since betting on black is almost a break-even wager (the mean is quite close to 0), the typical difference between an actual return X and the average return μ_X is roughly one dollar. ∎

A natural probability question arises: how often does X fall within this "typical distance of the mean"? That is, what's the chance that a rv X lies between $\mu_X - \sigma_X$ and $\mu_X + \sigma_X$? What about the likelihood that X is within two standard deviations of its mean? There are no universal answers: for different pmfs, varying amounts of probability may lie within one (or two or three) standard deviation(s) of the expected value. That said, the following theorem, due to Russian mathematician Pafnuty Chebyshev, partially addresses questions of this sort.

CHEBYSHEV'S INEQUALITY

Let X be a discrete rv with mean μ and standard deviation σ. Then, for any $k \geq 1$,

$$P(|X - \mu| \geq k\sigma) \leq \frac{1}{k^2}$$

That is, the probability X is at least k standard deviations away from its mean is at most $1/k^2$.

An equivalent statement to Chebyshev's inequality is that every random variable has a probability of at least $1 - 1/k^2$ to fall within k standard deviations of its mean.

Proof Let A denote the event $|X - \mu| \geq k\sigma$; or, equivalently, the set of values $\{x : |x - \mu| \geq k\sigma\}$. Begin by writing out the definition of $\text{Var}(X)$:

$$\text{Var}(X) = \sum_D \left[(x - \mu)^2 \cdot p(x) \right]$$

$$= \sum_A \left[(x - \mu)^2 \cdot p(x) \right] + \sum_{A'} \left[(x - \mu)^2 \cdot p(x) \right]$$

$$\geq \sum_A \left[(x - \mu)^2 \cdot p(x) \right] \quad \text{because the discarded term is} \geq 0$$

$$\geq \sum_A \left[(k\sigma)^2 \cdot p(x) \right] \quad \text{because } (x - \mu)^2 \geq (k\sigma)^2 \text{ on the set } A$$

$$= (k\sigma)^2 \sum_A p(x) = (k\sigma)^2 P(A) = k^2 \sigma^2 P\big(|X - \mu| \geq k\sigma\big)$$

The $\text{Var}(X)$ term on the left-hand side is the same as the σ^2 term on the right-hand side; cancelling the two, we are left with $1 \geq k^2 P(|X - \mu| \geq k\sigma)$, and Chebyshev's inequality follows. ∎

For $k = 1$, Chebyshev's inequality states that $P(|X - \mu| \geq \sigma) \leq 1$, which isn't very informative since all probabilities are bounded above by 1. In fact, distributions can be constructed for which 100% of the distribution is at least 1 standard deviation from the mean, so that the rv X has probability 0 of falling less than one standard deviation from its mean (see Exercise 47). Substituting $k = 2$, Chebyshev's inequality states that the chance any rv is at least 2 standard deviations from its mean cannot exceed $1/2^2 = .25 = 25\%$. Equivalently, *every* distribution has the property that at least 75% of its "mass" lies within 2 standard deviations of its mean value (in fact, for many distributions, the exact probability is more).

2.3.4 Properties of Variance

An alternative to the defining formula for $\text{Var}(X)$ reduces the computational burden.

PROPOSITION

$$\text{Var}(X) = \sigma^2 = E\big(X^2\big) - \mu^2$$

This equation is referred to as the *variance shortcut formula*.

In using this formula, $E(X^2)$ is computed first without any subtraction; then μ is computed, squared, and subtracted (once) from $E(X^2)$. This formula is more efficient because it entails only one subtraction, and $E(X^2)$ does not require calculating squared deviations from μ.

Example 2.25   Referring back to the Apgar score scenario of Examples 2.16 and 2.23,

$$E(X^2) = \sum_{x=1}^{10} x^2 \cdot p(x) = (0^2)(.002) + (1^2)(.001) + \cdots + (10^2)(.01) = 52.704$$

Thus, $\sigma^2 = 52.704 - (7.15)^2 = 1.5815$ as before, and again $\sigma = 1.26$.   ∎

Proof of the Variance Shortcut Formula   Expand $(X - \mu)^2$ in the definition of Var(X), and then apply linearity of expectation:

$$\begin{aligned}
\mathrm{Var}(X) = E\big[(X - \mu)^2\big] &= E\big[X^2 - 2\mu X + \mu^2\big] \\
&= E(X^2) - 2\mu E(X) + \mu^2 \qquad \text{by linearity of expectation} \\
&= E(X^2) - 2\mu \cdot \mu + \mu^2 = E(X^2) - 2\mu^2 + \mu^2 \\
&= E(X^2) - \mu^2
\end{aligned}$$

∎

The quantity $E(X^2)$ in the variance shortcut formula is called the **mean-square value** of the random variable X. Engineers may be familiar with the root-mean-square, or RMS, which is the square root of $E(X^2)$. Do not confuse this with the square of the mean of X, i.e., μ^2! For example, if X has a mean of 7.15, the mean-square value of X is *not* $(7.15)^2$, because $h(x) = x^2$ is not linear. (In Example 2.25, the mean-square value of X is 52.704.) It helps to look at the two formulas side-by-side:

$$E(X^2) = \sum_{D} x^2 \cdot p(x) \quad \text{versus} \quad \mu^2 = \left(\sum_{D} x \cdot p(x) \right)^2$$

The order of operations is clearly different. In fact, it can be shown (see Exercise 46) that $E(X^2) \geq \mu^2$ for every random variable, with equality if and only if X is constant.

The variance of a function $h(X)$ is the expected value of the squared difference between $h(X)$ and its expected value:

$$\mathrm{Var}[h(X)] = \sigma_{h(X)}^2 = \sum_{D} \left[\left(h(x) - \mu_{h(X)} \right)^2 \cdot p(x) \right]$$

$$= \left[\sum_{D} h^2(x) \cdot p(x) \right] - \left[\sum_{D} h(x) \cdot p(x) \right]^2$$

When $h(x)$ is a linear function, Var[$h(X)$] has a much simpler expression (see Exercise 43 for a proof).

PROPOSITION

$$\text{Var}(aX + b) = \sigma^2_{aX+b} = a^2 \cdot \sigma^2_X \quad \text{and} \quad \sigma_{aX+b} = |a| \cdot \sigma_X \qquad (2.13)$$

In particular,

$$\sigma_{aX} = |a| \cdot \sigma_X \quad \text{and} \quad \sigma_{X+b} = \sigma_X$$

The absolute value is necessary because a might be negative, yet a standard deviation cannot be. Usually multiplication by a corresponds to a change in the unit of measurement (e.g., kg to lb or dollars to euros); the sd in the new unit is just the original sd multiplied by the conversion factor. On the other hand, the addition of the constant b does not affect the variance, which is intuitive, because the addition of b changes the location (mean value) but not the spread of values. Together, Eqs. (2.12) and (2.13) comprise the *rescaling properties* of mean and standard deviation.

Example 2.26 In the computer sales scenario of Example 2.22, $E(X) = 2$ and

$$E(X^2) = (0^2)(.1) + (1^2)(.2) + (2^2)(.3) + (3^2)(.4) = 5$$

so $\text{Var}(X) = 5 - (2)^2 = 1$. The profit function $Y = h(X) = 800X - 900$ is linear, so Eq. (2.13) applies with $a = 800$ and $b = -900$. Hence Y has variance $a^2\sigma_X^2 = (800)^2(1) = 640,000$ and standard deviation $\$800$. ∎

2.3.5 Exercises: Section 2.3 (29–48)

29. The pmf of the amount of memory X (GB) in a purchased flash drive was given in Example 2.11 as

x	1	2	4	8	16
$p(x)$	.05	.10	.35	.40	.10

(a) Compute and interpret $E(X)$.
(b) Compute $\text{Var}(X)$ directly from the definition.
(c) Obtain and interpret the standard deviation of X.
(d) Compute $\text{Var}(X)$ using the shortcut formula.

30. An individual who has automobile insurance from a company is randomly selected. Let Y be the number of moving violations for which the individual was cited during the last 3 years. The pmf of Y is

y	0	1	2	3
$p(y)$	.60	.25	.10	.05

(a) Compute $E(Y)$.
(b) Suppose an individual with Y violations incurs a surcharge of $\$100Y^2$. Calculate the expected amount of the surcharge.

31. Refer to Exercise 12 and calculate $Var(Y)$ and σ_Y. Then determine the probability that Y is within 1 standard deviation of its mean value.

32. An appliance dealer sells three different models of upright freezers having 13.5, 15.9, and 19.1 cubic feet of storage space, respectively. Let $X =$ the amount of storage space purchased by the next customer to buy a freezer. Suppose that X has pmf

x	13.5	15.9	19.1
$p(x)$	.2	.5	.3

(a) Compute $E(X)$, $E(X^2)$, and $Var(X)$.
(b) If the price of a freezer having capacity X cubic feet is $17X + 180$, what is the expected price paid by the next customer to buy a freezer?
(c) What is the standard deviation of the price $17X + 180$ paid by the next customer?
(d) Suppose that although the rated capacity of a freezer is X, the actual capacity is $h(X) = X - .01X^2$. What is the expected actual capacity of the freezer purchased by the next customer?

33. Let X be a Bernoulli rv with pmf as in Example 2.17.
(a) Compute $E(X^2)$.
(b) Show that $Var(X) = p(1 - p)$.
(c) Compute $E(X^{79})$.

34. Suppose that the number of plants of a particular type found in a rectangular sampling region (called a quadrat by ecologists) in a certain geographic area is an rv X with pmf

$$p(x) = \begin{cases} c/x^3 & x = 1, 2, 3, \ldots \\ 0 & \text{otherwise} \end{cases}$$

Is $E(X)$ finite? Justify your answer. (This is another distribution that statisticians would call heavy-tailed.)

35. A small market orders copies of a certain magazine for its magazine rack each week. Let $X =$ demand for the magazine, with pmf

x	1	2	3	4	5	6
$p(x)$	$\dfrac{1}{15}$	$\dfrac{2}{15}$	$\dfrac{3}{15}$	$\dfrac{4}{15}$	$\dfrac{3}{15}$	$\dfrac{2}{15}$

Suppose the store owner actually pays $2.00 for each copy of the magazine and the price to customers is $4.00. If magazines left at the end of the week have no salvage value, is it better to order three or four copies of the magazine? [*Hint:* For both three and four copies ordered, express net revenue as a function of demand X, and then compute the expected revenue.]

36. Let X be the damage incurred (in $) in a certain type of accident during a given year. Possible X values are 0, 1000, 5000, and 10,000, with probabilities .8, .1, .08, and .02, respectively. A particular company offers a $500 deductible policy. If the company wishes its expected profit to be $100, what premium amount should it charge?

37. The n candidates for a job have been ranked 1, 2, 3, ..., n. Let $X =$ the rank of a randomly selected candidate, so that X has pmf

$$p(x) = \begin{cases} 1/n & x = 1, 2, 3, \ldots, n \\ 0 & \text{otherwise} \end{cases}$$

(this is called the *discrete uniform distribution*). Compute $E(X)$ and $\text{Var}(X)$ using the shortcut formula. [*Hint:* The sum of the first n positive integers is $n(n + 1)/2$, whereas the sum of their squares is $n(n + 1)(2n + 1)/6$.]

38. Let $X =$ the outcome when a fair die is rolled once. If before the die is rolled you are offered either $100 dollars or $h(X) = 350/X$ dollars, would you accept the guaranteed amount or would you gamble? [*Hint:* Determine $E[h(X)]$, but be careful: the mean of $350/X$ is not $350/\mu$.]

39. In the popular game Plinko on *The Price Is Right*, contestants drop a circular disk (a "chip") down a pegged board; the chip bounces down the board and lands in a slot corresponding to one of five dollar mounts. The random variable $X =$ winnings from one chip dropped from the middle slot has roughly the following distribution.

x	$0	$100	$500	$1000	$10,000
$p(x)$	.39	.03	.11	.24	.23

(a) Graph the probability mass function of X.
(b) What is the probability a contestant makes money on a chip?
(c) What is the probability a contestant makes at least $1000 on a chip?
(d) Determine the expected winnings. Interpret this number.
(e) Determine the corresponding standard deviation.

40. A supply company currently has in stock 500 lb of fertilizer, which it sells to customers in 10-lb bags. Let X equal the number of bags purchased by a randomly selected customer. Sales data shows that X has the following pmf:

x	1	2	3	4
$p(x)$	.2	.4	.3	.1

(a) Compute the average number of bags bought per customer.

(b) Determine the standard deviation for the number of bags bought per customer.

(c) Define Y to be the amount of fertilizer left in stock, in pounds, after the first customer. Construct the pmf of Y.

(d) Use the pmf of Y to find the expected amount of fertilizer left in stock, in pounds, after the first customer.

(e) Write Y as a linear function of X. Then use rescaling properties to find the mean and standard deviation of Y.

(f) The supply company offers a discount to each customer based on the formula $W = (X - 1)^2$. Determine the expected discount for a customer.

(g) Does your answer in part (f) equal $(\mu_X - 1)^2$? Why or why not?

(h) Calculate the standard deviation of W.

41. Refer back to the roulette scenario in Examples 2.18 and 2.24. Two other ways to wager at roulette are betting on a single number, or on a four-number "square." The pmfs for the returns on a $1 wager on a number and a square are displayed below. (Payoffs for winning are always based on the odds of losing a wager under the assumption the two green spaces didn't exist.)

Single number:

x	$-\$1$	$+\$35$
$p(x)$	37/38	1/38

Square:

x	$-\$1$	$+\$8$
$p(x)$	34/38	4/38

(a) Determine the expected return from a $1 wager on a single number, and then on a square.

(b) Compare your answers from (a) to Example 2.18. What can be said about the expected return for a $1 wager? Based on this, does expected return reflect most players' intuition that betting on black is "safer" and betting on a single number is "riskier"?

(c) Now calculate the standard deviations for the two pmfs above.

(d) How do the standard deviations of the three betting schemes (color, single number, square) compare? How do these values appear to relate to players' intuitive sense of risk?

42. (a) Draw a line graph of the pmf of X in Exercise 35. Then determine the pmf of $-X$ and draw its line graph. From these two pictures, what can you say about $\text{Var}(X)$ and $\text{Var}(-X)$?
    (b) Use the proposition involving $\text{Var}(aX + b)$ to establish a general relationship between $\text{Var}(X)$ and $\text{Var}(-X)$.

43. Use the definition of variance to prove that $\text{Var}(aX+b)=a^2\sigma_X^2$. [*Hint:* From Eq. (2.12), $\mu_{aX+b} = a\mu_X + b$.]

44. Suppose $E(X) = 5$ and $E[X(X-1)] = 27.5$.
    (a) Determine $E(X^2)$. [*Hint:* $E[X(X-1)] = E(X^2 - X) = E(X^2) - E(X)$.]
    (b) What is $\text{Var}(X)$?
    (c) What is the general relationship among the quantities $E(X)$, $E[X(X-1)]$, and $\text{Var}(X)$?

45. Write a general rule for $E(X-c)$ where c is a constant. What happens when you let $c = \mu$, the expected value of X?

46. Let X be a rv with mean μ. Show that $E(X^2) \geq \mu^2$, and that $E(X^2) > \mu^2$ unless X is a constant. [*Hint:* Consider variance.]

47. Refer to Chebyshev's inequality in this section.
    (a) What is the value of the upper bound for $k = 2$? $k = 3$? $k = 4$? $k = 5$? $k = 10$?
    (b) Compute μ and σ for the distribution of Exercise 13. Then evaluate for the values of k given in part (a). What does this suggest about the upper bound relative to the corresponding probability?
    (c) Suppose you will win d if a fair coin flips heads and lose d if it lands tails. Let X be the amount you get from a single coin flip. Compute $E(X)$ and $SD(X)$. What is the probability X will be less than one standard deviation from its mean value?
    (d) Let X have three possible values, $-1, 0$, and 1, with probabilities $\frac{1}{18}, \frac{8}{9}$, and $\frac{1}{18}$ respectively. What is $P(|X - \mu| \geq 3\sigma)$, and how does it compare to the corresponding Chebyshev bound?
    (e) Give a distribution for which $P(|X - \mu| \geq 5\sigma) = .04$.

48. For a discrete rv X taking values in $\{0, 1, 2, 3, \ldots\}$, we shall derive the following alternative formula for the mean:

$$\mu_X = \sum_{x=0}^{\infty} [1 - F(x)]$$

    (a) Suppose for now the range of X is $\{0, 1, \ldots N\}$ for some positive integer N. By regrouping terms, show that

$$\sum_{x=0}^{N}[x \cdot p(x)] = p(1) + p(2) + p(3) + \cdots + p(N)$$
$$+ p(2) + p(3) + \cdots + p(N)$$
$$+ p(3) + \cdots + p(N)$$
$$\vdots$$
$$+ p(N)$$

(b) Rewrite each row in the above expression in terms of the cdf of X, and use this to establish that

$$\sum_{x=0}^{N}[x \cdot p(x)] = \sum_{x=0}^{N-1}[1 - F(x)]$$

(c) Let $N \to \infty$ in part (b) to establish the desired result, and explain why the resulting formula works even if the maximum value of X is finite. [*Hint:* If the largest possible value of X is N, what does $1 - F(x)$ equal for $x \geq N$?] (This derivation also implies that a discrete rv X has a finite mean iff the series $\sum[1 - F(x)]$ converges.)

(d) Let X have the pmf from Examples 2.10 and 2.19. Use the cdf of X and the alternative mean formula just derived to determine μ_X.

2.4   The Binomial Distribution

Many experiments conform either exactly or approximately to the following list of requirements:

1. The experiment consists of a sequence of n smaller experiments called *trials*, where n is fixed in advance of the experiment.
2. Each trial can result in one of the same two possible outcomes (dichotomous trials), which we denote by success (S) or failure (F).
3. The trials are independent, so that the outcome on any particular trial does not influence the outcome on any other trial.
4. The probability of success is constant from trial to trial (homogeneous trials); we denote this probability by p.

DEFINITION

An experiment for which Conditions 1–4 are satisfied—a fixed number of dichotomous, independent, homogeneous trials—is called **a binomial experiment**.

Example 2.27 The same coin is tossed successively and independently n times. We arbitrarily use S to denote the outcome H (heads) and F to denote the outcome

T (tails). Then this experiment satisfies Conditions 1–4. Tossing a thumbtack n times, with S = point up and F = point down, also results in a binomial experiment. ■

Some experiments involve a sequence of independent trials for which there are more than two possible outcomes on any one trial. A binomial experiment can then be created by dividing the possible outcomes into two groups.

Example 2.28 The color of pea seeds is determined by a single genetic locus. If the two alleles at this locus are AA or Aa (the genotype), then the pea will be yellow (the phenotype), and if the allele is aa, the pea will be green. Suppose we pair off 20 Aa seeds and cross the two seeds in each of the ten pairs to obtain ten new genotypes. Call each new genotype a success S if it is aa and a failure otherwise. Then with this identification of S and F, the experiment is binomial with $n = 10$ and $p = P(\text{aa genotype})$. If each member of the pair is equally likely to contribute a or A, then $p = P(\text{a}) \cdot P(\text{a}) = (1/2)(1/2) = .25$. ■

Example 2.29 A student acquaintance of yours has an iPod playlist containing 50 songs, of which 35 were recorded prior to the year 2010 and the other 15 were recorded more recently. Suppose the random play function is used to select five from among these 50 songs for listening during a walk between classes. Each selection of a song constitutes a trial; regard a trial as a success if the selected song was recorded before 2010. Then

$$P(S \text{ on first trial}) = \frac{35}{50} = .70$$

and

$$P(S \text{ on second trial}) = P(SS) + P(FS)$$
$$= P(\text{second } S | \text{first } S)P(\text{first } S)$$
$$+ P(\text{second } S | \text{first } F)P(\text{first } F)$$
$$= \frac{34}{49} \cdot \frac{35}{50} + \frac{35}{49} \cdot \frac{15}{50} = \frac{35}{50}\left(\frac{34}{49} + \frac{15}{49}\right) = \frac{35}{50} = .70$$

Similarly, it can be shown that $P(S \text{ on } i\text{th trial}) = .70$ for $i = 3, 4, 5$, so the trials are homogeneous. However,

$$P(S \text{ on fifth trial} | SSSS) = \frac{31}{46} = .67$$

whereas

$$P\left(S \text{ on fifth trial} \middle| FFFF\right) = \frac{35}{46} = .76$$

The experiment is *not* binomial because the trials are not independent. In general, if sampling is without replacement, the experiment will not yield independent trials. If songs had been selected *with* replacement, then trials would have been independent, but this might have resulted in the same song being listened to more than once. ∎

Example 2.30 Suppose a state has 500,000 licensed drivers, of whom 400,000 are insured. A sample of 10 drivers is chosen without replacement. The ith trial is labeled S if the ith driver chosen is insured. Although this situation would seem identical to that of Example 2.29, the important difference is that the size of the population being sampled is very large relative to the sample size. In this case

$$P\left(S \text{ on } 2 \middle| S \text{ on } 1\right) = \frac{399,999}{499,999} \approx .80000$$

and

$$P\left(S \text{ on } 10 \middle| S \text{ on first } 9\right) = \frac{399,991}{499,991} = .799996 \approx .80000$$

These calculations suggest that although the trials are not exactly independent, the conditional probabilities differ so slightly from one another that for practical purposes the trials can be regarded as independent with constant $P(S) = .8$. Thus, to a very good approximation, the experiment is binomial with $n = 10$ and $p = .8$. ∎

We will use the following convention in deciding whether a "without-replacement" experiment can be treated as being (approximately) binomial.

RULE
Consider sampling without replacement from a dichotomous population of size N. If the sample size (number of trials) n is at most 5% of the population size, the experiment can be analyzed as though it were exactly a binomial experiment.

By "analyzed," we mean that probabilities based on the binomial experiment assumptions will be quite close to the actual "without-replacement" probabilities, which are typically more difficult to calculate. In Example 2.29, $n/N = 5/50 = .1 > .05$, so the binomial experiment is not a good approximation, but in Example 2.30, $n/N = 10/500,000 < .05$.

2.4.1   The Binomial Random Variable and Distribution

In most binomial experiments, it is the total number of successes, rather than knowledge of exactly which trials yielded successes, that is of interest.

DEFINITION
Given a binomial experiment consisting of n trials, the **binomial random variable** X associated with this experiment is defined as
$$X = \text{the number of successes among the } n \text{ trials}$$

Suppose, for example, that $n = 3$. Then there are eight possible outcomes for the experiment:
$$SSS \; SSF \; SFS \; SFF \; FSS \; FSF \; FFS \; FFF$$

From the definition of X, $X(SSF) = 2$, $X(SFF) = 1$, and so on. Possible values for X in an n-trial experiment are $x = 0, 1, 2, \ldots, n$.

NOTATION
We will write $X \sim \text{Bin}(n, p)$ to indicate that X is a binomial rv based on n trials with success probability p. Because the pmf of a binomial rv X depends on the two parameters n and p, we denote the pmf by $b(x; n, p)$.

Our next goal is to derive a formula for the binomial pmf. Consider first the case $n = 4$ for which each outcome, its probability, and corresponding x value are listed in Table 2.1. For example,

$$
\begin{aligned}
P(SSFS) &= P(S) \cdot P(S) \cdot P(F) \cdot P(S) & \text{independent trials} \\
&= p \cdot p \cdot (1 - p) \cdot p & \text{constant } P(S) \\
&= p^3 \cdot (1 - p)
\end{aligned}
$$

In this special case, we wish to determine $b(x; 4, p)$ for $x = 0, 1, 2, 3,$ and 4. For $b(3; 4, p)$, we identify which of the 16 outcomes yield an x value of 3 and sum the probabilities associated with each such outcome:

$$b(3; 4, p) = P(FSSS) + P(SFSS) + P(SSFS) + P(SSSF) = 4p^3(1 - p)$$

There are four outcomes with $x = 3$ and each has probability $p^3(1 - p)$; the probability depends only on the number of S's, *not* the order of S's and F's. So

$$b(3; 4, p) = \left\{ \begin{array}{l} \text{number of outcomes} \\ \text{with } X = 3 \end{array} \right\} \cdot \left\{ \begin{array}{l} \text{probability of any particular} \\ \text{outcome with } X = 3 \end{array} \right\}$$

Table 2.1 Outcomes and probabilities for a binomial experiment with four trials

Outcome	x	Probability	Outcome	x	Probability
SSSS	4	p^4	FSSS	3	$p^3(1-p)$
SSSF	3	$p^3(1-p)$	FSSF	2	$p^2(1-p)^2$
SSFS	3	$p^3(1-p)$	FSFS	2	$p^2(1-p)^2$
SSFF	2	$p^2(1-p)^2$	FSFF	1	$p(1-p)^3$
SFSS	3	$p^3(1-p)$	FFSS	2	$p^2(1-p)^2$
SFSF	2	$p^2(1-p)^2$	FFSF	1	$p(1-p)^3$
SFFS	2	$p^2(1-p)^2$	FFFS	1	$p(1-p)^3$
SFFF	1	$p(1-p)^3$	FFFF	0	$(1-p)^4$

Similarly, $b(2; 4, p) = 6p^2(1-p)^2$, which is also the product of the number of outcomes with $X = 2$ and the probability of any such outcome.

In general,

$$b(x; n, p) = \left\{ \begin{array}{c} \text{number of sequences of} \\ \text{length } n \text{ consisting of } x \text{ } S\text{'s} \end{array} \right\} \cdot \left\{ \begin{array}{c} \text{probability of any} \\ \text{particular such sequence} \end{array} \right\}$$

Since the ordering of S's and F's is not important, the second factor in the previous equation is $p^x(1-p)^{n-x}$ (for example, the first x trials resulting in S and the last $n-x$ resulting in F). The first factor is the number of ways of choosing x of the n trials to be S's—that is, the number of combinations of size x that can be constructed from n distinct objects (trials here).

THEOREM

$$b(x; n, p) = \left\{ \begin{array}{ll} \binom{n}{x} p^x (1-p)^{n-x} & x = 0, 1, 2, \ldots, n \\ 0 & \text{otherwise} \end{array} \right.$$

Example 2.31 Each of six randomly selected cola drinkers is given a glass containing cola S and one containing cola F. The glasses are identical in appearance except for a code on the bottom to identify the cola. Suppose there is actually no tendency among cola drinkers to prefer one cola to the other. Then $p = P(\text{a selected individual prefers } S) = .5$, so with $X =$ the number among the six who prefer S, $X \sim \text{Bin}(6, .5)$.

Thus

$$P(X = 3) = b(3; 6, .5) = \binom{6}{3}(.5)^3(.5)^3 = 20(.5)^6 = .313$$

The probability that at least three prefer S is

$$P(X \geq 3) = \sum_{x=3}^{6} b(x; 6, .5) = \sum_{x=3}^{6} \binom{6}{x}(.5)^x(.5)^{6-x} = .656$$

and the probability that at most one prefers S is

$$P(X \leq 1) = \sum_{x=0}^{1} b(x; 6, .5) = .109 \qquad \blacksquare$$

2.4.2   Computing Binomial Probabilities

Even for a relatively small value of n, the computation of binomial probabilities can be tedious. Software and statistical tables are both available for this purpose; both are often in terms of the cdf $F(x) = P(X \leq x)$ of the distribution, either in lieu of or in addition to the pmf. Various other probabilities can then be calculated using the proposition on cdfs from Sect. 2.2.

NOTATION
For $X \sim \text{Bin}(n, p)$, the cdf will be denoted by

$$B(x; n, p) = P(X \leq x) = \sum_{y=0}^{x} b(y; n, p) \qquad x = 0, 1, \ldots, n$$

Table 2.2 at the end of this section provides the code for performing binomial calculations in both Matlab and R. In addition, Appendix Table A.1 tabulates the binomial cdf for $n = 5, 10, 15, 20, 25$ in combination with selected values of p.

Example 2.32 Suppose that 20% of all copies of a particular textbook fail a binding strength test. Let X denote the number among 15 randomly selected copies that fail the test. Then X has a binomial distribution with $n = 15$ and $p = .2$.
(a) The probability that at most 8 fail the test is

$$P(X \leq 8) = \sum_{y=0}^{8} b(y; 15, .2) = B(8; 15, .2)$$

This is found at the intersection of the $p = .2$ column and $x = 8$ row in the $n = 15$ part of Table A.1: $B(8; 15, .2) = .999$. In Matlab, we may type `cdf('bin',8,15,.2)`; in R, `pbinom(8,15,.2)`.

(b) The probability that exactly 8 fail is $P(X = 8) = b(8; 15, .2) = \binom{15}{8}(.2)^8(.8)^7 = .0034$. We can evaluate this probability in Matlab or R with the calls `pdf('bin',8,15,.2)` and `dbinom(8,15,.2)`, respectively. To use Table A.1, write

$$P(X = 8) = P(X \le 8) - P(X \le 7) = B(8; 15, .2) - B(7; 15, .2)$$

which is the difference between two consecutive entries in the $p = .2$ column. The result is $.999 - .996 = .003$.

(c) The probability that at least 8 fail is $P(X \ge 8) = 1 - P(X \le 7) = 1 - B(7; 15, .2)$. The cdf may be evaluated using Matlab or R as above, or by looking up the entry in the $x = 7$ row of the $p = .2$ column in Table A.1. In any case, we find $P(X \ge 8) = 1 - .996 = .004$.

(d) Finally, the probability that between 4 and 7, inclusive, fail is

$$\begin{aligned}P(4 \le X \le 7) &= P(X = 4, 5, 6, \text{or } 7) = P(X \le 7) - P(X \le 3)\\ &= B(7; 15, .2) - B(3; 15, .2) = .996 - .648 = .348\end{aligned}$$

Notice that this latter probability is the difference between the cdf values at $x = 7$ and $x = 3$, *not* $x = 7$ and $x = 4$. ■

Example 2.33 An electronics manufacturer claims that at most 10% of its power supply units need service during the warranty period. To investigate this claim, technicians at a testing laboratory purchase 20 units and subject each one to accelerated testing to simulate use during the warranty period. Let p denote the probability that a power supply unit needs repair during the period (i.e., the proportion of *all* such units that need repair). The laboratory technicians must decide whether the data resulting from the experiment supports the claim that $p \le .10$. Let X denote the number among the 20 sampled that need repair, so $X \sim \text{Bin}(20, p)$. Consider the decision rule

Reject the claim that $p \le .10$ in favor of the conclusion that $p > .10$ if $x \ge 5$ (where x is the observed value of X), and consider the claim plausible if $x \le 4$

The probability that the claim is rejected when $p = .10$ (an incorrect conclusion) is

$$P(X \ge 5 \text{ when } p = .10) = 1 - B(4; 20, .1) = 1 - .957 = .043$$

The probability that the claim is not rejected when $p = .20$ (a different type of incorrect conclusion) is

$$P(X \leq 4 \text{ when } p = .2) = B(4; 20, .2) = .630$$

The first probability is rather small, but the second is intolerably large. When $p = .20$, so that the manufacturer has grossly understated the percentage of units that need service, and the stated decision rule is used, 63% of all samples of size 20 will result in the manufacturer's claim being judged plausible!

One might recognize that the probability of this second type of erroneous conclusion could be made smaller by changing the cutoff value 5 in the decision rule to something else. However, although replacing 5 by a smaller number would indeed yield a probability smaller than .630, the other probability would then increase. The only way to make both "error probabilities" small is to base the decision rule on an experiment involving many more units (i.e., to increase n). ∎

2.4.3   The Mean and Variance of a Binomial Random Variable

For $n = 1$, the binomial distribution becomes the Bernoulli distribution. From Example 2.17, the mean value of a Bernoulli variable is $\mu = p$, so the expected number of S's on any single trial is p. Since a binomial experiment consists of n trials, intuition suggests that for $X \sim \text{Bin}(n, p)$, $E(X) = np$, the product of the number of trials and the probability of success on a single trial. The expression for $\text{Var}(X)$ is not so obvious.

PROPOSITION

If $X \sim \text{Bin}(n, p)$, then $E(X) = np$, $\text{Var}(X) = np(1 - p) = npq$, and $SD(X) = \sqrt{npq}$ (where $q = 1 - p$).

Thus, calculating the mean and variance of a binomial rv does not necessitate evaluating summations of the sort we employed in Sect. 2.3. The proof of the result for $E(X)$ is sketched in Exercise 74.

Example 2.34 If 75% of all purchases at a store are made with a credit card and X is the number among ten randomly selected purchases made with a credit card, then $X \sim \text{Bin}(10, .75)$. Thus $E(X) = np = (10)(.75) = 7.5$, $\text{Var}(X) = np(1 - p) = 10(.75)(.25) = 1.875$, and $\sigma = \sqrt{1.875} = 1.37$. Again, even though X can take on only integer values, $E(X)$ need not be an integer. If we perform a large number of independent binomial experiments, each with $n = 10$ trials and $p = .75$, then the average number of S's per experiment will be close to 7.5. ∎

An important application of the binomial distribution is to estimating the precision of simulated probabilities, as in Sect. 1.6. The relative frequency definition of probability justified defining an estimate of a probability $P(A)$ by $\hat{P}(A) = X/n$, where n is the number of runs of the simulation program and X equals the number of runs in which event A occurred. Assuming the runs of our simulation are independent (and they usually are), the rv X has a binomial distribution with parameters n and $p = P(A)$. From the preceding proposition and the rescaling properties of mean and standard deviation, we have

$$E\left(\hat{P}(A)\right) = E\left(\frac{1}{n}X\right) = \frac{1}{n} \cdot E(X) = \frac{1}{n}(np) = p = P(A)$$

Thus we expect the value of our estimate to coincide with the probability being estimated, in the sense that there is no reason for $\hat{P}(A)$ to be systematically higher or lower than $P(A)$. Also,

$$SD\left(\hat{P}(A)\right) = SD\left(\frac{1}{n}X\right) = \left|\frac{1}{n}\right| \cdot SD(X)$$

$$= \frac{1}{n}\sqrt{np(1-p)} = \sqrt{\frac{p(1-p)}{n}} = \sqrt{\frac{P(A)[1-P(A)]}{n}} \qquad (2.14)$$

Expression (2.14) is called the **standard error** of $\hat{P}(A)$ (essentially a synonym for standard deviation) and indicates the amount by which an estimate $\hat{P}(A)$ "typically" varies from the true probability $P(A)$. However, this expression isn't of much use in practice: we most often simulate a probability when $P(A)$ is unknown, which prevents us from using Eq. (2.14). As a solution, we simply substitute the estimate $\hat{P} = \hat{P}(A)$ into this expression and get

$$SD\left(\hat{P}(A)\right) \approx \sqrt{\frac{\hat{P}(1-\hat{P})}{n}}$$

This is the estimated standard error formula (1.8) given in Sect. 1.6. Very importantly, this estimated standard error gets closer to 0 as the number of runs, n, in the simulation increases.

2.4.4   Binomial Calculations with Software

Many software packages, including Matlab and R, have built-in functions to evaluate both the pmf and cdf of the binomial distribution (and many other named distributions). Table 2.2 summarizes the relevant code in both packages. The use of these functions was illustrated in Example 2.32.

Table 2.2 Binomial probability calculations in Matlab and R

Function:	pmf	cdf
Notation:	$b(x; n, p)$	$B(x; n, p)$
Matlab:	pdf ('bin',x, n, p)	cdf ('bin',x, n, p)
R:	dbinom(x, n, p)	pbinom(x, n, p)

2.4.5   Exercises: Section 2.4 (49–74)

49. Determine whether each of the following rvs has a binomial distribution. If it does, identify the values of the parameters n and p (if possible).
    (a) $X =$ the number of ⚅s in 10 rolls of a fair die
    (b) $X =$ the number of multiple-choice questions a student gets right on a 40-question test, when each question has four choices and the student is completely guessing
    (c) $X =$ the same as (b), but half the questions have four choices and the other half have three
    (d) $X =$ the number of women in a random sample of 8 students, from a class comprising 20 women and 15 men
    (e) $X =$ the total weight of 15 randomly selected apples
    (f) $X =$ the number of apples, out of a random sample of 15, that weigh more than 150 g
50. Compute the following binomial probabilities directly from the formula for $b(x; n, p)$:
    (a) $b(3; 8, .6)$
    (b) $b(5; 8, .6)$
    (c) $P(3 \leq X \leq 5)$ when $n = 8$ and $p = .6$
    (d) $P(1 \leq X)$ when $n = 12$ and $p = .1$
51. Use Appendix Table A.1 or software to obtain the following probabilities:
    (a) $B(4; 10, .3)$
    (b) $b(4; 10, .3)$
    (c) $b(6; 10, .7)$
    (d) $P(2 \leq X \leq 4)$ when $X \sim \text{Bin}(10, .3)$
    (e) $P(2 \leq X)$ when $X \sim \text{Bin}(10, .3)$
    (f) $P(X \leq 1)$ when $X \sim \text{Bin}(10, .7)$
    (g) $P(2 < X < 6)$ when $X \sim \text{Bin}(10, .3)$
52. When circuit boards used in the manufacture of DVD players are tested, the long-run percentage of defectives is 5%. Let $X =$ the number of defective boards in a random sample of size $n = 25$, so $X \sim \text{Bin}(25, .05)$.
    (a) Determine $P(X \leq 2)$.
    (b) Determine $P(X \geq 5)$.
    (c) Determine $P(1 \leq X \leq 4)$.
    (d) What is the probability that none of the 25 boards is defective?
    (e) Calculate the expected value and standard deviation of X.

53. A company that produces fine crystal knows from experience that 10% of its goblets have cosmetic flaws and must be classified as "seconds."
    (a) Among six randomly selected goblets, how likely is it that only one is a second?
    (b) Among six randomly selected goblets, what is the probability that at least two are seconds?
    (c) If goblets are examined one by one, what is the probability that at most five must be selected to find four that are not seconds?

54. Suppose that only 25% of all drivers come to a complete stop at an intersection having flashing red lights in all directions when no other cars are visible. What is the probability that, of 20 randomly chosen drivers coming to an intersection under these conditions,
    (a) At most 6 will come to a complete stop?
    (b) Exactly 6 will come to a complete stop?
    (c) At least 6 will come to a complete stop?

55. Refer to the previous exercise.
    (a) What is the expected number of drivers among the 20 that come to a complete stop?
    (b) What is the standard deviation of the number of drivers among the 20 that come to a complete stop?
    (c) What is the probability that the number of drivers among these 20 that come to a complete stop differs from the expected number by more than 2 standard deviations?

56. Suppose that 30% of all students who have to buy a text for a particular course want a new copy (the successes!), whereas the other 70% want a used copy. Consider randomly selecting 25 purchasers.
    (a) What are the mean value and standard deviation of the number who want a new copy of the book?
    (b) What is the probability that the number who want new copies is more than two standard deviations away from the mean value?
    (c) The bookstore has 15 new copies and 15 used copies in stock. If 25 people come in one by one to purchase this text, what is the probability that all 25 will get the type of book they want from current stock? [*Hint:* Let $X =$ the number who want a new copy. For what values of X will all 25 get what they want?]
    (d) Suppose that new copies cost $100 and used copies cost $70. Assume the bookstore has 50 new copies and 50 used copies. What is the expected value of total revenue from the sale of the next 25 copies purchased? [*Hint:* Let $h(X) =$ the revenue when X of the 25 purchasers want new copies. Express this as a linear function.]

57. Exercise 30 (Sect. 2.3) gave the pmf of Y, the number of traffic citations for a randomly selected individual insured by a company. What is the probability that among 15 randomly chosen such individuals
    (a) At least 10 have no citations?
    (b) Fewer than half have at least one citation?
    (c) The number that have at least one citation is between 5 and 10, inclusive?

58. A particular type of tennis racket comes in a midsize version and an oversize version. Sixty percent of all customers at a store want the oversize version.
    (a) Among ten randomly selected customers who want this type of racket, what is the probability that at least six want the oversize version?
    (b) Among ten randomly selected customers, what is the probability that the number who want the oversize version is within 1 standard deviation of the mean value?
    (c) The store currently has seven rackets of each version. What is the probability that all of the next ten customers who want this racket can get the version they want from current stock?
59. Twenty percent of all telephones of a certain type are submitted for service while under warranty. Of these, 60% can be repaired, whereas the other 40% must be replaced with new units. If a company purchases ten of these telephones, what is the probability that exactly two will end up being replaced under warranty?
60. The College Board reports that 2% of the two million high school students who take the SAT each year receive special accommodations because of documented disabilities (*Los Angeles Times*, July 16, 2002). Consider a random sample of 25 students who have recently taken the test.
    (a) What is the probability that exactly 1 received a special accommodation?
    (b) What is the probability that at least 1 received a special accommodation?
    (c) What is the probability that at least 2 received a special accommodation?
    (d) What is the probability that the number among the 25 who received a special accommodation is within 2 standard deviations of the number you would expect to be accommodated?
    (e) Suppose that a student who does not receive a special accommodation is allowed 3 hours for the exam, whereas an accommodated student is allowed 4.5 hours. What would you expect the average time allowed the 25 selected students to be?
61. Suppose that 90% of all batteries from a supplier have acceptable voltages. A certain type of flashlight requires two type-D batteries, and the flashlight will work only if both its batteries have acceptable voltages. Among ten randomly selected flashlights, what is the probability that at least nine will work? What assumptions did you make in the course of answering the question posed?
62. A *k-out-of-n system* functions provided that at least k of the n components function. Consider independently operating components, each of which functions (for the needed duration) with probability .96.
    (a) In a 3-component system, what is the probability that exactly two components function?
    (b) What is the probability a 2-out-of-3 system works?
    (c) What is the probability a 3-out-of-5 system works?
    (d) What is the probability a 4-out-of-5 system works?
    (e) What does the component probability (previously .96) need to equal so that the 4-out-of-5 system will function with probability at least .9999?

63. Bit transmission errors between computers sometimes occur, where one computer sends a 0 but the other computer receives a 1 (or vice versa). Because of this, the computer sending a message repeats each bit three times, so a 0 is sent as 000 and a 1 as 111. The receiving computer "decodes" each triplet by majority rule: whichever number, 0 or 1, appears more often in a triplet is declared to be the intended bit. For example, both 000 and 100 are decoded as 0, while 101 and 011 are decoded as 1. Suppose that 6% of bits are switched (0 to 1, or 1 to 0) during transmission between two particular computers, and that these errors occur independently during transmission.
   (a) Find the probability that a triplet is decoded incorrectly by the receiving computer.
   (b) Using your answer to part (a), explain how using triplets reduces communication errors.
   (c) How does your answer to part (a) change if each bit is repeated five times (instead of three)?
   (d) Imagine a 25 kilobit message (i.e., one requiring 25,000 bits to send). What is the expected number of errors if there is no bit repetition implemented? If each bit is repeated three times?

64. A very large batch of components has arrived at a distributor. The batch can be characterized as acceptable only if the proportion of defective components is at most .10. The distributor decides to randomly select 10 components and to accept the batch only if the number of defective components in the sample is at most 2.
   (a) What is the probability that the batch will be accepted when the actual proportion of defectives is .01? .05? .10? .20? .25?
   (b) Let p denote the actual proportion of defectives in the batch. A graph of P(batch is accepted) as a function of p, with p on the horizontal axis and P(batch is accepted) on the vertical axis, is called the *operating characteristic curve* for the acceptance sampling plan. Use the results of part (a) to sketch this curve for $0 \leq p \leq 1$.
   (c) Repeat parts (a) and (b) with "1" replacing "2" in the acceptance sampling plan.
   (d) Repeat parts (a) and (b) with "15" replacing "10" in the acceptance sampling plan.
   (e) Which of the three sampling plans, that of part (a), (c), or (d), appears most satisfactory, and why?

65. An ordinance requiring that a smoke detector be installed in all previously constructed houses has been in effect in a city for 1 year. The fire department is concerned that many houses remain without detectors. Let p = the true proportion of such houses having detectors, and suppose that a random sample of 25 homes is inspected. If the sample strongly indicates that fewer than 80% of all houses have a detector, the fire department will campaign for a mandatory inspection program. Because of the costliness of the program, the department prefers not to call for such inspections unless sample evidence strongly argues for their necessity. Let X denote the number of homes with detectors among the 25 sampled. Consider rejecting the claim that $p \geq .8$ if $X \leq 15$.

(a) What is the probability that the claim is rejected when the actual value of p is .8?

(b) What is the probability of not rejecting the claim when $p = .7$? When $p = .6$?

(c) How do the "error probabilities" of parts (a) and (b) change if the value 15 in the decision rule is replaced by 14?

66. A toll bridge charges $1.00 for passenger cars and $2.50 for other vehicles. Suppose that during daytime hours, 60% of all vehicles are passenger cars. If 25 vehicles cross the bridge during a particular daytime period, what is the resulting expected toll revenue? [*Hint:* Let $X = $ the number of passenger cars; then the toll revenue $h(X)$ is a linear function of X.]

67. A student who is trying to write a paper for a course has a choice of two topics, A and B. If topic A is chosen, the student will order two books through interlibrary loan, whereas if topic B is chosen, the student will order four books. The student believes that a good paper necessitates receiving and using at least half the books ordered for either topic chosen. If the probability that a book ordered through interlibrary loan actually arrives in time is .9 and books arrive independently of one another, which topic should the student choose to maximize the probability of writing a good paper? What if the arrival probability is only .5 instead of .9?

68. Twelve jurors are randomly selected from a large population. Each juror arrives at her or his conclusion about the case before the jury independently of the other jurors.

(a) In a criminal case, all 12 jurors must agree on a verdict. Let p denote the probability that a randomly selected member of the population would reach a guilty verdict based on the evidence presented (so a proportion $1 - p$ would reach "not guilty"). What is the probability, in terms of p, that the jury reaches a unanimous verdict one way or the other?

(b) For what values of p is the probability in part (a) the highest? For what value of p is the probability in (a) the lowest? Explain why this makes sense.

(c) In most civil cases, only a nine-person majority is required to decide a verdict. That is, if nine or more jurors favor the plaintiff, then the plaintiff wins; if at least nine jurors side with the defendant, then the defendant wins. Let p denote the probability that someone would side with the plaintiff based on the evidence. What is the probability, in terms of p, that the jury reaches a verdict one way or the other? How does this compare with your answer to part (a)?

69. Customers at a gas station pay with a credit card (A), debit card (B), or cash (C). Assume that successive customers make independent choices, with $P(A) = .5$, $P(B) = .2$, and $P(C) = .3$.

(a) Among the next 100 customers, what are the mean and variance of the number who pay with a debit card? Explain your reasoning.

(b) Answer part (a) for the number among the 100 who don't pay with cash.

70. An airport limousine can accommodate up to four passengers on any one trip. The company will accept a maximum of six reservations for a trip, and a

passenger must have a reservation. From previous records, 20% of all those making reservations do not appear for the trip. In the following questions, assume independence, but explain why there could be dependence.

(a) If six reservations are made, what is the probability that at least one individual with a reservation cannot be accommodated on the trip?

(b) If six reservations are made, what is the expected number of available places when the limousine departs?

(c) Suppose the probability distribution of the number of reservations made is given in the accompanying table.

Number of reservations	3	4	5	6
Probability	.1	.2	.3	.4

Let X denote the number of passengers on a randomly selected trip. Obtain the probability mass function of X.

71. Let X be a binomial random variable with fixed n.
(a) Are there values of p $(0 \le p \le 1)$ for which $\text{Var}(X) = 0$? Explain why this is so.
(b) For what value of p is $\text{Var}(X)$ maximized? [*Hint*: Either graph $\text{Var}(X)$ as a function of p or else take a derivative.]

72. (a) Show that $b(x; n, 1 - p) = b(n - x; n, p)$.
(b) Show that $B(x; n, 1 - p) = 1 - B(n - x - 1; n, p)$. [*Hint*: At most x S's is equivalent to at least $(n - x)$ F's.]
(c) What do parts (a) and (b) imply about the necessity of including values of p greater than .5 in Table A.1?

73. Refer to Chebyshev's inequality given in Sect. 2.3. Calculate $P(|X - \mu| \ge k\sigma)$ for $k = 2$ and $k = 3$ when $X \sim \text{Bin}(20, .5)$, and compare to the corresponding upper bounds. Repeat for $X \sim \text{Bin}(20, .75)$.

74. Show that $E(X) = np$ when X is a binomial random variable. [*Hint*: Express $E(X)$ as a sum with lower limit $x = 1$. Then factor out np, let $y = x - 1$ so that the sum is from $y = 0$ to $y = n - 1$, and show that the sum equals 1.]

2.5   The Poisson Distribution

The binomial distribution was derived by starting with an experiment consisting of trials and applying the laws of probability to various outcomes of the experiment. There is no simple experiment on which the Poisson distribution is based, although we will shortly describe how it can be obtained from the binomial distribution by certain limiting operations.

> **DEFINITION**
>
> A random variable X is said to have a **Poisson distribution** with parameter μ $(\mu > 0)$ if the pmf of X is
>
> $$p(x; \mu) = \frac{e^{-\mu} \mu^x}{x!} \qquad x = 0, 1, 2, \ldots$$

We shall see shortly that μ is in fact the expected value of X, so the notation here is consistent with our previous use of the symbol μ. Because μ must be positive, $p(x; \mu) > 0$ for all possible x values. The fact that $\sum_{x=0}^{\infty} p(x; \mu) = 1$ is a consequence of the Maclaurin infinite series expansion of e^μ, which appears in most calculus texts:

$$e^\mu = 1 + \mu + \frac{\mu^2}{2!} + \frac{\mu^3}{3!} + \cdots = \sum_{x=0}^{\infty} \frac{\mu^x}{x!} \tag{2.15}$$

If the two extreme terms in Eq. (2.15) are multiplied by $e^{-\mu}$ and then $e^{-\mu}$ is placed inside the summation, the result is

$$1 = \sum_{x=0}^{\infty} \frac{e^{-\mu} \mu^x}{x!}$$

which shows that $p(x; \mu)$ fulfills the second condition necessary for specifying a pmf.

Example 2.35 Let X denote the number of creatures of a particular type captured in a trap during a given time period. Suppose that X has a Poisson distribution with $\mu = 4.5$, so on average traps will contain 4.5 creatures. [The article "Dispersal Dynamics of the Bivalve *Gemma gemma* in a Patchy Environment" (*Ecol. Monogr.*, 1995: 1–20) suggests this model; the bivalve *Gemma gemma* is a small clam.] The probability that a trap contains exactly five creatures is

$$P(X = 5) = \frac{e^{-4.5} (4.5)^5}{5!} = .1708$$

The probability that a trap has at most five creatures is

$$P(X \le 5) = \sum_{x=0}^{5} \frac{e^{-4.5} (4.5)^x}{x!} = e^{-4.5} \left[1 + 4.5 + \frac{4.5^2}{2!} + \cdots + \frac{4.5^5}{5!} \right] = .7029$$

∎

2.5.1 The Poisson Distribution as a Limit

The rationale for using the Poisson distribution in many situations is provided by the following proposition.

PROPOSITION

Suppose that in the binomial pmf $b(x; n, p)$ we let $n \to \infty$ and $p \to 0$ in such a way that np approaches a value $\mu > 0$. Then $b(x; n, p) \to p(x; \mu)$.

Proof Begin with the binomial pmf:

$$b(x; n, p) = \binom{n}{x} p^x (1-p)^{n-x} = \frac{n!}{x!(n-x)!} p^x (1-p)^{n-x}$$

$$= \frac{n \cdot (n-1) \cdot \cdots \cdot (n-x+1)}{x!} p^x (1-p)^{n-x}$$

Now multiply both the numerator and denominator by n^x:

$$b(x; n, p) = \frac{n}{n} \frac{n-1}{n} \cdots \frac{n-x+1}{n} \cdot \frac{(np)^x}{x!} \cdot \frac{(1-p)^n}{(1-p)^x}$$

Taking the limit as $n \to \infty$ and $p \to 0$ with $np \to \mu$,

$$\lim_{n \to \infty} b(x; n, p) = 1 \cdot 1 \cdots 1 \cdot \frac{\mu^x}{x!} \cdot \left(\lim_{n \to \infty} \frac{(1 - np/n)^n}{1} \right)$$

The limit on the right can be obtained from the calculus theorem that says the limit of $(1 - a_n/n)^n$ is e^{-a} if $a_n \to a$. Because $np \to \mu$,

$$\lim_{n \to \infty} b(x; n, p) = \frac{\mu^x}{x!} \cdot \lim_{n \to \infty} \left(1 - \frac{np}{n} \right)^n = \frac{\mu^x e^{-\mu}}{x!} = p(x; \mu) \qquad \blacksquare$$

According to the proposition, *in any binomial experiment for which n is large and p is small,* $b(x; n, p) \approx p(x; \mu)$ *where* $\mu = np$. It is interesting to note that Siméon Poisson discovered the distribution that bears his name by this approach in the 1830s.

Table 2.3 shows the Poisson distribution for $\mu = 3$ along with three binomial distributions with $np = 3$, and Fig. 2.8 (from R) plots the Poisson along with the first two binomial distributions. The approximation is of limited use for $n = 30$, but of course the accuracy is better for $n = 100$ and much better for $n = 300$.

Table 2.3 Comparing the Poisson and three binomial distributions

x	$n = 30, p = .1$	$n = 100, p = .03$	$n = 300, p = .01$	Poisson, $\mu = 3$
0	0.042391	0.047553	0.049041	0.049787
1	0.141304	0.147070	0.148609	0.149361
2	0.227656	0.225153	0.224414	0.224042
3	0.236088	0.227474	0.225170	0.224042
4	0.177066	0.170606	0.168877	0.168031
5	0.102305	0.101308	0.100985	0.100819
6	0.047363	0.049610	0.050153	0.050409
7	0.018043	0.020604	0.021277	0.021604
8	0.005764	0.007408	0.007871	0.008102
9	0.001565	0.002342	0.002580	0.002701
10	0.000365	0.000659	0.000758	0.000810

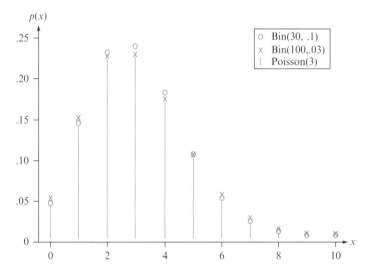

Fig. 2.8 Comparing a Poisson and two binomial distributions

Example 2.36 Suppose you have a 4-megabit modem (4,000,000 bits/s) with bit error probability 10^{-8}. Assume bit errors occur independently, and assume your bit rate stays constant at 4 Mbps. What is the probability of exactly 3 bit errors in the next minute? Of at most 3 bit errors in the next minute?

Define a random variable X = the number of bit errors in the next minute. From the description, X satisfies the conditions of a binomial distribution; specifically, since a constant bit rate of 4 Mbps equates to 240,000,000 bits transmitted per minute, $X \sim \text{Bin}(240000000, 10^{-8})$. Hence, the probability of exactly three bit errors in the next minute is

$$P(X = 3) = b\left(3; 240000000, 10^{-8}\right) = \binom{240000000}{3}\left(10^{-8}\right)^3\left(1 - 10^{-8}\right)^{239999997}$$

For a variety of reasons, some calculators will struggle with this computation. The expression for the chance of at most 3 bit errors, $P(X \leq 3)$, is even worse. (The inability to compute such expressions in the nineteenth century, even with modest values of n and p, was Poisson's motive to derive an easily computed approximation.)

We may approximate these probabilities using the Poisson distribution with $\mu = np = 240000000(10^{-8}) = 2.4$. Then

$$P(X = 3) \approx p(3; 2.4) = \frac{e^{-2.4}2.4^3}{3!} = .20901416$$

Similarly, the probability of at most 3 bit errors in the next minute is approximated by

$$P(X \leq 3) \approx \sum_{x=0}^{3} p(x, 2.4) = \sum_{x=0}^{3} \frac{e^{-2.4}2.4^x}{x!} = .77872291$$

Using modern software, the exact probabilities (i.e., using the binomial model) are .2090141655 and .7787229106, respectively. The Poisson approximations agree to eight decimal places and are clearly more computationally tractable. ∎

Many software packages will compute both $p(x; \mu)$ and the corresponding cdf $P(x; \mu)$ for specified values of x and μ upon request; the relevant Matlab and R functions appear in Table 2.4 at the end of this section. Appendix Table A.2 exhibits the cdf $P(x; \mu)$ for $\mu = .1, .2, \ldots, 1, 2, \ldots, 10, 15,$ and 20. For example, if $\mu = 2$, then $P(X \leq 3) = P(3; 2) = .857$, whereas $P(X = 3) = P(3; 2) - P(2; 2) = .180$.

2.5.2   The Mean and Variance of a Poisson Random Variable

Since $b(x; n, p) \to p(x; \mu)$ as $n \to \infty, p \to 0, np \to \mu$, one might guess that the mean and variance of a binomial variable approach those of a Poisson variable. These limits are $np \to \mu$ and $np(1 - p) \to \mu$.

> **PROPOSITION**
> If X has a Poisson distribution with parameter μ, then $E(X) = \text{Var}(X) = \mu$.

These results can also be derived directly from the definitions of mean and variance (see Exercise 88 for the mean).

Example 2.37 (Example 2.35 continued) Both the expected number of creatures trapped and the variance of the number trapped equal 4.5, and $\sigma_X = \sqrt{\mu} = \sqrt{4.5} = 2.12$. ∎

2.5.3 The Poisson Process

A very important application of the Poisson distribution arises in connection with the occurrence of events of a particular type over time. As an example, suppose that starting from a time point that we label $t = 0$, we are interested in counting the number of radioactive pulses recorded by a Geiger counter. If we make certain assumptions[2] about the way in which pulses occur—chiefly, that the number of pulses grows roughly linearly with time—then it can be shown that the number of pulses in any time interval of length t can be modeled by a Poisson distribution with mean $\mu = \lambda t$ for an appropriate positive constant λ. Since the expected number of pulses in an interval of length t is λt, the expected number in an interval of length 1 is λ. Thus λ is the long run number of pulses per unit of time.

If we replace "pulse" by "event," then the number of events occurring during a fixed time interval of length t has a Poisson distribution with parameter λt. Any process that has this distribution is called a **Poisson process**, and λ is called the *rate of the process*. Other examples of situations giving rise to a Poisson process include monitoring the status of a computer system over time, with breakdowns constituting the events of interest; recording the number of accidents in an industrial facility over time; answering 911 calls at a particular location; and observing the number of cosmic-ray showers from an observatory.

Example 2.36 hints at why this might be reasonable: if we "digitize" time—that is, divide time into discrete pieces, such as transmitted bits—and look at the number of the resulting time pieces that include an event, a binomial model is often applicable. If the number of time pieces is very large and the success probability close to zero, which would occur if we divided a fixed time frame into ever-smaller pieces, then we may invoke the Poisson approximation from earlier in this section.

Example 2.38 Suppose pulses arrive at the Geiger counter at an average rate of 6 per minute, so that $\lambda = 6$. To find the probability that in a 30-s interval at least one pulse is received, note that the number of pulses in such an interval has a Poisson distribution with parameter $\lambda t = 6(.5) = 3$ (.5 min is used because λ is expressed as a rate per minute). Then with $X =$ the number of pulses received in the 30-s interval,

$$P(X \geq 1) = 1 - P(X = 0) = 1 - \frac{e^{-3}3^0}{0!} = .950$$

In a 1-h interval ($t = 60$), the expected number of pulses is $\mu = \lambda t = 6(60) = 360$, with a standard deviation of $\sigma = \sqrt{\mu} = \sqrt{360} = 18.97$. According to this model, in a typical hour we will observe 360 ± 19 pulses arrive at the Geiger counter. ∎

---

[2] In Sect. 7.5, we present the formal assumptions required in this situation and derive the Poisson distribution that results from these assumptions.

Instead of observing events over time, consider observing events of some type that occur in a two- or three-dimensional region. For example, we might select on a map a certain region R of a forest, go to that region, and count the number of trees. Each tree would represent an event occurring at a particular point in space. Under appropriate assumptions (see Sect. 7.5), it can be shown that the number of events occurring in a region R has a Poisson distribution with parameter $\lambda \cdot a(R)$, where $a(R)$ is the area of R. The quantity λ is the expected number of events per unit area or volume.

2.5.4 Poisson Calculations with Software

Table 2.4 gives the Matlab and R commands for calculating Poisson probabilities.

Table 2.4 Poisson probability calculations	Function:	pmf	cdf
	Notation:	$p(x; \mu)$	$P(x; \mu)$
	Matlab:	pdf('pois',x,μ)	cdf('pois', x,μ)
	R:	dpois(x,μ)	ppois(x,μ)

2.5.5 Exercises: Section 2.5 (75–89)

75. Let X, the number of flaws on the surface of a randomly selected carpet of a particular type, have a Poisson distribution with parameter $\mu = 5$. Use software or Appendix Table A.2 to compute the following probabilities:
    (a) $P(X \le 8)$
    (b) $P(X = 8)$
    (c) $P(9 \le X)$
    (d) $P(5 \le X \le 8)$
    (e) $P(5 < X < 8)$

76. Let X be the number of material anomalies occurring in a particular region of an aircraft gas-turbine disk. The article "Methodology for Probabilistic Life Prediction of Multiple-Anomaly Materials" (*Amer. Inst. of Aeronautics and Astronautics J.*, 2006: 787-793) proposes a Poisson distribution for X. Suppose $\mu = 4$.
    (a) Compute both $P(X \le 4)$ and $P(X < 4)$.
    (b) Compute $P(4 \le X \le 8)$.
    (c) Compute $P(8 \le X)$.
    (d) What is the probability that the observed number of anomalies exceeds the expected number by no more than one standard deviation?

77. Suppose that the number of drivers who travel between a particular origin and destination during a designated time period has a Poisson distribution with parameter $\mu = 20$ (suggested in the article "Dynamic Ride Sharing: Theory

and Practice," *J. of Transp. Engr.*, 1997: 308–312). What is the probability that the number of drivers will

(a) Be at most 10?

(b) Exceed 20?

(c) Be between 10 and 20, inclusive? Be strictly between 10 and 20?

(d) Be within 2 standard deviations of the mean value?

78. Consider writing onto a computer disk and then sending it through a certifier that counts the number of missing pulses. Suppose this number X has a Poisson distribution with parameter $\mu = .2$. (Suggested in "Average Sample Number for Semi-Curtailed Sampling Using the Poisson Distribution," *J. Qual. Tech.*, 1983: 126–129.)

(a) What is the probability that a disk has exactly one missing pulse?

(b) What is the probability that a disk has at least two missing pulses?

(c) If two disks are independently selected, what is the probability that neither contains a missing pulse?

79. An article in the *Los Angeles Times* (Dec. 3, 1993) reports that 1 in 200 people carry the defective gene that causes inherited colon cancer. In a sample of 1000 individuals, what is the approximate distribution of the number who carry this gene? Use this distribution to calculate the approximate probability that

(a) Between 5 and 8 (inclusive) carry the gene.

(b) At least 8 carry the gene.

80. Suppose that only .10% of all computers of a certain type experience CPU failure during the warranty period. Consider a sample of 10,000 computers.

(a) What are the expected value and standard deviation of the number of computers in the sample that have the defect?

(b) What is the (approximate) probability that more than 10 sampled computers have the defect?

(c) What is the (approximate) probability that no sampled computers have the defect?

81. If a publisher of nontechnical books takes great pains to ensure that its books are free of typographical errors, so that the probability of any given page containing at least one such error is .005 and errors are independent from page to page, what is the probability that one of its 400-page novels will contain exactly one page with errors? At most three pages with errors?

82. In proof testing of circuit boards, the probability that any particular diode will fail is .01. Suppose a circuit board contains 200 diodes.

(a) How many diodes would you expect to fail, and what is the standard deviation of the number that are expected to fail?

(b) What is the (approximate) probability that at least four diodes will fail on a randomly selected board?

(c) If five boards are shipped to a particular customer, how likely is it that at least four of them will work properly? (A board works properly only if all its diodes work.)

83. The article "Expectation Analysis of the Probability of Failure for Water Supply Pipes" (*J. Pipeline Syst. Eng. Pract.* 2012.3:36-46) recommends using a

Poisson process to model the number of failures in commercial water pipes. The article also gives estimates of the failure rate λ, in units of failures per 100 miles of pipe per day, for four different types of pipe and for many different years.

(a) For PVC pipe in 2008, the authors estimate a failure rate of 0.0081 failures per 100 miles of pipe per day. Consider a 100-mile-long segment of such pipe. What is the expected number of failures in 1 year (365 days)? Based on this expectation, what is the probability of at least one failure along such a pipe in 1 year?

(b) For cast iron pipe in 2005, the authors' estimate is $\lambda = 0.0864$ failures per 100 miles per day. Suppose a town had 1500 miles of cast iron pipe underground in 2005. What is the probability of at least one failure somewhere along this pipe system on any given day?

84. Organisms are present in ballast water discharged from a ship according to a Poisson process with a concentration of 10 organisms/m^3 (the article "Counting at Low Concentrations: The Statistical Challenges of Verifying Ballast Water Discharge Standards" (*Ecological Applications*, 2013: 339–351) considers using the Poisson process for this purpose).

(a) What is the probability that one cubic meter of discharge contains at least 8 organisms?

(b) What is the probability that the number of organisms in 1.5 m^3 of discharge exceeds its mean value by more than one standard deviation?

(c) For what amount of discharge would the probability of containing at least one organism be .999?

85. Suppose small aircraft arrive at an airport according to a Poisson process with rate $\lambda = 8$ per hour, so that the number of arrivals during a time period of t hours is a Poisson rv with parameter $\mu = 8t$.

(a) What is the probability that exactly 6 small aircraft arrive during a 1-h period? At least 6? At least 10?

(b) What are the expected value and standard deviation of the number of small aircraft that arrive during a 90-min period?

(c) What is the probability that at least 20 small aircraft arrive during a 2.5-h period? That at most 10 arrive during this period?

86. The number of people arriving for treatment at an emergency room can be modeled by a Poisson process with a rate parameter of five per hour.

(a) What is the probability that exactly four arrivals occur during a particular hour?

(b) What is the probability that at least four people arrive during a particular hour?

(c) How many people do you expect to arrive during a 45-min period?

87. Suppose that trees are distributed in a forest according to a two-dimensional Poisson process with rate λ, the expected number of trees per acre, equal to 80.

(a) What is the probability that in a certain quarter-acre plot, there will be at most 16 trees?

(b) If the forest covers 85,000 acres, what is the expected number of trees in the forest?

(c) Suppose you select a point in the forest and construct a circle of radius .1 mile. Let $X =$ the number of trees within that circular region. What is the pmf of X? [*Hint*: 1 sq mile = 640 acres.]

88. Let X have a Poisson distribution with parameter μ. Show that $E(X) = \mu$ directly from the definition of expected value. [*Hint*: The first term in the sum equals 0, and then x can be canceled. Now factor out μ and show that what is left sums to 1.]

89. In some applications the distribution of a discrete rv X resembles the Poisson distribution except that zero is not a possible value of X. For example, let $X =$ the number of tattoos that an individual wants removed when s/he arrives at a tattoo removal facility. Suppose the pmf of X is

$$p(x) = k \frac{e^{-\theta}\theta^x}{x!} \quad x = 1, 2, 3, \ldots$$

(a) Determine the value of k. [*Hint:* The sum of all probabilities in the Poisson pmf is 1, and this pmf must also sum to 1.]

(b) If the mean value of X is 2.313035, what is the probability that an individual wants at most 5 tattoos removed?

(c) Determine the standard deviation of X when the mean value is as given in (b).

[*Note:* The article "An Exploratory Investigation of Identity Negotiation and Tattoo Removal" (*Academy of Marketing Science Review*, vol. 12, #6, 2008) gave a sample of 22 observations on the number of tattoos people wanted removed; estimates of μ and σ calculated from the data were 2.318182 and 1.249242, respectively.]

2.6   Other Discrete Distributions

The hypergeometric and negative binomial distributions are both closely related to the binomial distribution. Whereas the binomial distribution is the approximate probability model for sampling without replacement from a finite dichotomous (S-F) population, the hypergeometric distribution is the exact probability model for the number of S's in the sample. The binomial rv X is the number of S's when the number n of trials is fixed, whereas the negative binomial distribution arises from fixing the number of S's desired and letting the number of trials be random.

2.6.1   The Hypergeometric Distribution

The assumptions leading to the hypergeometric distribution are as follows:

1. The population or set to be sampled consists of N individuals, objects, or elements (a *finite* population).

2. Each individual can be characterized as a success (S) or a failure (F), and there are M successes in the population.
3. A sample of n individuals is selected without replacement in such a way that each subset of size n is equally likely to be chosen.

The random variable of interest is X = the number of S's in the sample. The probability distribution of X depends on the parameters n, M, and N, so we wish to obtain $P(X = x) = h(x; n, M, N)$.

Example 2.39 During a particular period a university's information technology office received 20 service orders for problems with laptops, of which 8 were Macs and 12 were PCs. A sample of five of these service orders is to be selected for inclusion in a customer satisfaction survey. Suppose that the five are selected in a completely random fashion, so that any particular subset of size 5 has the same chance of being selected as does any other subset (think of putting the numbers 1, 2, ... , 20 on 20 identical slips of paper, mixing up the slips, and choosing five of them). What then is the probability that exactly 2 of the selected service orders were for PC laptops?

In this example, the population size is $N = 20$, the sample size is $n = 5$, and the number of S's (PC = S) and F's (Mac = F) in the population are $M = 12$ and $N - M = 8$, respectively. Let X = the number of PCs among the five sampled service orders. Because all outcomes (each consisting of five particular orders) are equally likely,

$$P(X = 2) = h(2; 5, 12, 20) = \frac{\text{number of outcomes having } X = 2}{\text{number of possible outcomes}}$$

The number of possible outcomes in the experiment is the number of ways of selecting 5 from the 20 objects without regard to order—that is, $\binom{20}{5}$. To count the number of outcomes having $X = 2$, note that there are $\binom{12}{2}$ ways of selecting two of the PC orders, and for each such way there are $\binom{8}{3}$ ways of selecting the three Mac orders to fill out the sample. The Fundamental Counting Principle from Sect. 1.3 then gives $\binom{12}{2} \cdot \binom{8}{3}$ as the number of outcomes with $X = 2$, so

$$h(2; 5, 12, 20) = \frac{\binom{12}{2}\binom{8}{3}}{\binom{20}{5}} = \frac{77}{323} = .238$$

∎

In general, if the sample size n is smaller than the number of successes in the population (M), then the largest possible X value is n. However, if $M < n$ (e.g., a sample size of 25 and only 15 successes in the population), then X can be at most M.

Similarly, whenever the number of population failures $(N - M)$ exceeds the sample size, the smallest possible X value is 0 (since all sampled individuals might then be failures). However, if $N - M < n$, the smallest possible X value is $n - (N - M)$. Summarizing, the possible values of X satisfy the restriction $\max(0, n - N + M) \leq x \leq \min(n, M)$. An argument parallel to that of the previous example gives the pmf of X.

PROPOSITION

If X is the number of S's in a random sample of size n drawn from a population consisting of M S's and $(N - M)$ F's, then the probability distribution of X, called the **hypergeometric distribution**, is given by

$$P(X = x) = h(x; n, M, N) = \frac{\dbinom{M}{x}\dbinom{N - M}{n - x}}{\dbinom{N}{n}} \qquad (2.16)$$

for x an integer satisfying $\max(0, n - N + M) \leq x \leq \min(n, M)$.[3]

In Example 2.39, $n = 5$, $M = 12$, and $N = 20$, so $h(x; 5, 12, 20)$ for $x = 0, 1, 2, 3, 4, 5$ can be obtained by substituting these numbers into Eq. (2.16).

Example 2.40 *Capture–recapture.* Five individuals from an animal population thought to be near extinction in a region have been caught, tagged, and released to mix into the population. After they have had an opportunity to mix, a random sample of 10 of these animals is selected. Let $X =$ the number of tagged animals in the second sample. If there are actually 25 animals of this type in the region, what is the probability that (a) $X = 2$? (b) $X \leq 2$?

Application of the hypergeometric distribution here requires assuming that every subset of ten animals has the same chance of being captured. This in turn implies that released animals are no easier or harder to catch than are those not initially captured. Then the parameter values are $n = 10$, $M = 5$ (five tagged animals in the population), and $N = 25$, so

$$h(x; 10, 5, 25) = \frac{\dbinom{5}{x}\dbinom{20}{10 - x}}{\dbinom{25}{10}} \qquad x = 0, 1, 2, 3, 4, 5$$

---

[3] If we define $\dbinom{a}{b} = 0$ for $a < b$, then $h(x; n, M, N)$ may be applied for all integers $0 \leq x \leq n$.

For part (a),

$$P(X = 2) = h(2; 10, 5, 25) = \frac{\binom{5}{2}\binom{20}{8}}{\binom{25}{10}} = .385$$

For part (b),

$$P(X \leq 2) = P(X = 0, 1, \text{or } 2) = \sum_{x=0}^{2} h(x; 10, 5, 25)$$
$$= .057 + .257 + .385 = .699 \qquad \blacksquare$$

Matlab, R, and other software packages will easily generate hypergeometric probabilities; see Table 2.5 at the end of this section. Comprehensive tables of the hypergeometric distribution are available, but because the distribution has three parameters, these tables require much more space than tables for the binomial distribution.

As in the binomial case, there are simple expressions for $E(X)$ and $\text{Var}(X)$ for hypergeometric rvs.

PROPOSITION

The mean and variance of the hypergeometric rv X having pmf $h(x; n, M, N)$ are

$$E(X) = n \cdot \frac{M}{N} \quad \text{Var}(X) = \left(\frac{N-n}{N-1}\right) \cdot n \cdot \frac{M}{N}\left(1 - \frac{M}{N}\right)$$

The ratio M/N is the proportion of S's in the population. Replacing M/N by p in $E(X)$ and $\text{Var}(X)$ gives

$$E(X) = np \qquad\qquad (2.17)$$

$$\text{Var}(X) = \left(\frac{N-n}{N-1}\right) \cdot np(1-p)$$

Expression (2.17) shows that the means of the binomial and hypergeometric rvs are equal, whereas the variances of the two rvs differ by the factor $(N-n)/(N-1)$, often called the **finite population correction factor**. This factor is less than 1, so the hypergeometric variable has smaller variance than does the binomial rv. The correction factor can be written $(1 - n/N)/(1 - 1/N)$, which is approximately 1 when n is small relative to N.

Example 2.41 (Example 2.40 continued) In the animal-tagging example, $n = 10$, $M = 5$, and $N = 25$, so $p = \frac{5}{25} = .2$ and

$$E(X) = 10(.2) = 2$$

$$\text{Var}(X) = \frac{25 - 10}{25 - 1}(10)(.2)(.8) = (.625)(1.6) = 1$$

If the sampling were carried out *with* replacement, $\text{Var}(X) = 1.6$.

Suppose the population size N is not actually known, so the value x is observed and we wish to estimate N. It is reasonable to equate the observed sample proportion of S's, x/n, with the population proportion, M/N, giving the estimate

$$\hat{N} = \frac{M \cdot n}{x}$$

For example, if $M = 100$, $n = 40$, and $x = 16$, then $\hat{N} = 250$. ∎

Our rule in Sect. 2.4 stated that if sampling is without replacement but n/N is at most .05, then the binomial distribution can be used to compute approximate probabilities involving the number of S's in the sample. A more precise statement is as follows: Let the population size, N, and number of population S's, M, get large with the ratio M/N approaching p. Then $h(x; n, M, N)$ approaches $b(x; n, p)$; so for n/N small, the two are approximately equal provided that p is not too near either 0 or 1. This is the rationale for our rule.

2.6.2 The Negative Binomial and Geometric Distributions

The negative binomial distribution is based on an experiment satisfying the following conditions:
1. The experiment consists of a sequence of independent trials.
2. Each trial can result in either a success (S) or a failure (F).
3. The probability of success is constant from trial to trial, so $P(S \text{ on trial } i) = p$ for $i = 1, 2, 3 \ldots$.
4. The experiment continues (trials are performed) until a total of r successes has been observed, where r is a specified positive integer.

The random variable of interest is $X =$ the number of trials required to achieve the rth success, and X is called a **negative binomial random variable**. In contrast to the binomial rv, the number of *successes* is fixed and the number of *trials* is random. Possible values of X are $r, r + 1, r + 2, \ldots$, since it takes at least r trials to achieve r successes.

Let $nb(x; r, p)$ denote the pmf of X. The event $\{X = x\}$ is equivalent to $\{r - 1\ S$'s in the first $(x - 1)$ trials and an S on the xth trial$\}$, e.g., if $r = 5$ and $x = 15$, then there must be four S's in the first 14 trials and trial 15 must be an S. Since trials are independent,

$$nb(x; r, p) = P(X = x) = P(r - 1\ S\text{'s on the first } x - 1 \text{ trials}) \cdot P(S) \qquad (2.18)$$

The first probability on the far right of Eq. (2.18) is the binomial probability

$$\binom{x - 1}{r - 1} p^{r-1} (1 - p)^{(x-1)-(r-1)} \quad \text{where } P(S) = p$$

Simplifying and then multiplying by the extra factor of p at the end of Eq. (2.18) yields the following.

PROPOSITION

The pmf of the negative binomial rv X with parameters $r =$ desired number of S's and $p = P(S)$ is

$$nb(x; r, p) = \binom{x - 1}{r - 1} p^r (1 - p)^{x-r} \quad x = r, r + 1, r + 2, \ldots$$

Example 2.42 A pediatrician wishes to recruit four couples, each of whom is expecting their first child, to participate in a new natural childbirth regimen. Let $p = P($a randomly selected couple agrees to participate$)$. If $p = .2$, what is the probability that exactly 15 couples must be asked before 4 are found who agree to participate? Substituting $r = 4$, $p = .2$, and $x = 15$ into $nb(x; r, p)$ gives

$$nb(15; 4, 2) = \binom{15 - 1}{4 - 1} .2^4 .8^{11} = .050$$

The probability that at most 15 couples need to be asked is

$$P(X \le 15) = \sum_{x=4}^{15} nb(x; 4, .2) = \sum_{x=4}^{15} \binom{x - 1}{3} .2^4 .8^{x-4} = .352 \qquad \blacksquare$$

In the special case $r = 1$, the pmf is

$$nb(x; 1, p) = (1 - p)^{x-1} p \quad x = 1, 2, \ldots \qquad (2.19)$$

In Example 2.10, we derived the pmf for the number of trials necessary to obtain the first S, and the pmf there is identical to Eq. (2.19). The variable $X =$ number of trials required to achieve one success is referred to as a **geometric**

random variable, and the pmf in Eq. (2.19) is called the **geometric distribution**. The name is appropriate because the probabilities constitute a geometric series: $p, (1-p)p, (1-p)^2 p, \ldots$. To see that the sum of the probabilities is 1, recall that the sum of a geometric series is $a + ar + ar^2 + \ldots = a/(1-r)$ if $|r| < 1$, so for $p > 0$,

$$p + (1-p)p + (1-p)^2 p + \cdots = \frac{p}{1-(1-p)} = 1$$

In Example 2.19, the expected number of trials until the first S was shown to be $1/p$. Intuitively, we would then expect to need $r \cdot 1/p$ trials to achieve the rth S, and this is indeed $E(X)$. There is also a simple formula for $Var(X)$.

PROPOSITION

If X is a negative binomial rv with parameters r and p, then

$$E(X) = \frac{r}{p} \quad Var(X) = \frac{r(1-p)}{p^2}$$

Example 2.43 (Example 2.42 continued) With $p = .2$, the expected number of couples the doctor must speak to in order to find 4 that will agree to participate is $r/p = 4/.2 = 20$. This makes sense, since with $p = .2 = 1/5$ it will take five attempts, on average, to achieve one success. The corresponding variance is $4(1 - .2)/(.2)^2 = 80$, for a standard deviation of about 8.9. ∎

Since they are based on similar experiments, some caution must be taken to distinguish the binomial and negative binomial models, as seen in the next example.

Example 2.44 In many communication systems, a receiver will send a short signal back to the transmitter to indicate whether a message has been received correctly or with errors. (These signals are often called an *acknowledgement* and a *non-acknowledgement*, respectively. Bit sum checks and other tools are used by the receiver to determine the absence or presence of errors.) Assume we are using such a system in a noisy channel, so that each message is sent error-free with probability .86, independent of all other messages. What is the probability that in 10 transmissions, exactly 8 will succeed? What is the probability the system will require exactly 10 attempts to successfully transmit 8 messages?

While these two questions may sound similar, they require two different models for solution. To answer the first question, let $X =$ the number of successful transmissions out of 10. Then $X \sim Bin(10, .86)$, and the answer is

$$P(X = 8) = b(8; 10, .86) = \binom{10}{8}(.86)^8(.14)^2 = .2639$$

However, the event {exactly 10 attempts required to successfully transmit 8 messages} is more restrictive: not only must we observe 8 S's and 2 F's in 10 trials, but *the last trial must be a success*. Otherwise, it took fewer than 10 tries to send 8 messages successfully. Define a variable Y = the number of transmissions (trials) required to successfully transmit 8 messages. Then Y is negative binomial, with $r = 8$ and $p = .86$, and the answer to the second question is

$$P(Y = 10) = nb(10; 8, .86) = \binom{10 - 1}{8 - 1}(.86)^8(.14)^2 = .2111$$

Notice this is smaller than the answer to the first question, which makes sense because (as we noted) the second question imposes an additional constraint. In fact, you can think of the "-1" terms in the negative binomial pmf as accounting for this loss of flexibility in the placement of S's and F's.

Similarly, the expected number of successful transmissions in 10 attempts is $E(X) = np = 10(.86) = 8.6$, while the expected number of attempts required to successfully transmit 8 messages is $E(Y) = r/p = 8/.86 = 9.3$. In the first case, the number of trials ($n = 10$) is fixed, while in the second case the desired number of successes ($r = 8$) is fixed.   ∎

By expanding the binomial coefficient in front of $p^r(1 - p)^{x-r}$ and doing some cancellation, it can be seen that $nb(x; r, p)$ is well-defined even when r is not an integer. This *generalized negative binomial distribution* has been found to fit observed data quite well in a wide variety of applications.

2.6.3 Alternative Definition of the Negative Binomial Distribution

There is not universal agreement on the definition of a negative binomial random variable (or, by extension, a geometric rv). It is not uncommon in the literature, as well as in some textbooks, to see the number of *failures* preceding the rth success called "negative binomial"; in our notation, this simply equals $X - r$. Possible values of this "number of failures" variable are 0, 1, 2, ... Similarly, the geometric distribution is sometimes defined in terms of the number of failures preceding the first success in a sequence of independent and identical trials. If one uses these alternative definitions, then the pmf and mean formula must be adjusted accordingly. (The variance, however, will stay the same.)

The authors of Matlab and R are among those who have adopted this alternative definition; as a result, we must be careful with our inputs to the relevant software functions. The pmf syntax for the distributions in this section are cataloged in Table 2.5; cdfs may be invoked by changing pdf to cdf in Matlab or the initial letter d to p in R. Notice the input argument $x - r$ for the negative binomial functions: both software packages request the number of failures, rather than the number of trials.

Table 2.5 Matlab and R code for hypergeometric and negative binomial calculations

	Hypergeometric	Negative Binomial
Function:	pmf	pmf
Notation:	$h(x; n, M, N)$	$nb(x; r, p)$
Matlab:	pdf('hyge',x,N,M,n)	pdf('nbin',$x-r,r,p$)
R:	dhyper($x,M,N-M,n$)	dnbinom($x-r,r,p$)

For example, suppose X has a hypergeometric distribution with $n = 10$, $M = 5$, $N = 25$ as in Example 2.40. Using Matlab, we may calculate $P(X = 2) = $ pdf('hyge',2,25,5,10) and $P(X \leq 2) = $ cdf('hyge',2,25,5,10). The corresponding R function calls are dhyper(2,5,20,10) and phyper(2,5,20,10), respectively. If X is the negative binomial variable of Example 2.42 with parameters $r = 4$ and $p = .2$, then the chance of requiring 15 trials to achieve 4 successes (i.e., 11 total failures) can be found in Matlab with pdf('nbin',11,4,.2) and in R using the command dnbinom(11,4,.2).

2.6.4   Exercises: Section 2.6 (90–106)

90. An electronics store has received a shipment of 20 table radios that have connections for an iPod or iPhone. Twelve of these have two slots (so they can accommodate both devices), and the other eight have a single slot. Suppose that six of the 20 radios are randomly selected to be stored under a shelf where radios are displayed, and the remaining ones are placed in a storeroom. Let $X = $ the number among the radios stored under the display shelf that have two slots.
    (a) What kind of a distribution does X have (name and values of all parameters)?
    (b) Compute $P(X = 2)$, $P(X \leq 2)$, and $P(X \geq 2)$.
    (c) Calculate the mean value and standard deviation of X.

91. Each of 12 refrigerators has been returned to a distributor because of an audible, high-pitched, oscillating noise when the refrigerator is running. Suppose that 7 of these refrigerators have a defective compressor and the other 5 have less serious problems. If the refrigerators are examined in random order, let X be the number among the first 6 examined that have a defective compressor. Compute the following:
    (a) $P(X = 5)$
    (b) $P(X \leq 4)$
    (c) The probability that X exceeds its mean value by more than 1 standard deviation.
    (d) Consider a large shipment of 400 refrigerators, of which 40 have defective compressors. If X is the number among 15 randomly selected refrigerators

that have defective compressors, describe a less tedious way to calculate (at least approximately) $P(X \leq 5)$ than to use the hypergeometric pmf.

92. An instructor who taught two sections of statistics last term, the first with 20 students and the second with 30, decided to assign a term project. After all projects had been turned in, the instructor randomly ordered them before grading. Consider the first 15 graded projects.
    (a) What is the probability that exactly 10 of these are from the second section?
    (b) What is the probability that at least 10 of these are from the second section?
    (c) What is the probability that at least 10 of these are from the same section?
    (d) What are the mean and standard deviation of the number among these 15 that are from the second section?
    (e) What are the mean and standard deviation of the number of projects not among these first 15 that are from the second section?

93. A geologist has collected 10 specimens of basaltic rock and 10 specimens of granite. The geologist instructs a laboratory assistant to randomly select 15 of the specimens for analysis.
    (a) What is the pmf of the number of granite specimens selected for analysis?
    (b) What is the probability that all specimens of one of the two types of rock are selected for analysis?
    (c) What is the probability that the number of granite specimens selected for analysis is within 1 standard deviation of its mean value?

94. A personnel director interviewing 11 senior engineers for four job openings has scheduled six interviews for the first day and five for the second day of interviewing. Assume the candidates are interviewed in random order.
    (a) What is the probability that x of the top four candidates are interviewed on the first day?
    (b) How many of the top four candidates can be expected to be interviewed on the first day?

95. Twenty pairs of individuals playing in a bridge tournament have been seeded 1, ..., 20. In the first part of the tournament, the 20 are randomly divided into 10 east–west pairs and 10 north–south pairs.
    (a) What is the probability that x of the top 10 pairs end up playing east–west?
    (b) What is the probability that all of the top five pairs end up playing the same direction?
    (c) If there are $2n$ pairs, what is the pmf of $X =$ the number among the top n pairs who end up playing east–west? What are $E(X)$ and $\text{Var}(X)$?

96. A second-stage smog alert has been called in an area of Los Angeles County in which there are 50 industrial firms. An inspector will visit 10 randomly selected firms to check for violations of regulations.
    (a) If 15 of the firms are actually violating at least one regulation, what is the pmf of the number of firms visited by the inspector that are in violation of at least one regulation?
    (b) If there are 500 firms in the area, of which 150 are in violation, approximate the pmf of part (a) by a simpler pmf.

(c) For $X =$ the number among the 10 visited that are in violation, compute $E(X)$ and $\text{Var}(X)$ both for the exact pmf and the approximating pmf in part (b).

97. A shipment of 20 integrated circuits (ICs) arrives at an electronics manufacturing site. The site manager will randomly select 4 ICs and test them to see whether they are faulty. Unknown to the site manager, 5 of these 20 ICs are faulty.

(a) Suppose the shipment will be accepted if and only if none of the inspected ICs is faulty. What is the probability this shipment of 20 ICs will be accepted?

(b) Now suppose the shipment will be accepted if and only if at most one of the inspected ICs is faulty. What is the probability this shipment of 20 ICs will be accepted?

(c) How do your answers to (a) and (b) change if the number of faculty ICs in the shipment is 3 instead of 5? Recalculate (a) and (b) to verify your claim.

98. Suppose that 20% of all individuals have an adverse reaction to a particular drug. A medical researcher will administer the drug to one individual after another until the first adverse reaction occurs. Define an appropriate random variable and use its distribution to answer the following questions.

(a) What is the probability that when the experiment terminates, four individuals have not had adverse reactions?

(b) What is the probability that the drug is administered to exactly five individuals?

(c) What is the probability that at most four individuals do not have an adverse reaction?

(d) How many individuals would you expect to not have an adverse reaction, and how many individuals would you expect to be given the drug?

(e) What is the probability that the number of individuals given the drug is within one standard deviation of what you expect?

99. Suppose that $p = P(\text{female birth}) = .5$. A couple wishes to have exactly two female children in their family. They will have children until this condition is fulfilled.

(a) What is the probability that the family has x male children?

(b) What is the probability that the family has four children?

(c) What is the probability that the family has at most four children?

(d) How many children would you expect this family to have? How many male children would you expect this family to have?

100. A family decides to have children until it has three children of the same gender. Assuming $P(B) = P(G) = .5$, what is the pmf of $X =$ the number of children in the family?

101. Three brothers and their wives decide to have children until each family has two female children. Let $X =$ the total number of male children born to the brothers. What is $E(X)$, and how does it compare to the expected number of male children born to each brother?

102. According to the article "Characterizing the Severity and Risk of Drought in the Poudre River, Colorado" (*J. of Water Res. Planning and Mgmnt.*, 2005: 383-393), the drought length Y is the number of consecutive time intervals in which the water supply remains below a critical value y_0 (a deficit), preceded and followed by periods in which the supply exceeds this value (a surplus). The cited paper proposes a geometric distribution with $p = .409$ for this random variable.
    (a) What is the probability that a drought lasts exactly 3 intervals? At least 3 intervals?
    (b) What is the probability that the length of a drought exceeds its mean value by at least one standard deviation?

103. Individual A has a red die and B has a green die (both fair). If they each roll until they obtain five "doubles" (⊡⊡, ..., ⊞⊞), what is the pmf of X = the total number of times a die is rolled? What are $E(X)$ and $SD(X)$?

104. A carnival game consists of spinning a wheel with 10 slots, nine red and one blue. If you land on the blue slot, you win a prize. Suppose your significant other *really* wants that prize, so you will play until you win.
    (a) What is the probability you'll win on the first spin?
    (b) What is the probability you'll require exactly 5 spins? At least 5 spins? At most five spins?
    (c) What is the expected number of spins required for you to win the prize, and what is the corresponding standard deviation?

105. A kinesiology professor, requiring volunteers for her study, approaches students one by one at a campus hub. She will continue until she acquires 40 volunteers. Suppose that 25% of students are willing to volunteer for the study, that the professor's selections are random, and that the student population is large enough that individual "trials" (asking a student to participate) may be treated as independent.
    (a) What is the expected number of students the kinesiology professor will need to ask in order to get 40 volunteers? What is the standard deviation?
    (b) Determine the probability that the number of students the kinesiology professor will need to ask is within one standard deviation of the mean.

106. Refer back to the communication system of Example 2.44. Suppose a voice packet can be transmitted a maximum of 10 times, i.e., if the 10th attempt fails, no 11th attempt is made to retransmit the voice packet. Let X = the number of times a message is transmitted. Assuming each transmission succeeds with probability p, determine the pmf of X. Then obtain an expression for the expected number of times a packet is transmitted.

2.7 Moments and Moment Generating Functions

The expected values of integer powers of X and $X - \mu$ are often referred to as *moments*, terminology borrowed from physics. In this section, we'll discuss the general topic of moments and develop a shortcut for computing them.

> **DEFINITION**
> The **kth moment** of a random variable X is $E(X^k)$, while the **kth moment about the mean** (or **kth central moment**) of X is $E[(X - \mu)^k]$, where $\mu = E(X)$.

For example, $\mu = E(X)$ is the "first moment" of X and corresponds to the center of mass of the distribution of X. Similarly, $\text{Var}(X) = E[(X - \mu)^2]$ is the second moment of X about the mean, which is known in physics as the *moment of inertia*.

Example 2.45 A popular brand of dog food is sold in 5, 10, 15, and 20 lb bags. Let X be the weight of the next bag purchased, and suppose the pmf of X is

x	5	10	15	20
$p(x)$	.1	.2	.3	.4

The first moment of X is its mean:

$$\mu = E(X) = \sum_{x \in D} xp(x) = 5(.1) + 10(.2) + 15(.3) + 20(.4) = 15 \, \text{lbs}$$

The second moment about the mean is the variance:

$$\sigma^2 = E\left[(X - \mu)^2\right] = \sum_{x \in D} (x - \mu)^2 p(x)$$

$$= (5 - 15)^2(.1) + (10 - 15)^2(.2) + (15 - 15)^2(.3) + (20 - 15)^2(.4) = 25,$$

for a standard deviation of 5 lb. The third central moment of X is

$$E\left[(X - \mu)^3\right] = \sum_{x \in D} (x - \mu)^3 p(x)$$

$$= (5 - 15)^3(.1) + \left(10 - 15\right)^3(.2) + \left(15 - 15\right)^3(.3)$$

$$+ \left(20 - 15\right)^3(.4)$$

$$= -75$$

We'll discuss an interpretation of this last number shortly. ∎

It is not difficult to verify that the third moment about the mean is 0 if the pmf of X is symmetric. We would like to use $E[(X - \mu)^3]$ as a measure of lack of symmetry, but it depends on the scale of measurement. If we switch the unit of weight in Example 2.45 from pounds to ounces or kilograms, the value of the third moment about the mean (as well as the values of all the other moments) will change. But we can achieve scale independence by dividing the third moment about the mean by σ^3:

$$\frac{E\left[(X-\mu)^3\right]}{\sigma^3} = E\left[\left(\frac{X-\mu}{\sigma}\right)^3\right] \qquad (2.20)$$

Expression (2.20) is our measure of departure from symmetry, called the **skewness coefficient**. The skewness coefficient for a symmetric distribution is 0 because its third moment about the mean is 0. However, in the foregoing example the skewness coefficient is $E[(X-\mu)^3]/\sigma^3 = -75/5^3 = -0.6$. When the skewness coefficient is negative, as it is here, we say that the distribution is *negatively skewed* or that it is *skewed to the left*. Generally speaking, it means that the distribution stretches farther to the left of the mean than to the right.

If the skewness were positive, then we would say that the distribution is *positively skewed* or that it is *skewed to the right*. For example, reverse the order of the probabilities in the $p(x)$ table above, so the probabilities of the values 5, 10, 15, 20 are now .4, .3, .2, and .1, (customers now favor much smaller bags of dog food). Exercise 119 shows that this changes the sign but not the magnitude of the skewness coefficient, so it becomes +0.6 and the distribution is skewed right. Both distributions are illustrated in Fig. 2.9.

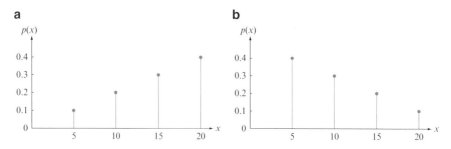

Fig. 2.9 Departures from symmetry: (**a**) skewness coefficient < 0 (skewed left); (**b**) skewness coefficient > 0 (skewed right)

2.7.1   The Moment Generating Function

Calculation of the mean, variance, skewness coefficient, etc. for a particular discrete rv requires extensive, sometimes tedious, summation. Mathematicians have developed a tool, the moment generating function, that will allow us to determine the moments of a distribution with less effort. Moreover, this function will allow us to derive properties of several of our major probability distributions here and in subsequent sections of the book.

DEFINITION
The **moment generating function (mgf)** of a discrete random variable X is defined to be

$$M_X(t) = E(e^{tX}) = \sum_{x \in D} e^{tx} p(x)$$

where D is the set of possible X values. The moment generating function exists iff $M_X(t)$ is defined for an interval that includes zero as well as positive and negative values of t.

For any random variable X, the mgf evaluated at $t = 0$ is

$$M_X(0) = E(e^{0X}) = \sum_{x \in D} e^{0x} p(x) = \sum_{x \in D} 1 p(x) = 1$$

That is, $M_X(0)$ is the sum of all the probabilities, so it must always be 1. However, in order for the mgf to be useful in generating moments, it will need to be defined for an interval of values of t including 0 in its interior. The moment generating function fails to exist in cases when moments themselves fail to exist (see Example 2.49 below).

Example 2.46 The simplest example of an mgf is for a Bernoulli distribution, where only the X values 0 and 1 receive positive probability. Let X be a Bernoulli random variable with $p(0) = 1/3$ and $p(1) = 2/3$. Then

$$M_X(t) = E(e^{tX}) = \sum_{x \in D} e^{tx} p(x) = e^{t \cdot 0} \cdot (1/3) + e^{t \cdot 1} \cdot (2/3) = (1/3) + (2/3)e^t$$

A Bernoulli random variable will always have an mgf of the form $p(0) + p(1)e^t$, a well-defined function for all values of t. ∎

A key property of the mgf is its "uniqueness," the fact that it completely characterizes the underlying distribution.

MGF UNIQUENESS THEOREM
If the mgf exists and is the same for two distributions, then the two distributions are identical. That is, the moment generating function uniquely specifies the probability distribution; there is a one-to-one correspondence between distributions and mgfs.

The proof of this theorem, originally due to Laplace, requires some sophisticated mathematics and is beyond the scope of this textbook.

Example 2.47 Let X, the number of claims submitted on a renter's insurance policy on a given year, have mgf $M_X(t) = .7 + .2e^t + .1e^{2t}$. It follows that X must have the pmf $p(0) = .7$, $p(1) = .2$, and $p(2) = .1$—because if we use this pmf to obtain the mgf, we get $M_X(t)$, and the distribution is uniquely determined by its mgf.   ∎

Example 2.48 Consider testing individuals' blood samples one by one in order to find someone whose blood type is Rh+. Suppose X, the number of tested samples, has a geometric distribution with $p = .85$:

$$p(x) = .85(.15)^{x-1} \text{ for } x = 1, 2, 3, \dots .$$

Determining the moment generating function here requires using the formula for the sum of a geometric series: $1 + r + r^2 + \cdots = 1/(1 - r)$ for $|r| < 1$. The moment generating function is

$$M_X(t) = E(e^{tX}) = \sum_{x \in D} e^{tx} p(x) = \sum_{x=1}^{\infty} e^{tx} .85(.15)^{x-1} = .85e^t \sum_{x=1}^{\infty} e^{t(x-1)} (.15)^{x-1}$$

$$= .85e^t \sum_{x=1}^{\infty} (.15e^t)^{x-1} = .85e^t \left[1 + .15e^t + (.15e^t)^2 + \cdots \right] = \frac{.85e^t}{1 - .15e^t}$$

The condition on r requires $|.15e^t| < 1$. Dividing by .15 and taking logs, this gives $t < -\ln(.15) \approx 1.90$, i.e., this function is defined in the interval $(-\infty, 1.90)$. The result is an interval of values that includes 0 in its interior, so the mgf exists. As a check, $M_X(0) = .85/(1 - .15) = 1$, as required.   ∎

Example 2.49 Reconsider Example 2.20, where $p(x) = k/x^2, x = 1, 2, 3, \dots$. Recall that $E(X)$ does not exist for this distribution, portending a problem for the existence of the mgf:

$$M_X(t) = E\left(e^{tX}\right) = \sum_{x=1}^{\infty} e^{tx} \frac{k}{x^2}$$

With the help of tests for convergence such as the ratio test, we find that the series converges if and only if $e^t \leq 1$, which means that $t \leq 0$, i.e., the mgf is only defined on the interval $(-\infty, 0]$. Because zero is on the *boundary* of this interval, not the interior of the interval (the interval must include both positive and negative values), the mgf of this distribution does not exist. In any case, it could not be useful for finding moments, because X does not have even a first moment (mean).   ∎

2.7.2 Obtaining Moments from the MGF

We now turn to the computation of moments from the mgf. For any positive integer r, let $M_X^{(r)}(t)$ denote the rth derivative of $M_X(t)$. By computing this and then setting $t = 0$, we get the rth moment about 0.

THEOREM

If the mgf of X exists, then $E(X^r)$ is finite for all positive integers r, and

$$E(X^r) = M_X^{(r)}(0) \qquad (2.21)$$

Proof The proof of the existence of all moments is beyond the scope of this book. We will show that Eq. (2.21) is true for $r = 1$ and $r = 2$. A proof by mathematical induction can be used for general r. Differentiate:

$$\frac{d}{dt}M_X(t) = \frac{d}{dt}\sum_{x \in D} e^{xt}p(x) = \sum_{x \in D}\frac{d}{dt}e^{xt}p(x) = \sum_{x \in D}xe^{xt}p(x)$$

where we have interchanged the order of summation and differentiation. (This is justified inside the interval of convergence, which includes 0 in its interior.) Next set $t = 0$ to obtain the first moment:

$$M_X'(0) = M_X^{(1)}(0) = \sum_{x \in D}xe^{x(0)}p(x) = \sum_{x \in D}xp(x) = E(X)$$

Differentiating a second time gives

$$\frac{d^2}{dt^2}M_X(t) = \frac{d}{dt}\sum_{x \in D}xe^{xt}p(x) = \sum_{x \in D}x\frac{d}{dt}e^{xt}p(x) = \sum_{x \in D}x^2e^{xt}p(x)$$

Set $t = 0$ to get the second moment:

$$M_X''(0) = M_X^{(2)}(0) = \sum_{x \in D}x^2p(x) = E(X^2) \qquad \blacksquare$$

For the pmfs in Examples 2.45 and 2.46, this may seem like needless work—after all, for simple distributions with just a few values, we can quickly determine the mean, variance, etc. The real utility of the mgf arises for more complicated distributions.

Example 2.50 (Example 2.48 continued) Recall that $p = .85$ is the probability of a person having Rh+ blood and we keep checking people until we find one with this blood type. If X is the number of people we need to check, then $p(x) = .85(.15)^{x-1}$, $x = 1, 2, 3, \ldots$, and the mgf is

$$M_X(t) = E(e^{tX}) = \frac{.85e^t}{1 - .15e^t}$$

Differentiating with the help of the quotient rule,

$$M_X'(t) = \frac{.85e^t}{(1 - .15e^t)^2}$$

Setting $t = 0$ then gives $\mu = E(X) = M_X'(0) = 1/.85 = 1.176$. This corresponds to the formula $1/p$ for a geometric distribution.

To get the second moment, differentiate again:

$$M_X''(t) = \frac{.85e^t(1 + .15e^t)}{(1 - .15e^t)^3}$$

Setting $t = 0$, $E(X^2) = M_X''(0) = \dfrac{1.15}{.85^2} = 1.15/.85^2$. Now use the variance shortcut formula:

$$\text{Var}(X) = \sigma^2 = E(X^2) - \mu^2 = \frac{1.15}{.85^2} - \left(\frac{1}{.85}\right)^2 = \frac{.15}{.85^2} = .2076$$

This matches the variance formula $(1 - p)/p^2$ given without proof toward the end of Sect. 2.6. ∎

As mentioned in Sect. 2.3, it is common to transform a rv X using a linear function $Y = aX + b$. What happens to the mgf when we do this?

PROPOSITION

Let X have the mgf $M_X(t)$ and let $Y = aX + b$. Then $M_Y(t) = e^{bt}M_X(at)$.

Example 2.51 Let X be a Bernoulli random variable with $p(0) = 20/38$ and $p(1) = 18/38$. Think of X as the number of wins, 0 or 1, in a single play of roulette. If you play roulette at an American casino and bet on red, then your chances of winning are 18/38 because 18 of the 38 possible outcomes are red. From Example 2.46, $M_X(t) = 20/38 + e^t(18/38)$. Suppose you bet \$5 on red, and let Y be your winnings. If $X = 0$ then $Y = -5$, and if $X = 1$ then $Y = 5$. The linear equation $Y = 10X - 5$ gives the appropriate relationship.

This equation is of the form $Y = aX + b$ with $a = 10$ and $b = -5$, so by the foregoing proposition

$$M_Y(t) = e^{bt}M_X(at) = e^{-5t}M_X(10t)$$

$$= e^{-5t}\left[\frac{20}{38} + e^{10t}\frac{18}{38}\right] = e^{-5t}\cdot\frac{20}{38} + e^{5t}\cdot\frac{18}{38}$$

This implies that the pmf of Y is $p(-5) = 20/38$ and $p(5) = 18/38$; moreover, we can compute the mean (and other moments) of Y directly from this mgf.   ∎

2.7.3   MGFs of Common Distributions

Several of the distributions presented in this chapter (binomial, Poisson, negative binomial) have fairly simple expressions for their moment generating functions. These mgfs, in turn, allow us to determine the means and variances of the distributions without some rather unpleasant summation. (Additionally, we will use these mgfs to prove some more advanced distributional properties in Chap. 4.)

To start, determining the moment generating function of a binomial rv requires use of the binomial theorem: $(a + b)^n = \sum_{x=0}^{n}\binom{n}{x}a^x b^{n-x}$. Then

$$M_X(t) = E(e^{tX}) = \sum_{x\in D}e^{tx}b(x; n, p) = \sum_{x=0}^{n}e^{tx}\binom{n}{x}p^x(1-p)^{n-x}$$

$$= \sum_{x=0}^{n}\binom{n}{x}(pe^t)^x(1-p)^{n-x} = (pe^t + 1 - p)^n$$

The mean and variance can be obtained by differentiating $M_X(t)$:

$$M_X'(t) = n(pe^t + 1 - p)^{n-1}pe^t \quad \Rightarrow \quad \mu = M_X'(0) = np;$$
$$M_X''(t) = n(n-1)(pe^t + 1 - p)^{n-2}pe^tpe^t + n(pe^t + 1 - p)^{n-1}pe^t \Rightarrow$$
$$E(X^2) = M_X''(0) = n(n-1)p^2 + np \Rightarrow$$
$$\sigma^2 = \text{Var}(X) = E(X^2) - \mu^2$$
$$= n(n-1)p^2 + np - n^2p^2 = np - np^2 = np(1-p),$$

in accord with the proposition in Sect. 2.4.

Derivation of the Poisson mgf utilizes the series expansion $\sum_{x=0}^{\infty}u^x/x! = e^u$:

$$M_X(t) = E(e^{tX}) = \sum_{x=0}^{\infty}e^{tx}e^{-\mu}\frac{\mu^x}{x!} = e^{-\mu}\sum_{x=0}^{\infty}\frac{(\mu e^t)^x}{x!} = e^{-\mu}e^{\mu e^t} = e^{\mu(e^t - 1)}$$

Successive differentiation then gives the mean and variance identified in Sect. 2.5 (see Exercise 127).

Finally, derivation of the negative binomial mgf is based on Newton's generalization of the binomial theorem. The result (see Exercise 124) is

$$M_X(t) = \left(\frac{pe^t}{1 - (1 - p)e^t} \right)^r$$

The geometric mgf is just the special case $r = 1$ (cf. Example 2.48 above). There is unfortunately no simple expression for the mgf of a hypergeometric rv.

2.7.4 Exercises: Section 2.7 (107–128)

107. For the entry-level employees of a certain fast food chain, the pmf of $X =$ highest grade level completed is specified by $p(9) = .01$, $p(10) = .05$, $p(11) = .16$, and $p(12) = .78$.
  (a) Determine the moment generating function of this distribution.
  (b) Use (a) to find $E(X)$ and $SD(X)$.
108. For a new car the number of defects X has the distribution given by the accompanying table. Find $M_X(t)$ and use it to find $E(X)$ and $SD(X)$.

x	0	1	2	3	4	5	6
$p(x)$	.04	.20	.34	.20	.15	.04	.03

109. In flipping a fair coin let X be the number of tosses to get the first head. Then $p(x) = .5^x$ for $x = 1, 2, 3, \ldots$. Find $M_X(t)$ and use it to get $E(X)$ and $SD(X)$.
110. If you toss a fair die with outcome X, $p(x) = 1/6$ for $x = 1, 2, 3, 4, 5, 6$. Find $M_X(t)$.
111. Find the skewness coefficients of the distributions in the previous four exercises. Do these agree with the "shape" of each distribution?
112. Given $M_X(t) = .2 + .3e^t + .5e^{3t}$, find $p(x)$, $E(X)$, $Var(X)$.
113. If $M_X(t) = 1/(1 - t^2)$, find $E(X)$ and $Var(X)$.
114. Show that $g(t) = te^t$ cannot be a moment generating function.
115. Using a calculation similar to the one in Example 2.48 show that, if X has a geometric distribution with parameter p, then its mgf is

$$M_X(t) = \frac{pe^t}{1 - (1 - p)e^t}$$

Assuming that Y has mgf $M_Y(t) = .75e^t/(1 - .25e^t)$, determine the probability mass function $p(y)$ with the help of the uniqueness property.
116. (a) Prove the result in the second proposition: $M_{aX+b}(t) = e^{bt}M_X(at)$.
  (b) Let $Y = aX + b$. Use (a) to establish the relationships between the means and variances of X and Y.

117. Let $M_X(t) = e^{5t+2t^2}$ and let $Y = (X - 5)/2$. Find $M_Y(t)$ and use it to find $E(Y)$ and Var(Y).

118. Let X have the moment generating function of Example 2.48 and let $Y = X - 1$. Recall that X is the number of people who need to be checked to get someone who is Rh+, so Y is the number of people checked before the first Rh+ person is found. Find $M_Y(t)$.

119. Let X be the number of points earned by a randomly selected student on a 10 point quiz, with possible values 0, 1, 2, ..., 10 and pmf $p(x)$, and suppose the distribution has a skewness coefficient of c. Now consider reversing the probabilities in the distribution, so that $p(0)$ is interchanged with $p(10)$, $p(1)$ is interchanged with $p(9)$, and so on. Show that the skewness coefficient of the resulting distribution is $-c$. [*Hint*: Let $Y = 10 - X$ and show that Y has the reversed distribution. Use this fact to determine μ_Y and then the value of skewness coefficient for the Y distribution.]

120. Let $M_X(t)$ be the moment generating function of a rv X, and define a new function by

$$L_X(t) = \ln[M_X(t)]$$

Show that (a) $L_X(0) = 0$, (b) $L_X'(0) = \mu$, and (c) $L_X''(0) = \sigma^2$.

121. Refer back to Exercise 120. If $M_X(t) = e^{5t+2t^2}$ then find $E(X)$ and Var(X) by differentiating
    (a) $M_X(t)$
    (b) $L_X(t)$

122. Refer back to Exercise 120. If $M_X(t) = e^{5(e^t-1)}$ then find $E(X)$ and Var(X) by differentiating
    (a) $M_X(t)$
    (b) $L_X(t)$

123. Obtain the moment generating function of the number of failures, $n - X$, in a binomial experiment, and use it to determine the expected number of failures and the variance of the number of failures. Are the expected value and variance intuitively consistent with the expressions for $E(X)$ and Var(X)? Explain.

124. Newton's generalization of the binomial theorem can be used to show that, for any positive integer r,

$$(1 - u)^{-r} = \sum_{k=0}^{\infty} \binom{r+k-1}{r-1} u^k$$

Use this to derive the negative binomial mgf presented in this section. Then obtain the mean and variance of a binomial rv using this mgf.

125. If X is a negative binomial rv, then $Y = X - r$ is the number of failures preceding the rth success. Obtain the mgf of Y and then its mean value and variance.

126. Refer back to Exercise 120. Obtain the negative binomial mean and variance from $L_X(t) = \ln[M_X(t)]$.
127. (a) Use derivatives of $M_X(t)$ to obtain the mean and variance for the Poisson distribution.
    (b) Obtain the Poisson mean and variance from $L_X(t) = \ln[M_X(t)]$. In terms of effort, how does this method compare with the one in part (a)?
128. Show that the binomial moment generating function converges to the Poisson moment generating function if we let $n \to \infty$ and $p \to 0$ in such a way that np approaches a value $\mu > 0$. [*Hint*: Use the calculus theorem that was used in showing that the binomial pmf converges to the Poisson pmf.] There is, in fact, a theorem saying that convergence of the mgf implies convergence of the probability distribution. In particular, convergence of the binomial mgf to the Poisson mgf implies $b(x; n, p) \to p(x; \mu)$.

2.8   Simulation of Discrete Random Variables

Probability calculations for complex systems often depend on the behavior of various random variables. When such calculations are difficult or impossible, simulation is the fallback strategy. In this section, we give a general method for simulating an arbitrary discrete random variable and consider implementations in existing software for simulating common discrete distributions.

Example 2.52 Refer back to the distribution of Example 2.11 for the random variable $X =$ the amount of memory (GB) in a purchased flash drive, and suppose we wish to simulate X. Recall from Sect. 1.6 that we begin with a "standard uniform" random number generator, i.e., a software function that generates evenly distributed numbers in the interval $[0, 1)$. Our goal is to convert these decimals into the values of X with the probabilities specified by its pmf: 5% 1s, 10% 2s, 35% 4s, and so on. To that end, we partition the interval $[0, 1)$ according to these percentages: $[0, .05)$ has probability .05; $[.05, .15)$ has probability .1, since the length of the interval is .1; $[.15, .5)$ has probability $.5 - .15 = .35$; etc. Proceed as follows: given a value u from the RNG,

– If $0 \le u < .05$, assign the value 1 to the variable x.
– If $.05 \le u < .15$, assign $x = 2$.
– If $.15 \le u < .50$, assign $x = 4$.
– If $.50 \le u < .90$, assign $x = 8$.
– If $.90 \le u < 1$, assign $x = 16$.

Repeating this algorithm n times gives n simulated values of X. Programs in Matlab and R that implement this algorithm appear in Fig. 2.10; both return a vector, x, containing $n = 10{,}000$ simulated values of the specified distribution.

Figure 2.11 shows a graph of the results of executing the code, in the form of a *histogram*: the height of each rectangle corresponds to the relative frequency of each x value in the simulation (i.e., the number of times that value occurred, divided

a
```
x=zeros(10000,1);
for i=1:10000
    u=rand;
    if u<.05
        x(i)=1;
    elseif u<.15
        x(i)=2;
    elseif u<.50
        x(i)=4;
    elseif u<.90
        x(i)=8;
    else
        x(i)=16;
    end
end
```

b
```
x <- NULL
for (i in 1:10000){
    u=runif(1)
    if (u<.05)
        x[i]<-1
    else if (u<.15)
        x[i]<-2
    else if (u<.50)
        x[i]<-4
    else if (u<.90)
        x[i]<-8
    else
        x[i]<-16
}
```

Fig. 2.10 Simulation code: (**a**) Matlab; (**b**) R

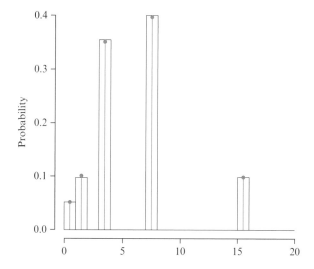

Fig. 2.11 Simulation and exact distribution for Example 2.52

by 10,000). The exact pmf of X is superimposed for comparison; as expected, simulation results are similar, but not identical, to the theoretical distribution.

Later in this section, we will present a faster, built-in way to simulate discrete distributions in Matlab and R. The method introduced here will, however, prove useful in adapting to the case of continuous random variables in Chap. 3.   ■

In the preceding example, the selected subintervals of [0, 1) were not our only choices—any five intervals with lengths .05, .10, .35, .40, and .10 would produce the desired result. However, those particular five subintervals have one desirable feature: the "cut points" for the intervals (i.e., 0, .05, .15, .50, .90, and 1) are precisely the possible heights of the graph of the cdf, $F(x)$. This permits a geometric

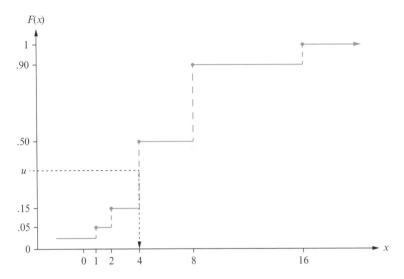

Fig. 2.12 The inverse cdf method for Example 2.52

interpretation of the algorithm, which can be seen in Fig. 2.12. The value u provided by the RNG corresponds to a position on the vertical axis between 0 and 1; we then "invert" the cdf by matching this u-value back to one of the gaps in the graph of $F(x)$, denoted by dotted lines in Fig. 2.12. If the gap occurs at horizontal position x, then x is our simulated value of the rv X for that run of the simulation. This is often referred to as the **inverse cdf method** for simulating discrete random variables. The general method is spelled out in the accompanying box.

Inverse cdf Method for Simulating Discrete Random Variables
Let X be a discrete random variable taking on values $x_1 < x_2 < \dots$ with corresponding probabilities p_1, p_2, ... Define $F_0 = 0$; $F_1 = F(x_1) = p_1$; $F_2 = F(x_2) = p_1 + p_2$; and, in general, $F_k = F(x_k) = p_1 + \dots + p_k = F_{k-1} + p_k$. To simulate a value of X, proceed as follows:
1. Use an RNG to produce a value, u, from $[0, 1)$.
2. If $F_{k-1} \leq u < F_k$, then assign $x = x_k$.

Example 2.53 (Example 2.52 continued): Suppose the prices for the flash drives, in increasing order of memory size, are \$10, \$15, \$20, \$25, and \$30. If the store sells 80 flash drives in a week, what's the probability they will make a gross profit of at least \$1800?

Let $Y =$ the amount spent on a flash drive, which has the following pmf:

y	10	15	20	25	30
$p(y)$	.05	.10	.35	.40	.10

The gross profit for 80 purchases is the sum of 80 values from this distribution. Let $A = \{$gross profit $\geq \$1800\}$. We can use simulation to estimate $P(A)$, as follows:

0. Set a counter for the number of times A occurs to zero.

Repeat n times:

1. Simulate 80 values $y_1, \ldots, y_{80}$ from the above pmf (using for example an inverse cdf program similar to those displayed in Fig. 2.10).
2. Compute the week's gross profit, $g = y_1 + \cdots + y_{80}$.
3. If $g \geq 1800$, add 1 to the count of occurrences for A.

Once the n runs are complete, then $\hat{P}(A) = $ (count of the occurrences of $A)/n$.

Figure 2.13 shows the resulting values of g for $n = 10{,}000$ simulations in R. In effect, our program is simulating a random variable $G = Y_1 + \ldots + Y_{80}$ whose pmf is not known (in light of all the possible G values, it would not be worthwhile to attempt to determine its pmf analytically). The highlighted bars in Fig. 2.13 correspond to g values of at least \$1800; in our simulation, such values occurred 1940 times. Thus, $\hat{P}(A) = 1940/10{,}000 = .194$, with an estimated standard error of $\sqrt{.194(1 - .194)/10{,}000} = .004$.

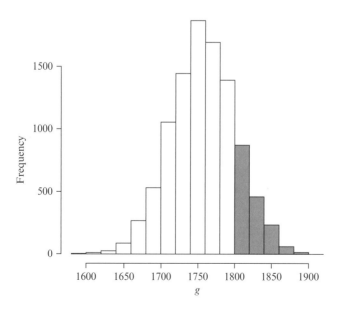

Fig. 2.13 Simulated distribution of weekly gross profit for Example 2.53   ■

2.8.1   Simulations Implemented in R and Matlab

Earlier in this section, we presented the inverse cdf method as a general way to simulate discrete distributions applicable in any software. In fact, one can simulate generic discrete rvs in both Matlab and R by clever use of the built-in `randsample` and `sample` functions, respectively. We saw these functions in the context of probability simulation in Chap. 1. Both are designed to generate a random sample from any selected set of values (even including text values, if desired); the "clever" part is that both can accommodate a set of weights. The following short example illustrates their use.

To simulate, say, 35 values from the pmf in Example 2.52, one can use the following code in Matlab:

```
randsample([1,2,4,8,16],35,true,[.05,.10,.35,.40,.10])
```

The function takes four arguments: the list of x-values, the desired number of simulated values (the "sample size"), whether to sample with replacement (here, `true`), and the list of probabilities in the same order as the x-values. The corresponding call in R is

```
sample(c(1,2,4,8,16),35,TRUE,c(.05,.10,.35,.40,.10))
```

Thanks to the ubiquity of the binomial, Poisson, and other distributions in probability modeling, many software packages have built-in tools for simulating values from these distributions. Table 2.6 summarizes the relevant functions in Matlab and R; the input argument *sampsize* refers to the desired number of simulated values of the distribution.

A word of warning (really, a reminder) about the way software treats the negative binomial distribution: both Matlab and R define a negative binomial rv as the number of failures preceding the rth success, which differs from our definition. Assuming you want to simulate the number of *trials* required to achieve r successes, execute the code in the last line of Table 2.6 and then add r to each value.

Table 2.6 Functions to simulate major discrete distributions in Matlab and R

Distribution	Matlab code	R code
Binomial	`random('bin',`$n,p,$ `[`*sampsize*`,1])`	`rbinom(`*sampsize*$,n,p)$
Poisson	`random('pois',`$\mu,$ `[`*sampsize*`,1])`	`rpois(`*sampsize*$,\mu)$
Hypergeometric	`random('hyge',`$N,M,n,$ `[`*sampsize*`,1])`	`rhyper(`*sampsize*$,M,$ $N-M,n)$
Negative binomial	`random('nbin',`$r,p,$ `[`*sampsize*`,1])`	`rnbinom(`*sampsize*$,r,p)$

Example 2.54 The number of customers shipping express mail packages at a certain store during any particular hour of the day is a Poisson rv with mean 5. Each such customer has 1, 2, 3, or 4 packages with probabilities .4, .3, .2, and .1, respectively. Let's carry out a simulation to estimate the probability that at most 10 packages are shipped during any particular hour.

Define an event $A = \{$ at most 10 packages shipped in an hour$\}$. Our simulation to estimate $P(A)$ proceeds as follows.

0. Set a counter for the number of times A occurs to zero.

   Repeat n times:
1. Simulate the number of customers in an hour, C, which is Poisson with $\mu = 5$.
2. For each of the C customers, simulate the number of packages shipped according to the pmf above.
3. If the total number of packages shipped is at most 10, add 1 to the counter for A.
   Matlab and R code to implement this simulation appear in Fig. 2.14.

a
```
A=0;
for i=1:10000
    c=random('pois',5,1);
    packages = randsample([1,2,3,4],c,
                true,[.4,.3,.2,.1]);
    if sum(packages)<=10
        A=A+1;
    end
end
```

b
```
A <- 0
for (i in 1:10000){
    c<-rpois(1,5)
    packages <- sample(c(1,2,3,4),c,
                TRUE,c(.4,.3,.2,.1))
    if (sum(packages)<=10){
        A<-A+1
    }
}
```

Fig. 2.14 Simulation code for Example 2.54: (**a**) Matlab; (**b**) R

In Matlab, 10,000 simulations resulted in 10 or fewer packages 5752 times, for an estimated probability of $\hat{P}(A) = .5752$, with an estimated standard error of $\sqrt{.5752(1 - .5752)/10000} = .0049$. ∎

2.8.2 Simulation Mean, Standard Deviation, and Precision

In Sect. 1.6 and in the preceding examples, we used simulation to estimate the probability of an event. But consider the "gross profit" variable G in Example 2.53: since we have 10,000 simulated values of this variable, we should be able to estimate its mean μ_G and its standard deviation σ_G. More generally, suppose we have simulated n values $x_1, \ldots, x_n$ of a random variable X. Then the following quantities based on our observed values serve as suitable estimates.

DEFINITION

For a set of numerical values $x_1, \ldots, x_n$, the **sample mean**, denoted by $\bar{x}$, is

$$\bar{x} = \frac{x_1 + \cdots + x_n}{n} = \frac{1}{n}\sum_{i=1}^{n} x_i$$

The **sample standard deviation** of these numerical values, denoted by s, is

$$s = \sqrt{\frac{1}{n-1}\sum_{i=1}^{n}(x_i - \bar{x})^2}$$

If $x_1, \ldots, x_n$ represent simulated values of a random variable X, then we may estimate the expected value and standard deviation of X by $\hat{\mu} = \bar{x}$ and $\hat{\sigma} = s$, respectively.

The justification for the use of the divisor $n - 1$ in s will be discussed in Chap. 5.

In Sect. 1.6, we introduced the standard error of an estimated probability, which quantifies the precision of a simulation result $\hat{P}(A)$ as an estimate of a "true" probability $P(A)$. By analogy, it is possible to quantify the amount by which a sample mean, $\bar{x}$, will generally differ from the corresponding expected value μ. For n simulated values of a random variable, with sample standard deviation s, the (**estimated**) **standard error of the mean** is

$$\frac{s}{\sqrt{n}} \tag{2.22}$$

Expression (2.22) will be derived in Chap. 4. As with an estimated probability, the formula indicates that the precision of $\bar{x}$ increases (i.e., its standard error *decreases*) as n increases, but not very quickly. To increase the precision of $\bar{x}$ as an estimate of μ by a factor of 10 (one decimal place) requires increasing the number of simulation runs, n, by a factor of 100. Unfortunately, there is no general formula for the standard error of s as an estimate of σ.

Example 2.55 (Ex. 2.53 continued) The 10,000 simulated values of the random variable G, which we denote by $g_1, \ldots, g_{10000}$, are displayed in the histogram in Fig. 2.13. From these simulated values, we can estimate both the expected value and standard deviation of G:

$$\hat{\mu}_G = \overline{g} = \frac{1}{10,000} \sum_{i=1}^{10,000} g_i = 1759.62$$

$$\hat{\sigma}_G = s = \sqrt{\frac{1}{10,000-1} \sum_{i=1}^{10,000} (g_i - \overline{g})^2} = \sqrt{\frac{1}{9999} \sum_{i=1}^{10,000} (g_i - 1759.62)^2} = 43.50$$

We estimate that the average weekly gross profit from flash drive sales is $1759.62, with a standard deviation of $43.50. Neither of these computations was performed by hand, of course: if the n simulated values of a variable are stored in a vector x, then mean(x) and sd(x) in R will provide the sample mean and standard deviation, respectively. In Matlab, the calls are mean(x) and std(x).

Applying Eq. (2.22), the (estimated) standard error of $\overline{g}$ is $s/\sqrt{n} = 43.50/\sqrt{10,000} = 0.435$. If 10,000 runs are used to simulate G, it's estimated that the resulting sample mean will differ from $E(G)$ by roughly 0.435. (In contrast, the sample standard deviation, s, estimates that gross profit for a single week—i.e., a single observation g—typically differs from $E(G)$ by about $43.50.)

In Chap. 4, we will see how the expected value and variance of random variables like G, that are sums of a fixed number of other rvs, can be obtained analytically. ∎

Example 2.56 The "help desk" at a university's computer center receives both hardware and software queries. Let X and Y be the number of hardware and software queries, respectively, in a given day. Each can be modeled by a Poisson distribution with mean 20. Because computer center employees need to be allocated efficiently, of interest is the *difference* between the sizes of the two queues: $D = |X - Y|$. Let's use simulation to estimate (1) the probability the queue sizes differ by more than 5; (2) the expected difference; (3) the standard deviation of the difference.

Figure 2.15 shows Matlab and R code to simulate this process. In both languages, the code exploits the built-in Poisson simulator, as well as the fact that 10,000 simulated values may be called simultaneously.

a
```
X=random('pois',20,[10000,1]);
Y=random('pois',20,[10000,1]);
D=abs(X-Y);
sum((D>5))
mean(D)
std(D)
```

b
```
X<-rpois(10000,20)
Y<-rpois(10000,20)
D<-abs(X-Y)
sum((D>5))
mean(D)
sd(D)
```

Fig. 2.15 Simulation code for Example 2.56: (**a**) Matlab; (**b**) R

The line sum((D>5)) performs two operations: first, (D>5) determines if each simulated d value exceeds 5, returning a logical vector of bits; second, sum() tallies the "success" bits (1s or TRUEs) and gives a count of the number of times the event $\{D > 5\}$ occurred in the 10,000 simulations. The results from one run in Matlab were

$$\hat{P}(D > 5) = \frac{3843}{10,000} = .3843 \quad \hat{\mu}_D = \bar{d} = 5.0380 \quad \hat{\sigma}_D = s = 3.8436$$

A histogram of the simulated values of D appears in Fig. 2.16.

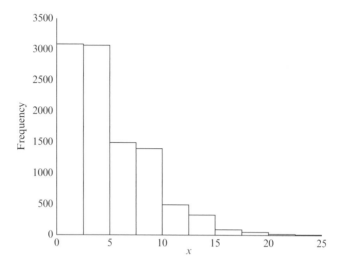

Fig. 2.16 Simulation histogram of D in Example 2.56   ■

2.8.3   Exercises: Section 2.8 (129–141)

129. Consider the pmf given in Exercise 30 for the random variable $Y =$ the number of moving violations for which the a randomly selected insured individual was cited during the last 3 years. Write a program to simulate this random variable, then use your simulation to estimate $E(Y)$ and $SD(Y)$. How do these compare to the exact values of $E(Y)$ and $SD(Y)$?

130. Consider the pmf given in Exercise 32 for the random variable $X =$ capacity of a purchased freezer. Write a program to simulate this random variable, then use your simulation to estimate $E(X)$ and $SD(X)$. How do these compare to the exact values of $E(X)$ and $SD(X)$?

131. Suppose person after person is tested for the presence of a certain characteristic. The probability that any individual tests positive is .75. Let $X =$ the number of people who must be tested to obtain five *consecutive* positive test results. Use simulation to estimate $P(X \le 25)$.

132. *The matching problem.* Suppose that N items labeled $1, 2, \ldots, N$ are shuffled so that they are in random order. Of interest is how many of these will be in their "correct" positions (e.g., item #5 situated at the 5th position in the sequence, etc.) after shuffling.

    (a) Write a program that simulates a permutation of the numbers 1 to N and then records the value of the variable $X =$ number of items in the correct position.

(b) Set $N = 5$ in your program, and use at least 10,000 simulations to estimate $E(X)$, the expected number of items in the correct position.

(c) Set $N = 52$ in your program (as if you were shuffling a deck of cards), and use at least 10,000 simulations to estimate $E(X)$. What do you discover? Is this surprising?

133. Exercise 109 of Chap. 1 referred to a multiple-choice exam in which 10 of the questions have two options, 13 have three options, 13 have four options, and the other 4 have five options. Let $X =$ the number of questions a student gets right, assuming s/he is completely guessing.

(a) Write a program to simulate X, and use your program to estimate the mean and standard deviation of X.

(b) Estimate the probability a student will score at least one standard deviation above the mean.

134. Example 2.53 of this section considered the gross profit G resulting from selling flash drives to 80 customers per week. Of course, it isn't realistic for the number of customers to remain fixed from week to week. So, instead, imagine the number of customers buying flash drives in a week follows a Poisson distribution with mean 80, and that the amount paid by each customer follows the distribution for Y provided in that example. Write a program to simulate the random variable G, and use your simulation to estimate

(a) The probability that weekly gross sales are at least \$1,800.

(b) The mean of G.

(c) The standard deviation of G.

135. Exercise 21 (Sect. 2.2) investigated Benford's law, a discrete distribution with pmf given by $p(x) = \log_{10}((x + 1)/x)$ for $x = 1, 2, \ldots, 9$. Use the inverse cdf method to write a program that simulates the Benford's law distribution. Then use your program to estimate the expected value and variance of this distribution.

136. Recall that a geometric rv has pmf $p(x) = p(1 - p)^{x-1}$ for $x = 1, 2, 3, \ldots$. In Example 2.12, it was shown that the cdf of this distribution is given by $F(x) = 1 - (1 - p)^x$ for positive integers x.

(a) Write a program that implements the inverse cdf method to simulate a geometric distribution. Your program should have as inputs the numerical value of p and the desired sample size.

(b) Use your program to simulate 10,000 values from a geometric rv X with $p = .85$. From these values, estimate each of the following: $P(X \le 2), E(X)$, $SD(X)$. How do these compare to the corresponding exact values?

137. Tickets for a particular flight are \$250 apiece. The plane seats 120 passengers, but the airline will knowingly overbook (i.e., sell more than 120 tickets), because not every paid passenger shows up. Let t denote the number of tickets the airline sells for this flight, and assume the number of passengers that actually show up for the flight, X, follows a Bin(t, .85) distribution.

Let $B =$ the number of paid passengers who show up at the airport but are denied a seat on the plane, so $B = X - 120$ if $X > 120$ and $B = 0$ otherwise. If the airline must compensate these passengers with \$500 apiece, then the profit the airline makes on this flight is $250t - 500B$. (Notice t is fixed, but B is random.)

(a) Write a program to simulate this scenario. Specifically, your program should take in t as an input and return many values of the profit variable $250t - 500B$.

(b) The airline wishes to determine the optimal value of t, i.e., the number of tickets to sell that will maximize their expected profit. Run your program for $t = 140, 141, \ldots, 150$, and record the average profit from many runs under each of these settings. What value of t appears to return the largest value? [*Note:* If a clear winner does not emerge, you might need to increase the number of runs for each t value!]

138. Imagine the following simple game: flip a fair coin repeatedly, winning $1 for every head and losing $1 for every tail. Your net winnings will potentially oscillate between positive and negative numbers as play continues. How many times do you think net winnings will *change signs* in, say, 1000 coin flips? 5000 flips?

(a) Let $X =$ the number of sign changes in 1000 coin flips. Write a program to simulate X, and use your program to estimate the probability of at least 10 sign changes.

(b) Use your program to estimate $E(X)$ and SD(X). Does your estimate for $E(X)$ match your intuition for the number of sign changes?

(c) Repeat parts (a)-(b) with 5000 flips.

139. Exercise 39 (Sect. 2.3) describes the game Plinko from *The Price is Right*. Each contestant drops between one and 5 chips down the Plinko board, depending on how well s/he prices several small items. Suppose the random variable $C =$ number of chips earned by a contestant has the following distribution:

c	1	2	3	4	5
$p(c)$	.03	.15	.35	.34	.13

The winnings from each chip follow the distribution presented in Exercise 39. Write a program to simulate Plinko; you will need to consider both the number of chips a contestant earns and how much money is won on each of those chips. Use your simulation estimate the answers to the following questions:

(a) What is the probability a contestant wins more than $11,000?

(b) What is a contestant's expected winnings?

(c) What is the corresponding standard deviation?

(d) In fact, a player gets one Plinko chip for free and can earn the other four by guessing the prices of small items (waffle irons, alarm clocks, etc.). Assume the player has a 50–50 chance of getting each price correct, so we may write $C = 1 + R$, where $R \sim \text{Bin}(4, .5)$. Use this revised model for C to estimate the answers to (a)-(c).

140. Recall the *Coupon Collector's Problem* described in the book's Introduction and again in Exercise 114 of Chap. 1. Let X = the number of cereal boxes purchased in order to obtain all 10 coupons.
   (a) Use a simulation program to estimate $E(X)$ and SD(X). Also compute the estimated standard error of your sample mean.
   (b) How does your estimate of $E(X)$ compare to the theoretical answer given in the Introduction?
   (c) Repeat (a) with 20 coupons required instead of 10. Does it appear to take roughly twice as long to collect 20 coupons as 10? More than twice as long? Less?

141. A small high school holds its graduation ceremony in the gym. Because of seating constraints, students are limited to a maximum of four tickets to graduation for family and friends. Suppose 30% of students want four tickets, 25% want three, 25% want two, 15% want one, and 5% want none.
   (a) Write a simulation for 150 graduates requesting tickets, where students' requests follow the distribution described above. In particular, keep track of the variable T = the total number of tickets requested by these 150 students.
   (b) The gym can seat a maximum of 410 guests. Based on your simulation, estimate the probability that all students' requests can be accommodated.

2.9   Supplementary Exercises (142–170)

142. Consider a deck consisting of seven cards, marked 1, 2, ..., 7. Three of these cards are selected at random. Define an rv W by W = the sum of the resulting numbers, and compute the pmf of W. Then compute $E(W)$ and Var(W). [*Hint*: Consider outcomes as unordered, so that (1, 3, 7) and (3, 1, 7) are not different outcomes. Then there are 35 outcomes, and they can be listed.] (This type of rv actually arises in connection with *Wilcoxon's rank-sum test*, in which there is an x sample and a y sample and W is the sum of the ranks of the x's in the combined sample.)

143. After shuffling a deck of 52 cards, a dealer deals out 5. Let X = the number of suits represented in the five-card hand.
   (a) Show that the pmf of X is

x	1	2	3	4
$p(x)$	.002	.146	.588	.264

   [*Hint*: $p(1) = 4P$(all are spades), $p(2) = 6P$(only spades and hearts with at least one of each), and $p(4) = 4P$(2 spades $\cap$ one of each other suit).]
   (b) Compute $E(X)$ and SD(X).

144. The negative binomial rv X was defined as the number of trials necessary to obtain the rth S. Let Y = the number of F's preceding the rth S. In the same manner in which the pmf of X was derived, derive the pmf of Y.

145. Of all customers purchasing automatic garage-door openers, 75% purchase a chain-driven model. Let X = the number among the next 15 purchasers who select the chain-driven model.
    (a) What is the pmf of X?
    (b) Compute $P(X > 10)$.
    (c) Compute $P(6 \leq X \leq 10)$.
    (d) Compute $E(X)$ and $SD(X)$.
    (e) If the store currently has in stock 10 chain-driven models and 8 shaft-driven models, what is the probability that the requests of these 15 customers can all be met from existing stock?

146. A friend recently planned a camping trip. He has two flashlights, one that required a single 6-V battery and another that used two size-D batteries. He had previously packed two 6-V and four size-D batteries in his camper. Suppose the probability that any particular battery works is p and that batteries work or fail independently of one another. Our friend wants to take just one flashlight. For what values of p should he take the 6-V flashlight?

147. Binary data are transmitted over a noisy communication channel. The probability that a received binary digit is in error due to channel noise is 0.05. Assume that such errors occur independently within the bit stream.
    (a) What is the probability that the 3rd error occurs on the 50th transmitted bit?
    (b) On average, how many bits will be transmitted correctly *before* the first error?
    (c) Consider a 32-bit "word." What is the probability of exactly 2 errors in this word?
    (d) Consider the next 10,000 bits. What approximating model could we use for X = the number of errors in these 10,000 bits? Give both the name of the model and the value(s) of the parameter(s).

148. A manufacturer of flashlight batteries wishes to control the quality of its product by rejecting any lot in which the proportion of batteries having unacceptable voltage appears to be too high. To this end, out of each large lot (10,000 batteries), 25 will be selected and tested. If at least 5 of these generate an unacceptable voltage, the entire lot will be rejected. What is the probability that a lot will be rejected if
    (a) 5% of the batteries in the lot have unacceptable voltages?
    (b) 10% of the batteries in the lot have unacceptable voltages?
    (c) 20% of the batteries in the lot have unacceptable voltages?
    (d) What would happen to the probabilities in parts (a)–(c) if the critical rejection number were increased from 5 to 6?

149. Of the people passing through an airport metal detector, .5% activate it; let X = the number among a randomly selected group of 500 who activate the detector.

   (a) What is the (approximate) pmf of X?
   (b) Compute $P(X = 5)$.
   (c) Compute $P(X \geq 5)$.

150. An educational consulting firm is trying to decide whether high school students who have never before used a handheld calculator can solve a certain type of problem more easily with a calculator that uses reverse Polish logic or one that does not use this logic. A sample of 25 students is selected and allowed to practice on both calculators. Then each student is asked to work one problem on the reverse Polish calculator and a similar problem on the other. Let $p = P(S)$, where S indicates that a student worked the problem more quickly using reverse Polish logic than without, and let $X =$ number of S's.
   (a) If $p = .5$, what is $P(7 \leq X \leq 18)$?
   (b) If $p = .8$, what is $P(7 \leq X \leq 18)$?
   (c) If the claim that $p = .5$ is to be rejected when either $X \leq 7$ or $X \geq 18$, what is the probability of rejecting the claim when it is actually correct?
   (d) If the decision to reject the claim $p = .5$ is made as in part (c), what is the probability that the claim is not rejected when $p = .6$? When $p = .8$?
   (e) What decision rule would you choose for rejecting the claim $p = .5$ if you wanted the probability in part (c) to be at most .01?

151. Consider a disease whose presence can be identified by carrying out a blood test. Let p denote the probability that a randomly selected individual has the disease. Suppose n individuals are independently selected for testing. One way to proceed is to carry out a separate test on each of the n blood samples. A potentially more economical approach, group testing, was introduced during World War II to identify syphilitic men among army inductees. First, take a part of each blood sample, combine these specimens, and carry out a single test. If no one has the disease, the result will be negative, and only the one test is required. If at least one individual is diseased, the test on the combined sample will yield a positive result, in which case the n individual tests are then carried out. If $p = .1$ and $n = 3$, what is the expected number of tests using this procedure? What is the expected number when $n = 5$? [The article "Random Multiple-Access Communication and Group Testing" (*IEEE Trans. Commun.*, 1984: 769–774) applied these ideas to a communication system in which the dichotomy was active/idle user rather than diseased/nondiseased.]

152. Let p_1 denote the probability that any particular code symbol is erroneously transmitted through a communication system. Assume that on different symbols, errors occur independently of one another. Suppose also that with probability p_2 an erroneous symbol is corrected upon receipt. Let X denote the number of correct symbols in a message block consisting of n symbols (after the correction process has ended). What is the probability distribution of X?

153. The purchaser of a power-generating unit requires c consecutive successful start-ups before the unit will be accepted. Assume that the outcomes of individual start-ups are independent of one another. Let p denote the probability that any particular start-up is successful. The random variable of interest is $X =$ the

number of start-ups that must be made prior to acceptance. Give the pmf of X for the case $c = 2$. If $p = .9$, what is $P(X \le 8)$? [*Hint*: For $x \ge 5$, express $p(x)$ "recursively" in terms of the pmf evaluated at the smaller values $x - 3$, $x - 4, \ldots, 2$.] (This problem was suggested by the article "Evaluation of a Start-Up Demonstration Test," *J. Qual. Tech.*, 1983: 103–106.)

154. A plan for an executive travelers' club has been developed by an airline on the premise that 10% of its current customers would qualify for membership.

   (a) Assuming the validity of this premise, among 25 randomly selected current customers, what is the probability that between 2 and 6 (inclusive) qualify for membership?

   (b) Again assuming the validity of the premise, what are the expected number of customers who qualify and the standard deviation of the number who qualify in a random sample of 100 current customers?

   (c) Let X denote the number in a random sample of 25 current customers who qualify for membership. Consider rejecting the company's premise in favor of the claim that $p > .10$ if $x \ge 7$. What is the probability that the company's premise is rejected when it is actually valid?

   (d) Refer to the decision rule introduced in part (c). What is the probability that the company's premise is not rejected even though $p = .20$ (i.e., 20% qualify)?

155. Forty percent of seeds from maize (modern-day corn) ears carry single spikelets, and the other 60% carry paired spikelets. A seed with single spikelets will produce an ear with single spikelets 29% of the time, whereas a seed with paired spikelets will produce an ear with single spikelets 26% of the time. Consider randomly selecting ten seeds.

   (a) What is the probability that exactly five of these seeds carry a single spikelet and produce an ear with a single spikelet?

   (b) What is the probability that exactly five of the ears produced by these seeds have single spikelets? What is the probability that at most five ears have single spikelets?

156. A trial has just resulted in a hung jury because eight members of the jury were in favor of a guilty verdict and the other four were for acquittal. If the jurors leave the jury room in random order and each of the first four leaving the room is accosted by a reporter in quest of an interview, what is the pmf of $X = $ the number of jurors favoring acquittal among those interviewed? How many of those favoring acquittal do you expect to be interviewed?

157. A reservation service employs five information operators who receive requests for information independently of one another, each according to a Poisson process with rate $\lambda = 2$ per minute.

   (a) What is the probability that during a given 1-min period, the first operator receives no requests?

   (b) What is the probability that during a given 1-min period, exactly four of the five operators receive no requests?

   (c) Write an expression for the probability that during a given 1-min period, all of the operators receive exactly the same number of requests.

158. Grasshoppers are distributed at random in a large field according to a Poisson process with parameter $\lambda = 2$ per square yard. How large should the radius r of a circular sampling region be taken so that the probability of finding at least one grasshopper in the region equals .99?

159. A newsstand has ordered five copies of a certain issue of a photography magazine. Let $X =$ the number of individuals who come in to purchase this magazine. If X has a Poisson distribution with parameter $\mu = 4$, what is the expected number of copies that are sold?

160. Individuals A and B begin to play a sequence of chess games. Let $S = \{$A wins a game$\}$, and suppose that outcomes of successive games are independent with $P(S) = p$ and $P(F) = 1 - p$ (they never draw). They will play until one of them wins ten games. Let $X =$ the number of games played (with possible values 10, 11, ..., 19).
    (a) For $x = 10, 11, \ldots, 19$, obtain an expression for $p(x) = P(X = x)$.
    (b) If a draw is possible, with $p = P(S)$, $q = P(F)$, $1 - p - q = P(\text{draw})$, what are the possible values of X? What is $P(20 \leq X)$? [*Hint*: $P(20 \leq X) = 1 - P(X < 20)$.]

161. A test for the presence of a disease has probability .20 of giving a false-positive reading (indicating that an individual has the disease when this is not the case) and probability .10 of giving a false-negative result. Suppose that ten individuals are tested, five of whom have the disease and five of whom do not. Let $X =$ the number of positive readings that result.
    (a) Does X have a binomial distribution? Explain your reasoning.
    (b) What is the probability that exactly three of the ten test results are positive?

162. The generalized negative binomial pmf is given by

$$nb(x; r, p) = k(r, x) \times p^r (1 - p)^x \quad x = 0, 1, 2, \ldots$$

    where

$$k(r, x) = \begin{cases} \dfrac{(x + r - 1)(x + r - 2) \ldots (x + r - x)}{x!} & x = 1, 2, \ldots \\ 1 & x = 0 \end{cases}$$

    Let X, the number of plants of a certain species found in a particular region, have this distribution with $p = .3$ and $r = 2.5$. What is $P(X = 4)$? What is the probability that at least one plant is found?

163. There are two certified public accountants (CPAs) in a particular office who prepare tax returns for clients. Suppose that for one type of complex tax form, the number of errors made by the first preparer has a Poisson distribution with mean μ_1, the number of errors made by the second preparer has a Poisson distribution with mean μ_2, and that each CPA prepares the same number of forms of this type. Then if one such form is randomly selected, the function

$$p(x; \mu_1, \mu_2) = .5e^{-\mu_1}\frac{\mu_1^x}{x!} + .5e^{-\mu_2}\frac{\mu_2^x}{x!} \quad x = 0, 1, 2, \ldots$$

gives the pmf of X = the number of errors in the selected form.
(a) Verify that $p(x; \mu_1, \mu_2)$ is a legitimate pmf (≥ 0 and sums to 1).
(b) What is the expected number of errors on the selected form?
(c) What is the standard deviation of the number of errors on the selected form?
(d) How does the pmf change if the first CPA prepares 60% of all such forms and the second prepares 40%?

164. The *mode* of a discrete random variable X with pmf $p(x)$ is that value x^* for which $p(x)$ is largest (the most probable x value).
(a) Let $X \sim \text{Bin}(n, p)$. By considering the ratio $b(x + 1; n, p)/b(x; n, p)$, show that $b(x; n, p)$ increases with x as long as $x < np - (1 - p)$. Conclude that the mode x^* is the integer satisfying $(n + 1)p - 1 \leq x^* \leq (n + 1)p$.
(b) Show that if X has a Poisson distribution with parameter μ, the mode is the largest integer less than μ. If μ is an integer, show that both $\mu - 1$ and μ are modes.

165. For a particular insurance policy the number of claims by a policy holder in 5 years is Poisson distributed. If the filing of one claim is four times as likely as the filing of two claims, find the expected number of claims.

166. If X is a hypergeometric rv, show directly from the definition that $E(X) = nM/N$ (consider only the case $n < M$). [*Hint:* Factor nM/N out of the sum for $E(X)$, and show that the terms inside the sum are a match to the pmf $h(y; n - 1, M - 1, N - 1)$, where $y = x - 1$.]

167. Suppose a store sells two different coffee makers of a particular brand, a basic model selling for \$30 and a fancy one selling for \$50. Let X be the number of people among the next 25 purchasing this brand who choose the fancy one. Then $h(X)$ = revenue = $50X + 30(25 - X) = 20X + 750$, a linear function. If the choices are independent and have the same probability, then how is X distributed? Find the mean and standard deviation of $h(X)$. Explain why the choices might not be independent with the same probability.

168. Let X be a discrete rv with possible values $0, 1, 2, \ldots$ or some subset of these. The function $\psi(s) = E(s^X) = \sum_{x=0}^{\infty} s^x \cdot p(x)$ is called the **probability generating function (pgf)** of X.
(a) Suppose X is the number of children born to a family, and $p(0) = .2$, $p(1) = .5$, and $p(2) = .3$. Determine the pgf of X.
(b) Determine the pgf when X has a Poisson distribution with parameter μ.
(c) Show that $\psi(1) = 1$.
(d) Show that $\psi'(0) = p(1)$. (You'll need to assume that the derivative can be brought inside the summation, which is justified.) What results from taking the second derivative with respect to s and evaluating at $s = 0$? The third derivative? Explain how successive differentiation of $\psi(s)$ and

evaluation at $s = 0$ "generates the probabilities in the distribution." Use this to recapture the probabilities of (a) from the pgf. [*Note*: This shows that the pgf contains all the information about the distribution—knowing $\psi(s)$ is equivalent to knowing $p(x)$.]

169. Consider a collection $A_1, \ldots, A_k$ of mutually exclusive and exhaustive events (a partition) and a random variable X whose distribution depends on which of the A_is occurs. (e.g., a commuter might select one of three possible routes from home to work, with X representing commute time.) Let $E(X \mid A_i)$ denote the expected value of X given that event A_i occurs. Then, analogous to the Law of Total Probability, it can be shown that the overall mean of X is given by the weighted average $E(X) = \sum E(X|A_i)P(A_i)$

   (a) The expected duration of a voice call to a particular office telephone number is 3 min, whereas the expected duration of a data call to that same number is 1 min. If 75% of all calls are voice calls, what is the expected duration of the next call?

   (b) A bakery sells three different types of chocolate chip cookies. The number of chocolate chips on a type i cookie has a Poisson distribution with mean $\mu_i = i + 1$ ($i = 1, 2, 3$). If 20% of all customers select a cookie of the first type, 50% choose the second type, and 30% opt for the third type, what is the expected number of chocolate chips in the next customer's cookie?

170. Consider a sequence of identical and independent trials, each of which will be a success S or failure F. Let $p = P(S)$ and $q = P(F)$.

   (a) Let $X =$ the number of trials necessary to obtain the first S, a geometric rv. Here is an alternative approach to determining $E(X)$. Apply the weighted average formula from the previous exercise with $k = 2$, $A_1 = \{S \text{ on 1st trial}\}$, and $A_2 = A'$. Show that $E(X) = 1/p$. [*Hint*: Denote $E(X)$ by μ. Given that the first trial is a failure, one trial has been performed and, starting from the 2nd trial, we are still looking for the first S. This implies that $E(X|A') = 1 + \mu$.]

   (b) Now let $Y =$ the number of trials necessary to obtain two *consecutive* S's. It is not possible to determine $E(Y)$ directly from the definition of expected value, because there is no formula for the pmf of Y; the complication is the word *consecutive*. Use the weighted average formula to determine $E(Y)$. [*Hint*: Consider the partition with $k = 3$ and $A_1 = \{F\}$, $A_2 = \{SS\}$, $A_3 = \{SF\}$.]

Continuous Random Variables and Probability Distributions

3

As emphasized at the beginning of Chap. 2, the two important types of random variables are discrete and continuous. In this chapter, we study the second general type of random variable that arises in many applied problems. Sections 3.1 and 3.2 present the basic definitions and properties of continuous random variables, their probability distributions, and their various expected values. The normal distribution, arguably the most important and useful model in all of probability and statistics, is introduced in Sect. 3.3. Sections 3.4 and 3.5 discuss some other continuous distributions that are often used in applied work. In Sect. 3.6, we introduce a method for assessing whether given sample data is consistent with a specified distribution. Section 3.7 presents methods for obtaining the distribution of a rv Y from the distribution of X when the two are related by some equation $Y = g(X)$. The last section of this chapter is dedicated to the simulation of continuous rvs.

3.1 Probability Density Functions and Cumulative Distribution Functions

A discrete random variable (rv) is one whose possible values either constitute a finite set or else can be listed in an infinite sequence (a list in which there is a first element, a second element, etc.). A random variable whose set of possible values is an entire interval of numbers is not discrete.

Recall from the beginning of Chap. 2 that a random variable X is **continuous** if (1) its possible values comprise either a single interval on the number line (for some $A < B$, any number x between A and B is a possible value) or a union of disjoint intervals, and (2) $P(X = c) = 0$ for any number c that is a possible value of X.

Example 3.1 If in the study of the ecology of a lake, we make depth measurements at randomly chosen locations, then $X =$ the depth at such a location is a continuous rv. Here A is the minimum depth in the region being sampled, and B is the maximum depth. ∎

M.A. Carlton and J.L. Devore, *Probability with Applications in Engineering, Science, and Technology*, Springer Texts in Statistics, DOI 10.1007/978-1-4939-0395-5_3, © Springer Science+Business Media New York 2014

Example 3.2 If a chemical compound is randomly selected and its pH X is determined, then X is a continuous rv because any pH value between 0 and 14 is possible. If more is known about the compound selected for analysis, then the set of possible values might be a subinterval of [0, 14], such as $5.5 \leq x \leq 6.5$, but X would still be continuous.   ■

Example 3.3 Let X represent the amount of time a randomly selected customer spends waiting for a haircut. Your first thought might be that X is a continuous random variable, since a measurement is required to determine its value. However, there are customers lucky enough to have no wait whatsoever before climbing into the barber or stylist's chair. So it must be the case that $P(X = 0) > 0$. Conditional on no chairs being empty, however, the waiting time will be continuous since X could then assume any value between some minimum possible time A and a maximum possible time B. This random variable is neither purely discrete nor purely continuous but instead is a mixture of the two types.   ■

One might argue that although in principle variables such as height, weight, and temperature are continuous, in practice the limitations of our measuring instruments restrict us to a discrete (though sometimes very finely subdivided) world. However, continuous models often approximate real-world situations very well, and continuous mathematics (the calculus) is frequently easier to work with than the mathematics of discrete variables and distributions.

3.1.1   Probability Distributions for Continuous Variables

Suppose the variable X of interest is the depth of a lake at a randomly chosen point on the surface. Let $M =$ the maximum depth (in meters), so that any number in the interval [0, M] is a possible value of X. If we "discretize" X by measuring depth to the nearest meter, then possible values are nonnegative integers less than or equal to M. The resulting discrete distribution of depth can be pictured using a probability histogram. If we draw the histogram so that the area of the rectangle above any possible integer k is the proportion of the lake whose depth is (to the nearest meter) k, then the total area of all rectangles is 1. A possible histogram appears in Fig. 3.1a.

If depth is measured much more precisely and the same measurement axis as in Fig. 3.1a is used, each rectangle in the resulting probability histogram is much narrower, although the total area of all rectangles is still 1. A possible histogram is pictured in Fig. 3.1b; it has a much smoother appearance than the histogram in Fig. 3.1a. If we continue in this way to measure depth more and more finely, the resulting sequence of histograms approaches a smooth curve, as pictured in Fig. 3.1c. Because for each histogram the total area of all rectangles equals 1, the total area under the smooth curve is also 1. The probability that the depth at a randomly chosen point is between a and b is just the area under the smooth curve between a and b. It is exactly a smooth curve of the type pictured in Fig. 3.1c that specifies a continuous probability distribution.

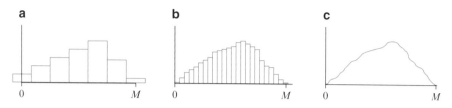

Fig. 3.1 (a) Probability histogram of depth measured to the nearest meter; (b) probability histogram of depth measured to the nearest centimeter; (c) a limit of a sequence of discrete histograms

DEFINITION

Let X be a continuous rv. Then a **probability distribution** or **probability density function** (pdf) of X is a function $f(x)$ such that for any two numbers a and b with $a \leq b$,

$$P(a \leq X \leq b) = \int_a^b f(x)dx$$

That is, the probability that X takes on a value in the interval $[a, b]$ is the area above this interval and under the graph of the density function, as illustrated in Fig. 3.2. The graph of $f(x)$ is often referred to as the *density curve*.

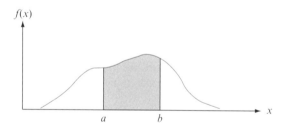

Fig. 3.2 $P(a \leq X \leq b) =$ the area under the density curve between a and b

For $f(x)$ to be a legitimate pdf, it must satisfy the following two conditions:

1. $f(x) \geq 0$ for all x
2. $\int_{-\infty}^{\infty} f(x)dx = $ [area under the entire graph of $f(x)$] $= 1$

Example 3.4 The direction of an imperfection with respect to a reference line on a circular object such as a tire, brake rotor, or flywheel is often subject to uncertainty. Consider the reference line connecting the valve stem on a tire to the center point, and let X be the angle measured clockwise to the location of an imperfection. One possible pdf for X is

$$f(x) = \begin{cases} \dfrac{1}{360} & 0 \le x < 360 \\[2mm] 0 & \text{otherwise} \end{cases}$$

The pdf is graphed in Fig. 3.3. Clearly $f(x) \ge 0$. The area under the density curve is just the area of a rectangle: $(\text{height})(\text{base}) = \left(\frac{1}{360}\right)(360) = 1$. The probability that the angle is between $90°$ and $180°$ is

$$P(90 \le X \le 180) = \int_{90}^{180} \frac{1}{360} dx = \frac{x}{360}\Big|_{x=90}^{x=180} = \frac{1}{4} = .25$$

The probability that the angle of occurrence is within $90°$ of the reference line is

$$P(0 \le X \le 90) + P(270 \le X < 360) = .25 + .25 = .50$$

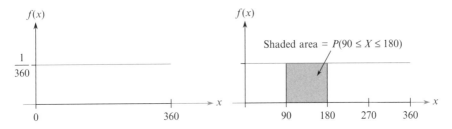

Fig. 3.3 The pdf and probability for Example 3.4   ■

Because whenever $0 \le a \le b \le 360$ in Example 3.4, $P(a \le X \le b)$ depends only on the width $b - a$ of the interval, X is said to have a *uniform* distribution.

DEFINITION

A continuous rv X is said to have a **uniform distribution** on the interval $[A, B]$ if the pdf of X is

$$f(x; A, B) = \begin{cases} \dfrac{1}{B - A} & A \le x \le B \\[2mm] 0 & \text{otherwise} \end{cases}$$

The statement that X has a uniform distribution on $[A, B]$ will be denoted $X \sim \text{Unif}[A, B]$.

The graph of any uniform pdf looks like the graph in Fig. 3.3 except that the interval of positive density is $[A, B]$ rather than $[0, 360)$.

In the discrete case, a probability mass function (pmf) tells us how little "blobs" of probability mass of various magnitudes are distributed along the measurement axis. In the continuous case, probability density is "smeared" in a continuous fashion along the interval of possible values. When density is smeared evenly over the interval, a uniform pdf, as in Fig. 3.3, results.

When X is a discrete random variable, each possible value is assigned positive probability. This is not true of a continuous random variable, because the area under a density curve that lies above any single value is zero:

$$P(X = c) = P(c \leq X \leq c) = \int_c^c f(x)\, dx = 0$$

The fact that $P(X = c) = 0$ when X is continuous has an important practical consequence: The probability that X lies in some interval between a and b does not depend on whether the lower limit a or the upper limit b is included in the probability calculation:

$$P(a \leq X \leq b) = P(a < X < b) = P(a < X \leq b) = P(a \leq X < b) \qquad (3.1)$$

In contrast, if X were discrete and both a and b are possible values of X (e.g., $X \sim \text{Bin}(20, .3)$ and $a = 5$, $b = 10$), then all four of the probabilities in Eq. (3.1) would be different. This also means that whether we include the endpoints of the range of values for a continuous rv X is somewhat arbitrary; for example, the pdf in Example 3.4 could be defined to be positive on $(0, 360)$ or $[0, 360]$ rather than $[0, 360)$, and the same applies for a uniform distribution on $[A, B]$ in general.

The zero probability condition has a physical analog. Consider a solid circular rod (with cross-sectional area of 1 in^2 for simplicity). Place the rod alongside a measurement axis and suppose that the density of the rod at any point x is given by the value $f(x)$ of a density function. Then if the rod is sliced at points a and b and this segment is removed, the amount of mass removed is $\int_a^b f(x)dx$; however, if the rod is sliced just at the point c, no mass is removed. Mass is assigned to interval segments of the rod but not to individual points.

So, if $P(X = c) = 0$ when X is a continuous rv, then what does $f(c)$ represent? After all, if X were discrete, its pmf evaluated at $x = c$, $p(c)$, would indicate the probability that X equals c. To help understand what $f(c)$ means, consider a small window near $x = c$—say, $[c, c + \Delta x]$. Using a rectangle to approximate the area under $f(x)$ between c and $c + \Delta x$ (the usual "Riemann approximation" idea from calculus), one obtains $\int_c^{c+\Delta x} f(x)dx \approx \Delta x \cdot f(c)$, from which

$$f(c) \approx \frac{\displaystyle\int_c^{c+\Delta x} f(x)dx}{\Delta x} = \frac{P(c \leq X \leq c + \Delta x)}{\Delta x}$$

This indicates that $f(c)$ is not a probability, but rather roughly the probability of an interval *divided by the length of the chosen interval*. If we associate mass with probability and remember that interval length is the one-dimensional analog of volume, then f represents their quotient, mass per volume, more commonly known as *density* (hence, the name pdf). The height of the function $f(x)$ at a particular point reflects how "dense" the values of X are near that point—taller sections of $f(x)$ contain more probability within a fixed interval length than do shorter sections.

Example 3.5 "Time headway" in traffic flow is the elapsed time between the time that one car finishes passing a fixed point and the instant that the next car begins to pass that point. Let $X =$ the time headway for two randomly chosen consecutive cars on a freeway during a period of heavy flow. The following pdf of X is essentially the one suggested in "The Statistical Properties of Freeway Traffic" (*Transp. Res.*, 11: 221–228):

$$f(x) = \begin{cases} .15e^{-.15(x-.5)} & x \geq .5 \\ 0 & \text{otherwise} \end{cases}$$

The graph of $f(x)$ is given in Fig. 3.4; there is no density associated with headway times less than .5, and headway density decreases rapidly (exponentially fast) as x increases from .5. The fact that the graph of $f(x)$ is taller near $x = .5$ and shorter near, say, $x = 10$ indicates that time headway values are more dense near the left boundary, i.e., there is a higher proportion of time headways in the interval [.5, 1.5] than in [10, 11], even though these two intervals have the same length.

Clearly, $f(x) \geq 0$; to show that $\int_{-\infty}^{\infty} f(x)dx = 1$ we use the calculus result $\int_{a}^{\infty} e^{-kx}dx = (1/k)e^{-ka}$. Then

$$\int_{-\infty}^{\infty} f(x)dx = \int_{-\infty}^{.5} 0\,dx + \int_{.5}^{\infty} .15e^{-.15(x-5)}dx$$
$$= .15e^{.075}\int_{.5}^{\infty} e^{-.15x}dx = .15e^{.075} \cdot \frac{1}{.15}e^{-.15(.5)} = 1$$

The probability that headway time is at most 5 s is

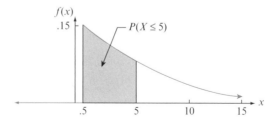

Fig. 3.4 The density curve for headway time in Example 3.5

$$P(X \leq 5) = \int_{-\infty}^{5} f(x)\,dx = \int_{.5}^{5} .15e^{-.15(x-.5)}\,dx = .15e^{.075} \int_{.5}^{5} e^{-.15x}\,dx$$

$$= .15e^{.075} \cdot \frac{-1}{.15} e^{-.15x} \Big|_{x=.5}^{x=5}$$

$$= e^{.075}\left(-e^{-.75} + e^{-.075}\right) = 1.078(-.472 + .928) = .491$$

Since X is a continuous rv, .491 also equals $P(X < 5)$, the probability that headway time is (strictly) less than 5 s. The difference between these two events is $\{X = 5\}$, i.e., that headway time is exactly 5 s, which has probability zero: $P(X = 5) = \int_{5}^{5} f(x)\,dx = 0$.

This last statement may feel uncomfortable to you: Is there really zero chance that the headway time between two cars is exactly 5 s? If time is treated as continuous, then "exactly 5 s" means $X = 5.000\ldots$, with an endless repetition of 0s. That is to say, X isn't rounded to the nearest second (or even tenth of a second); we are asking for the probability that X equals one specific number, $5.000\ldots$, out of the (uncountably) infinite collection of possible values of X. ∎

Unlike discrete distributions such as the binomial, hypergeometric, and negative binomial, the distribution of any given continuous rv cannot usually be derived using simple probabilistic arguments. Instead, one must make a judicious choice of pdf based on prior knowledge and available data. Fortunately, some general pdf families have been found to fit well in a wide variety of experimental situations; several of these are discussed later in the chapter.

Just as in the discrete case, it is often helpful to think of the population of interest as consisting of X values rather than individuals or objects. The pdf is then a model for the distribution of values in this numerical population, and from this model various population characteristics (such as the mean) can be calculated.

Several of the most important concepts introduced in the study of discrete distributions also play an important role for continuous distributions. Definitions analogous to those in Chap. 2 involve replacing summation by integration.

3.1.2 The Cumulative Distribution Function

The cumulative distribution function (cdf) $F(x)$ for a discrete rv X gives, for any specified number x, the probability $P(X \leq x)$. It is obtained by summing the pmf $p(y)$ over all possible values y satisfying $y \leq x$. The cdf of a continuous rv gives the same probabilities $P(X \leq x)$ and is obtained by integrating the pdf $f(y)$ between the limits $-\infty$ and x.

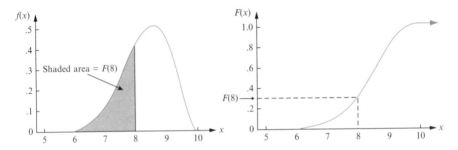

Fig. 3.5 A pdf and associated cdf

Example 3.6 Let X, the thickness of a membrane, have a uniform distribution on $[A, B]$. The density function is shown in Fig. 3.6.

For $x < A$, $F(x) = 0$, since there is no area under the graph of the density function to the left of such an x. For $x \ge B$, $F(x) = 1$, since all the area is accumulated to the left of such an x. Finally, for $A \le x < B$,

$$F(x) = \int_{-\infty}^{x} f(y)dy = \int_{A}^{x} \frac{1}{B-A}dy = \frac{1}{B-A} \cdot y \Big|_{y=A}^{y=x} = \frac{x-A}{B-A}$$

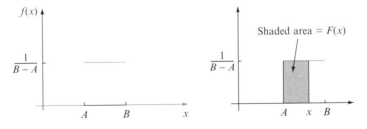

Fig. 3.6 The pdf for a uniform distribution

The entire cdf is

$$F(x) = \begin{cases} 0 & x < A \\ \dfrac{x - A}{B - A} & A \le x < B \\ 1 & x \ge B \end{cases}$$

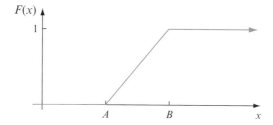

Fig. 3.7 The cdf for a uniform distribution ∎

The graph of this cdf appears in Fig. 3.7.

3.1.3 Using *F(x)* to Compute Probabilities

The importance of the cdf here, just as for discrete rvs, is that probabilities of various intervals can be computed from a formula or table for $F(x)$.

> **PROPOSITION**
>
> Let X be a continuous rv with pdf $f(x)$ and cdf $F(x)$. Then for any number a,
>
> $$P(X > a) = 1 - F(a)$$
>
> and for any two numbers a and b with $a < b$,
>
> $$P(a \le X \le b) = F(b) - F(a)$$

Figure 3.8 illustrates the second part of this proposition; the desired probability is the shaded area under the density curve between a and b, and it equals the difference between the two shaded cumulative areas. This is different from what is appropriate for a discrete integer-valued rv (e.g., binomial or Poisson): $P(a \le X \le b) = F(b) - F(a - 1)$ when a and b are integers.

Example 3.7 Suppose the pdf of the magnitude X of a dynamic load on a bridge (in newtons) is given by

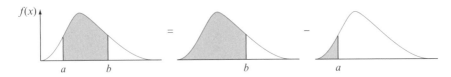

Fig. 3.8 Computing $P(a \leq X \leq b)$ from cumulative probabilities

$$f(x) = \begin{cases} \dfrac{1}{8} + \dfrac{3}{8}x & 0 \leq x \leq 2 \\ 0 & \text{otherwise} \end{cases}$$

For any number x between 0 and 2,

$$F(x) = \int_{-\infty}^{x} f(y)dy = \int_{0}^{x} \left(\frac{1}{8} + \frac{3}{8}y \right) dy = \frac{x}{8} + \frac{3x^2}{16}$$

Thus

$$F(x) = \begin{cases} 0 & x < 0 \\ \dfrac{x}{8} + \dfrac{3x^2}{16} & 0 \leq x \leq 2 \\ 1 & 2 < x \end{cases}$$

The graphs of $f(x)$ and $F(x)$ are shown in Fig. 3.9. The probability that the load is between 1 and 1.5 N is

$$P(1 \leq X \leq 1.5) = F(1.5) - F(1) = \left[\frac{1}{8}(1.5) + \frac{3}{16}(1.5)^2 \right] - \left[\frac{1}{8}(1) + \frac{3}{16}(1)^2 \right]$$

$$= \frac{19}{64} = .297$$

The probability that the load exceeds 1 N is

$$P(X > 1) = 1 - P(X \leq 1) = 1 - F(1) = 1 - \left[\frac{1}{8}(1) + \frac{3}{16}(1)^2 \right] = \frac{11}{16} = .688$$

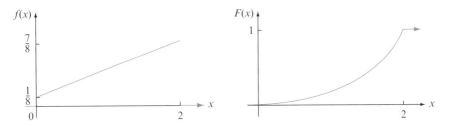

Fig. 3.9 The pdf and cdf for Example 3.7

The beauty of the cdf in the continuous case is that once it is available, any probability involving X can easily be calculated without any further integration.

3.1.4 Obtaining *f(x)* from *F(x)*

For X discrete, the pmf is obtained from the cdf by taking the difference between two $F(x)$ values. The continuous analog of a difference is a derivative. The following result is a consequence of the Fundamental Theorem of Calculus.

> **PROPOSITION**
>
> If X is a continuous rv with pdf $f(x)$ and cdf $F(x)$, then at every x at which the derivative $F'(x)$ exists, $F'(x) = f(x)$.

Example 3.8 (Example 3.6 continued) When $X \sim \text{Unif}[A, B]$, $F(x)$ is differentiable except at $x = A$ and $x = B$, where the graph of $F(x)$ has sharp corners. Since $F(x) = 0$ for $x < A$ and $F(x) = 1$ for $x > B$, $F'(x) = 0 = f(x)$ for such x. For $A < x < B$,

$$F'(x) = \frac{d}{dx}\left(\frac{x - A}{B - A}\right) = \frac{1}{B - A} = f(x)$$

∎

3.1.5 Percentiles of a Continuous Distribution

When we say that an individual's test score was at the 85th percentile of the population, we mean that 85% of all population scores were below that score and 15% were above. Similarly, the 40th percentile is the score that exceeds 40% of all scores and is exceeded by 60% of all scores.

> **DEFINITION**
>
> Let p be a number between 0 and 1. The **(100p)th percentile** of the distribution of a continuous rv X, denoted by η_p, is defined implicitly by the equation
>
> $$p = F(\eta_p) = \int_{-\infty}^{\eta_p} f(y)\,dy \tag{3.2}$$
>
> Assuming we can find the inverse of $F(x)$, this can also be written as
>
> $$\eta_p = F^{-1}(p)$$

In particular, the **median** of a continuous distribution is the 50th percentile, $\eta_{.5}$ or $F^{-1}(.5)$. That is, half the area under the density curve is to the left of the median and half is to the right of the median. We will occasionally denote the median of a distribution simply as η (i.e., without the .5 subscript).

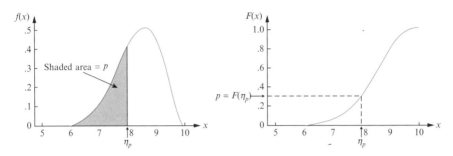

Fig. 3.10 The (100p)th percentile of a continuous distribution

According to Expression (3.2), η_p is that value on the measurement axis such that $100p\%$ of the area under the graph of $f(x)$ lies to the left of η_p and $100(1-p)\%$ lies to the right. Thus $\eta_{.75}$, the 75th percentile, is such that the area under the graph of $f(x)$ to the left of $\eta_{.75}$ is .75. Figure 3.10 illustrates the definition.

Example 3.9 The distribution of the amount of gravel (in tons) sold by a construction supply company in a given week is a continuous rv X with pdf

$$f(x) = \begin{cases} \dfrac{3}{2}\left(1 - x^2\right) & 0 \le x \le 1 \\ 0 & \text{otherwise} \end{cases}$$

The cdf of sales for any x between 0 and 1 is

$$F(x) = \int_0^x \frac{3}{2}\left(1 - y^2\right) dy = \frac{3}{2}\left(y - \frac{y^3}{3}\right)\Bigg|_{y=0}^{y=x} = \frac{3}{2}\left(x - \frac{x^3}{3}\right)$$

The graphs of both $f(x)$ and $F(x)$ appear in Fig. 3.11. The (100p)th percentile of this distribution satisfies the equation

$$p = F\left(\eta_p\right) = \frac{3}{2}\left(\eta_p - \frac{\eta_p^3}{3}\right)$$

that is,

$$\eta_p^3 - 3\eta_p + 2p = 0$$

For the median, $p = .5$ and the equation to be solved is $\eta^3 - 3\eta + 1 = 0$; the solution is $\eta = .347$. If the distribution remains the same from week to week, then in the long run 50% of all weeks will result in sales of less than .347 tons and 50% in more than .347 tons.

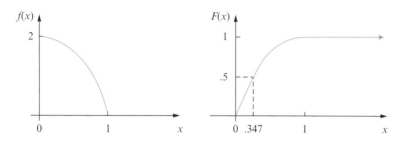

Fig. 3.11 The pdf and cdf for Example 3.9 ∎

A continuous distribution whose pdf is **symmetric**—which means that the graph of the pdf to the left of some point is a mirror image of the graph to the right of that point—has median η equal to the point of symmetry, since half the area under the curve lies to either side of this point. Figure 3.12 gives several examples. The amount of error in a measurement of a physical quantity is often assumed to have a symmetric distribution.

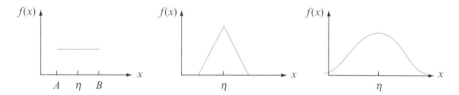

Fig. 3.12 Medians of symmetric distributions

3.1.6 Exercises: Section 3.1 (1–18)

1. The current in a certain circuit as measured by an ammeter is a continuous random variable X with the following density function:

$$f(x) = \begin{cases} .075x + 2 & 3 \leq x \leq 5 \\ 0 & \text{otherwise} \end{cases}$$

(a) Graph the pdf and verify that the total area under the density curve is indeed 1.
(b) Calculate $P(X \leq 4)$. How does this probability compare to $P(X < 4)$?
(c) Calculate $P(3.5 \leq X \leq 4.5)$ and $P(X > 4.5)$.
2. Suppose the reaction temperature X (in °C) in a chemical process has a uniform distribution with $A = -5$ and $B = 5$.
(a) Compute $P(X < 0)$.
(b) Compute $P(-2.5 < X < 2.5)$.
(c) Compute $P(-2 \leq X \leq 3)$.
(d) For k satisfying $-5 < k < k+4 < 5$, compute $P(k < X < k+4)$. Interpret this in words.

3. Suppose the error involved in making a measurement is a continuous rv X with pdf

$$f(x) = \begin{cases} .09375(4 - x^2) & -2 \le x \le 2 \\ 0 & \text{otherwise} \end{cases}$$

   (a) Sketch the graph of $f(x)$.
   (b) Compute $P(X > 0)$.
   (c) Compute $P(-1 < X < 1)$.
   (d) Compute $P(X < -.5 \text{ or } X > .5)$.

4. Let X denote the vibratory stress (psi) on a wind turbine blade at a particular wind speed in a wind tunnel. The article "Blade Fatigue Life Assessment with Application to VAWTS" (*J. Solar Energy Engr.*, 1982: 107–111) proposes the *Rayleigh* distribution, with pdf

$$f(x; \theta) = \begin{cases} \dfrac{x}{\theta^2} \cdot e^{-x^2/(2\theta^2)} & x > 0 \\ 0 & \text{otherwise} \end{cases}$$

   as a model for X, where θ is a positive constant.
   (a) Verify that $f(x; \theta)$ is a legitimate pdf.
   (b) Suppose $\theta = 100$ (a value suggested by a graph in the article). What is the probability that X is at most 200? Less than 200? At least 200?
   (c) What is the probability that X is between 100 and 200 (again assuming $\theta = 100$)?
   (d) Give an expression for the cdf of X.

5. A college professor never finishes his lecture before the end of the hour and always finishes his lectures within 2 min after the hour. Let $X =$ the time that elapses between the end of the hour and the end of the lecture and suppose the pdf of X is

$$f(x) = \begin{cases} kx^2 & 0 \le x \le 2 \\ 0 & \text{otherwise} \end{cases}$$

   (a) Find the value of k and draw the corresponding density curve. [*Hint*: Total area under the graph of $f(x)$ is 1.]
   (b) What is the probability that the lecture ends within 1 min of the end of the hour?
   (c) What is the probability that the lecture continues beyond the hour for between 60 and 90 s?
   (d) What is the probability that the lecture continues for at least 90 s beyond the end of the hour?

6. The actual tracking weight of a stereo cartridge that is set to track at 3 g on a particular changer can be regarded as a continuous rv X with pdf

$$f(x) = \begin{cases} k[1 - (x - 3)^2] & 2 \le x \le 4 \\ 0 & \text{otherwise} \end{cases}$$

(a) Sketch the graph of $f(x)$.

(b) Find the value of k.

(c) What is the probability that the actual tracking weight is greater than the prescribed weight?

(d) What is the probability that the actual weight is within .25 g of the prescribed weight?

(e) What is the probability that the actual weight differs from the prescribed weight by more than .5 g?

7. The article "Second Moment Reliability Evaluation vs. Monte Carlo Simulations for Weld Fatigue Strength" (*Quality and Reliability Engr. Intl.*, 2012: 887-896) considered the use of a uniform distribution with $A = .20$ and $B = 4.25$ for the diameter X of a certain type of weld (mm).

(a) Determine the pdf of X and graph it.

(b) What is the probability that diameter exceeds 3 mm?

(c) What is the probability that diameter is within 1 mm of the mean diameter?

(d) For any value a satisfying $.20 < a < a + 1 < 4.25$, what is $P(a < X < a + 1)$?

8. Commuting to work requires getting on a bus near home and then transferring to a second bus. If the waiting time (in minutes) at each stop has a Unif[0, 5] distribution, then it can be shown that the total waiting time Y has the pdf

$$f(y) = \begin{cases} \dfrac{1}{25}y & 0 \le y < 5 \\ \dfrac{2}{5} - \dfrac{1}{25}y & 5 \le y \le 10 \\ 0 & y < 0 \text{ or } y > 10 \end{cases}$$

(a) Sketch the pdf of Y.

(b) Verify that $\int_{-\infty}^{\infty} f(y)\,dy = 1$.

(c) What is the probability that total waiting time is at most 3 min?

(d) What is the probability that total waiting time is at most 8 min?

(e) What is the probability that total waiting time is between 3 and 8 min?

(f) What is the probability that total waiting time is either less than 2 min or more than 6 min?

9. Consider again the pdf of $X =$ time headway given in Example 3.5. What is the probability that time headway is

(a) At most 6 s?

(b) More than 6 s? At least 6 s?

(c) Between 5 and 6 s?

10. A family of pdfs that has been used to approximate the distribution of income, city population size, and size of firms is the *Pareto* family. The family has two parameters, k and θ, both > 0, and the pdf is

$$f(x; k, \theta) = \begin{cases} \dfrac{k \cdot \theta^k}{x^{k+1}} & x \geq \theta \\ 0 & x < \theta \end{cases}$$

(a) Sketch the graph of $f(x; k, \theta)$.
(b) Verify that the total area under the graph equals 1.
(c) If the rv X has pdf $f(x; k, \theta)$, obtain an expression for the cdf of X.
(d) For $\theta < a < b$, obtain an expression for the probability $P(a \leq X \leq b)$.
(e) Find an expression for the $(100p)$th percentile η_p.

11. Let X denote the amount of time a book on 2-h reserve is actually checked out, and suppose the cdf is

$$F(x) = \begin{cases} 0 & x < 0 \\ \dfrac{x^2}{4} & 0 \leq x < 2 \\ 1 & 2 \leq x \end{cases}$$

Use this to compute the following:
(a) $P(X \leq 1)$
(b) $P(.5 \leq X \leq 1)$
(c) $P(X > .5)$
(d) The median checkout duration η [*Hint*: Solve $F(\eta) = .5$.]
(e) $F'(x)$ to obtain the density function $f(x)$

12. The cdf for $X =$ measurement error of Exercise 3 is

$$F(x) = \begin{cases} 0 & x < -2 \\ \dfrac{1}{2} + \dfrac{3}{32}\left(4x - \dfrac{x^3}{3}\right) & -2 \leq x < 2 \\ 1 & 2 \leq x \end{cases}$$

(a) Compute $P(X < 0)$.
(b) Compute $P(-1 < X < 1)$.
(c) Compute $P(X > .5)$.
(d) Verify that $f(x)$ is as given in Exercise 3 by obtaining $F'(x)$.
(e) Verify that $\eta = 0$.

13. Example 3.5 introduced the concept of time headway in traffic flow and proposed a particular distribution for $X =$ the headway between two randomly selected consecutive car. Suppose that in a different traffic environment, the distribution of time headway has the form

$$f(x) = \begin{cases} \dfrac{k}{x^4} & x > 1 \\ 0 & x \le 1 \end{cases}$$

(a) Determine the value of k for which $f(x)$ is a legitimate pdf.
(b) Obtain the cumulative distribution function.
(c) Use the cdf from (b) to determine the probability that headway exceeds 2 s and also the probability that headway is between 2 and 3 s.

14. Let X denote the amount of space occupied by an article placed in a 1-ft^3 packing container. The pdf of X is

$$f(x) = \begin{cases} 90x^8(1-x) & 0 < x < 1 \\ 0 & \text{otherwise} \end{cases}$$

(a) Graph the pdf. Then obtain the cdf of X and graph it.
(b) What is $P(X \le .5)$ [i.e., $F(.5)$]?
(c) Using part (a), what is $P(.25 < X \le .5)$? What is $P(.25 \le X \le .5)$?
(d) What is the 75th percentile of the distribution?

15. Answer parts (a)–(d) of Exercise 14 for the random variable X, lecture time past the hour, given in Exercise 5.

16. The article "A Model of Pedestrians' Waiting Times for Street Crossings at Signalized Intersections" (*Transportation Research*, 2013: 17–28) suggested that under some circumstances the distribution of waiting time X could be modeled with the following pdf:

$$f(x; \theta, \tau) = \begin{cases} \dfrac{\theta}{\tau}(1 - x/\tau)^{\theta-1} & 0 \le x < \tau \\ 0 & \text{otherwise} \end{cases}$$

where $\theta, \tau > 0$.

(a) Graph $f(x; \theta, 80)$ for the three cases $\theta = 4$, 1, and .5 (these graphs appear in the cited article) and comment on their shapes.
(b) Obtain the cumulative distribution function of X.
(c) Obtain an expression for the median of the waiting time distribution.
(d) For the case $\theta = 4$ and $\tau = 80$, calculate $P(50 \le X \le 70)$ without doing any additional integration.

17. Let X be a continuous rv with cdf

$$F(x) = \begin{cases} 0 & x \le 0 \\ \dfrac{x}{4}\left[1 + \ln\left(\dfrac{4}{x}\right)\right] & 0 < x \le 4 \\ 1 & x > 4 \end{cases}$$

[This type of cdf is suggested in the article "Variability in Measured Bedload-Transport Rates" (*Water Resources Bull.*, 1985: 39–48) as a model for a hydrologic variable.] What is

(a) $P(X \leq 1)$?
(b) $P(1 \leq X \leq 3)$?
(c) The pdf of X?

18. Let X be the temperature in °C at which a chemical reaction takes place, and let Y be the temperature in °F (so $Y = 1.8X + 32$).

   (a) If the median of the X distribution is η, show that $1.8\eta + 32$ is the median of the Y distribution.
   (b) How is the 90th percentile of the Y distribution related to the 90th percentile of the X distribution? Verify your conjecture.
   (c) More generally, if $Y = aX + b$, how is any particular percentile of the Y distribution related to the corresponding percentile of the X distribution?

3.2   Expected Values and Moment Generating Functions

In Sect. 3.1 we saw that the transition from a discrete cdf to a continuous cdf entails replacing summation by integration. The same thing is true in moving from expected values of discrete variables to those of continuous variables.

3.2.1   Expected Values

For a discrete random variable X, the mean μ_X or $E(X)$ was defined as a weighted average and obtained by summing $x \cdot p(x)$ over possible X values. Here we replace summation by integration and the pmf by the pdf to get a continuous weighted average.

> **DEFINITION**
> The **expected value** or **mean value** of a continuous rv X with pdf $f(x)$ is
> $$\mu = \mu_X = E(X) = \int_{-\infty}^{\infty} x \cdot f(x)\, dx$$

Example 3.10 (Example 3.9 continued) The pdf of weekly gravel sales X was

$$f(x) = \begin{cases} \dfrac{3}{2}(1 - x^2) & 0 \leq x \leq 1 \\ 0 & \text{otherwise} \end{cases}$$

so

$$E(X) = \int_{-\infty}^{\infty} x \cdot f(x)dx = \int_0^1 x \cdot \frac{3}{2}(1 - x^2)dx = \frac{3}{2}\int_0^1 (x - x^3)dx = \frac{3}{2}\left(\frac{x^2}{2} - \frac{x^4}{4}\right)\Big|_{x=0}^{x=1} = \frac{3}{8}$$

If gravel sales are determined week after week according to the given pdf, then the long-run average value of sales per week will be .375 ton. ∎

Similar to the interpretation in the discrete case, the mean value μ can be regarded as the balance point (or fulcrum or center of mass) of a continuous distribution. In Example 3.10, if a piece of cardboard were cut out in the shape of the region under the density curve $f(x)$, then it would balance if supported at $\mu = 3/8$ along the bottom edge. When a pdf $f(x)$ is symmetric, then it will balance at its point of symmetry, which must be the mean μ. Recall from Sect. 3.1 that the median is also the point of symmetry; in general, if a distribution is symmetric and the mean exists, then it is equal to the median.

Often we wish to compute the expected value of some function $h(X)$ of the rv X. If we think of $h(X)$ as a new rv Y, methods from Sect. 3.7 can be used to derive the pdf of Y, and $E(Y)$ can be computed from the definition. Fortunately, as in the discrete case, there is an easier way to compute $E[h(X)]$.

PROPOSITION

If X is a continuous rv with pdf $f(x)$ and $h(X)$ is any function of X, then

$$\mu_{h(X)} = E[h(X)] = \int_{-\infty}^{\infty} h(x) \cdot f(x)\,dx$$

This is sometimes called the *Law of the Unconscious Statistician.*

Importantly, except in the cases where $h(x)$ is a linear function (see later in this section), $E[h(X)]$ is *not* equal to $h(\mu_X)$, the function h evaluated at the mean of X.

Example 3.11 The variation in a certain electrical current source X (in milliamps) can be modeled by the pdf

$$f(x) = \begin{cases} 1.25 - .25x & 2 \le x \le 4 \\ 0 & \text{otherwise} \end{cases}$$

The average current from this source is

$$E(X) = \int_2^4 x(1.25 - .25x)dx = \frac{17}{6} = 2.833\text{mA}$$

If this current passes through a 220-Ω resistor, the resulting power (in microwatts) is given by the expression $h(X) = (\text{current})^2(\text{resistance}) = 220X^2$. The expected power is given by

$$E(h(X)) = E(220X^2) = \int_2^4 220x^2(1.25 - .25x)dx = \frac{5500}{3} = 1833.3\mu\text{W}$$

Notice that the expected power is *not* equal to $220(2.833)^2$, a common error that results from substituting the mean current μ_X into the power formula. ∎

Example 3.12 Two species are competing in a region for control of a limited amount of a resource. Let $X =$ the proportion of the resource controlled by species 1 and suppose X has pdf

$$f(x) = \begin{cases} 1 & 0 \le x \le 1 \\ 0 & \text{otherwise} \end{cases}$$

which is a uniform distribution on [0, 1]. (In her book *Ecological Diversity*, E. C. Pielou calls this the "broken-stick" model for resource allocation, since it is analogous to breaking a stick at a randomly chosen point.) Then the species that controls the majority of this resource controls the amount

$$h(X) = \max(X, 1 - X) = \begin{cases} 1 - X & \text{if } 0 \le X < \frac{1}{2} \\ X & \text{if } \frac{1}{2} \le X \le 1 \end{cases}$$

The expected amount controlled by the species having majority control is then

$$\begin{aligned} E[h(X)] &= \int_{-\infty}^{\infty} \max(x, 1 - x) \cdot f(x)dx = \int_0^1 \max(x, 1 - x) \cdot 1 \ dx \\ &= \int_0^{1/2} (1 - x) \cdot 1 \ dx + \int_{1/2}^1 x \cdot 1 \ dx = \frac{3}{4} \end{aligned}$$ ∎

In the discrete case, the variance of X was defined as the expected squared deviation from μ and was calculated by summation. Here again integration replaces summation.

DEFINITION
The **variance** of a continuous random variable X with pdf $f(x)$ and mean value μ is

$$\sigma_X^2 = \text{Var}(X) = \int_{-\infty}^{\infty} (x - \mu)^2 \cdot f(x) \, dx = E\left[(X - \mu)^2\right]$$

The **standard deviation** of X is $\sigma_X = \text{SD}(X) = \sqrt{\text{Var}(X)}$.

As in the discrete case, σ_X^2 is the expected or average squared deviation about the mean μ, and σ_X can be interpreted roughly as the size of a representative deviation from the mean value μ. Note that σ_X has the same units as X itself.

Example 3.13 Let $X \sim \text{Unif}[A, B]$. Since a uniform distribution is symmetric, the mean of X is at the density curve's point of symmetry, which is clearly the midpoint $(A + B)/2$. This can be verified by integration:

$$\mu = \int_A^B x \cdot \frac{1}{B - A} dx = \frac{1}{B - A} \left. \frac{x^2}{2} \right|_A^B = \frac{1}{B - A} \frac{B^2 - A^2}{2} = \frac{A + B}{2}$$

The variance of X is then given by

$$\sigma^2 = \int_A^B (x - \mu)^2 \cdot \frac{1}{B - A} dx = \frac{1}{B - A} \int_A^B \left(x - \frac{A + B}{2} \right)^2 dx$$

$$= \frac{1}{B - A} \int_{-(B-A)/2}^{(B-A)/2} u^2 \, du \qquad \text{substitute } u = x - \frac{A + B}{2}$$

$$= \frac{2}{B - A} \int_0^{(B-A)/2} u^2 \, du \qquad \text{symmetry}$$

$$= \frac{2}{B - A} \left. \frac{u^3}{3} \right|_0^{(B-A)/2} = \frac{2}{B - A} \frac{(B - A)^3}{2^3 \cdot 3} = \frac{(B - A)^2}{12}$$

The standard deviation of X is the square root of the variance: $\sigma = (B - A)/\sqrt{12}$. Notice that the standard deviation of a $\text{Unif}[A, B]$ distribution is proportional to the length of the interval, $B - A$, which matches our intuitive notion that a larger standard deviation corresponds to greater "spread" in a distribution. ∎

Section 2.3 presented several properties of expected value, variance, and standard deviation for discrete random variables. Those same properties hold for the continuous case; proofs of these results are obtained by replacing summation with integration in the proofs presented in Chap. 2.

PROPOSITION

Let X be a continuous rv with pdf $f(x)$, mean μ, and standard deviation σ. Then the following properties hold.

1. (variance shortcut) $\text{Var}(X) = E(X^2) - \mu^2 = \int_{-\infty}^{\infty} x^2 \cdot f(x) dx - \left(\int_{-\infty}^{\infty} x \cdot f(x) dx \right)^2$

2. (Chebyshev's inequality) For any constant $k \geq 1$,

(continued)

$$P(|X - \mu| \geq k\sigma) \leq \frac{1}{k^2}$$

3. (linearity of expectation) For any functions $h_1(X)$ and $h_2(X)$ and any constants a_1, a_2, and b,

$$E[a_1 h_1(X) + a_2 h_2(X) + b] = a_1 E[h_1(X)] + a_2 E[h_2(X)] + b$$

4. (rescaling) For any constants a and b,

$$E(aX + b) = a\mu + b \qquad \text{Var}(aX + b) = a^2\sigma^2 \qquad \sigma_{aX+b} = |a|\sigma$$

Example 3.14 (Example 3.10 continued) For $X =$ weekly gravel sales, we computed $E(X) = 3/8$. Since

$$E(X^2) = \int_{-\infty}^{\infty} x^2 \cdot f(x)dx = \int_0^1 x^2 \cdot \frac{3}{2}(1 - x^2)dx = \frac{3}{2}\int_0^1 (x^2 - x^4)dx = \frac{1}{5},$$

$$\text{Var}(X) = \frac{1}{5} - \left(\frac{3}{8}\right)^2 = \frac{19}{320} = .059 \quad \text{and} \quad \sigma_X = .244$$

Suppose the amount of gravel actually received by customers in a week is $h(X) = X - .02X^2$; the second term accounts for the small amount that is lost in transport. Then the average weekly amount received by customers is

$$E(X - .02X^2) = E(X) - .02E(X^2) = \frac{3}{8} - .02 \cdot \frac{1}{5} = .371 \text{ tons} \qquad \blacksquare$$

Example 3.15 When a dart is thrown at a circular target, consider the location of the landing point relative to the bull's eye. Let X be the angle in degrees measured from the horizontal, and assume that $X \sim \text{Unif}[0, 360)$. By Example 3.13, $E(X) = 180$ and $\text{SD}(X) = 360/\sqrt{12}$. Define Y to be the angle measured in radians between $-\pi$ and π, so $Y = (2\pi/360)X - \pi$. Then, applying the rescaling properties with $a = 2\pi/360$ and $b = -\pi$,

$$E(Y) = \frac{2\pi}{360} \cdot E(X) - \pi = \frac{2\pi}{360}180 - \pi = 0$$

and

$$\sigma_Y = \left|\frac{2\pi}{360}\right| \cdot \sigma_X = \frac{2\pi}{360}\frac{360}{\sqrt{12}} = \frac{2\pi}{\sqrt{12}} \qquad \blacksquare$$

3.2.2 Moment Generating Functions

Moments and moment generating functions for discrete random variables were introduced in Sect. 2.7. These concepts carry over to the continuous case.

DEFINITION

The **moment generating function** (mgf) of a continuous random variable X is

$$M_X(t) = E(e^{tX}) = \int_{-\infty}^{\infty} e^{tx} f(x) dx.$$

As in the discrete case, the moment generating function exists iff $M_X(t)$ is defined for an interval that includes zero as well as positive and negative values of t.

Just as before, when $t = 0$ the value of the mgf is always 1:

$$M_X(0) = E(e^{0X}) = \int_{-\infty}^{\infty} e^{0x} f(x) dx = \int_{-\infty}^{\infty} f(x) dx = 1.$$

Example 3.16 At a store, the checkout time X in minutes has the pdf $f(x) = 2e^{-2x}$, $x \geq 0$; $f(x) = 0$ otherwise. Then

$$M_X(t) = \int_{-\infty}^{\infty} e^{tx} f(x) dx = \int_{0}^{\infty} e^{tx} (2e^{-2x}) dx = \int_{0}^{\infty} 2e^{-(2-t)x} dx$$

$$= -\frac{2}{2-t} e^{-(2-t)x} \Big|_{0}^{\infty} = \frac{2}{2-t} - \frac{2}{2-t} \lim_{x \to \infty} e^{-(2-t)x}.$$

The limit above exists (in fact, it equals zero) provided the coefficient on x is negative, i.e., $-(2-t) < 0$. This is equivalent to $t < 2$. The mgf exists because it is defined for an interval of values including 0 in its interior, specifically $(-\infty, 2)$. For t in that interval, the mgf of X is $M_X(t) = 2/(2-t)$.

Notice that $M_X(0) = 2/(2-0) = 1$. Of course, from the calculation preceding this example we know that $M_X(0) = 1$ must always be the case, but it is useful as a check to set $t = 0$ and see if the result is 1. ∎

Recall that in Sect. 2.7 we had a uniqueness property for the mgfs of discrete distributions. This proposition is equally valid in the continuous case: two distributions have the same pdf if and only if they have the same moment generating function, assuming that the mgf exists. For example, if a random variable X is known to have mgf $M_X(t) = 2/(2-t)$ for $t < 2$, then from Example 3.16 it must necessarily be the case that the pdf of X is $f(x) = 2e^{-2x}$ for $x \geq 0$ and $f(x) = 0$ otherwise.

In the discrete case we also had a theorem on how to get moments from the mgf, and this theorem applies also in the continuous case: the rth moment of a continuous rv with mgf $M_X(t)$ is given by

$$E(X^r) = M_X^{(r)}(0),$$

the rth derivative of the mgf with respect to t evaluated at $t = 0$, if the mgf exists.

Example 3.17 (Example 3.16 continued) The mgf of the rv $X =$ checkout time at the store was found to be $M_X(t) = 2/(2 - t) = 2(2 - t)^{-1}$ for $t < 2$. To find the mean and standard deviation, first compute the derivatives:

$$M_X'(t) = -2(2 - t)^{-2}(-1) = \frac{2}{(2 - t)^2}$$

$$M_X''(t) = \frac{d}{dt}\left[2(2 - t)^{-2}\right] = -4(2 - t)^{-3}(-1) = \frac{4}{(2 - t)^3}$$

Setting t to 0 in the first derivative gives the expected checkout time as

$$E(X) = M_X^{(1)}(0) = M_X'(0) = .5 \text{ min.}$$

Setting t to 0 in the second derivative gives the second moment

$$E(X^2) = M_X^{(2)}(0) = M_X''(0) = .5,$$

from which the variance of the checkout time is $\text{Var}(X) = \sigma^2 = E(X^2) - [E(X)]^2 = .5 - .5^2 = .25$ and the standard deviation is $\sigma = \sqrt{.25} = .5$ min. ∎

We will sometimes need to transform X using a linear function $Y = aX + b$. As discussed in the discrete case, if X has the mgf $M_X(t)$ and $Y = aX + b$, then $M_Y(t) = e^{bt}M_X(at)$.

Example 3.18 Let $X \sim \text{Unif}[A, B]$. As verified in Exercise 32, the moment generating function of X is

$$M_X(t) = \begin{cases} \dfrac{e^{Bt} - e^{At}}{(B - A)t} & t \neq 0 \\ 1 & t = 0 \end{cases}$$

In particular, consider the situation in Example 3.15. Let X, the angle measured in degrees, be uniform on [0, 360], so $A = 0$ and $B = 360$. Then

$$M_X(t) = \frac{e^{360t} - 1}{360t} \quad t \neq 0, \quad M_X(0) = 1$$

Now let $Y = (2\pi/360)X - \pi$, so Y is the angle measured in radians between $-\pi$ and π. Using the mgf rule for linear transformations with $a = 2\pi/360$ and $b = -\pi$, we get

$$M_Y(t) = e^{bt}M_X(at) = e^{-\pi t}M_X\left(\frac{2\pi}{360}\right)t$$

$$= e^{-\pi t}\frac{e^{360(2\pi/360)t} - 1}{360\left(\dfrac{2\pi}{360}\right)t}$$

$$= \frac{e^{\pi t} - e^{-\pi t}}{2\pi t} \quad t \neq 0, \qquad M_Y(0) = 1$$

This matches the general form of the moment generating function for a uniform random variable with $A = -\pi$ and $B = \pi$. Thus, by the mgf uniqueness property, $Y \sim \text{Unif}[-\pi, \pi]$.   ■

3.2.3   Exercises: Section 3.2 (19–38)

19. Reconsider the distribution of checkout duration X described in Exercise 11. Compute the following:
    (a) $E(X)$
    (b) $\text{Var}(X)$ and $\text{SD}(X)$
    (c) If the borrower is charged an amount $h(X) = X^2$ when checkout duration is X, compute the expected charge $E[h(X)]$.
20. The article "Modeling Sediment and Water Column Interactions for Hydrophobic Pollutants" (*Water Res.*, 1984: 1169–1174) suggests the uniform distribution on the interval [7.5, 20] as a model for depth (cm) of the bioturbation layer in sediment in a certain region.
    (a) What are the mean and variance of depth?
    (b) What is the cdf of depth?
    (c) What is the probability that observed depth is at most 10? Between 10 and 15?
    (d) What is the probability that the observed depth is within 1 standard deviation of the mean value?
    Within 2 standard deviations?
21. For the distribution of Exercise 14,
    (a) Compute $E(X)$ and $\text{SD}(X)$.
    (b) What is the probability that X is more than 1 standard deviation from its mean value?
22. Consider the pdf given in Exercise 6.
    (a) Obtain and graph the cdf of X.
    (b) From the graph of $f(x)$, what is the median, η?
    (c) Compute $E(X)$ and $\text{Var}(X)$.

23. Let $X \sim \text{Unif}[A, B]$.
    (a) Obtain an expression for the $(100p)$th percentile.
    (b) Obtain an expression for the median, η. How does this compare to the mean μ, and why does that make sense for this distribution?
    (c) For n a positive integer, compute $E(X^n)$.

24. Consider the pdf for total waiting time Y for two buses

$$f(y) = \begin{cases} \dfrac{1}{25}y & 0 \le y < 5 \\[2mm] \dfrac{2}{5} - \dfrac{1}{25}y & 5 \le y \le 10 \\[2mm] 0 & \text{otherwise} \end{cases}$$

introduced in Exercise 8.
    (a) Compute and sketch the cdf of Y. [*Hint*: Consider separately $0 \le y < 5$ and $5 \le y \le 10$ in computing $F(y)$. A graph of the pdf should be helpful.]
    (b) Obtain an expression for the $(100p)$th percentile. [*Hint*: Consider separately $0 < p < .5$ and $.5 \le p < 1$.]
    (c) Compute $E(Y)$ and $\text{Var}(Y)$. How do these compare with the expected waiting time and variance for a single bus when the time is uniformly distributed on $[0, 5]$?
    (d) Explain how symmetry can be used to obtain $E(Y)$.

25. An ecologist wishes to mark off a circular sampling region having radius 10 m. However, the radius of the resulting region is actually a random variable R with pdf

$$f(r) = \begin{cases} \dfrac{3}{4}[1 - (10 - r)^2] & 9 \le r \le 11 \\[2mm] 0 & \text{otherwise} \end{cases}$$

What is the expected area of the resulting circular region?

26. The weekly demand for propane gas (in 1000s of gallons) from a particular facility is an rv X with pdf

$$f(x) = \begin{cases} 2\left(1 - \dfrac{1}{x^2}\right) & 1 \le x \le 2 \\[2mm] 0 & \text{otherwise} \end{cases}$$

    (a) Compute the cdf of X.
    (b) Obtain an expression for the $(100p)$th percentile. What is the value of the median, η?
    (c) Compute $E(X)$. How do the mean and median of this distribution compare?
    (d) Compute $\text{Var}(X)$ and $\text{SD}(X)$.

(e) If 1.5 thousand gallons are in stock at the beginning of the week and no new supply is due in during the week, how much of the 1.5 thousand gallons is expected to be left at the end of the week? [*Hint:* Let $h(x) =$ amount left when demand is x.]

27. If the temperature at which a compound melts is a random variable with mean value 120°C and standard deviation 2°C, what are the mean temperature and standard deviation measured in °F? [*Hint:* °F $= 1.8$°C $+ 32$.]

28. Let X have the Pareto pdf introduced in Exercise 10:

$$f(x; k, \theta) = \begin{cases} \dfrac{k \cdot \theta^k}{x^{k+1}} & x \geq \theta \\ 0 & x < \theta \end{cases}$$

(a) If $k > 1$, compute $E(X)$.
(b) What can you say about $E(X)$ if $k = 1$?
(c) If $k > 2$, show that $\text{Var}(X) = k\theta^2(k-1)^{-2}(k-2)^{-1}$.
(d) If $k = 2$, what can you say about $\text{Var}(X)$?
(e) What conditions on k are necessary to ensure that $E(X^n)$ is finite?

29. The time (min) between successive visits to a particular Web site has pdf $f(x) = 4e^{-4x}, x \geq 0; f(x) = 0$ otherwise. Use integration by parts to obtain $E(X)$ and $SD(X)$.

30. Consider the weights, in grams, of walnuts harvested at a nearby farm. Suppose this weight distribution can be modeled by the following pdf:

$$f(x) = \begin{cases} .5 - \dfrac{x}{8} & 0 \leq x \leq 4 \\ 0 & \text{otherwise} \end{cases}$$

(a) Show that $E(X) = 4/3$ and $\text{Var}(X) = 8/9$.
(b) The *skewness coefficient* is defined as $E[(X - \mu)^3]/\sigma^3$. Show that its value for the given pdf is .566. What would the skewness be for a perfectly symmetric pdf?

31. The *delta method* provides approximations to the mean and variance of a nonlinear function $h(X)$ of a rv X. These approximations are based on a first-order Taylor series expansion of $h(x)$ about $x = \mu$, the mean of X:

$$h(X) \approx h_1(X) = h(\mu) + h'(\mu)(X - \mu)$$

(a) Show that $E[h_1(X)] = h(\mu)$. (This is the delta method approximation to $E[h(X)]$.)
(b) Show that $\text{Var}[h_1(X)] = [h'(\mu)]^2\text{Var}(X)$. (This is the delta method approximation to $\text{Var}[h(X)]$.)

(c) If the voltage v across a medium is fixed but current I is random, then resistance will also be a random variable related to I by $R = v/I$. If $\mu_I = 20$ and $\sigma_I = .5$, calculate approximations to μ_R and σ_R.

(d) Let R have the distribution in Exercise 25, whose mean and variance are 10 and 1/5, respectively. Let $h(R) = \pi R^2$, the area of the ecologist's sampling region. How does $E[h(R)]$ from Exercise 25 compare to the delta method approximation $h(10)$?

(e) It can be shown that $\mathrm{Var}[h(R)] = 14008\pi^2/175$. Compute the delta method approximation to $\mathrm{Var}[h(R)]$ using the formula in (b). How good is the approximation?

32. Let $X \sim \mathrm{Unif}[A, B]$, so its pdf is $f(x) = 1/(B - A)$, $A \leq x \leq B$, $f(x) = 0$ otherwise. Show that the moment generating function of X is

$$M_X(t) = \begin{cases} \dfrac{e^{Bt} - e^{At}}{(B - A)t} & t \neq 0 \\[2mm] 1 & t = 0 \end{cases}$$

33. Let $X \sim \mathrm{Unif}[0, 1]$. Find a linear function $Y = g(X)$ such that the interval $[0, 1]$ is transformed into $[-5, 5]$. Use the relationship for linear functions $M_{aX+b}(t) = e^{bt}M_X(at)$ to obtain the mgf of Y from the mgf of X. Compare your answer with the result of Exercise 32, and use this to obtain the pdf of Y.

34. If the pdf of a measurement error X is $f(x) = .5e^{-|x|}$, $-\infty < x < \infty$, show that $M_X(t) = 1/(1 - t^2)$ for $|t| < 1$.

35. Consider the rv $X = $ time headway in Example 3.5.

(a) Find the moment generating function and use it to find the mean and variance.

(b) Now consider a random variable whose pdf is

$$f(x) = \begin{cases} .15e^{-.15x} & x \geq 0 \\ 0 & \text{otherwise} \end{cases}$$

Find the moment generating function and use it to find the mean and variance. Compare with (a), and explain the similarities and differences.

(c) Let $Y = X - .5$ and use the relationship for linear functions $M_{aX+b}(t) = e^{bt}M_X(at)$ to obtain the mgf of Y from (a). Compare with the result of (b) and explain.

36. Define $L_X(t) = \ln[M_X(t)]$. It was shown in Exercise 120 of Chap. 2 that $L_X'(0) = E(X)$ and $L_X''(0) = \mathrm{Var}(X)$.

(a) Determine $M_X(t)$ for the pdf in Exercise 29, and use this mgf to obtain $E(X)$ and $\mathrm{Var}(X)$. How does this compare, in terms of difficulty, with the integration by parts required in that exercise?

(b) Determine $L_X(t)$ for this same distribution, and use $L_X(t)$ to obtain $E(X)$ and $\mathrm{Var}(X)$. How does the computational effort here compare with that of (a)?

37. Let X be a nonnegative, continuous rv with pdf $f(x)$ and cdf $F(x)$.

(a) Show that, for any constant $t > 0$,

$$\int_t^\infty x \cdot f(x)\,dx \ge t \cdot P(X > t) = t \cdot [1 - F(t)]$$

(b) Assume the mean of X is finite (i.e., the integral defining μ converges). Use part (a) to show that

$$\lim_{t \to \infty} t \cdot [1 - F(t)] = 0$$

[*Hint*: Write the integral for μ as the sum of two other integrals, one from 0 to t and another from t to ∞.]

38. Let X be a nonnegative, continuous rv with cdf $F(x)$.

(a) Assuming the mean μ of X is finite, show that

$$\mu = \int_0^\infty [1 - F(x)]\,dx$$

[*Hint*: Apply integration by parts to the integral above, and use the result of the previous exercise.] This is the continuous analog of the result established in Exercise 48 of Chap. 2.

(b) A similar argument can be used to show that the kth moment of X is given by

$$E(X^k) = k \int_0^\infty x^{k-1}[1 - F(x)]\,dx$$

and that $E(X^k)$ exists iff $t^k[1 - F(t)] \to 0$ as $t \to \infty$. (This was the topic of a 2012 article in *The American Statistician*.) Suppose the lifetime X, in weeks, of a low-grade transistor under continuous use has cdf $F(x) = 1 - (x + 1)^{-3}$ for $x > 0$. Without finding the pdf of X, determine its mean and its standard deviation.

3.3   The Normal (Gaussian) Distribution

The normal distribution, often called the Gaussian distribution by engineers, is the most important one in all of probability and statistics. Many numerical populations have distributions that can be fit very closely by an appropriate normal curve. Examples include heights, weights, and other physical characteristics, measurement errors in scientific experiments, measurements on fossils, reaction times in psychological experiments, measurements of intelligence and aptitude, scores on various tests, and numerous economic measures and indicators. Even when the underlying distribution is discrete, the normal curve often gives an excellent approximation. In addition, even when individual variables themselves are not normally distributed, sums and averages of the variables will, under suitable conditions, have approximately a normal distribution; this is the content of the Central Limit Theorem discussed in Chap. 4.

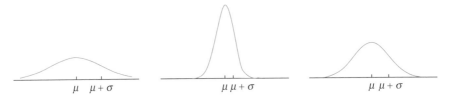

Fig. 3.13 Normal density curves

> **DEFINITION**
>
> A continuous rv X is said to have a **normal distribution** (or **Gaussian distribution**) with parameters μ and σ, where $-\infty < \mu < \infty$ and $\sigma > 0$, if the pdf of X is
>
> $$f(x; \mu, \sigma) = \frac{1}{\sigma\sqrt{2\pi}} e^{-(x-\mu)^2/(2\sigma^2)} \qquad -\infty < x < \infty \qquad (3.3)$$
>
> The statement that X is normally distributed with parameters μ and σ is often abbreviated $X \sim N(\mu, \sigma)$.

Figure 3.13 presents graphs of $f(x; \mu, \sigma)$ for several different (μ, σ) pairs. Each resulting density curve is symmetric about μ and bell-shaped, so the center of the bell (point of symmetry) is both the mean of the distribution and the median. The value of σ is the distance from μ to the inflection points of the curve (the points at which the curve changes between turning downward to turning upward). Large values of σ yield density curves that are quite spread out about μ, whereas small values of σ yield density curves with a high peak above μ and most of the area under the density curve quite close to μ. Thus a large σ implies that a value of X far from μ may well be observed, whereas such a value is quite unlikely when σ is small.

Clearly $f(x; \mu, \sigma) \geq 0$, but a somewhat complicated calculus argument is required to prove that $\int_{-\infty}^{\infty} f(x; \mu, \sigma) dx = 1$ (see Exercise 66). It can be shown using calculus (Exercise 67) or moment generating functions (Exercise 68) that $E(X) = \mu$ and $\text{Var}(X) = \sigma^2$, so the parameters μ and σ are the mean and the standard deviation, respectively, of X.

3.3.1 The Standard Normal Distribution

To compute $P(a \leq X \leq b)$ when $X \sim N(\mu, \sigma)$, we must evaluate

$$\int_a^b \frac{1}{\sigma\sqrt{2\pi}} e^{-(x-\mu)^2/(2\sigma^2)}\,dx \tag{3.4}$$

None of the standard integration techniques can be used here, and there is no closed-form expression for the integral. Table 3.1 at the end of this section provides the code for performing such normal distribution calculations in both Matlab and R. For the purpose of hand calculation of normal distribution probabilities, we now introduce a special normal distribution.

DEFINITION

The normal distribution with parameter values $\mu=0$ and $\sigma=1$ is called the **standard normal distribution**. A random variable that has a standard normal distribution is called a **standard normal random variable** and will be denoted by Z. The pdf of Z is

$$f(z;0,1) = \frac{1}{\sqrt{2\pi}} e^{-z^2/2} \qquad -\infty < z < \infty$$

The cdf of Z is $P(Z \le z) = \int_{-\infty}^{z} \frac{1}{\sqrt{2\pi}} e^{-y^2/2}\,dy$, which we will denote by $\Phi(z)$.

The standard normal distribution does not frequently serve as a model for a naturally arising population, since few variables have mean 0 and standard deviation 1. Instead, it is a reference distribution from which information about other normal distributions can be obtained. Appendix Table A.3 gives values of $\Phi(z)$ for $z = -3.49, -3.48, \ldots, 3.48, 3.49$ and is referred to as the *standard normal table* or *z table*. Figure 3.14 illustrates the type of cumulative area (probability) tabulated in Table A.3. From this table, various other probabilities involving Z can be calculated.

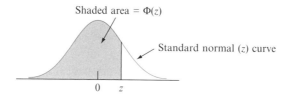

Shaded area = $\Phi(z)$

Standard normal (z) curve

0   z

Fig. 3.14 Standard normal cumulative areas tabulated in Appendix Table A.3

Example 3.19 Here we demonstrate how the z table is used to calculate various probabilities involving a standard normal rv.

(a) $P(Z \le 1.25) = \Phi(1.25)$, a probability that is tabulated in Table A.3 at the intersection of the row marked 1.2 and the column marked .05. The number there is .8944, so $P(Z \le 1.25) = .8944$. See Fig. 3.15a. In Matlab, we may type

a   **b**

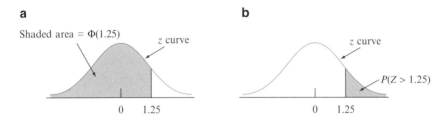

Fig. 3.15 Normal curve areas (probabilities) for Example 3.19

```
cdf('norm',1.25,0,1);  in  R,  use  pnorm(1.25,0,1)  or  just
pnorm(1.25).
```
(b) $P(Z > 1.25) = 1 - P(Z \le 1.25) = 1 - \Phi(1.25)$, the area under the standard normal curve to the right of 1.25 (an upper-tail area). Since $\Phi(1.25) = .8944$, it follows that $P(Z > 1.25) = .1056$. Since Z is a continuous rv, $P(Z \ge 1.25)$ also equals .1056. See Fig. 3.15b.
(c) $P(Z \le -1.25) = \Phi(-1.25)$, a lower-tail area. Directly from the z table, $\Phi(-1.25) = .1056$. By symmetry of the normal curve, this is identical to the probability in (b).
(d) $P(-.38 \le Z \le 1.25)$ is the area under the standard normal curve above the interval whose left endpoint is $-.38$ and whose right endpoint is 1.25. From Sect. 3.1, if Z is a continuous rv with cdf $F(z)$, then $P(a \le Z \le b) = F(b) - F(a)$. This gives $P(-.38 \le Z \le 1.25) = \Phi(1.25) - \Phi(-.38) = .8944 - .3520 = .5424$. (See Fig. 3.16.) To evaluate this probability in Matlab, type `cdf('norm',1.25,0,1)-cdf('norm',-.38,0,1)`; in R, type `pnorm(1.25,0,1)-pnorm(-.38,0,1)` or just `pnorm(1.25)-pnorm(-.38)`.

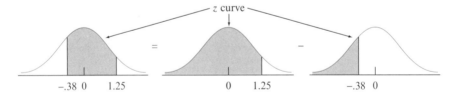

Fig. 3.16 $P(-.38 \le Z \le 1.25)$ as the difference between two cumulative areas   ■

From Sect. 3.1, we have that the $(100p)$th percentile of the standard normal distribution, for any p between 0 and 1, is the solution to the equation $\Phi(z) = p$. So, we may write the $(100p)$th percentile of the standard normal distribution as $\eta_p = \Phi^{-1}(p)$. Matlab, R, or the z table can be used to obtain this percentile.

Example 3.20 The 99th percentile of the standard normal distribution, $\Phi^{-1}(.99)$, is the value on the horizontal axis such that the area under the curve to the left of the value is .9900, as illustrated in Fig. 3.17. To solve the "inverse" problem $\Phi(z) = p$, the standard normal table is used in an inverse fashion: Find in the middle of the table

.9900; the row and column in which it lies identify the 99th z percentile. Here .9901 lies in the row marked 2.3 and column marked .03, so $\Phi(2.33) = .9901 \approx .99$ and the 99th percentile is approximately $z = 2.33$. By symmetry, the first percentile is the negative of the 99th percentile, so it equals -2.33 (1% lies below the first and above the 99th). See Fig. 3.18.

Fig. 3.17 Finding the 99th percentile

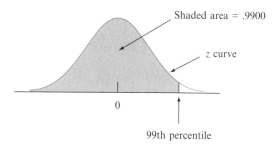

Fig. 3.18 The relationship between the 1st and 99th percentiles

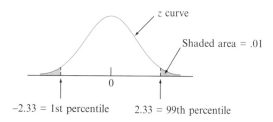

To find the 99th percentile in Matlab, use the command `icdf ('norm',.99,0,1)`; "icdf" stands for "inverse cumulative distribution function," meaning Φ^{-1}. In R, `qnorm(.99,0,1)` or just `qnorm(.99)` produces that same value of roughly $z = 2.33$. ∎

3.3.2 Non-standardized Normal Distributions

When $X \sim N(\mu, \sigma)$, probabilities involving X may be computed by "standardizing." A **standardized variable** has the form $(X - \mu)/\sigma$. Subtracting μ shifts the mean from μ to zero, and then dividing by σ scales the variable so that the standard deviation is 1 rather than σ.

PROPOSITION

If $X \sim N(\mu, \sigma)$, then the "standardized" rv Z defined by

$$Z = \frac{X - \mu}{\sigma}$$

(continued)

has a standard normal distribution. Thus

$$P(a \leq X \leq b) = P\left(\frac{a-\mu}{\sigma} \leq Z \leq \frac{b-\mu}{\sigma}\right) = \Phi\left(\frac{b-\mu}{\sigma}\right) - \Phi\left(\frac{a-\mu}{\sigma}\right),$$

$$P(X \leq a) = \Phi\left(\frac{a-\mu}{\sigma}\right), \qquad P(X \geq b) = 1 - \Phi\left(\frac{b-\mu}{\sigma}\right),$$

and the $(100p)$th percentile of the $N(\mu, \sigma)$ distribution is given by

$$\eta_p = \mu + \Phi^{-1}(p) \cdot \sigma.$$

Conversely, if $Z \sim N(0, 1)$ and μ and σ are constants (with $\sigma > 0$), then the "un-standardized" rv $X = \mu + \sigma Z$ has a normal distribution with mean μ and standard deviation σ.

Proof Let $X \sim N(\mu, \sigma)$ and define $Z = (X - \mu)/\sigma$ as in the statement of the proposition. Then the cdf of Z is given by

$$F_z(z) = P(Z \leq z)$$

$$= P\left(\frac{X - \mu}{\sigma} \leq z\right)$$

$$= P(X \leq \mu + z\sigma) = \int_{-\infty}^{\mu+z\sigma} f(x; \mu, \sigma)dx = \int_{-\infty}^{\mu+z\sigma} \frac{1}{\sigma\sqrt{2\pi}} e^{-(x-\mu)^2/(2\sigma^2)} dx$$

Now make the substitution $u = (x - \mu)/\sigma$. The new limits of integration become $-\infty$ to z, and the differential dx is replaced by $\sigma\, du$, resulting in

$$F_z(z) = \int_{-\infty}^{z} \frac{1}{\sigma\sqrt{2\pi}} e^{-u^2/2}\sigma du = \int_{-\infty}^{z} \frac{1}{\sqrt{2\pi}} e^{-u^2/2} du = \Phi(z)$$

Thus, the cdf of $(X - \mu)/\sigma$ is the standard normal cdf, which establishes that $(X - \mu)/\sigma \sim N(0, 1)$.

The probability formulas in the statement of the proposition follow directly from this main result, as does the formula for the $(100p)$th percentile:

$$p = P(X \leq \eta_p) = P\left(\frac{X-\mu}{\sigma} \leq \frac{\eta_p - \mu}{\sigma}\right) = \Phi\left(\frac{\eta_p - \mu}{\sigma}\right) \Rightarrow \frac{\eta_p - \mu}{\sigma} = \Phi^{-1}(p) \Rightarrow$$

$$\eta_p = \mu + \Phi^{-1}(p) \cdot \sigma$$

The converse statement $Z \sim N(0, 1) \Rightarrow \mu + \sigma Z \sim N(\mu, \sigma)$ is derived similarly. ∎

The key idea of this proposition is that by standardizing, any probability involving X can be expressed as a probability involving a standard normal rv Z, so that the z table can be used. This is illustrated in Fig. 3.19.

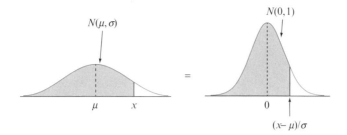

Fig. 3.19 Equality of nonstandard and standard normal curve areas

Software eliminates the need for standardizing X, although the standard normal distribution is still important in its own right. Table 3.1 at the end of this section details the relevant R and Matlab commands, which are also illustrated in the following examples.

Example 3.21 The time that it takes a driver to react to the brake lights on a decelerating vehicle is critical in avoiding rear-end collisions. The article "Fast-Rise Brake Lamp as a Collision-Prevention Device" (*Ergonomics*, 1993: 391–395) suggests that reaction time for an in-traffic response to a brake signal from standard brake lights can be modeled with a normal distribution having mean value 1.25 s and standard deviation of .46 s. What is the probability that reaction time is between 1.00 s and 1.75 s? If we let X denote reaction time, then standardizing gives $1.00 \leq X \leq 1.75$ if and only if

$$\frac{1.00 - 1.25}{.46} \leq \frac{X - 1.25}{.46} \leq \frac{1.75 - 1.25}{.46}$$

The middle expression, by the previous proposition, is a standard normal rv. Thus

$$P(1.00 \leq X \leq 1.75) = P\left(\frac{1.00 - 1.25}{.46} \leq Z \leq \frac{1.75 - 1.25}{.46}\right)$$
$$= P(-.54 \leq Z \leq 1.09) = \Phi(1.09) - \Phi(-.54)$$
$$= .8621 - .2946 = .5675$$

This is illustrated in Fig. 3.20. The same answer may be produced in Matlab with the command cdf('norm',1.75,1.25,.46)-cdf('norm',1.00, 1.25,.46); Matlab gives the answer .5681, which is more accurate than the value .5675 above (due to rounding the z-values to two decimal places). The analogous R command is pnorm(1.75,1.25,.46)-pnorm(1.00,1.25,.46).

Similarly, if we view 2 s as a critically long reaction time, the probability that actual reaction time will exceed this value is

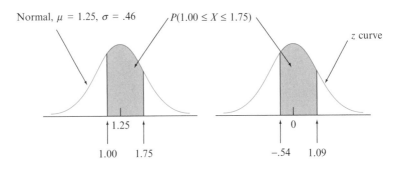

Fig. 3.20 Normal curves for Example 3.21

$$P(X > 2) = P\left(Z > \frac{2 - 1.25}{.46}\right) = P(Z > 1.63) = 1 - \Phi(1.63) = .0516$$

This probability is determined in Matlab and R with the commands 1-cdf ('norm',2,1.25,.46) and 1-pnorm(2,1.25,.46), respectively. ■

Example 3.22 The amount of distilled water dispensed by a machine is normally distributed with mean value 64 oz and standard deviation .78 oz. What container size c will ensure that overflow occurs only .5% of the time? If X denotes the amount dispensed, the desired condition is that $P(X > c) = .005$, or, equivalently, that $P(X \leq c) = .995$. Thus c is the 99.5th percentile of the normal distribution with $\mu = 64$ and $\sigma = .78$. The 99.5th percentile of the standard normal distribution is $\Phi^{-1}(.995) \approx 2.58$, so
$$c = \eta_{.995} = 64 + (2.58)(.78) = 64 + 2.0 = 66.0 \text{ oz}$$

This is illustrated in Fig. 3.21.

Fig. 3.21 Distribution of amount dispensed for Example 3.22

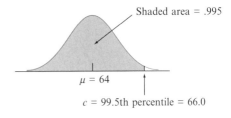

The relevant Matlab and R commands are icdf('norm',.995,64,.78) and qnorm(.995,64,.78), respectively. ■

Standardizing amounts to nothing more than calculating a distance from the mean and then reexpressing the distance as some number of standard deviations. For example, if $\mu = 100$ and $\sigma = 15$, then $x = 130$ corresponds to $z = (130 - 100)/15 = 30/15 = 2.00$. That is, 130 is 2 standard deviations above (to the right of) the mean value. Similarly, standardizing 85 gives $(85 - 100)/15 = -1.00$, so 85 is 1 standard deviation below the mean.

The z table applies to *any* normal distribution provided that we think in terms of number of standard deviations away from the mean value.

Example 3.23 The return on a diversified investment portfolio is normally distributed. What is the probability that the return is within 1 standard deviation of its mean value? This question can be answered without knowing either μ or σ, as long as the distribution is known to be normal; in other words, the answer is the same for *any* normal distribution. Going one standard deviation below μ lands us at $\mu - \sigma$, while $\mu + \sigma$ is one standard deviation above the mean. Thus

$$P\left(\begin{array}{c}X \text{ is within one standard}\\ \text{deviation of its mean}\end{array}\right) = P(\mu - \sigma \le X \le \mu + \sigma)$$
$$= P\left(\frac{\mu - \sigma - \mu}{\sigma} \le Z \le \frac{\mu + \sigma - \mu}{\sigma}\right)$$
$$= P(-1 \le Z \le 1)$$
$$= \Phi(1) - \Phi(-1) = .6826$$

The probability that X is within 2 standard deviations of the mean is $P(-2 \le Z \le 2) = .9544$ and the probability that X is within 3 standard deviations of the mean is $P(-3 \le Z \le 3) = .9973$. ∎

The results of Example 3.23 are often reported in percentage form and referred to as the *empirical rule* (because empirical evidence has shown that histograms of real data can very frequently be approximated by normal curves).

EMPIRICAL RULE
If the population distribution of a variable is (approximately) normal, then
1. Roughly 68% of the values are within 1 SD of the mean.
2. Roughly 95% of the values are within 2 SDs of the mean.
3. Roughly 99.7% of the values are within 3 SDs of the mean.

3.3.3   The Normal MGF

The moment generating function provides a straightforward way to establish several important results concerning the normal distribution.

PROPOSITION
The moment generating function of a normally distributed random variable X is
$$M_X(t) = e^{\mu t + \sigma^2 t^2 / 2}$$

Proof Consider first the special case of a standard normal rv Z. Then

$$M_Z(t) = E\left(e^{tZ}\right) = \int_{-\infty}^{\infty} e^{tz}\frac{1}{\sqrt{2\pi}}e^{-z^2/2}dz = \int_{-\infty}^{\infty} \frac{1}{\sqrt{2\pi}}e^{-\left(z^2-2tz\right)/2}dz$$

Completing the square in the exponent, we have

$$M_Z(t) = e^{t^2/2}\int_{-\infty}^{\infty}\frac{1}{\sqrt{2\pi}}e^{-\left(z^2-2tz+t^2\right)/2}dz = e^{t^2/2}\int_{-\infty}^{\infty}\frac{1}{\sqrt{2\pi}}e^{-(z-t)^2/2}dz$$

The last integral is the area under a normal density curve with mean t and standard deviation 1, so the value of the integral is 1. Therefore, $M_Z(t) = e^{t^2/2}$.

Now let X be any normal rv with mean μ and standard deviation σ. Then, by the proposition earlier in this section, $(X - \mu)/\sigma = Z$, where Z is standard normal. Rewrite this relationship as $X = \mu + \sigma Z$, and use the property $M_{aY+b}(t) = e^{bt}M_Y(at)$:

$$M_X(t) = M_{\mu+\sigma Z}(t) = e^{\mu t}M_Z(\sigma t) = e^{\mu t}e^{\sigma^2 t^2/2} = e^{\mu t+\sigma^2 t^2/2} \qquad \blacksquare$$

The normal mgf can be used to establish that μ and σ are indeed the mean and standard deviation of X, as claimed earlier (Exercise 68). Also, by the mgf uniqueness property, any rv X whose moment generating function has the form specified above is necessarily normally distributed. For example, if it is known that the mgf of X is $M_X(t) = e^{8t^2}$, then X must be a normal rv with mean $\mu = 0$ and standard deviation $\sigma = 4$ (since the $N(0, 4)$ distribution has e^{8t^2} as its mgf).

It was established earlier in this section that if $X \sim N(\mu, \sigma)$ and $Z = (X - \mu)/\sigma$, then $Z \sim N(0, 1)$, and vice versa. This standardizing transformation is actually a special case of a much more general property.

PROPOSITION

Let $X \sim N(\mu, \sigma)$. Then for any constants a and b with $a \neq 0$, $aX + b$ is also normally distributed. That is, any linear rescaling of a normal rv is normal.

The proof of this proposition uses mgfs and is left as an exercise (Exercise 70). This proposition provides a much easier proof of the earlier relationship between X and Z. The rescaling formulas and this proposition combine to give the following statement: if X is normally distributed and $Y = aX + b$ $(a \neq 0)$, then Y is also normal, with mean $\mu_Y = a\mu_X + b$ and standard deviation $\sigma_Y = |a|\sigma_X$.

3.3.4   The Normal Distribution and Discrete Populations

The normal distribution is often used as an approximation to the distribution of values in a discrete population. In such situations, extra care must be taken to ensure that probabilities are computed in an accurate manner.

Example 3.24 IQ (as measured by a standard test) is known to be approximately normally distributed with $\mu = 100$ and $\sigma = 15$. What is the probability that a randomly selected individual has an IQ of at least 125? Letting $X =$ the IQ of a randomly chosen person, we wish $P(X \geq 125)$. The temptation here is to standardize $X \geq 125$ immediately as in previous examples. However, the IQ population is actually discrete, since IQs are integer-valued, so the normal curve is an approximation to a discrete probability histogram, as pictured in Fig. 3.22.

The rectangles of the histogram are *centered* at integers, so IQs of at least 125 correspond to rectangles beginning at 124.5, as shaded in Fig. 3.22. Thus we really want the area under the approximating normal curve to the right of 124.5. Standardizing this value gives $P(Z \geq 1.63) = .0516$. If we had standardized $X \geq 125$, we would have obtained $P(Z \geq 1.67) = .0475$. The difference is not great, but the answer .0516 is more accurate. Similarly, $P(X = 125)$ would be approximated by the area between 124.5 and 125.5, since the area under the normal curve above the single value 125 is zero.

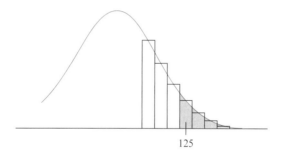

125

Fig. 3.22 A normal approximation to a discrete distribution   ∎

The correction for discreteness of the underlying distribution in Example 3.24 is often called a **continuity correction**; it adjusts for the use of a continuous distribution in approximating a probability involving a discrete rv. It is useful in the following application of the normal distribution to the computation of binomial probabilities. The normal distribution was actually created as an approximation to the binomial distribution (by Abraham de Moivre in the 1730s).

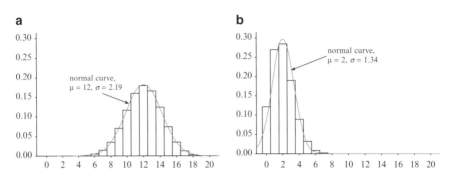

Fig. 3.23 Binomial probability histograms with normal approximation curves superimposed: (**a**) $n = 20$ and $p = .6$ (a good fit); (**b**) $n = 20$ and $p = .1$ (a poor fit)

3.3.5 Approximating the Binomial Distribution

Recall that the mean value and standard deviation of a binomial random variable X are $\mu = np$ and $\sigma = \sqrt{npq}$, respectively. Figure 3.23a displays a probability histogram for the binomial distribution with $n = 20$, $p = .6$ [so $\mu = 20(.6) = 12$ and $\sigma = \sqrt{20(.6)(.4)} = 2.19$]. A normal curve with mean value and standard deviation equal to the corresponding values for the binomial distribution has been superimposed on the probability histogram. Although the probability histogram is a bit skewed (because $p \neq .5$), the normal curve gives a very good approximation, especially in the middle part of the picture. The area of any rectangle (probability of any particular X value) except those in the extreme tails can be accurately approximated by the corresponding normal curve area. For example, $P(X = 10) = \binom{20}{10}(.6)^{10}(.4)^{10} = .117$, whereas the area under the normal curve between 9.5 and 10.5 is $P(-1.14 \leq Z \leq -.68) = .120$.

On the other hand, a normal distribution is a poor approximation to a discrete distribution that is heavily skewed. For example, Figure 3.23b shows a probability histogram for the Bin(20, .1) distribution and the normal pdf with the same mean and standard deviation ($\mu = 2$ and $\sigma = 1.34$). Clearly, we would not want to use this normal curve to approximate binomial probabilities, even with a continuity correction.

PROPOSITION

Let X be a binomial rv based on n trials with success probability p. Then if the binomial probability histogram is not too skewed, X has approximately a normal distribution with $\mu = np$ and $\sigma = \sqrt{npq}$. In particular, for $x = a$ possible value of X,

(continued)

$$P(X \leq x) = B(x; n, p) \approx \text{(area under the normal curve to the left of } x + .5)$$

$$= \Phi\left(\frac{x + .5 - np}{\sqrt{npq}}\right)$$

In practice, the approximation is adequate provided that both $np \geq 10$ and $nq \geq 10$.

If either $np < 10$ or $nq < 10$, the binomial distribution may be too skewed for the (symmetric) normal curve to give accurate approximations.

Example 3.25 Suppose that 25% of all licensed drivers in a state do not have insurance. Let X be the number of uninsured drivers in a random sample of size 50 (somewhat perversely, a success is an uninsured driver), so that $p = .25$. Then $\mu = 12.5$ and $\sigma = 3.062$. Since $np = 50(.25) = 12.5 \geq 10$ and $nq = 37.5 \geq 10$, the approximation can safely be applied:

$$P(X \leq 10) = B(10; 50, .25) \approx \Phi\left(\frac{10 + .5 - 12.5}{3.062}\right)$$

$$= \Phi(-.6532) = .2568$$

Similarly, the probability that between 5 and 15 (inclusive) of the selected drivers are uninsured is

$$P(5 \leq X \leq 15) = B(15; 50, .25) - B(4; 50, .25)$$
$$\approx \Phi\left(\frac{15.5 - 12.5}{3.062}\right) - \Phi\left(\frac{4.5 - 12.5}{3.062}\right) = .8319$$

The exact probabilities are .2622 and .8348, respectively, so the approximations are quite good. In the last calculation, the probability $P(5 \leq X \leq 15)$ is being approximated by the area under the normal curve between 4.5 and 15.5—the continuity correction is used for both the upper and lower limits.   ∎

The wide availability of software for doing binomial probability calculations, even for large values of n, has considerably diminished the importance of the normal approximation. However, it is important for another reason. When the objective of an investigation is to make an inference about a population proportion p, interest will focus on the sample proportion of successes $\hat{P} = X/n$ rather than on X itself. Because this proportion is just X multiplied by the constant $1/n$, the earlier rescaling proposition tells us it will also have approximately a normal distribution (with mean $\mu = p$ and standard deviation $\sigma = \sqrt{pq/n}$) provided that both $np \geq 10$ and $nq \geq 10$. This normal approximation is the basis for several inferential procedures to be discussed in Chap. 5.

It is quite difficult to give a direct proof of the validity of this normal approximation (the first one goes back about 270 years to de Moivre). In Chap. 4, we'll see that it is a consequence of an important general result called the Central Limit Theorem.

3.3.6 Normal Distribution Calculations with Software

Many software packages, including Matlab and R, have built-in functions to determine both probabilities under a normal curve and quantiles (aka percentiles) of any given normal distribution. Table 3.1 summarizes the relevant code in both packages.

Table 3.1 Normal probability and quantile calculations in Matlab and R

Function:	cdf	quantile, i.e., the $(100p)$th percentile
Notation:	$\Phi\left(\dfrac{x-\mu}{\sigma}\right)$	$\eta_p = \mu + \Phi^{-1}(p) \cdot \sigma$
Matlab:	$\texttt{cdf('norm'},x,\mu,\sigma)$	$\texttt{icdf('norm'},p,\mu,\sigma)$
R:	$\texttt{pnorm}(x,\mu,\sigma)$	$\texttt{qnorm}(p,\mu,\sigma)$

In the special case of a standard normal distribution, R (but not Matlab) will allow the user to drop the last two arguments, μ and σ. That is, the R commands $\texttt{pnorm(x)}$ and $\texttt{pnorm(x,0,1)}$ yield the same result for any number x, and a similar comment applies to qnorm. Both software packages also have built-in function calls for the normal pdf: $\texttt{pdf('norm'},x,\mu,\sigma)$ and $\texttt{dnorm}(x,\mu,\sigma)$, respectively. However, these two commands are generally only used when one desires to *graph* a normal density curve (x vs. $f(x; \mu, \sigma)$), since the pdf evaluated at particular x does not represent a probability, as discussed in Sect. 3.1.

3.3.7 Exercises: Section 3.3 (39–70)

39. Let Z be a standard normal random variable and obtain each of the following probabilities, drawing pictures wherever appropriate.
    (a) $P(0 \le Z \le 2.17)$
    (b) $P(0 \le Z \le 1)$
    (c) $P(-2.50 \le Z \le 0)$
    (d) $P(-2.50 \le Z \le 2.50)$
    (e) $P(Z \le 1.37)$
    (f) $P(-1.75 \le Z)$
    (g) $P(-1.50 \le Z \le 2.00)$
    (h) $P(1.37 \le Z \le 2.50)$
    (i) $P(1.50 \le Z)$
    (j) $P(|Z| \le 2.50)$
40. In each case, determine the value of the constant c that makes the probability statement correct.
    (a) $\Phi(c) = .9838$
    (b) $P(0 \le Z \le c) = .291$
    (c) $P(c \le Z) = .121$
    (d) $P(-c \le Z \le c) = .668$
    (e) $P(c \le |Z|) = .016$

41. Find the following percentiles for the standard normal distribution. Interpolate where appropriate.
    (a) 91st
    (b) 9th
    (c) 75th
    (d) 25th
    (e) 6th

42. Suppose the force acting on a column that helps to support a building is a normally distributed random variable X with mean value 15.0 kips and standard deviation 1.25 kips. Compute the following probabilities.
    (a) $P(X \leq 15)$
    (b) $P(X \leq 17.5)$
    (c) $P(X \geq 10)$
    (d) $P(14 \leq X \leq 18)$
    (e) $P(|X - 15| \leq 3)$

43. Mopeds (small motorcycles with an engine capacity below 50 cc) are very popular in Europe because of their mobility, ease of operation, and low cost. The article "Procedure to Verify the Maximum Speed of Automatic Transmission Mopeds in Periodic Motor Vehicle Inspections" (*J. of Automobile Engr.*, 2008: 1615-1623) described a rolling bench test for determining maximum vehicle speed. A normal distribution with mean value 46.8 km/h and standard deviation 1.75 km/h is postulated. Consider randomly selecting a single such moped.
    (a) What is the probability that maximum speed is at most 50 km/h?
    (b) What is the probability that maximum speed is at least 48 km/h?
    (c) What is the probability that maximum speed differs from the mean value by at most 1.5 standard deviations?

44. Let X be the birth weight, in grams, of a randomly selected full-term baby. The article "Fetal Growth Parameters and Birth Weight: Their Relationship to Neonatal Body Composition" (*Ultrasound in Obstetrics and Gynecology*, 2009: 441–446) suggests that X is normally distributed with mean 3500 and standard deviation 600.
    (a) Sketch the relevant density curve, including tick marks on the horizontal scale.
    (b) What is $P(3000 < X < 4500)$, and how does this compare to $P(3000 \leq X \leq 4500)$?
    (c) What is the probability that the weight of such a newborn is less than 2500 g?
    (d) What is the probability that the weight of such a newborn exceeds 6000 g (roughly 13.2 lb)?
    (e) How would you characterize the most extreme .1% of all birth weights?
    (f) Use the rescaling proposition from this section to determine the distribution of birth weight expressed in pounds (shape, mean, and standard deviation), and then recalculate the probability from part (c). How does this compare to your previous answer?

45. Based on extensive data from an urban freeway near Toronto, Canada, "it is assumed that free speeds can best be represented by a normal distribution" ("Impact of Driver Compliance on the Safety and Operational Impacts of Freeway Variable Speed Limit Systems," *J. of Transp. Engr.*, 2011: 260–268). The mean and standard deviation reported in the article were 119 km/h and 13.1 km/h, respectively.
    (a) What is the probability that the speed of a randomly selected vehicle is between 100 and 120 km/h?
    (b) What speed characterizes the fastest 10% of all speeds?
    (c) The posted speed limit was 100 km/h. What percentage of vehicles was traveling at speeds exceeding this posted limit?
    (d) If five vehicles are randomly and independently selected, what is the probability that at least one is not exceeding the posted speed limit?
    (e) What is the probability that the speed of a randomly selected vehicle exceeds 70 miles/h?

46. The defect length of a corrosion defect in a pressurized steel pipe is normally distributed with mean value 30 mm and standard deviation 7.8 mm (suggested in the article "Reliability Evaluation of Corroding Pipelines Considering Multiple Failure Modes and Time-Dependent Internal Pressure," *J. of Infrastructure Systems*, 2011: 216–224).
    (a) What is the probability that defect length is at most 20 mm? Less than 20 mm?
    (b) What is the 75th percentile of the defect length distribution, i.e., the value that separates the smallest 75% of all lengths from the largest 25%?
    (c) What is the 15th percentile of the defect length distribution?
    (d) What values separate the middle 80% of the defect length distribution from the smallest 10% and the largest 10%?

47. The plasma cholesterol level (mg/dL) for patients with no prior evidence of heart disease who experience chest pain is normally distributed with mean 200 and standard deviation 35. Consider randomly selecting an individual of this type. What is the probability that the plasma cholesterol level
    (a) Is at most 250?
    (b) Is between 300 and 400?
    (c) Differs from the mean by at least 1.5 standard deviations?

48. Suppose the diameter at breast height (in.) of trees of a certain type is normally distributed with $\mu = 8.8$ and $\sigma = 2.8$, as suggested in the article "Simulating a Harvester-Forwarder Softwood Thinning" (*Forest Products J.,* May 1997: 36–41).
    (a) What is the probability that the diameter of a randomly selected tree will be at least 10 in.? Will exceed 10 in.?
    (b) What is the probability that the diameter of a randomly selected tree will exceed 20 in.?
    (c) What is the probability that the diameter of a randomly selected tree will be between 5 and 10 in.?
    (d) What value c is such that the interval $(8.8 - c, 8.8 + c)$ includes 98% of all diameter values?

(e) If four trees are independently selected, what is the probability that at least one has a diameter exceeding 10 in.?

49. There are two machines available for cutting corks intended for use in wine bottles. The first produces corks with diameters that are normally distributed with mean 3 cm and standard deviation .1 cm. The second machine produces corks with diameters that have a normal distribution with mean 3.04 cm and standard deviation .02 cm. Acceptable corks have diameters between 2.9 and 3.1 cm. Which machine is more likely to produce an acceptable cork?

50. Human body temperatures for healthy individuals have approximately a normal distribution with mean 98.25 °F and standard deviation .75 °F. (The past accepted value of 98.6 °F was obtained by converting the Celsius value of 37°, which is correct to the nearest integer.)
(a) Find the 90th percentile of the distribution.
(b) Find the 5th percentile of the distribution.
(c) What temperature separates the coolest 25% from the others?

51. The article "Monte Carlo Simulation—Tool for Better Understanding of LRFD" (*J. Struct. Engr.*, 1993: 1586–1599) suggests that yield strength (ksi) for A36 grade steel is normally distributed with $\mu = 43$ and $\sigma = 4.5$.
(a) What is the probability that yield strength is at most 40? Greater than 60?
(b) What yield strength value separates the strongest 75% from the others?

52. The automatic opening device of a military cargo parachute has been designed to open when the parachute is 200 m above the ground. Suppose opening altitude actually has a normal distribution with mean value 200 m and standard deviation 30 m. Equipment damage will occur if the parachute opens at an altitude of less than 100 m. What is the probability that there is equipment damage to the payload of at least one of five independently dropped parachutes?

53. The temperature reading from a thermocouple placed in a constant-temperature medium is normally distributed with mean μ, the actual temperature of the medium, and standard deviation σ. What would the value of σ have to be to ensure that 95% of all readings are within .1° of μ?

54. Vehicle speed on a particular bridge in China can be modeled as normally distributed ("Fatigue Reliability Assessment for Long-Span Bridges under Combined Dynamic Loads from Winds and Vehicles," *J. of Bridge Engr.*, 2013: 735–747).
(a) If 5% of all vehicles travel less than 39.12 mph and 10% travel more than 73.24 mph, what are the mean and standard deviation of vehicle speed? [*Note*: The resulting values should agree with those given in the cited article.]
(b) What is the probability that a randomly selected vehicle's speed is between 50 and 65 mph?
(c) What is the probability that a randomly selected vehicle's speed exceeds the speed limit of 70 mph?

55. If adult female heights are normally distributed, what is the probability that the height of a randomly selected woman is
(a) Within 1.5 SDs of its mean value?
(b) Farther than 2.5 SDs from its mean value?

(c) Between 1 and 2 SDs from its mean value?

56. A machine that produces ball bearings has initially been set so that the true average diameter of the bearings it produces is .500 in. A bearing is acceptable if its diameter is within .004 in. of this target value. Suppose, however, that the setting has changed during the course of production, so that the bearings have normally distributed diameters with mean value .499 in. and standard deviation .002 in. What percentage of the bearings produced will not be acceptable?

57. The Rockwell hardness of a metal is determined by impressing a hardened point into the surface of the metal and then measuring the depth of penetration of the point. Suppose the Rockwell hardness of an alloy is normally distributed with mean 70 and standard deviation 3. (Rockwell hardness is measured on a continuous scale.)

(a) If a specimen is acceptable only if its hardness is between 67 and 75, what is the probability that a randomly chosen specimen has an acceptable hardness?

(b) If the acceptable range of hardness is $(70 - c, 70 + c)$, for what value of c would 95% of all specimens have acceptable hardness?

(c) If the acceptable range is as in part (a) and the hardness of each of ten randomly selected specimens is independently determined, what is the expected number of acceptable specimens among the ten?

(d) What is the probability that at most eight of ten independently selected specimens have a hardness of less than 73.84? [*Hint*: $Y =$ the number among the ten specimens with hardness less than 73.84 is a binomial variable; what is p?]

58. The weight distribution of parcels sent in a certain manner is normal with mean value 12 lb and standard deviation 3.5 lb. The parcel service wishes to establish a weight value c beyond which there will be a surcharge. What value of c is such that 99% of all parcels are at least 1 lb under the surcharge weight?

59. Suppose Appendix Table A.3 contained $\Phi(z)$ only for $z \geq 0$. Explain how you could still compute
(a) $P(-1.72 \leq Z \leq -.55)$
(b) $P(-1.72 \leq Z \leq .55)$
Is it necessary to table $\Phi(z)$ for z negative? What property of the standard normal curve justifies your answer?

60. Chebyshev's inequality (Sect. 3.2) states that for any number k satisfying $k \geq 1$, $P(|X - \mu| \geq k\sigma)$ is no more than $1/k^2$. Obtain this probability in the case of a normal distribution for $k = 1, 2$, and 3, and compare to Chebyshev's upper bound.

61. Let X denote the number of flaws along a 100-m reel of magnetic tape (an integer-valued variable). Suppose X has approximately a normal distribution with $\mu = 25$ and $\sigma = 5$. Use the continuity correction to calculate the probability that the number of flaws is
(a) Between 20 and 30, inclusive.
(b) At most 30. Less than 30.

62. Let X have a binomial distribution with parameters $n = 25$ and p. Calculate each of the following probabilities using the normal approximation (with the

continuity correction) for the cases $p = .5, .6$, and $.8$ and compare to the exact probabilities calculated from Appendix Table A.1.

(a) $P(15 \leq X \leq 20)$

(b) $P(X \leq 15)$

(c) $P(20 \leq X)$

63. Suppose that 10% of all steel shafts produced by a process are nonconforming but can be reworked (rather than having to be scrapped). Consider a random sample of 200 shafts, and let X denote the number among these that are nonconforming and can be reworked. What is the (approximate) probability that X is

(a) At most 30?

(b) Less than 30?

(c) Between 15 and 25 (inclusive)?

64. Suppose only 70% of all drivers in a state regularly wear a seat belt. A random sample of 500 drivers is selected. What is the probability that

(a) Between 320 and 370 (inclusive) of the drivers in the sample regularly wear a seat belt?

(b) Fewer than 325 of those in the sample regularly wear a seat belt? Fewer than 315?

65. In response to concerns about nutritional contents of fast foods, McDonald's announced that it would use a new cooking oil for its french fries that would decrease substantially trans fatty acid levels and increase the amount of more beneficial polyunsaturated fat. The company claimed that 97 out of 100 people cannot detect a difference in taste between the new and old oils. Assuming that this figure is correct (as a long-run proportion), what is the approximate probability that in a random sample of 1000 individuals who have purchased fries at McDonald's,

(a) At least 40 can taste the difference between the two oils?

(b) At most 5% can taste the difference between the two oils?

66. The following proof that the normal pdf integrates to 1 comes courtesy of Professor Robert Young, Oberlin College. Let $f(z)$ denote the standard normal pdf, and consider the function of two variables

$$g(x, y) = f(x) \cdot f(y) = \frac{1}{\sqrt{2\pi}} e^{-x^2/2} \frac{1}{\sqrt{2\pi}} e^{-y^2/2} = \frac{1}{2\pi} e^{-(x^2+y^2)/2}$$

Let V denote the volume under $g(x, y)$ above the xy-plane.

(a) Let A denote the area under the standard normal curve. By setting up the double integral for the volume underneath $g(x, y)$, show that $V = A^2$.

(b) Using the rotational symmetry of $g(x, y)$, V can be determined by adding up the volumes of shells from rotation about the y-axis:

$$V = \int_0^\infty 2\pi r \cdot \frac{1}{2\pi} e^{-r^2/2} dr$$

Show this integral equals 1, then use (a) to establish that the area under the standard normal curve is 1.

(c) Show that $\int_{-\infty}^{\infty} f(x; \mu, \sigma)dx = 1$. [*Hint*: Write out the integral, and then make a substitution to reduce it to the standard normal case. Then invoke (b).]

67. Suppose $X \sim N(\mu, \sigma)$.

(a) Show via integration that $E(X) = \mu$. [*Hint*: Make the substitution $u = (x - \mu)/\sigma$, which will create two integrals. For one, use the symmetry of the pdf; for the other, use the fact that the standard normal pdf integrates to 1.]

(b) Show via integration that $Var(X) = \sigma^2$. [*Hint*: Evaluate the integral for $E[(X-\mu)^2]$ rather than using the variance shortcut formula. Use the same substitution as in part (a).]

68. The moment generating function can be used to find the mean and variance of the normal distribution.

(a) Use derivatives of $M_X(t)$ to verify that $E(X) = \mu$ and $Var(X) = \sigma^2$.

(b) Repeat (a) using $L_X(t) = \ln[M_X(t)]$, and compare with part (a) in terms of effort. (Refer back to Exercise 36 for properties of the function $L_X(t)$.)

69. There is no nice formula for the standard normal cdf $\Phi(z)$, but several good approximations have been published in articles. The following is from "Approximations for Hand Calculators Using Small Integer Coefficients" (*Math. Comput.*, 1977: 214–222). For $0 < z \leq 5.5$,

$$P(Z \geq z) = 1 - \Phi(z) \approx .5\exp\left\{-\left[\frac{(83z + 351)z + 562}{(703/z) + 165}\right]\right\}$$

The relative error of this approximation is less than .042%. Use this to calculate approximations to the following probabilities, and compare whenever possible to the probabilities obtained from Appendix Table A.3.

(a) $P(Z \geq 1)$

(b) $P(Z < -3)$

(c) $P(-4 < Z < 4)$

(d) $P(Z > 5)$

70. (a) Use mgfs to show that if X has a normal distribution with parameters μ and σ, then $Y = aX + b$ (a linear function of X) also has a normal distribution. What are the parameters of the distribution of Y [i.e., $E(Y)$ and $SD(Y)$]?

(b) If when measured in °C, temperature is normally distributed with mean 115 and standard deviation 2, what can be said about the distribution of temperature measured in °F?

3.4   The Exponential and Gamma Distributions

The graph of any normal pdf is bell-shaped and thus symmetric. In many practical situations, the variable of interest to the experimenter might have a skewed distribution. A family of pdfs that yields a wide variety of skewed distributional shapes is the gamma family. We first consider a special case, the exponential distribution, and then generalize later in the section.

3.4.1   The Exponential Distribution

The family of exponential distributions provides probability models that are widely used in engineering and science disciplines.

> **DEFINITION**
> X is said to have an **exponential distribution** with parameter λ ($\lambda > 0$) if the pdf of X is
> $$f(x; \lambda) = \begin{cases} \lambda e^{-\lambda x} & x > 0 \\ 0 & \text{otherwise} \end{cases}$$

Some sources write the exponential pdf in the form $(1/\beta)e^{-x/\beta}$, so that $\beta = 1/\lambda$. Graphs of several exponential pdfs appear in Fig. 3.24.

The expected value of an exponentially distributed random variable X is

$$E(X) = \int_0^\infty x \cdot \lambda e^{-\lambda x} dx$$

Obtaining this expected value requires integration by parts. The variance of X can be computed using the shortcut formula $\text{Var}(X) = E(X^2) - [E(X)]^2$; evaluating $E(X^2)$ uses integration by parts twice in succession. In contrast, the exponential cdf is easily obtained by integrating the pdf. The results of these integrations are as follows.

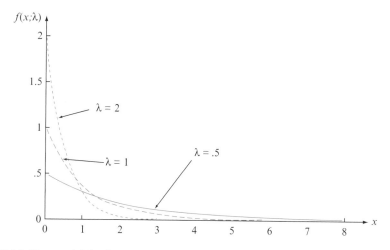

Fig. 3.24 Exponential density curves

PROPOSITION

Let X be an exponential variable with parameter λ. Then the cdf of X is

$$F(x; \lambda) = \begin{cases} 0 & x \leq 0 \\ 1 - e^{-\lambda x} & x > 0 \end{cases}$$

The mean and standard deviation of X are both equal to $1/\lambda$.

Under the alternative parameterization, the exponential cdf becomes $1 - e^{-x/\beta}$ for $x > 0$, and the mean and standard deviation are both equal to β.

Example 3.23 The response time X at an on-line computer terminal (the elapsed time between the end of a user's inquiry and the beginning of the system's response to that inquiry) has an exponential distribution with expected response time equal to 5 s. Then $E(X) = 1/\lambda = 5$, so $\lambda = .2$. The probability that the response time is at most 10 s is

$$P(X \leq 10) = F(10; 2) = 1 - e^{-(.2)(10)} = 1 - e^{-2} = 1 - .135 = .865$$

The probability that response time is between 5 and 10 s is

$$P(5 \leq X \leq 10) = F(10; 2) - F(5; 2) = \left(1 - e^{-2}\right) - \left(1 - e^{-1}\right) = .233 \quad \blacksquare$$

The exponential distribution is frequently used as a model for the distribution of times between the occurrence of successive events, such as customers arriving at a service facility or calls coming in to a call center. The reason for this is that the exponential distribution is closely related to the Poisson distribution introduced in Chap. 2. We will explore this relationship fully in Sect. 7.5 (Poisson Processes).

Another important application of the exponential distribution is to model the distribution of component lifetimes. A partial reason for the popularity of such applications is the "**memoryless**" **property** of the exponential distribution. Suppose component lifetime is exponentially distributed with parameter λ. After putting the component into service, we leave for a period of t_0 hours and then return to find the component still working; what now is the probability that it lasts at least an additional t hours? In symbols, we wish $P(X \geq t + t_0 \mid X \geq t_0)$. By the definition of conditional probability,

$$P\left(X \geq t + t_0 \middle| X \geq t_0\right) = \frac{P[(X \geq t + t_0) \cap (X \geq t_0)]}{P(X \geq t_0)}$$

But the event $X \geq t_0$ in the numerator is redundant, since both events can occur if and only if $X \geq t + t_0$. Therefore,

$$P\left(X \geq t + t_0 \middle| X \geq t_0\right) = \frac{P(X \geq t + t_0)}{P(X \geq t_0)} = \frac{1 - F(t + t_0; \lambda)}{1 - F(t_0; \lambda)} = \frac{e^{-\lambda(t + t_0)}}{e^{-\lambda t_0}} = e^{-\lambda t}$$

This conditional probability is identical to the original probability $P(X \geq t)$ that the component lasted t hours. Thus *the distribution of additional lifetime is exactly the same as the original distribution of lifetime*, so at each point in time the component shows no effect of wear. In other words, the distribution of remaining lifetime is independent of current age (we wish that were true of us!).

Although the memoryless property can be justified at least approximately in many applied problems, in other situations components deteriorate with age or occasionally improve with age (at least up to a certain point). More general lifetime models are then furnished by the gamma, Weibull, and lognormal distributions (the latter two are discussed in the next section). Lifetime distributions are at the heart of *reliability* models, which we'll consider in depth in Sect. 4.8.

3.4.2   The Gamma Distribution

To define the family of gamma distributions, which generalizes the exponential distribution, we first need to introduce a function that plays an important role in many branches of mathematics.

DEFINITION

For $\alpha > 0$, the **gamma function** $\Gamma(\alpha)$ is defined by

$$\Gamma(\alpha) = \int_0^\infty x^{\alpha-1}e^{-x}dx$$

The most important properties of the gamma function are the following:
1. For any $\alpha > 1$, $\Gamma(\alpha) = (\alpha - 1) \cdot \Gamma(\alpha - 1)$ (via integration by parts)
2. For any positive integer n, $\Gamma(n) = (n - 1)!$
3. $\Gamma(\frac{1}{2}) = \sqrt{\pi}$

The following proposition will prove useful for several computations that follow.

PROPOSITION

For any $\alpha, \beta > 0$,

$$\int_0^\infty x^{\alpha-1}e^{-x/\beta}dx = \beta^\alpha\Gamma(\alpha) \tag{3.5}$$

Proof Make the substitution $u = x/\beta$, so that $x = \beta u$ and $dx = \beta\,du$:

$$\int_0^\infty x^{\alpha-1}e^{-x/\beta}dx = \int_0^\infty (\beta u)^{\alpha-1}e^{-u}\beta du = \beta^\alpha \int_0^\infty u^{\alpha-1}e^{-u}du = \beta^\alpha\Gamma(\alpha)$$

The last equality comes from the definition of the gamma function.   ∎

With the preceding proposition in mind, we make the following definition.

DEFINITION

A continuous random variable X is said to have a **gamma distribution** if the pdf of X is

$$f(x; \alpha, \beta) = \begin{cases} \dfrac{1}{\beta^{\alpha}\Gamma(\alpha)}x^{\alpha-1}e^{-x/\beta} & x > 0 \\ \qquad 0 & \text{otherwise} \end{cases} \tag{3.6}$$

where the parameters α and β satisfy $\alpha > 0$, $\beta > 0$. When $\beta = 1$, X is said to have a **standard gamma distribution**, and its pdf may be denoted $f(x; \alpha)$.

The exponential distribution results from taking $\alpha = 1$ and $\beta = 1/\lambda$.

It's clear that $f(x; \alpha, \beta) \geq 0$ for all x; the previous proposition guarantees that this function integrates to 1, as required. Figure 3.25a illustrates the graphs of the gamma pdf for several (α, β) pairs, whereas Fig. 3.25b presents graphs of the standard gamma pdf. For the standard pdf, when $\alpha \leq 1$, $f(x; \alpha)$ is strictly decreasing as x increases; when $\alpha > 1$, $f(x; \alpha)$ rises to a maximum and then decreases. Because of this difference, α is referred to as a *shape parameter*. The parameter β in Eq. (3.6) is called the *scale parameter* because values other than 1 either stretch or compress the pdf in the x direction.

The mean and variance of a gamma distribution are

$$E(X) = \mu = \alpha\beta \qquad \text{Var}(X) = \sigma^2 = \alpha\beta^2$$

These can be calculated directly from the gamma pdf using integration by parts, or by employing properties of the gamma function along with Expression (3.5); see Exercise 83. Notice these are consistent with the aforementioned mean and variance of the exponential distribution: with $\alpha = 1$ and $\beta = 1/\lambda$ we obtain $E(X) = 1(1/\lambda) = 1/\lambda$ and $\text{Var}(X) = 1(1/\lambda)^2 = 1/\lambda^2$.

In the special case where the shape parameter α is a positive integer, n, the gamma distribution is sometimes rewritten with the substitution $\lambda = 1/\beta$, and the resulting pdf is

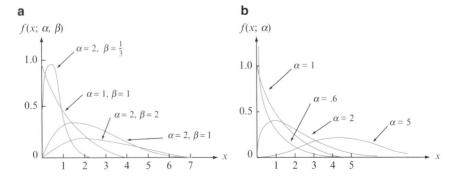

Fig. 3.25 (**a**) Gamma density curves; (**b**) standard gamma density curves

$$f(x; n, 1/\lambda) = \frac{\lambda^n}{(n-1)!} x^{n-1} e^{-\lambda x}, \qquad x > 0$$

This is often called an *Erlang distribution*, and it plays a central role in the study of Poisson processes (again, see Sect. 7.5; notice that the $n = 1$ case of the Erlang distribution is actually the exponential pdf). In Chap. 4, it will be shown that the sum of n independent exponential rvs follows this Erlang distribution.

When X is a standard gamma rv, the cdf of X, which for $x > 0$ is

$$G(x; \alpha) = P(X \leq x) = \int_0^x \frac{1}{\Gamma(\alpha)} y^{\alpha-1} e^{-y} dy \qquad (3.7)$$

is called the **incomplete gamma function** (in mathematics literature, the incomplete gamma function sometimes refers to Eq. (3.7) without the denominator $\Gamma(\alpha)$ in the integrand]. In Appendix Table A.4, we present a small tabulation of $G(x; \alpha)$ for $\alpha = 1, 2, \ldots, 10$ and $x = 1, 2, \ldots, 15$. Table 3.2 at the end of this section provides the Matlab and R commands related to the gamma cdf, which are illustrated in the following examples.

Example 3.27 Suppose the reaction time X (in seconds) of a randomly selected individual to a certain stimulus has a standard gamma distribution with $\alpha = 2$. Since X is continuous,

$$P(3 \leq X \leq 5) = P(X \leq 5) - P(X \leq 3) = G(5; 2) - G(3; 2) = .960 - .801 = .159$$

This probability can be obtained in Matlab with cdf('gamma',5,2,1)-cdf('gamma',3,2,1) and in R with pgamma(5,2)-pgamma(3,2).

The probability that the reaction time is more than 4 s is

$$P(X > 4) = 1 - P(X \leq 4) = 1 - G(4; 2) = 1 - .908 = .092 \qquad \blacksquare$$

The incomplete gamma function can also be used to compute probabilities involving nonstandard gamma distributions.

PROPOSITION

Let X have a gamma distribution with parameters α and β. Then for any $x > 0$, the cdf of X is given by

$$P(X \leq x) = G\left(\frac{x}{\beta}; \alpha\right),$$

the incomplete gamma function evaluated at x/β.

The proof is similar to that of Eq. (3.5).

Example 3.28 Suppose the survival time X in weeks of a randomly selected male mouse exposed to 240 rads of gamma radiation has, rather fittingly, a gamma distribution with $\alpha = 8$ and $\beta = 15$. (Data in *Survival Distributions: Reliability Applications in the Biomedical Services*, by A. J. Gross and V. Clark, suggest $\alpha \approx 8.5$ and $\beta \approx 13.3$.) The expected survival time is $E(X) = (8)(15) = 120$ weeks, whereas $\mathrm{SD}(X) = \sqrt{(8)(15)^2} = \sqrt{1800} = 42.43$ weeks. The probability that a mouse survives between 60 and 120 weeks is

$$
\begin{aligned}
P(60 \le X \le 120) &= P(X \le 120) - P(X \le 60) \\
&= G(120/15; 8) - G(60/15; 8) \\
&= G(8; 8) - G(4; 8) = .547 - .051 = .496
\end{aligned}
$$

In Matlab, the command `cdf('gamma',120,8,15)-cdf('gamma', 60,8,15)` yields the desired probability; the corresponding R code is `pgamma (120,8,1/15)-pgamma(60,8,1/15)`.

The probability that a mouse survives at least 30 weeks is

$$
P(X \ge 30) = 1 - P(X < 30) = 1 - P(X \le 30) = 1 - G(30/15; 8) = .999 \quad \blacksquare
$$

3.4.3 The Gamma MGF

The integral proposition earlier in this section made it easy to determine the mean and variance of a gamma rv. However, the moment generating function of the gamma distribution — and, as a special case, of the exponential model — will prove critical in establishing some of the more advanced properties of these distributions in Chap. 4.

> **PROPOSITION**
> The moment generating function of a gamma random variable is
> $$
> M_X(t) = \frac{1}{(1 - \beta t)^\alpha} \qquad t < 1/\beta
> $$

Proof By definition, the mgf is

$$
M_X(t) = E\left(e^{tX}\right) = \int_0^\infty e^{tx} \frac{x^{\alpha-1}}{\Gamma(\alpha)\beta^\alpha} e^{-x/\beta} dx = \frac{1}{\Gamma(\alpha)\beta^\alpha} \int_0^\infty x^{\alpha-1} e^{-(-t+1/\beta)x} dx
$$

Now use Expression (3.5): provided $-t + 1/\beta > 0$, i.e., $t < 1/\beta$,

$$
\frac{1}{\Gamma(\alpha)\beta^\alpha} \int_0^\infty x^{\alpha-1} e^{-(-t+1/\beta)x} dx = \frac{1}{\Gamma(\alpha)\beta^\alpha} \cdot \Gamma(\alpha) \left(\frac{1}{-t+1/\beta}\right)^\alpha = \frac{1}{(1-\beta t)^\alpha} \quad \blacksquare
$$

The exponential mgf can then be determined with the substitution $\alpha = 1$, $\beta = 1/\lambda$:

$$M_X(t) = \frac{1}{(1 - (1/\lambda)t)^1} = \frac{\lambda}{\lambda - t} \qquad t < \lambda$$

3.4.4 Gamma and Exponential Calculations with Software

Table 3.2 summarizes the syntax for gamma and exponential probability calculations in Matlab and R, which follows the pattern of the other distributions. In a sense, the exponential commands are redundant, since they are just a special case ($\alpha = 1$) of the gamma distribution.

Notice that Matlab and R parameterize the distributions differently: in Matlab, both the gamma and exponential functions require β (that is, $1/\lambda$) as the last input, whereas the R functions take as their last input the "rate" parameter $\lambda = 1/\beta$. So, for the gamma rv with parameters $\alpha = 8$ and $\beta = 15$ from Example 3.28, the probability $P(X \leq 30)$ would be evaluated as cdf('gamma',30,8,15) in Matlab but pgamma(30,8,1/15) in R. This inconsistency of gamma inputs can be remedied by using a name assignment in the last argument in R; specifically, pgamma(30,8,scale=15) will instruct R to use $\beta = 15$ in its gamma probability calculation and produce the same answer as the previous expressions. Interestingly, as of this writing the same option does not exist in the pexp function.

To graph gamma or exponential distributions, one can request their pdfs by replacing cdf with pdf (in Matlab) or the leading letter p with d (in R). To find quantiles of either of these distributions, the appropriate replacements are icdf and q, respectively. For example, the 75th percentile of the gamma distribution from Example 3.28 can be determined with icdf('gamma',.75,8,15) in Matlab or qgamma(.75,8,scale=15) in R (both give 145.2665 weeks).

Table 3.2 Matlab and R code for gamma and exponential calculations

	Gamma	Exponential
Function:	cdf	cdf
Notation:	$G(x/\beta; \alpha)$	$F(x; \lambda) = 1 - e^{-\lambda x}$
Matlab:	cdf('gamma',x,α,β)	cdf('exp',x,1/λ)
R:	pgamma(x,α,1/β)	pexp(x,λ)

3.4.5 Exercises: Section 3.4 (71–83)

71. Let $X =$ the time between two successive arrivals at the drive-up window of a local bank. If X has an exponential distribution with $\lambda = 1$, compute the following:
    (a) The expected time between two successive arrivals
    (b) The standard deviation of the time between successive arrivals
    (c) $P(X \leq 4)$
    (d) $P(2 \leq X \leq 5)$

72. Let X denote the distance (m) that an animal moves from its birth site to the first territorial vacancy it encounters. Suppose that for banner-tailed kangaroo rats, X has an exponential distribution with parameter $\lambda = .01386$ (as suggested in the article "Competition and Dispersal from Multiple Nests," *Ecology*, 1997: 873–883).
    (a) What is the probability that the distance is at most 100 m? At most 200 m? Between 100 and 200 m?
    (b) What is the probability that distance exceeds the mean distance by more than 2 standard deviations?
    (c) What is the value of the median distance?

73. In studies of anticancer drugs it was found that if mice are injected with cancer cells, the survival time can be modeled with the exponential distribution. Without treatment the expected survival time was 10 h. What is the probability that
    (a) A randomly selected mouse will survive at least 8 h? At most 12 h? Between 8 and 12 h?
    (b) The survival time of a mouse exceeds the mean value by more than 2 standard deviations? More than 3 standard deviations?

74. Data collected at Toronto Pearson International Airport suggests that an exponential distribution with mean value 2.725 h is a good model for rainfall duration (*Urban Stormwater Management Planning with Analytical Probabilistic Models*, 2000, p.69).
    (a) What is the probability that the duration of a particular rainfall event at this location is at least 2 h? At most 3 h? Between 2 and 3 h?
    (b) What is the probability that rainfall duration exceeds the mean value by more than 2 standard deviations? What is the probability that it is less than the mean value by more than one standard deviation?

75. Evaluate the following:
    (a) $\Gamma(6)$
    (b) $\Gamma(5/2)$
    (c) $G(4; 5)$ (the incomplete gamma function)
    (d) $G(5; 4)$
    (e) $G(0; 4)$

76. Let X have a standard gamma distribution with $\alpha = 7$. Evaluate the following:
    (a) $P(X \le 5)$
    (b) $P(X < 5)$
    (c) $P(X > 8)$
    (d) $P(3 \le X \le 8)$
    (e) $P(3 < X < 8)$
    (f) $P(X < 4 \text{ or } X > 6)$

77. Suppose that when a type of transistor is subjected to an accelerated life test, the lifetime X (in weeks) has a gamma distribution with mean 24 weeks and standard deviation 12 weeks.
    (a) What is the probability that a transistor will last between 12 and 24 weeks?
    (b) What is the probability that a transistor will last at most 24 weeks? Is the median of the lifetime distribution less than 24? Why or why not?

(c) What is the 99th percentile of the lifetime distribution?

(d) Suppose the test will actually be terminated after t weeks. What value of t is such that only .5% of all transistors would still be operating at termination?

78. The two-parameter gamma distribution can be generalized by introducing a third parameter γ, called a *threshold* or *location* parameter: replace x in Eq. (3.6) by $x - \gamma$ and $x \geq 0$ by $x \geq \gamma$. This amounts to shifting the density curves in Fig. 3.25 so that they begin their ascent or descent at γ rather than 0. The article "Bivariate Flood Frequency Analysis with Historical Information Based on Copulas" (*J. of Hydrologic Engr.*, 2013: 1018–1030) employs this distribution to model $X = $ 3-day flood volume (10^8 m^3). Suppose that values of the parameters are $\alpha = 12, \beta = 7, \gamma = 40$ (very close to estimates in the cited article based on past data).

(a) What are the mean value and standard deviation of X?

(b) What is the probability that flood volume is between 100 and 150?

(c) What is the probability that flood volume exceeds its mean value by more than one standard deviation?

(d) What is the 95th percentile of the flood volume distribution?

79. If X has an exponential distribution with parameter λ, derive a general expression for the $(100p)$th percentile of the distribution. Then specialize to obtain the median.

80. A system consists of five identical components connected in series as shown:

As soon as one component fails, the entire system will fail. Suppose each component has a lifetime that is exponentially distributed with $\lambda = .01$ and that components fail independently of one another. Define events $A_i = \{i$th component lasts at least t hours$\}$, $i = 1, \ldots, 5$, so that the A_is are independent events. Let $X = $ the time at which the system fails—that is, the shortest (minimum) lifetime among the five components.

(a) The event $\{X \geq t\}$ is equivalent to what event involving $A_1, \ldots, A_5$?

(b) Using the independence of the five A_is, compute $P(X \geq t)$. Then obtain $F(t) = P(X \leq t)$ and the pdf of X. What type of distribution does X have?

(c) Suppose there are n components, each having exponential lifetime with parameter λ. What type of distribution does X have?

81. Based on an analysis of sample data, the article "Pedestrians' Crossing Behaviors and Safety at Unmarked Roadways in China" (*Accident Analysis and Prevention*, 2011: 1927–1936) proposed the pdf $f(x) = .15e^{-.15(x-1)}$ when $x \geq 1$ as a model for the distribution of $X = $ time (sec) spent at the median line. This is an example of a *shifted exponential* distribution, i.e., an exponential model beginning at an x-value other than 0.

(a) What is the probability that waiting time is at most 5 s? More than 5 s?

(b) What is the probability that waiting time is between 2 and 5 s?

(c) What is the mean waiting time?

(d) What is the standard deviation of waiting times?
[*Hint*: For (c) and (d), you can either use integration or write $X = Y + 1$, where Y has an exponential distribution with parameter $\lambda = .15$. Then, apply rescaling properties of mean and standard deviation.]

82. The *double exponential distribution* has pdf

$$f(x) = .5\lambda e^{-\lambda|x|} \quad \text{for } -\infty < x < \infty$$

The article "Microwave Observations of Daily Antarctic Sea-Ice Edge Expansion and Contribution Rates" (*IEEE Geosci. and Remote Sensing Letters*, 2006: 54-58) states that "the distribution of the daily sea-ice advance/retreat from each sensor is similar and is approximately double exponential." The standard deviation is given as 40.9 km.

(a) What is the mean of the double exponential distribution? [*Hint:* Draw a picture of the density curve.]
(b) What is the value of the parameter λ?
(c) What is the probability that the extent of daily sea-ice change is within 1 standard deviation of the mean value?

83. (a) Find the mean and variance of the gamma distribution using integration and Expression (3.5) to obtain $E(X)$ and $E(X^2)$.
(b) Use the gamma mgf to find the mean and variance.

3.5   Other Continuous Distributions

The normal, gamma (including exponential), and uniform families of distributions provide a wide variety of probability models for continuous variables, but there are many practical situations in which no member of these families fits a set of observed data very well. Statisticians and other investigators have developed other families of distributions that are often appropriate in practice.

3.5.1 The Weibull Distribution

The family of Weibull distributions was introduced by the Swedish physicist Waloddi Weibull in 1939; his 1951 article "A Statistical Distribution Function of Wide Applicability" (*J. Appl. Mech.*, 18: 293–297) discusses a number of applications.

DEFINITION

A random variable X is said to have a **Weibull distribution** with parameters α and β ($\alpha > 0$, $\beta > 0$) if the pdf of X is

$$f(x; \alpha, \beta) = \begin{cases} \dfrac{\alpha}{\beta^\alpha} x^{\alpha-1} e^{-(x/\beta)^\alpha} & x \geq 0 \\ 0 & x < 0 \end{cases} \tag{3.8}$$

In some situations there are theoretical justifications for the appropriateness of the Weibull distribution, but in many applications $f(x;\ \alpha,\ \beta)$ simply provides a good fit to observed data for particular values of α and β. When $\alpha = 1$, the pdf reduces to the exponential distribution (with $\lambda = 1/\beta$), so the exponential distribution is a special case of both the gamma and Weibull distributions. However, there are gamma distributions that are not Weibull distributions and vice versa, so one family is not a subset of the other. Both α and β can be varied to obtain a number of different distributional shapes, as illustrated in Fig. 3.26. Note that β is a scale parameter, so different values stretch or compress the graph in the x-direction; α is referred to as a shape parameter. Integrating to obtain $E(X)$ and $E(X^2)$ yields

$$\mu = \beta\Gamma\left(1 + \frac{1}{\alpha}\right) \qquad \sigma^2 = \beta^2\left\{\Gamma\left(1 + \frac{2}{\alpha}\right) - \left[\Gamma\left(1 + \frac{1}{\alpha}\right)\right]^2\right\}$$

The computation of μ and σ^2 thus necessitate using the gamma function from Sect. 3.4. (The moment generating function of the Weibull distribution is very complicated, and so we do not include it here.) On the other hand, the integration $\int_0^x f(y;\alpha,\beta)dy$ is easily carried out to obtain the cdf of X:

$$F(x;\alpha,\beta) = \begin{cases} 0 & x < 0 \\ 1 - e^{-(x/\beta)^\alpha} & x \geq 0 \end{cases} \qquad (3.9)$$

Example 3.29 In recent years the Weibull distribution has been used to model engine emissions of various pollutants. Let X denote the amount of NO_x emission (g/gal) from a randomly selected four-stroke engine of a certain type, and suppose that X has a Weibull distribution with $\alpha = 2$ and $\beta = 10$ (suggested by information in the article "Quantification of Variability and Uncertainty in Lawn and Garden Equipment NO_x and Total Hydrocarbon Emission Factors," *J. Air Waste Manag. Assoc.*, 2002: 435–448). The corresponding density curve looks exactly like the one in Fig. 3.26 for $\alpha = 2$, $\beta = 1$ except that now the values 50 and 100 replace 5 and 10 on the horizontal axis (because β is a "scale parameter"). Then

$$P(X \leq 10) = F(10; 2, 10) = 1 - e^{-(10/10)^2} = 1 - e^{-1} = .632$$

Similarly, $P(X \leq 25) = .998$, so the distribution is almost entirely concentrated on values between 0 g/gal and 25 g/gal. The value c which separates the 5% of all engines having the largest amounts of NO_x emissions from the remaining 95%, satisfies

$$.95 = F(c; 2, 10) = 1 - e^{-(c/10)^2}$$

Isolating the exponential term on one side, taking logarithms, and solving the resulting equation gives $c \approx 17.3$ g/gal as the 95th percentile of the emission distribution. ∎

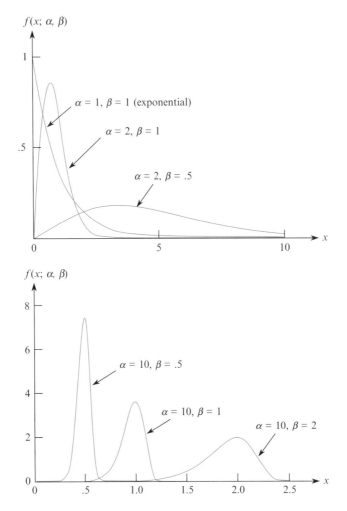

Fig. 3.26 Weibull density curves

Frequently, in practical situations, a Weibull model may be reasonable except that the smallest possible X value may be some value γ not assumed to be zero (this would also apply to a gamma model). The quantity γ can then be regarded as a third parameter of the distribution, which is what Weibull did in his original work. For, say, $\gamma = 3$, all curves in Fig. 3.26 would be shifted 3 units to the right. This is equivalent to saying that $X - \gamma$ has the pdf Eq. (3.8), so that the cdf of X is obtained by replacing x in Eq. (3.9) by $x - \gamma$.

Example 3.30 An understanding of the volumetric properties of asphalt is important in designing mixtures that will result in high-durability pavement. The article "Is a Normal Distribution the Most Appropriate Statistical Distribution for

Volumetric Properties in Asphalt Mixtures" (*J. of Testing and Evaluation*, Sept. 2009: 1–11) used the analysis of some sample data to recommend that for a particular mixture, $X =$ air void volume (%) be modeled with a three-parameter Weibull distribution. Suppose the values of the parameters are $\gamma = 4$, $\alpha = 1.3$, and $\beta = .8$, which are quite close to estimates given in the article.

For $x \geq 4$, the cumulative distribution function is

$$F(x; \alpha, \beta, \gamma) = F(x; 1.3, .8, 4) = 1 - e^{-[(x-4)/.8]^{1.3}}$$

The probability that the air void volume of a specimen is between 5% and 6% is

$$P(5 \leq X \leq 6) = F\left(6; 1.3, .8, 4\right) - F\left(5; 1.3, .8, 4\right) = e^{-[(5-4)/.8]^{1.3}} - e^{-[(6-4)/.8]^{1.3}}$$
$$= .263 - .037 = .226$$

Figure 3.27 shows a graph of the corresponding Weibull density function, in which the shaded area corresponds to the probability just calculated.

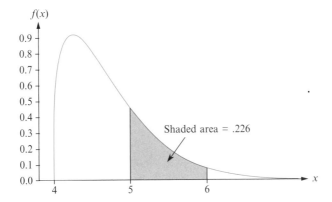

Fig. 3.27 Weibull density curve with threshold $= 4$, shape $= 1.3$, scale $= .8$ ∎

3.5.2 The Lognormal Distribution

Lognormal distributions have been used extensively in engineering, medicine, and more recently, finance.

DEFINITION

A nonnegative rv X is said to have a **lognormal distribution** if the rv $Y = \ln(X)$ has a normal distribution. The resulting pdf of a lognormal rv when $\ln(X)$ is normally distributed with parameters μ and σ is

$$f(x; \mu, \sigma) = \begin{cases} \dfrac{1}{\sqrt{2\pi}} \sigma x e^{-[\ln(x)-\mu]^2/(2\sigma^2)} & x \geq 0 \\ 0 & x < 0 \end{cases}$$

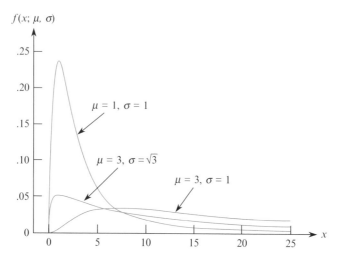

Fig. 3.28 Lognormal density curves

Be careful here: the parameters μ and σ are not the mean and standard deviation of X but of $\ln(X)$. The mean and variance of a lognormal random variable can be shown to be

$$E(X) = e^{\mu + \sigma^2/2} \qquad \text{Var}(X) = e^{2\mu + \sigma^2} \cdot \left(e^{\sigma^2} - 1 \right)$$

In Chap. 4, we will present a theoretical justification for this distribution in connection with the Central Limit Theorem, but as with other distributions, the lognormal can be used as a model even in the absence of such justification. Figure 3.28 illustrates graphs of the lognormal pdf; although a normal curve is symmetric, a lognormal curve has a positive skew.

Because $\ln(X)$ has a normal distribution, the cdf of X can be expressed in terms of the cdf $\Phi(z)$ of a standard normal rv Z. For $x \geq 0$,

$$F(x; \mu, \sigma) = P(X \leq x) = P[\ln(X) \leq \ln(x)] = P\left[\frac{\ln(X) - \mu}{\sigma} \leq \frac{\ln(x) - \mu}{\sigma} \right]$$

$$= P\left[Z \leq \frac{\ln(x) - \mu}{\sigma} \right] = \Phi\left[\frac{\ln(x) - \mu}{\sigma} \right] \tag{3.10}$$

Differentiating $F(x; \mu, \sigma)$ with respect to x gives the pdf $f(x; \mu, \sigma)$ above.

Example 3.31 According to the article "Predictive Model for Pitting Corrosion in Buried Oil and Gas Pipelines" (*Corrosion*, 2009: 332–342), the lognormal distribution has been reported as the best option for describing the distribution of maximum pit depth data from cast iron pipes in soil. The authors suggest that a lognormal distribution with $\mu = .353$ and $\sigma = .754$ is appropriate for maximum pit depth

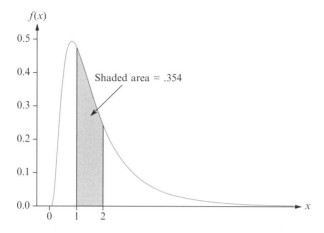

Fig. 3.29 Lognormal density curve with $\mu = .353$ and $\sigma = .754$

(mm) of buried pipelines. For this distribution, the mean value and variance of pit depth are

$$E(X) = e^{.353+(.754)^2/2} = e^{.6383} = 1.893$$
$$\text{Var}(X) = e^{2(.353)+(.754)^2} \cdot \left(e^{(.754)^2} - 1\right) = (3.57697)(.765645) = 2.7387$$

The probability that maximum pit depth is between 1 and 2 mm is
$$P(1 \leq X \leq 2) = P\left(\ln(1) \leq \ln(X) \leq \ln(2)\right) = P(0 \leq \ln(X) \leq .693)$$
$$= P\left(\frac{0 - .353}{.754} \leq Z \leq \frac{.693 - .353}{.754}\right) = \Phi(.45) - \Phi(-.47) = .354$$

This probability is illustrated in Fig. 3.29.

What value c is such that only 1% of all specimens have a maximum pit depth exceeding c? The desired value satisfies

$$.99 = P(X \leq c) = \Phi\left(\frac{\ln(c) - .353}{.754}\right)$$

Appendix Table A.3 indicates that $z = 2.33$ is the 99th percentile of the standard normal distribution, which implies that

$$\frac{\ln(c) - .353}{.754} = 2.33$$

Solving for c gives $\ln(c) = 2.1098$ and $c = 8.247$. Thus 8.247 mm is the 99th percentile of the maximum pit depth distribution. ∎

As with the Weibull distribution, a third parameter γ can be introduced so that the domain of the distribution is $x > \gamma$ rather than $x > 0$.

3.5.3   The Beta Distribution

All families of continuous distributions discussed so far except for the uniform distribution have positive density over an infinite interval (although typically the density function decreases rapidly to zero beyond a few standard deviations from the mean). The beta distribution provides positive density only for X in an interval of finite length.

DEFINITION

A random variable X is said to have a **beta distribution** with parameters α, β (both positive), A, and B if the pdf of X is

$$f(x; \alpha, \beta, A, B) = \begin{cases} \dfrac{1}{B-A} \cdot \dfrac{\Gamma(\alpha+\beta)}{\Gamma(\alpha) \cdot \Gamma(\beta)} \left(\dfrac{x-A}{B-A}\right)^{\alpha-1} \left(\dfrac{B-x}{B-A}\right)^{\beta-1} & A \leq x \leq B \\ \\ 0 & \text{otherwise} \end{cases}$$

The case $A = 0, B = 1$ gives the **standard beta distribution**.

Figure 3.30 illustrates several standard beta pdfs. Graphs of the general pdf are similar, except they are shifted and then stretched or compressed to fit over $[A, B]$. Unless α and β are integers, integration of the pdf to calculate probabilities is difficult, so either a table of the incomplete beta function or software is generally used.

The standard beta distribution is commonly used to model variation in the proportion or percentage of a quantity occurring in different samples, such as the proportion of a 24-h day that an individual is asleep or the proportion of a certain element in a chemical compound.

Fig. 3.30 Standard beta density curves

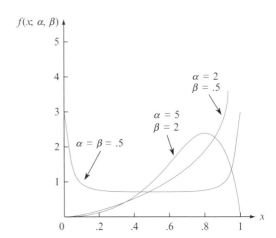

The mean and variance of X are

$$\mu = A + (B - A) \cdot \frac{\alpha}{\alpha + \beta} \qquad \sigma^2 = \frac{(B - A)^2 \alpha \beta}{(\alpha + \beta)^2 (\alpha + \beta + 1)}$$

The moment generating function of the beta distribution is too complicated to be useful.

Example 3.32 Project managers often use a method labeled PERT—for program evaluation and review technique—to coordinate the various activities making up a large project. (One successful application was in the construction of the *Apollo* spacecraft.) A standard assumption in PERT analysis is that the time necessary to complete any particular activity once it has been started has a beta distribution with $A =$ the optimistic time (if everything goes well) and $B =$ the pessimistic time (if everything goes badly). Suppose that in constructing a single-family house, the time X (in days) necessary for laying the foundation has a beta distribution with $A = 2$, $B = 5$, $\alpha = 2$, and $\beta = 3$. Then $\alpha/(\alpha + \beta) = .4$, so $E(X) = 2 + (3)(.4) = 3.2$. For these values of α and β, the pdf of X is a simple polynomial function. The probability that it takes at most 3 days to lay the foundation is

$$P(X \le 3) = \int_2^3 \frac{1}{3} \cdot \frac{4!}{1! \cdot 2!} \left(\frac{x - 2}{3}\right) \left(\frac{5 - x}{3}\right)^2 dx$$

$$= \frac{4}{27} \int_2^3 (x - 2)(5 - x)^2 dx = \frac{4}{27} \cdot \frac{11}{4} = \frac{11}{27} = .407 \qquad \blacksquare$$

Software, including Matlab and R, can be used to perform probability calculations for the Weibull, lognormal, and beta distributions. Interested readers should consult the help menus in those packages.

3.5.4   Exercises: Section 3.5 (84–100)

84. The lifetime X (in hundreds of hours) of a type of vacuum tube has a Weibull distribution with parameters $\alpha = 2$ and $\beta = 3$. Compute the following:
    (a) $E(X)$ and $\text{Var}(X)$
    (b) $P(X \le 6)$
    (c) $P(1.5 \le X \le 6)$
    (This Weibull distribution is suggested as a model for time in service in "On the Assessment of Equipment Reliability: Trading Data Collection Costs for Precision," *J. Engrg. Manuf.*, 1991: 105–109.)

85. The authors of the article "A Probabilistic Insulation Life Model for Combined Thermal-Electrical Stresses" (*IEEE Trans. Electr. Insul.*, 1985: 519–522) state that "the Weibull distribution is widely used in statistical problems relating to aging of solid insulating materials subjected to aging and stress." They propose the use of the distribution as a model for time (in hours) to

failure of solid insulating specimens subjected to ac voltage. The values of the parameters depend on the voltage and temperature; suppose $\alpha = 2.5$ and $\beta = 200$ (values suggested by data in the article).

(a) What is the probability that a specimen's lifetime is at most 250? Less than 250? More than 300?

(b) What is the probability that a specimen's lifetime is between 100 and 250?

(c) What value is such that exactly 50% of all specimens have lifetimes exceeding that value?

86. Let $X =$ the time (in 10^{-1} weeks) from shipment of a defective product until the customer returns the product. Suppose that the minimum return time is $\gamma = 3.5$ and that the excess $X = 3.5$ over the minimum has a Weibull distribution with parameters $\alpha = 2$ and $\beta = 1.5$ (see the article "Practical Applications of the Weibull Distribution," *Indust. Qual. Control*, 1964: 71–78).

(a) What is the cdf of X?

(b) What are the expected return time and variance of return time? [*Hint*: First obtain $E(X - 3.5)$ and $\mathrm{Var}(X - 3.5)$.]

(c) Compute $P(X > 5)$.

(d) Compute $P(5 \le X \le 8)$.

87. Let X have a Weibull distribution with the pdf from Expression (3.8). Verify that $\mu = \beta \Gamma(1 + 1/\alpha)$. [*Hint*: In the integral for $E(X)$, make the change of variable $y = (x/\beta)^\alpha$, so that $x = \beta y^{1/\alpha}$.]

88. (a) In Exercise 84, what is the median lifetime of such tubes? [*Hint*: Use Expression (3.9).]

(b) In Exercise 86, what is the median return time?

(c) If X has a Weibull distribution with the cdf from Expression (3.9), obtain a general expression for the $(100p)$th percentile of the distribution.

(d) In Exercise 86, the company wants to refuse to accept returns after t weeks. For what value of t will only 10% of all returns be refused?

89. Let X denote the ultimate tensile strength (ksi) at $-200°$ of a randomly selected steel specimen of a certain type that exhibits "cold brittleness" at low temperatures. Suppose that X has a Weibull distribution with $\alpha = 20$ and $\beta = 100$.

(a) What is the probability that X is at most 105 ksi?

(b) If specimen after specimen is selected, what is the long-run proportion having strength values between 100 and 105 ksi?

(c) What is the median of the strength distribution?

90. The article "On Assessing the Accuracy of Offshore Wind Turbine Reliability-Based Design Loads from the Environmental Contour Method" (*Intl. J. of Offshore and Polar Engr.*, 2005: 132–140) proposes the Weibull distribution with $\alpha = 1.817$ and $\beta = .863$ as a model for 1-h significant wave height (m) at a certain site.

(a) What is the probability that wave height is at most .5 m?

(b) What is the probability that wave height exceeds its mean value by more than one standard deviation?

(c) What is the median of the wave-height distribution?

(d) For $0 < p < 1$, give a general expression for the $100p$th percentile of the wave-height distribution.

91. Nonpoint source loads are chemical masses that travel to the main stem of a river and its tributaries in flows that are distributed over relatively long stream reaches, in contrast to those that enter at well-defined and regulated points. The article "Assessing Uncertainty in Mass Balance Calculation of River Nonpoint Source Loads" (*J. of Envir. Engr.*, 2008: 247–258) suggested that for a certain time period and location, nonpoint source load of total dissolved solids could be modeled with a lognormal distribution having mean value 10,281 kg/day/km and a coefficient of variation $CV = .40$ ($CV = \sigma_X/\mu_X$).
    (a) What are the mean value and standard deviation of $\ln(X)$?
    (b) What is the probability that X is at most 15,000 kg/day/km?
    (c) What is the probability that X exceeds its mean value, and why is this probability not .5?
    (d) Is 17,000 the 95th percentile of the distribution?

92. The authors of the article "Study on the Life Distribution of Microdrills" (*J. of Engr. Manufacture*, 2002: 301-305) suggested that a reasonable probability model for drill lifetime was a lognormal distribution with $\mu = 4.5$ and $\sigma = .8$.
    (a) What are the mean value and standard deviation of lifetime?
    (b) What is the probability that lifetime is at most 100?
    (c) What is the probability that lifetime is at least 200? Greater than 200?

93. Use Equation (3.10) to write a formula for the median η of the lognormal distribution. What is the median for the load distribution of Exercise 91?

94. As in the case of the Weibull distribution, the lognormal distribution can be modified by the introduction of a third parameter γ such that the pdf is shifted to be positive only for $x > \gamma$. The article cited in Exercise 46 suggested that a shifted lognormal distribution with shift $= 1.0$, mean value $= 2.16$, and standard deviation $= 1.03$ would be an appropriate model for the rv $X = maximum-to-average$ depth ratio of a corrosion defect in pressurized steel.
    (a) What are the values of μ and σ for the proposed distribution?
    (b) What is the probability that depth ratio exceeds 2?
    (c) What is the median of the depth ratio distribution?
    (d) What is the 99th percentile of the depth ratio distribution?

95. Sales delay is the elapsed time between the manufacture of a product and its sale. According to the article "Warranty Claims Data Analysis Considering Sales Delay" (*Quality and Reliability Engr. Intl.*, 2013: 113–123), it is quite common for investigators to model sales delay using a lognormal distribution. For a particular product, the cited article proposes this distribution with parameter values $\mu = 2.05$ and $\sigma^2 = .06$ (here the unit for delay is months).
    (a) What are the variance and standard deviation of delay time?
    (b) What is the probability that delay time exceeds 12 months?
    (c) What is the probability that delay time is within one standard deviation of its mean value?
    (d) What is the median of the delay time distribution?
    (e) What is the 99th percentile of the delay time distribution?

   (f) Among 10 randomly selected such items, how many would you expect to
      have a delay time exceeding 8 months?
96. The article "The Statistics of Phytotoxic Air Pollutants" (*J. Roy. Statist Soc.*,
   1989: 183–198) suggests the lognormal distribution as a model for SO_2
   concentration above a forest. Suppose the parameter values are $\mu = 1.9$ and
   $\sigma = .9$.
   (a) What are the mean value and standard deviation of concentration?
   (b) What is the probability that concentration is at most 10? Between 5 and 10?
97. What condition on α and β is necessary for the standard beta pdf to be
   symmetric?
98. Suppose the proportion X of surface area in a randomly selected quadrat that is
   covered by a certain plant has a standard beta distribution with $\alpha = 5$ and
   $\beta = 2$.
   (a) Compute $E(X)$ and $Var(X)$.
   (b) Compute $P(X \leq .2)$.
   (c) Compute $P(.2 \leq X \leq .4)$.
   (d) What is the expected proportion of the sampling region not covered by the
      plant?
99. Let X have a standard beta density with parameters α and β.
   (a) Verify the formula for $E(X)$ given in the section.
   (b) Compute $E[(1 - X)^m]$. If X represents the proportion of a substance
      consisting of a particular ingredient, what is the expected proportion
      that does not consist of this ingredient?
100. Stress is applied to a 20-in. steel bar that is clamped in a fixed position at each
   end. Let $Y =$ the distance from the left end at which the bar snaps. Suppose
   $Y/20$ has a standard beta distribution with $E(Y) = 10$ and $Var(Y) = 100/7$.
   (a) What are the parameters of the relevant standard beta distribution?
   (b) Compute $P(8 \leq Y \leq 12)$.
   (c) Compute the probability that the bar snaps more than 2 in. from where you
      expect it to snap.

3.6 Probability Plots

An investigator will often have obtained a numerical sample consisting of
n observations and wish to know whether it is plausible that this sample came
from a population distribution of some particular type (e.g., from a normal distribution). For one thing, many formal procedures from statistical inference (Chap. 5)
are based on the assumption that the population distribution is of a specified type.
The use of such a procedure is inappropriate if the actual underlying probability
distribution differs greatly from the assumed type. Additionally, understanding the
underlying distribution can sometimes give insight into the physical mechanisms
involved in generating the data. An effective way to check a distributional assumption is to construct what is called a **probability plot**. The basis for our construction
is a comparison between percentiles of the sample data and the corresponding
percentiles of the assumed underlying distribution.

3.6.1 Sample Percentiles

The details involved in constructing probability plots differ a bit from source to source. Roughly speaking, sample percentiles are defined in the same way that percentiles of a population distribution are defined. The sample 50th percentile (i.e., the sample median) should separate the smallest 50% of the sample from the largest 50%, the sample 90th percentile should be such that 90% of the sample lies below that value and 10% lies above, and so on. Unfortunately, we run into problems when we actually try to compute the sample percentiles for a particular sample of n observations. If, for example, $n = 10$, we can split off 20% or 30% of the data, but there is no value that will split off exactly 23% of these ten observations. To proceed further, we need an operational definition of sample percentiles (this is one place where different people and different software packages do slightly different things).

Statistical convention states that when n is odd, the sample median is the middle value in the ordered list of sample observations, for example, the sixth-largest value when $n = 11$. This amounts to regarding the middle observation as being half in the lower half of the data and half in the upper half. Similarly, suppose $n = 10$. Then if we call the third-smallest value the 25th percentile, we are regarding that value as being half in the lower group (consisting of the two smallest observations) and half in the upper group (the seven largest observations). This leads to the following general definition of sample percentiles.

> **DEFINITION**
> Order the n sample observations from smallest to largest. Then the ith-smallest observation in the list is taken to be the **[$100(i - .5)/n$]th sample percentile**.

For example, if $n = 10$, the percentages corresponding to the ordered sample observations are $100(1 - .5)/10 = 5\%$, $100(2 - .5)/10 = 15\%$, 25%, ..., and $100(10 - .5)/10 = 95\%$. All other percentiles could then be determined by interpolation, e.g., the sample 10th percentile would then be halfway between the 5th percentile (smallest sample observation) and the 15th percentile (second smallest observation) of the $n = 10$ values. For the purposes of a probability plot, such interpolation will not be necessary, because a probability plot will be based only on the percentages $100(i - .5)/n$ corresponding to the n sample observations.

3.6.2 A Probability Plot

We now wish to determine whether the sample data could plausibly have come from some particular population distribution (e.g., a normal distribution with $\mu = 10$ and $\sigma = 3$). If the sample was actually selected from the specified distribution, the sample percentiles (ordered sample observations) should be reasonably close to the corresponding population distribution percentiles. That is, for $i = 1, 2, \ldots, n$ there should be reasonable agreement between the ith-smallest sample observation and the theoretical $[100(i - .5)/n]$th percentile for the specified distribution. Consider the (sample percentile, population percentile) pairs—that is, the pairs

$$\left(\begin{matrix} i\text{th smallest sample} \\ \text{observation} \end{matrix} \; , \; \begin{matrix} [100(i - .5)/n]\text{th percentile} \\ \text{of the population distribution} \end{matrix} \right)$$

for $i = 1, \ldots, n$. Each such pair can be plotted as a point on a two-dimensional coordinate system. If the sample percentiles are close to the corresponding population distribution percentiles, the first number in each pair will be roughly equal to the second number, and the plotted points will then fall close to a 45° line. Substantial deviations of the plotted points from a 45° line suggest that the assumed distribution might be wrong.

Example 3.33 The value of a physical constant is known to an experimenter. The experimenter makes $n = 10$ independent measurements of this value using a measurement device and records the resulting measurement errors (error = observed value − true value). These observations appear in the accompanying table. Is it plausible that the random variable *measurement error* has a standard normal distribution? The needed standard normal (z) percentiles are also displayed in the table and were determined as follows: the 5th percentile of the distribution under consideration, $N(0,1)$, is given by $\Phi(z) = .05$. From software or Appendix Table A.3, the solution is roughly $z = -1.645$. The other nine population (z) percentiles were found in a similar fashion.

Percentage	5	15	25	35	45
Sample observation	−1.91	−1.25	−.75	−.53	.20
z percentile	−1.645	−1.037	−.675	−.385	−.126
Percentage	55	65	75	85	95
Sample observation	.35	.72	.87	1.40	1.56
z percentile	.126	.385	.675	1.037	1.645

Thus the points in the probability plot are $(-1.91, -1.645)$, $(-1.25, -1.037)$, ..., and $(1.56, 1.645)$. Figure 3.31 shows the resulting plot. Although the points deviate a bit from the 45° line, the predominant impression is that this line fits the points very well. The plot suggests that the standard normal distribution is a reasonable probability model for measurement error.

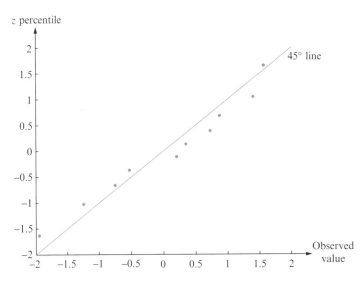

Fig. 3.31 Plots of pairs (observed value, z percentile) for the data of Example 3.33   ∎

An investigator is typically not interested in knowing whether a completely specified probability distribution, such as the standard normal distribution (normal with $\mu = 0$ and $\sigma = 1$) or the exponential distribution with $\lambda = .1$, is a plausible model for the population distribution from which the sample was selected. Instead, the investigator will want to know whether *some* member of a family of probability distributions specifies a plausible model—the family of normal distributions, the family of exponential distributions, the family of Weibull distributions, and so on. The values of the parameters of a distribution are usually not specified at the outset. If the family of Weibull distributions is under consideration as a model for lifetime data, the issue is whether there are *any* values of the parameters α and β for which the corresponding Weibull distribution gives a good fit to the data. Fortunately, it is almost always the case that just one probability plot will suffice for assessing the plausibility of an entire family. If the plot deviates substantially from a straight line, but not necessarily the 45° line, no member of the family is plausible.

To see why, let's focus on a plot for checking normality. As mentioned earlier, such a plot can be very useful in applied work because many formal statistical procedures are appropriate (give accurate inferences) only when the population distribution is at least approximately normal. These procedures should generally not be used if the normal probability plot shows a very pronounced departure from linearity. The key to constructing an omnibus normal probability plot is the relationship between standard normal (z) percentiles and those for any other normal distribution, which was presented in Sect. 3.3:

$$\begin{array}{c} \text{percentile for a} \\ N(\mu,\sigma) \text{ distribution} \end{array} = \mu + \sigma \cdot (\text{corresponding } z \text{ percentile})$$

Consider first the case $\mu = 0$. Then if each observation is exactly equal to the corresponding normal percentile for a particular value of σ, the pairs (observation,

$\sigma \cdot [z$ percentile]) fall on a 45° line, which has slope 1. This implies that the pairs (observation, z percentile) fall on a line passing through (0, 0) but having slope σ rather than 1. Similarly, the effect of a nonzero value of μ is simply to change the y-intercept from 0 to μ.

DEFINITION

A plot of the n pairs

(ith-smallest observation, $[100(i - .5)/n]$th z percentile)

on a two-dimensional coordinate system is called a **normal probability plot**. If the sample observations are in fact drawn from a normal distribution with mean value μ and standard deviation σ, the points should fall close to a straight line with slope σ and intercept μ. Thus a plot for which the points fall close to *some* straight line suggests that the assumption of a normal population distribution is plausible.

Example 3.34 The accompanying sample consisting of $n = 20$ observations on dielectric breakdown voltage of a piece of epoxy resin appeared in the article "Maximum Likelihood Estimation in the 3-Parameter Weibull Distribution" (*IEEE Trans. Dielectrics Electr. Insul.*, 1996: 43–55). Values of $(i - .5)/n$ for which z percentiles are needed are $(1 - .5)/20 = .025$, $(2 - .5)/20 = .075$, ..., and .975.

Observation	24.46	25.61	26.25	26.42	26.66	27.15	27.31	27.54	27.74	27.94
z percentile	−1.96	−1.44	−1.15	−.93	−.76	−.60	−.45	−.32	−.19	−.06

Observation	27.98	28.04	28.28	28.49	28.50	28.87	29.11	29.13	29.50	30.88
z percentile	.06	.19	.32	.45	.60	.76	.93	1.15	1.44	1.96

Figure 3.32 shows the resulting normal probability plot. The pattern in the plot is quite straight, indicating it is plausible that the population distribution of dielectric breakdown voltage is normal.

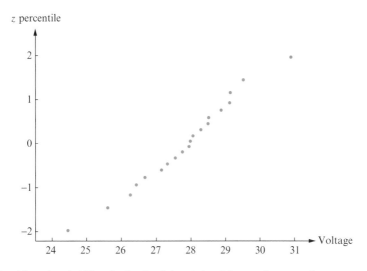

Fig. 3.32 Normal probability plot for the dielectric breakdown voltage sample   ■

There is an alternative version of a normal probability plot in which the z percentile axis is replaced by a nonlinear probability axis. The scaling on this axis is constructed so that plotted points should again fall close to a line when the sampled distribution is normal. Figure 3.33 shows such a plot from Matlab,

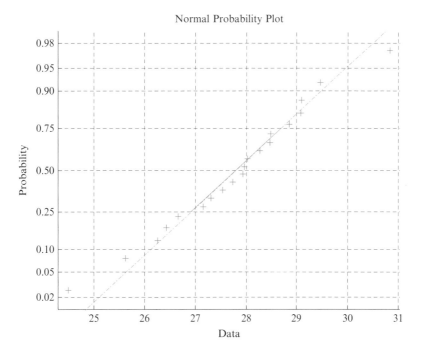

Fig. 3.33 Normal probability plot of the breakdown voltage data from Matlab

obtained using the `normplot` command, for the breakdown voltage data of Example 3.34. The plot remains essentially the same, and it is just the labeling of the axis that changes.

3.6.3   Departures from Normality

A nonnormal population distribution can often be placed in one of the following three categories:

1. It is symmetric and has "lighter tails" than does a normal distribution; that is, the density curve declines more rapidly out in the tails than does a normal curve.
2. It is symmetric and heavy-tailed compared to a normal distribution.
3. It is skewed; that is, the distribution is not symmetric, but rather tapers off more in one direction than the other.

A uniform distribution is light-tailed, since its density function drops to zero outside a finite interval. The density function $f(x) = 1/[\pi(1+x^2)]$, for $-\infty < x < \infty$, is one example of a heavy-tailed distribution, since $1/(1+x^2)$ declines much less rapidly than does $e^{-x^2/2}$. Lognormal and Weibull distributions are among those that are skewed. When the points in a normal probability plot do not adhere to a straight line, the pattern will frequently suggest that the population distribution is in a particular one of these three categories.

Figure 3.34 illustrates typical normal probability plots corresponding to three situations above. If the sample was selected from a light-tailed distribution, the largest and smallest observations are usually not as extreme as would be expected from a normal random sample. Visualize a straight line drawn through the middle part of the plot; points on the far right tend to be above the line (z percentile > observed value), whereas points on the left end of the plot tend to fall below the straight line (z percentile < observed value). The result is an S-shaped pattern of the type pictured in Fig. 3.34a. For sample observations from a heavy-tailed distribution, the opposite effect will occur, and a normal probability plot will have an S shape with the opposite orientation, as in Fig. 3.34b. If the underlying distribution is positively skewed (a short left tail and a long right tail), the smallest sample observations will be larger than expected from a normal sample and so will the largest observations. In this case, points on both ends of the plot will fall below a straight line through the middle part, yielding a curved pattern, as illustrated in Fig. 3.34c. For example, a sample from a lognormal distribution will usually produce such a pattern; a plot of (ln(observation), z percentile) pairs should then resemble a straight line.

Even when the population distribution is normal, the sample percentiles will not coincide exactly with the theoretical percentiles because of sampling variability. How much can the points in the probability plot deviate from a straight-line pattern before the assumption of population normality is no longer plausible? This is not an easy question to answer. Generally speaking, a small sample from a normal distribution is more likely to yield a plot with a nonlinear pattern than is a large sample. The book *Fitting Equations to Data* by Daniel Cuthbert and Fred Wood presents the results of a simulation study in which numerous samples of different

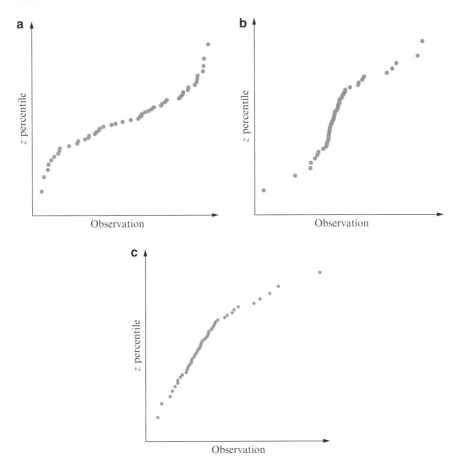

Fig. 3.34 Probability plots that suggest a non-normal distribution: (**a**) a plot consistent with a light-tailed distribution; (**b**) a plot consistent with a heavy-tailed distribution; (**c**) a plot consistent with a (positively) skewed distribution

sizes were selected from normal distributions. The authors concluded that there is typically greater variation in the appearance of the probability plot for sample sizes smaller than 30, and only for much larger sample sizes does a linear pattern generally predominate. When a plot is based on a small sample size, only a very substantial departure from linearity should be taken as conclusive evidence of nonnormality. A similar comment applies to probability plots for checking the plausibility of other types of distributions.

3.6.4 Beyond Normality

Consider a generic family of probability distributions involving two parameters, θ_1 and θ_2, and let $F(x; \theta_1, \theta_2)$ denote the corresponding cdf. The family of normal distributions is one such family, with $\theta_1 = \mu$, $\theta_2 = \sigma$, and $F(x; \mu, \sigma) = \Phi[(x - \mu)/\sigma]$. Another example is the Weibull family, with $\theta_1 = \alpha$, $\theta_2 = \beta$, and

$$F(x; \alpha, \beta) = 1 - e^{-(x/\beta)^\alpha}$$

Still another family of this type is the gamma family, for which the cdf is an integral involving the incomplete gamma function that cannot be expressed in any simpler form.

The parameters θ_1 and θ_2 are said to be **location** and **scale parameters**, respectively, if $F(x; \theta_1, \theta_2)$ is a function of $(x - \theta_1)/\theta_2$. The parameters μ and σ of the normal family are location and scale parameters, respectively. Changing μ shifts the location of the bell-shaped density curve to the right or left, and changing σ amounts to stretching or compressing the measurement scale (the scale on the horizontal axis when the density function is graphed). Another example is given by the cdf

$$F(x; \theta_1, \theta_2) = 1 - e^{-e^{(x-\theta_1)/\theta_2}} \qquad -\infty < x < \infty$$

A random variable with this cdf is said to have an *extreme value distribution*. It is used in applications involving component lifetime and material strength.

The parameter β of the Weibull distribution is a scale parameter. However, α is not a location parameter but instead is called a **shape parameter**. The same is true for the parameters α and β of the gamma distribution. In the usual form, the density function for any member of either the gamma or Weibull distribution is positive for $x > 0$ and zero otherwise. A location (or shift) parameter can be introduced as a third parameter γ (we did this for the Weibull distribution in Sect. 3.5) to shift the density function so that it is positive if $x > \gamma$ and zero otherwise.

When the family under consideration has only location and scale parameters, the issue of whether any member of the family is a plausible population distribution can be addressed via a single, easily constructed probability plot. One first obtains the percentiles of the *standard distribution*, the one with $\theta_1 = 0$ and $\theta_2 = 1$, for percentages $100(i - .5)/n$ ($i = 1, \ldots, n$). The n (observation, standardized percentile) pairs give the points in the plot. This is, of course, exactly what we did to obtain an omnibus normal probability plot.

Somewhat surprisingly, this methodology can be applied to yield an omnibus Weibull probability plot. The key result is that if X has a Weibull distribution with shape parameter α and scale parameter β, then the transformed variable $\ln(X)$ has an extreme value distribution with location parameter $\theta_1 = \ln(\beta)$ and scale parameter $\theta_2 = 1/\alpha$ (see Exercise 169). Thus a plot of the

(ln(observation), extreme value standardized percentile)

pairs that shows a strong linear pattern provides support for choosing the Weibull distribution as a population model.

Example 3.35 The accompanying observations are on lifetime (in hours) of power apparatus insulation when thermal and electrical stress acceleration were fixed at particular values ("On the Estimation of Life of Power Apparatus Insulation Under Combined Electrical and Thermal Stress," *IEEE Trans. Electr. Insul.*, 1985:

70–78). A Weibull probability plot necessitates first computing the 5th, 15th, ...,
and 95th percentiles of the standard extreme value distribution. The $(100p)$th
percentile η_p satisfies

$$p = F(\eta_p) = 1 - e^{-e^{\eta_p}}$$

from which $\eta_p = \ln(-\ln(1-p))$.

Observation	282	501	741	851	1072	1122	1202	1585	1905	2138
ln(observation)	5.64	6.22	6.61	6.75	6.98	7.02	7.09	7.37	7.55	7.67
Percentile	−2.97	−1.82	−1.25	−.84	−.51	−.23	.05	.33	.64	1.10

The pairs $(5.64, -2.97)$, $(6.22, -1.82)$, ..., $(7.67, 1.10)$ are plotted as points in
Fig. 3.35. The straightness of the plot argues strongly for using the Weibull
distribution as a model for insulation life, a conclusion also reached by the author
of the cited article.

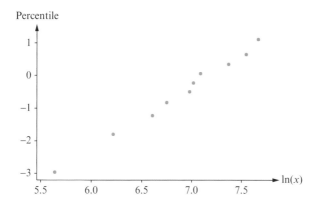

Fig. 3.35 A Weibull probability plot of the insulation lifetime data   ∎

The gamma distribution is an example of a family involving a shape parameter
for which there is no transformation into a distribution that depends only on
location and scale parameters. Construction of a probability plot necessitates first
estimating the shape parameter from sample data (some general methods for doing
this are described in Chap. 5).

Sometimes an investigator wishes to know whether the transformed variable X^θ
has a normal distribution for some value of θ (by convention, $\theta = 0$ is identified with
the logarithmic transformation, in which case X has a lognormal distribution). The
book *Graphical Methods for Data Analysis* by John Chambers et al. discusses this
type of problem as well as other refinements of probability plotting.

3.6.5   Probability Plots in Matlab and R

Matlab, along with many statistical software packages (including R), have built-in
probability plotting commands that vitiate the need for manual calculation of

percentiles from the assumed population distribution. In Matlab, the `normplot(x)` command will produce a graph like the one seen in Fig. 3.33, assuming the vector `x` contains the observed data. The R command `qqnorm(x)` creates a similar graph, except that the axes are transposed (ordered observations on the vertical axis, theoretical quantiles on the horizontal). Both Matlab and R have a package called `probplot` that, with appropriate specifications of the inputs, can create probability plots for distributions besides normal (e.g., Weibull, exponential, extreme value). Refer to the help documentation in those languages for more information.

3.6.6 Exercises: Section 3.6 (101–111)

101. The accompanying normal probability plot was constructed from a sample of 30 readings on tension for mesh screens behind the surface of video display tubes. Does it appear plausible that the tension distribution is normal?

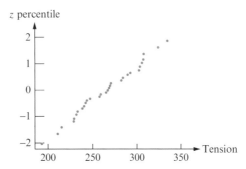

102. A sample of 15 female collegiate golfers was selected and the clubhead velocity (km/h) while swinging a driver was determined for each one, resulting in the following data ("Hip Rotational Velocities during the Full Golf Swing," *J. of Sports Science and Medicine*, 2009: 296-299):

69.0	69.7	72.7	80.3	81.0
85.0	86.0	86.3	86.7	87.7
89.3	90.7	91.0	92.5	93.0

The corresponding z percentiles are

−1.83	−1.28	−0.97	−0.73	−0.52
−0.34	−0.17	0.0	0.17	0.34
0.52	0.73	0.97	1.28	1.83

Construct a normal probability plot. Is it plausible that the population distribution is normal?

103. Construct a normal probability plot for the following sample of observations on coating thickness for low-viscosity paint ("Achieving a Target Value for a

Manufacturing Process: A Case Study," *J. Qual. Tech.*, 1992: 22–26). Would you feel comfortable estimating population mean thickness using a method that assumed a normal population distribution?

.83	.88	.88	1.04	1.09	1.12	1.29	1.31
1.48	1.49	1.59	1.62	1.65	1.71	1.76	1.83

104. The article "A Probabilistic Model of Fracture in Concrete and Size Effects on Fracture Toughness" (*Mag. Concrete Res.*, 1996: 311–320) gives arguments for why fracture toughness in concrete specimens should have a Weibull distribution and presents several histograms of data that appear well fit by superimposed Weibull curves. Consider the following sample of size $n = 18$ observations on toughness for high-strength concrete (consistent with one of the histograms); values of $p_i = (i - .5)/18$ are also given.

Observation	.47	.58	.65	.69	.72	.74
p_i	.0278	.0833	.1389	.1944	.2500	.3056
Observation	.77	.79	.80	.81	.82	.84
p_i	.3611	.4167	.4722	.5278	.5833	.6389
Observation	.86	.89	.91	.95	1.01	1.04
p_i	.6944	.7500	.8056	.8611	.9167	.9722

Construct a Weibull probability plot and comment.

105. The propagation of fatigue cracks in various aircraft parts has been the subject of extensive study. The accompanying data consists of propagation lives (flight hours/10^4) to reach a given crack size in fastener holes for use in military aircraft ("Statistical Crack Propagation in Fastener Holes Under Spectrum Loading," *J. Aircraft*, 1983: 1028-1032):

.736	.863	.865	.913	.915	.937	.983	1.007
1.011	1.064	1.109	1.132	1.140	1.153	1.253	1.394

Construct a normal probability plot for this data. Does it appear plausible that propagation life has a normal distribution? Explain.

106. The article "The Load-Life Relationship for M50 Bearings with Silicon Nitride Ceramic Balls" (*Lubricat. Engrg.*, 1984: 153–159) reports the accompanying data on bearing load life (million revs.) for bearings tested at a 6.45 kN load.

47.1	68.1	68.1	90.8	103.6	106.0	115.0
126.0	146.6	229.0	240.0	240.0	278.0	278.0
289.0	289.0	367.0	385.9	392.0	505.0	

(a) Construct a normal probability plot. Is normality plausible?

(b) Construct a Weibull probability plot. Is the Weibull distribution family plausible?

107. The accompanying data on rainfall (acre-feet) from 26 seed clouds is taken from the article "A Bayesian Analysis of a Multiplicative Treatment Effect in Weather Modification" (*Technometrics*, 1975: 161-166). Construct a probability plot that will allow you to assess the plausibility of the lognormal distribution as a model for the rainfall data, and comment on what you find.

4.1	7.7	17.5	31.4	32.7	40.6	92.4
115.3	118.3	119.0	129.6	198.6	200.7	242.5
255.0	274.7	274.7	302.8	334.1	430.0	489.1
703.4	978.0	1656.0	1697.8	2745.6		

108. The accompanying observations are precipitation values during March over a 30-year period in Minneapolis–St. Paul.

.77	1.20	3.00	1.62	2.81	2.48
1.74	.47	3.09	1.31	1.87	.96
.81	1.43	1.51	.32	1.18	1.89
1.20	3.37	2.10	.59	1.35	.90
1.95	2.20	.52	.81	4.75	2.05

(a) Construct and interpret a normal probability plot for this data set.
(b) Calculate the square root of each value and then construct a normal probability plot based on this transformed data. Does it seem plausible that the square root of precipitation is normally distributed?
(c) Repeat part (b) after transforming by cube roots.

109. Allowable mechanical properties for structural design of metallic aerospace vehicles requires an approval method for statistically analyzing empirical test data. The article "Establishing Mechanical Property Allowables for Metals" (*J. of Testing and Evaluation*, 1998: 293-299) used the accompanying data on tensile ultimate strength (ksi) as a basis for addressing the difficulties in developing such a method.

122.2	124.2	124.3	125.6	126.3	126.5	126.5	127.2	127.3
127.5	127.9	128.6	128.8	129.0	129.2	129.4	129.6	130.2
130.4	130.8	131.3	131.4	131.4	131.5	131.6	131.6	131.8
131.8	132.3	132.4	132.4	132.5	132.5	132.5	132.5	132.6
132.7	132.9	133.0	133.1	133.1	133.1	133.1	133.2	133.2
133.2	133.3	133.3	133.5	133.5	133.5	133.8	133.9	134.0
134.0	134.0	134.0	134.1	134.2	134.3	134.4	134.4	134.6
134.7	134.7	134.7	134.8	134.8	134.8	134.9	134.9	135.2
135.2	135.2	135.3	135.3	135.4	135.5	135.5	135.6	135.6
135.7	135.8	135.8	135.8	135.8	135.8	135.9	135.9	135.9
135.9	136.0	136.0	136.1	136.2	136.2	136.3	136.4	136.4
136.6	136.8	136.9	136.9	137.0	137.1	137.2	137.6	137.6

(continued)

137.8	137.8	137.8	137.9	137.9	138.2	138.2	138.3	138.3
138.4	138.4	138.4	138.5	138.5	138.6	138.7	138.7	139.0
139.1	139.5	139.6	139.8	139.8	140.0	140.0	140.7	140.7
140.9	140.9	141.2	141.4	141.5	141.6	142.9	143.4	143.5
143.6	143.8	143.8	143.9	144.1	144.5	144.5	147.7	147.7

Use software to construct a normal probability plot of this data, and comment.

110. Let the ordered sample observations be denoted by $y_1, y_2, \ldots, y_n$ (y_1 being the smallest and y_n the largest). Our suggested check for normality is to plot the $(y_i, \Phi^{-1}[(i - .5)/n])$ pairs. Suppose we believe that the observations come from a distribution with mean 0, and let $w_1, \ldots, w_n$ be the ordered *absolute values* of the observed data. A **half-normal plot** is a probability plot of the w_is. More specifically, since $P(|Z| \leq w) = P(-w \leq Z \leq w) = 2\Phi(w) - 1$, a half-normal plot is a plot of the $(w_i, \Phi^{-1}[(p_i + 1)/2])$ pairs, where $p_i = (i - .5)/n$. The virtue of this plot is that small or large outliers in the original sample will now appear only at the upper end of the plot rather than at both ends. Construct a half-normal plot for the following sample of measurement errors, and comment:

$-3.78, -1.27, 1.44, -.39, 12.38, -43.40, 1.15, -3.96, -2.34, 30.84.$

111. The following failure time observations (1000s of hours) resulted from accelerated life testing of 16 integrated circuit chips of a certain type:

82.8	11.6	359.5	502.5	307.8	179.7
242.0	26.5	244.8	304.3	379.1	212.6
229.9	558.9	366.7	203.6		

Use the corresponding percentiles of the exponential distribution with $\lambda = 1$ to construct a probability plot. Then explain why the plot assesses the plausibility of the sample having been generated from *any* exponential distribution.

3.7   Transformations of a Random Variable

Often we need to deal with a transformation $Y = g(X)$ of the random variable X. Here $g(X)$ could be a simple change of time scale. If X is the time to complete a task in minutes, then $Y = 60X$ is the completion time expressed in seconds. How can we get the pdf of Y from the pdf of X? Consider first a simple example.

Example 3.36 The interval X in minutes between calls to a 911 center is exponentially distributed with mean 2 min, so its pdf $f_X(x) = .5e^{-.5x}$ for $x > 0$. In order to get the pdf of $Y = 60X$, we first obtain its cdf:

$$F_Y(y) = P(Y \le y) = P(60X \le y) = P(X \le y/60) = F_X(y/60) = \int_0^{y/60} .5e^{-.5x} dx$$

$$= 1 - e^{-y/120}$$

Differentiating this with respect to y gives $f_Y(y) = (1/120)e^{-y/120}$ for $y > 0$. We see that the distribution of Y is exponential with mean 120 s (2 min).

There is nothing special here about the mean 2 and the multiplier 60. It should be clear that if we multiply an exponential random variable with mean μ by a positive constant c we get another exponential random variable with mean $c\mu$. ∎

Sometimes it isn't possible to evaluate the cdf in closed form. Could we still find the pdf of Y without evaluating the integral? Yes, thanks to the following theorem.

TRANSFORMATION THEOREM

Let X have pdf $f_X(x)$ and let $Y = g(X)$, where g is monotonic (either strictly increasing or strictly decreasing) on the set of all possible values of X, so it has an inverse function $X = h(Y)$. Assume that h has a derivative $h'(y)$. Then

$$f_Y(y) = f_X(h(y)) \, |h'(y)| \qquad (3.11)$$

Proof Here is the proof assuming that g is monotonically increasing. The proof for g monotonically decreasing is similar. First find the cdf of Y:

$$F_Y(y) = P(Y \le y) = P(g(X) \le y) = P(X \le h(y)) = F_X(h(y))$$

The third equality above, wherein $g(X) \le y$ is true iff $X \le g^{-1}(y) = h(y)$, relies on g being a monotonically increasing function. Now differentiate the cdf with respect to y, using the Chain Rule:

$$f_Y(y) = \frac{d}{dy} F_Y(y) = \frac{d}{dy} F_X(h(y)) = F_X'(h(y)) \cdot h'(y) = f_X(h(y)) \cdot h'(y)$$

The absolute value on the derivative in Eq. (3.11) is needed only in the other case where g is decreasing. The set of possible values for Y is obtained by applying g to the set of possible values for X. ∎

Example 3.37 Let's apply the Transformation Theorem to the situation introduced in Example 3.36. There $Y = g(X) = 60X$ and $X = h(Y) = Y/60$.

$$f_Y(y) = f_X(h(y)) |h'(y)| = .5e^{-.5x} \left| \frac{1}{60} \right| = \frac{1}{120} e^{-y/120} \qquad y > 0$$

This matches the pdf of Y derived through the cdf in Example 3.36. ∎

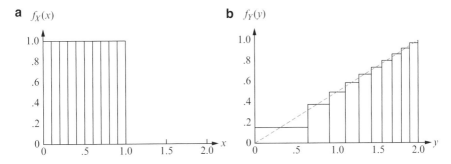

a $f_X(x)$ **b** $f_Y(y)$

Fig. 3.36 The effect on the pdf if X is uniform on $[0, 1]$ and $Y = 2\sqrt{X}$

Example 3.38 Let $X \sim \text{Unif}[0, 1]$, so $f_X(x) = 1$ for $0 \le x \le 1$, and define a new variable $Y = 2\sqrt{X}$. The function $g(x) = 2\sqrt{x}$ is monotone on $[0, 1]$, with inverse $x = h(y) = y^2/4$. Apply the Transformation Theorem:

$$f_Y(y) = f_X(h(y))\left|h'(y)\right| = (1)\left|\frac{2y}{4}\right| = \frac{y}{2} \qquad 0 \le y \le 2$$

The range $0 \le y \le 2$ comes from the fact that $y = 2\sqrt{x}$ maps $[0, 1]$ to $[0, 2]$. A graphical representation may help in understanding why the transform $Y = 2\sqrt{X}$ yields $f_Y(y) = y/2$ if $X \sim \text{Unif}[0, 1]$. Figure 3.36a shows the uniform distribution with $[0, 1]$ partitioned into ten subintervals. In Fig. 3.36b the endpoints of these intervals are shown after transforming according to $y = 2\sqrt{x}$. The heights of the rectangles are arranged so each rectangle still has area .1, and therefore the probability in each interval is preserved. Notice the close fit of the dashed line, which has the equation $f_Y(y) = y/2$. ∎

Example 3.39 The variation in a certain electrical current source X (in milliamps) can be modeled by the pdf

$$f_X(x) = \begin{cases} 1.25 - .25x & 2 \le x \le 4 \\ 0 & \text{otherwise} \end{cases}$$

If this current passes through a 220-Ω resistor, the resulting power Y (in microwatts) is given by the expression $Y = 220X^2$. The function $y = g(x) = 220x^2$ is monotonically increasing on the range of X, the interval $[2, 4]$, and has inverse function $x = h(y) = g^{-1}(y) = \sqrt{y/220}$. (Notice that $g(x)$ is a parabola and thus not monotone on the entire real number line, but for the purposes of the theorem $g(x)$ only needs to be monotone on the range of the rv X.) Apply Eq. (3.11):

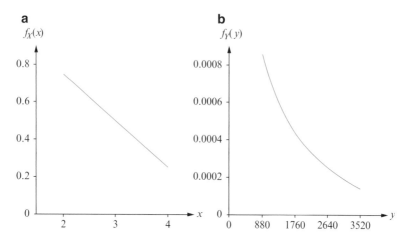

Fig. 3.37 pdfs from Example 3.39: (**a**) pdf of X; (**b**) pdf of Y ∎

$$f_Y(y) = f_X(h(y)) \cdot \left|h'(y)\right|$$

$$= f_X\left(\sqrt{y/220}\right) \cdot \left|\frac{d}{dy}\sqrt{y/220}\right|$$

$$= \left(1.25 - .25\sqrt{y/220}\right) \cdot \frac{1}{2\sqrt{220y}} = \frac{5}{8\sqrt{220y}} - \frac{1}{1760}$$

The set of possible Y-values is determined by substituting $x = 2$ and $x = 4$ into $g(x) = 220x^2$; the resulting range for Y is [880, 3520]. Therefore, the pdf of $Y = 220X^2$ is

$$f_Y(y) = \begin{cases} \dfrac{5}{8\sqrt{220y}} - \dfrac{1}{1760} & 880 \le y \le 3520 \\ 0 & \text{otherwise} \end{cases}$$

The pdfs of X and Y appear in Fig. 3.37.

The Transformation Theorem requires a monotonic transformation, but there are important applications in which the transformation is not monotone. Nevertheless, it may be possible to use the theorem anyway with a little trickery.

Example 3.40 In this example, we start with a standard normal random variable Z, and we transform to $Y = Z^2$. Of course, this is not monotonic over the interval for Z, $(-\infty, \infty)$. However, consider the transformation $U = |Z|$. Because Z has a symmetric distribution, the pdf of U is $f_U(u) = f_Z(u) + f_Z(-u) = 2f_Z(u)$. (Don't despair if this is not intuitively clear, because we'll verify it shortly. For the time being, assume it to be true.) Then $Y = Z^2 = |Z|^2 = U^2$, and the transformation in terms of U is monotonic because its set of possible values is [0, ∞). Thus we can use the Transformation Theorem with $h(y) = y^{1/2}$:

$$f_Y(y) = f_U[h(y)] \, |h'(y)| = 2f_X[h(y)] \, |h'(y)|$$

$$= \frac{2}{\sqrt{2\pi}} e^{-.5(y^{1/2})^2} \left|\frac{1}{2}y^{-1/2}\right| = \frac{1}{\sqrt{2\pi y}} e^{-y/2} \quad y > 0$$

This distribution is known as the *chi-squared distribution with one degree of freedom*. Chi-squared distributions arise frequently in statistical inference procedures, such as those in Chap. 5.

You were asked to believe intuitively that $f_U(u) = 2f_Z(u)$. Here is a little derivation that works as long as $f_Z(z)$ is an even function, [i.e., $f_Z(-z) = f_Z(z)$]. If $u > 0$,

$$F_U(u) = P(U \le u) = P(|Z| \le u) = P(-u \le Z \le u) = 2P(0 \le Z \le u)$$

$$= 2[F_Z(u) - F_Z(0)].$$

Differentiating this with respect to u gives $f_U(u) = 2\,f_Z(u)$.   ∎

Example 3.41 Sometimes the Transformation Theorem cannot be used at all, and you need to use the cdf. Let $f_X(x) = (x+1)/8$, $-1 < x < 3$, and $Y = X^2$. The transformation is not monotonic on $(-1, 3)$ and $f_X(x)$ is not an even function. Possible values of Y are $\{y: 0 \le y \le 9\}$. Considering first $0 \le y \le 1$,

$$F_Y(y) = P(Y \le y) = P(X^2 \le y) = P(-\sqrt{y} \le X \le \sqrt{y}) = \int_{-\sqrt{y}}^{\sqrt{y}} \frac{u+1}{8} du = \frac{\sqrt{y}}{4}$$

Then, on the other subinterval, $1 < y \le 9$,

$$F_Y(y) = P(Y \le y) = P(X^2 \le y) = P(-\sqrt{y} \le X \le \sqrt{y}) = P(-1 \le X \le \sqrt{y})$$

$$= \int_{-1}^{\sqrt{y}} \frac{u+1}{8} du = (1 + y + 2\sqrt{y})/16$$

Differentiating, we get

$$f_Y(y) = \begin{cases} \dfrac{1}{8\sqrt{y}} & 0 < y < 1 \\[2mm] \dfrac{y + \sqrt{y}}{16y} & 1 < y < 9 \\[2mm] 0 & \text{otherwise} \end{cases}$$

Figure 3.38 shows the pdfs of both X and Y.

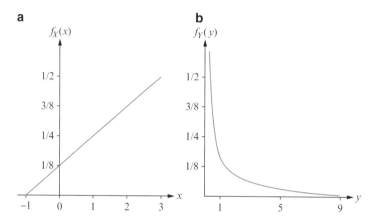

Fig. 3.38 pdfs from Example 3.41: (**a**) pdf of X; (**b**) pdf of Y　　　　　■

3.7.1 Exercises: Section 3.7 (112–128)

112. Relative to the winning time, the time X of another runner in a ten kilometer race has pdf $f_X(x) = 2/x^3$, $x > 1$. The reciprocal $Y = 1/X$ represents the ratio of the time for the winner divided by the time of the other runner. Find the pdf of Y. Explain why Y also represents the speed of the other runner relative to the winner.

113. Let X be the fuel efficiency in miles per gallon of an extremely inefficient vehicle (a military tank, perhaps?), and suppose X has the pdf $f_X(x) = 2x$, $0 < x < 1$. Determine the pdf of $Y = 1/X$, which is fuel efficiency in gallons per mile. [*Note*: The distribution of Y is a special case of the *Pareto distribution* (see Exercise 10).]

114. Let X have the pdf $f_X(x) = 2/x^3$, $x > 1$. Find the pdf of $Y = \sqrt{X}$.

115. Let X have an exponential distribution with mean 2, so $f_X(x) = \frac{1}{2} e^{-x/2}$, $x > 0$. Find the pdf of $Y = \sqrt{X}$. [*Note*: Suppose you choose a point in two dimensions randomly, with the horizontal and vertical coordinates chosen independently from the standard normal distribution. Then X has the distribution of the squared distance from the origin and Y has the distribution of the distance from the origin. Y has a *Rayleigh distribution* (see Exercise 4).]

116. If X is distributed as $N(\mu, \sigma)$, find the pdf of $Y = e^X$. Verify that the distribution of Y matches the lognormal pdf provided in Sect. 3.5.

117. If the side of a square X is random with the pdf $f_X(x) = x/8$, $0 < x < 4$, and Y is the area of the square, find the pdf of Y.

118. Let $X \sim \text{Unif}[0, 1]$. Find the pdf of $Y = -\ln(X)$.

119. Let $X \sim \text{Unif}[0, 1]$. Find the pdf of $Y = \tan[\pi(X - .5)]$. [*Note*: The random variable Y has the *Cauchy distribution*, named after the famous mathematician.]

120. If $X \sim \text{Unif}[0, 1]$, find a linear transformation $Y = cX + d$ such that Y is uniformly distributed on $[A, B]$, where A and B are any two numbers such that $A < B$. Is there any other solution? Explain.

121. If X has the pdf $f_X(x) = x/8$, $0 < x < 4$, find a transformation $Y = g(X)$ such that $Y \sim \text{Unif}[0, 1]$. [*Hint*: The target is to achieve $f_Y(y) = 1$ for $0 \le y \le 1$. The Transformation Theorem will allow you to find $h(y)$, from which $g(x)$ can be obtained.]

122. If a measurement error X is uniformly distributed on $[-1, 1]$, find the pdf of $Y = |X|$, which is the magnitude of the measurement error.

123. If $X \sim \text{Unif}[-1, 1]$, find the pdf of $Y = X^2$.

124. Ann is expected at 7:00 pm after an all-day drive. She may be as much as 1 h early or as much as 3 h late. Assuming that her arrival time X is uniformly distributed over that interval, find the pdf of $|X - 7|$, the unsigned difference between her actual and predicted arrival times.

125. If $X \sim \text{Unif}[-1, 3]$, find the pdf of $Y = X^2$.

126. If a measurement error X is distributed as $N(0, 1)$, find the pdf of $|X|$, which is the magnitude of the measurement error.

127. A circular target has radius 1 foot. Assume that you hit the target (we shall ignore misses) and that the probability of hitting any region of the target is proportional to the region's area. If you hit the target at a distance Y from the center, then let $X = \pi Y^2$ be the corresponding area. Show that
    (a) X is uniformly distributed on $[0, \pi]$. [*Hint*: Show that $F_X(x) = P(X \le x) = x/\pi$.]
    (b) Y has pdf $f_Y(y) = 2y$, $0 < y < 1$.

128. In Exercise 127 suppose instead that Y is uniformly distributed on $[0, 1]$. Find the pdf of $X = \pi Y^2$. Geometrically speaking, why should X have a pdf that is unbounded near 0?

3.8 Simulation of Continuous Random Variables

In Sects. 1.6 and 2.8, we discussed the need for simulation of random events and discrete random variables in situations where an "analytic" solution is very difficult or simply not possible. This section presents methods for simulating continuous random variables, including some of the built-in simulation tools of Matlab and R.

3.8.1 The Inverse CDF Method

Section 2.8 introduced the inverse cdf method for simulating discrete random variables. The basic idea was this: generate a Unif[0, 1) random number and align it with the cdf of the random variable X we want to simulate. Then, determine which X value corresponds to that cdf value. We now extend this methodology to the simulation of values from a continuous distribution; the heart of the algorithm relies on the following theorem, often called the *probability integral transform.*

THEOREM

Consider any continuous distribution with pdf f and cdf F. Let $U \sim \text{Unif}[0, 1)$, and define a random variable X by

$$X = F^{-1}(U) \tag{3.12}$$

Then the pdf of X is f.

Before proving this theorem, let's consider its practical usage: Suppose we want to simulate a continuous rv whose pdf is $f(x)$, i.e., obtain successive values of X having pdf $f(x)$. If we can compute the corresponding cdf $F(x)$ and apply its inverse F^{-1} to standard uniform variates $u_1, \ldots, u_n$, the theorem states that the resulting variates $x_1 = F^{-1}(u_1), \ldots, x_n = F^{-1}(u_n)$ will follow the desired distribution f. (We'll discuss the practical difficulties of implementing this method a little later.) A graphical description of the algorithm appears in Fig. 3.39.

Proof Apply the Transformation Theorem (Sect. 3.7) with $f_U(u) = 1$ for $0 \le u < 1$, $X = g(U) = F^{-1}(U)$, and thus $U = h(X) = g^{-1}(X) = F(X)$. The pdf of the transformed variable X is

$$f_X(x) = f_U(h(x)) \cdot \left| h'(x) \right| = f_U(F(x)) \cdot \left| F'(x) \right| = 1 \cdot |f(x)| = f(x)$$

In the last step, the absolute values may be removed because a pdf is always nonnegative.   ∎

The following box explains the implementation of the inverse cdf method justified by the preceding theorem.

Fig. 3.39 The inverse cdf method, illustrated

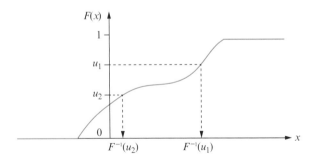

> **INVERSE CDF METHOD**
> It is desired to simulate n values from a distribution pdf $f(x)$. Let $F(x)$ be the corresponding cdf. Repeat n times:
> 1. Use a random-number generator (RNG) to produce a value, u, from $[0, 1)$.
> 2. Assign $x = F^{-1}(u)$.
> The resulting values $x_1, \ldots, x_n$ form a simulation of a random variable with the original pdf, $f(x)$.

Example 3.42 Consider the electrical current distribution model of Example 3.11, where the pdf of X is given by $f(x) = 1.25 - .25x$ for $2 \leq x \leq 4$. Suppose a simulation of X is required as part of some larger system analysis. To implement the above method, the inverse of the cdf of X is required. First, compute the cdf:

$$F(x) = P(X \leq x) = \int_2^x f(y)dy$$

$$= \int_2^x (1.25 - .25y)dy = -0.125x^2 + 1.25x - 2, \quad 2 \leq x \leq 4$$

To find the probability integral transform Eq. (3.12), set $u = F(x)$ and solve for x:

$$u = F(x) = -0.125x^2 + 1.25x - 2 \Rightarrow x = F^{-1}(u) = 5 - \sqrt{9 - 8u}$$

The equation above has been solved using the quadratic formula; care must be taken to select the solution whose values lie in the interval $[2, 4]$ (the other solution, $x = 5 + \sqrt{9 - 8u}$, does not have that feature). Beginning with the usual Unif$[0, 1)$ RNG, the algorithm for simulating X is the following: given a value u from the RNG, assign $x = 5 - \sqrt{9 - 8u}$. Repeating this algorithm n times gives n simulated values of X. Programs in Matlab and R that implement this algorithm appear in Fig. 3.40; both return a vector, x, containing $n = 10{,}000$ simulated values of the specified distribution.

a
```
x=zeros(10000,1);
for i=1:10000
    u=rand;
    x(i)=5-sqrt(9-8*u);
end
```

b
```
x <- NULL
for (i in 1:10000){
    u<-runif(1)
    x[i]<-5-sqrt(9-8*u)
}
```

Fig. 3.40 Simulation code for Example 3.42: (**a**) Matlab; (**b**) R

As discussed in Chap. 1, both of these programs can be accelerated by "vectorizing" the operations rather than using a for loop. In fact, a single line of code in either language can produce the desired result:

in Matlab: `x=5-sqrt(9-8*rand(10000,1))`
in R: `x<-5-sqrt(9-8*runif(10000))`

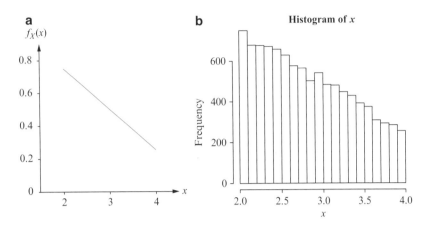

Fig. 3.41 (a) Theoretical pdf and (b) R simulation results for Example 3.42

The pdf of the rv X and a histogram of simulation results from R appear in Fig. 3.41. ∎

Example 3.43 The lifetime of a certain type of drill bit has an exponential distribution with mean 100 h. An analysis of a large manufacturing process that uses these drill bits requires the simulation of this lifetime distribution, which can be achieved through the inverse cdf method. From Sect. 3.4, the cdf of this exponential distribution is $F(x) = 1 - e^{-.01x}$, and so the inverse cdf is $x = F^{-1}(u) = -100\ln(1 - u)$. Applying this function to Unif[0, 1) random numbers will generate the desired simulation. (Don't let the negative sign at the front worry you: since $0 \leq u < 1$, $1 - u$ lies between 0 and 1, and so its logarithm is negative and the resulting value of x is actually positive.)

As a check, the code `x=-100*log(1-rand(10000,1))` was submitted to Matlab and the resulting sample mean and sd were obtained using `mean(x)` and `std(x)`. Exponentially distributed rvs have standard deviation equal to the mean, so the theoretical answers are $\mu = 100$ and $\sigma = 100$. The Matlab simulation yielded $\bar{x} = 99.3724$ and $s = 100.8908$, both of which are reasonably close to 100 and validate the inverse cdf formula.

In general, an exponential distribution with mean μ (equivalently, parameter $\lambda = 1/\mu$) can be simulated using the transform $x = -\mu\ln(1 - u)$. ∎

The preceding two examples illustrated the inverse cdf method for fairly simple density functions: a linear polynomial and an exponential function. In practice, the algebraic complexity of $f(x)$ can often be a barrier to implementing this simulation technique. After all, the algorithm requires that we can (1) obtain the cdf $F(x)$ in closed form and (2) find the inverse function of F in closed form. Consider, for example, attempting to simulate values from the $N(0, 1)$ distribution: its cdf is the function denoted $\Phi(z)$ and given by the integral expression

$(1/\sqrt{2\pi}) \int_{-\infty}^{z} e^{-u^2/2} du$. There is no closed-form expression for this integral, let alone a method to solve $u = \Phi(z)$ for z and implement Eq. (3.12). (As a reminder, the lack of a closed-form expression for $\Phi(z)$ is the reason that software or tables are always required for calculations involving normal probabilities.) Thankfully, most software packages, including Matlab and R, have built-in tools to simulate normally distributed variates (using a very clever algorithm called the *Box-Muller method*; see Sect. 4.6). We'll discuss built-in simulation tools at the end of this section.

As the next example illustrates, even when $F(x)$ can be determined in closed form we cannot necessarily implement the inverse cdf method, because $F(x)$ cannot always be inverted. This difficulty surfaces in practice when attempting to simulate values from a gamma distribution.

Example 3.44 The measurement error X (in mV) of a particular volt-meter has the following distribution: $f(x) = (4 - x^2)/9$ for $-1 \le x \le 2$ (and $f(x) = 0$ otherwise). To use the inverse cdf method to simulate X, begin by calculating its cdf:

$$F(x) = \int_{-1}^{x} (4 - y^2)/9 dy = \frac{-x^3 + 12x + 11}{27}$$

To implement step 2 of the inverse cdf method requires solving $F(x) = u$ for x; since $F(x)$ is a cubic polynomial, this is not a simple task. Advanced computer algebra systems can solve this equation, though the general solution is unwieldy (and such a solution doesn't exist at all for 5th-degree and higher polynomials). Readers familiar with numerical analysis methods may recognize that, for any specified numerical value of u, a root-finding algorithm (such as Newton–Raphson) can be implemented to *approximate* the solution x. This latter method, however, is computationally intensive, especially if it's desirable to generate 10,000 or more simulated values of x. ∎

The preceding example suggests that the inverse cdf method is insufficient for simulating all continuous distributions in practice. We next consider an alternative algorithm that, while less efficient, has a broader scope.

3.8.2 The Accept–Reject Method

When the inverse cdf method of simulation cannot be implemented, the *accept–reject method* provides an alternative. The downside of the accept–reject method, as will be explained below, is that only some of the random numbers generated by software will be used ("accepted"), while others will be "rejected." As a result, one needs to create more—sometimes, many more—random variates than the desired number of simulated values.

Suppose we wish to simulate a random variable X, whose pdf is $f(x)$. The key to the accept–reject method is to begin with a *different* pdf, call it $g(x)$, that satisfies two properties: (1) we can already simulate values from $g(x)$, so g is either algebraically simple or else built into our software package; (2) the set of possible x-values for the distribution specified by $g(x)$ equals (or exceeds) that of $f(x)$. For example, to simulate the distribution in Example 3.44, whose range of x-values is $[-1, 2]$, one might select for $g(x)$ the uniform distribution on $[-1, 2]$, i.e., $g(x) = 1/3$ for $-1 \leq x \leq 2$. If X takes on values across $[0, \infty)$, then an exponential pdf would be a logical choice for $g(x)$.

ACCEPT–REJECT METHOD

It is desired to simulate n values from a distribution pdf $f(x)$. Let $g(x)$ be some other pdf such that the ratio f/g is bounded, i.e., there exists a constant c such that $f(x)/g(x) \leq c$ for all x. (The constant c is sometimes called the *majorization constant*.) Proceed as follows:

1. Generate a variate, y, from the distribution g. This value y is called a *candidate*.
2. Generate a standard uniform variate, u.
3. If $u \cdot c \cdot g(y) \leq f(y)$, then assign $x = y$ (i.e., "accept" the candidate). Otherwise, discard ("reject") y and return to step 1.

These steps are repeated until n candidate values have been accepted. The resulting accepted values $x_1, \ldots, x_n$ form a simulation of a random variable with the original pdf, $f(x)$.

A proof that the method works—i.e., that the resulting values really do simulate the target distribution $f(x)$—requires material from Chap. 4 (see Exercise 22 at the end of Sect. 4.1).

Fig. 3.42 The accept–reject method

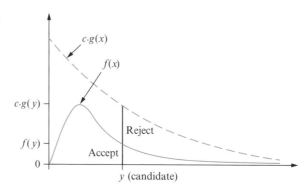

Figure 3.42 illustrates the key step in this algorithm. A candidate y has been generated on the common interval of the pdfs f and g. Given y, the left-hand side of the inequality in step 3, $u \cdot c \cdot g(y)$, is uniformly distributed on the interval from 0 to $c \cdot g(y)$ (since u itself is standard uniform). If $u \cdot c \cdot g(y)$ falls between

0 and $f(y)$, i.e., lies underneath the target pdf f, then that y-value is accepted as coming from f; otherwise, y is rejected.

As a corollary to proving the validity of the accept–reject method, it can also be shown that the probability any particular candidate y is accepted equals $1/c$. (The value of c must always exceed 1; can you see why?) Since successive candidates are independent, it follows that the number of candidates required to generate a single acceptable value has a geometric distribution, and the expected number of candidates to generate one x from $f(x)$ is $1/(1/c)=c$. By extension, the expected number of candidates required to generate our simulation sample of size n is cn. Consequently, the majorization constant c should always be made as small as possible, i.e., we should find the *smallest* value c such that $f(x)/g(x) \le c$ for all x under consideration.

Example 3.45 (Example 3.44 continued) In order to simulate 10,000 values from the pdf $f(x)=(4-x^2)/9$, $-1 \le x \le 2$, we will rely on our ability to generate variates from $g(x)=1/3$ on $-1 \le x \le 2$, the uniform pdf. To implement the accept–

a
```
x=zeros(10000,1);
i=0;
while i<10000
    y=-1+3*rand;
    u=rand;
    if u*4/3*1/3<=(4-y^2)/9
        i=i+1;
        x(i)=y;
    end
end
```

b
```
x <- NULL
i <- 0
while (i<10000){
    y <- -1+3*runif(1)
    u <- runif(1)
    if (u*4/3*1/3<=(4-y^2)/9){
        i <- i+1
        x[i] <- y
    }
}
```

Fig. 3.43 Simulation code for Example 3.45: (**a**) Matlab; (**b**) R

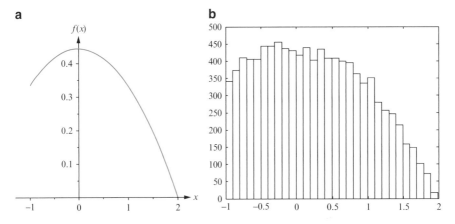

Fig. 3.44 pdf and histogram of simulated values for Example 3.45

reject method, we must determine the majorization constant, c, by looking at the ratio f/g:

$$\frac{f(x)}{g(x)} = \frac{(4 - x^2)/9}{1/3} = \frac{4 - x^2}{3} \leq \frac{4 - 0^2}{3} = \frac{4}{3} \qquad \text{for} - 1 \leq x \leq 2$$

The expression $4 - x^2$ represents a downward-facing parabola with vertex at $x = 0$, so it is clearly maximized at 0. We conclude that $c = 4/3$ is the smallest possible majorization constant, and that is what we shall use. To create the desired simulation, the following steps are repeated until 10,000 values are accepted in step 3.

1. Generate y from the uniform distribution on $[-1, 2]$.
2. Generate u from the standard uniform RNG.
3. If $u \cdot \dfrac{4}{3} \cdot \dfrac{1}{3} \leq \dfrac{4 - y^2}{9}$, assign $x = y$; otherwise, discard y and return to step 1.

Figure 3.43 shows the preceding algorithm implemented in Matlab and R. Both programs result in a vector of 10,000 simulated values from the pdf $f(x)$. Figure 3.44 shows $f(x)$ alongside the simulated values from Matlab. Since $c = 4/3$, it's expected to require $4/3(10,000) = 13,333$ iterations of the while loop to create the desired simulation size; by adding a counter to the program, one run of the Matlab code was found to use 13,303 candidates.

You may have noticed that step 3 may be simplified: the inequality $u \leq (4 - x^2)/4$ would be equivalent to the one presented. In fact, it is very common to see this final step of the accept–reject algorithm written as "accept y iff $u \leq f(y)/[c \cdot g(y)]$." ∎

For more information on the accept–reject method and selection of a sensible "candidate" distribution $g(x)$ consult the text *Simulation* by Ross listed in the references.

3.8.3   Built-In Simulation Packages for Matlab and R

As was true for the most common discrete distributions, many software packages have built-in tools for simulating values from the continuous models named in this chapter. Table 3.3 summarizes the relevant functions in Matlab and R for the uniform, normal, gamma, and exponential distributions; the variable n refers to the desired number of simulated values of the distribution. Both packages include similar commands for the Weibull, lognormal, and beta distributions.

As was the case with the cdf commands discussed in Sect. 3.4, Matlab and R parameterize the gamma and exponential distributions differently: Matlab always requires the "scale" parameter $\beta = 1/\lambda$, while R takes in the "rate" parameter $\lambda = 1/\beta$. (In the gamma simulation command, this can be overridden by naming the final argument scale, as in rgamma(n, α, scale = β).) In R, the command rnorm(n) will generate standard normal variates (i.e., with $\mu = 0$ and $\sigma = 1$), but

Table 3.3 Functions to simulate major continuous distributions in Matlab and R

Distribution	Matlab code	R code
Unif[A, B]	`random('unif',A,B,[n,1])`	`runif(n,A,B)`
$N(\mu, \sigma)$	`random('norm',`μ`,` σ`,[n,1])`	`rnorm(n,`μ`,`σ`)`
Gamma(α, β)	`random('gam',`α`,`β`,[n,1])`	`rgamma(n,`α`,1/`β`)`
Exponential(λ)	`random('exp',1/`λ`,[n,1])`	`rexp(n,`λ`)`

those arguments are required in Matlab. Similarly, R will generate standard uniform variates ($A = 0$ and $B = 1$), the basis for many of our simulation methods, with the command `runif(n)`. Matlab's corresponding syntax is `rand([n,1])`; if you type `rand(100)` instead of `rand([100,1])`, you will receive a 100-by-100 *matrix* of Unif[0, 1) values.

3.8.4 Precision of Simulation Results

Sect. 2.8 discusses in detail the precision of estimates associated with simulating discrete random variables. The same results apply in the continuous case. In particular, the estimated standard error in using a sample proportion $\hat{p}$ to estimate the true probability of an event is still $\sqrt{\hat{p}(1 - \hat{p})/n}$, where n is the simulation size. Also, the estimated standard error in using a sample mean, $\bar{x}$, to estimate the true expected value μ of a (continuous) rv X is $s/\sqrt{n}$, where s is the sample standard deviation of the simulated values of X. Refer back to Sect. 2.8 for more details.

3.8.5 Exercises: Section 3.8 (129–139)

129. The amount of time (hours) required to complete an unusually short statistics homework assignment is modeled by the pdf $f(x) = x/2$ for $0 < x < 2$ (and $= 0$ otherwise).
    (a) Obtain the cdf and then its inverse.
    (b) Write a program to simulate 10,000 values from this distribution.
    (c) Compare the sample mean and standard deviation of your 10,000 simulated values to the theoretical mean and sd of this distribution (which you can determine by calculating the appropriate integrals).
130. The Weibull distribution was introduced in Sect. 3.5.
    (a) Find the inverse cdf for the Weibull distribution.
    (b) Write a program to simulate n values from a Weibull distribution. Your program should have three inputs: the desired number of simulated values n and the two parameters α and β. It should have a single output: an $n \times 1$ vector of simulated values.

(c) Use your program from part (b) to simulate 10,000 values from a Weibull (4, 6) distribution and estimate the mean of this distribution. The correct value of the mean is $6\Gamma(5/4) \approx 5.438$; how close is your sample mean?

131. Consider the pdf for the rv X = magnitude (in newtons) of a dynamic load on a bridge, given in Example 3.7:

$$f(x) = \begin{cases} \dfrac{1}{8} + \dfrac{3}{8}x & 0 \le x \le 2 \\ 0 & \text{otherwise} \end{cases}$$

Write a program to simulate values from this distribution using the inverse cdf method.

132. In distributed computing, any given task is split into smaller sub-tasks which are handled by separate processors (which are then recombined by a multiplexer). Consider a distributed computing system with 4 processors, and suppose for one particular purpose that pdf of completion time for a particular sub-task (microseconds) on any one of the processors is given by

$$f(x) = \begin{cases} \dfrac{20}{3x^2} & 4 \le x \le 10 \\ 0 & \text{otherwise} \end{cases}$$

That is, the sub-task completion times X_1, X_2, X_3, X_4 of the four processors each have the above pdf.

(a) Write a program to simulate the above pdf using the inverse cdf method.

(b) The overall time to complete any task is the largest of the four sub-task completion times: if we call this variable Y, then $Y = \max(X_1, X_2, X_3, X_4)$. (We assume that the multiplexing time is negligible). Use your program in part (a) to simulate 10,000 values of the rv Y. Create a histogram of the simulated values of Y, and also use your simulation to estimate both $E(Y)$ and $SD(Y)$.

133. Exercise 16 in Sect. 3.1 introduced the following model for wait times at street crossings:

$$f(x; \theta, \tau) = \begin{cases} \dfrac{\theta}{\tau}(1 - x/\tau)^{\theta - 1} & 0 \le x < \tau \\ 0 & \text{otherwise} \end{cases}$$

where $\theta > 0$ and $\tau > 0$ are the parameters of the model.

(a) Write a function to simulate values from this distribution, implementing the inverse cdf method. Your function should have three inputs: the desired number of simulated values n and values for the two parameters for θ and τ.

(b) Use your function in part (a) to simulate 10,000 values from this wait time distribution with $\theta = 4$ and $\tau = 80$. Estimate $E(X)$ under these parameter settings. How close is your estimate to the correct value of 16?

134. Explain why the transformation $x = -\mu\ln(u)$ may be used to simulate values from an exponential distribution with mean μ. (This expression is slightly simpler than the one established in this section.)

135. Recall the rv X = amount of gravel (in tons) sold by a construction supply company in a given week from Example 3.9, whose pdf is

$$f(x) = \begin{cases} \dfrac{3}{2}(1 - x^2) & 0 \leq x \leq 1 \\ 0 & \text{otherwise} \end{cases}$$

Consider simulating values from this distribution using the accept–reject method with a Unif[0, 1] "candidate" distribution, i.e., $g(x) = 1$ for $0 \leq x \leq 1$.

(a) Find the smallest majorization constant c so that $f(x)/g(x) \leq c$ for all x in $[0, 1]$.

(b) Write a program to simulate values from this distribution.

(c) On the average, how many candidate values must your program generate in order to create 10,000 "accepted" values?

(d) Simulate 10,000 values from this distribution, and use these to estimate the mean μ of this distribution. How close is your sample mean to the true value of μ (which you can determine using the appropriate integral)?

(e) The supply company's management looks at quarterly data for X, i.e., values $X_1, \ldots, X_{13}$ for 13 weeks (one quarter of a year). Of particular interest is the variable $M = \min(X_1, \ldots, X_{13})$, the least amount of gravel sold in one week during a quarter. Use your program in (b) to simulate the rv M, and use the results of at least 10,000 simulated values of M to estimate $P(M < .1)$, the chance that the worst sales week in a quarter saw less than .1 tons of gravel sold. [*Hint*: Simulate each X_i 10,000 times for $i = 1, \ldots, 13$, and then compute the minimum of each set of 13 values to create a value for M.]

136. The time required to complete a 3-h final exam is modeled by the following pdf:

$$f(x) = \begin{cases} \dfrac{4}{27}x^2(3 - x) & 0 \leq x \leq 3 \\ 0 & \text{otherwise} \end{cases}$$

Consider simulating values from this distribution using the accept–reject method with a uniform "candidate" distribution on the interval [0, 3].

(a) Find the smallest majorization constant c so that $f(x)/g(x) \leq c$ for all x in $[0, 3]$. [*Hint*: What is the pdf of the uniform distribution on [0, 3]?]

(b) Write a program to simulate values from this distribution.

(c) On the average, how many candidate values must your program generate in order to create 10,000 "accepted" values?

(d) A professor has 20 students taking her class (lucky professor!). Assume her 20 students' completion times on the final exam can be modeled as 20 independent observations from the above pdf. The professor must stay at the final exam until all 20 students are finished (i.e., until the last student leaves). Use your program in (b) to simulate the rv $L = $ time, in hours, that the professor sits proctoring her final exam to 20 students. Use your simulation to estimate $P(L \geq 35/12)$, the probability she will have to stay into the last 5 min of the final exam period.

137. The *half-normal* distribution has the following pdf:

$$f(x) = \begin{cases} \sqrt{\frac{2}{\pi}} \cdot e^{-x^2/2} & x \geq 0 \\ 0 & \text{otherwise} \end{cases}$$

This is the distribution of $|Z|$, where $Z \sim N(0, 1)$; equivalently, it's the pdf that arises by "folding" the standard normal distribution in half along its line of symmetry. Consider simulating values from this distribution using the accept–reject method with a candidate distribution $g(x) = e^{-x}$ for $x \geq 0$ (i.e., an exponential pdf with $\lambda = 1$).

(a) Find the inverse cdf corresponding to $g(x)$. (This will allow us to simulate values from the candidate distribution.)
(b) Find the smallest majorization constant c so that $f(x)/g(x) \leq c$ for all $x \geq 0$. [*Hint*: Use calculus to determine where the ratio $f(x)/g(x)$ is maximized.]
(c) On the average, how many candidate values will be required to generate 10,000 "accepted" values?
(d) Write a program to construct 10,000 values from a half-normal distribution.

138. As discussed previously, the normal distribution cannot be simulated using the inverse cdf method. One possibility for simulating from a standard normal distribution is to employ the accept–reject method with candidate distribution

$$g(x) = \frac{1}{\pi(1 + x^2)} \quad -\infty < x < \infty$$

(This is the *Cauchy* distribution.)
(a) Find the cdf and inverse cdf corresponding to $g(x)$. (This will allow us to simulate values from the candidate distribution.)
(b) Find the smallest majorization constant c so that $f(x)/g(x) \leq c$ for all x, where $f(x)$ is the standard normal pdf. [*Hint*: Use calculus to determine where the ratio $f(x)/g(x)$ is maximized.]
(c) On the average, how many candidate values will be required to generate 10,000 "accepted" values?
(d) Write a program to construct 10,000 values from a standard normal distribution.
(e) Suppose that you now wish to simulate from a $N(\mu, \sigma)$ distribution. How would you modify your program in part (d)?

139. Explain why the majorization constant c in the accept–reject algorithm must be ≥ 1. [*Hint*: If $c < 1$, then $f(x) < g(x)$ for all x. Why is this bad?]

3.9 Supplementary Exercises (140–172)

140. An insurance company issues a policy covering losses up to 5 (in thousands of dollars). The loss, X, follows a distribution with density function:

$$f(x) = \begin{cases} \dfrac{3}{x^4} & x \geq 1 \\ 0 & x < 1 \end{cases}$$

What is the expected value of the amount paid under the policy?

141. Let $X =$ the time it takes a read/write head to locate a desired record on a computer disk memory device once the head has been positioned over the correct track. If the disks rotate once every 25 msec, a reasonable assumption is that X is uniformly distributed on the interval $[0, 25]$.
   (a) Compute $P(10 \leq X \leq 20)$.
   (b) Compute $P(X \geq 10)$.
   (c) Obtain the cdf $F(X)$.
   (d) Compute $E(X)$ and SD(X).

142. A 12-in. bar clamped at both ends is subjected to an increasing amount of stress until it snaps. Let $Y =$ the distance from the left end at which the break occurs. Suppose Y has pdf

$$f(y) = \begin{cases} \dfrac{y}{24}\left(1 - \dfrac{y}{12}\right) & 0 \leq y \leq 12 \\ 0 & \text{otherwise} \end{cases}$$

Compute the following:
   (a) The cdf of Y, and graph it.
   (b) $P(Y \leq 4)$, $P(Y > 6)$, and $P(4 \leq Y \leq 6)$.
   (c) $E(Y)$, $E(Y^2)$, and SD(Y).
   (d) The probability that the break point occurs more than 2 in. from the expected break point.
   (e) The expected length of the shorter segment when the break occurs.

143. Let X denote the time to failure (in years) of a hydraulic component. Suppose the pdf of X is $f(x) = 32/(x+4)^3$ for $x > 0$.
   (a) Verify that $f(x)$ is a legitimate pdf.
   (b) Determine the cdf.
   (c) Use the result of part (b) to calculate the probability that time to failure is between 2 and 5 years.
   (d) What is the expected time to failure?

(e) If the component has a salvage value equal to $100/(4+x)$ when its time to failure is x, what is the expected salvage value?

144. The completion time X for a task has cdf $F(x)$ given by

$$
\begin{cases}
0 & x < 0 \\[2mm]
\dfrac{x^3}{3} & 0 \le x < 1 \\[4mm]
1 - \dfrac{1}{2}\left(\dfrac{7}{3} - x\right)\left(\dfrac{7}{4} - \dfrac{3}{4}x\right) & 1 \le x \le \dfrac{7}{3} \\[4mm]
1 & x \ge \dfrac{7}{3}
\end{cases}
$$

(a) Obtain the pdf $f(x)$ and sketch its graph.
(b) Compute $P(.5 \le X \le 2)$.
(c) Compute $E(X)$.

145. The breakdown voltage of a randomly chosen diode of a certain type is known to be normally distributed with mean value 40 V and standard deviation 1.5 V.
(a) What is the probability that the voltage of a single diode is between 39 and 42?
(b) What value is such that only 15% of all diodes have voltages exceeding that value?
(c) If four diodes are independently selected, what is the probability that at least one has a voltage exceeding 42?

146. The article "Computer Assisted Net Weight Control" (*Qual. Prog.*, 1983: 22–25) suggests a normal distribution with mean 137.2 oz and standard deviation 1.6 oz, for the actual contents of jars of a certain type. The stated contents was 135 oz.
(a) What is the probability that a single jar contains more than the stated contents?
(b) Among ten randomly selected jars, what is the probability that at least eight contain more than the stated contents?
(c) Assuming that the mean remains at 137.2, to what value would the standard deviation have to be changed so that 95% of all jars contain more than the stated contents?

147. When circuit boards used in the manufacture of compact disk players are tested, the long-run percentage of defectives is 5%. Suppose that a batch of 250 boards has been received and that the condition of any particular board is independent of that of any other board.
(a) What is the approximate probability that at least 10% of the boards in the batch are defective?
(b) What is the approximate probability that there are exactly 10 defectives in the batch?

148. The article "Reliability of Domestic-Waste Biofilm Reactors" (*J. Envir. Engr.*, 1995: 785–790) suggests that substrate concentration (mg/cm^3) of influent to a reactor is normally distributed with $\mu = .30$ and $\sigma = .06$.
   (a) What is the probability that the concentration exceeds .25?
   (b) What is the probability that the concentration is at most .10?
   (c) How would you characterize the largest 5% of all concentration values?

149. Let $X =$ the hourly median power (in decibels) of received radio signals transmitted between two cities. The authors of the article "Families of Distributions for Hourly Median Power and Instantaneous Power of Received Radio Signals" (*J. Res. Nat. Bureau Standards*, vol. 67D, 1963: 753–762) argue that the lognormal distribution provides a reasonable probability model for X. If the parameter values are $\mu = 3.5$ and $\sigma = 1.2$, calculate the following:
   (a) The mean value and standard deviation of received power.
   (b) The probability that received power is between 50 and 250 dB.
   (c) The probability that X is less than its mean value. Why is this probability not .5?

150. Let X be a nonnegative continuous random variable with cdf $F(x)$ and mean $E(X)$.
   (a) The definition of expected value is $E(X) = \displaystyle\int_0^\infty xf(x)dx$. Replace the first x inside the integral with $\displaystyle\int_0^x 1\, dy$ to create a double integral expression for $E(X)$. [The "order of integration" should be $dy\, dx$.]
   (b) Rearrange the order of integration, keeping track of the revised limits of integration, to show that
$$E(X) = \int_0^\infty \int_y^\infty f(x)dxdy$$
   (c) Evaluate the dx integral in (b) to show that $E(X) = \displaystyle\int_0^\infty [1 - F(y)]dy$. (This provides an alternate derivation of the formula established in Exercise 38.)
   (d) Use the result of (c) to verify that the expected value of an exponentially distributed rv with parameter λ is $1/\lambda$.

151. The reaction time (in seconds) to a stimulus is a continuous random variable with pdf
$$f(x) = \begin{cases} \dfrac{3}{2x^2} & 1 \le x \le 3 \\ 0 & \text{otherwise} \end{cases}$$

   (a) Obtain the cdf.
   (b) What is the probability that reaction time is at most 2.5 s? Between 1.5 and 2.5 s?
   (c) Compute the expected reaction time.

(d) Compute the standard deviation of reaction time.

(e) If an individual takes more than 1.5 s to react, a light comes on and stays on either until one further second has elapsed or until the person reacts (whichever happens first). Determine the expected amount of time that the light remains lit. [*Hint:* Let $h(X) =$ the time that the light is on as a function of reaction time X.]

152. The article "Characterization of Room Temperature Damping in Aluminum-Indium Alloys" (*Metallurgical Trans.*, 1993: 1611-1619) suggests that aluminum matrix grain size (μm) for an alloy consisting of 2% indium could be modeled with a normal distribution with mean 96 and standard deviation 14.

    (a) What is the probability that grain size exceeds 100 μm?

    (b) What is the probability that grain size is between 50 and 80 μm?

    (c) What interval (a, b) includes the central 90% of all grain sizes (so that 5% are below a and 5% are above b)?

153. The article "Determination of the MTF of Positive Photoresists Using the Monte Carlo Method" (*Photographic Sci. Engrg.*, 1983: 254–260) proposes the exponential distribution with parameter $\lambda = .93$ as a model for the distribution of a photon's free path length (mm) under certain circumstances. Suppose this is the correct model.

    (a) What is the expected path length, and what is the standard deviation of path length?

    (b) What is the probability that path length exceeds 3.0? What is the probability that path length is between 1.0 and 3.0?

    (c) What value is exceeded by only 10% of all path lengths?

154. The article "The Prediction of Corrosion by Statistical Analysis of Corrosion Profiles" (*Corrosion Sci.*, 1985: 305–315) suggests the following cdf for the depth X of the deepest pit in an experiment involving the exposure of carbon manganese steel to acidified seawater:

$$F(x; \theta_1, \theta_2) = e^{-e^{-(x-\theta_1)/\theta_2}} \qquad -\infty < x < \infty$$

(This is called the *largest extreme value distribution* or *Gumbel distribution*.) The investigators proposed the values $\theta_1 = 150$ and $\theta_2 = 90$. Assume this to be the correct model.

    (a) What is the probability that the depth of the deepest pit is at most 150? At most 300? Between 150 and 300?

    (b) Below what value will the depth of the maximum pit be observed in 90% of all such experiments?

    (c) What is the density function of X?

    (d) The density function can be shown to be unimodal (a single peak). Above what value on the measurement axis does this peak occur? (This value is the mode.)

    (e) It can be shown that $E(X) \approx .5772\theta_2 + \theta_1$. What is the mean for the given values of θ_1 and θ_2, and how does it compare to the median and mode? Sketch the graph of the density function.

155. Let $t =$ the amount of sales tax a retailer owes the government for a certain period. The article "Statistical Sampling in Tax Audits" (*Statistics and the Law*, 2008: 320–343) proposes modeling the uncertainty in t by regarding it as a normally distributed random variable with mean value μ and standard deviation σ (in the article, these two parameters are estimated from the results of a tax audit involving n sampled transactions). If a represents the amount the retailer is assessed, then an underassessment results if $t > a$ and an overassessment if $a > t$. We can express this in terms of a *loss function*, a function that shows zero loss if $t = a$ but increases as the gap between t and a increases. The proposed loss function is $L(a, t) = t - a$ if $t > a$ and $= k(a - t)$ if $t \leq a$ ($k > 1$ is suggested to incorporate the idea that over-assessment is more serious than under-assessment).

   (a) Show that $a* = \mu + \sigma\Phi^{-1}(1/(k + 1))$ is the value of a that minimizes the expected loss, where Φ^{-1} is the inverse function of the standard normal cdf.

   (b) If $k = 2$ (suggested in the article), $\mu = \$100,000$, and $\sigma = \$10,000$, what is the optimal value of a, and what is the resulting probability of over-assessment?

156. A *mode* of a continuous distribution is a value $x*$ that maximizes $f(x)$.

   (a) What is the mode of a normal distribution with parameters μ and σ?

   (b) Does the uniform distribution with parameters A and B have a single mode? Why or why not?

   (c) What is the mode of an exponential distribution with parameter λ? (Draw a picture.)

   (d) If X has a gamma distribution with parameters α and β, and $\alpha > 1$, find the mode. [*Hint*: $\ln[f(x)]$ will be maximized if and only if $f(x)$ is, and it may be simpler to take the derivative of $\ln[f(x)]$.]

157. The article "Error Distribution in Navigation" (*J. Institut. Navigation*, 1971: 429–442) suggests that the frequency distribution of positive errors (magnitudes of errors) is well approximated by an exponential distribution. Let $X =$ the lateral position error (nautical miles), which can be either negative or positive. Suppose the pdf of X is

$$f(x) = .1e^{-.2|x|} \qquad -\infty < x < \infty$$

   (a) Sketch a graph of $f(x)$ and verify that $f(x)$ is a legitimate pdf (show that it integrates to 1).

   (b) Obtain the cdf of X and sketch it.

   (c) Compute $P(X \leq 0)$, $P(X \leq 2)$, $P(-1 \leq X \leq 2)$, and the probability that an error of more than 2 miles is made.

158. The article "Statistical Behavior Modeling for Driver-Adaptive Precrash Systems" (*IEEE Trans. on Intelligent Transp. Systems*, 2013: 1-9) proposed the following distribution for modeling the behavior of what the authors called "the criticality level of a situation" X.

$$f(x; \lambda_1, \lambda_2, p) = \begin{cases} p\lambda_1 e^{-\lambda_1 x} + (1 - p)\lambda_2 e^{-\lambda_2 x} & x \geq 0 \\ 0 & \text{otherwise} \end{cases}$$

This is often called the *hyperexponential* or *mixed exponential distribution*.

(a) What is the cdf $F(x; \lambda_1, \lambda_2, p)$?
(b) If $p = .5$, $\lambda_1 = 40$, $\lambda_2 = 200$ (values of the λs suggested in the cited article), calculate $P(X > .01)$.
(c) If X has $f(x; \lambda_1, \lambda_2, p)$ as its pdf, what is $E(X)$?
(d) Using the fact that $E(X^2) = 2/\lambda^2$ when X has an exponential distribution with parameter λ, compute $E(X^2)$ when X has pdf $f(x; \lambda_1, \lambda_2, p)$. Then compute $\text{Var}(X)$.
(e) The *coefficient of variation* of a random variable (or distribution) is $CV = \sigma/\mu$. What is the CV for an exponential rv? What can you say about the value of CV when X has a hyperexponential distribution?
(f) What is the CV for an Erlang distribution with parameters λ and n as defined in Sect. 3.4? [*Note*: In applied work, the sample CV is used to decide which of the three distributions might be appropriate.]
(g) For the parameter values given in (b), calculate the probability that X is within one standard deviation of its mean value. Does this probability depend upon the values of the λs (it does not depend on λ when X has an exponential distribution)?

159. Suppose a state allows individuals filing tax returns to itemize deductions only if the total of all itemized deductions is at least $5,000. Let X (in 1000s of dollars) be the total of itemized deductions on a randomly chosen form. Assume that X has the pdf

$$f(x; \alpha) = \begin{cases} k/x^\alpha & x \geq 5 \\ 0 & \text{otherwise} \end{cases}$$

(a) Find the value of k. What restriction on α is necessary?
(b) What is the cdf of X?
(c) What is the expected total deduction on a randomly chosen form? What restriction on α is necessary for $E(X)$ to be finite?
(d) Show that $\ln(X/5)$ has an exponential distribution with parameter $\alpha - 1$.

160. Let I_i be the input current to a transistor and I_o be the output current. Then the current gain is proportional to $\ln(I_o/I_i)$. Suppose the constant of proportionality is 1 (which amounts to choosing a particular unit of measurement), so that current gain $= X = \ln(I_o/I_i)$. Assume X is normally distributed with $\mu = 1$ and $\sigma = .05$.

(a) What type of distribution does the ratio I_o/I_i have?
(b) What is the probability that the output current is more than twice the input current?
(c) What are the expected value and variance of the ratio of output to input current?

161. The article "Response of SiC_f/Si_3N_4 Composites Under Static and Cyclic Loading—An Experimental and Statistical Analysis" (*J. Engr. Materials Tech.*, 1997: 186–193) suggests that tensile strength (MPa) of composites under specified conditions can be modeled by a Weibull distribution with $\alpha = 9$ and $\beta = 180$.
    (a) Sketch a graph of the density function.
    (b) What is the probability that the strength of a randomly selected specimen will exceed 175? Will be between 150 and 175?
    (c) If two randomly selected specimens are chosen and their strengths are independent of each other, what is the probability that at least one has strength between 150 and 175?
    (d) What strength value separates the weakest 10% of all specimens from the remaining 90%?

162. (a) Suppose the lifetime X of a component, when measured in hours, has a gamma distribution with parameters α and β. Let $Y =$ lifetime measured in minutes. Derive the pdf of Y.
    (b) If X has a gamma distribution with parameters α and β, what is the probability distribution of $Y = cX$?

163. Based on data from a dart-throwing experiment, the article "Shooting Darts" (*Chance*, Summer 1997: 16–19) proposed that the horizontal and vertical errors from aiming at a point target should be independent of each other, each with a normal distribution having mean 0 and standard deviation σ. It can then be shown that the pdf of the distance V from the target to the landing point is

$$f(v) = \frac{v}{\sigma^2} \cdot e^{-v^2/(2\sigma^2)} \qquad v > 0$$

    (a) This pdf is a member of what family introduced in this chapter?
    (b) If $\sigma = 20$ mm (close to the value suggested in the paper), what is the probability that a dart will land within 25 mm (roughly 1 in.) of the target?

164. The article "Three Sisters Give Birth on the Same Day" (*Chance*, Spring 2001: 23–25) used the fact that three Utah sisters had all given birth on March 11, 1998, as a basis for posing some interesting questions regarding birth coincidences.
    (a) Disregarding leap year and assuming that the other 365 days are equally likely, what is the probability that three randomly selected births all occur on March 11? Be sure to indicate what, if any, extra assumptions you are making.
    (b) With the assumptions used in part (a), what is the probability that three randomly selected births all occur on the same day?
    (c) The author suggested that, based on extensive data, the length of gestation (time between conception and birth) could be modeled as having a normal distribution with mean value 280 days and standard deviation 19.88 days. The due dates for the three Utah sisters were March 15, April 1, and April 4, respectively. Assuming that all three due dates are at the mean of the distribution, what is the probability that all births occurred on March 11?

[*Hint*: The deviation of birth date from due date is normally distributed with mean 0.]

(d) Explain how you would use the information in part (c) to calculate the probability of a common birth date.

165. Exercise 49 introduced two machines that produce wine corks, the first one having a normal diameter distribution with mean value 3 cm and standard deviation .1 cm and the second having a normal diameter distribution with mean value 3.04 cm and standard deviation .02 cm. Acceptable corks have diameters between 2.9 and 3.1 cm. If 60% of all corks used come from the first machine and a randomly selected cork is found to be acceptable, what is the probability that it was produced by the first machine?

166. A function $g(x)$ is *convex* if the chord connecting any two points on the function's graph lies above the graph. When $g(x)$ is differentiable, an equivalent condition is that for every x, the tangent line at x lies entirely on or below the graph. (See the figure below.) How does $g(\mu) = g[E(X)]$ compare to the expected value $E[g(X)]$? [*Hint*: The equation of the tangent line at $x = \mu$ is $y = g(\mu) + g'(\mu) \cdot (x - \mu)$. Use the condition of convexity, substitute X for x, and take expected values. *Note*: Unless $g(x)$ is linear, the resulting inequality (usually called Jensen's inequality) is strict ($<$ rather than $\leq$); it is valid for both continuous and discrete rvs.]

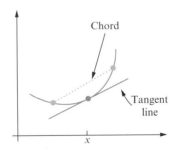

167. Let X have a Weibull distribution with parameters $\alpha = 2$ and β. Show that $Y = 2X^2/\beta^2$ has an exponential distribution with $\lambda = 1/2$.

168. Let X have the pdf $f(x) = 1/[\pi(1 + x^2)]$ for $-\infty < x < \infty$ (a central Cauchy distribution), and show that $Y = 1/X$ has the same distribution. [*Hint*: Consider $P(|Y| \leq y)$, the cdf of $|Y|$, then obtain its pdf and show it is identical to the pdf of $|X|$.]

169. Let X have a Weibull distribution with shape parameter α and scale parameter β. Show that the transformed variable $Y = \ln(X)$ has an extreme value distribution as defined in Section 3.6, with $\theta_1 = \ln(\beta)$ and $\theta_2 = 1/\alpha$.

170. A store will order q gallons of a liquid product to meet demand during a particular time period. This product can be dispensed to customers in any amount desired, so demand during the period is a continuous random variable X with cdf $F(x)$. There is a fixed cost c_0 for ordering the product plus a cost of c_1 per gallon purchased. The per-gallon sale price of the product is d. Liquid

left unsold at the end of the time period has a salvage value of e per gallon. Finally, if demand exceeds q, there will be a shortage cost for loss of goodwill and future business; this cost is f per gallon of unfulfilled demand. Show that the value of q that maximizes expected profit, denoted by q^*, satisfies

$$P(\text{satisfying demand}) = F(q^*) = \frac{d - c_1 + f}{d - e + f}$$

Then determine the value of $F(q^*)$ if $d = \$35$, $c_0 = \$25$, $c_1 = \$15$, $e = \$5$, and $f = \$25$. [*Hint*: Let x denote a particular value of X. Develop an expression for profit when $x \leq q$ and another expression for profit when $x > q$. Now write an integral expression for expected profit (as a function of q) and differentiate.]

171. An individual's credit score is a number calculated based on that person's credit history that helps a lender determine how much s/he should be loaned or what credit limit should be established for a credit card. An article in the *Los Angeles Times* gave data which suggested that a beta distribution with parameters $A = 150$, $B = 850$, $\alpha = 8$, $\beta = 2$ would provide a reasonable approximation to the distribution of American credit scores. [*Note*: credit scores are integer-valued.]

   (a) Let X represent a randomly selected American credit score. What are the mean and standard deviation of this random variable? What is the probability that X is within 1 standard deviation of its mean?

   (b) What is the approximate probability that a randomly selected score will exceed 750 (which lenders consider a very good score)?

172. Let V denote rainfall volume and W denote runoff volume (both in mm). According to the article "Runoff Quality Analysis of Urban Catchments with Analytical Probability Models" (*J. of Water Resource Planning and Management*, 2006: 4–14), the runoff volume will be 0 if $V \leq v_d$ and will be $k(V - v_d)$ if $V > v_d$. Here v_d is the volume of depression storage (a constant), and k (also a constant) is the runoff coefficient. The cited article proposes an exponential distribution with parameter λ for V.

   (a) Obtain an expression for the cdf of W. [*Note*: W is neither purely continuous nor purely discrete; instead it has a "mixed" distribution with a discrete component at 0 and is continuous for values $w > 0$.]

   (b) What is the pdf of W for $w > 0$? Use this to obtain an expression for the expected value of runoff volume.

Joint Probability Distributions and Their Applications

4

In Chaps. 2 and 3, we studied probability models for a single random variable. Many problems in probability and statistics lead to models involving several random variables simultaneously. For example, we might consider randomly selecting a college student and defining $X =$ the student's high school GPA and $Y =$ the student's college GPA. In this chapter, we first discuss probability models for the joint behavior of several random variables, putting special emphasis on the case in which the variables are independent of each other. We then study expected values of functions of several random variables, including *covariance* and *correlation* as measures of the degree of association between two variables.

Many problem scenarios involve linear combinations of random variables. For example, suppose an investor owns 100 share of one stock and 200 shares of another. If X_1 and X_2 are the prices per share of the two stocks, then the value of investor's portfolio is $100X_1 + 200X_2$. Sections 4.3 and 4.5 enumerate the properties of linear combinations of random variables, including the celebrated *Central Limit Theorem* (CLT), which characterizes the behavior of a sum $X_1 + X_2 + \ldots + X_n$ as n increases.

The fifth section considers conditional distributions, the distributions of some random variables given the values of other random variables, e.g., the distribution of fuel efficiency conditional on the weight of a vehicle.

In Sect. 3.7, we developed methods for obtaining the distribution of some function $g(X)$ of a random variable. Section 4.6 extends these ideas to transformations of two or more rvs. For example, if X and Y are the scores on a two-part exam, we might be interested in the total score $X + Y$ and also $X/(X + Y)$, the proportion of total points achieved on the first part.

The chapter ends with sections on the bivariate normal distribution (Sect. 4.9), the reliability of devices and systems (Sect. 4.8), "order statistics" such as the median and range obtained by ordering sample observations from smallest to largest (Sect. 4.9), and simulation techniques for jointly distributed random variables (Sect. 4.10).

M.A. Carlton and J.L. Devore, *Probability with Applications in Engineering, Science, and Technology*, Springer Texts in Statistics, DOI 10.1007/978-1-4939-0395-5_4, © Springer Science+Business Media New York 2014

4.1   Jointly Distributed Random Variables

There are many experimental situations in which more than one random variable (rv) will be of interest to an investigator. For example X might be the number of books checked out from a public library on a particular day and Y the number of videos checked out on the same day. Or X and Y might be the height and weight, respectively, of a randomly selected adult. In general, the two rvs of interest could both be discrete, both be continuous, or one could be discrete and the other continuous. In practice, the two "pure" cases—both of the same type—predominate. We shall first consider joint probability distributions for two discrete rvs, then for two continuous variables, and finally for more than two variables.

4.1.1   The Joint Probability Mass Function for Two Discrete Random Variables

The probability mass function (pmf) of a single discrete rv X specifies how much probability mass is placed on each possible X value. The joint pmf of two discrete rvs X and Y describes how much probability mass is placed on each possible pair of values (x, y).

DEFINITION
Let X and Y be two discrete rvs defined on the sample space $\mathscr{S}$ of an experiment. The **joint probability mass function** $p(x, y)$ is defined for each pair of numbers (x, y) by

$$p(x,y) = P(X = x \text{ and } Y = y)$$

A function $p(x, y)$ can be used as a joint pmf provided that $p(x, y) \geq 0$ for all x and y and $\sum_x \sum_y p(x,y) = 1$. Let A be any set consisting of pairs of (x, y) values, such as $\{(x, y): x+y < 10\}$. Then the probability that the random pair (X, Y) lies in A is obtained by summing the joint pmf over pairs in A:

$$P((X,Y) \in A) = \sum_{(x,y) \in A} \sum p(x,y)$$

Example 4.1 A large insurance agency services a number of customers who have purchased both a homeowner's policy and an automobile policy from the agency. For each type of policy, a deductible amount must be specified. For an automobile policy, the choices are $100 and $250, whereas for a homeowner's policy, the choices are 0, $100, and $200. Suppose an individual with both types of policy is selected at random from the agency's files. Let $X =$ the deductible amount on the auto policy and $Y =$ the deductible amount on the homeowner's policy. Possible (X, Y) pairs are then (100, 0), (100, 100), (100, 200), (250, 0), (250, 100), and (250, 200); the joint pmf specifies the probability associated with each one of these pairs, with any other pair having probability zero. Suppose the joint pmf is as given in the accompanying **joint probability table**:

	$p(x, y)$	0	100	200
x	100	.20	.10	.20
	250	.05	.15	.30

Then $p(100,\ 100) = P(X = 100$ and $Y = 100) = P(\$100$ deductible on both policies$) = .10$. The probability $P(Y \geq 100)$ is computed by summing probabilities of all (x, y) pairs for which $y \geq 100$:

$$P(Y \geq 100) = p(100, 100) + p(250, 100) + p(100, 200) + p(250, 200) = .75$$

∎

It should be obvious from the preceding example that a probability such as $P(Y = 0)$, i.e., $p_Y(0)$, results from summing $p(x, 0)$ over all possibly x values. More generally the pmf of Y is obtained by fixing the value of y in turn at each possible value and summing $p(x, y)$ over all values of x. The pmf of X can be obtained by analogous summation. The result is called a *marginal pmf*, because when the $p(x, y)$ values appear in a rectangular table, the sums are just marginal (row or column) totals.

> **DEFINITION**
>
> The **marginal probability mass functions** of X and of Y, denoted by $p_X(x)$ and $p_Y(y)$, respectively, are given by
>
> $$p_X(x) = \sum_y p(x, y) \qquad p_Y(y) = \sum_x p(x, y)$$

Thus to obtain the marginal pmf of X evaluated at, say, $x = 100$, the probabilities $p(100, y)$ are added over all possible y values. Doing this for each possible X value gives the marginal pmf of X alone (i.e., without reference to Y). From the marginal pmfs, probabilities of events involving only X or only Y can be computed.

Example 4.2 (Example 4.1 continued) The possible X values are $x = 100$ and $x = 250$, so computing row totals in the joint probability table yields

$$p_X(100) = p(100, 0) + p(100, 100) + p(100, 200) = .50$$

And

$$p_X(250) = p(250, 0) + p(250, 100) + p(250, 200) = .50$$

The marginal pmf of X is then

$$p_X(x) = \begin{cases} .5 & x = 100, 250 \\ 0 & \text{otherwise} \end{cases}$$

Similarly, the marginal pmf of Y is obtained from column totals as

$$p_Y(y) = \begin{cases} .25 & y = 0, 100 \\ .50 & y = 200 \\ 0 & \text{otherwise} \end{cases}$$

so $P(Y \geq 100) = p_Y(100) + p_Y(200) = .75$ as before.   ∎

4.1.2   The Joint Probability Density Function for Two Continuous Random Variables

The probability that the observed value of a continuous rv X lies in a one-dimensional set A (such as an interval) is obtained by integrating the pdf $f(x)$ over the set A. Similarly, the probability that the pair (X, Y) of continuous rvs falls in a two-dimensional set A (such as a rectangle) is obtained by integrating a function called the *joint density function*.

DEFINITION

Let X and Y be continuous rvs. Then $f(x, y)$ is the **joint probability density function** for X and Y if for any two-dimensional set A,

$$P((X, Y) \in A) = \iint\limits_A f(x, y) dx dy$$

In particular, if A is the two-dimensional rectangle $\{(x, y): a \leq x \leq b, c \leq y \leq d\}$, then

$$P((X, Y) \in A) = P(a \leq X \leq b, c \leq Y \leq d) = \int_a^b \int_c^d f(x, y) dy dx$$

Fig. 4.1 $P((X, Y) \in A) =$ volume under density surface above A

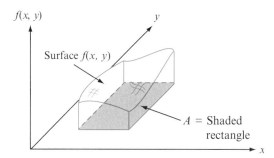

For $f(x, y)$ to be a candidate for a joint pdf, it must satisfy $f(x, y) \geq 0$ and $\int_{-\infty}^{\infty} \int_{-\infty}^{\infty} f(x, y)dxdy = 1$. We can think of $f(x, y)$ as specifying a surface at height $f(x, y)$ above the point (x, y) in a three-dimensional coordinate system. Then $P((X, Y) \in A)$ is the volume underneath this surface and above the region A, analogous to the area under a curve in the one-dimensional case. This is illustrated in Fig. 4.1.

Example 4.3 A bank operates both a drive-up facility and a walk-up window. On a randomly selected day, let $X =$ the proportion of time that the drive-up facility is in use (at least one customer is being served or waiting to be served) and $Y =$ the proportion of time that the walk-up window is in use. Then the set of possible values for (X, Y) is the rectangle $\{(x, y): 0 \leq x \leq 1, 0 \leq y \leq 1\}$. Suppose the joint pdf of (X, Y) is given by

$$f(x, y) = \begin{cases} \dfrac{6}{5}(x + y^2) & 0 \leq x \leq 1, \ 0 \leq y \leq 1 \\ \\ 0 & \text{otherwise} \end{cases}$$

To verify that this is a legitimate pdf, note that $f(x, y) \geq 0$ and

$$\int_{-\infty}^{\infty} \int_{-\infty}^{\infty} f(x, y)dxdy = \int_{0}^{1} \int_{0}^{1} \frac{6}{5}(x + y^2)dxdy = \int_{0}^{1} \int_{0}^{1} \frac{6}{5}xdxdy + \int_{0}^{1} \int_{0}^{1} \frac{6}{5}y^2dxdy$$

$$= \int_{0}^{1} \frac{6}{5}xdx + \int_{0}^{1} \frac{6}{5}y^2dy = \frac{6}{10} + \frac{6}{15} = 1$$

The probability that neither facility is busy more than one-quarter of the time is

$$P\left(0 \leq X \leq \frac{1}{4}, 0 \leq Y \leq \frac{1}{4}\right) = \int_0^{1/4} \int_0^{1/4} \frac{6}{5}(x+y^2)dxdy$$

$$= \frac{6}{5}\int_0^{1/4} \int_0^{1/4} xdxdy + \frac{6}{5}\int_0^{1/4} \int_0^{1/4} y^2 dxdy$$

$$= \frac{6}{20} \cdot \frac{x^2}{2}\Big|_{x=0}^{x=1/4} + \frac{6}{20} \cdot \frac{y^3}{3}\Big|_{y=0}^{y=1/4} = \frac{7}{640} = .0109$$

∎

The marginal pmf of one discrete variable results from summing the joint pmf over all values of the *other* variable. Similarly, the marginal pdf of one continuous variable is obtained by integrating the joint pdf over all values of the other variable.

DEFINITION

The **marginal probability density functions** of X and Y, denoted by $f_X(x)$ and $f_Y(y)$, respectively, are given by

$$f_X(x) = \int_{-\infty}^{\infty} f(x,y)dy \qquad \text{for } -\infty < x < \infty$$

$$f_Y(y) = \int_{-\infty}^{\infty} f(x,y)dx \qquad \text{for } -\infty < y < \infty$$

Example 4.4 (Example 4.3 continued) The marginal pdf of X, which gives the probability distribution of busy time for the drive-up facility without reference to the walk-up window, is

$$f_X(x) = \int_{-\infty}^{\infty} f(x,y)dy = \int_0^1 \frac{6}{5}(x+y^2)dy = \frac{6}{5}x + \frac{2}{5}$$

for $0 \leq x \leq 1$ and 0 otherwise. The marginal pdf of Y is

$$f_Y(y) = \begin{cases} \frac{6}{5}y^2 + \frac{3}{5} & 0 \leq y \leq 1 \\ 0 & \text{otherwise} \end{cases}$$

Then, for example,

$$P\left(\frac{1}{4} \le Y \le \frac{3}{4}\right) = \int_{1/4}^{3/4}\left(\frac{6}{5}y^2 + \frac{3}{5}\right)dy = \frac{37}{80} = .4625$$ ∎

In Example 4.3, the region of positive joint density was a rectangle, which made computation of the marginal pdfs relatively easy. Consider now an example in which the region of positive density is a more complicated figure.

Example 4.5 A nut company markets cans of deluxe mixed nuts containing almonds, cashews, and peanuts. Suppose the net weight of each can is exactly 1 lb, but the weight contribution of each type of nut is random. Because the three weights sum to 1, a joint probability model for any two gives all necessary information about the weight of the third type. Let $X =$ the weight of almonds in a selected can and $Y =$ the weight of cashews. Then the region of positive density is $D = \{(x, y): 0 \le x \le 1, 0 \le y \le 1, x + y \le 1\}$, the shaded region pictured in Fig. 4.2.
Now let the joint pdf for (X, Y) be

$$f(x, y) = \begin{cases} 24xy & 0 \le x \le 1, \quad 0 \le y \le 1, \quad x + y \le 1 \\ 0 & \text{otherwise} \end{cases}$$

For any fixed $x, f(x, y)$ increases with y; for fixed $y, f(x, y)$ increases with x. This is appropriate because the word *deluxe* implies that most of the can should consist of almonds and cashews rather than peanuts, so that the density function should be large near the upper boundary and small near the origin. The surface determined by $f(x, y)$ slopes upward from zero as (x, y) moves away from either axis.
   Clearly, $f(x, y) \ge 0$. To verify the second condition on a joint pdf, recall that a double integral is computed as an iterated integral by holding one variable fixed (such as x as in Fig. 4.2), integrating over values of the other variable lying along the straight line passing through the value of the fixed variable, and finally integrating over all possible values of the fixed variable. Thus

Fig. 4.2 Region of positive density for Example 4.5

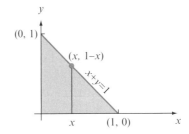

Fig. 4.3 Computing
$P((X, Y) \in A)$ for
Example 4.5

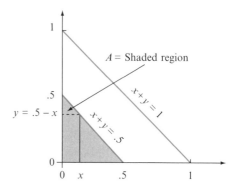

$$\int_{-\infty}^{\infty} \int_{-\infty}^{\infty} f(x,y)dydx = \int \int_{D} f(x,y)dydx = \int_{0}^{1} \left\{ \int_{0}^{1-x} 24xydy \right\} dx$$

$$= \int_{0}^{1} 24x \left\{ \frac{y^2}{2} \Big|_{y=0}^{y=1-x} \right\} dx = \int_{0}^{1} 12x(1-x)^2 dx = 1$$

To compute the probability that the two types of nuts together make up at most 50% of the can, let $A = \{(x, y): 0 \le x \le 1, 0 \le y \le 1, \text{ and } x + y \le .5\}$, as shown in Fig. 4.3. Then

$$P((X,Y) \in A) = \int \int_{A} f(x,y)dxdy = \int_{0}^{.5} \int_{0}^{.5-x} 24xydydx = .0625$$

The marginal pdf for almonds is obtained by holding X fixed at x (again, as in Fig. 4.2) and integrating $f(x, y)$ along the vertical line through x:

$$f_X(x) = \int_{-\infty}^{\infty} f(x,y)dy = \begin{cases} \int_{0}^{1-x} 24xydy = 12x(1-x)^2 & 0 \le x \le 1 \\ 0 & \text{otherwise} \end{cases}$$

By symmetry of $f(x, y)$ and the region D, the marginal pdf of Y is obtained by replacing x and X in $f_X(x)$ by y and Y, respectively.   ■

4.1.3   Independent Random Variables

In many situations, information about the observed value of one of the two variables X and Y gives information about the value of the other variable. In Example 4.1, the marginal probability of X at $x = 250$ was .5, as was the probability that $X = 100$. If, however, we are told that the selected individual had $Y = 0$, then $X = 100$ is four times as likely as $X = 250$. Thus there is a dependence between the two variables.

In Chap. 1 we pointed out that one way of defining independence of two events is to say that A and B are independent if $P(A \cap B) = P(A) \cdot P(B)$. Here is an analogous definition for the independence of two rvs.

DEFINITION

Two random variables X and Y are said to be **independent** if for every pair of x and y values,

$$p(x, y) = p_X(x) \cdot p_Y(y) \text{ when } X \text{ and } Y \text{ are discrete}$$

or   (4.1)

$$f(x, y) = f_X(x) \cdot f_Y(y) \text{ when } X \text{ and } Y \text{ are continuous}$$

If Eq. (4.1) is not satisfied for all (x, y), then X and Y are said to be **dependent**.

The definition says that two variables are independent if their joint pmf or pdf is the product of the two marginal pmfs or pdfs.

Example 4.6 In the insurance situation of Examples 4.1 and 4.2,

$$p(100, 100) = .10 \neq (.5)(.25) = p_X(100) \cdot p_Y(100)$$

so X and Y are not independent. Independence of X and Y requires that *every* entry in the joint probability table be the product of the corresponding row and column marginal probabilities. ∎

Example 4.7 (Example 4.5 continued) Because $f(x, y)$ in the nut scenario has the form of a product, X and Y would appear to be independent. However, although $f_X\left(\frac{3}{4}\right) = f_Y\left(\frac{3}{4}\right) = \frac{9}{16}$, $f\left(\frac{3}{4}, \frac{3}{4}\right) = 0 \neq \frac{9}{16} \cdot \frac{9}{16}$, so the variables are not in fact independent. To be independent, $f(x, y)$ must have the form $g(x) \cdot h(y)$ *and* the region of positive density must be a rectangle whose sides are parallel to the coordinate axes. ∎

Independence of two random variables most often arises when the description of the experiment under study tells us that X and Y have no effect on each other. Then once the marginal pmfs or pdfs have been specified, the joint pmf or pdf is simply the product of the two marginal functions. It follows that

$$P(a \leq X \leq b, c \leq Y \leq d) = P(a \leq X \leq b) \cdot P(c \leq Y \leq d)$$

Example 4.8 Suppose that the lifetimes of two components are independent of each other and that the first lifetime, X_1, has an exponential distribution with parameter λ_1 whereas the second, X_2, has an exponential distribution with parameter λ_2. Then the joint pdf is

$$f(x_1, x_2) = f_{X_1}(x_1) \cdot f_{X_2}(x_2)$$
$$= \begin{cases} \lambda_1 e^{-\lambda_1 x_1} \cdot \lambda_2 e^{-\lambda_2 x_2} = \lambda_1 \lambda_2 e^{-\lambda_1 x_1 - \lambda_2 x_2} & x_1 > 0, x_2 > 0 \\ 0 & \text{otherwise} \end{cases}$$

Let $\lambda_1 = 1/1000$ and $\lambda_2 = 1/1200$, so that the expected lifetimes are 1000 h and 1200 h, respectively. The probability that both component lifetimes are at least 1500 h is

$$P(X_1 \geq 1500, X_2 \geq 1500) = P(X_1 \geq 1500) \cdot P(X_2 \geq 1500)$$
$$= \int_{1500}^{\infty} \lambda_1 e^{-\lambda_1 x_1} dx_1 \cdot \int_{1500}^{\infty} \lambda_2 e^{-\lambda_2 x_2} dx_2$$
$$= e^{-\lambda_1(1500)} \cdot e^{-\lambda_2(1500)} = (.2231)(.2865) = .0639$$

The probability that the sum of their lifetimes, $X_1 + X_2$, is at most 3000 h requires a double integral of the joint pdf:

$$P(X_1 + X_2 \leq 3000) = P(X_1 \leq 3000 - X_2) = \int_0^{3000} \int_0^{3000-x_2} f(x_1, x_2) dx_1 dx_2$$
$$= \int_0^{3000} \int_0^{3000-X_2} \lambda_1 \lambda_2 e^{-\lambda_1 x_1 - \lambda_2 x_2} dx_1 dx_2 = \int_0^{3000} \lambda_2 e^{-\lambda_2 x_2} \left[-e^{\lambda_1 x_1} \right]_0^{3000-x_2} dx_2$$
$$= \int_0^{3000} \left[\lambda_2 e^{-\lambda_2 x_2} - \lambda_2 e^{-3000\lambda_1} e^{(\lambda_1 - \lambda_2)x_2} \right] dx_2 = .7564 \qquad \blacksquare$$

4.1.4   More Than Two Random Variables

To model the joint behavior of more than two random variables, we extend the concept of a joint distribution of two variables.

> **DEFINITION**
>
> If X_1, X_2, ..., X_n are all discrete random variables, the **joint pmf** of the variables is the function
>
> $$p(x_1, x_2, \ldots, x_n) = P(X_1 = x_1 \cap X_2 = x_2 \cap \ldots \cap X_n = x_n)$$
>
> If the variables are continuous, the **joint pdf** of X_1, X_2, ..., X_n is the function $f(x_1, x_2, \ldots, x_n)$ such that for any n intervals $[a_1, b_1]$, ..., $[a_n, b_n]$,
>
> $$P(a_1 \leq X_1 \leq b_1, \ldots, a_n \leq X_n \leq b_n) = \int_{a_1}^{b_1} \cdots \int_{a_n}^{b_n} f(x_1, \ldots, x_n) dx_n \ldots dx_1$$

Example 4.9 A binomial experiment consists of n dichotomous (success-failure), homogenous (constant success probability) independent trials. Now consider a *trinomial experiment* in which each of the n trials can result in one of *three* possible outcomes. For example, each successive customer at a store might pay with cash, a credit card, or a debit card. The trials are assumed independent. Let $p_1 = P(\text{trial}$ results in a type 1 outcome) and define p_2 and p_3 analogously for type 2 and type 3 outcomes. The random variables of interest here are $X_i =$ the number of trials that result in a type i outcome for $i = 1, 2, 3$.

In $n = 10$ trials, the probability that the first five are type 1 outcomes, the next three are type 2, and the last two are type 3—i.e., the probability of the experimental outcome 1111122233—is $p_1^5 \cdot p_2^3 \cdot p_3^2$. This is also the probability of the outcome 1122311123, and in fact the probability of any outcome that has exactly five 1s, three 2s, and two 3s. Now to determine the probability $P(X_1 = 5, X_2 = 3, \text{and}$ $X_3 = 2)$, we have to count the number of outcomes that have exactly five 1s, three 2s, and two 3s. First, there are $\binom{10}{5}$ ways to choose five of the trials to be the type 1 outcomes. Now from the remaining five trials, we choose three to be the type 2 outcomes, which can be done in $\binom{5}{3}$ ways. This determines the remaining two trials which consist of type 3 outcomes. So the total number of ways of choosing five 1s, three 2s, and two 3s is

$$\binom{10}{5} \cdot \binom{5}{3} = \frac{10!}{5!5!} \cdot \frac{5!}{3!2!} = \frac{10!}{5!3!2!} = 2520$$

Thus we see that $P(X_1 = 5, X_2 = 3, X_3 = 2) = 2520 \, p_1^5 \cdot p_2^3 \cdot p_3^2$. Generalizing this to n trials gives

$$p(x_1, x_2, x_3) = P(X_1 = x_1, x_2, X_2 = x_2, X_3 = x_3) = \frac{n!}{x_1! x_2! x_3!} p_1^{x_1} p_2^{x_2} p_3^{x_3}$$

for $x_1 = 0, 1, 2, \ldots$; $x_2 = 0, 1, 2, \ldots$; $x_3 = 0, 1, 2, \ldots$ such that $x_1 + x_2 + x_3 = n$. Notice that whereas there are three random variables here, the third variable X_3 is actually redundant, because for example in the case $n = 10$, having $X_1 = 5$ and $X_2 = 3$ implies that $X_3 = 2$ (just as in a binomial experiment there are actually two rvs—the number of successes and number of failures—but the latter is redundant).

As an example, the genotype of a pea section can be either AA, Aa, or aa. A simple genetic model specifies $P(AA) = .25$, $P(Aa) = .50$, and $P(aa) = .25$. If the alleles of ten independently obtained sections are determined, the probability that exactly five of these are Aa and two are AA is

$$p(2, 5, 3) = \frac{10!}{2!5!3!}(.25)^2(.50)^5(.25)^3 = .0769 \qquad \blacksquare$$

The trinomial scenario of Example 4.9 can be generalized by considering a **multinomial experiment** consisting of n independent and identical trials, in which each trial can result in any one of r possible outcomes. Let $p_i = P(\text{outcome } i \text{ on any particular trial})$, and define random variables by $X_i = $ the number of trials resulting in outcome i ($i = 1, \ldots, r$). The joint pmf of $X_1, \ldots, X_r$ is called the **multinomial distribution**. An argument analogous to what was done in Example 4.9 gives the joint pmf of $X_1, \ldots, X_r$:

$$p(x_1, \ldots, x_r) = \begin{cases} \dfrac{n!}{x_1! x_2! \cdots x_r!} p_1^{x_1} \cdots \cdot p_r^{x_r} & \text{for } x_i = 0, 1, 2, \ldots \text{ with } x_1 + \cdots + x_r = n \\ 0 & \text{otherwise} \end{cases}$$

The case $r = 2$ reduces to the binomial distribution, with $X_1 = $ number of successes and $X_2 = n - X_1 = $ number of failures.

Example 4.10 When a certain method is used to collect a fixed volume of rock samples in a region, there are four resulting rock types. Let X_1, X_2, and X_3 denote the proportion by volume of rock types 1, 2, and 3 in a randomly selected sample (the proportion of rock type 4 is $1 - X_1 - X_2 - X_3$, so a variable X_4 would be redundant). If the joint pdf of X_1, X_2, X_3 is

$$f(x_1, x_2, x_3) = \begin{cases} kx_1 x_2 (1 - x_3) & 0 \le x_1 \le 1, \ 0 \le x_2 \le 1, \ 0 \le x_3 \le 1, \ x_1 + x_2 + x_3 \le 1 \\ 0 & \text{otherwise} \end{cases}$$

then k is determined by

$$1 = \int_{-\infty}^{\infty} \int_{-\infty}^{\infty} \int_{-\infty}^{\infty} f(x_1, x_2, x_3) dx_3 dx_2 dx_1$$

$$= \int_{0}^{1} \left\{ \int_{0}^{1-x_1} \left[\int_{0}^{1-x_1-x_2} kx_1 x_2 (1-x_3) dx_3 \right] dx_2 \right\} dx_1$$

This iterated integral has value $k/144$, so $k = 144$. The probability that rocks of types 1 and 2 together account for at most 50% of the sample is

$$P(X_1 + X_2 \le .5) = \iiint_{\left\{\begin{array}{l} 0 \le x_i \le 1 \text{ for } i = 1,2,3 \\ x_1 + x_2 + x_3 \le 1, \ x_1 + x_2 \le .5 \end{array}\right\}} f(x_1, x_2, x_3) dx_3 dx_2 dx_1$$

$$= \int_{0}^{.5} \left\{ \int_{0}^{.5-x_1} \left[\int_{0}^{1-x_1-x_2} 144 x_1 x_2 (1-x_3) dx_3 \right] dx_2 \right\} dx_1 = .6066$$

∎

The notion of independence of more than two random variables is similar to the notion of independence of more than two events. Random variables $X_1, X_2, \ldots, X_n$ are said to be independent if for every subset $X_{i_1}, X_{i_2}, \ldots, X_{i_k}$ of the variables (each pair, each triple, and so on), the joint pmf or pdf of the subset is equal to the product of the marginal pmfs or pdfs. Thus if the variables are independent with $n = 4$, then the joint pmf or pdf of any two variables is the product of the two marginals, and similarly for any three variables and all four variables together. Most important, once we are told that n variables are independent, then the joint pmf or pdf is the product of the n marginals.

Example 4.11 If $X_1, \ldots, X_n$ represent the lifetimes of n components, the components operate independently of each other, and each lifetime is exponentially distributed with parameter λ, then

$$f(x_1, x_2, \ldots, x_n) = \left(\lambda e^{-\lambda x_1}\right) \cdot \left(\lambda e^{-\lambda x_2}\right) \cdots \cdots \left(\lambda e^{-\lambda x_n}\right)$$
$$= \begin{cases} \lambda^n e^{-\lambda \Sigma x_i} & x_1 > 0, x_2 > 0, \ldots, x_n > 0 \\ 0 & \text{otherwise} \end{cases}$$

If these n components are connected in series, so that the system will fail as soon as a single component fails, then the probability that the system lasts past time t is

$$P(X_1 > t, \ldots, X_n > t) = \int_{t}^{\infty} \cdots \int_{t}^{\infty} f(x_1, \ldots, x_n) dx_1 \ldots dx_n$$
$$= \left(\int_{t}^{\infty} \lambda e^{-\lambda x_1} dx_1\right) \cdots \left(\int_{t}^{\infty} \lambda e^{-\lambda x_n} dx_n\right) = \left(e^{-\lambda t}\right)^n = e^{-n\lambda t}$$

Therefore,

$$P(\text{system lifetime} \le t) = 1 - e^{-n\lambda t} \text{ for } t \ge 0$$

which shows that *system* lifetime has an exponential distribution with parameter $n\lambda$; the expected value of system lifetime is $1/(n\lambda)$.

A variation on the foregoing scenario appeared in the article "A Method for Correlating Field Life Degradation with Reliability Prediction for Electronic Modules" (*Quality and Reliability Engr. Intl.*, 2005: 715–726). The investigators considered a circuit card with n soldered chip resistors. The failure time of a card is the minimum of the individual solder connection failure times (mileages here). It was assumed that the solder connection failure mileages were independent, that failure mileage would exceed t if and only if the shear strength of a connection exceeded a threshold d, and that each shear strength was normally distributed with a mean value and standard deviation that depended on the value of mileage t: $\mu(t) = a_1 - a_2 t$ and $\sigma(t) = a_3 + a_4 t$ (a weld's shear strength typically deteriorates and becomes more variable as mileage increases). Then the probability that the failure mileage of a card exceeds t is

$$P(T > t) = \left(1 - \Phi \left(\frac{d - (a_1 - a_2 t)}{a_3 + a_4 t} \right) \right)^n$$

The cited article suggested values for d and the a_is based on data. In contrast to the exponential scenario, normality of individual lifetimes does not imply normality of system lifetime.   ∎

Example 4.11 gives you a taste of the sub-field of probability called *reliability*, the study of how long devices and/or systems operate; see Exercises 16 and 17 as well. We will explore reliability in great depth in Sect. 4.8.

4.1.5   Exercises: Section 4.1 (1–22)

1. A service station has both self-service and full-service islands. On each island, there is a single regular unleaded pump with two hoses. Let X denote the number of hoses being used on the self-service island at a particular time, and let Y denote the number of hoses on the full-service island in use at that time. The joint pmf of X and Y appears in the accompanying table.

		y		
$p(x, y)$		0	1	2
x	0	.10	.04	.02
	1	.08	.20	.06
	2	.06	.14	.30

(a) What is $P(X = 1 \text{ and } Y = 1)$?
(b) Compute $P(X \le 1 \text{ and } Y \le 1)$.

(c) Give a word description of the event $\{X \neq 0 \text{ and } Y \neq 0\}$, and compute the probability of this event.

(d) Compute the marginal pmf of X and of Y. Using $p_X(x)$, what is $P(X \leq 1)$?

(e) Are X and Y independent rvs? Explain.

2. A large but sparsely populated county has two small hospitals, one at the south end of the county and the other at the north end. The south hospital's emergency room has 4 beds, whereas the north hospital's emergency room has only 3 beds. Let X denote the number of south beds occupied at a particular time on a given day, and let Y denote the number of north beds occupied at the same time on the same day. Suppose that these two rvs are independent, that the pmf of X puts probability masses .1, .2, .3, .3, and .2 on the x values 0, 1, 2, 3, and 4, respectively, and that the pmf of Y distributes probabilities .1, .3, .4, and .2 on the y values 0, 1, 2, and 3, respectively.

(a) Display the joint pmf of X and Y in a joint probability table.

(b) Compute $P(X \leq 1 \text{ and } Y \leq 1)$ by adding probabilities from the joint pmf, and verify that this equals the product of $P(X \leq 1)$ and $P(Y \leq 1)$.

(c) Express the event that the total number of beds occupied at the two hospitals combined is at most 1 in terms of X and Y, and then calculate this probability.

(d) What is the probability that at least one of the two hospitals has no beds occupied?

3. A market has both an express checkout line and a superexpress checkout line. Let X_1 denote the number of customers in line at the express checkout at a particular time of day, and let X_2 denote the number of customers in line at the superexpress checkout at the same time. Suppose the joint pmf of X_1 and X_2 is as given in the accompanying table.

		x_2			
		0	1	2	3
	0	.08	.07	.04	.00
	1	.06	.15	.05	.04
x_1	2	.05	.04	.10	.06
	3	.00	.03	.04	.07
	4	.00	.01	.05	.06

(a) What is $P(X_1 = 1, X_2 = 1)$, that is, the probability that there is exactly one customer in each line?

(b) What is $P(X_1 = X_2)$, that is, the probability that the numbers of customers in the two lines are identical?

(c) Let A denote the event that there are at least two more customers in one line than in the other line. Express A in terms of X_1 and X_2, and calculate the probability of this event.

(d) What is the probability that the total number of customers in the two lines is exactly four? At least four?

(e) Determine the marginal pmf of X_1, and then calculate the expected number of customers in line at the express checkout.

(f) Determine the marginal pmf of X_2.

(g) By inspection of the probabilities $P(X_1 = 4)$, $P(X_2 = 0)$, and $P(X_1 = 4, X_2 = 0)$, are X_1 and X_2 independent random variables? Explain.

4. Suppose 51% of the individuals in a certain population have brown eyes, 32% have blue eyes, and the remainder have green eyes. Consider a random sample of 10 people from this population.

(a) What is the probability that 5 of the 10 people have brown eyes, 3 of 10 have blue eyes, and the other 2 have green eyes?

(b) What is the probability that exactly one person in the sample has blue eyes and exactly one has green eyes?

(c) What is the probability that at least 7 of the 10 people have brown eyes? [*Hint:* Think of brown as a success and all other eye colors as failures.]

5. At a certain university, 20% of all students are freshmen, 18% are sophomores, 21% are juniors, and 41% are seniors. As part of a promotion, the university bookstore is running a raffle for which all students are eligible. Ten students will be randomly selected to receive prizes (in the form of textbooks for the term).

(a) What is the probability the winners consist of two freshmen, two sophomores, two juniors, and four seniors?

(b) What is the probability the winners are split equally among underclassmen (freshmen and sophomores) and upperclassmen (juniors and seniors)?

(c) The raffle resulted in no freshmen being selected. The freshman class president complained that something must be amiss for this to occur. Do you agree? Explain.

6. According to the Mars Candy Company, the long-run percentages of various colors of M&M's milk chocolate candies are as follows:

   Blue: 24% Orange: 20% Green: 16% Yellow: 14% Red: 13% Brown: 13%

(a) In a random sample of 12 candies, what is the probability that there are exactly two of each color?

(b) In a random sample of 6 candies, what is the probability that at least one color is not included?

(c) In a random sample of 10 candies, what is the probability that there are exactly 3 blue candies and exactly 2 orange candies?

(d) In a random sample of 10 candies, what is the probability that there are at most 3 orange candies? [*Hint*: Think of an orange candy as a success and any other color as a failure.]

(e) In a random sample of 10 candies, what is the probability that at least 7 are either blue, orange, or green?

7. The number of customers waiting for gift-wrap service at a department store is an rv X with possible values 0, 1, 2, 3, 4 and corresponding probabilities .1, .2, .3, .25, .15. A randomly selected customer will have 1, 2, or 3 packages for

wrapping with probabilities .6, .3, and .1, respectively. Let $Y =$ the total number of packages to be wrapped for the customers waiting in line (assume that the number of packages submitted by one customer is independent of the number submitted by any other customer).

(a) Determine $P(X = 3, Y = 3)$, that is, $p(3, 3)$.

(b) Determine $p(4, 11)$.

8. Let X denote the number of Canon digital cameras sold during a particular week by a certain store. The pmf of X is

x	0	1	2	3	4
$p_X(x)$	.1	.2	.3	.25	.15

Sixty percent of all customers who purchase these cameras also buy an extended warranty. Let Y denote the number of purchasers during this week who buy an extended warranty.

(a) What is $P(X = 4, Y = 2)$? [*Hint*: This probability equals $P(Y = 2 | X = 4) \cdot P(X = 4)$; now think of the four purchases as four trials of a binomial experiment, with success on a trial corresponding to buying an extended warranty.]

(b) Calculate $P(X = Y)$.

(c) Determine the joint pmf of X and Y and then the marginal pmf of Y.

9. The joint probability distribution of the number X of cars and the number Y of buses per signal cycle at a proposed left-turn lane is displayed in the accompanying joint probability table.

		y	
$p(x, y)$	0	1	2
0	.025	.015	.010
1	.050	.030	.020
2	.125	.075	.050
x \hspace{6pt} 3	.150	.090	.060
4	.100	.060	.040
5	.050	.030	.020

(a) What is the probability that there is exactly one car and exactly one bus during a cycle?

(b) What is the probability that there is at most one car and at most one bus during a cycle?

(c) What is the probability that there is exactly one car during a cycle? Exactly one bus?

(d) Suppose the left-turn lane is to have a capacity of five cars, and one bus is equivalent to three cars. What is the probability of an overflow during a cycle?

(e) Are X and Y independent rvs? Explain.

10. A stockroom currently has 30 components of a certain type, of which 8 were provided by supplier 1, 10 by supplier 2, and 12 by supplier 3. Six of these are to be randomly selected for a particular assembly. Let $X =$ the number of supplier 1's components selected, $Y =$ the number of supplier 2's components selected, and $p(x, y)$ denote the joint pmf of X and Y.
   (a) What is $p(3, 2)$? [*Hint*: Each sample of size 6 is equally likely to be selected. Therefore, $p(3, 2) =$ (number of outcomes with $X = 3$ and $Y = 2$)/(total number of outcomes). Now use the product rule for counting to obtain the numerator and denominator.]
   (b) Using the logic of part (a), obtain $p(x, y)$. (This can be thought of as a multivariate hypergeometric distribution—sampling without replacement from a finite population consisting of more than two categories.)

11. Each front tire of a vehicle is supposed to be filled to a pressure of 26 psi. Suppose the actual air pressure in each tire is a random variable—X for the right tire and Y for the left tire, with joint pdf

$$f(x, y) = \begin{cases} k(x^2 + y^2) & 20 \le x \le 30, \quad 20 \le y \le 30 \\ 0 & \text{otherwise} \end{cases}$$

   (a) What is the value of k?
   (b) What is the probability that both tires are underfilled?
   (c) What is the probability that the difference in air pressure between the two tires is at most 2 psi?
   (d) Determine the (marginal) distribution of air pressure in the right tire alone.
   (e) Are X and Y independent rvs?

12. Annie and Alvie have agreed to meet between 5:00 and 6:00 p.m. for dinner at a local health-food restaurant. Let $X =$ Annie's arrival time and $Y =$ Alvie's arrival time. Suppose X and Y are independent with each uniformly distributed on the interval $[5, 6]$.
   (a) What is the joint pdf of X and Y?
   (b) What is the probability that they both arrive between 5:15 and 5:45?
   (c) If the first one to arrive will wait only 10 min before leaving to eat elsewhere, what is the probability that they have dinner at the health-food restaurant? [*Hint*: The event of interest is $A = \{(x, y) : | x - y | \le \frac{1}{6}\}$.]

13. Two different professors have just submitted final exams for duplication. Let X denote the number of typographical errors on the first professor's exam and Y denote the number of such errors on the second exam. Suppose X has a Poisson distribution with parameter μ_1, Y has a Poisson distribution with parameter μ_2, and X and Y are independent.
   (a) What is the joint pmf of X and Y?
   (b) What is the probability that at most one error is made on both exams combined?

(c) Obtain a general expression for the probability that the total number of errors in the two exams is m (where m is a nonnegative integer). [*Hint*: $A = \{(x, y): x + y = m\} = \{(m, 0), (m - 1, 1), \ldots, (1, m - 1), (0, m)\}$. Now sum the joint pmf over $(x, y) \in A$ and use the binomial theorem, which says that

$$\sum_{k=0}^{m} \binom{m}{k} a^k b^{m-k} = (a + b)^m$$

for any a, b.]

14. Two components of a computer have the following joint pdf for their useful lifetimes X and Y:

$$f(x, y) = \begin{cases} xe^{-x(1+y)} & x \geq 0 \quad \text{and} \quad y \geq 0 \\ 0 & \text{otherwise} \end{cases}$$

(a) What is the probability that the lifetime X of the first component exceeds 3?
(b) What are the marginal pdfs of X and Y? Are the two lifetimes independent? Explain.
(c) What is the probability that the lifetime of at least one component exceeds 3?

15. You have two lightbulbs for a particular lamp. Let $X =$ the lifetime of the first bulb and $Y =$ the lifetime of the second bulb (both in thousands of hours). Suppose that X and Y are independent and that each has an exponential distribution with parameter $\lambda = 1$.
(a) What is the joint pdf of X and Y?
(b) What is the probability that each bulb lasts at most 1000 h (i.e., $X \leq 1$ and $Y \leq 1$)?
(c) What is the probability that the total lifetime of the two bulbs is at most 2? [*Hint*: Draw a picture of the region $A = \{(x, y): x \geq 0, y \geq 0, x + y \leq 2\}$ before integrating.]
(d) What is the probability that the total lifetime is between 1 and 2?

16. Suppose that you have ten lightbulbs, that the lifetime of each is independent of all the other lifetimes, and that each lifetime has an exponential distribution with parameter λ.
(a) What is the probability that all ten bulbs fail before time t?
(b) What is the probability that exactly k of the ten bulbs fail before time t?
(c) Suppose that nine of the bulbs have lifetimes that are exponentially distributed with parameter λ and that the remaining bulb has a lifetime that is exponentially distributed with parameter θ (it is made by another manufacturer). What is the probability that exactly five of the ten bulbs fail before time t?

17. Consider a system consisting of three components as pictured. The system will continue to function as long as the first component functions and either component 2 or component 3 functions. Let $X_1, X_2,$ and X_3 denote the lifetimes

of components 1, 2, and 3, respectively. Suppose the X_is are independent of each other and each X_i has an exponential distribution with parameter λ.

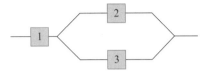

(a) Let Y denote the system lifetime. Obtain the cumulative distribution function of Y and differentiate to obtain the pdf. [*Hint*: $F(y) = P(Y \le y)$; express the event $\{Y \le y\}$ in terms of unions and/or intersections of the three events $\{X_1 \le y\}$, $\{X_2 \le y\}$, and $\{X_3 \le y\}$.]

(b) Compute the expected system lifetime.

18. (a) For $f(x_1, x_2, x_3)$ as given in Example 4.10, compute the **joint marginal density function** of X_1 and X_3 alone (by integrating over x_2).

   (b) What is the probability that rocks of types 1 and 3 together make up at most 50% of the sample? [*Hint*: Use the result of part (a).]

   (c) Compute the marginal pdf of X_1 alone. [*Hint*: Use the result of part (a).]

19. An ecologist selects a point inside a circular sampling region according to a uniform distribution. Let $X =$ the x coordinate of the point selected and $Y =$ the y coordinate of the point selected. If the circle is centered at $(0, 0)$ and has radius r, then the joint pdf of X and Y is

$$f(x, y) = \begin{cases} \dfrac{1}{\pi r^2} & x^2 + y^2 \le r^2 \\[2mm] 0 & \text{otherwise} \end{cases}$$

(a) What is the probability that the selected point is within $r/2$ of the center of the circular region? [*Hint*: Draw a picture of the region of positive density D. Because $f(x, y)$ is constant on D, computing a probability reduces to computing an area.]

(b) What is the probability that both X and Y differ from 0 by at most $r/2$?

(c) Answer part (b) for $r/\sqrt{2}$ replacing $r/2$.

(d) What is the marginal pdf of X? Of Y? Are X and Y independent?

20. Each customer making a particular Internet purchase must pay with one of three types of credit cards (think Visa, MasterCard, AmEx). Let A_i ($i = 1, 2, 3$) be the event that a type i credit card is used, with $P(A_1) = .5$, $P(A_2) = .3$, $P(A_3) = .2$. Suppose that the number of customers who make a purchase on a given day, N, is a Poisson rv with parameter μ. Define rvs X_1, X_2, X_3 by $X_i =$ the number among the N customers who use a type i card ($i = 1, 2, 3$). Show that these three rvs are independent with Poisson distributions having parameters

.5μ, .3μ, and .2μ, respectively. [*Hint*: For non-negative integers x_1, x_2, x_3, let $n = x_1 + x_2 + x_3$. Then $P(X_1 = x_1, X_2 = x_2, X_3 = x_3) = P(X_1 = x_1, X_2 = x_2, X_3 = x_3, N = n)$. Now condition on $N = n$, in which case the three X_is have a trinomial distribution (multinomial with 3 categories) with category probabilities .5, .3, and .2.]

21. Consider randomly selecting two points A and B on the circumference of a circle by selecting their angles of rotation, in degrees, independently from a uniform distribution on the interval [0, 360]. Connect points A and B with a straight line segment. What is the probability that this random chord is longer than the side of an equilateral triangle inscribed inside the circle?

    (This is called *Bertrand's Chord Problem* in the probability literature. There are other ways of randomly selecting a chord that give different answers from the one appropriate here.) [*Hint:* Place one of the vertices of the inscribed triangle at A. You should then be able to intuit the answer visually without having to do any integration.]

22. Section 3.8 introduced the accept–reject method for simulating continuous rvs. Refer back to that algorithm in order to answer the questions below.
    (a) Show that the probability a candidate value is "accepted" equals $1/c$. [*Hint:* According to the algorithm, this occurs iff $U \leq f(Y)/cg(Y)$, where $U \sim \text{Uniform}[0, 1)$ and $Y \sim g$. Compute the relevant double integral.]
    (b) Argue that the average number of candidates required to generate a single accepted value is c.
    (c) Show that the accept–reject method does result in an observation from the pdf f by showing that $P(\text{accepted value} \leq x) = F(x)$, where F is the cdf corresponding to f. [*Hint:* Let X denote the accepted value. Then $P(X \leq x) = P(Y \leq x | Y \text{ accepted}) = P(Y \leq x \cap Y \text{ accepted})/P(Y \text{ accepted}).$]

4.2   Expected Values, Covariance, and Correlation

We previously saw that any function $h(X)$ of a single rv X is itself a random variable. However, to compute $E[h(X)]$, it was not necessary to obtain the probability distribution of $h(X)$; instead, $E[h(X)]$ was computed as a weighted average of $h(X)$ values, where the weight function was the pmf $p(x)$ or pdf $f(x)$ of X. A similar result holds for a function $h(X, Y)$ of two jointly distributed random variables.

PROPOSITION

Let X and Y be jointly distributed rvs with pmf $p(x, y)$ or pdf $f(x, y)$ according to whether the variables are discrete or continuous. Then the expected value of a function $h(X, Y)$, denoted by $E[h(X, Y)]$ or $\mu_{h(X,Y)}$, is given by

$$E[h(X,Y)] = \begin{cases} \sum_x \sum_y h(x,y) \cdot p(x,y) & \text{if } X \text{ and } Y \text{ are discrete} \\ \int_{-\infty}^{\infty} \int_{-\infty}^{\infty} h(x,y) \cdot f(x,y)\, dx\, dy & \text{if } X \text{ and } Y \text{ are continuous} \end{cases} \tag{4.2}$$

The method of computing the expected value of a function $h(X_1, \ldots, X_n)$ of n random variables is similar to Eq. (4.2). If the X_is are discrete, $E[h(X_1, \ldots, X_n)]$ is an n-dimensional sum; if the X_is are continuous, it is an n-dimensional integral.

Example 4.12 Five friends have purchased tickets to a concert. If the tickets are for seats 1–5 in a particular row and the tickets are randomly distributed among the five, what is the expected number of seats separating any particular two of the five friends? Let X and Y denote the seat numbers of the first and second individuals, respectively. Possible (X, Y) pairs are $\{(1, 2), (1, 3), \ldots, (5, 4)\}$,and the joint pmf of (X, Y) is

$$p(x,y) = \begin{cases} \dfrac{1}{20} & x = 1, \ldots, 5; \quad y = 1, \ldots, 5; \quad x \neq y \\ 0 & \text{otherwise} \end{cases}$$

The number of seats separating the two individuals is $h(X, Y) = |X - Y| - 1$. The accompanying table gives $h(x, y)$ for each possible (x, y) pair.

	$h(x, y)$	x = 1	2	3	4	5
	1	–	0	1	2	3
	2	0	–	0	1	2
y	3	1	0	–	0	1
	4	2	1	0	–	0
	5	3	2	1	0	–

Thus

$$E[h(X,Y)] = \sum\sum_{(x,\,y)} h(x,y) \cdot p(x,y) = \sum_{x=1}^{5}\sum_{\substack{y=1 \\ x \neq y}}^{5} (|x - y| - 1) \cdot \frac{1}{20} = 1$$

■

Example 4.13 In Example 4.5, the joint pdf of the amount X of almonds and amount Y of cashews in a 1-lb can of nuts was

$$f(x,y) = \begin{cases} 24xy & 0 \leq x \leq 1, \quad 0 \leq y \leq 1, \quad x + y \leq 1 \\ 0 & \text{otherwise} \end{cases}$$

If 1 lb of almonds costs the company $6.00, 1 lb of cashews costs $10.00, and 1 lb of peanuts costs $3.50, then the total cost of the contents of a can is

$$h(X,Y) = 6X + 10Y + 3.5(1 - X - Y) = 3.5 + 2.5X + 6.5Y$$

(since $1 - X - Y$ of the weight consists of peanuts). The expected total cost is

$$E[h(X,Y)] = \int_{-\infty}^{\infty}\int_{-\infty}^{\infty} h(x,y) \cdot f(x,y)dxdy$$

$$= \int_{0}^{1}\int_{0}^{1-x} (3.5 + 2.5x + 6.5y) \cdot 24xy \; dydx = \$7.10$$

■

4.2.1 Properties of Expected Value

In Chaps. 2 and 3, we saw that expected values can be distributed across addition, subtraction, and multiplication by constants. In the language of mathematics, expected value is a *linear operator*. This was a simple consequence of expectation being a sum or an integral, both of which are linear. This obvious but important property, linearity of expectation, extends to more than one variable.

LINEARITY OF EXPECTATION

Let X and Y be random variables. Then, for any functions h_1, h_2 and any constants a_1, a_2, b,

$$E[a_1h_1(X,Y) + a_2h_2(X,Y) + b] = a_1E[h_1(X,Y)] + a_2E[h_2(X,Y)] + b$$

In the previous example, $E(3.5 + 2.5X + 6.5Y)$ can be rewritten as $3.5 + 2.5E(X) + 6.5E(Y)$; the means of X and Y can be computed either by using Eq. (4.2) or by first finding the marginal pdfs of X and Y and then computing the appropriate single integrals.

As another illustration, linearity of expectation tells us that for any two rvs X and Y,

$$E\left(5XY^2 - 4XY + e^X + 12\right) = 5E\left(XY^2\right) - 4E(XY) + E\left(e^X\right) + 12 \qquad (4.3)$$

In general, we cannot distribute the expected value operation any further. But when $h(X, Y)$ is a product of a function of X and a function of Y, the expected value simplifies in the case of independence.

THEOREM

Let X and Y be *independent* random variables. If $h(X, Y) = g_1(X) \cdot g_2(Y)$, then

$$E[h(X, Y)] = E[g_1(X) \cdot g_2(Y)] = E[g_1(X)] \cdot E[g_2(Y)]$$

Proof We present the proof here for two continuous rvs; the discrete case is similar. Apply Eq. (4.2):

$$E\left[h(X,Y)\right] = E\left[g_1(X) \cdot g_2(Y)\right] = \int_{-\infty}^{\infty} \int_{-\infty}^{\infty} g_1(x) \cdot g_2(y) \cdot f(x,y) \, dx \, dy \quad \text{by } (4.2)$$

$$= \int_{-\infty}^{\infty} \int_{-\infty}^{\infty} g_1(x) \cdot g_2(y) \cdot f_X(x) \cdot f_Y(y) \, dx \, dy \quad \text{because } X \text{ and } Y \text{ are independent}$$

$$= \left(\int_{-\infty}^{\infty} g_1(x) \cdot f_X(x) \, dx\right) \left(\int_{-\infty}^{\infty} g_2(y) \cdot f_Y(y) \, dy\right) = E\left[g_1(X)\right] E\left[g_2(Y)\right] \quad \blacksquare$$

So, if X and Y are independent, Eq. (4.3) simplifies further, to $5E(X)E(Y^2) - 4E(X)E(Y) + E(e^X) + 12$. Not surprisingly, both linearity of expectation and the foregoing corollary can be extended to more than two random variables.

4.2.2   Covariance

When two random variables X and Y are not independent, it is frequently of interest to assess how strongly they are related to each other.

DEFINITION

The **covariance** between two rvs X and Y is

$$\text{Cov}(X,Y) = E\left[(X - \mu_X)(Y - \mu_Y)\right]$$

$$= \begin{cases} \displaystyle\sum_x \sum_y (x - \mu_X)(y - \mu_Y)p(x,y) & \text{if } X \text{ and } Y \text{ are discrete} \\[2mm] \displaystyle\int_{-\infty}^{\infty} \int_{-\infty}^{\infty} (x - \mu_X)(y - \mu_Y)f(x,y)\, dx\, dy & \text{if } X \text{ and } Y \text{ are continuous} \end{cases}$$

The rationale for the definition is as follows. Suppose X and Y have a strong positive relationship to each other, by which we mean that large values of X tend to occur with large values of Y and small values of X with small values of Y (e.g., $X =$ height and $Y =$ weight). Then most of the probability mass or density will be associated with $(x - \mu_X)$ and $(y - \mu_Y)$ either both positive (both X and Y above their respective means) or both negative, so the product $(x - \mu_X)(y - \mu_Y)$ will tend to be positive. Thus for a strong positive relationship, $\text{Cov}(X, Y)$ should be quite positive. For a strong negative relationship, the signs of $(x - \mu_X)$ and $(y - \mu_Y)$ will tend to be opposite, resulting in a negative product. Thus for a strong negative relationship, $\text{Cov}(X, Y)$ should be quite negative. If X and Y are not strongly related, positive and negative products will tend to cancel each other, yielding a covariance near 0. Figure 4.4 illustrates the different possibilities. The covariance depends on *both* the set of possible pairs and the probabilities. In Fig. 4.4, the probabilities could be changed without altering the set of possible pairs, and this could drastically change the value of $\text{Cov}(X, Y)$.

Example 4.14 The joint and marginal pmfs for $X =$ automobile policy deductible amount and $Y =$ homeowner policy deductible amount in Example 4.1 were

		y		
$p(x, y)$		0	100	200
	100	.20	.10	.20
x	250	.05	.15	.30

x	100	250		y	0	100	200
$p_X(x)$	.5	.5		$p_Y(y)$	.25	.25	.50

from which $\mu_X = \sum x \cdot p_X(x) = 175$ and $\mu_Y = 125$. Therefore,

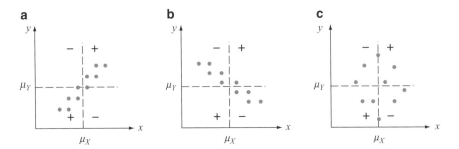

Fig. 4.4 $p(x,\ y) = .10$ for each of ten pairs corresponding to indicated points; (**a**) positive covariance; (**b**) negative covariance; (**c**) covariance near zero

$$\mathrm{Cov}(X, Y) = \sum_{(x,y)} \sum (x - 175)(y - 125)p(x, y)$$
$$= (100 - 175)(0 - 125)(.20) + \cdots$$
$$+ (250 - 175)(200 - 125)(.30) = 1875 \qquad \blacksquare$$

The following proposition summarizes some important properties of covariance.

PROPOSITION

For any two random variables X and Y,
1. $\mathrm{Cov}(X,\ Y) = \mathrm{Cov}(Y,\ X)$
2. $\mathrm{Cov}(X,\ X) = \mathrm{Var}(X)$
3. (Covariance shortcut formula) $\mathrm{Cov}(X,\ Y) = E(XY) - \mu_X \cdot \mu_Y$
4. (Distributive property of covariance) For any rv Z and any constants, a, b, c,

$$\mathrm{Cov}(aX + bY + c, Z) = a\mathrm{Cov}(X, Z) + b\mathrm{Cov}(Y, Z)$$

Proof Property 1 is obvious from the definition of covariance. To establish property 2, replace Y with X in the definition:

$$\mathrm{Cov}(X, X) = E[(X - \mu_X)(X - \mu_X)] = E\left[(X - \mu_X)^2\right] = \mathrm{Var}(X)$$

To prove property 3, apply linearity of expectation:

$$\mathrm{Cov}(X, Y) = E[(X - \mu_X)(Y - \mu_Y)]$$
$$= E(XY - \mu_X Y - \mu_Y X + \mu_X \mu_Y)$$
$$= E(XY) - \mu_X E(Y) - \mu_Y E(X) + \mu_X \mu_Y$$
$$= E(XY) - \mu_X \mu_Y - \mu_X \mu_Y + \mu_X \mu_Y = E(XY) - \mu_X \mu_Y$$

Property 4 also follows from linearity of expectation (Exercise 39). $\blacksquare$

According to property 3, the covariance shortcut, no intermediate subtractions are necessary to calculate covariance; only at the end of the computation is $\mu_X \cdot \mu_Y$ subtracted from $E(XY)$.

Example 4.15 (Example 4.5 continued) The joint and marginal pdfs of $X =$ amount of almonds and $Y =$ amount of cashews were

$$f(x, y) = \begin{cases} 24xy & 0 \leq x \leq 1, \quad 0 \leq y \leq 1, \quad x + y \leq 1 \\ 0 & \text{otherwise} \end{cases}$$

$$f_X(x) = \begin{cases} 12x(1 - x)^2 & 0 \leq x \leq 1 \\ 0 & \text{otherwise} \end{cases}$$

with $f_Y(y)$ obtained by replacing x by y in $f_X(x)$. It is easily verified that $\mu_X = \mu_Y = \frac{2}{5}$, and

$$E(XY) = \int_{-\infty}^{\infty} \int_{-\infty}^{\infty} xy f(x, y) dx dy = \int_0^1 \int_0^{1-x} xy \cdot 24xy \; dy dx = 8 \int_0^1 x^2 (1 - x)^3 dx$$

$$= \frac{2}{15}$$

Thus $\text{Cov}(X, Y) = \frac{2}{15} - \left(\frac{2}{5}\right)\left(\frac{2}{5}\right) = \frac{2}{15} - \frac{4}{25} = -\frac{2}{75}$. A negative covariance is reasonable here because more almonds in the can implies fewer cashews.   ■

4.2.3   Correlation

It would appear that the relationship in the insurance example is quite strong since $\text{Cov}(X, Y) = 1875$, whereas in the nut example $\text{Cov}(X, Y) = -2/75$ would seem to imply quite a weak relationship. Unfortunately, the covariance has a serious defect that makes it impossible to interpret a computed value of the covariance. In the insurance example, suppose we had expressed the deductible amount in cents rather than in dollars. Then $100X$ would replace X, $100Y$ would replace Y, and the resulting covariance would be $\text{Cov}(100X, 100Y) = (100)(100)\text{Cov}(X, Y) = 18,750,000$. [To see why, apply properties 1 and 4 of the previous proposition.] If, on the other hand, the deductible amounts had been expressed in hundreds of dollars, the computed covariance would have been $(.01)(.01)(1875) = .1875$. *The defect of covariance is that its computed value depends critically on the units of measurement.* Ideally, the choice of units should have no effect on a measure of strength of relationship. This is achieved by scaling the covariance.

> **DEFINITION**
> The **correlation coefficient** of X and Y, denoted by $\text{Corr}(X, Y)$, or $\rho_{X,Y}$, or just ρ, is defined by
> $$\rho_{X,Y} = \frac{\text{Cov}(X, Y)}{\sigma_X \cdot \sigma_Y}$$

Example 4.16 It is easily verified that in the insurance scenario of Example 4.14, $E(X^2) = 36{,}250$, $\sigma_X^2 = 36{,}250 - (175)^2 = 5625$, $\sigma_X = 75$, $E(Y^2) = 22{,}500$, $\sigma_Y^2 = 6875$, and $\sigma_Y = 82.92$. This gives

$$\rho = \frac{1875}{(75)(82.92)} = .301 \qquad \blacksquare$$

The following proposition shows that ρ remedies the defect of $\text{Cov}(X, Y)$ and also suggests how to recognize the existence of a strong (linear) relationship.

> **PROPOSITION**
> For any two rvs X and Y,
> 1. $\text{Corr}(X, Y) = \text{Corr}(Y, X)$
> 2. $\text{Corr}(X, X) = 1$
> 3. (Scale invariance property) If a, b, c, d are constants and $ac > 0$,
> $$\text{Corr}(aX + b, cY + d) = \text{Corr}(X, Y)$$
> 4. $-1 \leq \text{Corr}(X, Y) \leq 1$

Proof Property 1 is clear from the definition of correlation and the corresponding property of covariance. For Property 2, write $\text{Corr}(X, X) = \text{Cov}(X, X)/[\sigma_X \cdot \sigma_X] = \text{Var}(X)/\sigma_X^2 = 1$. The second-to-last step uses Property 2 of covariance. The proofs of Properties 3 and 4 appear as exercises. $\blacksquare$

Property 3 (scale invariance) says precisely that the correlation coefficient is not affected by a linear change in the units of measurement. If, say, $Y =$ completion time for a chemical reaction in seconds and $X =$ temperature in °C, then $Y/60 =$ time in minutes and $1.8X + 32 =$ temperature in °F, but $\text{Corr}(X, Y)$ will be exactly the same as $\text{Corr}(1.8X + 32, Y/60)$.

According to Properties 2 and 4, the strongest possible positive relationship is evidenced by $\rho = +1$, whereas the strongest possible negative relationship corresponds to $\rho = -1$. Therefore, the correlation coefficient provides information about both the nature and strength of the relationship between X and Y: the sign of ρ indicates whether X and Y are positively or negatively related, and the magnitude of ρ describes the strength of that relationship on an absolute 0–1 scale.

While superior to covariance, the correlation coefficient ρ is actually not a completely general measure of the strength of a relationship.

PROPOSITION

1. If X and Y are independent, then $\rho = 0$, but $\rho = 0$ does not imply independence.
2. $\rho = 1$ or -1 iff $Y = aX + b$ for some numbers a and b with $a \neq 0$.

Exercise 38 and Example 4.17 relate to Statement 1, and Statement 2 is investigated in Exercises 41 and 42(d).

This proposition says that ρ is a measure of the degree of *linear* relationship between X and Y, and only when the two variables are perfectly related in a linear manner will ρ be as positive or negative as it can be. A ρ less than 1 in absolute value indicates only that the relationship is not completely linear, but there may still be a very strong nonlinear relation. Also, $\rho = 0$ does not imply that X and Y are independent, but only that there is complete absence of a linear relationship. When $\rho = 0$, X and Y are said to be **uncorrelated**. Two variables could be uncorrelated yet highly dependent because of a strong nonlinear relationship, so be careful not to conclude too much from knowing that $\rho = 0$.

Example 4.17 Let X and Y be discrete rvs with joint pmf

$$p(x, y) = \begin{cases} .25 & (x, y) = (-4, 1), (4, -1), (2, 2), (-2, -2) \\ 0 & \text{otherwise} \end{cases}$$

The points that receive positive probability mass are identified on the (x, y) coordinate system in Fig. 4.5. It is evident from the figure that the value of X is completely determined by the value of Y and vice versa, so the two variables are completely dependent. However, by symmetry $\mu_X = \mu_Y = 0$ and $E(XY) = (-4)(.25) + (-4)(.25) + (4)(.25) + (4)(.25) = 0$, so $\text{Cov}(X, Y) = E(XY) - \mu_X \cdot \mu_Y = 0$ and thus $\rho_{X,Y} = 0$. Although there is perfect dependence, there is also complete absence of any linear relationship!

Fig. 4.5 The population of
pairs for Example 4.17

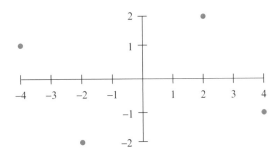

The next result provides an alternative view of zero correlation.

PROPOSITION

Two rvs X and Y are uncorrelated if, and only if, $E[XY] = \mu_X \cdot \mu_Y$.

Proof By its definition, $\mathrm{Corr}(X, Y) = 0$ iff $\mathrm{Cov}(X, Y) = 0$. Apply the covariance shortcut formula:

$$\rho = 0 \Leftrightarrow \mathrm{Cov}(X, Y) = 0 \Leftrightarrow E[XY] - \mu_X \cdot \mu_Y = 0 \Leftrightarrow E[XY] = \mu_X \cdot \mu_Y \qquad \blacksquare$$

Contrast this with an earlier proposition from this section: if X and Y are *independent*, then $E[g_1(X)g_2(Y)] = E[g_1(X)] \cdot E[g_2(Y)]$ for *all* functions g_1 and g_2. Thus, independence is stronger than zero correlation, the latter just being the special case corresponding to $g_1(X) = X$ and $g_2(Y) = Y$.

4.2.4 Correlation Versus Causation

A value of ρ near 1 does not necessarily imply that increasing the value of X *causes* Y to increase. It implies only that large X values are *associated* with large Y values. For example, in the population of children, vocabulary size and number of cavities are quite positively correlated, but it is certainly not true that cavities cause vocabulary to grow. Instead, the values of both these variables tend to increase as the value of age, a third variable, increases. For children of a fixed age, there is probably a very low correlation between number of cavities and vocabulary size. In summary, association (a high correlation) is not the same as causation.

4.2.5   Exercises: Section 4.2 (23–42)

23. The two most common types of errors made by programmers are syntax errors and logic errors. Let X denote the number of syntax errors and Y the number of logic errors on the first run of a program. Suppose X and Y have the following joint pmf for a particular programming assignment:

<table>
<tr><td colspan="5" align="center">x</td></tr>
<tr><td>$p(x,y)$</td><td>0</td><td>1</td><td>2</td><td>3</td></tr>
<tr><td>0</td><td>.71</td><td>.03</td><td>.02</td><td>.01</td></tr>
<tr><td>y 1</td><td>.04</td><td>.06</td><td>.03</td><td>.01</td></tr>
<tr><td>2</td><td>.03</td><td>.03</td><td>.02</td><td>.01</td></tr>
</table>

(a) What is the probability a program has more syntax errors than logic errors on the first run?

(b) Find the marginal pmfs of X and Y.

(c) Are X and Y independent? How can you tell?

(d) What is the average number of syntax errors in the first run of a program? What is the average number of logic errors?

(e) Suppose an evaluator assigns points to each program with the formula $100 - 4X - 9Y$. What is the expected point score for a randomly selected program?

24. An instructor has given a short quiz consisting of two parts. For a randomly selected student, let $X =$ the number of points earned on the first part and $Y =$ the number of points earned on the second part. Suppose that the joint pmf of X and Y is given in the accompanying table.

<table>
<tr><td colspan="5" align="center">y</td></tr>
<tr><td>$p(x, y)$</td><td>0</td><td>5</td><td>10</td><td>15</td></tr>
<tr><td>0</td><td>.02</td><td>.06</td><td>.02</td><td>.10</td></tr>
<tr><td>x 5</td><td>.04</td><td>.15</td><td>.20</td><td>.10</td></tr>
<tr><td>10</td><td>.01</td><td>.15</td><td>.14</td><td>.01</td></tr>
</table>

(a) If the score recorded in the grade book is the total number of points earned on the two parts, what is the expected recorded score $E(X+Y)$?

(b) If the maximum of the two scores is recorded, what is the expected recorded score?

25. The difference between the number of customers in line at the express checkout and the number in line at the superexpress checkout in Exercise 3 is $X_1 - X_2$. Calculate the expected difference.

26. Six individuals, including A and B, take seats around a circular table in a completely random fashion. Suppose the seats are numbered 1, ..., 6. Let $X = $ A's seat number and $Y = $ B's seat number. If A sends a written message around the table to B in the direction in which they are closest, how many individuals (including A and B) would you expect to handle the message?

27. A surveyor wishes to lay out a square region with each side having length L. However, because of measurement error, he instead lays out a rectangle in which the north–south sides both have length X and the east–west sides both have length Y. Suppose that X and Y are independent and that each is uniformly distributed on the interval $[L - A, L + A]$ (where $0 < A < L$). What is the expected area of the resulting rectangle?

28. Consider a small ferry that can accommodate cars and buses. The toll for cars is $3, and the toll for buses is $10. Let X and Y denote the number of cars and buses, respectively, carried on a single trip. Suppose the joint distribution of X and Y is as given in the table of Exercise 9. Compute the expected revenue from a single trip.

29. Annie and Alvie have agreed to meet for lunch between noon (0:00 p.m.) and 1:00 p.m. Denote Annie's arrival time by X, Alvie's by Y, and suppose X and Y are independent with pdfs

$$f_X(x) = \begin{cases} 3x^2 & 0 \le x \le 1 \\ 0 & \text{otherwise} \end{cases}$$

$$f_Y(y) = \begin{cases} 2y & 0 \le y \le 1 \\ 0 & \text{otherwise} \end{cases}$$

What is the expected amount of time that the one who arrives first must wait for the other person? [*Hint*: $h(X, Y) = |X - Y|$.]

30. Suppose that X and Y are independent rvs with moment generating functions $M_X(t)$ and $M_Y(t)$, respectively. If $Z = X + Y$, show that $M_Z(t) = M_X(t) \cdot M_Y(t)$. [*Hint*: Use the proposition on the expected value of a product.]

31. Compute the correlation coefficient ρ for X and Y of Example 4.15 (the covariance has already been computed).

32. (a) Compute the covariance for X and Y in Exercise 24.
    (b) Compute ρ for X and Y in the same exercise.

33. (a) Compute the covariance between X and Y in Exercise 11.
    (b) Compute the correlation coefficient ρ for this X and Y.

34. Reconsider the computer component lifetimes X and Y as described in Exercise 14. Determine $E(XY)$. What can be said about $Cov(X, Y)$ and ρ?

35. Refer back to Exercise 23.

    (a) Calculate the covariance of X and Y.

    (b) Calculate the correlation coefficient of X and Y. Interpret this value.

36. In practice, it is often desired to predict the value of a variable Y from the known value of some other variable, X. For example, a doctor might wish to predict the lifespan Y of someone who smokes X cigarettes a day, or an engineer may require predictions of the tensile strength Y of steel made with concentration X of a certain additive. A *linear predictor* of Y is anything of the form $\hat{Y} = a + bX$; the "hat" ^ on Y indicates prediction.

    A common measure of the quality of a predictor is given by the *mean square prediction error*:

    $$E\left[(Y - \hat{Y})^2\right]$$

    (a) Show that the choices of a and b that minimize mean square prediction error are

    $$b = \rho \cdot \frac{\sigma_Y}{\sigma_X} \qquad a = \mu_Y - b \cdot \mu_X$$

    where $\rho = Corr(X, Y)$. The resulting expression for $\hat{Y}$ is often called the *best linear predictor* of Y, given X. [*Hint:* Expand the expression for mean square prediction error, apply linearity of expectation, and then use calculus.]

    (b) Determine the mean square prediction error for the best linear predictor. How does the value of ρ affect this quantity?

37. (a) Recalling the definition of σ^2 for a single rv X, write a formula that would be appropriate for computing the variance of a function $h(X, Y)$ of two random variables. [*Hint:* Remember that variance is just a special expected value.]

    (b) Use this formula to compute the variance of the recorded score $h(X, Y)$ [$=\max(X, Y)$] in part (b) of Exercise 24.

38. Show that when X and Y are independent, $Cov(X, Y) = Corr(X, Y) = 0$.

39. Use linearity of expectation to establish the covariance property

    $$Cov(aX + bY + c, Z) = aCov(X, Z) + bCov(Y, Z)$$

40. (a) Use the properties of covariance to show that $Cov(aX + b, cY + d) = acCov(X, Y)$.

(b) Use part (a) along with the rescaling property of standard deviation to show that $\mathrm{Corr}(aX + b, cY + d) = \mathrm{Corr}(X, Y)$ when $ac > 0$ (this is the scale invariance property of correlation).

(c) What happens if a and c have opposite signs, so $ac < 0$?

41. Show that if $Y = aX + b$ $(a \neq 0)$, then $\mathrm{Corr}(X, Y) = +1$ or -1. Under what conditions will $\rho = +1$?

42. Let Z_X be the standardized X, $Z_X = (X - \mu_X)/\sigma_X$, let Z_Y be the standardized Y, $Z_Y = (Y - \mu_Y)/\sigma_Y$, and let $\rho = \mathrm{Corr}(X, Y)$.

(a) Use properties of covariance and correlation to show that $\mathrm{Corr}(X, Y) = \mathrm{Cov}(Z_X, Z_Y) = E(Z_X Z_Y)$.

(b) Use the linearity of expectation along with part (a), to show that $E[(Z_Y - \rho Z_X)^2] = 1 - \rho^2$. [*Hint:* If Z is a standardized rv, what are its mean and variance, and how can you use those to determine $E(Z^2)$?]

(c) Use part (b) to show that $-1 \leq \rho \leq 1$.

(d) Use part (b) to show that $\rho = 1$ implies that $Y = aX + b$ where $a > 0$, and $\rho = -1$ implies that $Y = aX + b$ where $a < 0$.

4.3   Properties of Linear Combinations

A **linear combination** of random variables refers to anything of the form $a_1 X_1 + \cdots + a_n X_n + b$, where the X_is are random variables and the a_is and b are numerical constants. (Some sources do not include the constant b in the definition.) For example, suppose your investment portfolio with a particular financial institution includes 100 shares of stock #1, 200 shares of stock #2, and 500 shares of stock #3. Let X_1, X_2, and X_3 denote the share prices of these three stocks at the end of the current fiscal year. Suppose also that the financial institution will levy a management fee of \$150. Then the value of your investments with this institution at the end of the year is $100X_1 + 200X_2 + 500X_3 - 150$, which is a particular linear combination. Important special cases include the total $X_1 + \cdots + X_n$ (take $a_1 = \cdots = a_n = 1$, $b = 0$), the difference of two rvs $X_1 - X_2$ ($n = 2, a_1 = 1, a_2 = -1$), and anything of the form $aX + b$ (take $n = 1$ or, equivalently, set $a_2 = \ldots = a_n = 0$). Another very important linear combination is the *sample mean* $(X_1 + \cdots + X_n)/n$, conventionally denoted $\bar{X}$; just take $a_1 = \cdots = a_n = 1/n$ and $b = 0$.

Notice that we are not requiring the X_is to be independent or to have the same probability distribution. All the X_is could have different distributions and therefore different mean values and standard deviations. In this section, we investigate the general properties of linear combinations. Section 4.5 will explore some special properties of the total and the sample mean under additional assumptions.

We first consider the expected value and variance of a linear combination.

THEOREM

Let the rvs $X_1, X_2, \ldots, X_n$ have mean values $\mu_1, \ldots, \mu_n$ and standard deviations $\sigma_1, \ldots, \sigma_n$, respectively.

1. Whether or not the X_is are independent,

$$
\begin{aligned}
E(a_1 X_1 + \cdots + a_n X_n + b) &= a_1 E(X_1) + \cdots + a_n E(X_n) + b \\
&= a_1 \mu_1 + \cdots + a_n \mu_n + b
\end{aligned}
\tag{4.4}
$$

and

$$
\begin{aligned}
\mathrm{Var}(a_1 X_1 + \cdots + a_n X_n + b) &= \sum_{i=1}^{n} \sum_{j=1}^{n} a_i a_j \mathrm{Cov}(X_i, X_j) \\
&= \sum_{i=1}^{n} a_i^2 \sigma_i^2 + 2 \sum \sum_{i<j} a_i a_j \mathrm{Cov}(X_i, X_j)
\end{aligned}
\tag{4.5}
$$

2. If $X_1, \ldots, X_n$ are independent,

$$
\begin{aligned}
\mathrm{Var}(a_1 X_1 + \cdots + a_n X_n + b) &= a_1^2 \mathrm{Var}(X_1) + \cdots + a_n^2 \mathrm{Var}(X_n) \\
&= a_1^2 \sigma_1^2 + \cdots + a_n^2 \sigma_n^2
\end{aligned}
\tag{4.6}
$$

and

$$
\mathrm{SD}(a_1 X_1 + \cdots + a_n X_n + b) = \sqrt{a_1^2 \sigma_1^2 + \cdots + a_n^2 \sigma_n^2}
$$

A paraphrase of Eq. (4.4) is that the expected value of a linear combination is the same linear combination of the expected values—for example, $E(2X_1 + 5X_2) = 2\mu_1 + 5\mu_2$. Equation (4.6) in Statement 2 is a special case of Eq. (4.5) in Statement 1: when the X_is are independent, $\mathrm{Cov}(X_i, X_j) = 0$ for $i \neq j$ (this simplification actually occurs when the X_is are uncorrelated, a weaker condition than independence).

Proofs for the Case $n = 2$ To establish Eq. (4.4), we could invoke linearity of expectation from Sect. 4.2, but we present an independent proof here. Suppose that X_1 and X_2 are continuous with joint pdf $f(x_1, x_2)$. Then

$$E(a_1X_1 + a_2X_2 + b) = \int_{-\infty}^{\infty}\int_{-\infty}^{\infty} (a_1x_1 + a_2x_2 + b)f(x_1,x_2)dx_1dx_2$$

$$= a_1 \int_{-\infty}^{\infty}\int_{-\infty}^{\infty} x_1f(x_1,x_2)dx_2dx_1 + a_2 \int_{-\infty}^{\infty}\int_{-\infty}^{\infty} x_2f(x_1,x_2)dx_1dx_2$$

$$+ b\int_{-\infty}^{\infty}\int_{-\infty}^{\infty} f(x_1,x_2)dx_1dx_2$$

$$= a_1 \int_{-\infty}^{\infty} x_1 \left[\int_{-\infty}^{\infty} f(x_1,x_2)dx_2\right]dx_1$$

$$+ a_2 \int_{-\infty}^{\infty} x_2 \left[\int_{-\infty}^{\infty} f(x_1,x_2)dx_1\right]dx_2 + b(1)$$

$$= a_1 \int_{-\infty}^{\infty} x_1f_{X_1}(x_1)dx_1 + a_2 \int_{-\infty}^{\infty} x_2f_{X_2}(x_2)dx_2 + b$$

$$= a_1E(X_1) + a_2E(X_2) + b$$

Summation replaces integration in the discrete case. The argument for Eq. (4.5) does not require specifying whether either variable is discrete or continuous. Recalling that $\mathrm{Var}(Y) = E[(Y-\mu_Y)^2]$,

$$\mathrm{Var}(a_1X_1 + a_2X_2 + b) = E\left[(a_1X_1 + a_2X_2 + b - (a_1\mu_1 + a_2\mu_2 + b))^2\right]$$

$$= E\left[(a_1X_1 - a_1\mu_1 + a_2X_2 - a_2\mu_2)^2\right]$$

$$= E\left[a_1^2(X_1 - \mu_1)^2 + a_2^2(X_2 - \mu_2)^2\right.$$

$$\left. + 2a_1a_2(X_1 - \mu_1)(X_2 - \mu_2)\right]$$

$$= a_1^2E\left[(X_1 - \mu_1)^2\right] + a_2^2E\left[(X_2 - \mu_2)^2\right]$$

$$+ 2a_1a_2E\left[(X_1 - \mu_1)(X_2 - \mu_2)\right]$$

where the last equality comes from linearity of expectation. We recognize the terms in this last expression as variances and covariance, all together $a_1^2\mathrm{Var}(X_1) + a_2^2\mathrm{Var}(X_2) + 2a_1a_2\mathrm{Cov}(X_1,X_2)$, as required. ∎

Example 4.18 A gas station sells three grades of gasoline: regular, plus, and premium. These are priced at \$3.50, \$3.65, and \$3.80 per gallon, respectively. Let X_1, X_2, and X_3 denote the amounts of these grades purchased (gallons) on a particular day. Suppose the X_is are independent with $\mu_1 = 1000$, $\mu_2 = 500$, $\mu_3 = 300$, $\sigma_1 = 100$, $\sigma_2 = 80$, and $\sigma_3 = 50$. The revenue from sales is $Y = 3.5X_1 + 3.65X_2 + 3.8X_3$, and

$$E(Y) = 3.5\mu_1 + 3.65\mu_2 + 3.8\mu_3 = \$6465$$

$$\mathrm{Var}(Y) = 3.5^2\sigma_1^2 + 3.65^2\sigma_2^2 + 3.8^2\sigma_3^2 = 243,864$$

$$\mathrm{SD}(Y) = \sqrt{243,864} = \$493.83$$

∎

Example 4.19 Recall that a hypergeometric rv X is the number of successes in a random sample of size n selected without replacement from a population of size N consisting of M successes and $N - M$ failures. It is tricky to obtain the mean value and variance of X directly from the pmf, and the hypergeometric moment generating function is very complicated. We now show how the foregoing proposition on linear combinations can be used to accomplish this task.

To this end, let $X_1 = 1$ if the first individual or object selected is a success and $X_1 = 0$ if it is a failure; define $X_2, X_3, \ldots, X_n$ analogously for the second selection, third selection, and so on. Each X_i is a Bernoulli rv, and each has the same marginal distribution: $p(1) = M/N$ and $p(0) = 1 - M/N$ (this is obvious for X_1, which is based on the very first draw from the population, and can be verified for the other draws as well). Thus $E(X_i) = 0(1 - M/N) + 1(M/N) = M/N$. The total number of successes in the sample is $X = X_1 + \ldots + X_n$ (a 1 is added in for each success and a 0 for each failure), so

$$E(X) = E(X_1) + \ldots + E(X_n) = M/N + M/N + \ldots + M/N = n(M/N) = np$$

where p denotes the success probability on any particular draw (trial). That is, just as in the case of a binomial rv, the expected value of a hypergeometric rv is the success probability on any trial multiplied by the number of trials. Notice that we were able to apply Statement 1 of the foregoing theorem, even though the X_i are not independent.

However, the variance of X here is not the same as the binomial variance because the successive draws are not independent. Consider the joint distribution of X_1 and X_2:

$$p(1,1) = \frac{M}{N}\left(\frac{M-1}{N-1}\right), \quad p(0,0) = \left(\frac{N-M}{N}\right)\left(\frac{N-M-1}{N-1}\right),$$
$$p(1,0) = p(0,1) = \frac{M}{N}\left(\frac{N-M}{N-1}\right)$$

This is also the joint pmf of any pair X_i, X_j. A slightly tedious calculation then results in

$$\mathrm{Cov}(X_i, X_j) = -\frac{p(1-p)}{N-1} \quad (i \neq j)$$

Applying the variance formula from statement 1 of the theorem eventually yields

$$\mathrm{Var}(X) = \mathrm{Var}(X_1 + \ldots + X_n) = n\mathrm{Var}(X_1) + n(n-1)\mathrm{Cov}(X_1, X_2)$$
$$= np(1-p)\left(\frac{N-n}{N-1}\right)$$

This is quite close to the binomial variance provided that n is much smaller than N so that the last term in parentheses is close to 1. ∎

The following corollary expresses the $n = 2$ case of the main theorem for ease of use, including the important special cases of the sum and the difference of two random variables.

COROLLARY

For any two rvs X_1 and X_2, and any constants a_1, a_2, b,

$$E(a_1X_1 + a_2X_2 + b) = a_1E(X_1) + a_2E(X_2) + b$$

and

$$\text{Var}(a_1X_1 + a_2X_2 + b) = a_1^2\text{Var}(X_1) + a_2^2\text{Var}(X_2) + 2a_1a_2\text{Cov}(X_1, X_2)$$

In particular, $E(X_1 + X_2) = E(X_1) + E(X_2)$ and, if X_1 and X_2 are independent, $\text{Var}(X_1 + X_2) = \text{Var}(X_1) + \text{Var}(X_2)$.[1] Also, $E(X_1 - X_2) = E(X_1) - E(X_2)$ and, if X_1 and X_2 are independent,

$$\text{Var}(X_1 - X_2) = \text{Var}(X_1) + \text{Var}(X_2).$$

The expected value of a difference is the difference of the two expected values, but the variance of a difference between two independent variables is the *sum*, not the difference, of the two variances. There is just as much variability in $X_1 - X_2$ as in $X_1 + X_2$: writing $X_1 - X_2 = X_1 + (-1)X_2$, $(-1)X_2$ has the same amount of variability as X_2 itself.

Example 4.20 An automobile manufacturer equips a particular model with either a six-cylinder engine or a four-cylinder engine. Let X_1 and X_2 be fuel efficiencies for independently and randomly selected six-cylinder and four-cylinder cars, respectively. With $\mu_1 = 22$, $\mu_2 = 26$, $\sigma_1 = 1.2$, and $\sigma_2 = 1.5$,

$$E(X_1 - X_2) = \mu_1 - \mu_2 = 22 - 26 = -4$$
$$\text{Var}(X_1 - X_2) = \sigma_1^2 + \sigma_2^2 = 1.2^2 + 1.5^2 = 3.69$$
$$\text{SD}(X_1 - X_2) = \sqrt{3.69} = 1.92$$

If we relabel so that X_1 refers to the four-cylinder car, then $E(X_1 - X_2) = 26 - 22 = 4$, but the variance of the difference is still 3.69. ∎

---

[1] This property of independent rvs can also be written as $\text{SD}(X_1)^2 + \text{SD}(X_2)^2 = \text{SD}(X_1 + X_2)^2$. In part because the formula has the format $a^2 + b^2 = c^2$, statisticians sometimes call this property the *Pythagorean Theorem*.

4.3.1 The PDF of a Sum

Generally speaking, knowing the mean and standard deviation of a random variable W is not enough to specify its probability distribution and thus be able to compute probabilities such as $P(W > 10)$ or $P(W \le -2)$. In the case of independent rvs, a general method exists for determining the pdf of the sum $X_1 + \cdots + X_n$ from their marginal pdfs. We present first the result for two random variables.

> **THEOREM**
>
> Suppose X and Y are independent, continuous rvs with marginal pdfs $f_X(x)$ and $f_Y(y)$, respectively. Then the pdf of the rv $W = X + Y$ is given by
>
> $$f_W(w) = \int_{-\infty}^{\infty} f_X(x) f_Y(w - x) dx$$
>
> [In mathematics, this integral operation is known as the **convolution** of $f_X(x)$ and $f_Y(y)$ and is sometimes denoted $f_W = f_X \star f_Y$.] The limits of integration are determined by which x values make both $f_X(x) > 0$ and $f_Y(w - x) > 0$.

Proof Since X and Y are independent, their joint pdf is given by $f_X(x) \cdot f_Y(y)$. The cdf of W is then

$$F_W(w) = P(W \le w) = P(X + Y \le w)$$

To calculate $P(X + Y \le w)$, we must integrate over the set of numbers $\{(x, y): x + y \le w\}$, which is the shaded region indicated in Fig. 4.6.

The resulting limits of integration are $-\infty < x < \infty$ and $-\infty < y \le w - x$, and so

Fig. 4.6 Region of integration for $P(X + Y \le w)$

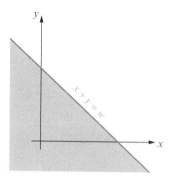

$$F_W(w) = P(X + Y \leq w)$$
$$= \int_{-\infty}^{\infty} \int_{-\infty}^{w-x} f_X(x) f_Y(y) dy dx = \int_{-\infty}^{\infty} f_X(x) \int_{-\infty}^{w-x} f_Y(y) dy dx$$
$$= \int_{-\infty}^{\infty} f_X(x) F_Y(w - x) dx$$

The pdf of W is the derivative of this expression with respect to w; taking the derivative underneath the integral sign yields the desired result.   ∎

By a similar argument, the pdf of $W = X + Y$ can be determined even when X and Y are not independent. Assuming X and Y have joint pdf $f(x, y)$,

$$f_W(w) = \int_{-\infty}^{\infty} f(x, w - x) dx$$

Example 4.21 In a *standby system*, a component is used until it wears out and is then immediately replaced by another, not necessarily identical, component. (The second component is said to be "in standby mode," i.e., waiting to be used.) The overall lifetime of a standby system is just the sum of the lifetimes of its individual components. Let X and Y denote the lifetimes of the two components of a standby system, and suppose X and Y are independent exponentially distributed random variables with expected lifetimes 3 weeks and 4 weeks, respectively. Let $W = X + Y$, the lifetime of the standby system.

Using the first theorem of this section, the expected lifetime of the standby system is $E(W) = E(X) + E(Y) = 3 + 4 = 7$ weeks. Since X and Y are exponential, the variance of each one is the square of its mean (9 and 16, respectively); since they are also independent,

$$\text{Var}(W) = \text{Var}(X) + \text{Var}(Y) = 3^2 + 4^2 = 25$$

It follows that $\text{SD}(W) = 5$ weeks. Since $\mu_W \neq \sigma_W$, W cannot itself be exponentially distributed, but we can use the previous theorem to find its pdf.

The marginal pdfs of X and Y are $f_X(x) = (1/3)e^{-x/3}$ for $x > 0$ and $f_Y(y) = (1/4)e^{-y/4}$ for $y > 0$. Substituting $y = w - x$, the inequalities $x > 0$ and $w - x > 0$ imply $0 < x < w$, which are the limits of integration of the convolution integral:

$$f_W(w) = \int_{-\infty}^{\infty} f_X(x) f_Y(w - x) dx = \int_0^w (1/3)e^{-x/3}(1/4)e^{-(w-x)/4} dx$$
$$= \frac{1}{12} e^{-w/4} \int_0^w e^{-x/12} dx$$
$$= e^{-w/4}\left(1 - e^{-w/12}\right), \quad w > 0$$

The pdf of W appears in Fig. 4.7. As a check, the mean and variance of W can be verified directly from its pdf.

The probability the standby system lasts more than its expected lifetime of 7 weeks is given by

Fig. 4.7 The pdf of
$W = X + Y$ for Example 4.21

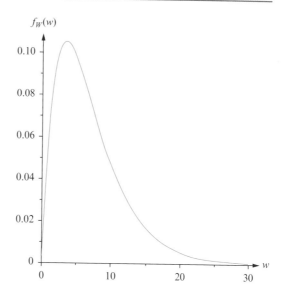

$$P(W > 7) = \int_7^\infty f_W(w)\,dw = \int_7^\infty e^{-w/4}\left(1 - e^{-w/12}\right)dw = .4042 \qquad \blacksquare$$

As a generalization of the previous proposition, the pdf of the sum $W = X_1 + \cdots + X_n$ of n independent, continuous rvs can be determined by successive convolution: $f_W = f_1 \star \cdots \star f_n$. In most situations, it isn't practical to evaluate such a complicated object. Thankfully, as we'll see next, such tedious computations can sometimes be avoided with the use of moment generating functions.

4.3.2 Moment Generating Functions for Linear Combinations

A corollary in Sect. 4.2 stated that the expected value of a product of functions of independent random variables is the product of the individual expected values. We now use this to formulate the moment generating function of a linear combination of independent random variables.

PROPOSITION

Let $X_1, X_2, \ldots, X_n$ be independent random variables with moment generating functions $M_{X_1}(t), M_{X_2}(t), \ldots, M_{X_n}(t)$, respectively. Then the moment generating function of the linear combination $Y = a_1X_1 + a_2X_2 + \cdots + a_nX_n + b$ is

(continued)

$$M_Y(t) = e^{bt} M_{X_1}(a_1 t) \cdot M_{X_2}(a_2 t) \cdot \ \cdots \ \cdot M_{X_n}(a_n t)$$

In the special case that $a_1 = a_2 = \cdots = a_n = 1$ and $b = 0$, so $Y = X_1 + \cdots + X_n$,

$$M_Y(t) = M_{X_1}(t) \cdot M_{X_2}(t) \cdot \ \cdots \ \cdot M_{X_n}(t)$$

That is, the mgf of a sum of independent rvs is the *product* of the individual mgfs.

Proof First, we write the moment generating function of Y as the expected value of a product.

$$
\begin{aligned}
M_Y(t) = E\!\left[e^{tY}\right] &= E\!\left[e^{t(a_1 X_1 + a_2 X_2 + \cdots + a_n X_n + b)}\right] \\
&= E\!\left[e^{ta_1 X_1 + ta_2 X_2 + \cdots + ta_n X_n + tb}\right] = e^{bt} E\!\left[e^{a_1 t X_1} \cdot e^{a_2 t X_2} \cdot \ \cdots \ \cdot e^{a_n t X_n}\right]
\end{aligned}
$$

The last expression inside brackets is the product of functions of $X_1, X_2, \ldots, X_n$. Since the X_i are independent, the expected value can be distributed across this product:

$$
\begin{aligned}
e^{bt} E\!\left[e^{a_1 t X_1} \cdot e^{a_2 t X_2} \cdot \ \cdots \ \cdot e^{a_n t X_n}\right] &= e^{bt} E\!\left[e^{a_1 t X_1}\right] \cdot E\!\left[e^{a_2 t X_2}\right] \cdot \ \cdots \ \cdot E\!\left[e^{a_n t X_n}\right] \\
&= e^{bt} M_{X_1}(a_1 t) \cdot M_{X_2}(a_2 t) \cdot \ \cdots \ \cdot M_{X_n}(a_n t) \quad \blacksquare
\end{aligned}
$$

Now suppose we wish to determine the pdf of some linear combination of independent rvs. Provided we have their mgfs, the previous proposition makes it easy to determine the mgf of the linear combination. Then, if we can recognize this mgf as belonging to some known distributional family (binomial, exponential, etc.), the uniqueness property of mgfs guarantees our linear combination has that particular distribution. The next several propositions illustrate this technique.

PROPOSITION

If $X_1, X_2, \ldots, X_n$ are independent, normally distributed rvs (with possibly different means and/or sds), then any linear combination of the X_is also has a normal distribution. In particular, the sum of independent normally distributed rvs itself has a normal distribution, and the difference $X_1 - X_2$ between two independent, normally distributed variables is itself normally distributed.

Proof Let $Y = a_1 X_1 + a_2 X_2 + \cdots + a_n X_n + b$, where X_i is normally distributed with mean μ_i and standard deviation σ_i, and the X_i are independent. From Sect. 3.3, $M_{X_i}(t) = e^{\mu_i t + \sigma_i^2 t^2/2}$. Therefore,

$$M_Y(t) = e^{bt}M_{X_1}(a_1t) \cdot M_{X_2}(a_2t) \cdot \cdots \cdot M_{X_n}(a_nt)$$
$$= e^{bt}e^{\mu_1 a_1 t + \sigma_1^2 a_1^2 t^2/2}e^{\mu_2 a_2 t + \sigma_2^2 a_2^2 t^2/2} \cdots \cdot e^{\mu_n a_n t + \sigma_n^2 a_n^2 t^2/2}$$
$$= e^{(\mu_1 a_1 + \mu_2 a_2 + \cdots + \mu_n a_n + b)t + (\sigma_1^2 a_1^2 + \sigma_2^2 a_2^2 + \cdots + \sigma_n^2 a_n^2)t^2/2}$$
$$= e^{\mu t + \sigma^2 t^2/2}$$

where $\mu = a_1\mu_1 + a_2\mu_2 + \cdots + a_n\mu_n + b$ and $\sigma^2 = a_1^2\sigma_1^2 + a_2^2\sigma_2^2 + \cdots + a_n^2\sigma_n^2$. We recognize this function as the mgf of a normal random variable, and it follows by the uniqueness property of mgfs that Y is normally distributed. Notice that the mean and variance are in agreement with the first proposition of this section. ∎

Example 4.22 (Example 4.18 continued) The total revenue from the sale of the three grades of gasoline on a particular day was $Y = 3.5X_1 + 3.65X_2 + 3.8X_3$, and we calculated $\mu_Y = 6465$ and (assuming independence) $\sigma_Y = 493.83$. If the X_is are normally distributed, the probability that revenue exceeds 5000 is

$$P(Y > 5000) = P\left(Z > \frac{5000 - 6465}{493.83}\right) = P(Z > -2.967) = 1 - \Phi(-2.967) = .9985$$

∎

This same method may be applied to Poisson rvs, as the next proposition indicates.

> **PROPOSITION**
>
> Suppose $X_1, \ldots, X_n$ are independent Poisson random variables, where X_i has mean μ_i. Then $Y = X_1 + \cdots + X_n$ also has a Poisson distribution, with mean $\mu_1 + \cdots + \mu_n$.

Proof From Sect. 2.7, the mgf of a Poisson rv with mean μ is $e^{\mu(e^t - 1)}$. Since Y is the sum of the X_is, and the X_is are independent,

$$M_Y(t) = M_{X_1}(t) \cdots M_{X_n}(t) = e^{\mu_1(e^t - 1)} \cdots e^{\mu_n(e^t - 1)} = e^{(\mu_1 + \cdots + \mu_n)(e^t - 1)}$$

This is the mgf of a Poisson rv with mean $\mu_1 + \cdots + \mu_n$. Therefore, by the uniqueness property of mgfs, Y has a Poisson distribution with mean $\mu_1 + \cdots + \mu_n$. ∎

Example 4.23 During the open enrollment period at a large university, the number of freshmen registering for classes through the online registration system in 1 h follows a Poisson distribution with mean 80 students; denote this rv by X_1. Define X_2, X_3, and X_4 similarly for sophomores, juniors, and seniors, and suppose the corresponding means are 125, 118, and 140, respectively. Assume these four counts are independent. The rv $Y = X_1 + X_2 + X_3 + X_4$ represents the total number of undergraduate students registering in 1 h; by the preceding proposition, Y is also a Poisson rv, but with mean $80 + 125 + 118 + 140 = 463$ students and standard deviation $\sqrt{463} = 21.5$ students. The probability that more than 500 students enroll during 1 h, exceeding the registration system's capacity, is then $P(Y > 500) = 1 - P(Y \leq 500) = .042$ (software was used to perform the calculation).   ■

Because of the properties stated in the preceding two propositions, both the normal and Poisson models are sometimes called *additive distributions*, meaning that the sum of independent rvs from that family (normal or Poisson) will also belong to that family. The next proposition shows that not all of the major probability distributions are additive; its proof is left as an exercise (Exercise 65).

> **PROPOSITION**
>
> Suppose $X_1, \ldots, X_n$ are independent exponential random variables with common parameter λ. Then $Y = X_1 + \cdots + X_n$ has a gamma distribution, with parameters $\alpha = n$ and $\beta = 1/\lambda$ (aka the Erlang distribution).

Notice this proposition requires the X_i to have the same "rate" parameter λ, i.e., the X_i must be independent *and* identically distributed. As we saw in Example 4.21, the sum of two independent exponential rvs with different parameters does not follow an exponential distribution.

4.3.3   Exercises: Section 4.3 (43–65)

43. A shipping company handles containers in three different sizes: (1) 27 ft^3 ($3 \times 3 \times 3$), (2) 125 ft^3, and (3) 512 ft^3. Let X_i ($i = 1, 2, 3$) denote the number of type i containers shipped during a given week. With $\mu_i = E(X_i)$ and $\sigma_i = SD(X_i)$, suppose the mean values and standard deviations are as follows:

$$\mu_1 = 200 \qquad \mu_2 = 250 \qquad \mu_3 = 100$$
$$\sigma_1 = 10 \qquad \sigma_2 = 12 \qquad \sigma_3 = 8$$

(a) Assuming that X_1, X_2, X_3 are independent, calculate the expected value and standard deviation of the total volume shipped. [*Hint:* Volume $= 27X_1 + 125X_2 + 512X_3$.]

(b) Would your calculations necessarily be correct if the X_is were not independent? Explain.

(c) Suppose that the X_is are independent with each one having a normal distribution. What is the probability that the total volume shipped is at most 100,000 ft^3?

44. Let X_1, X_2, and X_3 represent the times necessary to perform three successive repair tasks at a service facility. Suppose they are independent, normal rvs with expected values μ_1, μ_2, and μ_3 and variances σ_1^2, σ_2^2, and σ_3^2, respectively.
   (a) If $\mu_1 = \mu_2 = \mu_3 = 60$ and $\sigma_1^2 = \sigma_2^2 = \sigma_3^2 = 15$, calculate $P(X_1 + X_2 + X_3 \leq 200)$.
   (b) Using the μ_is and σ_is given in part (a), what is $P(150 \leq X_1 + X_2 + X_3 \leq 200)$?
   (c) Using the μ_is and σ_is given in part (a), calculate $P(55 \leq \overline{X})$ and $P(58 \leq \overline{X} \leq 62)$. [As noted at the beginning of this section, $\overline{X}$ denotes the sample mean, so here $\overline{X} = (X_1 + X_2 + X_3)/3$.]
   (d) Using the μ_is and σ_is given in part (a), calculate $P(-10 \leq X_1 - .5X_2 - .5X_3 \leq 5)$.
   (e) If $\mu_1 = 40$, $\mu_2 = 50$, $\mu_3 = 60$, $\sigma_1^2 = 10$, $\sigma_2^2 = 12$, and $\sigma_3^2 = 14$, calculate $P(X_1 + X_2 + X_3 \leq 160)$ and $P(X_1 + X_2 \geq 2X_3)$.

45. Five automobiles of the same type are to be driven on a 300-mile trip. The first two have six-cylinder engines, and the other three have four-cylinder engines. Let X_1, X_2, X_3, X_4, and X_5 be the observed fuel efficiencies (mpg) for the five cars. Suppose these variables are independent and normally distributed with $\mu_1 = \mu_2 = 20$, $\mu_3 = \mu_4 = \mu_5 = 22$, and $\sigma^2 = 4$ for the smaller engines and 3.5 for the larger engines. Define an rv Y by

$$Y = \frac{X_1 + X_2}{2} - \frac{X_3 + X_4 + X_5}{3}$$

so that Y is a measure of the difference in efficiency between the six-cylinder and four-cylinder engines. Compute $P(0 \leq Y)$ and $P(-1 \leq Y \leq 1)$. [Hint: $Y = a_1 X_1 + \cdots + a_5 X_5$, with $a_1 = \frac{1}{2}, \ldots, a_5 = -\frac{1}{3}$.]

46. Exercise 28 introduced random variables X and Y, the number of cars and buses, respectively, carried by a ferry on a single trip. The joint pmf of X and Y is given in the table in Exercise 9. It is readily verified that X and Y are independent.
   (a) Compute the expected value, variance, and standard deviation of the total number of vehicles on a single trip.
   (b) If each car is charged $3 and each bus $10, compute the expected value, variance, and standard deviation of the revenue resulting from a single trip.

47. A concert has three pieces of music to be played before intermission. The time taken to play each piece has a normal distribution. Assume that the three times are independent of each other. The mean times are 15, 30, and 20 min, respectively, and the standard deviations are 1, 2, and 1.5 min, respectively. What is the probability that this part of the concert takes at most 1 h? Are there reasons to question the independence assumption? Explain.

48. Refer to Exercise 3.
   (a) Calculate the covariance between $X_1 =$ the number of customers in the express checkout and $X_2 =$ the number of customers in the superexpress checkout.
   (b) Calculate $\text{Var}(X_1 + X_2)$. How does this compare to $\text{Var}(X_1) + \text{Var}(X_2)$?

49. Suppose your waiting time for a bus in the morning is uniformly distributed on [0, 8], whereas waiting time in the evening is uniformly distributed on [0, 10] independent of morning waiting time.

    (a) If you take the bus each morning and evening for a week, what is your total expected waiting time? [*Hint*: Define rvs $X_1, \ldots, X_{10}$ and use a rule of expected value.]

    (b) What is the variance of your total waiting time?

    (c) What are the expected value and variance of the difference between morning and evening waiting times on a given day?

    (d) What are the expected value and variance of the difference between total morning waiting time and total evening waiting time for a particular week?

50. An insurance office buys paper by the ream (500 sheets) for use in the copier, fax, and printer. Each ream lasts an average of 4 days, with standard deviation 1 day. The distribution is normal, independent of previous reams.

    (a) Find the probability that the next ream outlasts the present one by more than 2 days.

    (b) How many reams must be purchased if they are to last at least 60 days with probability at least 80%?

51. If two loads are applied to a cantilever beam as shown in the accompanying drawing, the bending moment at 0 due to the loads is $a_1X_1 + a_2X_2$.

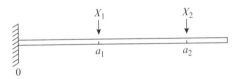

    (a) Suppose that X_1 and X_2 are independent rvs with means 2 and 4 kips, respectively, and standard deviations .5 and 1.0 kip, respectively. If $a_1 = 5$ ft and $a_2 = 10$ ft, what is the expected bending moment and what is the standard deviation of the bending moment?

    (b) If X_1 and X_2 are normally distributed, what is the probability that the bending moment will exceed 75 kip-ft?

    (c) Suppose the positions of the two loads are random variables. Denoting them by A_1 and A_2, assume that these variables have means of 5 and 10 ft, respectively, that each has a standard deviation of .5, and that all A_is and X_is are independent of each other. What is the expected moment now?

    (d) For the situation of part (c), what is the variance of the bending moment?

    (e) If the situation is as described in part (a) except that $\mathrm{Corr}(X_1, X_2) = .5$ (so that the two loads are not independent), what is the variance of the bending moment?

52. One piece of PVC pipe is to be inserted inside another piece. The length of the first piece is normally distributed with mean value 20 in. and standard deviation .5 in. The length of the second piece is a normal rv with mean and standard

deviation 15 in. and .4 in., respectively. The amount of overlap is normally distributed with mean value 1 in. and standard deviation .1 in. Assuming that the lengths and amount of overlap are independent of each other, what is the probability that the total length after insertion is between 34.5 and 35 in.?

53. Two airplanes are flying in the same direction in adjacent parallel corridors. At time $t = 0$, the first airplane is 10 km ahead of the second one. Suppose the speed of the first plane (km/h) is normally distributed with mean 520 and standard deviation 10 and the second plane's speed, independent of the first, is also normally distributed with mean and standard deviation 500 and 10, respectively.
    (a) What is the probability that after 2 h of flying, the second plane has not caught up to the first plane?
    (b) Determine the probability that the planes are separated by at most 10 km after 2 h.

54. Three different roads feed into a particular freeway entrance. Suppose that during a fixed time period, the number of cars coming from each road onto the freeway is a random variable, with expected value and standard deviation as given in the table.

	Road 1	Road 2	Road 3
Expected value	800	1000	600
Standard deviation	16	25	18

   (a) What is the expected total number of cars entering the freeway at this point during the period? [*Hint:* Let $X_i =$ the number from road i.]
   (b) What is the standard deviation of the total number of entering cars? Have you made any assumptions about the relationship between the numbers of cars on the different roads?
   (c) With X_i denoting the number of cars entering from road i during the period, suppose that $\mathrm{Cov}(X_1, X_2) = 80$, $\mathrm{Cov}(X_1, X_3) = 90$, and $\mathrm{Cov}(X_2, X_3) = 100$ (so that the three streams of traffic are not independent). Compute the expected total number of entering cars and the standard deviation of the total.

55. Suppose we take a random sample of size n from a continuous distribution having median 0 so that the probability of any one observation being positive is .5. We now disregard the signs of the observations, rank them from smallest to largest in absolute value, and then let $W =$ the sum of the ranks of the observations having positive signs. For example, if the observations are $-.3$, $+.7$, $+2.1$, and -2.5, then the ranks of positive observations are 2 and 3, so $W = 5$. In statistics literature, W is called *Wilcoxon's signed-rank statistic*. W can be represented as follows:

$$W = 1 \cdot Y_1 + 2 \cdot Y_2 + 3 \cdot Y_3 + \cdots + n \cdot Y_n = \sum_{i=1}^{n} i \cdot Y_i$$

where the Y_is are independent Bernoulli rvs, each with $p = .5$ ($Y_i = 1$ corresponds to the observation with rank i being positive). Compute the following:

(a) $E(Y_i)$ and then $E(W)$ using the equation for W [*Hint*: The first n positive integers sum to $n(n + 1)/2$.]

(b) $\text{Var}(Y_i)$ and then $\text{Var}(W)$ [*Hint*: The sum of the squares of the first n positive integers is $n(n + 1)(2n + 1)/6$.]

56. In Exercise 51, the weight of the beam itself contributes to the bending moment. Assume that the beam is of uniform thickness and density so that the resulting load is uniformly distributed on the beam. If the weight of the beam is random, the resulting load from the weight is also random; denote this load by W (kip-ft).

(a) If the beam is 12 ft long, W has mean 1.5 and standard deviation .25, and the fixed loads are as described in part (a) of Exercise 51, what are the expected value and variance of the bending moment? [*Hint:* If the load due to the beam were w kip-ft, the contribution to the bending moment would be $w \int_0^{12} x dx$.]

(b) If all three variables (X_1, X_2, and W) are normally distributed, what is the probability that the bending moment will be at most 200 kip-ft?

57. A professor has three errands to take care of in the Administration Building. Let X_i = the time that it takes for the ith errand ($i = 1, 2, 3$), and let X_4 = the total time in minutes that she spends walking to and from the building and between each errand. Suppose the X_is are independent, normally distributed, with the following means and standard deviations: $\mu_1 = 15$, $\sigma_1 = 4$, $\mu_2 = 5$, $\sigma_2 = 1$, $\mu_3 = 8$, $\sigma_3 = 2$, $\mu_4 = 12$, $\sigma_4 = 3$. She plans to leave her office at precisely 10:00 a.m. and wishes to post a note on her door that reads, "I will return by t a.m." What time t should she write down if she wants the probability of her arriving after t to be .01?

58. In an area having sandy soil, 50 small trees of a certain type were planted, and another 50 trees were planted in an area having clay soil. Let X = the number of trees planted in sandy soil that survive 1 year and Y = the number of trees planted in clay soil that survive 1 year. If the probability that a tree planted in sandy soil will survive 1 year is .7 and the probability of 1-year survival in clay soil is .6, compute an approximation to $P(-5 \leq X - Y \leq 5)$. [*Hint:* Use a normal approximation from Sect. 3.3. Do not bother with the continuity correction.]

59. Let X and Y be independent rvs, with $X \sim N(0, 1)$ and $Y \sim N(0, 1)$.

(a) Use convolution to show that $X + Y$ is also normal, and identify its mean and standard deviation.

(b) Use the additive property of the normal distribution presented in this section to verify your answer to part (a).

60. Karen throws two darts at a board with radius 10 in.; let X and Y denote the distances of the two darts from the center of the board. Under the system Karen uses, the score she receives depends upon $W = X + Y$, the sum of these two distances. Assume X and Y are independent.

(a) Suppose X and Y are both uniform on the interval [0, 10]. Use convolution to determine the pdf of $W = X + Y$. Be very careful with your limits of integration!

(b) Based on the pdf in part (a), calculate $P(X + Y \leq 5)$.

(c) If Karen's darts are equally likely to land anywhere on the board, it can be shown that the pdfs of X and Y are $f_X(x) = x/50$ for $0 \leq x \leq 10$ and $f_Y(y) = y/50$ for $0 \leq y \leq 10$. Use convolution to determine the pdf of $W = X + Y$. Again, be very careful with your limits of integration.

(d) Based on the pdf in part (c), calculate $P(X + Y \leq 5)$.

61. Siblings Matt and Liz both enjoy playing roulette. One day, Matt brought $10 to the local casino and Liz brought $15. They sat at different tables, and each made $1 wagers on red on consecutive spins (10 spins for Matt, 15 for Liz). Let $X =$ the number of times Matt won and $Y =$ the number of times Liz won.

(a) What is a reasonable probability model for X? [*Hint:* Successive spins of a roulette wheel are independent, and P(land on red) $= 18/38$.]

(b) What is a reasonable probability model for Y?

(c) What is a reasonable probability model for $X + Y$, the total number of times Matt and Liz win that day? Explain. [*Hint:* Since the siblings sat at different table, their gambling results are independent.]

(d) Use moment-generating functions, along with your answers to (a) and (b), to show that your answer to part (c) is correct.

(e) Generalize part (d): If $X_1, \ldots, X_k$ are independent binomial rvs, with $X_i \sim \text{Bin}(n_i, p)$, show that their sum is also binomially distributed.

(f) Does the result of part (e) hold if the probability parameter p is different for each X_i (e.g., if Matt bets on red but Liz bets on the number 27)?

62. The children attending Milena's birthday party are enjoying taking swings at a piñata. Let $X =$ the number of swings it takes Milena to hit the piñata once (since she's the birthday girl, she goes first), and let $Y =$ the number of swings it takes her brother Lucas to hit the piñata once (he goes second). Assume the results of successive swings are independent (the children don't improve, since they're blindfolded), and that each child has a .2 probability of hitting the piñata on any attempt.

(a) What is a reasonable probability model for X?

(b) What is a reasonable probability model for Y?

(c) What is a reasonable probability model for $X + Y$, the total number of swings taken by Milena and Lucas? Explain. (Assume Milena's and Lucas' results are independent.)

(d) Use moment-generating functions, along with your answers to (a) and (b), to show that $X + Y$ has a negative binomial distribution.

(e) Generalize part (d): If $X_1, \ldots, X_r$ are independent geometric rvs with common parameter p, show that their sum has a negative binomial distribution.

(f) Does the result of part (e) hold if the probability parameter p is different for each X_i (e.g., if Milena has probability .4 on each attempt while Lucas' success probability is only .1)?

63. Let $X_1, \ldots, X_n$ be independent rvs, with X_i having a negative binomial distribution with parameters r_i and p $(i = 1, \ldots, n)$. Use moment generating functions to show that $X_1 + \cdots + X_n$ has a negative binomial distribution, and identify the

parameters of this distribution. Explain why this answer makes sense, based on the negative binomial model. [*Note:* Each X_i may have a different parameter r_i, but all have the same p parameter.]

64. Let X and Y be independent gamma random variables, both with the same scale parameter β. The value of the shape parameter is α_1 for X and α_2 for Y. Use moment generating functions to show that $X + Y$ is also gamma distributed, with shape parameter $\alpha_1 + \alpha_2$ and scale parameter β. Is $X + Y$ gamma distributed if the scale parameters are different? Explain.

65. Let X and Y be independent exponential random variables with common parameter λ.

   (a) Use convolution to show that $X + Y$ has a gamma distribution, and identify the parameters of that gamma distribution.

   (b) Use the previous exercise to establish the same result.

   (c) Generalize part (b): If $X_1, \ldots, X_n$ are independent exponential rvs with common parameter λ, what is the distribution of their sum?

4.4   Conditional Distributions and Conditional Expectation

The distribution of Y can depend strongly on the value of another variable X. For example, if X is height and Y is weight, the distribution of weight for men who are 6 ft tall is very different from the distribution of weight for short men. The *conditional distribution* of Y given $X = x$ describes for each possible x value how probability is distributed over the set of y values. We define below the conditional distribution of Y given X, but the conditional distribution of X given Y can be obtained by just reversing the roles of X and Y. Both definitions are analogous to that of the conditional probability, $P(A|B)$, as the ratio $P(A \cap B)/P(B)$.

DEFINITION

Let X and Y be two discrete random variables with joint pmf $p(x,y)$ and marginal X pmf $p_X(x)$. Then for any x value such that $p_X(x) > 0$, the **conditional probability mass function of Y given $X = x$** is

$$p_{Y|X}(y|x) = \frac{p(x,y)}{p_X(x)}$$

An analogous formula holds in the continuous case. Let X and Y be two continuous random variables with joint pdf $f(x,y)$ and marginal X pdf $f_X(x)$. Then for any x value such that $f_X(x) > 0$, the **conditional probability density function of Y given $X = x$** is

$$f_{Y|X}(y|x) = \frac{f(x,y)}{f_X(x)}$$

Example 4.24 For a discrete example, reconsider Example 4.1, where X represents the deductible amount on an automobile policy and Y represents the deductible amount on a homeowner's policy. Here is the joint distribution again.

		y		
$p(x, y)$	0	100	200	
100	.20	.10	.20	.50
250	.05	.15	.30	.50
	.25	.25	.50	

(x labels the rows 100 and 250.)

The distribution of Y depends on X. In particular, let's find the conditional probability that Y is 200, given that X is 250, first using the definition of conditional probability from Sect. 1.4:

$$P(Y = 200\,|\,X = 250) = \frac{P(Y = 200 \cap X = 250)}{P(X = 250)} = \frac{.30}{.05 + .15 + .30} = .6$$

With our new definition we obtain the same result:

$$p_{Y|X}(200\,|\,250) = \frac{p(250, 200)}{p_X(250)} = \frac{.30}{.50} = .6$$

The conditional probabilities for the two other possible values of Y are

$$p_{Y|X}(0\,|\,250) = \frac{p(250, 0)}{p_X(250)} = \frac{.05}{.50} = .1$$

$$p_{Y|X}(100\,|\,250) = \frac{p(250, 100)}{p_X(250)} = \frac{.15}{.50} = .3$$

Notice that $p_{Y|X}(0|250) + p_{Y|X}(100|250) + p_{Y|X}(200|250) = .1 + .3 + .6 = 1$. This is no coincidence: conditional probabilities satisfy the properties of ordinary probabilities (i.e., they are nonnegative and they sum to 1). Essentially, the denominator in the definition of conditional probability is designed to make the total be 1.

Reversing the roles of X and Y, we find the conditional distribution for X, given that $Y = 0$:

$$p_{X|Y}(100\,|\,0) = \frac{p(100, 0)}{p_Y(0)} = \frac{.20}{.20 + .05} = .8$$

$$p_{X|Y}(250\,|\,0) = \frac{p(250, 0)}{p_Y(0)} = \frac{.05}{.20 + .05} = .2$$

Again, the conditional probabilities add to 1. ∎

Example 4.25 For a continuous example, recall Example 4.5, where X is the weight of almonds and Y is the weight of cashews in a can of mixed nuts. The sum of $X + Y$ is at most 1 lb, the total weight of the can of nuts. The joint pdf of X and Y is

$$f(x,y) = \begin{cases} 24xy & 0 \leq x \leq 1, \ 0 \leq y \leq 1, \ x+y \leq 1 \\ 0 & \text{otherwise} \end{cases}$$

and in Example 4.5 it was shown that

$$f_X(x) = \begin{cases} 12x(1-x)^2 & 0 \leq x \leq 1 \\ 0 & \text{otherwise} \end{cases}$$

Thus, the conditional pdf of Y given that $X = x$ is

$$f_{Y|X}(y \mid x) = \frac{f(x,y)}{f_X(x)} = \frac{24xy}{12x(1-x)^2} = \frac{2y}{(1-x)^2} \quad 0 \leq y \leq 1-x$$

This can be used to get conditional probabilities for Y. For example,

$$P(Y \leq .25 \mid X = .5) = \int_{-\infty}^{.25} f_{Y|X}(y \mid .5)\,dy = \int_0^{.25} \frac{2y}{(1-.5)^2}\,dy = \left[4y^2\right]_0^{.25} = .25$$

Given that the weight of almonds (X) is .5 lb, the probability is .25 for the weight of cashews (Y) to be less than .25 lb.

Just as in the discrete case, the conditional distribution assigns a total probability of 1 to the set of all possible Y values. That is, integrating the conditional density over its set of possible values should yield 1:

$$\int_{-\infty}^{\infty} f_{Y|X}(y \mid x)\,dy = \int_0^{1-x} \frac{2y}{(1-x)^2}\,dy = \left[\frac{y^2}{(1-x)^2}\right]_0^{1-x} = 1$$

Whenever you calculate a conditional density, we recommend doing this integration as a validity check.   ∎

4.4.1   Conditional Distributions and Independence

Recall that in Sect. 4.1 two random variables were defined to be independent if their joint pmf or pdf factors into the product of the marginal pmfs or pdfs. We can understand this definition better with the help of conditional distributions. For example, suppose there is independence in the discrete case. Then

$$p_{Y|X}(y \mid x) = \frac{p(x,y)}{p_X(x)} = \frac{p_X(x)p_Y(y)}{p_X(x)} = p_Y(y)$$

That is, independence implies that the conditional distribution of Y is the same as the unconditional (i.e., marginal) distribution, and that this is true no matter the value of X. The implication works in the other direction, too. If $p_{Y|X}(y|x) = p_Y(y)$, then

$$\frac{p(x,y)}{p_X(x)} = p_Y(y)$$

so $p(x, y) = p_X(x)\, p_Y(y)$, and therefore X and Y are independent.

In Example 4.7 we said that independence necessitates the region of positive density being a rectangle (possibly infinite in extent). In terms of conditional distributions, this region tells us the domain of Y for each possible x value. For independence we need to have the domain of Y not be dependent on X, so the interval of positive density must be the same for each x, implying a rectangular region.

4.4.2 Conditional Expectation and Variance

Because the conditional distribution is a valid probability distribution, it makes sense to define the conditional mean and variance.

DEFINITION

Let X and Y be two discrete random variables with conditional probability mass function $p_{Y|X}(y|x)$. Then the **conditional expectation** (or **conditional mean**) of Y given $X = x$ is

$$\mu_{Y|X=x} = E(Y \mid X = x) = \sum_y y \cdot p_{Y|X}(y \mid x)$$

Analogously, for two continuous rvs X and Y with conditional probability density function $f_{Y|X}(y|x)$,

$$\mu_{Y|X=x} = E(Y \mid X = x) = \int_{-\infty}^{\infty} y \cdot f_{Y|X}(y \mid x)\, dy$$

More generally, the conditional mean of any function $h(Y)$ is given by

(continued)

$$E(h(Y)|X = x) = \begin{cases} \sum_y h(y) \cdot p_{Y|X}(y|x) & \text{(discrete case)} \\ \int_{-\infty}^{\infty} h(y) \cdot f_{Y|X}(y|x) dy & \text{(continous case)} \end{cases}$$

In particular, the **conditional variance of Y given** $X = x$ is

$$\sigma_{Y|X=x}^2 = \text{Var}(Y|X = x) = E[(Y - \mu_{Y|X=x})^2|X = x]$$
$$= E(Y^2|X = x) - \mu_{Y|X=x}^2$$

Example 4.26 Having previously found the conditional distribution of Y given $X = 250$ in Example 4.24, we now compute the conditional mean and variance.

$$\mu_{Y|X=250} = E(Y|X=250) = 0 \cdot p_{Y|X}(0|250) + 100 \cdot p_{Y|X}(100|250) + 200 \cdot p_{Y|X}(200|250)$$
$$= 0(.1) + 100(.3) + 200(.6) = 150$$

The average homeowner's policy deductible, among customers with a $50 auto deductible, is $150. Given that the possibilities for Y are 0, 100, and 200 and most of the probability is on the latter two values, it is reasonable that the conditional mean should be between 100 and 200.

Using the alternative (shortcut) formula for the conditional variance requires first obtaining the conditional expectation of Y^2:

$$E(Y^2|X = 250) = 0^2 p_{Y|X}(0|250) + 100^2 p_{Y|X}(100|250) + 200^2 p_{Y|X}(200|250)$$
$$= 0^2(.1) + 100^2(.3) + 200^2(.6) = 27,000$$

Thus,

$$\sigma_{Y|X=250}^2 = \text{Var}(Y|X = 250) = E(Y^2|X = 250) - \mu_{Y|X=250}^2 = 27,000 - 150^2 = 4500$$

Taking the square root gives $\sigma_{Y|X=250} = \$67.08$, which is in the right ballpark when we recall that the possible values of Y are 0, 100, and 200. ∎

Example 4.27 (Example 4.25 continued) Suppose a 1-lb can of mixed nuts contains .1 lbs of almonds (i.e., we know that $X = .1$). Given this information, the amount of cashews Y in the can is constrained by $0 \le y \le 1 - x = .9$, and the expected amount of cashews in such a can is

$$E(Y|X = .1) = \int_0^{.9} y \cdot f_{Y|X}(y|.1) dy = \int_0^{.9} y \cdot \frac{2y}{(1 - .1)^2} dy = .6$$

The conditional variance of Y given that $X = .1$ is

$$\text{Var}(Y \mid X = .1) = \int_0^{.9} (y - .6)^2 \cdot f_{Y \mid X}(y \mid .1) dy = \int_0^{.9} (y - .6)^2 \cdot \frac{2y}{(1 - .1)^2} dy = .045$$

Using the aforementioned shortcut, this can also be calculated in two steps:

$$E(Y^2 \mid X = .1) = \int_0^{.9} y^2 \cdot f_{Y \mid X}(y \mid .1) dy = \int_0^{.9} y^2 \cdot \frac{2y}{(1 - .1)^2} dy = .405$$

$$\Rightarrow \text{Var}(Y \mid X = .1) = .405 - (.6)^2 = .045$$

More generally, conditional on $X = x$ lbs (where $0 < x < 1$), integrals similar to those above can be used to show that the conditional mean amount of cashews is $2(1 - x)/3$, and the corresponding conditional variance is $(1 - x)^2/18$. This formula implies that the variance gets smaller as the weight of almonds in a can approaches 1 lb. Does this make sense? When the weight of almonds is 1 lb, the weight of cashews is *guaranteed* to be 0, implying that the variance is 0. Indeed, Fig. 4.2 shows that the set of possible y-values narrows to 0 as x approaches 1. ∎

4.4.3 The Laws of Total Expectation and Variance

By the definition of conditional expectation, the rv Y has a conditional mean for every possible value x of the variable X. In Example 4.26, we determined the mean of Y given that $X = 250$, but a different mean would result if we conditioned on $X = 100$. For the continuous rvs in Example 4.27, every value x between 0 and 1 yielded a different conditional mean of Y (and, in fact, we even found a general formula for this conditional expectation). As it turns out, these conditional means can be related back to the *unconditional* mean of Y, i.e., μ_Y. Our next example illustrates the connection.

Example 4.28 Apartments in a certain city have $x = 0, 1, 2,$ or 3 bedrooms (0 for a studio apartment), and $y = 1, 1.5,$ or 2 bathrooms. The accompanying table gives the proportions of apartments for the various number of bedroom/number of bathroom combinations.

		y			
$p(x, y)$		1	1.5	2	
	0	.10	.00	.00	.1
x	1	.20	.08	.02	.3
	2	.15	.10	.15	.4
	3	.05	.05	.10	.2
		.50	.23	.27	

Let X and Y denote the number of bedrooms and bathrooms, respectively, in a randomly selected apartment in this city. The marginal distribution of Y comes from the column totals in the joint probability table, from which it is easily verified that $E(Y) = 1.385$ and $\text{Var}(Y) = .179275$. The conditional distributions (pmfs) of Y given that $X = x$ for $x = 0$, 1, 2, and 3 are as follows:

$$x = 0 : \quad p_{Y|X=0}(1) = 1 \qquad \text{(all studio apartments have one bathroom)}$$

$$x = 1 : \quad p_{Y|X=1}(1) = .667, \quad p_{Y|X=1}(1.5) = .267, \quad p_{Y|X=1}(2) = .067$$

$$x = 2 : \quad p_{Y|X=2}(1) = .375, \quad p_{Y|X=2}(1.5) = .25, \quad p_{Y|X=2}(2) = .375$$

$$x = 3 : \quad p_{Y|X=3}(1) = .25, \quad p_{Y|X=3}(1.5) = .25, \quad p_{Y|X=3}(2) = .50$$

From these conditional pmfs, we obtain the expected value of Y given $X = x$ for each of the four possible x values:

$$E(Y|X = 0) = 1, \; E(Y|X = 1) = 1.2, \; E(Y|X = 2) = 1.5, \; E(Y|X = 3) = 1.625$$

So, on the average, studio apartments have 1 bathroom, one-bedroom apartments have 1.2 bathrooms, 2-bedrooms have 1.5 baths, and luxurious 3-bedroom apartments have 1.625 baths.

Now, instead of writing $E(Y|X = x)$ for some specific value x, let's consider the expected number of bathrooms for an apartment of *randomly selected* size, X. This expectation, denoted $E(Y|X)$, is itself a random variable, since it is a function of the random quantity X. Its smallest possible value is 1, which occurs when $X = 0$, and that happens with probability .1 (the sum of probabilities in the first row of the joint probability table). Similarly, the random variable $E(Y|X)$ takes on the value 1.2 with probability $p_X(1) = .3$. Continuing in this manner, the probability distribution of the rv $E(Y|X)$ is as follows:

Value of $E(Y|X)$	1	1.2	1.5	1.625
Probability of value	.1	.3	.4	.2

(4.7)

The expected value of this random variable, denoted $E[E(Y|X)]$, is computed by taking the weighted average of the four values of $E(Y|X = x)$ against the probabilities specified by $p_X(x)$, as suggested by (4.7):

$$E[E(Y|X)] = 1(.1) + 1.2(.3) + 1.5(.4) + 1.625(.2) = 1.385$$

But this is exactly $E(Y)$, the expected number of bathrooms. ∎

LAW OF TOTAL EXPECTATION

For any two random variables X and Y,

$$E[E(Y|X)] = E(Y)$$

(This is sometimes referred to as computing $E(Y)$ by means of *iterated expectation*.)

The Law of Total Expectation says that $E(Y)$ is a weighted average of the conditional means $E(Y|X=x)$, where the weights are given by the pmf or pdf of X. It is analogous to the Law of Total Probability, which describes how to find $P(B)$ as a weighted average of conditional probabilities $P(B|A_i)$.

Proof Here is the proof when both rvs are discrete; in the jointly continuous case, simply replace summation by integration and pmfs by pdfs.

$$
\begin{aligned}
E[E(Y \mid X)] &= \sum_{x \in D_X} E(Y|X = x)p_X(x) = \sum_{x \in D_X} \sum_{y \in D_Y} y p_{Y|X}(y \mid x)p_X(x) \\
&= \sum_{x \in D_X} \sum_{y \in D_Y} y \frac{p(x, y)}{p_X(x)} p_X(x) = \sum_{y \in D_Y} y \sum_{x \in D_X} p(x, y) \\
&= \sum_{y \in D_Y} y p_Y(y) = E(Y)
\end{aligned}
$$

∎

In Example 4.28, the use of iterated expectation to compute $E(Y)$ is unnecessarily cumbersome; working from the marginal pmf of Y is more straightforward. However, there are many situations in which the distribution of a variable Y is only expressed conditional on the value of another variable X. For these so-called *hierarchical models*, the Law of Total Expectation proves very useful.

Example 4.29 A ferry goes from the left bank of a small river to the right bank once an hour. The ferry can accommodate at most two vehicles. The probability that no vehicles show up is .1, than exactly one shows up is .7, and that two or more show up is .2 (but only two can be transported). The fare paid for a vehicle depends upon its weight, and the average fare per vehicle is $25. What is the expected fare for a single trip made by this ferry?

Let X represent the number of vehicles that show up, and let Y denote the total fare for a single trip. The conditional mean of Y, given X, is given by $E(Y|X) = 25X$. So, by the Law of Total Expectation,

$$E(Y) = E[E(Y|X)] = E[25X] = \sum_{x=0}^{2}[25x \cdot p_X(x)]$$
$$= (0)(.1) + (25)(.7) + (50)(.2) = 27.50$$

∎

The next theorem provides a way to compute the variance of Y by conditioning on the value of X. There are two contributions to Var(Y). The first part is the variance of the random variable $E(Y|X)$. The second part involves the random variable Var($Y|X$)—the variance of Y as a function of X—and in particular the expected value of this random variable.

LAW OF TOTAL VARIANCE
For any two random variables X and Y,
$$\text{Var}(Y) = \text{Var}[E(Y|X)] + E[\text{Var}(Y|X)]$$

Proving the Law of Total Variance requires some more sophisticated algebra; see Exercise 84. For those familiar with statistical methods, the Law of Total Variance is analogous to the famous ANOVA identity, wherein the total variability in a response variable Y can be decomposed into the differences between group means (here, the term Var[$E(Y|X)$]) and the variation of responses within groups (represented by $E[\text{Var}(Y|X)]$ above).

Example 4.30 Let's verify the Law of Total Variance for the apartment scenario of Example 4.28. The pmf of the rv $E(Y|X)$ appears in (4.7), from which its variance is given by

$$\text{Var}[E(Y|X)] = (1 - 1.385)^2(.1) + (1.2 - 1.385)^2(.3)$$
$$+ (1.5 - 1.385)^2(.4) + (1.625 - 1.385)^2(.2)$$
$$= 0.0419$$

(Recall that 1.385 is the mean of the rv $E(Y|X)$, which, by the Law of Total Expectation, is also $E(Y)$.) The second term in the Law of Total Variance involves the variable Var($Y|X$), which requires determining the conditional variance of Y given $X=x$ for $x=0, 1, 2, 3$. Using the four conditional distributions displayed in Example 4.28, these are

$$\text{Var}(Y \mid X = 0) = 0; \quad \text{Var}(Y \mid X = 1) = .0933$$
$$\text{Var}(Y \mid X = 2) = .1875; \quad \text{Var}(Y \mid X = 3) = .171875$$

The rv Var($Y|X$) takes on these four values with probabilities .1, .4, .3, and .2, respectively (again, these are inherited from the distribution of X). Thus,

$$E[\text{Var}(Y \mid X)] = 0(.1) + .0933(.3) + .1875(.4) + .171875(.2) = .137375$$

Combining, $\text{Var}[E(Y|X)] + E[\text{Var}(Y|X)] = .0419 + .137375 = .179275$

This is exactly Var(Y) computed using the marginal pmf of Y in Example 4.28, and the Law of Total Variance is verified for this example.   ■

The computation of Var(Y) in Example 4.30 is clearly not efficient; it is much easier, given the joint pmf of X and Y, to determine the variance of Y from its marginal pmf. As with the Law of Total Expectation, the real worth of the Law of Total Variance comes from its application to hierarchical models, where the distribution of one variable (Y, say) is only known conditional on the distribution of another rv.

Example 4.31 In the manufacture of ceramic tiles used for heat shielding, the proportion of tiles that meet the required thermal specifications varies from day to day. Let P denote the proportion of tiles meeting specifications on a randomly selected day, and suppose P can be modeled by the following pdf:

$$f(p) = 9p^8 \qquad 0 < p < 1$$

At the end of each day, a random sample of $n = 20$ tiles is selected and each tile is tested. Let Y denote the number of tiles among the 20 that meet specifications; conditional on $P = p$, $Y \sim \text{Bin}(20, p)$. Find the expected number of tiles meeting thermal specifications in a daily sample of 20, and find the corresponding standard deviation.

From the properties of the binomial distribution, we know that $E(Y|P = p) = np = 20p$, so $E(Y|P) = 20P$. Applying the Law of Total Expectation,

$$E(Y) = E[E(Y \mid P)] = E[20P] = \int_0^1 20p \cdot f(p)dp = \int_0^1 180p^9 dp = 18$$

This is reasonable: since $E(P) = .9$ by integration, the expected proportion of good tiles is 90%, and thus the expected number of good tiles in a random sample of 20 tiles is 18.

Determining the standard deviation of Y requires the two pieces of the Law of Total Variance. First, using the rescaling property of variance,

$$\mathrm{Var}[E(Y\,|\,P)] = \mathrm{Var}(20P) = 20^2\mathrm{Var}(P) = 400\mathrm{Var}(P)$$

The variance of P can be determined directly from the pdf of P via integration. The result is $\mathrm{Var}(P) = 9/1100$, so $\mathrm{Var}[E(Y|P)] = 400(9/1100) = 36/11$. Second, the binomial variance formula $np(1-p)$ implies that the conditional variance of Y given P is $\mathrm{Var}(Y|P) = 20P(1-P)$, so

$$E[\mathrm{Var}(Y\,|\,P)] = E[20P(1-P)] = \int_0^1 20p(1-p)\cdot 9p^8 dp = \frac{18}{11}$$

Therefore, by the Law of Total Variance,

$$\mathrm{Var}(Y) = \mathrm{Var}[E(Y\,|\,P)] + E[\mathrm{Var}(Y\,|\,P)] = \frac{36}{11} + \frac{18}{11} = \frac{54}{11} = 4.909,$$

and the standard deviation of Y is $\sigma_Y = \sqrt{4.909} = 2.22$. This "total" standard deviation accounts for two effects: day-to-day variation in quality as modeled by P (the first term in the variance expression), and random variation in the number of observed good tiles as modeled by the binomial distribution (the second term). ∎

Here is an example where the Laws of Total Expectation and Variance are helpful in finding the mean and variance of a random variable that is neither discrete nor continuous.

Example 4.32 The probability of a claim being filed on an insurance policy is .1, and only one claim can be filed. If a claim is filed, the amount is exponentially distributed with mean $1,000. Recall from Sect. 3.4 that the mean and standard deviation of the exponential distribution are the same, so the variance is the square of this value. We want to find the mean and variance of the amount paid. Let X be the number of claims (0 or 1) and let Y be the payment. We know that $E(Y|X=0) = 0$ and $E(Y|X=1) = 1000$. Also, $\mathrm{Var}(Y|X=0) = 0$ and $\mathrm{Var}(Y|X=1) = 1000^2 = 1,000,000$. Here is a table for the both the distribution of $E(Y|X=x)$ and that of $\mathrm{Var}(Y|X=x)$:

x	$P(X=x)$	$E(Y|X=x)$	$\mathrm{Var}(Y|X=x)$
0	.9	0	0
1	.1	1000	1,000,000

Therefore

$$E(Y) = E[E(Y\,|\,X)] = E(Y\,|\,X=0)P(X=0) + E(Y\,|\,X=1)P(X=1)$$
$$= 0(.9) + 1000(.1) = 100$$

The average claim amount across all customers is $100. Next, the variance of the conditional mean is

$$\text{Var}[E(Y|X)] = (0 - 100)^2(.9) + (1000 - 100)^2(.1) = 90,000,$$

and the expected value of the conditional variance is

$$E[\text{Var}(Y|X)] = 0(.9) + 1,000,000(.1) = 100,000$$

Apply the Law of Total Variance to get $\text{Var}(Y)$:

$$\text{Var}(Y) = \text{Var}[E(Y|X)] + E[\text{Var}(Y|X)] = 90,000 + 100,000 = 190,000$$

Taking the square root gives the standard deviation, $\sigma_Y = \$435.89$.

Suppose that we want to compute the mean and variance of Y directly. Notice that X is discrete, but the conditional distribution of Y given $X = 1$ is continuous. The random variable Y itself is neither discrete nor continuous, because it has probability .9 of being 0, but the other .1 of its probability is spread out from 0 to ∞. Such "mixed" distributions may require a little extra effort to evaluate means and variances, although it is not especially hard in this case (because the discrete mass is at 0 and doesn't contribute to expectations).

$$E(Y) = (.1)\int_0^\infty y\frac{1}{1000}e^{-y/1000}dy = (.1)(1000) = 100$$

$$E(Y^2) = (.1)\int_0^\infty y^2\frac{1}{1000}e^{-y/1000}dy = (.1)2(1000^2) = 200,000$$

$$\text{Var}(Y) = E(Y^2) - [E(Y)]^2 = 200,000 - 10,000 = 190,000$$

These agree with what we found using the theorems. ∎

4.4.4 Exercises: Section 4.4 (66–84)

66. Refer back to Exercise 1 of this chapter.
    (a) Given that $X = 1$, determine the conditional pmf of Y—that is, $p_{Y|X}(0|1)$, $p_{Y|X}(1|1)$, and $p_{Y|X}(2|1)$.
    (b) Given that two hoses are in use at the self-service island, what is the conditional pmf of the number of hoses in use on the full-service island?
    (c) Use the result of part (b) to calculate the conditional probability $P(Y \le 1| X = 2)$.
    (d) Given that two hoses are in use at the full-service island, what is the conditional pmf of the number in use at the self-service island?
67. A system consists of two components. Suppose the joint pdf of the lifetimes of the two components in a system is given by $f(x, y) = c[10 - (x+y)]$ for $x > 0$, $y > 0$, $x + y < 10$, where x and y are in months.

(a) If the first component functions for exactly 3 months, what is the probability that the second functions for more than 2 months?

(b) Suppose the system will continue to work only as long as both components function. Among 20 of these systems that operate independently of each other, what is the probability that at least half work for more than 3 months?

68. The joint pdf of pressures for right and left front tires is given in Exercise 11.

(a) Determine the conditional pdf of Y given that $X = x$ and the conditional pdf of X given that $Y = y$.

(b) If the pressure in the right tire is found to be 22 psi, what is the probability that the left tire has a pressure of at least 25 psi? Compare this to $P(Y \geq 25)$.

(c) If the pressure in the right tire is found to be 22 psi, what is the expected pressure in the left tire, and what is the standard deviation of pressure in this tire?

69. Suppose that X is uniformly distributed between 0 and 1. Given $X = x$, Y is uniformly distributed between 0 and x^2.

(a) Determine $E(Y|X = x)$ and then $\text{Var}(Y|X = x)$.

(b) Determine $f(x,y)$ using $f_X(x)$ and $f_{Y|X}(y|x)$.

(c) Determine $f_Y(y)$.

70. Consider three Ping-Pong balls numbered 1, 2, and 3. Two balls are randomly selected with replacement. If the sum of the two resulting numbers exceeds 4, two balls are again selected. This process continues until the sum is at most 4. Let X and Y denote the last two numbers selected. Possible (X, Y) pairs are $\{(1, 1), (1, 2), (1, 3), (2, 1), (2, 2), (3, 1)\}$.

(a) Determine $p_{X,Y}(x,y)$.

(b) Determine $p_{Y|X}(y|x)$.

(c) Determine $E(Y|X = x)$.

(d) Determine $E(X|Y = y)$. What special property of $p(x, y)$ allows us to get this from (c)?

(e) Determine $\text{Var}(Y|X = x)$.

71. Let X be a random digit (0, 1, 2, ..., 9 are equally likely) and let Y be a random digit not equal to X. That is, the nine digits other than X are equally likely for Y.

(a) Determine $p_X(x)$, $p_{Y|X}(y|x)$, $p_{X,Y}(x,y)$.

(b) Determine a formula for $E(Y|X = x)$.

72. A pizza delivery business has two phones. On each phone the waiting time until the first call is exponentially distributed with mean 1 min. Each phone is not influenced by the other. Let X be the shorter of the two waiting times and let Y be the longer. Using techniques from Sect. 4.9, it can be shown that the joint pdf of X and Y is $f(x, y) = 2e^{-(x+y)}$ for $0 < x < y < \infty$.

(a) Determine the marginal density of X.

(b) Determine the conditional density of Y given $X = x$.

(c) Determine the probability that Y is greater than 2, given that $X = 1$.

(d) Are X and Y independent? Explain.

(e) Determine the conditional mean of Y given $X = x$.

(f) Determine the conditional variance of Y given $X = x$.

73. Teresa and Allison each have arrival times uniformly distributed between 12:00 and 1:00. Their times do not influence each other. If Y is the first of the two times and X is the second, on a scale of 0 to 1, it can be shown that the joint pdf of X and Y is $f(x, y) = 2$ for $0 < y < x < 1$.

(a) Determine the marginal density of X.

(b) Determine the conditional density of Y given $X = x$.

(c) Determine the conditional probability that Y is between 0 and .3, given that X is .5.

(d) Are X and Y independent? Explain.

(e) Determine the conditional mean of Y given $X = x$.

(f) Determine the conditional variance of Y given $X = x$.

74. Refer back to the previous exercise.

(a) Determine the marginal density of Y.

(b) Determine the conditional density of X given $Y = y$.

(c) Determine the conditional mean of X given $Y = y$.

(d) Determine the conditional variance of X given $Y = y$.

75. According to an article in the August 30, 2002 issue of the *Chronicle of Higher Education*, 30% of first-year college students are liberals, 20% are conservatives, and 50% characterize themselves as middle-of-the-road. Choose two students at random, let X be the number of liberals among the two, and let Y be the number of conservatives among the two.

(a) Using the multinomial distribution from Sect. 4.1, give the joint probability mass function $p(x, y)$ of X and Y and the corresponding joint probability table.

(b) Determine the marginal probability mass functions by summing $p(x, y)$ numerically. How could these be obtained directly? [*Hint*: What are the univariate distributions of X and Y?]

(c) Determine the conditional probability mass function of Y given $X = x$ for $x = 0$, 1, 2. Compare this to the binomial distribution with $n = 2 - x$ and $p = .2/(.2 + .5)$. Why should this work?

(d) Are X and Y independent? Explain.

(e) Find $E(Y|X = x)$ for $x = 0$, 1, 2. Do this numerically and then compare with the use of the formula for the binomial mean, using the binomial distribution given in part (c).

(f) Determine $\mathrm{Var}(Y|X = x)$ for $x = 0$, 1, 2. Do this numerically and then compare with the use of the formula for the binomial variance, using the binomial distribution given in part (c).

76. A class has 10 mathematics majors, 6 computer science majors, and 4 statistics majors. Two of these students are randomly selected to make a presentation. Let X be the number of mathematics majors and let Y be the number of computer science majors chosen.

(a) Determine the joint probability mass function $p(x,y)$. This generalizes the hypergeometric distribution studied in Sect. 2.6. Give the joint probability table showing all nine values, of which three should be 0.

(b) Determine the marginal probability mass functions by summing numerically. How could these be obtained directly? [*Hint*: What type of rv is X?
Y?]

(c) Determine the conditional probability mass function of Y given $X = x$ for
$x = 0$, 1, 2. Compare with the $h(y; 2 - x, 6, 10)$ distribution. Intuitively,
why should this work?

(d) Are X and Y independent? Explain.

(e) Determine $E(Y|X = x)$, $x = 0$, 1, 2. Do this numerically and then compare
with the use of the formula for the hypergeometric mean, using the
hypergeometric distribution given in part (c).

(f) Determine $\mathrm{Var}(Y|X = x)$, $x = 0, 1, 2$. Do this numerically and then compare
with the use of the formula for the hypergeometric variance, using the
hypergeometric distribution given in part (c).

77. A 1-ft-long stick is broken at a point X (measured from the left end) chosen
randomly uniformly along its length. Then the left part is broken at a point
Y chosen randomly uniformly along its length. In other words, X is uniformly
distributed between 0 and 1 and, given $X = x$, Y is uniformly distributed
between 0 and x.

(a) Determine $E(Y|X = x)$ and then $\mathrm{Var}(Y|X = x)$.

(b) Determine $f(x,y)$ using $f_X(x)$ and $f_{Y|X}(y|x)$.

(c) Determine $f_Y(y)$.

(d) Use $f_Y(y)$ from (c) to get $E(Y)$ and $\mathrm{Var}(Y)$.

(e) Use (a) and the Laws of Total Expectation and Variance to get $E(Y)$ and
$\mathrm{Var}(Y)$.

78. Consider the situation in Example 4.29, and suppose further that the standard
deviation for fares per car is \$4.

(a) Find the variance of the rv $E(Y|X)$.

(b) Using Expression (4.6) from the previous section, the conditional variance
of Y given $X = x$ is $4^2 x = 16x$. Determine the mean of the rv $\mathrm{Var}(Y|X)$.

(c) Use the Law of Total Variance to find σ_Y, the unconditional standard
deviation of Y.

79. This week the number X of claims coming into an insurance office has a
Poisson distribution with mean 100. The probability that any particular claim
relates to automobile insurance is .6, independent of any other claim. If Y is
the number of automobile claims, then Y is binomial with X trials, each with
"success" probability .6.

(a) Determine $E(Y|X = x)$ and $\mathrm{Var}(Y|X = x)$.

(b) Use part (a) to find $E(Y)$.

(c) Use part (a) to find $\mathrm{Var}(Y)$.

80. In the previous exercise, show that the distribution of Y is Poisson with mean
60. [You will need to recognize the Maclaurin series expansion for the
exponential function.] Use the knowledge that Y is Poisson with mean 60 to
find $E(Y)$ and $\mathrm{Var}(Y)$.

81. The heights of American men follow a normal distribution with mean 70 in.
and standard deviation 3 in. Suppose that the weight distribution (lbs) for men

that are x inches tall also has a normal distribution, but with mean $4x - 104$ and standard deviation $.3x - 17$. Let Y denote the weight of a randomly selected American man. Find the (unconditional) mean and standard deviation of Y.

82. A statistician is waiting behind one person to check out at a store. The check-out time for the first person, X, can be modeled by an exponential distribution with some parameter $\lambda > 0$. The statistician observes the first person's check-out time, x; being a statistician, she surmises that her check-out time Y will follow an exponential distribution with mean x.
    (a) Determine $E(Y|X=x)$ and $\text{Var}(Y|X=x)$.
    (b) Use the Laws of Total Expectation and Variance to find $E(Y)$ and $\text{Var}(Y)$.
    (c) Write out the joint pdf of X and Y. [*Hint:* You have $f_X(x)$ and $f_{Y|X}(y|x)$.] Then write an integral expression for the marginal pdf of Y (from which, at least in theory, one could determine the mean and variance of Y). What happens?

83. In the game Plinko on the television game show *The Price is Right*, contestants have the opportunity to earn "chips" (flat, circular disks) that can be dropped down a peg board into slots labeled with cash amounts. Every contestant is given one chip automatically and can earn up to four more chips by correctly guessing the prices of certain small items. If we let p denote the probability a contestant correctly guesses the price of a prize, then the number of chips a contestant earns, X, can be modeled as $X = 1 + N$, where $N \sim \text{Bin}(4, p)$.
    (a) Determine $E(X)$ and $\text{Var}(X)$.
    (b) For each chip, the amount of money won on the Plinko board has the following distribution:

Value	$0	$100	$500	$1000	$10,000
Probability	.39	.03	.11	.24	.23

    Determine the mean and variance of the winnings from a single chip.
    (c) Let Y denote the total winnings of a randomly selected contestant. Using results from the previous section, the conditional mean and variance of Y, given a player gets x chips, are μx and $\sigma^2 x$, respectively, where μ and σ^2 are the mean and variance for a single chip computed in (b). Find expressions for the (unconditional) mean and standard deviation of Y. [*Note:* Your answers will be functions of p.]
    (d) Evaluate your answers to part (c) for $p = 0, .5,$ and 1. Do these answers make sense? Explain.

84. Let X and Y be any two random variables.
    (a) Show that $E[\text{Var}(Y|X)] = E[Y^2] - E[\mu_{Y|X}^2]$. [*Hint:* Use the variance shortcut formula and apply the Law of Total Expectation to the first term.]
    (b) Show that $\text{Var}(E[Y|X]) = E[\mu_{Y|X}^2] - (E[Y])^2$. [*Hint:* Use the variance shortcut formula again; this time, apply the Law of Total Expectation to the second term.]
    (c) Combine the previous two results to establish the Law of Total Variance.

4.5   Limit Theorems (What Happens as *n* Gets Large)

Many problems in probability and statistics involve either a sum or an average of random variables. In this section we consider what happens as n, the number of variables in such sums and averages, gets large. The most important result of this type is the celebrated Central Limit Theorem, according to which the approximate distribution is normal when n is sufficiently large.

4.5.1   Random Samples

The random variables from which our sums and averages will be created must satisfy two general conditions.

> **DEFINITION**
> The rvs X_1, X_2, ..., X_n are said to be **independent and identically distributed (iid)** if
> 1. The X_is are independent rvs.
> 2. Every X_i has the same probability distribution.
> Such a collection of rvs is also called a (simple) **random sample** of size n.

For example, $X_1, X_2, \ldots X_n$ might be a random sample from a normal distribution with mean 100 and standard deviation 15; then the X_is are independent and each one has the specified normal distribution. Similarly, for these variables to constitute a random sample from an exponential distribution, they must be independent and the value of the exponential parameter λ must be the same for each variable.

The notion of iid rvs is meant to resemble (simple) random sampling from a population: X_1 is the value of some variable for the first individual or object selected, X_2 is the value of that same variable for the second selected individual or object, and so on. If sampling is either with replacement or from a (potentially) infinite population, Conditions 1 and 2 are satisfied exactly. These conditions will be approximately satisfied if sampling is without replacement, yet the sample size n is much smaller than the population size N. In practice, if $n/N \leq .05$ (at most 5% of the population is sampled), we proceed as if the X_is form a random sample.

Throughout this section, we will be primarily interested in the properties of two particular rvs derived from random samples: the **sample total** T and the **sample mean** $\overline{X}$:

$$T = X_1 + \cdots + X_n = \sum_{i=1}^{n} X_i, \qquad \overline{X} = \frac{X_1 + \cdots + X_n}{n} = \frac{T}{n}.$$

Note that both T and $\overline{X}$ are linear combinations of the X_is.

> **PROPOSITION**
>
> Suppose $X_1, X_2, \ldots, X_n$ are iid with common mean μ and common standard deviation σ. T and $\overline{X}$ have the following properties:
>
> 1. $E(T) = n\mu$ 1. $E(\overline{X}) = \mu$
>
> 2. $\operatorname{Var}(T) = n\sigma^2$ and $\operatorname{SD}(T) = \sqrt{n}\sigma$ 2. $\operatorname{Var}(\overline{X}) = \dfrac{\sigma^2}{n}$ and $\operatorname{SD}(\overline{X}) = \dfrac{\sigma}{\sqrt{n}}$
>
> 3. If the X_is are normally distributed, then T is also normally distributed.   3. If the X_is are normally distributed, then $\overline{X}$ is also normally distributed.

Proof Recall from the main theorem of Sect. 4.3 that the expected value of a sum is the sum of individual expected values; moreover, when the variables in the sum are independent, the variance of the sum is the sum of the individual variances:

$$E(T) = E(X_1 + \cdots + X_n) = E(X_1) + \cdots + E(X_n) = \mu + \cdots + \mu = n\mu$$
$$\operatorname{Var}(T) = \operatorname{Var}(X_1 + \cdots + X_n) = \operatorname{Var}(X_1) + \cdots + \operatorname{Var}(X_n) = \sigma^2 + \cdots + \sigma^2 = n\sigma^2$$
$$\operatorname{SD}(T) = \sqrt{n\sigma^2} = \sqrt{n}\sigma$$

The corresponding results for $\overline{X}$ can be derived by writing $\overline{X} = \frac{1}{n} \cdot T$ and using basic rescaling properties, such as $E(cY) = cE(Y)$. Property 3 is a consequence of the more general result from Sect. 4.3 that any linear combination of independent normal rvs is normal.   ∎

According to Property 1, the distribution of $\overline{X}$ is centered precisely at the mean of the population from which the sample has been selected. If the sample mean is used to compute an estimate (educated guess) of the population mean μ, there will be no systematic tendency for the estimate to be too large or too small.

Property 2 shows that the $\overline{X}$ distribution becomes more concentrated about μ as the sample size n increases, because its standard deviation decreases. In marked contrast, the distribution of T becomes more spread out as n increases. Averaging moves probability in toward the middle, whereas totaling spreads probability out over a wider and wider range of values. The expression $\sigma/\sqrt{n}$ for the standard deviation of $\overline{X}$ is called the **standard error of the mean**, and it indicates the typical amount by which a value of $\overline{X}$ will deviate from the true mean, μ (in contrast, σ itself represents the typical difference between an *individual* X_i and μ).

When σ is unknown, as is usually the case when μ is unknown and we are trying to estimate it, we may substitute the sample standard deviation, s, of our sample into the standard error formula and say that an observed value of $\overline{X}$ will typically differ by about $s/\sqrt{n}$ from μ. This is the estimated standard error formula presented in Sects. 2.8 and 3.8.

Fig. 4.8 A normal
population distribution and $\overline{X}$
sampling distributions

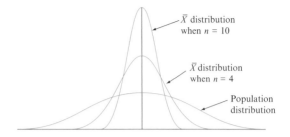

Finally, Property 3 says that we know everything there is to know about the $\overline{X}$ and T distributions when the population distribution is normal. In particular, probabilities such as $P\left(a \leq \overline{X} \leq b\right)$ and $P(c \leq T \leq d)$ can be obtained simply by standardizing. Figure 4.8 illustrates the $\overline{X}$ part of the proposition.

Example 4.33 The amount of time that a patient undergoing a particular procedure spends in a certain outpatient surgery center is a random variable with a mean value of 4.5 h and a standard deviation of 1.4 h. Let $X_1, \ldots, X_{25}$ be the times for a random sample of 25 patients. Then the expected total time for the 25 patients is $E(T) = n\mu = 25(4.5) = 112.5$ h, whereas the expected sample mean amount of time is $E\left(\overline{X}\right) = \mu = 4.5$ hours. The standard deviations of T and $\overline{X}$ are

$$\sigma_T = \sqrt{n}\sigma = \sqrt{25}(1.4) = 7 \text{ hours}$$

$$\sigma_{\overline{X}} = \frac{\sigma}{\sqrt{n}} = \frac{1.4}{\sqrt{25}} = .28 \text{ hours}$$

Suppose further that such patient times follow a normal distribution, i.e., $X_i \sim N(4.5, 1.4)$. Then the total time spent by 25 randomly selected patients in this center is also normal: $T \sim N(112.5, 7)$. The probability their total time exceeds 5 days (120 h) is

$$P(T > 120) = 1 - P(T \leq 120) = 1 - \Phi\left(\frac{120 - 112.5}{7}\right) = 1 - \Phi(1.07) = .8577$$

This same probability can be reframed in terms of $\overline{X}$: for 25 patients, a total time of 120 h equates to an average time of $120/25 = 4.8$ h, and since $\overline{X} \sim N(4.5, .28)$,

$$P\left(\overline{X} > 4.8\right) = 1 - \Phi\left(\frac{4.8 - 4.5}{.28}\right) = 1 - \Phi(1.07) = .8577 \qquad \blacksquare$$

Example 4.34 Resistors used in electronics manufacturing are labeled with a "nominal" resistance as well as a percentage tolerance. For example, a 330-ohm resistor with a 5% tolerance is anticipated to have an actual resistance between 313.5 Ω and 346.5 Ω. Consider five such resistors, randomly selected from the population of all resistors with those specifications, and model the resistance of each by a uniform distribution on $[313.5, 346.5]$. If these are connected in series, the resistance R of the system is given by $R = X_1 + \cdots + X_5$, where the X_i are the iid uniform resistances.

A random variable uniformly distributed on $[A, B]$ has mean $(A + B)/2$ and standard deviation $(B - A)/\sqrt{12}$. For our uniform model, the mean resistance is $E(X_i) = (313.5 + 346.5)/2 = 330 \ \Omega$, the nominal resistance, with a standard deviation of $(346.5 - 313.5)/\sqrt{12} = 9.526 \ \Omega$. The system's resistance has mean and standard deviation

$$E(R) = n\mu = 5(330) = 1650\Omega, \qquad SD(R) = \sqrt{n}\sigma = \sqrt{5}(9.526) = 21.3\Omega$$

But what is the probability distribution of R? Is R also uniformly distributed? Determining the exact pdf of R is difficult (it requires four convolutions). And the mgf of R, while easy to obtain, is not recognizable as coming from any particular family of known distributions. Instead, we resort to a simulation of R, the results of which appear in Fig. 4.9. For 10,000 iterations in R (appropriately), five independent uniform variates on $[313.5, 346.5]$ were created and summed; see Sect. 3.8 for information on simulating a uniform distribution. The histogram in Fig. 4.9 clearly indicates that R is not uniform; in fact, if anything, R appears (from the simulation, anyway) to be approximately normal!

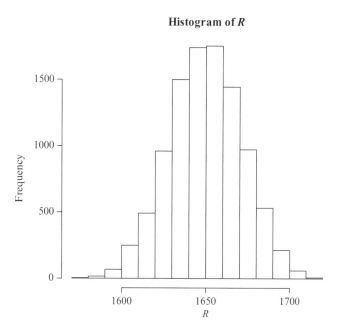

Fig. 4.9 Simulated distribution of the random variable R in Example 4.34   ∎

4.5.2   The Central Limit Theorem

When iid X_is are normally distributed, so are T and $\overline{X}$ for every sample size n. The simulation results from Example 4.34 suggest that even when the population distribution is not normal, summing (or averaging) produces a distribution more bell-shaped than the one being sampled. Upon reflection, this is quite intuitive: in order for R to be near $5(346.5) = 1732.5$, its theoretical maximum, all five randomly selected resistors would have to exert resistances at the high end of their common range (i.e., every X_i would have to be near 346.5). Thus, R-values near 1732.5 are unlikely, and the same applies to R's theoretical minimum of $5(313.5) = 1567.5$. On the other hand, there are many ways for R to be near the mean value of 1650: all five resistances in the middle, two low and one middle and two high, and so on. Thus, R is more likely to be "centrally" located than out at the extremes. (This is analogous to the well-known fact that rolling a pair of dice is far more likely to result in a sum of 7 than 2 or 12, because there are more ways to obtain 7.)

This general pattern of behavior for sample totals and sample means is formalized by the most important theorem of probability, the *Central Limit Theorem* (CLT). A proof of this theorem is beyond the scope of this book, but interested readers may consult the text by Devore and Berk listed in the references.

CENTRAL LIMIT THEOREM
Let $X_1, X_2, \ldots, X_n$ be a random sample from a distribution with mean μ and standard deviation σ. Then, in the limit as $n \rightarrow \infty$, the standardized versions of T and $\overline{X}$ have the standard normal distribution. That is,

$$\lim_{n \to \infty} P\left(\frac{T - n\mu}{\sqrt{n}\sigma} \le z\right) = P(Z \le z) = \Phi(z)$$

and

$$\lim_{n \to \infty} P\left(\frac{\overline{X} - \mu}{\sigma/\sqrt{n}} \le z\right) = P(Z \le z) = \Phi(z)$$

where Z is a standard normal rv. It is customary to say that T and $\overline{X}$ are **asymptotically normal**. Thus when n is sufficiently large, the sample total T has approximately a normal distribution with mean $\mu_T = n\mu$ and standard deviation $\sigma_T = \sqrt{n}\sigma$. Equivalently, for large n the sample mean $\overline{X}$ has approximately a normal distribution with mean $\mu_{\overline{X}} = \mu$ and standard deviation $\sigma_{\overline{X}} = \sigma/\sqrt{n}$.

Figure 4.10 illustrates the Central Limit Theorem for $\overline{X}$. According to the CLT, when n is large and we wish to calculate a probability such as $P(a \le \overline{X} \le b)$ or $P(c \le T \le d)$, we need only "pretend" that $\overline{X}$ or T is normal, standardize it, and use software or the standard normal table. The resulting answer will be approximately

Fig. 4.10 The Central Limit Theorem for $\overline{X}$ illustrated

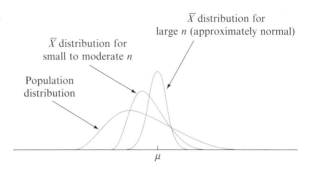

correct. The exact answer could be obtained only by first finding the distribution of *T* or $\overline{X}$, so the CLT provides a truly impressive shortcut.

A practical difficulty in applying the CLT is in knowing when *n* is "sufficiently large." The problem is that the accuracy of the approximation for a particular *n* depends on the shape of the original underlying distribution being sampled. If the underlying distribution is symmetric and there is not much probability far out in the tails, then the approximation will be good even for a small *n*, whereas if it is highly skewed or has "heavy" tails, then a large *n* will be required. For example, if the distribution is uniform on an interval, then it is symmetric with no probability in the tails, and the normal approximation is very good for *n* as small as 10 (in Example 4.34, even for $n = 5$, the distribution of the sample total appeared rather bell-shaped). However, at the other extreme, a distribution can have such fat tails that its mean fails to exist and the Central Limit Theorem does not apply, so no *n* is big enough. A popular, although frequently somewhat conservative, convention is that the Central Limit Theorem may be safely applied when $n > 30$. Of course, there are exceptions, but this rule applies to most distributions of real data.

Example 4.35 When a batch of a certain chemical product is prepared, the amount of a particular impurity in the batch is a random variable with mean value 4.0 g and standard deviation 1.5 g. If 50 batches are independently prepared, what is the (approximate) probability that the total amount of impurity is between 175 and 190 g? According to the convention mentioned above, $n = 50$ is large enough for the CLT to be applicable. The total *T* then has approximately a normal distribution with mean value $\mu_T = 50(4.0) = 200$ g and standard deviation $\sigma_T = \sqrt{50}(1.5) = 10.6066$ g. So, with *Z* denoting a standard normal rv,

$$P(175 \leq T \leq 190) \approx P\left(\frac{175 - 200}{10.6066} \leq Z \leq \frac{190 - 200}{10.6066}\right) = \Phi(-.94) - \Phi(-2.36)$$

$$= .1645$$

Notice that nothing was said initially about the shape of the underlying impurity distribution. It could be normally distributed, or uniform, or positively skewed—regardless, the CLT ensures that the distribution of their total, *T*, is approximately normal. ∎

Example 4.36 Suppose the number of times a randomly selected customer of a
large bank uses the bank's ATM during a particular period is a random variable
with a mean value of 3.2 and a standard deviation of 2.4. Among 100 randomly
selected customers, how likely is it that the sample mean number of times the
bank's ATM is used exceeds 4? Let X_i denote the number of times the ith customer
in the sample uses the bank's ATM. Notice that X_i is a discrete rv, but the CLT is not
limited to continuous random variables. Also, although the fact that the standard
deviation of this nonnegative variable is quite large relative to the mean value
suggests that its distribution is positively skewed, the large sample size implies that
$\overline{X}$ does have approximately a normal distribution. Using $\mu_{\overline{X}} = \mu = 3.2$ and
$\sigma_{\overline{X}} = \sigma/\sqrt{n} = 2.4/\sqrt{100} = .24$,

$$P\left(\overline{X} > 4\right) \approx P\left(Z > \frac{4 - 3.2}{.24}\right) = 1 - \Phi(3.33) = .0004 \qquad \blacksquare$$

Example 4.37 Consider the distribution shown in Fig. 4.11 for the amount purchased (rounded to the nearest dollar) by a randomly selected customer at a
particular gas station (a similar distribution for purchases in Britain (in £) appeared
in the article "Data Mining for Fun and Profit," *Statistical Science*, 2000:
$111 - 131$; there were big spikes at the values 10, 15, 20, 25, and 30). The
distribution is obviously quite non-normal.

   We asked Matlab to select 1000 different samples, each consisting of $n = 15$
observations, and calculate the value of the sample mean for each one. Figure 4.12
is a histogram of the resulting 1000 values; this is the approximate distribution of $\overline{X}$
under the specified circumstances. This distribution is clearly approximately

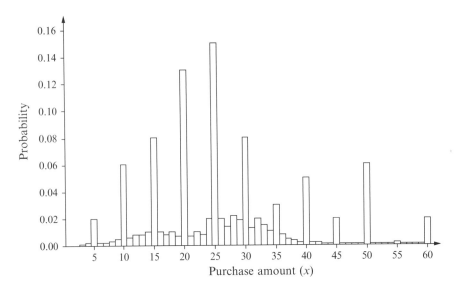

Fig. 4.11 Probability distribution of $X =$ amount of gasoline purchased (\$) in Example 4.37

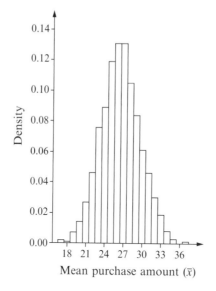

Fig. 4.12   Approximate sampling distribution of the sample mean amount purchased when $n = 15$ and the population distribution is as shown in Fig. 4.11

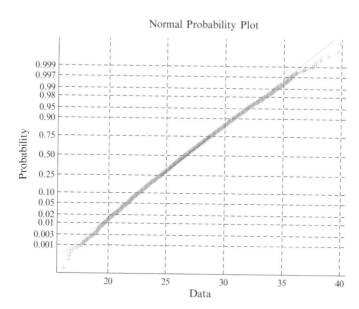

Fig. 4.13   Normal probability plot from Matlab of the 1000 $\bar{x}$ values based on samples of size $n = 15$

normal even though the sample size is not all that large. As further evidence for normality, Fig. 4.13 shows a normal probability plot of the 1000 $\bar{x}$ values; the linear pattern is very prominent. It is typically not non-normality in the central part of the population distribution that causes the CLT to fail, but instead very substantial skewness or heavy tails.   ■

The CLT can also be generalized so it applies to non-identically distributed independent random variables and certain linear combinations. Roughly speaking, if n is large and no individual term is likely to contribute too much to the overall value, then asymptotic normality prevails (see Exercise 190). It can also be generalized to sums of variables which are not independent provided the extent of dependence between most pairs of variables is not too strong.

4.5.3   Other Applications of the Central Limit Theorem

The CLT can be used to justify the normal approximation to the binomial distribution discussed in Sect. 3.3. Recall that a binomial variable X is the number of successes in a binomial experiment consisting of n independent success/failure trials with $p = P(\text{success})$ for any particular trial. Define new rvs $X_1, X_2, \ldots, X_n$ by

$$X_i = \begin{cases} 1 & \text{if the } i\text{th trial results in a success} \\ 0 & \text{if the } i\text{th trial results in a failure} \end{cases} \qquad (i = 1, \ldots, n)$$

Because the trials are independent and $P(\text{success})$ is constant from trial to trial, the X_is are iid (a random sample from a Bernoulli distribution). When the X_is are summed, a 1 is added for every success that occurs and a 0 for every failure, so $X = X_1 + \cdots + X_n$, their total. The sample mean of the X_is is $\overline{X} = X/n$, the sample proportion of successes, which in previous discussions we have denoted $\hat{P}$. The Central Limit Theorem then implies that if n is sufficiently large, both X and $\hat{P}$ are approximately normal when n is large. We summarize properties of the $\hat{P}$ distribution in the following corollary; Statements 1 and 2 were derived in Sect. 2.4.

COROLLARY

Consider an event A in the sample space of some experiment with $p = P(A)$. Let $X =$ the number of times A occurs when the experiment is repeated n independent times, and define

$$\hat{P} = \hat{P}(A) = \frac{X}{n}$$

Then

1. $$\mu_{\hat{P}} = E(\hat{P}) = p$$

(continued)

2.
$$\sigma_{\hat{P}} = \text{SD}(\hat{P}) = \sqrt{\frac{p(1-p)}{n}}$$

3. As n increases, the distribution of $\hat{P}$ approaches a normal distribution. In practice, Property 3 is taken to say that $\hat{P}$ is approximately normal, provided that $np \geq 10$ and $n(1-p) \geq 10$.

The necessary sample size for this approximation depends on the value of p: when p is close to .5, the distribution of each X_i is reasonably symmetric (see Fig. 4.14), whereas the distribution is quite skewed when p is near 0 or 1. Using the approximation only if both $np \geq 10$ and $n(1-p) \geq 10$ ensures that n is large enough to overcome any skewness in the underlying Bernoulli distribution.

Fig. 4.14 Two Bernoulli distributions: (**a**) $p = .4$ (reasonably symmetric); (**b**) $p = .1$ (*very* skewed)

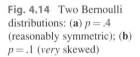
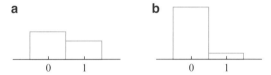

Example 4.38 A computer simulation in the style of Sect. 1.6 is used to determine the probability that a complex system of components operates properly throughout the warranty period. Unknown to the investigator, the true probability is $P(A) = .18$. If 10,000 simulations of the underlying process are run, what is the chance the estimated probability $\hat{P}(A)$ will lie within .01 of the true probability $P(A)$?

Apply the preceding corollary, with $n = 10,000$ and $p = P(A) = .18$. The expected value of the estimator $\hat{P}(A)$ is $p = .18$, and the standard deviation is $\sigma_{\hat{P}} = \sqrt{.18(.82)/10,000} = .00384$. Since $np = 1800 \geq 10$ and $n(1-p) = 8200 \geq 10$, a normal distribution can safely be used to approximate the distribution of $\hat{P}(A)$. This sample proportion is within .01 of the true probability, .18, iff $.17 < \hat{P}(A) < .19$, so the desired likelihood is approximately

$$P(.17 < \hat{P} < .19) \approx P\left(\frac{.17 - .18}{.00384} < Z < \frac{.19 - .18}{.00384}\right) = \Phi(2.60) - \Phi(-2.60)$$

$$= .9906 \qquad \blacksquare$$

The normal distribution serves as a reasonable approximation to the binomial pmf when n is large because the binomial distribution is *additive*, i.e., a binomial rv can be expressed as the sum of other, iid rvs. Other additive distributions include the

Poisson, negative binomial, gamma, and (of course) normal distributions; some of these were discussed at the end of Sect. 4.3. In particular, CLT justifies normal approximations to the following distributions:

- Poisson, when μ is large
- Negative binomial, when r is large
- Gamma, when α is large

As a final application of the CLT, first recall from Sect. 3.5 that X has a lognormal distribution if $\ln(X)$ has a normal distribution.

PROPOSITION

Let $X_1, X_2, \ldots, X_n$ be a random sample from a distribution for which only positive values are possible $[P(X_i > 0) = 1]$. Then if n is sufficiently large, the product $Y = X_1 X_2 \cdots X_n$ has approximately a lognormal distribution; that is, $\ln(Y)$ has a normal distribution.

To verify this, note that

$$\ln(Y) = \ln(X_1) + \ln(X_2) + \cdots + \ln(X_n)$$

Since $\ln(Y)$ is a sum of independent and identically distributed rvs [the $\ln(X_i)$s], it is approximately normal when n is large, so Y itself has approximately a lognormal distribution. As an example of the applicability of this result, it has been argued that the damage process in plastic flow and crack propagation is a multiplicative process, so that variables such as percentage elongation and rupture strength have approximately lognormal distributions.

4.5.4 The Law of Large Numbers

In the simulation sections of Chaps. 1–3, we described how a sample proportion $\hat{P}$ could estimate a true probability p, and a sample mean $\overline{X}$ served to approximate a theoretical expected value μ. Moreover, in both cases the precision of the estimation improves as the number of simulation runs, n, increases. We would like to be able to say that our estimates "converge" to the correct values in some sense. Such a convergence statement is justified by another important theoretical result, called the *Law of Large Numbers*.

To begin, recall the first proposition in this section: If $X_1, X_2, \ldots, X_n$ is a random sample from a distribution with mean μ and standard deviation σ, then $E(\overline{X}) = \mu$ and $\text{Var}(\overline{X}) = \sigma^2/n$. As n increases, the expected value of $\overline{X}$ remains at μ but the variance approaches zero; that is, $E\left[(\overline{X} - \mu)\right]^2 = \text{Var}(\overline{X}) = \sigma^2/n \to 0$. We say that

$\overline{X}$ converges *in mean square* to μ because the mean of the squared difference between $\overline{X}$ and μ goes to zero. This is one form of the Law of Large Numbers.

Another form of convergence states that as the sample size n increases, $\overline{X}$ is increasingly unlikely to differ by any set amount from μ. More precisely, let ε be a positive number close to 0, such as .01 or .001, and consider $P(|\overline{X} - \mu| \geq \varepsilon)$, the probability that $\overline{X}$ differs from μ by at least ε (at least .01, at least .001, etc.). We will prove shortly with the help of Chebyshev's inequality that, no matter how small the value of ε, this probability will approach zero as $n \to \infty$. Because of this, statisticians say that $\overline{X}$ *converges to μ in probability*.

The two forms of the Law of Large Numbers are summarized in the following theorem.

LAW OF LARGE NUMBERS
If $X_1, X_2, \ldots, X_n$ is a random sample from a distribution with mean μ, then $\overline{X}$ converges to μ

1. In mean square: $E\left[(\overline{X} - \mu)^2\right] \to 0$ as $n \to \infty$
2. In probability: $P(|\overline{X} - \mu| \geq \varepsilon) \to 0$ as $n \to \infty$ for any $\varepsilon > 0$

Proof   The proof of Statement 1 appears a few paragraphs above. For Statement 2, recall Chebyshev's inequality, which states that for any rv Y, $P(|Y - \mu_Y| \geq k\sigma_Y) \leq 1/k^2$ for any $k \geq 1$ (i.e., the probability that Y is at least k standard deviations away from its mean is at most $1/k^2$). Let $Y = \overline{X}$, so $\mu_Y = E(\overline{X}) = \mu$ and $\sigma_Y = \mathrm{SD}(\overline{X}) = \sigma/\sqrt{n}$. Now, for any $\varepsilon > 0$, determine the value of k such that $\varepsilon = k\sigma_Y = k\sigma/\sqrt{n}$. Solving for k yields $k = \varepsilon\sqrt{n}/\sigma$, which for sufficiently large n will exceed 1. Apply Chebyshev's inequality:

$$P(|Y - \mu_Y| \geq k\sigma_Y) \leq \frac{1}{k^2} \Rightarrow P\left(|\overline{X} - \mu| \geq \frac{\varepsilon\sqrt{n}}{\sigma} \cdot \frac{\sigma}{\sqrt{n}}\right) \leq \frac{1}{(\varepsilon\sqrt{n}/\sigma)^2} \Rightarrow$$

$$P(|\overline{X} - \mu| \geq \varepsilon) \leq \frac{\sigma^2}{\varepsilon^2 n} \to 0 \text{ as } n \to \infty$$

That is, $P(|\overline{X} - \mu| \geq \varepsilon) \to 0$ as $n \to \infty$ for any $\varepsilon > 0$.   ∎

Convergence of $\overline{X}$ to μ in probability actually holds even if the variance σ^2 does not exist (a heavy-tailed distribution) as long as μ is finite. But then Chebyshev's inequality cannot be used, and the proof is much more complicated.

An analogous result holds for proportions. If the X_i are iid Bernoulli(p) rvs, then similar to the discussion earlier in this section we may write $\overline{X}$ as $\hat{P}$, and $\mu = E(X_i) = p$. It follows that the sample proportion $\hat{P}$ converges to the "true" proportion p

1. In mean square: $E\left[(\hat{P} - p)^2\right] \to 0$ as $n \to \infty$, and

2. In probability: $P(|\hat{P} - p| \geq \varepsilon) \to 0$ as $n \to \infty$ for any $\varepsilon > 0$.

In statistical language, the Law of Large Numbers states that $\overline{X}$ is a *consistent* estimator of μ, and $\hat{P}$ is a consistent estimator of p. This consistency property also applies to other estimators. For example, it can be shown that the sample variance $S^2 = \sum (X_i - \overline{X})^2 / (n-1)$ converges in probability to the population variance σ^2.

4.5.5   Exercises: Section 4.5 (85–102)

85. The inside diameter of a randomly selected piston ring is a random variable with mean value 12 cm and standard deviation .04 cm.
    (a) If $\overline{X}$ is the sample mean diameter for a random sample of $n = 16$ rings, where is the sampling distribution of $\overline{X}$ centered, and what is the standard deviation of the $\overline{X}$ distribution?
    (b) Answer the questions posed in part (a) for a sample size of $n = 64$ rings.
    (c) For which of the two random samples, the one of part (a) or the one of part (b), is $\overline{X}$ more likely to be within .01 cm of 12 cm? Explain your reasoning.

86. Refer to the previous exercise. Suppose the distribution of diameter is normal.
    (a) Calculate $P(11.99 \leq \overline{X} \leq 12.01)$ when $n = 16$.
    (b) How likely is it that the sample mean diameter exceeds 12.01 when $n = 25$?

87. Suppose that the fracture angle under pure compression of a randomly selected specimen of fiber reinforced polymer-matrix composite material is normally distributed with mean value 53 and standard deviation 1 (suggested in the article "Stochastic Failure Modelling of Unidirectional Composite Ply Failure," *Reliability Engr. and System Safety*, 2012: 1–9; this type of material is used extensively in the aerospace industry).
    (a) If a random sample of 4 specimens is selected, what is the probability that the sample mean fracture angle is at most 54? Between 53 and 54?
    (b) How many such specimens would be required to ensure that the first probability in (a) is at least .999?

88. The time taken by a randomly selected applicant for a mortgage to fill out a certain form has a normal distribution with mean value 10 min and standard deviation 2 min. If five individuals fill out a form on 1 day and six on another, what is the probability that the sample average amount of time taken on each day is at most 11 min?

89. The lifetime of a type of battery is normally distributed with mean value 10 h and standard deviation 1 h. There are four batteries in a package. What lifetime value is such that the total lifetime of all batteries in a package exceeds that value for only 5% of all packages?

90. The National Health Statistics Reports dated Oct. 22, 2008 stated that for a sample size of 277 18-year-old American males, the sample mean waist circumference was 86.3 cm. A somewhat complicated method was used to estimate various population percentiles, resulting in the following values:

5th	10th	25th	50th	75th	90th	95th
69.6	70.9	75.2	81.3	95.4	107.1	116.4

   (a) Is it plausible that the waist size distribution is at least approximately normal? Explain your reasoning. If your answer is no, conjecture the shape of the population distribution.

   (b) Suppose that the population mean waist size is 85 cm and that the population standard deviation is 15 cm. How likely is it that a random sample of 277 individuals will result in a sample mean waist size of at least 86.3 cm?

   (c) Referring back to (b), suppose now that the population mean waist size is 82 cm (closer to the median than the mean). Now what is the (approximate) probability that the sample mean will be at least 86.3? In light of this calculation, do you think that 82 is a reasonable value for μ?

91. A friend commutes by bus to and from work 6 days per week. Suppose that waiting time is uniformly distributed between 0 and 10 min, and that waiting times going and returning on various days are independent of each other. What is the approximate probability that total waiting time for an entire week is at most 75 min?

92. There are 40 students in an elementary statistics class. On the basis of years of experience, the instructor knows that the time needed to grade a randomly chosen paper from the first exam is a random variable with an expected value of 6 min and a standard deviation of 6 min.

   (a) If grading times are independent and the instructor begins grading at 6:50 p.m. and grades continuously, what is the (approximate) probability that he is through grading before the 11:00 p.m. TV news begins?

   (b) If the sports report begins at 11:10, what is the probability that he misses part of the report if he waits until grading is done before turning on the TV?

93. The tip percentage at a restaurant has a mean value of 18% and a standard deviation of 6%.

   (a) What is the approximate probability that the sample mean tip percentage for a random sample of 40 bills is between 16 and 19%?

   (b) If the sample size had been 15 rather than 40, could the probability requested in part (a) be calculated from the given information?

94. A small high school holds its graduation ceremony in the gym. Because of seating constraints, students are limited to a maximum of four tickets to graduation for family and friends. The vice principal knows that historically 30% of students want four tickets, 25% want three, 25% want two, 15% want one, and 5% want none.
    (a) Let $X =$ the number of tickets requested by a randomly selected graduating student, and assume the historical distribution applies to this rv. Find the mean and standard deviation of X.
    (b) Let $T =$ the total number of tickets requested by the 150 students graduating this year. Assuming all 150 students' requests are independent, determine the mean and standard deviation of T.
    (c) The gym can seat a maximum of 500 guests. Calculate the (approximate) probability that all students' requests can be accommodated. [*Hint:* Express this probability in terms of T. What distribution does T have?]

95. Let X represent the amount of gasoline (gallons) purchased by a randomly selected customer at a gas station. Suppose that the mean and standard deviation of X are 11.5 and 4.0, respectively.
    (a) In a sample of 50 randomly selected customers, what is the approximate probability that the sample mean amount purchased is at least 12 gallons?
    (b) In a sample of 50 randomly selected customers, what is the approximate probability that the total amount of gasoline purchased is at most 600 gallons?
    (c) What is the approximate value of the 95th percentile for the total amount purchased by 50 randomly selected customers?

96. For males the expected pulse rate is 70 per second and the standard deviation is 10 per second. For women the expected pulse rate is 77 per second and the standard deviation is 12 per second. Let $\overline{X} =$ the sample average pulse rate for a random sample of 40 men and let $\overline{Y} =$ the sample average pulse rate for a random sample of 36 women.
    (a) What is the approximate distribution of $\overline{X}$? Of $\overline{Y}$?
    (b) What is the approximate distribution of $\overline{X} - \overline{Y}$? Justify your answer.
    (c) Calculate (approximately) the probability $P\left(-2 \le \overline{X} - \overline{Y} \le 1\right)$.
    (d) Calculate (approximately) $P\left(\overline{X} - \overline{Y} \le -15\right)$. If you actually observed $\overline{X} - \overline{Y} \le -15$, would you doubt that $\mu_1 - \mu_2 = -7$? Explain.

97. The first assignment in a statistical computing class involves running a short program. If past experience indicates that 40% of all students will make no programming errors, use an appropriate normal approximation to compute the probability that in a class of 50 students
    (a) At least 25 will make no errors.
    (b) Between 15 and 25 (inclusive) will make no errors.

98. The number of parking tickets issued in a certain city on any given weekday has a Poisson distribution with parameter $\mu = 50$. What is the approximate probability that

(a) Between 35 and 70 tickets are given out on a particular day?

(b) The total number of tickets given out during a 5-day week is between 225 and 275? [For parts (a) and (b), use an appropriate CLT approximation.]

(c) Use software to obtain the exact probabilities in (a) and (b), and compare to the approximations.

99. Suppose the distribution of the time X (in hours) spent by students at a certain university on a particular project is gamma with parameters $\alpha = 50$ and $\beta = 2$. Use CLT to compute the (approximate) probability that a randomly selected student spends at most 125 h on the project.

100. The Central Limit Theorem says that $\overline{X}$ is approximately normal if the sample size is large. More specifically, the theorem states that the standardized $\overline{X}$ has a limiting standard normal distribution. That is, $(\overline{X} - \mu)/(\sigma/\sqrt{n})$ has a distribution approaching the standard normal. Can you reconcile this with the Law of Large Numbers?

101. It can be shown that if Y_n converges in probability to a constant τ, then $h(Y_n)$ converges to $h(\tau)$ for any function h that is continuous at τ. Use this to obtain a consistent estimator for the rate parameter λ of an exponential distribution. [*Hint*: How does μ for an exponential distribution relate to the exponential parameter λ?]

102. Let $X_1, \ldots, X_n$ be a random sample from the uniform distribution on $[0, \theta]$. Let Y_n be the maximum of these observations: $Y_n = \max(X_1, \ldots, X_n)$. Show that Y_n converges in probability to θ, that is, that $P(|Y_n - \theta| \geq \varepsilon) \to 0$ as n approaches ∞. [*Hint*: We shall show in Sect. 4.9 that the pdf of Y_n is $f(y) = ny^{n-1}/\theta^n$ for $0 \leq y \leq \theta$.]

4.6   Transformations of Jointly Distributed Random Variables

In the previous chapter we discussed the problem of starting with a single random variable X, forming some function of X, such as $Y = X^2$ or $Y = e^X$, and investigating the distribution of this new random variable Y. We now generalize this scenario by starting with more than a single random variable. Consider as an example a system having a component that can be replaced just once before the system itself expires. Let X_1 denote the lifetime of the original component and X_2 the lifetime of the replacement component. Then any of the following functions of X_1 and X_2 may be of interest to an investigator:

1. The total lifetime, $X_1 + X_2$.

2. The ratio of lifetimes X_1/X_2 (for example, if the value of this ratio is 2, the original component lasted twice as long as its replacement).

3. The ratio $X_1/(X_1 + X_2)$, which represents the proportion of system lifetime during which the original component operated.

4.6.1   The Joint Distribution of Two New Random Variables

Given two random variables X_1 and X_2, consider forming two new random variables $Y_1 = u_1(X_1, X_2)$ and $Y_2 = u_2(X_1, X_2)$. Our focus is on finding the joint distribution of these two new variables. Since most applications assume that the X_is are continuous, we restrict ourselves to that case. Some notation is needed before a general result can be given. Let

$f(x_1, x_2) = $ the joint pdf of the two original variables
$g(y_1, y_2) = $ the joint pdf of the two new variables

The $u_1(\,\cdot\,)$ and $u_2(\,\cdot\,)$ functions express the new variables in terms of the original ones. The general result presumes that these functions can be inverted to solve for the original variables in terms of the new ones:

$$X_1 = v_1(Y_1, Y_2), \quad X_2 = v_2(Y_1, Y_2)$$

For example, if

$$y_1 = x_1 + x_2 \quad \text{and } y_2 = \frac{x_1}{x_1 + x_2}$$

then multiplying y_2 by y_1 gives an expression for x_1, and then we can substitute this into the expression for y_1 and solve for x_2:

$$x_1 = y_1 y_2 = v_1(y_1, y_2) \qquad\qquad x_2 = y_1(1 - y_2) = v_2(y_1, y_2)$$

In a final burst of notation, let

$$S = \{(x_1, x_2) : f(x_1, x_2) > 0\} \quad T = \{(y_1, y_2) : g(y_1, y_2) > 0\}$$

That is, S is the region of positive density for the original variables and T is the region of positive density for the new variables; T is the "image" of S under the transformation.

TRANSFORMATION THEOREM (bivariate case)
Suppose that the partial derivative of each $v_i(y_1, y_2)$ with respect to both y_1 and y_2 exists and is continuous for every $(y_1, y_2) \in T$. Form the 2×2 matrix

$$\mathbf{M} = \begin{pmatrix} \dfrac{\partial v_1(y_1, y_2)}{\partial y_1} & \dfrac{\partial v_1(y_1, y_2)}{\partial y_2} \\[3mm] \dfrac{\partial v_2(y_1, y_2)}{\partial y_1} & \dfrac{\partial v_2(y_1, y_2)}{\partial y_2} \end{pmatrix}$$

The determinant of this matrix, called the *Jacobian*, is

(continued)

$$\det(\mathbf{M}) = \frac{\partial v_1}{\partial y_1} \cdot \frac{\partial v_2}{\partial y_2} - \frac{\partial v_1}{\partial y_2} \cdot \frac{\partial v_2}{\partial y_1}$$

The joint pdf for the new variables then results from taking the joint pdf $f(x_1, x_2)$ for the original variables, replacing x_1 and x_2 by their expressions in terms of y_1 and y_2, and finally multiplying this by the absolute value of the Jacobian:

$$g(y_1, y_2) = f(v_1(y_1, y_2), v_2(y_1, y_2)) \cdot |\det(\mathbf{M})| \qquad (y_1, y_2) \in T$$

The theorem can be rewritten slightly by using the notation

$$\det(\mathbf{M}) = \left| \frac{\partial(x_1, x_2)}{\partial(y_1, y_2)} \right|$$

Then we have

$$g(y_1, y_2) = f(x_1, x_2) \left| \frac{\partial(x_1, x_2)}{\partial(y_1, y_2)} \right|.$$

which is the natural extension of the univariate transformation theorem $g(y) = f(x) \cdot |dx/dy|$ discussed in Chap. 3.

Example 4.39 Continuing with the component lifetime situation, suppose that X_1 and X_2 are independent, each having an exponential distribution with parameter λ. Let's determine the joint pdf of

$$Y_1 = u_1(X_1, X_2) = X_1 + X_2 \quad \text{and} \quad Y_2 = u_2(X_1, X_2) = \frac{X_1}{X_1 + X_2}.$$

We have already inverted this transformation:

$$x_1 = v_1(y_1, y_2) = y_1 y_2 \qquad x_2 = v_2(y_1, y_2) = y_1(1 - y_2)$$

The image of the transformation, i.e., the set of (y_1, y_2) pairs with positive density, is $y_1 > 0$ and $0 < y_2 < 1$. The four relevant partial derivatives are

$$\frac{\partial v_1}{\partial y_1} = y_2 \qquad \frac{\partial v_1}{\partial y_2} = y_1 \qquad \frac{\partial v_2}{\partial y_1} = 1 - y_2 \qquad \frac{\partial v_2}{\partial y_2} = -y_1$$

from which the Jacobian is $-y_1 y_2 - y_1(1 - y_2) = -y_1$.

Since the joint pdf of X_1 and X_2 is

$$f(x_1, x_2) = \lambda e^{-\lambda x_1} \cdot \lambda e^{-\lambda x_2} = \lambda^2 e^{-\lambda(x_1 + x_2)} \qquad x_1 > 0, x_2 > 0$$

we have

$$g(y_1, y_2) = \lambda^2 e^{-\lambda y_1} \cdot y_1 = \lambda^2 y_1 e^{-\lambda y_1} \cdot 1 \qquad y_1 > 0, \quad 0 < y_2 < 1$$

The joint pdf thus factors into two parts. The first part is a gamma pdf with parameters $\alpha = 2$ and $\beta = 1/\lambda$, and the second part is a uniform pdf on (0, 1). Since the pdf factors and the region of positive density is rectangular, we have demonstrated that

1. The distribution of system lifetime $X_1 + X_2$ is gamma (with $\alpha = 2$, $\beta = 1/\lambda$)
2. The distribution of the proportion of system lifetime during which the original component functions is uniform on (0, 1)
3. $Y_1 = X_1 + X_2$ and $Y_2 = X_1/(X_1 + X_2)$ are independent of each other ∎

In the foregoing example, because the joint pdf factored into one pdf involving y_1 alone and another pdf involving y_2 alone, the individual (i.e., marginal) pdfs of the two new variables were obtained from the joint pdf without any further effort. Often this will not be the case—that is, Y_1 and Y_2 will not be independent. Then to obtain the marginal pdf of Y_1, the joint pdf must be integrated over all values of the second variable. In fact, in many applications an investigator wishes to obtain the distribution of a single function $u_1(X_1, X_2)$ of the original variables. To accomplish this, a second function $u_2(X_1, X_2)$ is selected, the joint pdf is obtained, and then y_2 integrated out. There are of course many ways to select the second function. The choice should be made so that the transformation can be easily inverted *and* the integration in the last step is straightforward.

Example 4.40 Consider a rectangular coordinate system with a horizontal x_1 axis and a vertical x_2 axis as shown in Fig. 4.15a. First a point (X_1, X_2) is randomly selected, where the joint pdf of X_1, X_2 is

$$f(x_1, x_2) = \begin{cases} x_1 + x_2 & 0 < x_1 < 1, \quad 0 < x_2 < 1 \\ 0 & \text{otherwise} \end{cases}$$

Then a rectangle with vertices $(0, 0)$, $(X_1, 0)$, $(0, X_2)$, and (X_1, X_2) is formed as shown in Fig. 4.15a. What is the distribution of $X_1 X_2$, the area of this rectangle? To answer this question, let

$$Y_1 = X_1 X_2 \qquad\qquad Y_2 = X_2$$

so

$$y_1 = u_1(x_1, x_2) = x_1 x_2 \qquad y_2 = u_2(x_1, x_2) = x_2$$

Then

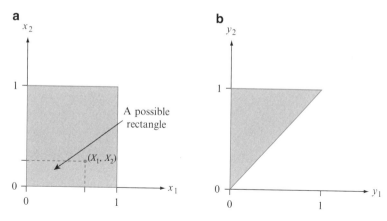

Fig. 4.15 Regions of positive density for Example 4.40

$$x_1 = v_1(y_1, y_2) = \frac{y_1}{y_2} \qquad x_2 = v_2(y_1, y_2) = y_2$$

Notice that because $x_2 \, (= y_2)$ is between 0 and 1 and y_1 is the product of the two x_is, it must be the case that $0 < y_1 < y_2$. The region of positive density for the new variables is then

$$T = \{(y_1, y_2) : 0 < y_1 < y_2, 0 < y_2 < 1\}$$

which is the triangular region shown in Fig. 4.15b.

Since $\partial v_2 / \partial y_1 = 0$, the product of the two off-diagonal elements in the matrix **M** will be 0, so only the two diagonal elements contribute to the Jacobian:

$$\mathbf{M} = \begin{pmatrix} \dfrac{1}{y_2} & ? \\ 0 & 1 \end{pmatrix}, \qquad |\det(\mathbf{M})| = \frac{1}{y_2}$$

The joint pdf of the two new variables is now

$$g(y_1, y_2) = f\left(\frac{y_1}{y_2}, y_2\right) \cdot |\det(\mathbf{M})| =$$

$$\begin{cases} \left(\dfrac{y_1}{y_2} + y_2\right) \cdot \dfrac{1}{y_2} & 0 < y_1 < y_2 < 1 \\ 0 & \text{otherwise} \end{cases}$$

To obtain the marginal pdf of Y_1 alone, we must now fix y_1 at some arbitrary value between 0 and 1, and integrate out y_2. Figure 4.15b shows that we must integrate along the vertical line segment passing through y_1 whose lower limit is y_1 and whose upper limit is 1:

$$g_1(y_1) = \int_{y_1}^1 \left(\frac{y_1}{y_2} + y_2\right) \cdot \frac{1}{y_2} \, dy_2 = 2(1 - y_1) \quad 0 < y_1 < 1$$

This marginal pdf can now be integrated to obtain any desired probability involving the area. For example, integrating from 0 to .5 gives $P(\text{area} < .5) = .75$. ∎

4.6.2   The Joint Distribution of More Than Two New Variables

Consider now starting with three random variables X_1, X_2, and X_3, and forming three new variables Y_1, Y_2, and Y_3. Suppose again that the transformation can be inverted to express the original variables in terms of the new ones:

$$x_1 = v_1(y_1, y_2, y_3), \quad x_2 = v_2(y_1, y_2, y_3), \quad x_3 = v_3(y_1, y_2, y_3)$$

Then the foregoing theorem can be extended to this new situation. The Jacobian matrix has dimension 3×3, with the entry in the ith row and jth column being $\partial v_i/\partial y_j$. The joint pdf of the new variables results from replacing each x_i in the original pdf $f(\cdot)$ by its expression in terms of the y_js and multiplying by the absolute value of the Jacobian.

Example 4.41 Consider $n = 3$ identical components with independent lifetimes X_1, X_2, X_3, each having an exponential distribution with parameter λ. If the first component is used until it fails, replaced by the second one which remains in service until it fails, and finally the third component is used until failure, then the total lifetime of these components is $Y_3 = X_1 + X_2 + X_3$. (This design structure, where one component is replaced by the next in succession, is called a *standby system*.) To find the distribution of total lifetime, let's first define two other new variables: $Y_1 = X_1$ and $Y_2 = X_1 + X_2$ (so that $Y_1 < Y_2 < Y_3$). After finding the joint pdf of all three variables, we integrate out the first two variables to obtain the desired information. Solving for the old variables in terms of the new gives

$$x_1 = y_1 \qquad\qquad x_2 = y_2 - y_1 \qquad\qquad x_3 = y_3 - y_2$$

It is obvious by inspection of these expressions that the three diagonal elements of the Jacobian matrix are all 1s and that the elements above the diagonal are all 0s, so the determinant is 1, the product of the diagonal elements. Since

$$f(x_1, x_2, x_3) = \lambda^3 e^{-\lambda(x_1 + x_2 + x_3)} \qquad x_1 > 0, x_2 > 0, x_3 > 0$$

by substitution,

$$g(y_1, y_2, y_3) = \lambda^3 e^{-\lambda y_3} \qquad 0 < y_1 < y_2 < y_3$$

Integrating this joint pdf first with respect to y_1 between 0 and y_2 and then with respect to y_2 between 0 and y_3 (try it!) gives

$$g_3(y_3) = \frac{\lambda^3}{2} y_3^2 e^{-\lambda y_3} \qquad y_3 > 0$$

which is the gamma pdf with $\alpha = 3$ and $\beta = 1/\lambda$. This result and Example 3.39 are both special cases of a proposition from Sect. 4.3, stating that the sum of n iid exponential rvs has a gamma distribution with $\alpha = n$. ∎

4.6.3 Exercises: Section 4.6 (103–110)

103. Let X_1 and X_2 be independent, standard normal rvs.
    (a) Define $Y_1 = X_1 + X_2$ and $Y_2 = X_1 - X_2$. Determine the joint pdf of Y_1 and Y_2.
    (b) Determine the marginal pdf of Y_1. [*Note:* We know the sum of two independent normal rvs is normal, so you can check your answer against the appropriate normal pdf.]
    (c) Are Y_1 and Y_2 independent?
104. Consider two components whose lifetimes X_1 and X_2 are independent and exponentially distributed with parameters λ_1 and λ_2, respectively. Obtain the joint pdf of total lifetime $X_1 + X_2$ and the proportion of total lifetime $X_1/(X_1 + X_2)$ during which the first component operates.
105. Let X_1 denote the time (hr) it takes to perform a first task and X_2 denote the time it takes to perform a second one. The second task always takes at least as long to perform as the first task. The joint pdf of these variables is

$$f(x_1, x_2) = \begin{cases} 2(x_1 + x_2) & 0 \le x_1 \le x_2 \le 1 \\ 0 & \text{otherwise} \end{cases}$$

    (a) Obtain the pdf of the total completion time for the two tasks.
    (b) Obtain the pdf of the difference $X_2 - X_1$ between the longer completion time and the shorter time.
106. An exam consists of a problem section and a short-answer section. Let X_1 denote the amount of time (h) that a student spends on the problem section and X_2 represent the amount of time the same student spends on the short-answer section. Suppose the joint pdf of these two times is

$$f(x_1, x_2) = \begin{cases} c x_1 x_2 & \frac{x_1}{3} < x_2 < \frac{x_1}{2}, \quad 0 < x_1 < 1 \\ 0 & \text{otherwise} \end{cases}$$

    (a) What is the value of c?

(b) If the student spends exactly .25 h on the short-answer section, what is the probability that at most .60 h was spent on the problem section? [*Hint:* First obtain the relevant conditional distribution.]

(c) What is the probability that the amount of time spent on the problem part of the exam exceeds the amount of time spent on the short-answer part by at least .5 h?

(d) Obtain the joint distribution of $Y_1 = X_2/X_1$, the ratio of the two times, and $Y_2 = X_2$. Then obtain the marginal distribution of the ratio.

107. Consider randomly selecting a point (X_1, X_2, X_3) in the unit cube $\{(x_1, x_2, x_3): 0 < x_1 < 1, 0 < x_2 < 1, 0 < x_3 < 1\}$ according to the joint pdf

$$f(x_1, x_2, x_3) = \begin{cases} 8x_1x_2x_3 & 0 < x_1 < 1, \quad 0 < x_2 < 1, \quad 0 < x_3 < 1 \\ 0 & \text{otherwise} \end{cases}$$

(so the three variables are independent). Then form a rectangular solid whose vertices are $(0, 0, 0)$, $(X_1, 0, 0)$, $(0, X_2, 0)$, $(X_1, X_2, 0)$, $(0, 0, X_3)$, $(X_1, 0, X_3)$, $(0, X_2, X_3)$, and (X_1, X_2, X_3). The volume of this cube is $Y_3 = X_1X_2X_3$. Obtain the pdf of this volume. [*Hint:* Let $Y_1 = X_1$ and $Y_2 = X_1X_2$.]

108. Let X_1 and X_2 be independent, each having a standard normal distribution. The pair (X_1, X_2) corresponds to a point in a two-dimensional coordinate system. Consider now changing to polar coordinates via the transformation,

$$Y_1 = X_1^2 + X_2^2$$

$$Y_2 = \begin{cases} \arctan\left(\dfrac{X_2}{X_1}\right) & X_1 > 0, X_2 \geq 0 \\[2mm] \arctan\left(\dfrac{X_2}{X_1}\right) + 2\pi & X_1 > 0, X_2 < 0 \\[2mm] \arctan\left(\dfrac{X_2}{X_1}\right) + \pi & X_1 < 0 \\[2mm] 0 & X_1 = 0 \end{cases}$$

from which $X_1 = \sqrt{Y_1}\cos(Y_2)$, $X_2 = \sqrt{Y_1}\sin(Y_2)$. Obtain the joint pdf of the new variables and then the marginal distribution of each one. [*Note:* It would be nice if we could simply let $Y_2 = \arctan(X_2/X_1)$, but in order to insure invertibility of the arctan function, it is defined to take on values only between $-\pi/2$ and $\pi/2$. Our specification of Y_2 allows it to assume any value between 0 and 2π.]

109. The result of the previous exercise suggests how observed values of two independent standard normal variables can be generated by first generating their polar coordinates with an exponential rv with $\lambda = \frac{1}{2}$ and an independent Unif$(0, 2\pi)$ rv: Let U_1 and U_2 be independent Unif$(0, 1)$ rvs, and then let

$$Y_1 = -2\ln(U_1) \qquad\qquad Y_2 = 2\pi U_2,$$
$$Z_1 = \sqrt{Y_1}\cos(Y_2) \qquad\qquad Z_2 = \sqrt{Y_1}\sin(Y_2)$$

Show that the Z_is are independent standard normal. [*Note*: This is called the *Box-Muller transformation* after the two individuals who discovered it. Now that statistical software packages will generate almost instantaneously observations from a normal distribution with any mean and variance, it is thankfully no longer necessary for people like you and us to carry out the transformations just described—let the software do it!]

110. Let X_1 and X_2 be independent random variables, each having a standard normal distribution. Show that the pdf of the ratio $Y = X_1/X_2$ is given by $f(y) = 1/[\pi(1 + y^2)]$ for $-\infty < y < \infty$. (This is called the *standard Cauchy distribution*; its density curve is bell-shaped, but the tails are so heavy that μ does not exist.)

4.7   The Bivariate Normal Distribution

Perhaps the most useful joint distribution is the bivariate normal. Although the formula may seem rather complicated, it is based on a simple quadratic expression in the standardized variables (subtract the mean and then divide by the standard deviation). The bivariate normal density is

$$f(x,y) = \frac{1}{2\pi\sigma_1\sigma_2\sqrt{1-\rho^2}}\exp\left(-\frac{1}{2(1-\rho^2)}\left[\left(\frac{x-\mu_1}{\sigma_1}\right)^2 - 2\rho\left(\frac{x-\mu_1}{\sigma_1}\right)\left(\frac{y-\mu_2}{\sigma_2}\right) + \left(\frac{y-\mu_2}{\sigma_2}\right)^2\right]\right)$$

The notation used here for the five parameters reflects the roles they play. Some tedious integration shows that μ_1 and σ_1 are the mean and standard deviation, respectively, of X, μ_2 and σ_2 are the mean and standard deviation, respectively, of Y, and ρ is the correlation coefficient between the two variables. The integration required to do bivariate normal probability calculations is quite difficult. Computer code is available for calculating $P(X \le x, Y \le y)$ approximately using numerical integration, and some software packages, including Matlab and R, incorporate this feature (see the end of this section).

The density surface in three dimensions looks like a mountain with elliptical cross-sections, as shown in Fig. 4.16a. The vertical cross-sections are all proportional to normal densities. If we set $f(x, y) = c$ to investigate the contours (curves along which the density is constant), this amounts to equating the exponent of the joint pdf to a constant. The contours are then concentric ellipses centered at $(x, y) = (\mu_1, \mu_2)$, as shown in Fig. 4.16b.

If $\rho = 0$, then $f(x,y) = f_X(x) f_Y(y)$, where X is normal with mean μ_1 and standard deviation σ_1, and Y is normal with mean μ_2 and standard deviation σ_2. That is, X and Y have independent normal distributions. In this case the elliptical contours reduce to circles. Recall that in Sect. 4.2 we emphasized that independence of X and Y implies $\rho = 0$ but, in general, $\rho = 0$ does not imply independence. However, we

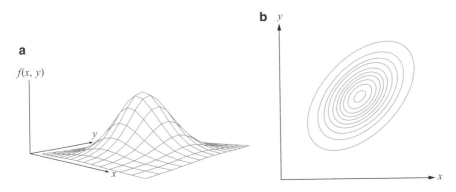

Fig. 4.16 (a) A graph of the bivariate normal pdf; (b) contours of the bivariate normal pdf

have just seen that when X and Y are bivariate normal $\rho = 0$ does imply independence. Therefore, in the bivariate normal case $\rho = 0$ if and only if the two rvs are independent.

Regardless of whether or not $\rho = 0$, the marginal distribution $f_X(x)$ is just a normal pdf with mean μ_1 and standard deviation σ_1:

$$f_X(x) = \frac{1}{\sigma_1\sqrt{2\pi}}e^{-(x-\mu_1)^2/(2\sigma_1^2)}$$

The integration to show this [integrating $f(x,y)$ on y from $-\infty$ to ∞] is rather messy. Likewise, the marginal distribution of Y is $N(\mu_2, \sigma_2)$. These two marginal pdfs are, in fact, just special cases of a much stronger result.

> **PROPOSITION**
> If X and Y have a bivariate normal distribution, then any linear combination of X and Y is also normal. That is, for any constants a, b, c, the random variable $aX + bY + c$ has a normal distribution.

This proposition can be proved using the transformation techniques of Sect. 4.6 along with some extremely tedious algebra. Setting $a = 1$ and $b = c = 0$, we have that X is normally distributed; $a = 0$, $b = 1$, $c = 0$ yields the same result for Y. To find the mean and standard deviation of a general linear combination, one can use the rules for linear combinations established in Sect. 4.3.

Example 4.42 Many students applying for college take the SAT, which consists of three components: Critical Reading, Mathematics, and Writing. While some colleges use all three components to determine admission, many only look at the first two (reading and math). Let X and Y denote the Critical Reading and Mathematics scores, respectively, for a randomly selected student. According to the

College Board Web site, the population of students taking the exam in Fall 2012 had the following results:

$$\mu_1 = 496, \sigma_1 = 114, \mu_2 = 514, \sigma_2 = 117$$

Suppose that X and Y have approximately (because both X and Y are discrete) a bivariate normal distribution with correlation coefficient $\rho = .25$. Let's determine the probability that a student's total score across these two components exceeds 1250, the minimum admission score for a particular university.

Our goal is to calculate $P(X + Y > 1250)$. Using the bivariate normal pdf, the desired probability is a daunting double integral:

$$\frac{1}{2\pi(114)(117)\sqrt{1 - .25^2}} \int_{-\infty}^{\infty} \int_{1250-y}^{\infty}$$

$$e^{-\left\{[(x-496)/114]^2 - 2(.25)(x-496)(y-514)/(114)(117) + [(y-514)/117]^2\right\}/\left[2\left(1-.25^2\right)\right]} dx dy$$

This is not a practical way to solve this problem! Instead, recognize $X + Y$ as a linear combination of X and Y; by the preceding proposition, $X + Y$ has a normal distribution. The mean and variance of $X + Y$ are calculated using the formulas from Sect. 4.5:

$$E(X + Y) = E(X) + E(Y) = \mu_1 + \mu_2 = 496 + 514 = 1010$$
$$\mathrm{Var}(X + Y) = \mathrm{Var}(X) + \mathrm{Var}(Y) + 2\mathrm{Cov}(X, Y)$$
$$= \sigma_1^2 + \sigma_2^2 + 2\rho\sigma_1\sigma_2 = 114^2 + 117^2 + 2(.25)(114)(117) = 33{,}354$$

Therefore,

$$P(X + Y > 1250) = 1 - \Phi\left(\frac{1250 - 1010}{\sqrt{33{,}354}}\right) = 1 - \Phi(1.31) = .0951.$$

Suppose instead we wish to determine $P(X < Y)$, the probability a student scores better on math than on reading. If we rewrite this probability as $P(X - Y < 0)$, then we may apply the preceding proposition to the linear combination $X - Y$. With $E(X - Y) = -18$ and $\mathrm{Var}(X - Y) = 20{,}016$,

$$P(X < Y) = P(X - Y < 0) = \Phi\left(\frac{0 - (-18)}{\sqrt{20{,}016}}\right) = \Phi(0.13) = .5517. \quad \blacksquare$$

4.7.1 Conditional Distributions of X and Y

As in Sect. 4.4, the conditional density of Y given $X = x$ results from dividing the marginal density of X into $f(x,y)$. The algebra is again a mess, but the result is fairly simple.

PROPOSITION
Let X and Y have a bivariate normal distribution. Then the conditional distribution of Y, given $X = x$, is normal with mean and variance

$$\mu_{Y|X=x} = E(Y \mid X = x) = \mu_2 + \rho\sigma_2 \frac{x - \mu_1}{\sigma_1}$$

$$\sigma^2_{Y|X=x} = \text{Var}(Y \mid X = x) = \sigma_2^2(1 - \rho^2)$$

Notice that the conditional mean of Y is a linear function of x, and the conditional variance of Y doesn't depend on x at all. When $\rho = 0$, the conditional mean is the mean of Y, μ_2, and the conditional variance is just the variance of Y, σ_2^2. In other words, if $\rho = 0$, then the conditional distribution of Y is the same as the unconditional distribution of Y. When ρ is close to 1 or -1 the conditional variance will be much smaller than $\text{Var}(Y)$, which says that knowledge of X will be very helpful in predicting Y. If ρ is near 0 then X and Y are nearly independent and knowledge of X is not very useful in predicting Y.

Example 4.43 Let X and Y be the heights of a randomly selected mother and her daughter, respectively. A similar situation was one of the first applications of the bivariate normal distribution, by Francis Galton in 1886, and the data was found to fit the distribution very well. Suppose a bivariate normal distribution with mean $\mu_1 = 64$ in. and standard deviation $\sigma_1 = 3$ in. for X and mean $\mu_2 = 65$ in. and standard deviation $\sigma_2 = 3$ in. for Y. Here $\mu_2 > \mu_1$, which is in accord with the increase in height from one generation to the next. Assume $\rho = .4$. Then

$$\mu_{Y|X=x} = \mu_2 + \rho\sigma_2 \frac{x - \mu_1}{\sigma_1} = 65 + .4(3)\frac{x - 64}{3} = 65 + .4\,(x - 64) = .4x + 39.4$$

$$\sigma^2_{Y|X=x} = \text{Var}(Y \mid X = x) = \sigma_2^2(1 - \rho^2) = 9(1 - .4^2) = 7.56 \quad \text{and} \quad \sigma_{Y|X=x} = 2.75$$

Notice that the conditional variance is 16% less than the variance of Y. Squaring the correlation gives the percentage by which the conditional variance is reduced relative to the variance of Y.   ∎

4.7.2   Regression to the Mean

The formula for the conditional mean can be reexpressed as

$$\frac{\mu_{Y|X=x} - \mu_2}{\sigma_2} = \rho \cdot \frac{x - \mu_1}{\sigma_1}$$

In words, when the formula is expressed in terms of standardized quantities, the standardized conditional mean is just ρ times the standardized x. In particular, for the height scenario

$$\frac{\mu_{Y|X=x} - 65}{3} = .4 \cdot \frac{x - 64}{3}$$

If the mother is 5 in. above the mean of 64 in. for mothers, then the daughter's conditional expected height is just 2 in. above the mean for daughters. In this example, with equal standard deviations for Y and X, the daughter's conditional expected height is always closer to its mean than the mother's height is to its mean. One can think of the conditional expectation as falling back toward the mean, and that is why Galton called this *regression to the mean*.

Regression to the mean occurs in many contexts. For example, let X be a baseball player's average for the first half of the season and let Y be the average for the second half. Most of the players with a high X (above .300) will not have such a high Y. The same kind of reasoning applies to the "sophomore jinx," which says that if a player has a very good first season, then the player is unlikely to do as well in the second season.

4.7.3 The Multivariate Normal Distribution

The multivariate normal distribution extends the bivariate normal distribution to situations involving models for n random variables $X_1, X_2, \ldots X_n$ with $n > 2$. The joint density function is quite complicated; the only way to express it compactly is to make use of matrix algebra notation. And probability calculations based on this distribution are extremely complex.

Here are some of the most important properties of the distribution:

- The distribution of any linear combination of $X_1, X_2, \ldots X_n$ is normal
- The marginal distribution of any X_i is normal
- The joint distribution of any pair X_i, X_j is bivariate normal
- The conditional distribution of any X_i given values of the other $n - 1$ variables is normal

Many procedures for the analysis of multivariate data (observations simultaneously on three or more variables) are based on assuming that the data was selected from a multivariate normal distribution. We recommend *Methods of Multivariate Analysis*, 3rd ed., by Rencher for more information on multivariate analysis and the multivariate normal distribution.

4.7.4 Bivariate Normal Calculations with Software

Matlab will compute probabilities under the bivariate normal pdf using the `mvncdf` command ("mvn" abbreviates multivariate normal). This function is illustrated in the next example.

Example 4.44 Consider the SAT reading/math scenario of Example 4.42. What is the probability that a randomly selected student scored at most 650 on both components, i.e., what is $P(X \leq 650 \cap Y \leq 650)$?

The desired probability cannot be expressed in terms of a linear combination of X and Y, and so the technique of the earlier example does not apply. Figure 4.17 shows the required Matlab code. The first two inputs are the desired cdf values $[x, y] = [650, 650]$ and the means $[\mu_1, \mu_2] = [496, 514]$, respectively. The third input is called the **covariance matrix** of X and Y, defined by

$$\mathbf{C}(X, Y) = \begin{bmatrix} \mathrm{Var}(X) & \mathrm{Cov}(X, Y) \\ \mathrm{Cov}(X, Y) & \mathrm{Var}(Y) \end{bmatrix} = \begin{bmatrix} \sigma_1^2 & \rho\sigma_1\sigma_2 \\ \rho\sigma_1\sigma_2 & \sigma_2^2 \end{bmatrix}$$

Fig. 4.17 Matlab code for Example 4.44

```
mu=[496, 514];
C=[114^2, .25*114*117; .25*114*117, 117^2];
mvncdf([650, 650],mu,C)
```

Matlab returns an answer of .8097, so for X and Y having a bivariate normal distribution with the parameters specified in Example 4.42, $P(X \leq 650 \cap Y \leq 650) = .8097$. About 81% of students scored 650 or below on both the Critical Reading and Mathematics components, according to this model. ∎

The `pmvnorm` function in R will perform the same calculation with the same inputs (the covariance matrix is labeled `sigma`). Users must install the `mvtnorm` package to access this function.

4.7.5 Exercises: Section 4.7 (111–120)

111. Example 4.42 introduced a bivariate normal model for $X =$ SAT Critical Reading score and $Y =$ SAT Mathematics score. Let $W =$ SAT Writing score (the third component of the SAT), which has mean 488 and standard deviation 114. Suppose X and W have a bivariate normal distribution with $\rho_{X,W} = \mathrm{Corr}(X, W) = .5$.
    (a) An English department plans to use $X + W$, a student's total score on the non-math sections of the SAT, to help determine admission. Determine the distribution of $X + W$.
    (b) Calculate $P(X + W > 1200)$.

(c) Suppose the English department wishes to admit only those students who score in the top 10% on this Critical Reading + Writing criterion. What combined score separates the top 10% of students from the rest?

112. In the context of the previous exercise, let $T = X + Y + W$, a student's grand total score on the three components of the SAT.
    (a) Find the expected value of T.
    (b) Assume $\text{Corr}(Y, W) = .2$. Find the variance of T. [*Hint:* Use Expression (4.5) from Sect. 4.3.]
    (c) Suppose X, Y, W have a multivariate normal distribution, in which case T is also normally distributed. Determine $P(T > 2000)$.
    (d) What is the 99th percentile of SAT grand total scores, according to this model?

113. Let X = height (inches) and Y = weight (lbs) for an American male. Suppose X and Y have a bivariate normal distribution, the mean and sd of heights are 70 in and 3 in. the mean and sd of weights are 170 lbs and 20 lbs, and the correlation coefficient is $\rho = .9$.
    (a) Determine the distribution of Y given $X = 68$, i.e., the weight distribution for 5'8" American males.
    (b) Determine the distribution of Y given $X = 70$, i.e., the weight distribution for 5'10" American males. In what ways is this distribution similar to that of part (a), and how are they different?
    (c) Calculate $P(Y < 180 | X = 72)$, the probability that a 6-ft-tall American male weighs less than 180 lb.

114. In electrical engineering, the unwanted "noise" in voltage or current signals is often modeled by a Gaussian (i.e., normal) distribution. Suppose that the noise in a particular voltage signal has a constant mean of 0.9 V, and that two noise instances sampled τ seconds apart have a bivariate normal distribution with covariance equal to $0.04e^{-|\tau|/10}$. Let X and Y denote the noise at times 3 s and 8 s, respectively.
    (a) Determine $\text{Cov}(X, Y)$.
    (b) Determine σ_X and σ_Y. [*Hint:* $\text{Var}(X) = \text{Cov}(X, X)$.]
    (c) Determine $\text{Corr}(X, Y)$.
    (d) Find the probability we observe greater voltage noise at time 3 s than at time 8 s.
    (e) Find the probability that the voltage noise at time 3 s is more than 1 V above the voltage noise at time 8 s.

115. For a Calculus I class, the final exam score Y and the average X of the four earlier tests have a bivariate normal distribution with mean $\mu_1 = 73$, standard deviation $\sigma_1 = 12$, mean $\mu_2 = 70$, standard deviation $\sigma_2 = 15$. The correlation is $\rho = .71$. Determine
    (a) $\mu_{Y|X=x}$
    (b) $\sigma^2_{Y|X=x}$
    (c) $\sigma_{Y|X=x}$

(d) $P(Y > 90|X = 80)$, i.e., the probability that the final exam score exceeds 90 given that the average of the four earlier tests is 80

116. Refer to the previous exercise. Suppose a student's Calculus I grade is determined by $4X + Y$, the total score across five tests.
(a) Find the mean and standard deviation of $4X + Y$.
(b) Determine $P(4X + Y < 320)$.
(c) Suppose the instructor sets the curve in such a way that the top 15% of students, based on total score across the five tests, will receive As. What point total is required to get an A in Calculus I?

117. Let X and Y, reaction times (sec) to two different stimuli, have a bivariate normal distribution with mean $\mu_1 = 20$ and standard deviation $\sigma_1 = 2$ for X and mean $\mu_2 = 30$ and standard deviation $\sigma_2 = 5$ for Y. Assume $\rho = .8$. Determine
(a) $\mu_{Y|X=x}$
(b) $\sigma^2_{Y|X=x}$
(c) $\sigma_{Y|X=x}$
(d) $P(Y > 46 | X = 25)$

118. Refer to the previous exercise.
(a) One researcher is interested in $X + Y$, the total reaction time to the two stimuli. Determine the mean and standard deviation of $X + Y$.
(b) If X and Y were independent, what would be the standard deviation of $X + Y$? Explain why it makes sense that the sd in part (a) is much larger than this.
(c) Another researcher is interested in $Y - X$, the difference in the reaction times to the two stimuli. Determine the mean and standard deviation of $Y - X$.
(d) If X and Y were independent, what would be the standard deviation of $Y - X$? Explain why it makes sense that the sd in part (c) is much smaller than this.

119. Let X and Y be the times for a randomly selected individual to complete two different tasks, and assume that (X, Y) has a bivariate normal distribution with $\mu_1 = 100$, $\sigma_1 = 50$, $\mu_2 = 25$, $\sigma_2 = 5$, $\rho = .4$. From statistical software we obtain $P(X < 100, \ Y < 25) = .3333$, $P(X < 50, \ Y < 20) = .0625$, $P(X < 50, Y < 25) = .1274$, and $P(X < 100, Y < 20) = .1274$.
(a) Determine $P(50 < X < 100, 20 < Y < 25)$.
(b) Leave the other parameters the same but change the correlation to $\rho = 0$ (independence). Now recompute the probability in part (a). Intuitively, why should the original be larger?

120. One of the propositions of this section gives an expression for $E(Y|X = x)$.
(a) By reversing the roles of X and Y give a similar formula for $E(X|Y = y)$.
(b) Both $E(Y|X = x)$ and $E(X|Y = y)$ are linear functions. Show that the product of the two slopes is ρ^2.

4.8 Reliability

Reliability theory is the branch of statistics and operations research devoted to studying how long systems will function properly. A "system" can refer to a single device, such as a DVR, or a network of devices or objects connected together (e.g., electronic components or stages in an assembly line). For any given system, the primary variable of interest is $T =$ the system's lifetime, i.e., the duration of time until the system fails (either permanently or until repairs/upgrades are made). Since T measures time, we always have $T \geq 0$. Most often, T is modeled as a continuous rv on $(0, \infty)$, though occasionally lifetimes are modeled as discrete or, at least, having positive probability of equaling zero (such as a light bulb that never turns on). The probability distribution of T is often described in terms of its reliability function.

4.8.1 The Reliability Function

> **DEFINITION**
> Let T denote the lifetime (i.e., the time to failure) of some system. The **reliability function** of T (or of the system), denoted by $R(t)$, is defined for $t \geq 0$ by
>
> $$R(t) = P(T > t) = 1 - F(t),$$
>
> where $F(t)$ is the cdf of T. That is, $R(t)$ is the probability that the system lasts more that t time units. The reliability function is sometimes also called the *survival function* of T.

Properties of $F(t)$ and the relation $R(t) = 1 - F(t)$ imply that
1. If T is a continuous rv on $[0, \infty)$, then $R(0) = 1$.
2. $R(t)$ is a non-increasing function of t.
3. $R(t) \to 0$ as $t \to \infty$.

Example 4.45 The exponential distribution serves as one of the most common lifetime models in engineering practice. Suppose the lifetime T, in hours, of a certain drill bit is exponential with parameter $\lambda = .01$ (equivalently, mean 100). From Sect. 3.4, we know that T has cdf $F(t) = 1 - e^{-.01t}$, so the reliability function of T is

$$R(t) = 1 - F(t) = e^{-.01t} \qquad t \geq 0$$

This function satisfies properties 1–3 above. A graph of $R(t)$ appears in Fig. 4.18a.

Now suppose instead that 5% of these drill bits shatter upon initial use, so that $P(T=0) = .05$, while the remaining 95% of such drill bits follow the aforementioned exponential distribution. Since T cannot be negative, $R(0) = P(T > 0) = 1 - P(T=0) = .95$. For $t > 0$, the reliability function of T is determined as follows:

$$R(t) = P(T > t)$$
$$= P(\text{bit doesn't shatter})P(T > t \mid \text{bit doesn't shatter})$$
$$= (.95)(e^{-.01t}) = .95e^{-.01t}$$

The expression $e^{-.01t}$ comes from the previous reliability function calculation. Since this expression for $R(t)$ equals .95 at $t = 0$, we have for all $t \geq 0$ that $R(t) = .95e^{-.01t}$ (see Fig. 4.18b). This, too, is a non-increasing function of t with $R(t) \rightarrow 0$ as $t \rightarrow \infty$, but property 1 does not hold because T is not a continuous rv (it has a "mass" of .05 at $t = 0$).

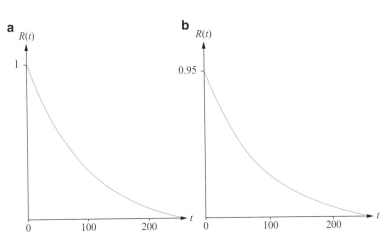

Fig. 4.18 Reliability functions: (**a**) a continuous lifetime distribution; (**b**) lifetime with positive probability of failure at $t = 0$ ∎

Example 4.46 The Weibull family of distributions offers a broader class of models than does the exponential family. Recall from Sect. 3.5 that the cdf of a Weibull rv is given by $F(x) = 1 - \exp(-(x/\beta)^{\alpha})$, where α is the shape parameter and β is the scale parameter (both > 0). If a system's time to failure follows a Weibull distribution, then the reliability function is

$$R(t) = 1 - F(t) = \exp(-(t/\beta)^{\alpha})$$

Several examples of Weibull reliability functions are illustrated in Fig. 4.19. The $\alpha = 1$ case corresponds to an exponential distribution with $\lambda = 1/\beta$. Interestingly, models with larger values of α have higher reliability for small values of t (to be precise, $t < \beta$) but lower reliability for larger t than do Weibull models with small α parameter values.

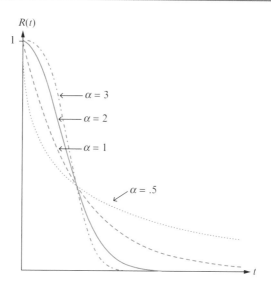

Fig. 4.19 Reliability functions for Weibull lifetime distributions

4.8.2   Series and Parallel Designs

Now consider assessing the reliability of systems configured in series and/or parallel designs. Figure 4.20 illustrates the two basic designs: a **series system** works if and only if all of its components work, while a **parallel system** continues to function as long as at least one of its components is still functioning. Let $T_1, \ldots,$ T_n denote the n component lifetimes and let $R_i(t) = P(T_i > t)$ be the reliability function of the ith component. A standard assumption in reliability theory is that the n components operate *independently*, i.e., that the T_is are independent rvs.

Let T denote the lifetime of the series system depicted in Fig. 4.20a. Under the assumption of component independence, the system reliability function is

$$
\begin{aligned}
R(t) = P(T > t) &= P(\text{the system's lifetime exceeds } t) \\
&= P(\text{all } n \text{ component lifetimes exceed } t) \qquad \text{series system} \\
&= P(T_1 > t \cap \ldots \cap T_n > t) \\
&= P(T_1 > t) \cdot \ldots \cdot P(T_n > t) \qquad \text{by independence} \\
&= R_1(t) \cdot \ldots \cdot R_n(t)
\end{aligned}
$$

That is, for a series design, the system reliability function equals the product of the component reliability functions. On the other hand, the reliability function for the parallel system in Fig. 4.20b is given by

Fig. 4.20 Basic system
designs: (a) series
connection; (b) parallel
connection

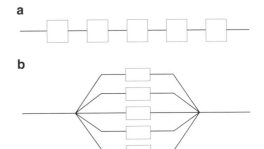

$$R(t) = P(\text{the system's lifetime exceeds } t)$$
$$= P(\text{at least one component lifetime exceeds } t) \qquad \text{parallel system}$$
$$= 1 - P(\text{all component lifetimes are } \leq t)$$
$$= 1 - P(T_1 \leq t \cap \ldots \cap T_n \leq t)$$
$$= 1 - P(T_1 \leq t) \cdot \ldots \cdot P(T_n \leq t) \qquad \text{by independence}$$
$$= 1 - [1 - R_1(t)] \cdot \ldots \cdot [1 - R_n(t)]$$

These two results are summarized in the following proposition.

PROPOSITION

Suppose a system consists of n independent components with reliability functions $R_1(t), \ldots, R_n(t)$.

1. If the n components are connected in series, the system reliability function is

$$R(t) = \prod_{i=1}^{n} R_i(t)$$

2. If the n components are connected in parallel, the system reliability function is

$$R(t) = 1 - \prod_{i=1}^{n} [1 - R_i(t)]$$

Example 4.47 Consider three independently operating devices, each of whose lifetime (in hours) is exponentially distributed with mean 100. From the previous example, $R_1(t) = R_2(t) = R_3(t) = e^{-.01t}$. If these three devices are connected in series, the reliability function of the resulting system is

$$R(t) = \prod_{i=1}^{3} R_i(t) = (e^{-.01t})(e^{-.01t})(e^{-.01t}) = e^{-.03t}$$

In contrast, a parallel system using these three devices as its components has reliability function

$$R(t) = 1 - \prod_{i=1}^{3} [1 - R_i(t)] = 1 - \left(1 - e^{-.01t}\right)^3$$

These two reliability functions are graphed on the same set of axes in Fig. 4.21. Both functions obey properties 1–3 from p. 383, but for any $t > 0$ the parallel system reliability exceeds that of the series system, as it logically should. For example, the probability the series system's lifetime exceeds 100 h (the expected lifetime of a single component) is $R(100) = e^{-.03(100)} = e^{-3} = .0498$, while the corresponding reliability for the parallel system is $R(100) = 1 - (1 - e^{-.01(100)})^3 = 1 - (1 - e^{-1})^3 = .7474$.

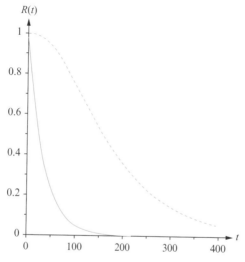

Fig. 4.21 Reliability functions for the series (*solid*) and parallel (*dashed*) systems of Example 4.47

Example 4.48 Consider the system depicted below, which consists of a combination of series and parallel elements. Using previous notation and assuming component lifetimes are independent, let's determine the reliability function of this system. More than one method may be applied here; we will rely on the Addition Rule:

$$R(t) = P\left(\left[T_1 > t \cap T_2 > t\right] \cup T_3 > t\right)$$
$$= P(T_1 > t \cap T_2 > t) + P(T_3 > t) - P(T_1 > t \cap T_2 > t \cap T_3 > t) \quad \text{Addition Rule}$$
$$= P(T_1 > t)P(T_2 > t) + P(T_3 > t) - P(T_1 > t)P(T_2 > t)P(T_3 > t) \quad \text{independence}$$
$$= R_1(t)R_2(t) + R_3(t) - R_1(t)R_2(t)R_3(t)$$

It can be shown that this reliability function satisfies properties 1–3 (the first and last are quite easy). If all three components have common reliability function $= R_i(t) = e^{-.01t}$ as in Example 4.47, the system reliability function becomes $R(t) = e^{-.02t} + e^{-.01t} - e^{-.03t}$, which lies in between the two reliability functions of Example 4.47 for all $t > 0$.   ∎

4.8.3   Mean Time to Failure

If T denotes a system's lifetime, i.e., its time until failure, then the **mean time to failure** (mttf) of the system is simply $E(T)$. The following proposition relates mean time to failure to the reliability function.

> **PROPOSITION**
>
> Suppose a system has reliability function $R(t)$ for $t \geq 0$. Then the system's mean time to failure is given by
>
> $$\mu_T = \int_0^\infty [1 - F(t)]dt = \int_0^\infty R(t)dt \qquad (4.8)$$

Expression (4.8) was established for all non-negative random variables in Exercises 38 and 150 of Chap. 3.

As a simple demonstration of this proposition, consider a single exponential lifetime with mean 100 h. We have already seen that $R(t) = e^{-.01t}$ for this particular lifetime model; integrating the reliability function yields

$$\int_0^\infty R(t)dt = \int_0^\infty e^{-.01t}dt = \left.\frac{e^{-.01t}}{-.01}\right|_0^\infty = 0 - \frac{1}{-.01} = 100$$

which is indeed the mean lifetime (aka mean time to failure) in this situation. The advantage of using Eq. (4.8) instead of the definition of $E(T)$ from Chap. 3 is that the former is usually an easier integral to calculate than the latter. Here, for example, direct computation of the mean time to failure would be

$$E(T) = \int_0^\infty t \cdot f(t)dt = \int_0^\infty .01te^{-.01t}dt,$$

which requires integration by parts (while the preceding computation did not).

Example 4.49 Consider again the series and parallel systems of Example 4.47. Using Eq. (4.8), the mttf of the series system is

$$\mu_T = \int_0^\infty R(t)dt = \int_0^\infty e^{-.03t}dt = \frac{1}{.03} \approx 33.33\,\text{hours}$$

More generally, if n independent components are connected in series, and each one has an exponentially distributed lifetime with common mean μ, then the system's mean time to failure is μ/n.

In contrast, mttf for the parallel system is given by

$$\int_0^\infty R(t)dt = \int_0^\infty \left(1 - [1 - e^{-.01t}]^3\right)dt = \int_0^\infty \left(3e^{-.01t} - 3e^{-.02t} + e^{-.03t}\right)dt$$

$$= \frac{3}{.01} - \frac{3}{.02} + \frac{1}{.03} = \frac{550}{3} \approx 183.33\,\text{hours}$$

There is no simple formula for the mttf of a parallel system, even when the components have identical exponential distributions. ∎

4.8.4 Hazard Functions

The reliability function of a system specifies the likelihood that the system will last beyond a prescribed time, t. An alternative characterization of reliability, called the hazard function, conveys information about the likelihood of imminent failure at any time t.

DEFINITION
Let T denote the lifetime of a system. If the rv T has pdf $f(t)$ and cdf $F(t)$, the **hazard function** is defined by

$$h(t) = \frac{f(t)}{1 - F(t)}$$

If T has reliability function $R(t)$, the hazard function may also be written as $h(t) = f(t)/R(t)$.

Since the pdf $f(t)$ is not a probability, neither is the hazard function $h(t)$. To get a sense of what the hazard function represents, consider the following question:

Given that the system has survived past time t, what is the probability the system will fail within the next Δt time units (an imminent failure)? Such a probability may be computed as follows:

$$P(T \le t + \Delta t \mid T > t) = \frac{P(T \le t + \Delta t \cap T > t)}{P(T > t)} = \frac{\int_t^{t+\Delta t} f(t)dt}{R(t)} \approx \frac{f(t) \cdot \Delta t}{R(t)}$$

$$= h(t) \cdot \Delta t$$

Rearranging, we have $h(t) \approx P(T \le t + \Delta t | T > t)/\Delta t$; more precisely, $h(t)$ is the limit of the right-hand side as $\Delta t \to 0$. This suggests that the hazard function $h(t)$ is a density function, like $f(t)$, except that $h(t)$ relates to the conditional probability that the system is about to fail.

Example 4.50 Once again, consider an exponentially distributed lifetime, T, but with arbitrary parameter λ. The pdf and reliability function of T are $\lambda e^{-\lambda t}$ and $e^{-\lambda t}$, respectively, so the hazard function of T is

$$h(t) = \frac{f(t)}{R(t)} = \frac{\lambda e^{-\lambda t}}{e^{-\lambda t}} = \lambda$$

In other words, a system whose time to failure follows an exponential distribution will have a constant hazard function. (The converse is true, too; we'll see how to recover $f(t)$ from $h(t)$ shortly.) This relates to the memoryless property of the exponential distribution: given the system has functioned for t hours thus far, the chance of surviving any additional amount of time is independent of t. As mentioned in Sect. 3.4, this suggests the system does not "wear out" as time progresses (which may be realistic for some devices in the short term, but not in the long term). ∎

Example 4.51 Suppose instead that we model a system's lifetime with a Weibull distribution. From the formulas presented in Sect. 3.5,

$$h(t) = \frac{f(t)}{1 - F(t)} = \frac{(\alpha/\beta^\alpha)t^{\alpha-1}e^{-(t/b)^\alpha}}{1 - \left[1 - e^{-(t/b)^\alpha}\right]} = \frac{\alpha}{\beta^\alpha}t^{\alpha-1}$$

For $\alpha = 1$, this is identical to the exponential distribution hazard function (with $\beta = 1/\lambda$). For $\alpha > 1$, $h(t)$ is an increasing function of t, meaning that we are more likely to observe an imminent failure as time progresses (this is equivalent to the system wearing out). For $0 < \alpha < 1$, $h(t)$ decreases with t, which would suggest that failures become *less* likely as t increases! This can actually be realistic for small values of t: for many devices, manufacturing flaws cause a handful to fail very early, and those that survive this initial "burn in" period are actually more likely to survive a while longer (since they presumably don't have severe faults). ∎

Fig. 4.22 A prototype hazard function

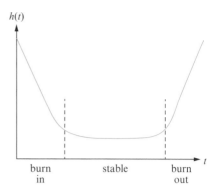

Figure 4.22 shows a prototype hazard function, popularly called a "bathtub" shape. The function can be divided into three time intervals: (1) a "burn in" period of early failures due to manufacturing errors; (2) a "stable" period where failures are due primarily to chance; and (3) a "burn out" period with an increasing failure rate due to devices wearing out. In practice, most hazard functions exhibit one or more of these behaviors.

There is a one-to-one correspondence between the pdf of a system lifetime, $f(t)$, and its hazard function, $h(t)$. The definition of the hazard function shows how one may derive $h(t)$ from $f(t)$; the following proposition reverses the process.

PROPOSITION
Suppose a system has a continuous lifetime distribution on $[0, \infty)$ with hazard function $h(t)$. Then its lifetime (aka time to failure) has reliability function $R(t)$ given by

$$R(t) = e^{-\int_0^t h(u)du}$$

and pdf $f(t)$ given by

$$f(t) = -R'(t) = h(t)e^{-\int_0^t h(u)du}$$

Proof Since $R(t) = 1 - F(t)$, $R'(t) = -f(t)$, and the hazard function may be written as $h(t) = -R'(t)/R(t)$. Now integrate both sides:

$$\int_0^t h(u)du = -\int_0^t \frac{R'(u)}{R(u)}du = -\ln[R(u)]\big|_0^t = -\ln[R(t)] + \ln[R(0)]$$

Since the system lifetime is assumed to be continuous on $[0, \infty)$, $R(0) = 1$ and $\ln[R(0)] = 0$. This leaves the equation

$$-\ln[R(t)] = \int_0^t h(u)du,$$

and the formula for $R(t)$ follows by solving for $R(t)$. The formula for $f(t)$ follows from the previous observation that $R'(t) = -f(t)$, so $f(t) = -R'(t)$, and then applying the chain rule:

$$f(t) = -R'(t) = -\frac{d}{dt}\left[e^{-\int_0^t h(u)du}\right] = -e^{-\int_0^t h(u)du} \cdot \frac{d}{dt}\left[-\int_0^t h(u)du\right]$$

$$= e^{-\int_0^t h(u)du} \cdot h(t)$$

The last step utilizes the Fundamental Theorem of Calculus.   ∎

The formulas for $R(t)$ and $f(t)$ in the preceding proposition can be easily modified for the case where $T = 0$ with some positive probability, and so $R(0) < 1$ (see Exercise 132).

Example 4.52 A certain type of high-quality transistors has hazard function $h(t) = 1 + t^6$ for $t \geq 0$, where t is measured in thousands of hours. This function is illustrated in Fig. 4.23a; notice there is no "burn in" period, but we see a fairly stable interval followed by burnout. The corresponding pdf for transistor lifetimes is

$$f(t) = h(t)e^{-\int_0^t h(u)du} = \left(1 + t^6\right)e^{-\int_0^t \left(1+u^6\right)du} = \left(1 + t^6\right)e^{-(t+t^7/7)}$$

This pdf appears in Fig. 4.23b.

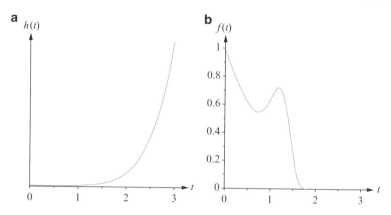

Fig. 4.23 (a) Hazard function and (b) pdf for Example 4.52

4.8.5 Exercises: Section 4.8 (121–132)

121. The lifetime, in thousands of hours, of the motor in a certain brand of kitchen blender has a Weibull distribution with $\alpha = 2$ and $\beta = 1$.
    (a) Determine the reliability function of such a motor and then graph it.
    (b) What is the probability a motor of this type will last more than 1,500 h?
    (c) Determine the hazard function of such a motor and then graph it.
    (d) Find the mean time to failure of such a motor. Compare your answer to the expected value of a Weibull distribution given in Sect. 3.5. [*Hint:* Let $u = x^2$, and apply the gamma integral formula (3.5) to the resulting integral.]

122. High-speed Internet customers are often frustrated by modem crashes. Suppose the time to "failure" for one particular brand of cable modem, measured in hundreds of hours, follows a gamma distribution with $\alpha = \beta = 2$.
    (a) Determine and graph the reliability function for this brand of cable modem.
    (b) What is the probability such a modem does not need to be refreshed for more than 300 h?
    (c) Find the mean time to "failure" for such a modem. Verify that your answer matches the formula for the mean of a gamma rv given in Sect. 3.4.
    (d) Determine and graph the hazard function for this type of modem.

123. Empirical evidence suggests that the electric ignition on a certain brand of gas stove has the following lifetime distribution, measured in thousands of days:

$$f(t) = \begin{cases} .375t^2 & 0 \le t \le 2 \\ 0 & \text{otherwise} \end{cases}$$

(Notice that the model indicates that all such ignitions expire within 2,000 days, a little less than 6 years.)

(a) Determine and graph the reliability function for this model, for all $t \ge 0$.

(b) Determine and graph the hazard function for $0 \le t \le 2$.

(c) What happens to the hazard function for $t > 2$?

124. The manufacture of a certain children's toy involves an assembly line with five stations. The lifetimes of the equipment at these stations are independent and all exponentially distributed; the mean time to failure at the first three stations (in hundreds of hours) is 1.5, while the mttf at the last two stations is 2.4.

(a) Determine the reliability function for each of the five individual stations.

(b) Determine the reliability function for the assembly line. [*Hint:* An assembly line is an example of what type of design?]

(c) Find the mean time to failure for the assembly line.

(d) Determine the hazard function for the assembly line.

125. A local bar owns four of the blenders described in Exercise 121, each having a Weibull(2, 1) lifetime distribution. During peak hours, these blenders are in continuous use, but the bartenders can keep making blended drinks (margaritas, etc.) provided that at least one of the four blenders is still functional. Define the "system" to be the four blenders under continuous use as described above, and define the lifetime of the system to be the length of time that at least one of the blenders is still functional. (Assume none of the blenders is replaced until all four have worn out.)

(a) What sort of system design do we have in this example?

(b) Find the reliability function of the system.

(c) Find the hazard function of the system.

(d) Find the mean time to failure of the system. [See the hint from Exercise 121(d).]

126. Consider the six-component system displayed below. Let $R_1(t)$, ..., $R_6(t)$ denote the reliability functions of the components. Assume the six components operate independently.

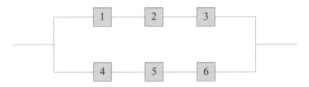

(a) Find the system reliability function.

(b) Assuming all six components have exponentially distributed lifetimes with mean 100 h, find the mean time to failure for the system.

127. Consider the six-component system displayed below. Let $R_1(t), \ldots, R_6(t)$ denote the component reliability functions. Assume the six components operate independently.

   (a) Find the system reliability function.
   (b) Assuming all six components have exponentially distributed lifetimes with mean 100 h, find the mean time to failure for the system.
128. A certain machine has the following hazard function:

$$h(t) = \begin{cases} .002 & 0 < t \le 200 \\ .001 & t > 200 \end{cases}$$

   This corresponds to a situation where a device with an exponentially distributed lifetime is replaced after 200 h of operation by another, better device also having an exponential lifetime distribution.
   (a) Determine and graph the reliability function.
   (b) Determine the probability density function of the machine's lifetime.
   (c) Find the mean time to failure.
129. Suppose the hazard function of a device is given by

$$h(t) = \begin{cases} \alpha\left(1 - \dfrac{t}{\beta}\right) & 0 \le t \le \beta \\ 0 & \text{otherwise} \end{cases}$$

   for some $\alpha, \beta > 0$. This model asserts that if a device lasts β hours, it will last forever (while seemingly unreasonable, this model can be used to study just "initial wearout").
   (a) Find the reliability function.
   (b) Find the pdf of device lifetime.
130. Suppose n independent devices are connected in series and that the ith device has an exponential lifetime distribution with parameter λ_i.
   (a) Find the reliability function of the series system.
   (b) Show that the system lifetime also has an exponential distribution, and identify its parameter in terms of $\lambda_1, \ldots, \lambda_n$.
   (c) If the mean lifetimes of the individual devices are $\mu_1, \ldots, \mu_n$, find an expression for the mean lifetime of the series system.

(d) If the same devices were connected in parallel, would the resulting system's lifetime also be exponentially distributed? How can you tell?

131. Show that a device whose hazard function is constant must have an exponential lifetime distribution.

132. Reconsider the drill bits described in Example 4.45, of which 5% shatter instantly (and so have lifetime $T = 0$). It was established that the reliability function for this scenario is $R(t) = .95e^{-.01t}$, $t \geq 0$.

(a) A generalized version of expected value that applies to distributions with both discrete and continuous elements can be used to show that the mean lifetime of these drill bits is $(.05)(0) + (.95)(100) = 95$ h. Verify that Eq. (4.8) applied to $R(t)$ gives the same answer. [This suggests that our proposition about mttf can be used even when the lifetime distribution assigns positive probability to 0.]

(b) For $t > 0$, the expression $h(t) = -R'(t)/R(t)$ is still valid. Find the hazard function for $t > 0$.

(c) Find a formula for $R(t)$ in terms of $h(t)$ that applies in situations where $R(0) < 1$. Verify that you recover $R(t) = .95e^{-.01t}$ when your formula is applied to $h(t)$ from part (b). [*Hint:* Look at the earlier proposition in this section. What one change needs to occur to accommodate $R(0) < 1$?]

4.9   Order Statistics

Many situations arise in practice that involve ordering sample observations from smallest to largest and then manipulating these ordered values in various ways. For example, once the bidding has closed in a hidden-bid auction (one in which bids are submitted independently of one another), the largest bid in the resulting sample is the amount paid for the item being auctioned, and the difference between the largest and second largest bids can be regarded as the amount that the successful bidder has overpaid.

Suppose that $X_1, X_2, \ldots, X_n$ is a random sample from a continuous distribution. Because of continuity, for any i, j with $i \neq j$, $P(X_i = X_j) = 0$. This implies that with probability 1, the n sample observations will all be different (of course, in practice all measuring instruments have accuracy limitations, so tied values may in fact result).

> **DEFINITION**
> The **order statistics** from a random sample are the random variables $Y_1, \ldots Y_n$ given by

(continued)

> $Y_1 =$ the smallest among $X_1, X_2, \ldots, X_n$ (i.e., the sample minimum)
> $Y_2 =$ the second smallest among $X_1, X_2, \ldots, X_n$
> $\vdots$
> $Y_n =$ the largest among $X_1, X_2, \ldots, X_n$ (the sample maximum)
>
> Thus, with probability 1, $Y_1 < Y_2 < \ldots < Y_{n-1} < Y_n$.

The *sample median* (i.e., the middle value in the ordered list) is then $Y_{(n+1)/2}$ when n is odd, while the *sample range* is $Y_n - Y_1$.

4.9.1   The Distributions of Y_n and Y_1

The key idea in obtaining the distribution of the sample maximum Y_n is the observation that Y_n is at most y if and only if every one of the X_is is at most y. Similarly, the distribution of Y_1 is based on the fact that it will exceed y if and only if all X_is exceed y.

Example 4.53 Consider 5 identical components connected in parallel as shown in Fig. 4.20b. Let X_i denote the lifetime, in hours, of the ith component ($i = 1, 2, 3, 4, 5$). Suppose that the X_is are independent and that each has an exponential distribution with $\lambda = .01$, so the expected lifetime of any particular component is $1/\lambda = 100$ h. Because of the parallel configuration, the system will continue to function as long as at least one component is still working, and will fail as soon as the last component functioning ceases to do so. That is, the system lifetime is Y_5, the largest order statistic in a sample of size 5 from the specified exponential distribution. Now Y_5 will be at most y if and only if every one of the five X_is is at most y. With $G_5(y)$ denoting the cdf of Y_5,

$$G_5(y) = P(Y_5 \leq y) = P(X_1 \leq y \cap X_2 \leq y \cap \ldots \cap X_5 \leq y)$$
$$= P(X_1 \leq y) \cdot P(X_2 \leq y) \cdot \ldots \cdot P(X_5 \leq y)$$

For every one of the X_is, $P(X_i \leq y) = F(y) = \int_0^y .01 e^{-.01x} dx = 1 - e^{-.01y}$; this is the common cdf of the X_is evaluated at y. Hence, $G_5(y) = (1 - e^{-.01y}) \cdots (1 - e^{-.01y}) = (1 - e^{-.01y})^5$. The pdf of Y_5 can now be obtained by differentiating the cdf with respect to y.

Suppose instead that the five components are connected in series rather than in parallel (Fig. 4.20a). In this case the system lifetime will be Y_1, the *smallest* of the five order statistics, since the system will crash as soon as a single one of the individual components fails. Note that system lifetime will exceed y hours if and only if the lifetime of *every* component exceeds y hours. Thus, the cdf of Y_1 is

$$G_1(y) = P(Y_1 \leq y) = 1 - P(Y_1 > y) = 1 - P(X_1 > y \cap X_2 > y \cap \ldots \cap X_5 > y)$$
$$= 1 - P(X_1 > y) \cdot P(X_2 > y) \cdots\cdots P(X_5 > y) = 1 - \left(e^{-.01y}\right)^5 = 1 - e^{-.05y}$$

This is the form of an exponential cdf with parameter .05. More generally, if the n components in a series connection have lifetimes that are independent, each exponentially distributed with the same parameter λ, then the system lifetime will be exponentially distributed with parameter $n\lambda$. We saw a similar result in Example 4.49. The expected system lifetime will then be $1/(n\lambda)$, much smaller than the expected lifetime of an individual component.   ∎

An argument parallel to that of the previous example for a general sample size n and an arbitrary pdf $f(x)$ gives the following general results.

PROPOSITION

Let Y_1 and Y_n denote the smallest and largest order statistics, respectively, based on a random sample from a continuous distribution with cdf $F(x)$ and pdf $f(x)$. Then the cdf and pdf of Y_n are

$$G_n(y) = [F(y)]^n, \quad g_n(y) = n[F(y)]^{n-1} \cdot f(y)$$

The cdf and pdf of Y_1 are

$$G_1(y) = 1 - \left[1 - F(y)\right]^n, \quad g_1(y) = n\left[1 - F(y)\right]^{n-1} \cdot f(y)$$

Example 4.54 Let X denote the contents of a one-gallon container, and suppose that its pdf is $f(x) = 2x$ for $0 \leq x \leq 1$ (and 0 otherwise) with corresponding cdf $F(x) = x^2$ in the interval of positive density. Consider a random sample of four such containers. The order statistics Y_1 and Y_4 represent the contents of the least-filled container and the most-filled container, respectively. The pdfs of Y_1 and Y_4 are

$$g_1(y) = 4\left(1 - y^2\right)^3 \cdot 2y = 8y\left[1 - y^2\right]^3 \quad 0 \leq y \leq 1$$
$$g_4(y) = 4\left(y^2\right)^3 \cdot 2y = 8y^7 \quad 0 \leq y \leq 1$$

The corresponding density curves appear in Fig. 4.24.

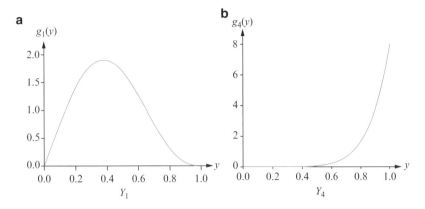

Fig. 4.24 Density curves for the order statistics (**a**) Y_1 and (**b**) Y_4 in Example 4.54

Let's determine the expected value of $Y_4 - Y_1$, the difference between the contents of the most-filled container and the least-filled container; $Y_4 - Y_1$ is just the sample range. Apply linearity of expectation:

$$E(Y_4 - Y_1) = E(Y_4) - E(Y_1) = \int_0^1 y \cdot 8y^7 dy - \int_0^1 y \cdot 8y(1 - y^2)^3 dy$$

$$= \frac{8}{9} - \frac{384}{945} = .889 - .406 = .483$$

If random samples of four containers were repeatedly selected and the sample range of contents determined for each one, the long run average value of the range would be .483. ∎

4.9.2 The Distribution of the *i*th Order Statistic

We have already obtained the (marginal) distribution of the largest order statistic Y_n and also that of the smallest order statistic Y_1. A generalization of the argument used previously results in the following proposition; Exercise 140 suggests how this result can be derived.

PROPOSITION

Suppose $X_1, X_2, \ldots, X_n$ is a random sample from a continuous distribution with cdf $F(x)$ and pdf $f(x)$. The pdf of the *i*th smallest order statistic Y_i is

$$g_i(y) = \frac{n!}{(i - 1)!(n - i)!}[F(y)]^{i-1}[1 - F(y)]^{n-i}f(y) \qquad (4.9)$$

An intuitive justification for Expression (4.9) will be given shortly. Notice that it is consistent with the pdf expressions for $g_1(y)$ and $g_n(y)$ given previously; just substitute $i = 1$ and $i = n$, respectively.

Example 4.55 Suppose that component lifetime is exponentially distributed with parameter λ. For a random sample of $n = 5$ components, the expected value of the sample median lifetime is

$$E(Y_3) = \int_0^\infty y \cdot g_3(y)dy = \int_0^\infty y \cdot \frac{5!}{2! \cdot 2!}\left(1 - e^{-\lambda y}\right)^2 \left(e^{-\lambda y}\right)^2 \cdot \lambda e^{-\lambda y}dy$$

Expanding out the integrand and integrating term by term, the expected value is $.783/\lambda$. The median of the exponential distribution is, from solving $F(\eta) = .5$, $\eta = -\ln(.5)/\lambda = .693/\lambda$. Thus if sample after sample of five components is selected, the long run average value of the sample median will be somewhat larger than the median value of the individual lifetime distribution. This is because the exponential distribution has a positive skew. ∎

Here is the promised intuitive derivation of Eq. (4.9). Let Δ be a number quite close to 0, and consider the three intervals $(-\infty, y]$, $(y, y + \Delta]$, and $(y + \Delta, \infty)$. For a single X, the probabilities of these three intervals are

$$p_1 = P(X \le y) = F(y) \qquad p_2 = P(y < X \le y + \Delta) = \int_y^{y+\Delta} f(x)dx \approx f(y) \cdot \Delta$$
$$p_3 = P(X > y + \Delta) = 1 - F(y + \Delta)$$

For a random sample of size n, it is very unlikely that two or more Xs will fall in the middle interval, since its width is only Δ. The probability that the ith order statistic falls in the middle interval is then approximately the probability that $i - 1$ of the Xs are in the first interval, one is in the middle, and the remaining $n - i$ are in the third. This is just a trinomial probability:

$$P(y < Y_i \le y + \Delta) \approx \frac{n!}{(i - 1)!1!(n - i)!}[F(y_i)]^{i-1} \cdot f(y) \cdot \Delta \cdot [1 - F(y + \Delta)]^{n-i}$$

Dividing both sides by Δ and taking the limit as $\Delta \to 0$ gives exactly Expression (4.9). That is, we may interpret the pdf $g_i(y)$ as loosely specifying that $i - 1$ of the original observations are below y, one is "at" y, and the other $n - i$ are above y.

Similar reasoning works to intuitively derive the joint pdf of Y_i and Y_j $(i < j)$. In this case there are five relevant intervals: $(-\infty, y_i]$, $(y_i, y_i + \Delta_1]$, $(y_i + \Delta_1, y_j]$, $(y_j, y_j + \Delta_2]$, and $(y_j + \Delta_2, \infty)$.

4.9.3   The Joint Distribution of the n Order Statistics

We now develop the joint pdf of $Y_1, Y_2, \ldots, Y_n$. Consider first a random sample X_1, X_2, X_3 of fuel efficiency measurements (mpg). The joint pdf of this random sample is

$$f(x_1, x_2, x_3) = f(x_1) \cdot f(x_2) \cdot f(x_3)$$

The joint pdf of Y_1, Y_2, Y_3 will be positive only for values of y_1, y_2, y_3 satisfying $y_1 < y_2 < y_3$. What is this joint pdf at the values $y_1 = 28.4$, $y_2 = 29.0$, $y_3 = 30.5$? There are six different ways to obtain these ordered values:

$$X_1 = 28.4, X_2 = 29.0, X_3 = 30.5$$
$$X_1 = 28.4, X_2 = 30.5, X_3 = 29.0$$
$$X_1 = 29.0, X_2 = 28.4, X_3 = 30.5$$
$$X_1 = 29.0, X_2 = 30.5, X_3 = 28.4$$
$$X_1 = 30.5, X_2 = 28.4, X_3 = 29.0$$
$$X_1 = 30.5, X_2 = 29.0, X_3 = 28.4$$

These six possibilities come from the 3! ways to order the three numerical observations once their values are fixed. Thus

$$g(28.4, 29.0, 30.5) = f(28.4) \cdot f(29.0) \cdot f(30.5) + \cdots + f(30.5) \cdot f(29.0) \cdot f(28.4)$$
$$= 3! f(28.4) \cdot f(29.0) \cdot f(30.5)$$

Analogous reasoning with a sample of size n yields the following result:

PROPOSITION

Let $g(y_1, y_2, \ldots, y_n)$ denote the joint pdf of the order statistics $Y_1, Y_2, \ldots, Y_n$ resulting from a random sample of X_is from a pdf $f(x)$. Then

$$g(y_1, y_2, \ldots, y_n) = \begin{cases} n! f(y_1) \cdot f(y_2) \cdots \cdots f(y_n) & y_1 < y_2 < \ldots < y_n \\ 0 & \text{otherwise} \end{cases}$$

For example, if we have a random sample of component lifetimes and the lifetime distribution is exponential with parameter λ, then the joint pdf of the order statistics is

$$g(y_1, \ldots, y_n) = n! \lambda^n e^{-\lambda(y_1 + \cdots + y_n)} \qquad 0 < y_1 < y_2 < \cdots < y_n$$

Example 4.56 Suppose X_1, X_2, X_3, and X_4 are independent random variables, each uniformly distributed on the interval from 0 to 1. The joint pdf of the four corresponding order statistics Y_1, Y_2, Y_3, and Y_4 is $g(y_1, y_2, y_3, y_4) = 4! \cdot 1$ for $0 < y_1 < y_2 < y_3 < y_4 < 1$. The probability that every pair of X_is is separated by more than .2 is the same as the probability that $Y_2 - Y_1 > .2$, $Y_3 - Y_2 > .2$, and

$Y_4 - Y_3 > .2$. This latter probability results from integrating the joint pdf of the Y_is over the region $.6 < y_4 < 1$, $.4 < y_3 < y_4 - .2$, $.2 < y_2 < y_3 - .2$, $0 < y_1 < y_2 - .2$:

$$P(Y_2 - Y_1 > .2, Y_3 - Y_2 > .2, Y_4 - Y_3 > .2) = \int_{.6}^{1} \int_{.4}^{y_4 - .2} \int_{.2}^{y_3 - .2} \int_{0}^{y_2 - .2} 4! \, dy_1 dy_2 dy_3 dy_4$$

The inner integration gives $4!(y_2 - .2)$, and this must then be integrated between $.2$ and $y_3 - .2$. Making the change of variable $z_2 = y_2 - .2$, the integration of z_2 is from 0 to $y_3 - .4$. The result of this integration is $4! \cdot (y_3 - .4)^2/2$. Continuing with the third and fourth integration, each time making an appropriate change of variable so that the lower limit of each integration becomes 0, the result is

$$P(Y_2 - Y_1 > .2, Y_3 - Y_2 > .2, Y_4 - Y_3 > .2) = .4^4 = .0256$$

A more general multiple integration argument for n independent uniform $[0, B]$ rvs shows that the probability that all values are separated by more than some distance d is

$$P(\text{all values are separated by more than } d) = \begin{cases} [1 - (n-1)d/B]^n & 0 \le d \le B/(n-1) \\ 0 & d > B/(n-1) \end{cases}$$

As an application, consider a year that has 365 days, and suppose that the birth time of someone born in that year is uniformly distributed throughout the 365-day period. Then in a group of 10 independently selected people born in that year, the probability that all of their birth times are separated by more than 24 h ($d = 1$ day) is $(1 - (10 - 1)(1)/365)^{10} = .779$. Thus the probability that at least two of the 10 birth times are separated by at most 24 h is .221. As the group size n increases, it becomes more likely that at least two people have birth times that are within 24 h of each other (but not necessarily on the same day). For $n = 16$, this probability is .467, and for $n = 17$ it is .533. So with as few as 17 people in the group, it is more likely than not that at least two of the people were born within 24 h of each other.

Coincidences such as this are not as surprising as one might think. The probability that at least two people are born on the same calendar day (assuming equally likely birthdays) is much easier to calculate than what we have shown here; see the Birthday Problem in Example 1.22.   ∎

4.9.4   Exercises: Section 4.9 (133–142)

133. A friend of ours takes the bus 5 days per week to her job. The five waiting times until she can board the bus are a random sample from a uniform distribution on the interval from 0 to 10 min.

(a) Determine the pdf and then the expected value of the largest of the five waiting times.

(b) Determine the expected value of the difference between the largest and smallest times.

(c) What is the expected value of the sample median waiting time?

(d) What is the standard deviation of the largest time?

134. Refer back to Example 4.54. Because $n=4$, the sample median is the average of the two middle order statistics, $(Y_2+Y_3)/2$. What is the expected value of the sample median, and how does it compare to the median of the population distribution?

135. An insurance policy issued to a boat owner has a deductible amount of \$1000, so the amount of damage claimed must exceed this deductible before there will be a payout. Suppose the amount (thousands of dollars) of a randomly selected claim is a continuous rv with pdf $f(x)=3/x^4$ for $x>1$. Consider a random sample of three claims.

(a) What is the probability that at least one of the claim amounts exceeds \$5000?

(b) What is the expected value of the largest amount claimed?

136. A store is expecting n deliveries between the hours of noon and 1 p.m. Suppose the arrival time of each delivery truck is uniformly distributed on this 1-h interval and that the times are independent of each other. What are the expected values of the ordered arrival times?

137. Let X be the amount of time an ATM is in use during a particular 1-h period, and suppose that X has the cdf $F(x)=x^\theta$ for $0<x<1$ (where $\theta>1$). Give expressions involving the gamma function for both the mean and variance of the ith smallest amount of time Y_i from a random sample of n such time periods.

138. The logistic pdf $f(x)=e^{-x}/(1+e^{-x})^2$ for $-\infty<x<\infty$ is sometimes used to describe the distribution of measurement errors.

(a) Graph the pdf. Does the appearance of the graph surprise you?

(b) For a random sample of size n, obtain an expression involving the gamma function for the moment generating function of the ith smallest order statistic Y_i. This expression can then be differentiated to obtain moments of the order statistics. [Hint: Set up the appropriate integral, and then let $u=1/(1+e^{-x})$.]

139. Let X represent a measurement error. It is natural to assume that the pdf $f(x)$ is symmetric about 0, so that the density at a value $-c$ is the same as the density at c (an error of a given magnitude is equally likely to be positive or negative). Consider a random sample of n measurements, where $n=2k+1$, so that Y_{k+1} is the sample median. What can be said about $E(Y_{k+1})$? If the X distribution is symmetric about some other value, so that value is the median of the distribution, what does this imply about $E(Y_{k+1})$? [Hints: For the first question, symmetry implies that $1-F(x)=P(X>x)=P(X<-x)=F(-x)$. For the second question, consider $W=X-\eta$; what is the median of the distribution of W?]

140. The pdf of the second-largest order statistic, Y_{n-1} can be obtained using reasoning analogous to how the pdf of Y_n was first obtained.
    (a) For any number y, $Y_{n-1} \le y$ if and only if *at least* $n-1$ of the original Xs are $\le y$. (Do you see why?) Use this fact to derive a formula for the cdf of Y_{n-1} in terms of F, the cdf of the Xs. [*Hint*: Separate "at least $n-1$" into two cases and apply the binomial distribution formula.]
    (b) Differentiate the cdf in part (a) to obtain the pdf of Y_{n-1}. Simplify and verify it matches the formula for $g_{n-1}(y)$ provided in this section.

141. Use the intuitive argument sketched in this section to obtain the following general formula for the joint pdf of two order statistics Y_i and Y_j with $i < j$:

$$g(y_i, y_j) = \frac{n!}{(i-1)!(j-i-1)!(n-j)!} F(y_i)^{i-1} \left[F(y_j) - F(y_i) \right]^{j-i-1} \left[1 - F(y_j) \right]^{n-j}.$$

$$f(y_i)f(y_j) \text{ for } y_i < y_j$$

142. Consider a sample of size $n = 3$ from the standard normal distribution, and obtain the expected value of the largest order statistic. What does this say about the expected value of the largest order statistic in a sample of this size from *any* normal distribution? [*Hint*: With $\phi(x)$ denoting the standard normal pdf, use the fact that $(d/dx)\phi(x) = -x\phi(x)$ along with integration by parts.]

4.10   Simulation of Joint Probability Distributions and System Reliability

In Chaps. 2 and 3, we saw several methods for simulating "generic" discrete and continuous distributions (in addition to built-in functions for binomial, Poisson, normal, etc.). Unfortunately, most of these general methods do not carry over easily to joint distributions or else require significant re-tooling. In this section, we briefly survey some simulation techniques for general bivariate discrete and continuous distributions and discuss how to simulate normal distributions in more than one dimension. We then consider simulations for the lifetimes of interconnected systems, in order to understand the reliability of such systems.

4.10.1 Simulating Values from a Joint PMF

Simulating two dependent discrete rvs X and Y can be rather tedious and is easier to understand with a specific example in mind. Suppose we desire to simulate (X, Y) values from the joint pmf in Example 4.1:

	$p(x, y)$	0	100	200
x	100	.20	.10	.20
	250	.05	.15	.30

The *exhaustive search* approach uses the inverse cdf method of Sect. 2.8 by reformatting the table as a single row of (x, y) pairs along with cumulative probabilities. Starting in the upper left corner and going across, create "cumulative" probabilities for the entire table:

		0	100	200
x	100	.20	.30	.50
	250	.55	.70	1

Be careful not to interpret these increasing decimals as cumulative probabilities in the traditional sense, e.g., it is *not* the case that .70 in the preceding table represents $P(X \le 250 \cap Y \le 100)$. For ease of reading, this same table has been rendered below as two parallel rows, one enumerating the (x, y) values and another with the corresponding cumulative probabilities.

(x, y)	(100, 0)	(100, 100)	(100, 200)	(250, 0)	(250, 100)	(250, 250)
cum. prob.	.20	.30	.50	.55	.70	1

Now the simulation proceeds similarly to those illustrated in Fig. 2.10 for simulating a single discrete random variable: use if-else statements, specifying the pair of values (x, y) for each range of standard uniform random numbers. Figure 4.25 provides the needed Matlab and R code.

In both languages, executing the code in Fig. 4.25 results in two vectors x and y that, when regarded as *paired* values, form a simulation of the original joint pmf. That is to say, if x and y were laid in parallel roughly 20% of the paired values would be (100, 0), about 10% would be (100, 100), and so on.

At the end of Sect. 2.8 we mentioned that both Matlab and R have built-in functions to speed up the inverse cdf method (randsample and sample, respectively) for a single discrete rv. Unfortunately, these are not designed to take pairs of values as an input, and so the lengthier code is required. You might be tempted to use these built-in functions to simulate the (marginal) pmfs of X and Y separately, but beware: by design, the resulting simulated values of X and Y would be independent, and the rvs displayed in the original joint pmf are clearly dependent. For example, (100, 0) ought to appear roughly 20% of the time in a simulation; however, separate simulations of X and Y will result in about 50% 100 s for X and 25% 0 s for Y, *independently*, meaning the pair (100, 0) will appear in approximately $(.5)(.25) = 12.5\%$ of simulated (X, Y) values.

It's worth noting that the choice to add across rows first was arbitrary. We could just as well have added down the left-most column ($Y = 0$) of the original joint pmf

a
```
x=zeros(10000,1); y=x;
for i=1:10000
    u=rand;
    if u<.2
        x(i)=100; y(i)=0;
    elseif u<.3
        x(i)=100; y(i)=100;
    elseif u<.5
        x(i)=100; y(i)=200;
    elseif u<.55
        x(i)=250; y(i)=0;
    elseif u<.7
        x(i)=250; y(i)=100;
    else
        x(i)=250; y(i)=200;
    end
end
```

b
```
x <- NULL; y <- NULL
for (i in 1:10000){
    u=runif(1)
    if (u<.2){
        x[i]<-100; y[i]<-0}
    else if (u<.3){
        x[i]<-100; y[i]<-1}
    else if (u<.5){
        x[i]<-100; y[i]<-2}
    else if (u<.55){
        x[i]<-250; y[i]<-0}
    else if (u<.7){
        x[i]<-250; y[i]<-100}
    else{
        x[i]<-250; y[i]<-200}
}
```

Fig. 4.25 The exhaustive search method for simulating two discrete rvs: (**a**) Matlab; (**b**) R

table, then the middle column, then the right column to create "cumulative" probabilities and then rewritten our code accordingly.

4.10.2 Simulating Values from a Joint PDF

As in the discrete case, a pair of *independent* continuous rvs X and Y can be simulated separately using any of the methods from Sect. 3.8 (inverse cdf, accept–reject). In the general case, however, the inverse cdf method breaks down in two or more dimensions, because we cannot "invert" the joint cdf of X and Y. Hence, we rely primarily on the accept–reject method. The following proposition repeats the algorithm from Sect. 3.8 but expands it to two dimensions; a simulation scheme for three or more dependent rvs would be analogous.

ACCEPT–REJECT METHOD (bivariate case)
It is desired to simulate n values from a joint pdf $f(x, y)$. Let $g_1(x)$ and $g_2(y)$ be two univariate pdfs such that the ratio $f/[g_1 g_2]$ is bounded, i.e., there exists a constant c such that $f(x, y)/[g_1(x)g_2(y)] \leq c$ for all x and y. Proceed as follows:
1. Generate a variate, x^*, from the distribution g_1; *independently*, generate a variate, y^*, from the distribution g_2. This pair (x^*, y^*) is our candidate.
2. Generate a standard uniform variate, u.
3. If $u \cdot c \cdot g_1(x^*)g_2(y^*) \leq f(x^*, y^*)$, then assign $(x, y) = (x^*, y^*)$, i.e., "accept" the candidate. Otherwise, reject (x^*, y^*) and return to step 1.
These steps are repeated until n candidate pairs have been accepted. The resulting accepted pairs $(x_1, y_1), \ldots, (x_n, y_n)$ constitute a simulation of a pair of random variables (X, Y) with the original joint pdf, $f(x, y)$.

a

```
x=zeros(10000,1); y=x;
i=0;
while i<10000
    xstar=random('unif',20,30);
    ystar=random('unif',20,30);
    u=rand;
    if u<=(xstar^2+ystar^2)/1800
        i=i+1;
        x(i)=xstar;
        y(i)=ystar;
    end
end
```

b

```
x <- NULL; y <- NULL
i <- 0
while (i <10000){
    xstar <- runif(1,20,30)
    ystar <- runif(1,20,30)
    u <- runif(1)
    if (u<=(xstar^2+ystar^2)/1800){
        i <- i+1
        x[i] <- xstar
        y[i] <- ystar
    }
}
```

Fig. 4.26 Simulation code for Example 4.57: (**a**) Matlab; (**b**) R

As in the one-dimensional case, the accept–reject method hinges on generating some other distribution on the same set of values as the "target" pdf. What's special here is that we leverage our ability to simulate univariate distributions—namely, $g_1(x)$ and $g_2(y)$—and create a candidate pair (x^*, y^*) from two independent rvs. In particular, the product $g_1(x^*)g_2(y^*)$ that appears in the algorithm is the joint pdf of two independent rvs having marginal distributions g_1 and g_2, respectively.

Example 4.57 It is desired to simulate values from the following joint pdf, introduced in Exercise 11:

$$f(x, y) = \begin{cases} k(x^2 + y^2) & 20 \leq x \leq 30, \quad 20 \leq y \leq 30 \\ 0 & \text{otherwise} \end{cases}$$

(Determining the constant of integration, k, won't be necessary.) Since both X and Y are bounded between 20 and 30, a sensible choice for both g_1 and g_2 is the uniform distribution on [20, 30]. That is, $g_1(x) = 1/(30-20) = .1$ for $20 \leq x \leq 30$, and $g_2(y) = g_1(y)$. The majorization constant c is determined by requiring

$$\frac{f(x, y)}{g_1(x)g_2(y)} = \frac{k(x^2 + y^2)}{(.1)(.1)} \leq c \quad \text{for } 20 \leq x \leq 30, 20 \leq y \leq 30$$

The left-hand expression is obviously maximized at $x = y = 30$, from which we have $c \geq k(30^2 + 30^2)/(.1)^2 = 180{,}000k$. Setting $c = 180{,}000k$, the accept–reject scheme for this joint pdf proceeds as follows:
1. Generate independent $x^* \sim \text{Unif}[20, 30]$ and $y^* \sim \text{Unif}[20, 30]$.
2. Generate a standard uniform variate, u.
3. Accept (x^*, y^*) iff $u \cdot c \cdot g_1(x^*)g_2(y^*) \leq f(x^*, y^*)$, i.e., $u \cdot 180{,}000k \cdot (.1)(.1) \leq k((x^*)^2 + (y^*)^2)$. This is algebraically equivalent to $u \leq ((x^*)^2 + (y^*)^2)/1800$.
Figure 4.26 provides Matlab and R code for this example. The output of either one is a pair of vectors, x and y, whose paired values simulate the original joint pdf.

Figure 4.27 shows the joint pdf $f(x, y)$ alongside a "three-dimensional histogram" of 10,000 (x, y) values simulated in Matlab (the latter was created using the `hist3` command). Observe that both show a slight rise as the x- or y-values increase from 20 to 30.

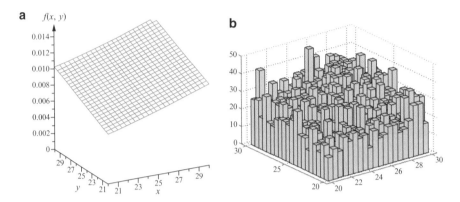

Fig. 4.27 Joint pdf (**a**) and histogram of simulated values (**b**) for Example 4.57   ∎

As indicated in Sect. 3.8, it can be shown that the majorization constant c is also the expected number of candidates required to generate a single accepted value (here, a pair). In the preceding example, the numerical value of c turns out to be $c = 27/19 \approx 1.421$, so we expect our programs to require about 14,210 iterations of the while loop to create 10,000 simulated valued of (X, Y).

As an alternative to the accept–reject method, a technique based on conditional distributions can be employed. The basic idea is this: suppose X has pdf $f(x)$ and, conditional on $X = x$, Y has conditional distribution $f(y|x)$. Then one can simulate (X, Y) by first simulating from $f(x)$ using the techniques of Sect. 3.8 and then, given the simulated value of x, simulating a value y from $f(y|x)$.

Example 4.58 Consider the following joint pdf, introduced in Exercise 14:

$$f(x, y) = \begin{cases} xe^{-x(1+y)} & x \geq 0 \quad \text{and} \quad y \geq 0 \\ 0 & \text{otherwise} \end{cases}$$

Straightforward integration shows the marginal pdf of X to be $f_X(x) = e^{-x}$, from which the conditional distribution of Y given $X = x$ is

$$f(y \mid x) = \frac{f(x, y)}{f_X(x)} = \frac{xe^{-x(1+y)}}{e^{-x}} = xe^{-xy}$$

Each of these has an algebraically simple cdf, so we will employ the inverse cdf method for each step. The cdf of X is $F(x) = 1 - e^{-x}$, whose inverse is given by $x = -\ln(1 - u)$. Similarly, the conditional cdf of Y given $X = x$ is $F(y|x) = 1 - e^{-xy}$, whose inverse function (with respect to y) is $y = -(1/x)\ln(1 - u)$. The resulting

a
```
x=zeros(10000,1); y=x;
for i=1:10000
    u=rand;
    x(i)=-log(1-u);
    v=rand;
    y(i)=-(1/x(i))*log(1-v);
end
```

b
```
x <- NULL; y <- NULL
for (i in 1:10000){
    u<-runif(1)
    x[i]<- -log(1-u)
    v<-runif(1)
    y[i]<- -(1/x[i])*log(1-v)
}
```

Fig. 4.28 Simulation code for Example 4.58: (**a**) Matlab; (**b**) R

simulation code, in Matlab and R, appears in Fig. 4.28. Notice in each program that *two* standard uniform variates, u and v, are required: one to simulate x, and another to simulate y given x.

Some simplifications can be made to the preceding code. As in many other simulations, the for loop can be vectorized (summoning all 10,000 simulated values at once). Additionally, you might recognize the pdfs under consideration: the marginal distribution of X is exponential with $\lambda = 1$, while Y given $X = x$ is exponential with parameter $\lambda = x$. Hence, we could exploit Matlab's or R's built-in exponential distribution simulator, rather than finding and inverting the cdfs. ∎

This method can also be extended to three or more variables, but finding the required conditional pdfs from the joint pdf can be difficult. This conditional distributions method is best suited to so-called *hierarchical* models, where the distribution of each rv is specified conditional on its predecessors, e.g., we are provided initially with $f(x)$, $f(y|x)$, $f(z|y,x)$, and so on.

The conditional distributions approach may also be implemented to simulate a joint discrete distribution; see Exercise 149.

4.10.3 Simulating a Bivariate Normal Distribution

The prevalence of normal distributions makes the ability to simulate both univariate and multivariate normal rvs especially important. A simple method exists for simulating pairs from an arbitrary bivariate normal distribution, as indicated in the following proposition.

PROPOSITION

Let Z_1 and Z_2 be independent standard normal rvs and let

$$W_1 = Z_1, \qquad W_2 = \rho \cdot Z_1 + \sqrt{1 - \rho^2} Z_2$$

Then W_1 and W_2 have a bivariate normal distribution, each having mean 0 and standard deviation 1, and $\mathrm{Corr}(W_1, W_2) = \rho$.

a

```
z1=random('norm',0,1,[10000 1]);
z2=random('norm',0,1,[10000 1]);
x=496+114*Z1;
y=514+117*(.25*z1+sqrt(1-.25^2)*z2);
```

b

```
z1 <- rnorm(10000)
z2 <- rnorm(10000)
x <- 496+114*z1
y <- 514+117*(.25*z1+sqrt(1-.25^2)*z2)
```

Fig. 4.29 Code for Example 4.59: (**a**) Matlab (**b**) R

This result can be proved using the transformation methods of Sect. 4.6. The means, variances, and correlation coefficient of W_1 and W_2 are established in Exercise 161.

Now suppose we wish to simulate from a bivariate normal distribution with an arbitrary set of parameters μ_1, σ_1, μ_2, σ_2, and ρ. Define X and Y by

$$X = \mu_1 + \sigma_1 W_1 = \mu_1 + \sigma_1 Z_1,$$

$$Y = \mu_2 + \sigma_2 W_2 = \mu_2 + \sigma_2 \left(\rho Z_1 + \sqrt{1 - \rho^2} Z_2 \right) \tag{4.10}$$

Since X and Y in Expression (4.10) are just linear functions of W_1 and W_2, it follows from Sect. 4.2 that Corr(X, Y) = Corr(W_1, W_2) = ρ. Moreover, since W_1 and W_2 have mean zero and standard deviation 1, these linear transformations give X and Y the desired means and standard deviations. So, to simulate a bivariate normal distribution, create a pair of independent standard normal variates z_1 and z_2, and then apply the formulas for X and Y in Eq. (4.10).

Example 4.59 Consider the joint distribution of SAT reading and math scores described in Example 4.42. Using the parameters from that example, Eq. (4.10) becomes

$$X = 496 + 114Z_1, \qquad Y = 514 + 117 \left(.25Z_1 + \sqrt{1 - .25^2} Z_2 \right)$$

Figure 4.29 shows this transformation implemented in Matlab and R; both programs have been vectorized and produce 10,000 (X, Y) pairs.

Now define a new rv $R = Y/X$, the ratio of a student's SAT Math and Critical Reading scores. Arguably, this measures a student's math ability *relative to* her or his reading skills. Determining the pdf of R is simply not feasible, especially since X and Y are dependent. But the above simulation, along with the command r=y/x (in R, or r=y./x in Matlab) gives us information about its distribution. A histogram of the simulated values of R appears in Fig. 4.30. For these 10,000 simulated values, the sample mean and standard deviation are $\bar{r} = 1.0161$ and $s = 0.1677$. So, we estimate the true expected ratio $E(R)$ for all students that took the SAT in Fall 2012 is 1.0161, with an estimated standard error of $s/\sqrt{n} = 0.1677/\sqrt{10,000} = .001677$.

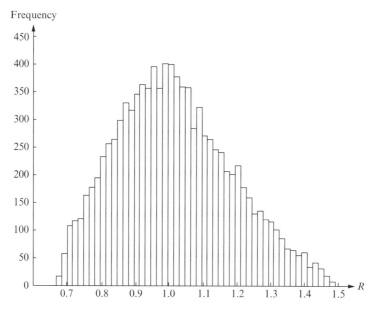

Fig. 4.30 Histogram of $R = Y/X$ from Example 4.59   ∎

Matlab and R also have built-in programs to simulate multivariate normal distributions that work for 2 or more dimensions and do not rely on the preceding proposition. In fact, users have created several such tools in R; we illustrate here the function available in the mvtnorm package. The mvnrnd function in Matlab and the rmvnorm function in R take three inputs: the desired number of simulated values (in the bivariate case, simulated pairs), a vector of means, and a covariance matrix (see the end of Sect. 4.7). Figure 4.31 illustrates these commands for the distribution specified in Example 4.59.

a

```
mu=[496, 514];
C=[114^2, .25*114*117;
      .25*114*117, 117^2];
x=mvnrnd(mu,C,10000)
```

b

```
mu <- c(496,514)
C <- matrix(c(114^2, .25*114*117,
.25*114*117, 117^2),2,2)
x <- rmvnorm(10000,mu,C)
```

Fig. 4.31 Built-in multivariate normal simulations: (**a**) Matlab; (**b**) R

4.10.4 Simulations Methods for Reliability

One area of application for the simulation methods presented in this section is to the lifetime distributions of complex systems. It can sometimes be difficult to derive the exact pdf of the lifetime of a system comprised of many components (in series and/or parallel), but simulation provides a way out.

Example 4.60 Consider the system described in Example 4.48; this is actually a comparatively simple configuration. Let T_1, T_2, and T_3 denote the lifetimes of the three components. Since components 1 and 2 are connected in series, the "1–2 subsystem" functions only as long as the smaller of T_1 and T_2, e.g., if $T_1 = 135$ h and $T_2 = 119$ h, then the lifetime of the 1–2 subsystem is 119 h. The lifetime of the 1–2 subsystem, therefore, can be expressed mathematically as $\min(T_1, T_2)$.

Similarly, the 1–2 subsystem is linked in parallel with component 3, and so the lifetime of the overall system is the *larger* of the lifetimes of the two pieces (the 1–2 subsystem and component 3). For example, if the lifetime of the 1–2 subsystem is 119 h and the lifetime of component 3 is 127 h, then the overall system lifetime is 127 h. If we let T_{sys} denote the system lifetime, then we have

$$T_{sys} = \max(1\text{–}2 \text{ subsystem lifetime}, \text{component 3 lifetime})$$
$$= \max\big(\min(T_1, T_2), T_3\big)$$

This expression combining max and min functions can be used to simulate the system lifetime, provided we have models (that we can simulate) for the lifetimes of the three individual components. Figure 4.32 shows example code for simulating the system lifetime assuming each of the three components has an exponentially distributed lifetime with mean 100 h. The 10,000 simulation runs have been vectorized to accelerate the process. Notice that the R code requires using the pmax and pmin commands, which treat their inputs as parallel vectors and find the "row-wise" maximum or minimum.

a

```
T1=-100*log(1-rand(10000,1));
T2=-100*log(1-rand(10000,1));
T3=-100*log(1-rand(10000,1));
Tsys=max(min(T1,T2),T3);
```

b

```
T1 <- -100*log(1-runif(10000))
T2 <- -100*log(1-runif(10000))
T3 <- -100*log(1-runif(10000))
Tsys <- pmax(pmin(T1,T2),T3)
```

Fig. 4.32 Simulation code for Example 4.60: (**a**) Matlab; (**b**) R

A histogram of 10,000 simulated values of T_{sys} from R appears in Fig. 4.33; this should look similar to the pdf of that rv. We can use these same simulated values to estimate the expectation and standard deviation of T_{sys}: for our run, the sample mean and standard deviation were 115.37 h and 93.49 h, respectively.

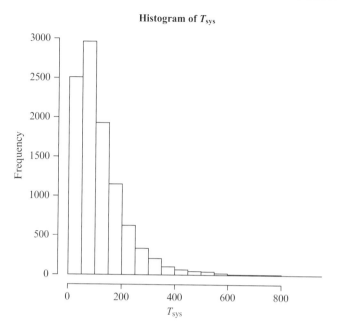

Fig. 4.33 A histogram of simulated values of T_{sys} in Example 4.60   ▪

4.10.5 Exercises: Section 4.10 (143–153)

143. Consider the service station scenario presented in Exercise 1 of this chapter.
    (a) Write a program to simulate the rvs (X, Y) described in that exercise.
    (b) Use your program to estimate $P(X \leq 1 \text{ and } Y \leq 1)$, and compare your estimate to the exact answer from the joint pmf. Use at least 10,000 simulation runs.
    (c) Define a new variable $D = |X - Y|$, the (absolute) difference in the number of hoses in use at the two gas pumps. Use your program (with at least 10,000 runs) to simulate D, and estimate both the mean and standard deviation of D.
144. Refer back to the quiz scenario of Exercise 24.
    (a) Write a program to simulate students' scores (X, Y) on the two parts of the quiz.
    (b) Use your program to estimate the probability that a student's total score is at least 20 points. How does your estimate compare to the exact answer from the joint pmf?

    (c) Define a new rv $M =$ the maximum of the two scores. Use your program
to simulate M, and estimate both the mean and standard deviation of M.

145. Consider the situation presented in Example 4.13: the joint pdf of the
amounts X and Y of almonds and cashews, respectively, in a 1-lb can of
nuts is

$$f(x, y) = \begin{cases} 24xy & 0 \le x \le 1, \quad 0 \le y \le 1, \quad x + y \le 1 \\ 0 & \text{otherwise} \end{cases}$$

    With the prices specified in that example, the total cost of the contents of
one can is $W = 3.5 + 2.5X + 6.5Y$.

    (a) Write a program implementing the accept–reject method of this section
to simulate (X, Y).

    (b) On the average, how many iterations will your program require to
generate 10,000 "accepted" (X, Y) pairs?

    (c) Use your program to simulate the rv W. Create a histogram of the
simulated values of W, and report estimates of the mean and standard
deviation of W. How close is your sample mean to the value $E(W) =$
\$7.10 determined in Example 4.13?

    (d) Use your simulation in part (c) to estimate the probability that the cost of
the contents of a can of nuts exceeds \$8.

146. Suppose a randomly chosen individual's verbal score X and quantitative
score Y on a nationally administered aptitude examination, each scaled down
to [0, 1], have joint pdf

$$f(x, y) = \begin{cases} \dfrac{2}{5}(2x + 3y) & 0 \le x \le 1, \quad 0 \le y \le 1 \\ 0 & \text{otherwise} \end{cases}$$

    (a) Write a program implementing the accept–reject method of this section
to simulate (X, Y).

    (b) The engineering school at a certain university uses a weighted total
$T = 3X + 7Y$ as part of its admission process. Use your program in part
(a) to simulate the rv T, and estimate $P(T \ge 9)$.

    (c) Suppose the engineering school decides to only admit students whose
weighted totals are above the 85th percentile for the national distribution. That is, if $\eta_{.85}$ is the 85th percentile of the distribution of T, a
student's weighted total must exceed $\eta_{.85}$ for admission. Use your
simulated values of T from part (b) to estimate $\eta_{.85}$. [*Hint:* $\eta_{.85}$ separates
the bottom 85% of the T distribution from the remaining 15%. What
value separates the lowest 85% of *simulated* T values from the rest?]

147. Refer back to Exercise 145.

    (a) Determine the marginal pdf of X and the conditional pdf of Y given
$X = x$.

(b) Write a program to simulate (X, Y) using the conditional distributions method presented in this section.

(c) What advantage does this method have over the accept–reject approach used in Exercise 145?

148. Consider the situation in Example 4.31: the proportion P of tiles meeting thermal specifications varies according to the pdf $f(p) = 9p^8$, $0 < p < 1$; conditional on $P = p$, the number of inspected tiles that meet specifications is a rv $Y \sim \text{Bin}(20, p)$.

(a) Write a program to simulate Y. Your program will first need to simulate a value of P, and then generate a variate from the appropriate binomial distribution. [*Hint:* Use your software's built-in binomial simulation tool.]

(b) Simulate (at least) 10,000 values of Y, and report estimates of both $E(Y)$ and $\text{Var}(Y)$. How do these compare to the exact answers found in Example 4.31?

(c) Use your simulation to estimate both $P(Y = 18)$ and $P(Y \geq 18)$.

149. The conditional distributions method described in this section can also be applied to joint discrete rvs. Refer back to the joint pmf presented in this section, which is originally from Example 4.1.

(a) Determine the marginal pmf of X. (This should be very easy.)

(b) Determine the conditional pmfs of Y given $X = 100$ and given $X = 250$.

(c) Write a program that first simulates X using its marginal pmf, then simulates Y via the appropriate conditional pmf. [*Hint:* For each stage, use your program's built-in discrete simulator (randsample in Matlab, sample in R).]

(d) Use your program in part (c) to simulate at least 10,000 (X, Y) pairs. Verify that the relative frequencies of the six possible pairs in your sample are close to the probabilities specified in the original joint pmf table.

150. Refer back to Exercise 113, which specifies a bivariate normal distribution for the rvs $X =$ height (inches) and $Y =$ weight (lbs) for American males. The parameters of that model were $\mu_1 = 70$, $\sigma_1 = 3$, $\mu_2 = 170$, $\sigma_2 = 20$, and $\rho = .9$.

(a) Use your software's built-in multivariate normal simulation function to generate (at least) 10,000 (X, Y) pairs according to this bivariate normal model.

(b) A person's body-mass index (BMI) is determined by the formula $703Y/X^2$. Use the result of part (a) to create a histogram of BMIs for the population of American males.

(c) BMI scores between 18.5 and 25 are considered healthy. By that criterion, what proportion of American males are healthy? Report both an estimate of this proportion and its estimated standard error.

151. The conditional distributions method of this section can be implemented to simulate a bivariate normal distribution, providing an alternative to built-in

multivariate simulation tools or Expression (4.10). Let X and Y have a bivariate normal distribution with parameters $\mu_1, \sigma_1, \mu_2, \sigma_2$, and ρ.

(a) What are the marginal distribution of X and the conditional distribution of Y given $X = x$? [*Hint:* Refer back to Sect. 4.7.]

(b) Write a program to simulate (X, Y) values from a bivariate normal distribution by first simulating X and then $Y|X = x$. The inputs to your program should be the five parameters and the desired number of simulated values; the outputs should be vectors containing the simulated values of X and Y.

(c) Use your program to simulate the height-weight distribution from the previous exercise. Verify that the sample mean and standard deviation of your simulated Y values are roughly 170 and 20, respectively.

152. Consider the system design illustrated in Exercise 126. Suppose that components 1, 2, and 3 have exponential lifetimes with mean 250 h, while components 4, 5, and 6 have exponential lifetimes with mean 300 h.

(a) Write a program to simulate the lifetime of the system.

(b) Let μ denote the true mean system lifetime. Provide an estimate of μ, along with its estimated standard error.

(c) Let p denote the true probability that the system lasts more than 200 h. Provide an estimate of p, along with its estimated standard error.

153. Consider the system design illustrated in Exercise 127. Suppose the odd-numbered components have exponential lifetimes with mean 250 h, while the even-numbered components have gamma lifetime distributions with $\alpha = 2$ and $\beta = 125$. (This second distribution also has mean 250 h.)

(a) Write a program to simulate the lifetime of the system. [You might want to use your software's built-in gamma random number generator.]

(b) Let μ denote the true mean system lifetime. Provide an estimate of μ, along with its estimated standard error.

(c) Let p denote the true probability that the system fails prior to 400 h. Provide an estimate of p, along with its estimated standard error.

4.11   Supplementary Exercises (154–192)

154. Suppose the amount of rainfall in one region during a particular month has an exponential distribution with mean value 3 in., the amount of rainfall in a second region during that same month has an exponential distribution with mean value 2 in., and the two amounts are independent of each other. What is the probability that the second region gets more rainfall during this month than does the first region?

155. Two messages are to be sent. The time (min) necessary to send each message has an exponential distribution with parameter $\lambda = 1$, and the two times are independent of each other. It costs \$2 per minute to send the first message and \$1 per minute to send the second. Obtain the density function of the total

cost of sending the two messages. [*Hint*: First obtain the cumulative distribution function of the total cost, which involves integrating the joint pdf.]

156. A restaurant serves three fixed-price dinners costing $20, $25, and $30. For a randomly selected couple dining at this restaurant, let $X =$ the cost of the man's dinner and $Y =$ the cost of the woman's dinner. The joint pmf of X and Y is given in the following table:

			y	
	$p(x, y)$	20	25	30
	20	.05	.05	.10
x	25	.05	.10	.35
	30	0	.20	.10

(a) Compute the marginal pmfs of X and Y.
(b) What is the probability that the man's and the woman's dinner cost at most $25 each?
(c) Are X and Y independent? Justify your answer.
(d) What is the expected total cost of the dinner for the two people?
(e) Suppose that when a couple opens fortune cookies at the conclusion of the meal, they find the message "You will receive as a refund the difference between the cost of the more expensive and the less expensive meal that you have chosen." How much does the restaurant expect to refund?

157. A health-food store stocks two different brands of a type of grain. Let $X =$ the amount (lb) of brand A on hand and $Y =$ the amount of brand B on hand. Suppose the joint pdf of X and Y is

$$f(x, y) = \begin{cases} kxy & x \geq 0, \quad y \geq 0, \quad 20 \leq x + y \leq 30 \\ 0 & \text{otherwise} \end{cases}$$

(a) Draw the region of positive density and determine the value of k.
(b) Are X and Y independent? Answer by first deriving the marginal pdf of each variable.
(c) Compute $P(X + Y \leq 25)$.
(d) What is the expected total amount of this grain on hand?
(e) Compute $\text{Cov}(X, Y)$ and $\text{Corr}(X, Y)$.
(f) What is the variance of the total amount of grain on hand?

158. Let $X_1, X_2, \ldots, X_n$ be random variables denoting n independent bids for an item that is for sale. Suppose each X_i is uniformly distributed on the interval $[100, 200]$. If the seller sells to the highest bidder, how much can he expect to earn on the sale? [*Hint*: Let $Y = \max(X_1, X_2, \ldots, X_n)$. Use the results of Sect. 4.9 to find $E(Y)$.]

159. Suppose a randomly chosen individual's verbal score X and quantitative score Y on a nationally administered aptitude examination have joint pdf

$$f(x,y) = \begin{cases} \dfrac{2}{5}(2x + 3y) & 0 \le x \le 1 \quad 0 \le y \le 1 \\ 0 & \text{otherwise} \end{cases}$$

You are asked to provide a prediction t of the individual's total score $X + Y$. The error of prediction is the mean squared error $E[(X + Y - t)^2]$. What value of t minimizes the error of prediction?

160. Let X_1 and X_2 be quantitative and verbal scores on one aptitude exam, and let Y_1 and Y_2 be the corresponding scores on another exam. If $\text{Cov}(X_1, Y_1) = 5$, $\text{Cov}(X_1, Y_2) = 1$, $\text{Cov}(X_2, Y_1) = 2$, and $\text{Cov}(X_2, Y_2) = 8$, what is the covariance between the two total scores $X_1 + X_2$ and $Y_1 + Y_2$?

161. Let Z_1 and Z_2 be independent standard normal rvs and let

$$W_1 = Z_1 \qquad W_2 = \rho \cdot Z_1 + \sqrt{1 - \rho^2} Z_2$$

(a) By definition, W_1 has mean 0 and standard deviation 1. Show that the same is true for W_2.
(b) Use the properties of covariance to show that $\text{Cov}(W_1, W_2) = \rho$.
(c) Show that $\text{Corr}(W_1, W_2) = \rho$.

162. You are driving on a highway at speed X_1. Cars entering this highway after you travel at speeds $X_2, X_3, \ldots$. Suppose these X_is are independent and identically distributed. Unfortunately there is no way for a faster car to pass a slower one—it will catch up to the slower one and then travel at the same speed. For example, if $X_1 = 52.3, X_2 = 37.5$, and $X_3 = 42.8$, then no car will catch up to yours, but the third car will catch up to the second. Let $N =$ the number of cars that ultimately travel at your speed (in your "cohort"), including your own car. Possible values of N are 1, 2, 3, $\ldots$. Show that the pmf of N is $p(n) = 1/[n(n+1)]$, and then determine the expected number of cars in your cohort. [*Hint*: $N = 3$ requires that $X_1 < X_2$, $X_1 < X_3, X_4 < X_1$.]

163. Suppose the number of children born to an individual has pmf $p(x)$. A *Galton–Watson branching process* unfolds as follows: At time $t = 0$, the population consists of a single individual. Just prior to time $t = 1$, this individual gives birth to X_1 individuals according to the pmf $p(x)$, so there are X_1 individuals in the first generation. Just prior to time $t = 2$, each of these X_1 individuals gives birth independently of the others according to the pmf $p(x)$, resulting in X_2 individuals in the second generation (e.g., if $X_1 = 3$, then $X_2 = Y_1 + Y_2 + Y_3$, where Y_i is the number of progeny of the ith individual in the first generation). This process then continues to yield a third generation of size X_3, and so on.
(a) If $X_1 = 3, Y_1 = 4, Y_2 = 0, Y_3 = 1$, draw a tree diagram with two generations of branches to represent this situation.
(b) Let A be the event that the process ultimately becomes extinct (one way for A to occur would be to have $X_1 = 3$ with none of these three second-generation individuals having any progeny) and let $p^* = P(A)$. Argue that p^* satisfies the equation

$$p* = \sum (p*)^x \cdot p(x)$$

[*Hint*: $A = \bigcup_{x=0}^{\infty} (A \cap X_1 = x)$, so the Law of Total Probability can be applied. Now given that $X_1 = 3$, A will occur if and only if each of the three separate branching processes starting from the first generation ultimately becomes extinct; what is the probability of this happening?]

(c) Verify that one solution to the equation in (b) is $p* = 1$. It can be shown that this equation has just one other solution, and that the probability of ultimate extinction is in fact the *smaller* of the two roots. If $p(0) = .3$, $p(1) = .5$, and $p(2) = .2$, what is $p*$? Is this consistent with the value of μ, the expected number of progeny from a single individual? What happens if $p(0) = .2$, $p(1) = .5$, and $p(2) = .3$?

164. Let $f(x)$ and $g(y)$ be pdfs with corresponding cdfs $F(x)$ and $G(y)$, respectively. With c denoting a numerical constant satisfying $|c| \le 1$, consider

$$f(x, y) = f(x)g(y)\{1 + c[2F(x) - 1][2G(y) - 1]\}$$

(a) Show that $f(x, y)$ satisfies the conditions necessary to specify a joint pdf for two continuous rvs.
(b) What is the marginal pdf of the first variable X? Of the second variable Y?
(c) For what values of c are X and Y independent?
(d) If $f(x)$ and $g(y)$ are normal pdfs, is the joint distribution of X and Y bivariate normal?

165. The **joint cumulative distribution function** of two random variables X and Y, denoted by $F(x, y)$, is defined by

$$F(x, y) = P[(X \le x) \cap (Y \le y)] \qquad -\infty < x < \infty, \qquad -\infty < y < \infty$$

(a) Suppose that X and Y are both continuous variables. Once the joint cdf is available, explain how it can be used to determine $P((X, Y) \in A)$, where A is the rectangular region $\{(x, y): a \le x \le b, c \le y \le d\}$.
(b) Suppose the only possible values of X and Y are 0, 1, 2, ... and consider the values $a = 5$, $b = 10$, $c = 2$, and $d = 6$ for the rectangle specified in (a). Describe how you would use the joint cdf to calculate the probability that the pair (X, Y) falls in the rectangle. More generally, how can the rectangular probability be calculated from the joint cdf if a, b, c, and d are all integers?
(c) Determine the joint cdf for the scenario of Example 4.1. [*Hint*: First determine $F(x, y)$ for $x = 100, 250$ and $y = 0, 100$, and 200. Then describe the joint cdf for various other (x, y) pairs.]
(d) Determine the joint cdf for the scenario of Example 4.3 and use it to calculate the probability that X and Y are both between .25 and .75. [*Hint*: For $0 \le x \le 1$ and $0 \le y \le 1$, $F(x, y) = \int_0^x \int_0^y f(u, v) dv du$.]
(e) Determine the joint cdf for the scenario of Example 4.4. [*Hint*: Proceed as in (d), but be careful about the order of integration and consider separately (x, y) points that lie inside the triangular region of positive density and then points that lie outside this region.]

166. A circular sampling region with radius X is chosen by a biologist, where X has an exponential distribution with mean value 10 ft. Plants of a certain type occur in this region according to a (spatial) Poisson process with "rate" .5 plant per square foot. Let Y denote the number of plants in the region.
   (a) Find $E(Y| X = x)$ and $\text{Var}(Y| X = x)$.
   (b) Use part (a) to find $E(Y)$.
   (c) Use part (a) to find $\text{Var}(Y)$.

167. The number of individuals arriving at a post office to mail packages during a certain period is a Poisson random variable X with mean value 20. Independently of each other, any particular customer will mail either 1, 2, 3, or 4 packages with probabilities .4, .3, .2, and .1, respectively. Let Y denote the total number of packages mailed during this time period.
   (a) Find $E(Y| X = x)$ and $\text{Var}(Y| X = x)$.
   (b) Use part (a) to find $E(Y)$.
   (c) Use part (a) to find $\text{Var}(Y)$.

168. Sandstone is mined from two different quarries. Let $X =$ the amount mined (in tons) from the first quarry in one day and $Y =$ the amount mined (in tons) from the second quarry in one day. The variables X and Y are independent, with $\mu_X = 12$, $\sigma_X = 4$, $\mu_Y = 10$, $\sigma_Y = 3$.
   (a) Find the mean and standard deviation of the variable $X + Y$, the total amount of sandstone mined in a day.
   (b) Find the mean and standard deviation of the variable $X - Y$, the difference in the mines' outputs in a day.
   (c) The manager of the first quarry sells sandstone at \$25/t, while the manager of the second quarry sells sandstone at \$28/t. Find the mean and standard deviation for the combined amount of money the quarries generate in a day.
   (d) Assuming X and Y are both normally distributed, find the probability the quarries generate more than \$750 revenue in a day.

169. The article "Stochastic Modeling for Pavement Warranty Cost Estimation" (*J. of Constr. Engr. and Mgmt.*, 2009: 352 –359) proposes the following model for the distribution of $Y =$ time to pavement failure. Let X_1 be the time to failure due to rutting, and X_2 be the time to failure due to transverse cracking; these two rvs are assumed independent. Then $Y = \min(X_1, X_2)$. The probability of failure due to either one of these distress modes is assumed to be an increasing function of time t. After making certain distributional assumptions, the following form of the cdf for each mode is obtained:

$$\Phi\left(\frac{a + bt}{\sqrt{c + dt + et^2}}\right)$$

where Φ is the standard normal cdf. Values of the five parameters a, b, c, d, and e are –25.49, 1.15, 4.45, –1.78, and .171 for cracking and –21.27, .0325, .972, –.00028, and .00022 for rutting. Determine the probability of pavement failure within $t = 5$ years and also $t = 10$ years.

170. Consider a sealed-bid auction in which each of the n bidders has his/her valuation (assessment of inherent worth) of the item being auctioned. The valuation of any particular bidder is not known to the other bidders. Suppose these valuations constitute a random sample $X_1, \ldots, X_n$ with corresponding order statistics $Y_1 \le Y_2 \le \cdots \le Y_n$. The *rent* of the winning bidder is the difference between the winner's valuation and the price. The article "Mean Sample Spacings, Sample Size and Variability in an Auction-Theoretic Framework" (*Oper. Res. Lett.*, 2004: 103–108) argues that the rent is just $Y_n - Y_{n-1}$ (do you see why?).
    (a) Suppose that the valuation distribution is uniform on [0, 100]. What is the expected rent when there are $n = 10$ bidders?
    (b) Referring back to (a), what happens when there are 11 bidders? More generally, what is the relationship between the expected rent for n bidders and for $n + 1$ bidders? Is this intuitive? [*Note*: The cited article presents a counterexample.]

171. Suppose two identical components are connected in parallel, so the system continues to function as long as at least one of the components does so. The two lifetimes are independent of each other, each having an exponential distribution with mean 1000 h. Let W denote system lifetime. Obtain the moment generating function of W, and use it to calculate the expected lifetime.

172. Let Y_0 denote the initial price of a particular security and Y_n denote the price at the end of n additional weeks for $n = 1, 2, 3, \ldots$. Assume that the successive price ratios $Y_1/Y_0, Y_2/Y_1, Y_3/Y_2, \ldots$ are independent of one another and that each ratio has a lognormal distribution with $\mu = .4$ and $\sigma = .8$ (the assumptions of independence and lognormality are common in such scenarios).
    (a) Calculate the probability that the security price will increase over the course of a week.
    (b) Calculate the probability that the security price will be higher at the end of the next week, be lower the week after that, and then be higher again at the end of the following week. [*Hint*: What does "higher" say about the ratio Y_{i+1}/Y_i?]
    (c) Calculate the probability that the security price will have increased by at least 20% over the course of a five-week period. [*Hint*: Consider the ratio Y_5/Y_0, and write this in terms of successive ratios Y_{i+1}/Y_i.]

173. In cost estimation, the total cost of a project is the sum of component task costs. Each of these costs is a random variable with a probability distribution. It is customary to obtain information about the total cost distribution by adding together characteristics of the individual component cost distributions—this is called the "roll-up" procedure. For example, $E(X_1 + \cdots + X_n) = E(X_1) + \cdots + E(X_n)$, so the roll-up procedure is valid for mean cost. Suppose that there are two component tasks and that X_1 and X_2 are independent, normally distributed random variables. Is the roll-up procedure valid for the 75th percentile? That is, is the 75th percentile of the distribution of $X_1 + X_2$ the same as the sum of the 75th percentiles of the two individual

distributions? If not, what is the relationship between the percentile of the sum and the sum of percentiles? For what percentiles is the roll-up procedure valid in this case?

174. Suppose that for a certain individual, calorie intake at breakfast is a random variable with expected value 500 and standard deviation 50, calorie intake at lunch is random with expected value 900 and standard deviation 100, and calorie intake at dinner is a random variable with expected value 2000 and standard deviation 180. Assuming that intakes at different meals are independent of each other, what is the probability that average calorie intake per day over the next (365-day) year is at most 3500? [*Hint*: Let X_i, Y_i, and Z_i denote the three calorie intakes on day i. Then total intake is given by $\sum(X_i + Y_i + Z_i)$.]

175. The mean weight of luggage checked by a randomly selected tourist-class passenger flying between two cities on a certain airline is 40 lb, and the standard deviation is 10 lb. The mean and standard deviation for a business-class passenger are 30 lb and 6 lb, respectively.

   (a) If there are 12 business-class passengers and 50 tourist-class passengers on a particular flight, what are the expected value of total luggage weight and the standard deviation of total luggage weight?

   (b) If individual luggage weights are independent, normally distributed rvs, what is the probability that total luggage weight is at most 2500 lb?

176. *Random sums.* If $X_1, X_2, \ldots, X_n$ are independent rvs, each with the same mean value μ and variance σ^2, then we have seen that $E(X_1 + X_2 + \cdots + X_n) = n\mu$ and $\text{Var}(X_1 + X_2 + \cdots + X_n) = n\sigma^2$. In some applications, the number of X_is under consideration is not a fixed number n but instead a rv N. For example, let N be the number of components of a certain type brought into a repair shop on a particular day and let X_i represent the repair time for the ith component. Then the total repair time is $T_N = X_1 + X_2 + \cdots + X_N$, the sum of a *random* number of rvs.

   (a) Suppose that N is independent of the X_is. Use the Law of Total Expectation to obtain an expression for $E(T_N)$ in terms of μ and $E(N)$.

   (b) Use the Law of Total Variance to obtain an expression for $\text{Var}(T_N)$ in terms of μ, σ^2, $E(N)$, and $\text{Var}(N)$.

   (c) Customers submit orders for stock purchases at a certain online site according to a Poisson process with a rate of 3 per hour. The amount purchased by any particular customer (in thousands of dollars) has an exponential distribution with mean 30. What is the expected total amount ($) purchased during a particular 4-h period, and what is the standard deviation of this total amount?

177. Suppose the proportion of rural voters in a certain state who favor a particular gubernatorial candidate is .45 and the proportion of suburban and urban voters favoring the candidate is .60. If a sample of 200 rural voters and 300 urban and suburban voters is obtained, what is the approximate probability that at least 250 of these voters favor this candidate?

178. Let μ denote the true pH of a chemical compound. A sequence of n independent sample pH determinations will be made. Suppose each sample pH is a random variable with expected value μ and standard deviation .1. How many determinations are required if we wish the probability that the sample average is within .02 of the true pH to be at least .95? What theorem justifies your probability calculation?

179. The amount of soft drink that Ann consumes on any given day is independent of consumption on any other day and is normally distributed with $\mu = 13$ oz and $\sigma = 2$. If she currently has two six-packs of 16-oz bottles, what is the probability that she still has some soft drink left at the end of 2 weeks (14 days)? Why should we worry about the validity of the independence assumption here?

180. A large university has 500 single employees who are covered by its dental plan. Suppose the number of claims filed during the next year by such an employee is a Poisson rv with mean value 2.3. Assuming that the number of claims filed by any such employee is independent of the number filed by any other employee, what is the approximate probability that the total number of claims filed is at least 1200?

181. A student has a class that is supposed to end at 9:00 a.m. and another that is supposed to begin at 9:10 a.m. Suppose the actual ending time of the 9 a.m. class is a normally distributed rv X_1 with mean 9:02 and standard deviation 1.5 min and that the starting time of the next class is also a normally distributed rv X_2 with mean 9:10 and standard deviation 1 min. Suppose also that the time necessary to get from one classroom to the other is a normally distributed rv X_3 with mean 6 min and standard deviation 1 min. What is the probability that the student makes it to the second class before the lecture starts? (Assume independence of X_1, X_2, and X_3, which is reasonable if the student pays no attention to the finishing time of the first class.)

182. This exercise provides an alternative approach to establishing the properties of correlation.
   (a) Use the general formula for the variance of a linear combination to write an expression for $\mathrm{Var}(aX + Y)$. Then let $a = \sigma_Y / \sigma_X$, and show that $\rho \geq -1$. [Hint: Variance is always ≥ 0, and $\mathrm{Cov}(X, Y) = \sigma_X \cdot \sigma_Y \cdot \rho$.]
   (b) By considering $\mathrm{Var}(aX - Y)$, conclude that $\rho \leq 1$.
   (c) Use the fact that $\mathrm{Var}(W) = 0$ only if W is a constant to show that $\rho = 1$ only if $Y = aX + b$.

183. A rock specimen from a particular area is randomly selected and weighed two different times. Let W denote the actual weight and X_1 and X_2 the two measured weights. Then $X_1 = W + E_1$ and $X_2 = W + E_2$, where E_1 and E_2 are the two measurement errors. Suppose that the E_is are independent of each other and of W and that $\mathrm{Var}(E_1) = \mathrm{Var}(E_2) = \sigma_E^2$.
   (a) Express ρ, the correlation coefficient between the two measured weights X_1 and X_2, in terms of σ_W^2, the variance of actual weight, and σ_X^2, the variance of measured weight.
   (b) Compute ρ when $\sigma_W = 1$ kg and $\sigma_E = .01$ kg.

184. Let A denote the percentage of one constituent in a randomly selected rock specimen, and let B denote the percentage of a second constituent in that same specimen. Suppose D and E are measurement errors in determining the values of A and B so that measured values are $X = A + D$ and $Y = B + E$, respectively. Assume that measurement errors are independent of each other and of actual values.

    (a) Show that

$$\text{Corr}(X, Y) = \text{Corr}(A, B) \cdot \sqrt{\text{Corr}(X_1, X_2)} \cdot \sqrt{\text{Corr}(Y_1, Y_2)}$$

    where X_1 and X_2 are replicate measurements on the value of A, and Y_1 and Y_2 are defined analogously with respect to B. What effect does the presence of measurement error have on the correlation?

    (b) What is the maximum value of $\text{Corr}(X, Y)$ when $\text{Corr}(X_1, X_2) = .8100$ and $\text{Corr}(Y_1, Y_2) = .9025$? Is this disturbing?

185. Let $X_1, \ldots, X_n$ be independent rvs with mean values $\mu_1, \ldots, \mu_n$ and variances $\sigma_1^2, \ldots, \sigma_n^2$. Consider a function $h(x_1, \ldots, x_n)$, and use it to define a new rv $Y = h(X_1, \ldots, X_n)$. Under rather general conditions on the h function, if the σ_is are all small relative to the corresponding μ_is, it can be shown that $E(Y) \approx h(\mu_1, \ldots, \mu_n)$ and

$$\text{Var}(Y) \approx \left(\frac{\partial h}{\partial x_1}\right)^2 \cdot \sigma_1^2 + \cdots + \left(\frac{\partial h}{\partial x_n}\right)^2 \cdot \sigma_n^2$$

    where each partial derivative is evaluated at $(x_1, \ldots, x_n) = (\mu_1, \ldots, \mu_n)$. Suppose three resistors with resistances X_1, X_2, X_3 are connected in parallel across a battery with voltage X_4. Then by Ohm's law, the current is

$$Y = X_4 \left(\frac{1}{X_1} + \frac{1}{X_2} + \frac{1}{X_3}\right)$$

    Let $\mu_1 = 10\,\Omega$, $\sigma_1 = 1.0\,\Omega$, $\mu_2 = 15\,\Omega$, $\sigma_2 = 1.0\,\Omega$, $\mu_3 = 20\,\Omega$, $\sigma_3 = 1.5\,\Omega$, $\mu_4 = 120$ V, $\sigma_4 = 4.0$ V. Calculate the approximate expected value and standard deviation of the current (suggested by "Random Samplings," *CHEMTECH*, 1984: 696–697).

186. A more accurate approximation to $E[h(X_1, \ldots, X_n)]$ in the previous exercise is

$$h(\mu_1, \ldots, \mu_n) + \frac{1}{2}\sigma_1^2\left(\frac{\partial^2 h}{\partial x_1^2}\right) + \cdots + \frac{1}{2}\sigma_n^2\left(\frac{\partial^2 h}{\partial x_n^2}\right)$$

    Compute this for $Y = h(X_1, X_2, X_3, X_4)$ given in the previous exercise, and compare it to the leading term $h(\mu_1, \ldots, \mu_n)$.

187. Let Y_1 and Y_n be the smallest and largest order statistics, respectively, from a random sample of size n.

(a) Use the result of Exercise 141 to determine the joint pdf of Y_1 and Y_n. (Your answer will include the pdf f and cdf F of the original random sample.)

(b) Let $W_1 = Y_1$ and $W_2 = Y_n - Y_1$ (the latter is the sample range). Use the method of Sect. 4.6 to obtain the joint pdf of W_1 and W_2, and then derive an expression involving an integral for the pdf of the sample range.

(c) For the case in which the random sample is from a uniform distribution on $[0, 1]$, carry out the integration of (b) to obtain an explicit formula for the pdf of the sample range. [*Hint:* For the Unif$[0, 1]$ distribution, what are f and F?]

188. Consider independent and identically distributed random variables X_1, X_2, X_3, ... where each X_i has a discrete uniform distribution on the integers 0, 1, 2, ..., 9; that is, $P(X_i = k) = 1/10$ for $k = 0, 1, 2, \ldots, 9$. Now form the sum

$$U_n = \sum_{i=1}^{n} \frac{1}{(10)^i} X_i = .1X_1 + .01X_2 + \cdots + (.1)^n X_n.$$

Intuitively, this is just the first n digits in the decimal expansion of a random number on the interval $[0, 1]$. Show that as $n \to \infty$, $P(U_n \leq u) \to P(U \leq u)$ where $U \sim$ Unif$[0, 1]$ (this is called *convergence in distribution*, the type of convergence involved in the CLT) by showing that the moment generating function of U_n converges to the moment generating function of U.

[The argument for this appears on p. 52 of the article "A Few Counter Examples Useful in Teaching Central Limit Theorems," *The American Statistician*, Feb. 2013.]

189. The following example is based on "Conditional Moments and Independence" (*The American Statistician*, 2008: 219). Consider the following joint pdf of two rvs X and Y:

$$f(x,y) = \frac{1}{2\pi} \frac{e^{-\left[(\ln x)^2 + (\ln y)^2\right]/2}}{xy} [1 + \sin(2\pi\ln x)\sin(2\pi\ln y)] \quad \text{for } x > 0, \ y > 0$$

(a) Show that the marginal distribution of each rv is lognormal. [*Hint:* When obtaining the marginal pdf of X, make the change of variable $u = \ln(y)$.]

(b) Obtain the conditional pdf of Y given that $X = x$. Then show that for every positive integer n, $E(Y^n | X = x) = E(Y^n)$. [*Hint:* Make the change of variable $\ln(y) = u + n$ in the second integrand.]

(c) Redo (b) with X and Y interchanged.

(d) The results of (b) and (c) suggest intuitively that X and Y are independent rvs. Are they in fact independent?

190. Let X_1, X_2, ... be a sequence of independent, but not necessarily identically distributed random variables, and let $T = X_1 + \cdots + X_n$. *Lyapunov's Theorem* states that the distribution of the standardized variable $(T - \mu_T)/\sigma_T$ converges to a $N(0, 1)$ distribution as $n \to \infty$, provided that

$$\lim_{n \to \infty} \frac{\sum_{i=1}^{n} E\left(|X_i - \mu_i|^3\right)}{\sigma_T^3} = 0$$

where $\mu_i = E(X_i)$. This limit is sometimes referred to as the *Lyapunov condition* for convergence.

(a) Assuming $E(X_i) = \mu_i$ and $\text{Var}(X_i) = \sigma_i^2$, write expressions for μ_T and σ_T.

(b) Show that the Lyapunov condition is automatically met when the X_is are iid. [*Hint:* Let $\tau = E(|X_i - \mu_i|^3)$, which we assume is finite, and observe that τ is the same for every X_i. Then simplify the limit.]

(c) Let X_1, X_2, ... be independent random variables, with X_i having an exponential distribution with mean i. Show that $X_1 + \cdots + X_n$ has an approximately normal distribution as n increases.

(d) An online trivia game presents progressively harder questions to players; specifically, the probability of answering the ith question correctly is $1/i$. Assume any player's successive answers are independent, and let T denote the number of questions a player has right out of the first n. Show that T has an approximately normal distribution for large n.

191. This exercise and the next complete our investigation of the Coupon Collector's Problem begun in the book's Introduction. A box of a certain brand of cereal marketed for children is equally likely to contain one of 10 different small toys. Suppose someone purchases boxes of this cereal one by one, stopping only when all 10 toys have been obtained.

(a) After obtaining a toy in the first box, let Y_2 be the subsequent number of boxes purchased until a toy different from the one in the first box is obtained. Argue that this rv has a geometric distribution, and determine its expected value.

(b) Let Y_3 be the number of additional boxes purchased to get a third type of toy once two types have been obtained. What kind of a distribution does this rv have, and what is its expected value?

(c) Analogous to Y_2 and Y_3, define Y_4, ..., Y_{10} as the numbers of additional boxes purchased to get a new type of toy. Express the total number of boxes purchased in terms of the Y_is and determine its expected value.

(d) Determine the standard deviation of the total number of boxes purchased. [*Hint:* The Y_is are independent.]

192. Return to the scenario described in the previous problem. Suppose an individual purchases 25 boxes of this cereal.

(a) Let $X_1 = 1$ if at least one type 1 toy is included in the 25 boxes and $X_1 = 0$ otherwise (a Bernoulli rv). Determine $E(X_1)$.

(b) Define X_2, ..., X_{10} analogously to X_1 for the other nine types of toys. Express the number of different toys obtained from the 25 boxes in terms of the X_is and determine its expected value.

(c) What happens to the expected value in (b) as the number of boxes purchased increases? As the number of different toys available increases?

(d) Show that, for $i \neq j$,

$$\mathrm{Cov}(X_i, X_j) = \left(\frac{8}{10}\right)^{25} - \left(\frac{9}{10}\right)^{50}.$$

Then determine the variance of the number of different toys obtained from the 25 boxes by applying Eq. (4.5) to the expression from part (b). [*Hint:* Refer back to Example 4.19 for the required method.]

The Basics of Statistical Inference

The overarching objective of *statistical inference* is to draw conclusions (make inferences) based on available sample data. In this chapter we generally assume that data have been acquired by observing the values of a random sample $X_1, X_2, \ldots, X_n$; recall from Sect. 4.5 that a random sample consists of rvs that are independent and have the same underlying probability distribution (what we also called iid). For example, highway fuel efficiency of a certain type of vehicle might have a normal distribution with mean μ and standard deviation σ. Then each observed fuel efficiency value would come from this normal distribution, with the various observed values obtained independently of one another—a normal random sample. Or the number of blemishes on a new type of DVD might have a Poisson distribution with mean value μ. If n of these disks were to be randomly selected and the number of blemishes on each one counted, the result would be data from a Poisson random sample. In either example, the values of the parameters would typically not be known to an investigator. The sample data would then be used to draw some type of conclusion about these values.

In this chapter we introduce several different inferential procedures. The first, *point estimation*, involves using the available data to obtain a single number that can be regarded as an educated guess for the value of some parameter ("point" refers to the fact that a single number corresponds to a single point on a number line). Thus we might offer up 31.2 mpg as a sensible estimate of population mean fuel efficiency, or 0.8 as an estimate of the true mean number of blemishes per DVD. Section 5.1 introduces some general concepts of point estimation and methods for assessing the quality of an estimate, while Sect. 5.2 discusses a popular method for producing point estimates.

A point estimate by itself, being a single number, does not provide any information as to how close the estimate might be to the value of the parameter being estimated. This deficiency can be remedied by calculating an entire set of plausible values for the parameter of interest, called a *confidence interval*. For example, it might be reported with a high degree of confidence—more precisely, a *confidence level* of 95%—that the true average breaking strength of hockey sticks made from a certain type of graphite-Kevlar composite is estimated to be between 459.5 and

M.A. Carlton and J.L. Devore, *Probability with Applications in Engineering, Science, and Technology*, Springer Texts in Statistics, DOI 10.1007/978-1-4939-0395-5_5, © Springer Science+Business Media New York 2014

466.2 N. Later in the chapter we consider confidence intervals for a population mean and also a population proportion (e.g., the proportion of all college students who regularly text during class).

Rather than estimating the value of some parameter, we may wish to decide which of two contradictory claims about the parameter is correct. Suppose, for example, that 1,000,000 signatures have been submitted in support of putting a particular initiative on a statewide ballot. State law requires that more than 500,000 of these signatures be valid. If we let p denote the proportion of valid signatures among those submitted, then the initiative qualifies if $p > .5$ and does not qualify if $p \leq .5$. Because it is extremely tedious and time consuming to check all one million signatures, it is customary to select a random sample, determine how many of those are valid, and then use the result as a basis for deciding between the two contradictory hypotheses $p > .5$ and $p \leq .5$. In this chapter we shall consider methods for "testing" hypotheses (that is, deciding which of two hypotheses is more plausible) about a population mean and also a population proportion.

Thus far our paradigm for inference has been to regard a parameter such as μ as having a fixed but unknown value. A different perspective, referred to as the *Bayesian* method, views any parameter whose value is unknown as being a random variable with some type of "prior" probability distribution. Once sample data is available, Bayes' theorem can be used to obtain the "posterior" distribution of the parameter conditional on the observed data. Adherents of the Bayesian method of inference then use this posterior distribution to draw some type of conclusion about the unknown parameter. The last section of this chapter introduces Bayesian methodology.

5.1   Point Estimation

Recall that a **parameter** is a numerical characteristic of a probability distribution. Often the distribution under consideration furnishes a model for how some variable is distributed in a population of interest. Examples include the distribution of yield strength values in a population of building-grade steel bars, or the distribution of time-to-recovery from a dental anesthetic in the conceptual population of all individuals given the anesthetic (we say "conceptual" here because the population includes both past and future recipients of the treatment). One parameter in the anesthetic scenario is the population mean recovery time μ, while in the steel bar population an investigator might focus on the parameter $\eta_{.95}$, the 95th percentile of the distribution (i.e., the yield strength that separates the strongest 5% of steel bars from the other 95%).

Statistical inference is frequently directed toward drawing some type of conclusions about one or more parameters. To do so requires that an investigator obtain sample data from the underlying distribution. If the sample consists of observations on some random variable X, we will denote the number of sample observations (the sample size) by n, the first observation by x_1, the second by x_2, and so on, with the last observation represented by x_n. The subscripts on x generally have no relationship to the magnitudes of the observations. They are often listed in the order in which they were acquired by an investigator. Conclusions (inferences) about the population distribution can then be based on the computed values of various sample quantities.

DEFINITION

A **statistic** is any random variable whose value can be computed from sample data.

Example 5.1 Zinfandel is a popular red wine varietal produced almost exclusively in California. It is rather controversial among wine connoisseurs because its alcohol content varies rather substantially from one producer to another. We went to the Web site klwines.com, randomly selected 10 from among the 325 available zinfandels, and obtained the following values of alcohol content (%):

$$x_1 = 14.8 \quad x_2 = 14.5 \quad x_3 = 16.1 \quad x_4 = 14.2 \quad x_5 = 15.9$$
$$x_6 = 13.7 \quad x_7 = 16.2 \quad x_8 = 14.6 \quad x_9 = 13.8 \quad x_{10} = 15.0$$

Here are examples of some statistics and their values calculated from the foregoing data:

(a) The **sample mean** $\overline{X}$, the arithmetic average of the n observations:

$$\overline{X} = \frac{X_1 + \cdots + X_n}{n} = \frac{\sum X_i}{n}$$

We encountered $\overline{X}$ previously in Sect. 4.5; this is the most frequently used measure of center for sample data. The calculated value of the sample mean for the given data is

$$\overline{x} = \frac{\sum x_i}{n} = \frac{14.8 + 14.5 + \ldots + 15.0}{10} = \frac{148.8}{10} = 14.88$$

Another sample of 10 such wines might yield $\overline{x} = 15.23$, and yet another give $\overline{x} = 14.70$. Prior to obtaining the data, there is uncertainty in what the value of the sample mean will be; hence we think of it as a random variable.

(b) The value of the sample mean can be unduly influenced by even a single unusually large or small observation, e.g., a sample of incomes that includes Bill Gates, or a sample of cities' populations that includes Shanghai. An alternative measure of center is the **sample median** $\tilde{X}$: list the n observations in increasing order from smallest to largest; then if n is an odd number, the median is the middle value in this ordered list [the $(n+1)/2$th value in from either end], and if n is even, the median is the average of the two middle values. Clearly several extreme values on either end of the ordered list will have no impact on the median. The ordered observations in our sample are

| 13.7 | 13.8 | 14.2 | 14.5 | 14.6 | 14.8 | 15.0 | 15.9 | 16.1 | 16.2 |

Because $n = 10$, the calculated value of the sample median is the average of the fifth and sixth largest values: $\tilde{x} = (14.6 + 14.8)/2 = 14.70$. Again, a second sample might result in $\tilde{x} = 15.15$, a third sample in $\tilde{x} = 14.95$, and so on. Prior to obtaining data, there is uncertainty in what value of the sample median will result, so the sample median is regarded as a random variable.

(c) The sample mean and sample median are both assessments of where the sample is centered—a typical or representative value. Another important characteristic of data is the extent to which the observations spread out about the center. The simplest measure of spread (dispersion, variability) is the **sample range W**: the difference between the largest and smallest observations. For our data, $w = 16.2 - 13.7 = 2.5$. A second sample might yield 14.1 and 15.8 as the smallest and largest observations, giving $w = 1.7$. Clearly the value of the sample range varies from one sample to another, so before data is available it is viewed as a random variable.

(d) In the context of simulation in Chaps. 2 and 3, we previously introduced another measure of variability, the **sample standard deviation S**:

$$S = \sqrt{\frac{1}{n-1} \sum_{i=1}^{n} (X_i - \bar{X})^2}$$

(The **sample variance** is defined as S^2.) For our data, the observed value of the sample standard deviation is

$$s = \sqrt{\frac{1}{10-1} \sum_{i=1}^{10} (x_i - 14.88)^2} = \sqrt{\frac{1}{9}\left[(14.8 - 14.88)^2 + \cdots + (15.0 - 14.88)^2\right]}$$

$$= 0.915$$

As with the previous examples of statistics, this value is particular to our data; a different random sample of 10 zinfandel wines might provide $s = 1.018$, another $s = 0.882$, and so on. Thus, prior to obtaining data, we regard S as a random variable.

(e) Finally, consider the random variable

$$Z = \frac{\bar{X} - \mu}{\sigma/\sqrt{n}},$$

which expresses the distance between the sample mean and its expected value μ in standard deviations (e.g., if $z = 3$, then the value of the sample mean is three standard deviations larger than would be expected). This rv is *not* a statistic unless the values of μ and σ are known; without those values, the sample does not provide enough information to calculate z. ∎

5.1.1   Estimates and Estimators

When discussing general concepts and methods of inference, it is convenient to have a generic symbol for the parameter of interest. We will use the Greek letter θ for this purpose. The objective of *point estimation* is to select a single number, based on sample data, that represents a sensible value for θ. Suppose, for example, that the parameter of interest is μ, the true average lifetime of batteries of a certain type. A random sample of $n = 3$ batteries might yield observed lifetimes (hours) $x_1 = 5.0$, $x_2 = 6.4$, $x_3 = 5.9$. The computed value of the sample mean lifetime is $\bar{x} = 5.77$, and it is reasonable to regard 5.77 h as a plausible value of μ, our "best guess" for the value of μ based on the available sample information.

> **DEFINITION**
>
> **A point estimate** of a parameter θ is a single number that can be regarded as a sensible value for θ. A point estimate is obtained by selecting a suitable statistic and computing its value from the given sample data. The selected statistic is called the **point estimator** of θ.

In the battery scenario just described, the point estimator (i.e., the statistic) used to obtain the point estimate of μ was $\bar{X}$, and the point estimate of μ was 5.77. If the three observed lifetimes had instead been $x_1 = 5.6$, $x_2 = 4.5$, and $x_3 = 6.1$, using the same *estimator* $\bar{X}$ would have resulted in a different *estimate*, $\bar{x} = (5.6 + 4.5 + 6.1)/3 = 5.40$ h.

The symbol $\hat{\theta}$ ("theta hat") is customarily used to denote the point estimate resulting from a given sample; we shall also use it to denote the estimator, as using an uppercase $\hat{\Theta}$ is somewhat awkward to write. Thus $\hat{\mu} = \bar{X}$ is read as "the point estimator of μ is the sample mean $\bar{X}$." The statement "the point estimate of μ is 5.77 h" can be written concisely as $\hat{\mu} = \bar{x} = 5.77$. Notice that in writing a statement such as $\hat{\theta} = 72.5$, there is no indication of how this point estimate was obtained (i.e., what statistic was used). It is recommended that both the estimator and the resulting estimate be reported.

Example 5.2   The National Health and Nutrition Examination Survey (NHANES) collects demographic, socioeconomic, dietary, and health-related information on an annual basis. Here is a sample of 20 observations on HDL-cholesterol level (mg/dl) obtained from the 2009–2010 survey (HDL is "good" cholesterol, and the higher the value, the lower the risk for heart disease):

35   49   52   54   65   51   51   47   86   36   46   33   39   45   39   63   95   35   30   48

Figure 5.1 shows both a normal probability plot and a brief descriptive summary of the data.

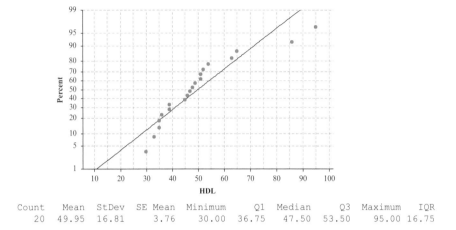

Count	Mean	StDev	SE Mean	Minimum	Q1	Median	Q3	Maximum	IQR
20	49.95	16.81	3.76	30.00	36.75	47.50	53.50	95.00	16.75

Fig. 5.1 Normal probability plot and descriptive summary of the HDL sample

(a) Let's first consider estimating the population mean HDL level μ. The natural estimator is of course the sample mean $\overline{X}$. The resulting point estimate is

$$\hat{\mu} = \overline{x} = \frac{\sum x_i}{n} = \frac{35 + 49 + \cdots + 48}{20} = 49.95$$

The NHANES data file contained 7846 HDL observations. We could regard our sample of size 20 as coming from the *population* consisting of these 7846 values. The population mean is then known to be $\mu = 52.6$ mg/dl, so our estimate of 49.95 is somewhat smaller than the value of the parameter we are trying to estimate. We extracted a second sample of size 20 from the population; for this sample, $\hat{\mu} = \overline{x} = 57.40$, a substantial overestimate of μ.

(b) Now let's consider estimating the population median η, the value that separates the smallest 50% of all HDL levels in the population from the largest 50%. The natural statistic for estimating this parameter is the sample median $\widetilde{X}$ described previously. The estimate here is the average of the 10th and 11th values in the ordered list of sample observations:

$$\hat{\eta} = \widetilde{x} = \frac{47 + 48}{2} = 47.5 \text{ mg/dl}$$

This is somewhat smaller than the sample mean because the sample has somewhat of a positive skew—values on the upper end stretch out more than do values on the lower end, and these pull the mean rightward compared to the median. If for the moment we regard the NHANES data set as constituting the population, the population median is 51.0 (again somewhat smaller than the population mean because of a positive skew). Our estimate of 47.5 is also smaller than what we are attempting to estimate (51.0). For the second sample alluded to in part (a), the sample median was 57.0, an overestimate of the population median.

(c) To estimate the HDL population standard deviation σ, it is natural to use the sample standard deviation S as our point estimator. The resulting point estimate of σ is

$$\hat{\sigma} = s = \sqrt{\frac{1}{20-1}\left[(35-49.95)^2 + \ldots + (48-49.95)^2\right]} = 16.81$$

Roughly speaking, the sample SD describes the size of a typical deviation within the sample from the sample mean. A second sample from the same population would almost surely give a somewhat different value of s, and thus a different point estimate of σ.

(d) An HDL level of at least 60 mg/dl is considered desirable, as it corresponds to a significantly lower risk of heart disease. How can we estimate the proportion p of the population having an HDL level of at least 60? If we think of a sample observation of at least 60 as being a "success," then a natural estimator of p is the *sample* proportion of successes:

$$\hat{P} = \frac{\# \text{ of successes in the sample}}{n}$$

(We have encountered $\hat{P}$ several times already, both in the context of simulation and of the Central Limit Theorem.) Four of the 20 sample observations are at least 60. Thus our point estimate is $\hat{p} = 4/20 = .20$. That is, we estimate that 20% of the individuals in the population have an HDL level of at least 60. If a second sample is selected, it may be that 7 of the 20 individuals have such a level. Use of the same estimator then gives the point estimate $7/20 = .35$. Just as with the other estimators proposed in this example, the value of the estimator $\hat{P}$ will in general vary from one sample to another. ∎

The foregoing example may have suggested that point estimation is deceptively straightforward: once the parameter to be estimated is identified, use intuition to specify a suitable estimator (statistic) and then just calculate. However, there are at least two major problems with this strategy. The first is that intuition may not be up to the task of identifying an estimator. For example, suppose a materials engineer is willing to assume (based on subject matter expertise and an appropriate probability plot) that the data she collected were sampled from a Weibull distribution. This distribution has two parameters, α and β, which appear in the Weibull pdf in a rather complicated way. Furthermore, the mean μ and standard deviation σ both involve the gamma function. So the sample mean and sample SD estimate complicated functions of the two parameters; it is not at all obvious how to sensibly estimate α

and β. In the next section we introduce a constructive method for producing estimators that will generally be reliable.

The second problem with relying solely on intuition is that, in many situations, there are two or more estimators for a particular parameter that could sensibly be used. For example, suppose an investigator is quite convinced (again, by a combination of subject matter expertise and a probability plot) that available data was generated by a normal distribution. A major objective of the investigation is to estimate the parameter μ. Since μ is the mean value of the normal population distribution, it certainly makes sense to use the sample mean $\overline{X}$ as its estimator. However, because any normal density curve is symmetric, μ is also the median of the normal population distribution. It is then sensible to think of using the sample median $\widetilde{X}$ as an estimator. Two other potential estimators of μ are the *mid-range*, the average of the largest and smallest observations, and a *trimmed mean*, obtained by eliminating a specified percentage of the values from each end of the ordered list and averaging those values that remain.

As a second example of competing estimators, consider data resulting from a Poisson random sample. This distribution has one parameter, μ, which is both the mean and the variance of the Poisson model. So one sensible estimator of μ is the sample mean, another is the sample variance, and a third is the average of these two. The choice between competing estimators such as these cannot usually be based on intuitive reasoning. Instead we need to introduce desirable properties for an estimator and then try to find one that satisfies the properties.

5.1.2   Assessing Estimators: Accuracy and Precision

When a particular statistic is selected to estimate an unknown parameter, two criteria often used to assess the quality of that estimator are its accuracy and its precision. Loosely speaking, an estimator is *accurate* if it has no systematic tendency, across repeated values of the estimator calculated from different samples, to overestimate or underestimate the value of the parameter. An estimator is *precise* if those same repeated values are "close together," so that two statisticians using the same estimator formula (but two different random samples) are liable to get similar point estimates.

The notions of accuracy and precision are made more rigorous by the following definitions.

> **DEFINITION**
>
> A point estimator $\hat{\theta}$ is said to be an **unbiased estimator** of θ if $E(\hat{\theta}) = \theta$ for every possible value of θ. If $\hat{\theta}$ is not unbiased, the difference $E(\hat{\theta}) - \theta$ is called the **bias** of $\hat{\theta}$.

(continued)

The **standard error** of $\hat{\theta}$ is its standard deviation, $\sigma_{\hat{\theta}} = \mathrm{SD}(\hat{\theta})$. If the standard error itself involves unknown parameters whose values can be estimated, substitution of these estimates into $\sigma_{\hat{\theta}}$ yields the **estimated standard error** of $\hat{\theta}$. The estimated standard error can be denoted by either $\hat{\sigma}_{\hat{\theta}}$ or by $s_{\hat{\theta}}$.

The bias of an estimator $\hat{\theta}$ quantifies its accuracy be measuring how far, on the average, $\hat{\theta}$ differs from θ. The standard error of $\hat{\theta}$ quantifies its precision by measuring the variability of $\hat{\theta}$ across different possible realizations (i.e., different random samples). It is important to note that both bias and standard error are properties of an *estimator* (the random variable), such as $\overline{X}$, and not of any specific value or *estimate*, $\bar{x}$.

Figure 5.2 illustrates bias and standard error for three potential estimators of a population parameter θ. Figure 5.2a shows the distribution of an estimator $\hat{\theta}_1$ whose expected value is very close to θ but whose distribution is quite dispersed. Hence, $\hat{\theta}_1$ has low bias but relatively high standard error. In contrast, the distribution of $\hat{\theta}_2$ displayed in Fig. 5.2b is very concentrated but is "off target": the values of $\hat{\theta}_2$ across different random samples will systematically over-estimate θ by a large amount. So, $\hat{\theta}_2$ has low standard error but high bias. The "ideal" estimator is illustrated in

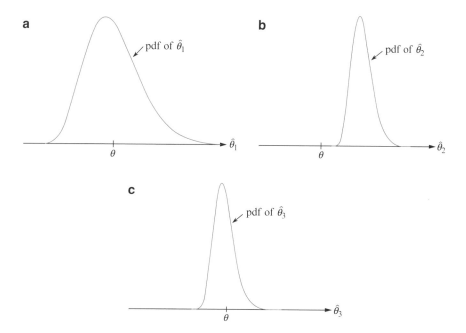

Fig. 5.2 Three potential types of estimators: (**a**) accurate, but not precise; (**b**) precise, but not accurate; (**c**) both accurate and precise

Fig. 5.2c: $\hat{\theta}_3$ has a mean roughly equal to θ, so it has low bias, and it also has a relatively small standard error.

Example 5.3 Consider the scenario of Example 5.1, wherein a sample mean $\overline{X}$ from a random sample of $n = 10$ observations will be used to estimate the population mean alcohol content μ of all zinfandel wines. In Sect. 4.5, we showed that the expected value and standard deviation of $\overline{X}$ are μ and $\sigma/\sqrt{n}$, respectively, where σ is the population standard deviation (i.e., the SD of the alcohol content of *all* zinfandel wines). Hence, the bias of $\overline{X}$ in estimating μ is

$$E(\overline{X}) - \mu = \mu - \mu = 0$$

That is, $\overline{X}$ is an unbiased estimator of μ. This is true for any random sample and for any sample size, n. The standard error of $\overline{X}$ is simply $\mathrm{SD}(\overline{X}) = \sigma/\sqrt{n} = \sigma/\sqrt{10}$; clearly the precision of $\overline{X}$ would be improved (i.e., the standard error reduced) by increasing the sample size n.

Since the value of σ is almost always unknown, we can estimate the standard error of $\overline{X}$ by $\hat{\sigma}_{\overline{X}} = s/\sqrt{n}$, where s denotes the sample standard deviation, as we did in the context of simulations in Sects. 2.8 and 3.8. For the random sample of 10 wines presented in Example 5.1, we have a point estimate $\hat{\mu} = \overline{x} = 14.88$ with an estimated standard error of $s/\sqrt{n} = 0.915/\sqrt{10} = 0.29$. The latter indicates that, based on the available data, we believe our estimate of μ is liable to differ by about ± 0.29 from the actual value of μ.   ∎

Example 5.4 Consider once again estimating a population proportion of "successes" p (for example, the proportion of all engineering graduates who have taken a statistics course, or the proportion of all vehicle accidents in which cell phone use was not a factor). The natural estimator of p is the sample proportion of successes $\hat{P} = X/n$, where X denotes the number of successes in the sample. Using the fact that $X \sim \mathrm{Bin}(n, p)$, we showed in Sect. 2.4 that the mean and standard error of $\hat{P}$ are

$$E(\hat{P}) = p \quad \text{and} \quad \mathrm{SD}(\hat{P}) = \sqrt{\frac{p(1-p)}{n}}$$

The first equation tells us that $\hat{P}$ is an unbiased estimator for p, and that this is true no matter the sample size. As for the standard error, since p is unknown (else why estimate?), we substitute $\hat{p} = x/n$ into $\sigma_{\hat{P}}$, yielding the estimated standard error $\hat{\sigma}_{\hat{P}} = \sqrt{\hat{p}(1-\hat{p})/n}$. This was used several times in the context of simulation in earlier chapters, and we will see this expression again in Sect. 5.5. When $n = 25$ and $\hat{p} = .6$, this gives $\hat{\sigma}_{\hat{P}} = \sqrt{(.6)(.4)/25} = .098$. Alternatively, since the largest value of $p(1-p)$ is attained when $p=.5$, an upper bound on the standard error is $\sqrt{(.5)(.5)/n} = 1/(2\sqrt{n})$. ∎

Example 5.5 The time a customer spends in service after waiting in a queue is often modeled with an exponential distribution. Recall that the exponential model has a single parameter, λ, and that the mean of the exponential distribution is $1/\lambda$. Thus, since $\lambda = 1/\mu$, a reasonable estimator of λ might be

$$\hat{\lambda} = \frac{1}{\overline{X}}$$

where $\overline{X}$ is the average of a random sample of wait times $X_1, \ldots, X_n$ from the aforementioned single-server queue. How accurate is $\hat{\lambda}$ as an estimator of λ? How precise is it?

It can be shown (Exercise 11) that the mean and variance of $\hat{\lambda} = 1/\overline{X}$ are

$$E(\hat{\lambda}) = \frac{n\lambda}{n-1} \quad \text{and} \quad \text{Var}(\hat{\lambda}) = \frac{n^2\lambda^2}{(n-1)^2(n-2)}$$

The bias of $\hat{\lambda}$ as an estimator of λ is therefore $E(\hat{\lambda}) - \lambda = \lambda/(n-1)$. We see that $\hat{\lambda}$ is not an unbiased estimator of λ; since $\lambda/(n-1) > 0$, we say that $\hat{\lambda}$ is *biased high*, meaning it will tend to systematically over-estimate λ. Clearly the bias approaches 0 as n increases.

The standard error of $\hat{\lambda}$ is the square root of the variance expression above. It can be estimated by replacing the unknown λ with the calculated value of $\hat{\lambda}$, $1/\overline{x}$, resulting in

$$\hat{\sigma}_{\hat{\lambda}} = \sqrt{\frac{n^2\hat{\lambda}^2}{(n-1)^2(n-2)}} = \sqrt{\frac{n^2}{(n-1)^2(n-2)\overline{x}^2}}$$ ∎

As mentioned before, in some situations more than one estimator might be proposed for the same parameter. It is sometimes the case in such scenarios that one estimator is more accurate (lower bias) while the other is more precise (smaller standard error). Which consideration should prevail?

Principle of Unbiased Estimation
When choosing among several different estimators of θ, select one that is unbiased.

According to this principle, the sample mean $\overline{X}$ would be selected as an estimator of a population mean μ over any biased estimator (see Example 5.3), and a sample proportion $\hat{P}$ is preferred over any biased estimator of a true proportion p (Example

5.4). In contrast, the estimator $\hat{\lambda}$ of Example 5.5 is *not* unbiased; if we can find some other estimator of λ which is unbiased, we would choose this latter estimator over $\hat{\lambda}$.

If two or more estimators of a parameter are unbiased, then naturally one selects the estimator among them with the smallest standard error. For example, we previously proposed several different estimators for the mean μ of a normal distribution. When the sampled distribution is continuous and symmetric, all four of the proposed estimators—the sample mean $\overline{X}$, the sample median $\widetilde{X}$, the midrange, and a trimmed mean—are unbiased estimators of μ (provided μ is finite). Using some sophisticated mathematics, it can be shown that when drawing from a normal distribution, $\overline{X}$ has the smallest standard error not only among these four estimators but in fact among *all* unbiased estimators of μ. For this reason, $\overline{X}$ is referred to as the **minimum variance unbiased estimator** (MVUE) of μ when sampling from a normally distributed population.

An alternative approach to the Principle of Unbiased Estimation is to combine the considerations of accuracy (bias) and precision (standard error) into a single measure, which can be achieved through the *mean squared error*; see Exercise 22. Under this method, the estimator with the smallest mean squared error is selected, even if it is biased and other estimators are not.

5.1.3   Exercises: Section 5.1 (1–23)

1. A study of children's intelligence and behavior included the following IQ data for 33 first-graders that participated in the study.

82	96	99	102	103	103	106	107	108	108	108
108	109	110	110	111	113	113	113	113	115	115
118	118	119	121	122	122	127	132	136	140	146

(a) Calculate a point estimate of the mean IQ for the conceptual population of all first graders in this school, and state which estimator you used.

(b) Calculate a point estimate of the IQ value that separates the lowest 50% of all such students from the highest 50%, and state which estimator you used.

(c) Calculate and interpret a point estimate of the population standard deviation σ. Which estimator did you use?

(d) Calculate a point estimate of the proportion of all such students whose IQ exceeds 100. [*Hint*: Think of an observation as a "success" if it exceeds 100.]

(e) Calculate a point estimate of the *population coefficient of variation*, $100\sigma/\mu$, and state what estimator you used.

2. A sample of 20 students who had recently taken elementary statistics yielded the following information on brand of calculator owned (T = Texas Instruments, H = Hewlett-Packard, C = Casio, S = Sharp):

T	T	H	T	C	T	T	S	C	H
S	S	T	H	C	T	T	T	H	T

(a) Estimate the true proportion of all such students who own a Texas Instruments calculator.

(b) Of the 10 students who owned a TI calculator, 4 had graphing calculators. Estimate the proportion of students who do not own a TI graphing calculator.

3. Consider the following sample of observations on coating thickness for low-viscosity paint ("Achieving a Target Value for a Manufacturing Process: A Case Study," *J. Qual. Technol.*, 1992: 22–26):

.83	.88	.88	1.04	1.09	1.12	1.29	1.31
1.48	1.49	1.59	1.62	1.65	1.71	1.76	1.83

Assume that the distribution of coating thickness is normal (a normal probability plot strongly supports this assumption).

(a) Calculate a point estimate of the mean value of coating thickness, and state which estimator you used.

(b) Calculate a point estimate of the median of the coating thickness distribution, and state which estimator you used.

(c) Calculate a point estimate of the value that separates the largest 10% of all values in the thickness distribution from the remaining 90%, and state which estimator you used. [*Hint*: Express what you are trying to estimate in terms of μ and σ.]

(d) Estimate $P(X < 1.5)$, i.e., the proportion of all thickness values less than 1.5. [*Hint*: If you knew the values of μ and σ, you could calculate this probability. These values are not available, but they can be estimated.]

(e) What is the estimated standard error of the estimator that you used in (b)?

4. The data set mentioned in Exercise 1 also includes these third-grade IQ observations for males:

117	103	121	112	120	132	113	117	132
149	125	131	136	107	108	113	136	114

and females:

114	102	113	131	124	117	120	90
114	109	102	114	127	127	103	

Prior to obtaining data, denote the male values by $X_1, \ldots, X_m$ and the female values by $Y_1, \ldots, Y_n$. Suppose that the X_is constitute a random sample from a distribution with mean μ_1 and standard deviation σ_1 and that the Y_is form a random sample (independent of the X_is) from another distribution with mean μ_2 and standard deviation σ_2.

(a) Show that $\overline{X} - \overline{Y}$ is an unbiased estimator of $\mu_1 - \mu_2$. Then calculate the estimate for the given data.

(b) Use rules of variance from Chap. 4 to obtain an expression for the standard error of the estimator in (a), and then compute the estimated standard error.

(c) Calculate a point estimate of the ratio σ_1/σ_2 of the two standard deviations.

(d) Suppose one male third-grader and one female third-grader are randomly selected. Calculate a point estimate of the variance of the difference $X - Y$ between their IQs.

5. As an example of a situation in which several different statistics could reasonably be used to calculate a point estimate, consider a population of N invoices. Associated with each invoice is its "book value," the recorded amount of that invoice. Let τ denote the total book value, a known amount. Some of these book values are erroneous. An audit will be carried out by randomly selecting n invoices and determining the audited (correct) value for each one. Suppose that the sample gives the following results (in dollars).

	Invoice				
	1	2	3	4	5
Book value	300	720	526	200	127
Audited value	300	520	526	200	157
Error	0	200	0	0	−30

Let $\overline{X}$ = the sample mean book value, $\overline{Y}$ = the sample mean audited value, and $\overline{D}$ = the sample mean error. Propose three different statistics for estimating the total audited (i.e., correct) value θ—one involving just N and $\overline{X}$, another involving N, τ, and $\overline{D}$, and the last involving τ and $\overline{X}/\overline{Y}$. Then calculate the resulting estimates when $N = 5000$ and $\tau = 1{,}761{,}300$ (The article "Statistical Models and Analysis in Auditing," *Statistical Science*, 1989: 2–33 discusses properties of these estimators.)

6. Consider the accompanying observations on stream flow (thousands of acre-feet) recorded at a station in Colorado for the period April 1–August 31 over a 31-year span (from an article in the 1974 volume of *Water Resources Res.*).

127.96	210.07	203.24	108.91	178.21
285.37	100.85	89.59	185.36	126.94
200.19	66.24	247.11	299.87	109.64
125.86	114.79	109.11	330.33	85.54
117.64	302.74	280.55	145.11	95.36
204.91	311.13	150.58	262.09	477.08
94.33				

An appropriate probability plot supports the use of the lognormal distribution (see Sect. 3.5) as a reasonable model for stream flow.

(a) Estimate the parameters of the distribution. [*Hint*: Remember that X has a lognormal distribution with parameters μ and σ if $\ln(X)$ is normally distributed with mean μ and standard deviation σ.]

(b) Use the estimates of part (a) to calculate an estimate of the expected value of stream flow. [*Hint*: What is the expression for $E(X)$?]

7. (a) A random sample of 10 houses in a particular area, each of which is heated with natural gas, is selected and the amount of gas (therms) used during the month of January is determined for each house. The resulting observations are 103, 156, 118, 89, 125, 147, 122, 109, 138, 99. Let μ denote the average gas usage during January by all houses in this area. Compute a point estimate of μ.

(b) Suppose there are 10,000 houses in this area that use natural gas for heating. Let τ denote the total amount of gas used by all of these houses during January. Estimate τ using the data of (a). What estimator did you use in computing your estimate?

(c) Use the data in (a) to estimate p, the proportion of all houses that used at least 100 therms.

(d) Give a point estimate of the population median usage based on the sample of (a). What estimator did you use?

8. In a random sample of 80 components of a certain type, 12 are found to be defective.

(a) Give a point estimate of the proportion of all such components that are *not* defective.

(b) A system is to be constructed by randomly selecting two of these components and connecting them in series, as shown here.

The series connection implies that the system will function if and only if neither component is defective (i.e., both components work properly). Estimate the proportion of all such systems that work properly. [*Hint*: If p denotes the probability that a component works properly, how can $P(\text{system works})$ be expressed in terms of p?]

(c) Let $\hat{p}$ be the sample proportion of successes. Is $\hat{p}^2$ an unbiased estimator for p^2? [*Hint*: For any rv Y, $E(Y^2) = \text{Var}(Y) + [E(Y)]^2$.]

9. Each of 150 newly manufactured items is examined and the number of scratches per item is recorded (the items are supposed to be free of scratches), yielding the following data:

Number of scratches per item	0	1	2	3	4	5	6	7
Observed frequency	18	37	42	30	13	7	2	1

Let $X =$ the number of scratches on a randomly chosen item, and assume that X has a Poisson distribution with parameter μ.

(a) Find an unbiased estimator of μ and compute the estimate for the data.

(b) What is the standard error of your estimator? Compute the estimated standard error. [*Hint*: $\sigma_X^2 = \mu$ for X Poisson.]

10. Let $X_1, \ldots, X_n$ be a random sample from a distribution with mean μ and variance σ^2.

(a) Show that $\sum (X_i - \overline{X})^2 = (\sum X_i^2) - n\overline{X}^2$.

(b) Show that $E(\sum X_i^2) = n(\mu^2 + \sigma^2)$. [*Hint*: Use linearity of expectation, along with the relation $E(Y^2) = \mathrm{Var}(Y) + [E(Y)]^2$.]

(c) Show that $E(n\overline{X}^2) = n\mu^2 + \sigma^2$. [*Hint*: Apply the relation given in the previous hint, but this time to $Y = \overline{X}$.]

(d) Combine parts (a)–(c) to show that S^2 is an unbiased estimator of σ^2.

(e) Does it follow that the sample standard deviation, S, of a random sample is an unbiased estimator of σ? Why or why not?

11. Example 5.5 considered the estimator $\hat{\lambda} = 1/\overline{X}$ for the unknown parameter λ of an exponential distribution, based on a random sample $X_1, X_2, \ldots, X_n$ from that distribution.

(a) Show using a moment generating function argument that $\overline{X}$ has a gamma distribution, with parameters $\alpha = n$ and $\beta = 1/(n\lambda)$.

(b) Find the expected value of $\hat{\lambda}$. [*Hint*: The goal is to find $E(1/Y)$, where $Y \sim \mathrm{gamma}(n, 1/(n\lambda))$. Use the gamma pdf to determine this expected value.]

(c) Find the variance of $\hat{\lambda}$. [*Hint*: Now find $E(1/Y^2)$. Then apply the variance shortcut formula.]

12. Using a long rod that has length μ, you are going to lay out a square plot in which the length of each side is μ. Thus the area of the plot will be μ^2. However, you do not know the value of μ, so you decide to make n independent measurements $X_1, X_2, \ldots X_n$ of the length. Assume that each X_i has mean μ (unbiased measurements) and variance σ^2.

(a) Show that $\overline{X}^2$ is not an unbiased estimator for μ^2. [*Hint*: Apply the hint from Exercises 8 and 10 with $Y = \overline{X}$.]

(b) For what value of k is the estimator $\overline{X}^2 - kS^2$ unbiased for μ^2? [*Hint*: Compute $E(\overline{X}^2 - kS^2)$, using the result of Exercise 10(d).]

13. Of n_1 randomly selected male smokers, X_1 smoked filter cigarettes, whereas of n_2 randomly selected female smokers, X_2 smoked filter cigarettes. Let p_1 and

p_2 denote the probabilities that a randomly selected male and female, respectively, smoke filter cigarettes.

(a) Show that $(X_1/n_1) - (X_2/n_2)$ is an unbiased estimator for $p_1 - p_2$. [*Hint:* What type of rvs are X_1 and X_2?]

(b) What is the standard error of the estimator in (a)?

(c) How would you use the observed values x_1 and x_2 to estimate the standard error of your estimator?

(d) If $n_1 = n_2 = 200$, $x_1 = 127$, and $x_2 = 176$, use the estimator of (a) to obtain an estimate of $p_1 - p_2$.

(e) Use the result of (c) and the data of (d) to estimate the standard error of the estimator.

14. Suppose a certain type of fertilizer has an expected yield per acre of μ_1 with variance σ^2, whereas the expected yield for a second type of fertilizer is μ_2 with the same variance σ^2. Let S_1^2 and S_2^2 denote the sample variances of yields based on sample sizes n_1 and n_2, respectively, of the two fertilizers. Use the result of Exercise 10(d) to show that the pooled (combined) estimator

$$\hat{\sigma}^2 = \frac{(n_1 - 1)S_1^2 + (n_2 - 1)S_2^2}{n_1 + n_2 - 2}$$

is an unbiased estimator of σ^2.

15. Consider a random sample $X_1, \ldots, X_n$ from the pdf

$$f(x; \theta) = .5(1 + \theta x) \qquad -1 \le x \le 1$$

where $-1 \le \theta \le 1$ (this distribution arises in particle physics). Show that $\hat{\theta} = 3\overline{X}$ is an unbiased estimator of θ. [*Hint:* First determine $\mu = E(X) = E(\overline{X})$.]

16. A sample of n captured jet fighters results in serial numbers $x_1, x_2, x_3, \ldots, x_n$. The CIA knows that the aircraft were numbered consecutively at the factory starting with α and ending with β, so that the total number of planes manufactured is $\beta - \alpha + 1$ (e.g., if $\alpha = 17$ and $\beta = 29$, then $29 - 17 + 1 = 13$ planes having serial numbers 17, 18, 19, $\ldots$, 28, 29 were manufactured). However, the CIA does not know the values of α or β. A CIA statistician suggests using the estimator $\max(X_i) - \min(X_i) + 1$ to estimate the total number of planes manufactured.

(a) If $n = 5$, $x_1 = 237$, $x_2 = 375$, $x_3 = 202$, $x_4 = 525$, and $x_5 = 418$, what is the corresponding estimate?

(b) Under what conditions on the sample will the value of the estimate be exactly equal to the true total number of planes? Will the estimate ever be larger than the true total? Do you think the estimator is unbiased for estimating $\beta - \alpha + 1$? Explain in one or two sentences.

(A similar method was used to estimate German tank production in World War II.)

17. Let $X_1, X_2, \ldots, X_n$ represent a random sample from a *Rayleigh distribution* with pdf

$$f(x;\theta) = \frac{x}{\theta}e^{-x^2/(2\theta)} \qquad x > 0$$

(a) It can be shown that $E(X^2) = 2\theta$. Use this fact to construct an unbiased estimator of θ based on $\sum X_i^2$ (and use rules of expected value to show that it is unbiased).

(b) Estimate θ from the following measurements of blood plasma beta concentration (in pmol/L) for $n = 10$ men.

16.88	10.23	4.59	6.66	13.68
14.23	19.87	9.40	6.51	10.95

18. Suppose the true average growth μ of one type of plant during a 1-year period is identical to that of a second type, but the variance of growth for the first type is σ^2, whereas for the second type the variance is $4\sigma^2$. Let $X_1, \ldots, X_m$ be m independent growth observations on the first type [so $E(X_i) = \mu$, $\text{Var}(X_i) = \sigma^2$], and let $Y_1, \ldots, Y_n$ be n independent growth observations on the second type [$E(Y_i) = \mu$, $\text{Var}(Y_i) = 4\sigma^2$].

Let c be a numerical constant and consider the estimator $\hat{\mu} = c\bar{X} + (1 - c)\bar{Y}$. For any c between 0 and 1, this is a weighted average of the two sample means, e.g., $.7\bar{X} + .3\bar{Y}$.

(a) Show that for any c the estimator is unbiased.

(b) For fixed m and n, what value c minimizes the standard error of $\hat{\mu}$? [*Hint*: The estimator is a linear combination of the two sample means and these means are independent. Once you have an expression for the variance, differentiate with respect to c.]

19. In Chap. 2, we defined a negative binomial rv as the number of trials required to achieve the rth success in a sequence of independent and identical success/ failure trials. The probability mass function (pmf) of X is

$$nb(x;r,p) = \begin{cases} \binom{x-1}{r-1} p^r (1-p)^{x-r} & x = r, r+1, r+2, \ldots \\ 0 & \text{otherwise} \end{cases}$$

(a) Suppose that $r \geq 2$. Show that

$$\hat{P} = (r - 1)/(X - 1)$$

is an unbiased estimator for p. [*Hint*: Write out $E(\hat{P})$ as a sum, then make the substitutions $y = x - 1$ and $s = r - 1$.]

(b) A reporter wishing to interview five individuals who support a certain candidate begins asking people whether (S) or not (F) they support the candidate. If the sequence of responses is $SFFSFFFSSS$, estimate $p=$ the true proportion who support the candidate.

20. Suppose that X, the reaction time to a stimulus, has a uniform distribution on the interval from 0 to an unknown upper limit θ. An investigator wants to estimate θ on the basis of a random sample $X_1, X_2, \ldots, X_n$ of reaction times. Consider two possible estimators:

$$\hat{\theta}_1 = \max(X_1, \ldots, X_n) \qquad \hat{\theta}_2 = 2\overline{X}$$

(a) The following observed reaction times, in seconds, are for a sample of $n=5$ subjects: $x_1=4.2$, $x_2=1.7$, $x_3=2.4$, $x_4=3.9$, $x_5=1.3$. Calculate a point estimate of θ based on $\hat{\theta}_1$ and a point estimate of θ based on $\hat{\theta}_2$.

(b) The techniques of Sect. 4.9 imply that the pdf of $\hat{\theta}_1$ is $f(y)=ny^{n-1}/\theta^n$ for $0 \le y \le \theta$ (we're using y as the argument instead of $\hat{\theta}_1$ so that the notation is less confusing). Use this to obtain the mean and variance of $\hat{\theta}_1$.

(c) Is $\hat{\theta}_1$ an unbiased estimator of θ? Explain why this is reasonable. [*Hint*: If the population maximum is θ, what must be true of the sample maximum?]

(d) The mean and variance of a uniform distribution on $[0, \theta]$ are $\theta/2$ and $\theta^2/12$, respectively. Use these and the properties of $\overline{X}$ to find the mean and variance of $\hat{\theta}_2$.

(e) If a statistician elected to apply the Principle of Unbiased Estimation, which estimator would she select? Why?

(f) Find a constant k such that $\hat{\theta}_3 = k \cdot \hat{\theta}_1$ is unbiased for θ, and compare the standard error of $\hat{\theta}_3$ to the standard error of $\hat{\theta}_2$.

21. An investigator wishes to estimate the proportion of students at a certain university who have violated the honor code. Having obtained a random sample of n students, she realizes that asking each, "Have you violated the honor code?" will probably result in some untruthful responses. Consider the following scheme, called a **randomized response** technique. The investigator makes up a deck of 100 cards, of which 50 are of Type I and 50 are of Type II.

   Type I: Have you violated the honor code (yes or no)?
   Type II: Is the last digit of your telephone number a 0, 1, or 2 (yes or no)?
   Each student in the random sample is asked to mix the deck, draw a card, and answer the resulting question truthfully. Because of the irrelevant question on Type II cards, a yes response no longer stigmatizes the respondent, so we assume that responses are truthful. Let p denote the proportion of honor-code

violators (i.e., the probability of a randomly selected student being a violator), and let $\lambda = P(\text{yes response})$. Then λ and p are related by $\lambda = .5p + (.5)(.3)$.

(a) Let Y denote the number of yes responses, so $Y \sim \text{Bin}(n, \lambda)$. Thus Y/n is an unbiased estimator of λ. Derive an estimator for p based on Y. If $n = 80$ and $y = 20$, what is your estimate? [*Hint*: Solve $\lambda = .5p + .15$ for p and then substitute Y/n for λ.]

(b) Use the fact that $E(Y/n) = \lambda$ to show that your estimator is unbiased for p.

(c) If there were 70 Type I and 30 Type II cards, what would be your estimator for p?

22. The **mean squared error** of an estimator $\hat{\theta}$ is defined by

$$\text{MSE}(\hat{\theta}) = E\left[(\hat{\theta} - \theta)^2\right]$$

(a) Show that $\text{MSE}(\hat{\theta}) = \left[E(\hat{\theta}) - \theta\right]^2 + \text{Var}(\hat{\theta})$ by expanding out the quadratic expression "inside" the expected value operation in the definition of MSE and then using linearity of expectation.

(b) If $\hat{\theta}$ is an unbiased estimator of the parameter θ, how does $\text{MSE}(\hat{\theta})$ simplify?

(c) Refer back to Example 5.5. Determine the mean squared error of the estimator $\hat{\lambda}$ using the mean and variance expressions provided in that example.

(d) Consider an alternative estimator, $\hat{\lambda}_a$, defined by

$$\hat{\lambda}_a = \frac{n-1}{\sum X_i} = \frac{n-1}{n} \cdot \frac{1}{\bar{X}} = \frac{n-1}{n} \hat{\lambda}$$

Obtain the mean, variance, and MSE of $\hat{\lambda}_a$. [*Hint*: Use rescaling properties.]

(e) Which of the two estimators, $\hat{\lambda}$ or $\hat{\lambda}_a$, is preferable? Explain your reasoning.

23. Return to the problem of estimating a population proportion p, and consider the following adjusted estimator:

$$\hat{P}_a = \frac{X + \sqrt{n/4}}{n + \sqrt{n}}$$

The justification for this estimator comes from the Bayesian approach to point estimation to be introduced in Sect. 5.6.

(a) Determine the mean, variance, and mean squared error of this estimator. What do you find interesting about this MSE?

(b) Compare the MSE of this estimator to the MSE of the usual estimator $\hat{P}$ (the sample proportion).

5.2 Maximum Likelihood Estimation

The point estimators introduced in Sect. 5.1 were obtained via intuition and/or educated guesswork. We now introduce a "constructive" method for obtaining point estimators: the method of maximum likelihood. By constructive we mean that the general definition of a maximum likelihood estimator suggests explicitly how to obtain the estimator in any specific problem.

The method of maximum likelihood was first introduced by R. A. Fisher, a geneticist and statistician, in the 1920s. Most statisticians recommend this method, at least when the sample size is large, since the resulting estimators have certain desirable efficiency properties (see the proposition on large-sample behavior toward the end of this section).

Example 5.6 The best protection against hacking into an online account is to use a password that has at least eight characters containing upper and lower case letters, a numeral, and a special character. Suppose that ten individuals who have a certain type of email account are selected, and it is found that the first, third, and tenth individuals have such strong protection, whereas the others do not (the January 2012 issue of *Consumer Reports* reported that only 25% of individuals surveyed used a strong password). Let $p = P(\text{strong protection})$, i.e., p is the proportion of all account holders having strong protection. Define Bernoulli random variables X_1, $X_2, \ldots, X_{10}$ by

$$X_i = \begin{cases} 1 \text{ if the } i\text{th person has strong protection} \\ 0 \text{ if not} \end{cases} \quad i = 1, 2, \ldots 10$$

Then for the obtained sample, $X_1 = X_3 = X_{10} = 1$ and the other seven X_is are all zero. The probability mass function of any particular X_i is $p^{x_i}(1-p)^{1-x_i}$, which becomes p if $x_i = 1$ and $1 - p$ when $x_i = 0$. Finally, the strengths of various passwords are presumably independent of one another, so that the X_is are independent and their joint probability mass function is the product of the individual pmfs. Thus the joint pmf evaluated at the observed X_is is

$$p \cdot (1-p) \cdot p \cdot (1-p) \cdot (1-p) \cdots p = p^3(1-p)^7 \qquad (5.1)$$

Suppose that $p = .25$. Then the probability of observing the sample that we actually obtained is $(.25)^3(.75)^7 = .002086$. If instead $p = .50$, then this probability is $(.50)^3(.50)^7 = .000977$. For what value of p is the obtained sample most likely to have occurred? That is, what value of p maximizes the joint pmf (Eq. 5.1)? Figure 5.3 shows a graph of the *likelihood* (Eq. 5.1) as a function of p. It appears

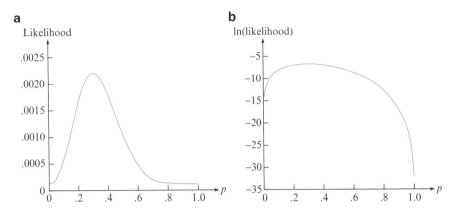

Fig. 5.3 Likelihood and log likelihood plotted against p

that the graph reaches its peak above $p = .3$, which is the proportion of strong passwords in the sample. The second figure shows a graph of the natural logarithm of (Eq. 5.1); since $\ln[g(u)]$ is a strictly increasing function of $g(u)$, finding u to maximize the function $g(u)$ is the same as finding u to maximize $\ln[g(u)]$.

We can verify our visual impression by using calculus to find the value of p that maximizes (Eq. 5.1). Working with the natural log of the joint pmf is often easier than working with the joint pmf itself, since the joint pmf is typically a product so its logarithm will be a sum. Here

$$\ln\left[p^3(1-p)^7\right] = 3\ln(p) + 7\ln(1-p)$$

Thus

$$\frac{d}{dp}\ln\left[p^3(1-p)^7\right] = \frac{d}{dp}\left[3\ln(p) + 7\ln(1-p)\right] = \frac{3}{p} + \frac{7}{1-p}(-1) = \frac{3}{p} - \frac{7}{1-p}$$

The (-1) comes from the chain rule in calculus. Equating this derivative to 0 and solving for p gives $3(1-p) = 7p$, from which $3 = 10p$ and so $p = 3/10 = .30$ as conjectured. That is, our point estimate is $\hat{p} = .30$. It is called the *maximum likelihood estimate* because it is the parameter value that maximizes the likelihood (joint pmf) of the observed sample. In general, the second derivative should be examined to make sure a maximum has been obtained, but here this is obvious from the figure.

Suppose that rather than being told the condition of every password, we had only been informed that three of the ten were strong. Then we would have the observed value of a binomial random variable $X =$ the number of strong passwords. The pmf of X is $\binom{10}{x}p^x(1-p)^{10-x}$. For $x = 3$, this becomes $\binom{10}{3}p^3(1-p)^7$. The binomial coefficient $\binom{10}{3}$ is irrelevant to the maximization, so the value of p that maximizes the likelihood of observing $X = 3$ is again $\hat{p} = .30$. ■

DEFINITION

Let $X_1, \ldots, X_n$ have a joint distribution (i.e., a joint pmf or pdf) that depends on a parameter θ whose value is unknown. Suppose that we observe $X_1 = x_1$, $X_2 = x_2, \ldots, X_n = x_n$. Substitute the observed data into the joint distribution and regard it as a function of θ, called the **likelihood function** and denoted by $L(\theta)$. Then the **maximum likelihood estimate** $\hat{\theta}$ is the value of θ that maximizes the likelihood, so that $L(\hat{\theta}) \geq L(\theta)$ for every possible value of θ. Replacing the x_is in $\hat{\theta}$ by X_is gives the **maximum likelihood estimator** (**mle**) of θ.

In Example 5.6, the joint pmf of $X_1, \ldots, X_{10}$ became $p^3(1-p)^7$ once the observed values of the X_is were substituted. So, the likelihood function would be written $L(p) = p^3(1-p)^7$. If we take the perspective that our data consists of a single binomial observation, then $L(p) = \binom{10}{3} p^3(1-p)^7$. In either case, the value of p that maximizes $L(p)$ is $\hat{p} = .3$.

The likelihood function tells us how likely the observed sample is as a function of the possible parameter values. Maximizing the likelihood gives the parameter value for which the observed sample is most likely to have been generated, that is, the parameter value that "agree most closely" with the observed data. Maximizing the likelihood is equivalent to maximizing the logarithm of the likelihood, and the latter is typically computationally more straightforward.

Example 5.77 Suppose $X_1, \ldots, X_n$ is a random sample from an exponential distribution with parameter λ. Because of independence, the likelihood function is a product of the individual pdfs:

$$f(x_1, \ldots, x_n; \lambda) = \left(\lambda e^{-\lambda x_1}\right) \cdots \cdots \left(\lambda e^{-\lambda x_n}\right) = \lambda^n e^{-\lambda \Sigma x_i} = L(\lambda)$$

The log of the likelihood function is

$$\ln[L(\lambda)] = n\ln(\lambda) - \lambda \sum x_i$$

Equating $(d/d\lambda)\ln[L(\lambda)]$ to zero results in $n/\lambda - \sum x_i = 0$, or $\lambda = n/\sum x_i = 1/\bar{x}$. Thus the mle is $\hat{\lambda} = 1/\bar{X}$. As we saw in Example 5.5, $\hat{\lambda}$ is unfortunately not an unbiased estimator, since $E(1/\bar{X}) \neq 1/E(\bar{X})$. ∎

Example 5.8 In Chap. 2, we indicated that the Poisson distribution could be used for modeling the number of events of some sort that occur in a two-dimensional region (e.g., the occurrence of tornadoes during a particular time period). Assume that when the region R being sampled has area $a(R)$, the number X of events occurring in R has a Poisson distribution with mean $\lambda \cdot a(R)$, where λ is the expected number of events per unit area, and that nonoverlapping regions yield independent Xs. (This is called a *spatial Poisson process*.)

Suppose an ecologist selects n nonoverlapping regions $R_1, \ldots, R_n$ and counts the number of plants of a certain species found in each region. The joint pmf (likelihood) is then

$$p(x_1, \ldots, x_n; \lambda) = \frac{[\lambda \cdot a(R_1)]^{x_1} e^{-\lambda \cdot a(R_1)}}{x_1!} \cdots \cdots \frac{[\lambda \cdot a(R_n)]^{x_n} e^{-\lambda \cdot a(R_n)}}{x_n!}$$

$$= \frac{[a(R_1)]^{x_1} \cdots \cdots [a(R_n)]^{x_n} \cdot \lambda^{\sum x_i} \cdot e^{-\lambda \sum a(R_i)}}{x_1! \cdots \cdots x_n!} = L(\lambda)$$

The log-likelihood is

$$\ln[L(\lambda)] = \sum \{x_i \ln[a(R_i)]\} + \ln(\lambda) \cdot \sum x_i - \lambda \sum a(R_i) - \sum \ln(x_i!)$$

Taking $(d/d\lambda)\ln[L(\lambda)]$ and equating it to zero yields

$$0 + \frac{\sum x_i}{\lambda} - \sum a(R_i) - 0 = 0 \Rightarrow \lambda = \frac{\sum x_i}{\sum a(R_i)}$$

The mle is then $\hat{\lambda} = \sum X_i / \sum a(R_i)$. This is intuitively reasonable because λ is the true density (plants per unit area), whereas $\hat{\lambda}$ is the sample density: $\sum X_i$ is the number of plants counted, and $\sum a(R_i)$ is just the total area sampled. Because $E(X_i) = \lambda \cdot a(R_i)$, the estimator is unbiased.

Sometimes an alternative sampling procedure is used. Instead of fixing regions to be sampled, the ecologist will select n points in the entire region of interest and let $y_i =$ the distance from the ith point to the nearest plant. The cdf of $Y =$ distance to the nearest plant is

$$F_Y(y) = P(Y \le y) = 1 - P(Y > y) = 1 - P\left(\begin{array}{c} \text{no plants in a} \\ \text{circle of radius } y \end{array}\right)$$

$$= 1 - \frac{e^{-\lambda \pi y^2} (\lambda \pi y^2)^0}{0!} = 1 - e^{-\lambda \pi y^2}$$

Taking the derivative of $F_Y(y)$ with respect to y yields

$$f_Y(y; \lambda) = \begin{cases} 2\pi\lambda y e^{-\lambda\pi y^2} & y \geq 0 \\ 0 & \text{otherwise} \end{cases}$$

If we now form the likelihood $L(\lambda) = f_Y(y_1; \lambda)\cdots f_Y(y_n; \lambda)$, differentiate $\ln[L(\lambda)]$, and so on, the resulting mle is

$$\hat{\lambda} = \frac{n}{\pi \sum Y_i^2} = \frac{\text{number of plants observed}}{\text{total area sampled}}$$

which is also a sample plant density. It can be shown that in a sparse environment (small λ), the distance method is in a certain sense better, whereas in a dense environment, the first sampling method is better.   ■

The definition of maximum likelihood estimates can be extended in the natural way to distributional families that include two or more parameters. The mles of parameters $\theta_1, \ldots, \theta_m$ are those values $\hat{\theta}_1, \ldots, \hat{\theta}_m$ that maximize the likelihood function $L(\theta_1, \ldots, \theta_m)$.

Example 5.9 Let $X_1, \ldots, X_n$ be a random sample from a normal distribution, which includes the two parameters μ and σ. The likelihood function is

$$f(x_1, \ldots, x_n; \mu, \sigma) = \frac{1}{\sqrt{2\pi\sigma^2}} e^{-(x_1-\mu)^2/(2\sigma^2)} \cdots \cdots \frac{1}{\sqrt{2\pi\sigma^2}} e^{-(x_n-\mu)^2/(2\sigma^2)}$$

$$= (2\pi\sigma^2)^{-n/2} e^{-\sum(x_i-\mu)^2/(2\sigma^2)} = L(\mu, \sigma)$$

so

$$\ln[L(\mu, \sigma)] = -\frac{n}{2}\ln(2\pi) - n\ln\sigma - \frac{1}{2\sigma^2}\sum(x_i - \mu)^2$$

To find the maximizing values of μ and σ, we must take the *partial* derivatives of $\ln(L)$ with respect to both μ and σ, equate them to zero, and solve the resulting two equations:

$$\frac{\partial}{\partial\mu}\ln[L(\mu, \sigma)] = -\frac{2}{2\sigma^2}\sum(x_i - \mu)(-1) = \frac{1}{\sigma^2}\sum(x_i - \mu) = 0$$

$$\frac{\partial}{\partial\sigma}\ln[L(\mu, \sigma)] = -\frac{n}{\sigma} + \frac{1}{\sigma^3}\sum(x_i - \mu)^2 = 0$$

The first equation implies that $\sum(x_i - \mu) = 0$, from which $\sum x_i - n\mu = 0$ and finally $\mu = \sum x_i/n = \bar{x}$. The mle of μ is the sample mean, independent of what the mle of σ turns out to be. Solving the second equation for σ yields $\sigma = \sqrt{\sum(x_i - \mu)^2/n}$; we must then substitute the solution from the first equation into this expression in order to get the simultaneous solution to the two partial derivative equations. Thus the maximum likelihood estimators of the two parameters are

$$\hat{\mu} = \overline{X} \qquad \hat{\sigma} = \sqrt{\frac{\sum (X_i - \overline{X})^2}{n}}$$

Notice that the mle of σ is *not* the sample standard deviation, S, since the denominator in the latter is $n - 1$ and not n.   ■

Example 5.10 Let $X_1, \ldots, X_n$ be a random sample from a Weibull pdf

$$f(x; \alpha, \beta) = \begin{cases} \dfrac{\alpha}{\beta^\alpha} \cdot x^{\alpha-1} \cdot e^{-(x/\beta)^\alpha} & x \geq 0 \\[2mm] 0 & \text{otherwise} \end{cases}$$

Writing the likelihood L and log-likelihood $\ln[L]$, then setting both partial derivatives $(\partial/\partial\alpha)[\ln(L)] = 0$ and $(\partial/\partial\beta)[\ln(L)] = 0$ yields the equations

$$\alpha = \left[\frac{\sum [x_i^\alpha \cdot \ln(x_i)]}{\sum x_i^\alpha} - \frac{\sum \ln(x_i)}{n} \right]^{-1} \qquad \beta = \left(\frac{\sum x_i^\alpha}{n} \right)^{1/\alpha}$$

These two equations cannot be solved explicitly to give general formulas for the mles $\hat{\alpha}$ and $\hat{\beta}$. Instead, for each sample $x_1, \ldots, x_n$, the equations must be solved using an iterative numerical procedure.

The iterative mle computations can be done using statistical software. In Matlab, the command wblfit(x) will return $\hat{\alpha}$ and $\hat{\beta}$ assuming the data is stored in the vector x. The corresponding R command is fitdistr(x,"weibull") performs the same estimation (the MASS package must be installed first). As an example, consider the following survival time data alluded to in Example 3.28:

152	115	109	94	88	137	152	77	160	165
125	40	128	123	136	101	62	153	83	69

A Weibull probability plot supports the plausibility of assuming that survival time has a Weibull distribution. The maximum likelihood estimates of the Weibull parameters are $\hat{\alpha} = 3.799$ and $\hat{\beta} = 125.88$. Figure 5.4 shows the Weibull log likelihood as a function of both α and β. The surface near the top has a rounded shape, allowing the maximum to be found easily, but for some distributions the surface can be much more irregular, and the maximum may be hard to find.

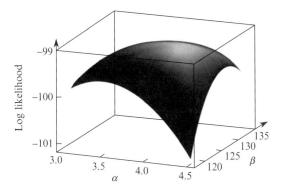

Fig. 5.4 Weibull log likelihood for Example 5.10   ■

Sometimes calculus cannot be used to obtain mles.

Example 5.11 Suppose the waiting time for a bus is uniformly distributed on $[0, \theta]$ and the results $x_1, \ldots, x_n$ of a random sample from this distribution have been observed. Since $f(x; \theta) = 1/\theta$ for $0 \leq x \leq \theta$ and 0 otherwise,

$$
f(x_1, \ldots, x_n; \theta) = \begin{cases} \dfrac{1}{\theta^n} & 0 \leq x_1 \leq \theta, \ldots, 0 \leq x_n \leq \theta \\ 0 & \text{otherwise} \end{cases}
$$

As long as $\max(x_i) \leq \theta$, the likelihood is $1/\theta^n$, which is positive, but as soon as $\theta < \max(x_i)$, the likelihood drops to 0. This is illustrated in Fig. 5.5. Calculus will not work because the maximum of the likelihood occurs at a point of discontinuity, but the figure shows that $\hat{\theta} = \max(x_i)$. Thus if the waiting times are 2.3, 3.7, 1.5, .4, and 3.2, then the mle is $\hat{\theta} = 3.7$. Note that the mle is biased (see Exercise 20(b)).

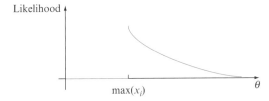

Fig. 5.5 The likelihood function for Example 5.11   ■

5.2.1 Some Properties of MLEs

In Example 5.9, we obtained the mle of σ when the underlying distribution is normal. The mle of σ^2, as well as many other mles, can be easily derived using the following proposition.

PROPOSITION (MLE INVARIANCE PRINCIPLE)

Let $\hat{\theta}_1, \hat{\theta}_2, \ldots, \hat{\theta}_m$ be the mles of the parameters $\theta_1, \theta_2, \ldots, \theta_m$. Then the mle of any function $h(\theta_1, \theta_2, \ldots, \theta_m)$ of these parameters is the function $h(\hat{\theta}_1, \hat{\theta}_2, \ldots, \hat{\theta}_m)$, of the mles.

Proof For an intuitive idea of the proof, consider the special case $m = 1$, with $\theta_1 = \theta$, and assume that $h(\cdot)$ is a one-to-one function. On the graph of the likelihood as a function of the parameter θ, the highest point occurs where $\theta = \hat{\theta}$. Now consider the graph of the likelihood as a function of $h(\theta)$. In the new graph the same heights occur, but the height that was previously plotted at $\theta = a$ is now plotted at $h(\theta) = h(a)$, and the highest point is now plotted at $h(\theta) = h(\hat{\theta})$. Thus, the maximum remains the same, but it now occurs at $h(\hat{\theta})$.   ∎

Example 5.12 (Example 5.9 continued) In the case of a random sample from a normal pdf, the mles of μ and σ are $\hat{\mu} = \overline{X}$ and $\hat{\sigma} = \sqrt{\sum (X_i - \overline{X})^2 / n}$. To obtain the mle of the function $h(\mu, \sigma) = \sigma^2$, substitute the mles into the function:

$$\widehat{\sigma^2} = \hat{\sigma}^2 = \frac{1}{n} \sum (X_i - \overline{X})^2$$

The mle of σ^2 is not the unbiased estimator (the sample variance S^2; see Exercise 10), although they are close unless n is quite small. Similarly, the mle of the population coefficient of variation, $h(\mu, \sigma) = 100\,\mu/\sigma$, is $100\hat{\mu}/\hat{\sigma}$.   ∎

Example 5.13 (Example 5.10 continued) The mean value of an rv X that has a Weibull distribution is

$$\mu = \beta \cdot \Gamma(1 + 1/\alpha)$$

The mle of μ is therefore $\hat{\mu} = \hat{\beta} \cdot \Gamma(1 + 1/\hat{\alpha})$, where $\hat{\alpha}$ and $\hat{\beta}$ are the mles of α and β. In particular, $\overline{X}$ is not the mle of μ, although it is an unbiased estimator. At least for large n, $\hat{\mu}$ is a better estimator than $\overline{X}$.   ∎

The method of maximum likelihood estimation has considerable intuitive appeal. The following proposition provides additional rationale for the use of mles.

> **PROPOSITION**
> Under very general conditions on the joint distribution of the sample, when the
> sample size is large, the maximum likelihood estimator of any particular θ
> - Is highly likely to be close to θ (consistency);
> - Is either unbiased or at least approximately unbiased $\left[E(\hat{\theta}) \approx \theta\right]$; and
> - Has variance that is either as small or nearly as small as can be achieved by
>   any unbiased estimator.

Because of this result and the fact that calculus-based techniques can usually be
used to derive the mles (although often numerical methods, such as Newton–
Raphson, are necessary), maximum likelihood estimation is the most widely used
estimation technique among statisticians. Obtaining an mle, however, does require
that the underlying distribution be specified. For example, the mle of the mean
value of a Weibull distribution is different from the mle of the mean value of a
Gamma distribution.

Suppose $X_1, X_2, \ldots, X_n$ is a random sample from a pdf $f(x; \theta)$ that is symmetric
about θ, but the investigator is unsure of the form of the f function. It is then
desirable to use an estimator that is *robust*, that is, one that performs well for a wide
variety of underlying pdfs. One such estimator, called an *M-estimator*, is based on a
generalization of maximum likelihood estimation. Instead of maximizing the
log-likelihood $\sum \ln[f(x; \theta)]$ for a specified f, one maximizes $\sum \psi(x_i; \theta)$, where
the "objective function" ψ is selected to yield an estimator with good robustness
properties. The book by David Hoaglin et al. (see the references) contains a good
exposition on this subject.

5.2.2   Exercises: Section 5.2 (24–36)

24. Let X represent the error in making a measurement of a physical characteristic
    or property (e.g., the boiling point of a particular liquid). It is often reasonable
    to assume that $E(X) = 0$ and that X has a normal distribution. Thus the pdf of
    any particular measurement error is

    $$f(x) = \frac{1}{\sqrt{2\pi\theta}} e^{-x^2/2\theta}$$

    where θ denotes the population variance. Now suppose that n independent
    measurements are made, resulting in measurement errors $X_1 = x_1, X_2 = x_2, \ldots,$
    $X_n = x_n$.
    (a) Determine the likelihood function of θ.
    (b) Find and simplify the log-likelihood function.
    (c) Differentiate (b) to determine the mle of θ.
    (d) The *precision* of a normal distribution is defined to be $\tau = 1/\theta$. Find the
        mle of τ.

25. A random sample of n bike helmets manufactured by a company is selected. Let $X =$ the number among the n that are flawed, and let $p = P(\text{flawed})$. Assume that only X is observed, rather than the sequence of Ss and Fs.
    (a) Derive the maximum likelihood estimator of p. If $n = 20$ and $x = 3$, what is the estimate?
    (b) Is the estimator of (a) unbiased?
    (c) If $n = 20$ and $x = 3$, what is the mle of the probability $(1 - p)^5$ that none of the next five helmets examined is flawed?

26. Let X denote the proportion of allotted time that a randomly selected student spends working on a certain aptitude test. Suppose the pdf of X is

$$f(x; \theta) = \begin{cases} (\theta + 1)x^\theta & 0 \le x \le 1 \\ 0 & \text{otherwise} \end{cases}$$

where $-1 < \theta$. A random sample of ten students yields data $x_1 = .92$, $x_2 = .79$, $x_3 = .90$, $x_4 = .65$, $x_5 = .86$, $x_6 = .47$, $x_7 = .73$, $x_8 = .97$, $x_9 = .94$, $x_{10} = .77$.

Obtain the maximum likelihood estimator of θ, and then compute the estimate for the given data.

27. Two different computer systems are monitored for a total of n weeks. Let X_i denote the number of breakdowns of the first system during the ith week, and suppose the X_is are independent and drawn from a Poisson distribution with parameter μ_1. Similarly, let Y_i denote the number of breakdowns of the second system during the ith week, and assume independence with each Y_i Poisson with parameter μ_2. Derive the mles of μ_1, μ_2, and $\mu_1 - \mu_2$. [*Hint:* Using independence, write the joint pmf (likelihood) of the X_is and Y_is together.]

28. Six Pepperidge Farm bagels were weighed, yielding the following data (grams):

| 117.6 | 109.5 | 111.6 | 109.2 | 119.1 | 110.8 |

    (a) Assuming that the six bagels are a random sample and that weights are normally distributed, estimate the true average weight and standard deviation of the weight using maximum likelihood.
    (b) Again assuming a normal distribution, estimate the weight below which 95% of all bagels will have their weights. [*Hint:* What is the 95th percentile in terms of μ and σ? Now use the invariance principle.]
    (c) Suppose we choose another bagel and weigh it. Let $X =$ weight of the bagel. Use the given data to obtain the mle of $P(X \le 113.4)$. [*Hint:* $P(X \le 113.4) = \Phi[(113.4 - \mu)/\sigma]$.]

29. Refer to Exercise 25. Instead of selecting $n = 20$ helmets to examine, suppose we examine helmets in succession until we have found $r = 3$ flawed ones. If the 20th helmet is the third flawed one, what is the mle of p? Is this the same as the estimate in Exercise 25? Why or why not?

30. Let $X_1, \ldots, X_n$ be a random sample from a gamma distribution with parameters α and β.
    (a) Derive the equations whose solution yields the maximum likelihood estimators of α and β. Do you think they can be solved explicitly?
    (b) Show that the mle of $\mu = \alpha\beta$ is $\hat{\mu} = \overline{X}$.

31. Let $X_1, X_2, \ldots, X_n$ represent a random sample from the Rayleigh distribution with density function given in Exercise 17.
    (a) Determine the maximum likelihood estimator of θ and then calculate the estimate for the vibratory stress data given in that exercise. Is this estimator the same as the unbiased estimator suggested in Exercise 17?
    (b) Determine the mle of the median of the vibratory stress distribution. [*Hint:* First express the median η in terms of θ.]

32. Consider a random sample $X_1, X_2, \ldots, X_n$ from the shifted exponential pdf

$$f(x; \lambda, \theta) = \begin{cases} \lambda e^{-\lambda(x-\theta)} & x \ge \theta \\ 0 & \text{otherwise} \end{cases}$$

Taking $\theta = 0$ gives the pdf of the exponential distribution considered previously (with positive density to the right of zero). An example of the shifted exponential distribution appeared in Example 3.5, in which the variable of interest was time headway in traffic flow and $\theta = .5$ was the minimum possible time headway.
    (a) Obtain the maximum likelihood estimators of θ and λ.
    (b) If $n = 10$ time headway observations are made, resulting in the values 3.11, .64, 2.55, 2.20, 5.44, 3.42, 10.39, 8.93, 17.82, and 1.30, calculate the estimates of θ and λ.

33. The article "A Model of Pedestrians' Waiting Times for Street Crossings at Signalized Intersections" (*Transportation Research*, 2013: 17–28) suggested that under some circumstances the distribution of waiting time X could be modeled with the following pdf:

$$f(x; \theta, \tau) = \begin{cases} \dfrac{\theta}{\tau}(1 - x/\tau)^{\theta-1} & 0 \le x < \tau \\ 0 & \text{otherwise} \end{cases} \quad \text{where } \theta > 0$$

(a) Suppose we observe a random sample of waiting times $X_1, \ldots, X_n$, and assume that the value of the parameter τ is known. Find the mle of θ.
(b) Suppose instead that θ is known but τ is unknown. Determine an equation whose solution is the mle of τ.

34. Twenty identical components are simultaneously tested. The lifetime distribution of each is exponential with parameter λ. The experimenter then leaves the test facility unmonitored. On his return 24 h later, the experimenter immediately terminates the test after noticing that $y = 15$ of the 20 components are still in operation (so 5 have failed). Derive the mle of λ. [*Hint:* Let $Y =$ the number that survive 24 h. Then $Y \sim \text{Bin}(n, p)$. What is the mle of p? Now

notice that $p = P(X_i \geq 24)$, where X_i is exponentially distributed. This relates λ to p, so the former can be estimated once the latter has been.]

35. Consider randomly selecting n segments of pipe and determining the corrosion loss (mm) in the wall thickness for each one. Denote these corrosion losses by $Y_1, \ldots, Y_n$. The article "A Probabilistic Model for A Gas Explosion Due to Leakages in the Grey Cast Iron Gas Mains" (*Reliability Engr. and System Safety* 2013:270–279) proposes a linear corrosion model $Y_i = t_i R_i$, where t_i is the age of the pipe and R_i, the corrosion rate, is exponentially distributed with parameter λ. Obtain the maximum likelihood estimator of λ (the resulting mle appears in the cited article). [*Hint:* First determine the pdf of Y_i.]

36. A method that is often used to estimate the size of a wildlife population involves performing a *capture/recapture* experiment. In this experiment, an initial sample of M animals is captured, each of these animals is tagged, and the animals are then returned to the population. After allowing enough time for the tagged individuals to mix into the population, another sample of size n is captured. With $X =$ the number of tagged animals in the second sample, the objective is to use the observed x to estimate the population size N.

   (a) What is the probability distribution of X?
   (b) Set $L(N)$ equal to the distribution specified in (a); this is the likelihood function. Since N can only assume integer values, using calculus to maximize $L(N)$ would present difficulties. Instead, determine the mle of N be considering the ratio $L(N)/L(N-1)$. [*Hint:* the mle can be found by determining when this ratio is greater than 1 or less than 1 (do you see why?).]
   (c) If 200 fish are taken from a lake and tagged, then subsequently 100 fish are recaptured and among the 100 there are 11 tagged fish, what is the mle of the size of the fish population in this lake? Does your answer make intuitive sense?

5.3   Confidence Intervals for a Population Mean

A point estimate, because it is a single number, by itself provides no information about the precision and reliability of estimation. Consider, for example, using the statistic $\overline{X}$ to calculate a point estimate for the true average breaking strength of paper towels of a certain brand, and suppose that a particular random sample yields $\overline{x} = 9322.7$ g. Because of sampling variability, it is virtually never the case that $\overline{x} = \mu$, and the point estimate alone says nothing about how close it might be to μ. An alternative to reporting a single sensible value for the parameter being estimated is to calculate and report an entire interval of plausible values—an *interval estimate* or *confidence interval* (CI).

A confidence interval is always calculated by first selecting a *confidence level*, which is a measure of the degree of reliability of the interval. A confidence interval with a 95% confidence level for the true average breaking strength might have a lower limit of 9162.5 g and an upper limit of 9482.9 g. Then at the "95% confidence level," any value of μ between 9162.5 and 9482.9 is plausible. A confidence level of 95% implies that 95% of all samples would give an interval that includes μ, or whatever other parameter is being estimated, and only 5% of all samples would yield an erroneous interval. The most frequently used confidence levels are 95, 99, and 90%. The higher the confidence level, the more strongly we believe that the value of the parameter being estimated lies within the interval.

Information about the precision of an interval estimate is conveyed by the width of the interval. If the confidence level is high and the resulting interval is quite narrow, our knowledge of the value of the parameter is reasonably precise. A very wide confidence interval, however, gives the message that there is a great deal of uncertainty concerning the value of what we are estimating. Figure 5.6 shows 95% confidence intervals for true average breaking strengths of two different brands of paper towels. One of these intervals suggests precise knowledge about μ, whereas the other suggests a very wide range of plausible values.

Fig. 5.6 Confidence intervals indicating precise (brand 1) and imprecise (brand 2) information about μ

5.3.1   A Confidence Interval for a Normal Population Mean

For much of this section we will assume that the available data results from a random sample $X_1, X_2, \ldots, X_n$ selected from a *normal* population distribution. The plausibility of assuming a normal population distribution can of course be checked by examining a normal probability plot of the data. Particularly when the sample size is small, the confidence interval to be developed here should not be used if the plot shows a substantial departure from a linear pattern. We'll comment later on what might be done in the presence of non-normality.

Recall from the previous chapter that as a consequence of our normality assumption, the sample mean $\overline{X}$ also is normally distributed, with mean value μ (the mean of the population from which the sample was selected) and standard deviation $\sigma/\sqrt{n}$. We now standardize $\overline{X}$ to obtain a random variable having a *standard* normal distribution:

$$Z = \frac{\overline{X} - \mu}{\sigma/\sqrt{n}}$$

Unfortunately this standardized variable cannot serve as a basis for deriving a confidence interval for μ unless the value of the population standard deviation σ happens to be known. So instead let's consider the standardized variable obtained by replacing σ in Z by the *sample* standard deviation S. Define a new random variable T by

$$T = \frac{\overline{X} - \mu}{S/\sqrt{n}}$$

It is important to contrast the behavior of Z in repeated sampling with that of T. The only variability in Z from one sample to another is because $\overline{X}$ in the numerator varies in value. However, there are two sources of sample-to-sample variability in T: both $\overline{X}$ in the numerator and S in the denominator. Because of this extra variation in T, it stands to reason that the distribution of T should be more spread out than that of Z. That is, the density curve for T should be more spread out than the standard normal curve.

At this point we need to introduce a new (to the reader) family of probability distributions that describes how T varies from one sample to another. This is the **family of t distributions**. The formula for the density function that specifies a t distribution is quite complicated (see the reference by Devore and Berk, where the formula and a derivation appear). Fortunately for our purpose we need only be acquainted with some general properties.

PROPERTIES OF T DISTRIBUTIONS

1. Any particular t distribution is obtained by specifying the value of a single parameter ν, called the *number of degrees of freedom* (df) of the distribution. Any positive integer is a possible value of ν, so there is a t distribution with 1 df, another with 2 df, and so on.
2. Each t_ν density curve is bell shaped and centered at 0, just like the standard normal (z) curve.
3. Each t_ν density curve is more spread out than the z curve.
4. As ν increases, the spread of the t_ν curve decreases (so the t_1 curve is the most spread out, the t_2 curve is next most spread out, and so on).
5. As $\nu \to \infty$, the sequence of t_ν curves approaches the z curve (for this reason, the z curve is often called the t curve with df $= \infty$).

Figure 5.7 shows several different t density curves and the z curve to illustrate how the curves compare and change as df increases.

Appendix Table A.5 displays what are called t *critical values*; these are numbers on the horizontal axis that capture certain central areas under t curves. For example, looking down the left column to $\nu = 24$ and then over to the column headed 95%, we learn that 95% of the area under the t curve with 24 df lies between -2.064 and 2.064. Notice that in any particular column of the table, the numbers decrease as we move down; this is because the spread of t curves decreases as df increases. And the numbers in any row increase from right to left because a larger central area is being captured. Also note that toward the bottom of the table df skips from 30 to 40 to 60 to 120 to ∞. Once past 30 df, the t curves do not change all that much, so it is not worth continuing to tabulate in increments of 1 df. For an intermediate number of

Fig. 5.7 Comparison of several t curves and the z curve

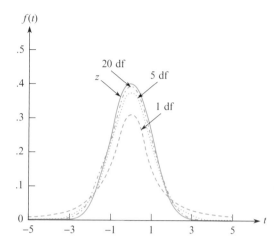

df, linear interpolation can be used to get a reasonable approximation, or appropriate software will produce an exact value. Lastly, the t distribution with an infinite number of df is actually the standard normal distribution. Thus the bottom row of the table contains *standard normal critical values*; for example, 95% of the area under the z curve lies between -1.96 and 1.96.

With information about t distributions in hand, we are now ready for the key theoretical result on which our confidence interval will be based. This result was originally discovered in 1908 by William Sealy Gosset, a statistician at the Guinness Brewery in Dublin, Ireland.

GOSSET'S THEOREM

Let $X_1, \ldots, X_n$ be a random sample from a normal population distribution having mean μ, with corresponding sample mean $\overline{X}$ and sample standard deviation S. Then the random variable

$$T = \frac{\overline{X} - \mu}{S/\sqrt{n}}$$

has a t distribution with $n - 1$ degrees of freedom.

An intuitive justification for degrees of freedom here is that although there are n deviations $X_1 - \overline{X}, X_2 - \overline{X}, \ldots, X_n - \overline{X}$ from the sample mean, it is easily verified that $\sum (X_i - \overline{X}) = 0$. This implies that any particular one of the deviations can be obtained from the other $n - 1$ deviations. For example, in the case $n = 5$, if the first four deviations are $-2, 5, 1,$ and -8, then the last one must be 4 to produce a sum of zero. The number of df here is the number of "freely varying" deviations that are inputs to the sample standard deviation.

Consider for the moment a sample size of $n = 25$, for which the standardized variable T is based on 24 df. Then 95% of the area under this t curve lies between -2.064 and 2.064. The foregoing theorem then allows us to make the following probability statement:

$$P\left(-2.064 < \frac{\overline{X} - \mu}{S/\sqrt{25}} < 2.064\right) = .95$$

Let's now manipulate the inequalities inside the parentheses to isolate μ in the middle. This requires three steps: (1) multiply all three terms by $S/\sqrt{n}$, (2) subtract $\overline{X}$ from all three terms, and (3) multiply through by -1 to eliminate the negative sign in front of μ. The last step will reverse the direction of each inequality, resulting in

$$\overline{X} + \frac{2.064S}{\sqrt{25}} > \mu > \overline{X} - \frac{2.064S}{\sqrt{25}}$$

These new inequalities are completely equivalent to those in the original probability statement, so

$$P\left(\overline{X} - \frac{2.064S}{\sqrt{25}} < \mu < \overline{X} + \frac{2.064S}{\sqrt{25}}\right) = .95$$

To interpret this latter probability, think of obtaining sample after sample of size 25 from a normal population distribution; calculate the sample mean and sample standard deviation for each one, and then form the lower limit $\overline{x} - 2.064s/\sqrt{25}$ and the upper limit $\overline{x} + 2.064s/\sqrt{25}$. Both the center of the interval ($\overline{x}$) and its width will vary from sample to sample. In the long run, 95% of such samples will result in the value of μ being captured in between the lower limit and the upper limit—the long-run capture percentage for the sequence of intervals is 95%. Any particular one of these intervals is called a *confidence interval for μ with confidence level 95%*.

Generalizing the foregoing derivation for an arbitrary sample size leads to the following confidence interval formula.

ONE-SAMPLE T CONFIDENCE INTERVAL
Let $\overline{x}$ and s be the sample mean and sample standard deviation of a random sample of size n selected from a normal population distribution. Then a confidence interval (interval of plausible values) for the population mean μ has endpoints

$$\overline{x} \pm t^* \cdot \frac{s}{\sqrt{n}}$$

where t^* is the appropriate t critical value with $n - 1$ df from Table A.5.

Example 5.14 Have you ever dreamed of owning a Porsche? Even though academic salaries leave little room for luxuries, the authors thought maybe the purchase of a used Boxster, the least expensive Porsche model, might be feasible. So on December 30, 2012 we went to www.cars.com to peruse prices. The news was discouraging, so we instead selected a random sample of 16 such vehicles and obtained the following odometer readings (miles):

| 1445 | 25,822 | 26,892 | 29,860 | 35,285 | 47,874 | 49,544 | 64,763 |
| 72,698 | 75,732 | 84,457 | 91,577 | 93,000 | 109,538 | 113,399 | 137,652 |

Figure 5.8 shows a normal probability plot of the data; this version includes a superimposed line which makes it easier to judge whether the pattern in the plot is reasonably linear. Very clearly that is the case. It is therefore quite plausible that the distribution of odometer readings is (at least approximately) normal, which validates the use of the one-sample t confidence interval to estimate the population mean odometer reading, μ.

The sample mean and sample standard deviation are 66,221.1 and 37,683.1672, respectively, and the (estimated) standard error of the mean is $s/\sqrt{n} = 9420.7918$. Table A.5 shows that the t critical value for a confidence level of 95% when df $= 16 - 1 = 15$ is $t^* = 2.131$. The confidence interval is then

$$\bar{x} \pm t^* \cdot \frac{s}{\sqrt{n}} = 66{,}221.1 \pm (2.131)(9420.7918) = 66{,}221.1 \pm 20{,}075.7$$

$$= (46{,}145.4, \ 86{,}296.8)$$

That is, we can say with a confidence level of 95% that $46{,}145.4 < \mu < 86{,}296.8$.

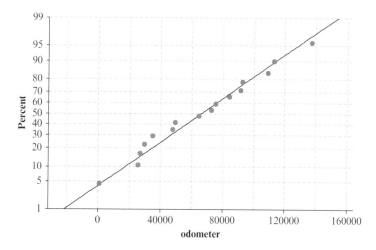

Fig. 5.8 Normal probability plot of the Boxster odometer reading data

Note that it is not correct at this point to write $P(46{,}145.4 < \mu < 86{,}296.8) = .95$, because nothing inside the parentheses is random. The interval we have calculated may or may not include the actual value of μ. If we were to obtain sample after sample of size 16 from this population distribution and for each one use the given formula with $t^* = 2.131$, in the long run 95% of the calculated CIs would include μ whereas 5% would not. Without knowing the value of μ, we can't know whether the *particular* interval we have calculated is one of the "good" 95% or the "bad" 5%.   ■

5.3.2   A Large-Sample Confidence Interval for μ

When the sample size n is sufficiently large, the Central Limit Theorem says that $\overline{X}$ has approximately a normal distribution even when the population distribution is not normal. Furthermore, it can be shown in this case that the standardized variable $(\overline{X} - \mu)/(S/\sqrt{n})$ has approximately a standard normal distribution; using S in place of σ in the denominator does not appreciably increase the variability of Z when n is large. This in turn implies that for large n, a legitimate confidence interval for a population mean μ is

$$\overline{x} \pm z^* \cdot \frac{s}{\sqrt{n}} \qquad\qquad (5.2)$$

where the z critical values for the most frequently employed confidence levels appear in the bottom row of Appendix Table A.5 (or can be extracted from the z table). For example, the z critical value for 95% confidence, the most common level used in practice, is $z^* = 1.96$.

Example 5.15 Magnetic resonance imaging is a commonly used noninvasive technique for assessing the extent of cartilage damage. However, there is concern that the MRI sizing of articular cartilage defects may not be accurate. The article "Preoperative MRI Underestimates Articular Cartilage Defect Size Compared with Findings at Arthroscopic Knee Surgery" (*Amer. J. of Sports Med.*, 2013: 590–595) reported on a study involving a sample of 92 cartilage defects. For each one, the size of the lesion area was determined by an MRI analysis and also during arthroscopic surgery. Each MRI value was then subtracted from the corresponding arthroscopic value to obtain a difference value; this is commonly referred to as "paired differ-ence" data. The sample mean difference was calculated to be 1.04 cm^2, with a sample standard deviation of 1.67. Let's now calculate a confidence interval using a confidence level of (at least approximately) 95% for μ_D, the mean difference for the population of all such defects (as did the authors of the cited article). Using the $z^* = 1.96$ and Expression (5.2), the CI is

$$1.04 \pm 1.96 \cdot \frac{1.67}{\sqrt{92}} = 1.04 \pm .34 = (.70, 1.38)$$

At the 95% confidence level, we conclude that $.70 < \mu_D < 1.38$. Perhaps the most important aspect of this interval is that 0 is not included; only certain positive values of μ_D are plausible. It is this fact that led the investigators to conclude that MRIs tend to underestimate defect size. ∎

Many statisticians do not use Expression (5.2) unless their sample size is extremely large, electing instead to use the one-sample t interval for virtually all cases. In Example 5.15, for instance, $z^* = 1.96$ would be replaced by the more conservative t critical value at 91 df, which happens to be 1.986. This would make very little practical change to the resulting interval. In the simulation sections of Chaps. 2–4, where the "sample" size was typically 10,000 or more, there would be no controversy in using $\bar{x} \pm 1.96s/\sqrt{n}$ as a 95% CI for the unknown mean μ of the rv being simulated.

When the sample size is small and the population distribution is substantially non-normal, neither the one sample t interval nor Expression (5.2) should be used. In this case there are other techniques for obtaining a valid CI. One relatively recent, computationally intensive such method is called a *bootstrap confidence interval*. This entails obtaining a large number of samples of size n by *resampling with replacement* from the sample that was actually obtained—e.g., if the sample size is 20, a bootstrap might be based on 1000 samples of size 20, each obtained with replacement from the original sample. Details can be found in the book by Devore and Berk listed in the references.

5.3.3 Software for Confidence Interval Calculation

It should be no surprise that modern software can compute confidence intervals automatically once we have supplied the software with our data. In R, the `t.test` function takes in a vector of data and returns, among other things, a one-sample t 95% confidence interval for the population mean μ. The optional argument `conf.level` can be used to select any other confidence level (the default is `conf.level=.95`). The analogous function in Matlab is `ttest`, although the inputs and outputs are managed differently. Both are illustrated in Fig. 5.9.

Notice that both R and Matlab give a CI of (46,147, 86,303) for the true mean odometer reading. This is roughly what we computed in Example 5.14, and the disparity is primarily due to rounding in the critical value t^*. The other information provided by R relates to *hypothesis testing*, which we will discuss in Sect. 5.4.

To simply find the t critical value for a particular df, the inverse cdf commands can be implemented, but with one proviso: for the *central* area of a t curve to equal some confidence level C, the cumulative probability from $-\infty$ to the critical value must be

$$C + \frac{1-C}{2} = \frac{1+C}{2}$$

e.g., for 95% confidence, $C = .95$, and the cumulative probability is $(1+.95)/2 = .975$. In Matlab, the command `icdf('t',.975,15)` returns 2.1314, the t critical

a
```
>> x=[1445,25882,26892,29860,35285,
47874,49544,64763,72698,75732,84457,
91577,93000,109538,113399,137652];
>> [~,~,CI]=ttest(x)

CI =
        46147           86303
```

b
```
> x<-c(1445,25882,26892,29860,35285,
47874,49544,64763,72698,75732,
84457,91577,93000,109538,113399,137652)
> t.test(x)

                  One Sample t-test
data:  x
t = 7.0305, df = 15,
p-value = 4.068e-06
alternative hypothesis: true mean is
not equal to 0
95 percent confidence interval:
 46147.22 86302.53
```

Fig. 5.9 One-sample t intervals for μ using the data in Example 5.14: (**a**) Matlab; (**b**) R

value at 15 df that we used in Example 5.14. In R, qt(.975,15) gives this same value.

To construct the interval (Eq. 5.2) for a population mean μ, use the command ztest in Matlab; the z-based CI for μ is not implemented in the R base package.

5.3.4 Exercises: Section 5.3 (37–50)

37. Determine the t critical value for a one-sample t confidence interval in each of the following situations.
    (a) Confidence level $= 95\%$, df $= 10$
    (b) Confidence level $= 95\%$, df $= 15$
    (c) Confidence level $= 99\%$, df $= 15$
    (d) Confidence level $= 99\%$, $n = 5$
    (e) Confidence level $= 98\%$, df $= 24$
    (f) Confidence level $= 99\%$, $n = 38$

38. According to the article "Fatigue Testing of Condoms" (*Polymer Testing*, 2009: 567–571), "tests currently used for condoms are surrogates for the challenges they face in use," including a test for holes, an inflation test, a package seal test, and tests of dimensions and lubricant quality (all fertile territory for the use of statistical methodology!). The investigators developed a new test that adds cyclic strain to a level well below breakage and determines the number of cycles to break. A sample of 20 condoms of one particular type resulted in a sample mean number of 1584 and a sample standard deviation of 607. Calculate and interpret a confidence interval at the 99% confidence level for the true average number of cycles to break. [*Note*: The article presented the results of hypothesis tests based on the t distribution; the validity of these depends on assuming normal population distributions.]

39. Here is a sample of ACT scores (average of the Math, English, Social Science, and Natural Science scores) for students taking college freshman calculus:

24.00	28.00	27.75	27.00	24.25	23.50	26.25
24.00	25.00	30.00	23.25	26.25	21.50	26.00
28.00	24.50	22.50	28.25	21.25	19.75	

(a) Using an appropriate graph, see if it is plausible that the observations were selected from a normal distribution.

(b) Calculate a 95% confidence interval for the population mean.

(c) The university ACT average for entering freshmen that year was about 21. Are the calculus students better than average, as measured by the ACT?

40. Even as traditional markets for sweetgum lumber have declined, large section solid timbers traditionally used for construction bridges and mats have become increasingly scarce. The article "Development of Novel Industrial Laminated Planks from Sweetgum Lumber" (*J. of Bridge Engr.*, 2008: 64–66) described the manufacturing and testing of composite beams designed to add value to low-grade sweetgum lumber. Here is data on the modulus of rupture (psi; the article contained summary data expressed in MPa):

6807.99	7637.06	6663.28	6165.03	6991.41	6992.23
6981.46	7569.75	7437.88	6872.39	7663.18	6032.28
6906.04	6617.17	6984.12	7093.71	7659.50	7378.61
7295.54	6702.76	7440.17	8053.26	8284.75	7347.95
7422.69	7886.87	6316.67	7713.65	7503.33	7674.99

(a) Verify the plausibility of assuming a normal population distribution.

(b) Estimate the true average modulus of rupture in a way that conveys information about precision and reliability.

41. A sample of 26 offshore oil workers took part in a simulated escape exercise, resulting in the accompanying data on time (seconds) to complete the escape ("Oxygen Consumption and Ventilation During Escape from an Offshore Platform," *Ergonomics*, 1997: 281–292):

389	356	359	363	375	424	325	394	402
373	373	370	364	366	364	325	339	393
392	369	374	359	356	403	334	397	

(a) Calculate a 99% confidence interval for the population mean escape time.

(b) Would a 90% CI based on the same data be wider or narrower? Explain.

42. A study of the ability of individuals to walk in a straight line ("Can We Really Walk Straight?" *Amer. J. Phys. Anthropol.*, 1992: 19–27) reported the accompanying data on cadence (strides per second) for a sample of $n = 20$ randomly selected healthy men.

.95	.85	.92	.95	.93	.86	1.00	.92	.85	.81
.78	.93	.93	1.05	.93	1.06	1.06	.96	.81	.96

A normal probability plot gives substantial support to the assumption that the population distribution of cadence is approximately normal. Calculate and interpret a 95% confidence interval for population mean cadence.

43. The article "Measuring and Understanding the Aging of Kraft Insulating Paper in Power Transformers" (*IEEE Electrical Insul. Mag.*, 1996: 28–34) contained the following observations on degree of polymerization for paper specimens for which viscosity times concentration fell in a certain middle range:

418	421	421	422	425	427	431
434	437	439	446	447	448	453
454	463	465				

(a) Is it plausible that the given sample observations were selected from a normal distribution?

(b) Calculate a 95% confidence interval for true average degree of polymerization (as did the authors of the article). Does the interval suggest that 440 is a plausible value for true average degree of polymerization? What about 450?

44. Silicone implant augmentation rhinoplasty is used to correct congenital nose deformities. The success of the procedure depends on various biomechanical properties of the human nasal periosteum and fascia. The article "Biomechanics in Augmentation Rhinoplasty" (*J. of Med. Engr. and Tech.*, 2005: 14–17) reported that for a sample of 15 (newly deceased) adults, the mean failure strain (%) was 25.0, and the standard deviation was 3.5. Assuming a normal distribution for failure strain, estimate true average strain in a way that conveys information about precision and reliability.

45. A more extensive tabulation of t critical values than what appears in this book shows that for the t distribution with 20 df, the areas to the right of the values .687, .860, and 1.064 are .25, .20, and .15, respectively. What is the confidence level for each of the following three confidence intervals for the mean μ of a normal population distribution? Which of the three intervals would you recommend be used, and why?

(a) $\left(\bar{x} - .687s/\sqrt{21}, \bar{x} + 1.725s/\sqrt{21}\right)$

(b) $\left(\bar{x} - .860s/\sqrt{21}, \bar{x} + 1.325s/\sqrt{21}\right)$

(c) $\left(\bar{x} - 1.064s/\sqrt{21}, \bar{x} + 1.064s/\sqrt{21}\right)$

46. In many applications, it suffices to have a reliable lower bound for the mean μ, because underestimating μ would be far more serious that overestimating it. This gives rise to the idea of a *lower confidence bound* for μ: a quantity L so that we can say with 95% confidence (for example) that $L < \mu$.

(a) Let t^* be a value such that

$$P\left(\frac{\overline{X} - \mu}{S/\sqrt{n}} < t^*\right) = .95$$

Manipulate the inequality inside the parentheses to isolate μ, and conclude that $L = \overline{x} - t^*(s/\sqrt{n})$ is a 95% lower confidence bound for μ.

(b) Notice that the expression given in (a) specifies that the area from $-\infty$ to t^* under the appropriate t curve is .95; equivalently, the upper tail area designated by t^* is .05. What is the appropriate t critical value for a 95% lower confidence bound with df $= 10$? df $= 15$? [*Hint:* Do not use the header row in the t table as a reference; those confidence levels refer to central areas, or equivalently two-sided confidence intervals.]

(c) A sample of 14 joint specimens of a particular type gave a sample mean proportional limit stress of 8.48 MPa and a sample standard deviation of .79 MPa ("Characterization of Bearing Strength Factors in Pegged Timber Connections," *J. Struct. Engr.*, 1997: 326–332). Assuming the data are drawn from a normally distributed population, calculate and interpret a 95% lower confidence bound for the true average proportional limit stress of all such joints.

47. An *upper confidence bound*, U, for μ is obtained by replacing the $-$ sign with a $+$ sign in the expression for L from the previous exercise: $U = \overline{x} + t^*(s/\sqrt{n})$. As in the previous exercise, the t critical value is determined by a one-tail area, not a central area.

Consider the following sample of fat content (in percentage) of $n = 10$ randomly selected hot dogs ("Sensory and Mechanical Assessment of the Quality of Frankfurters," *J. Texture Stud.*, 1990: 395-409):

| 25.2 | 21.3 | 22.8 | 17.0 | 29.8 | 21.0 | 25.5 | 16.0 | 20.9 | 19.5 |

Assuming that these were selected from a normal population distribution, calculate and interpret a 99% upper confidence bound for the population mean fat content.

48. When the sample size n is very large, lower and upper confidence bounds for μ can be obtained by replacing t^* with z^* in the expressions from the previous two exercises. For example, a large-sample lower confidence bound for μ is given by $L = \overline{x} - z^* s/\sqrt{n}$, where z^* satisfies the relation $P(Z < z^*) = c$ when $Z \sim N(0, 1)$ and $100c\%$ is the prescribed confidence level (e.g., $c = .95$).

(a) Show that the z critical value for a one-sided (i.e., upper or lower) confidence bound for μ, with confidence level $100c\%$, is given by $z^* = \Phi^{-1}(c)$.

(b) Find the one-sided z critical values for 90, 95, and 99% confidence.

(c) A certain random variable is simulated 10,000 times. The sample mean and standard deviation of the resulting 10,000 values are 41.63 and 8.05, respectively. Calculate and interpret a 95% lower confidence bound for the true expected value of this rv.

49. Often an investigator wishes to *predict* a single value of a variable to be observed at some future time, rather than to estimate the mean value of that variable. Suppose we will observe the values of a random sample $X_1, \ldots, X_n$ from a normal population with mean μ and standard deviation σ, and from these we wish to predict the value of a future independent observation X_{n+1}.
    (a) Show that

$$Z = \frac{\overline{X} - X_{n+1}}{\sigma\sqrt{1 + \dfrac{1}{n}}}$$

has a standard normal distribution, where $\overline{X}$ is the sample mean of $X_1, \ldots,$ X_n. [*Hint:* Since the population is normal, the linear combination $\overline{X} - X_{n+1}$ is normal. Show that $\overline{X} - X_{n+1}$ has mean 0 and variance $\sigma^2(1 + 1/n)$, then standardize.]

    (b) If we replace σ with the sample standard deviation S in the expression for Z from (a), it can be shown that the resulting quantity has a t distribution with $n - 1$ df. Use this fact and a derivation similar to the one presented in this section to show that a **prediction interval (PI)** for a single future observation X_{n+1} is given by

$$\overline{x} \pm t^* \cdot s\sqrt{1 + \frac{1}{n}}$$

    (c) Use the previous expression, along with the data in Exercise 47, to provide a 95% prediction interval for the fat content of a randomly selected hot dog you will consume at some future time.

50. Independent observations $X_1, \ldots, X_n \sim N(\mu_1, \sigma_1)$ and $Y_1, \ldots, Y_m \sim N(\mu_2, \sigma_2)$ will be taken. For example the heights of n men and m women might be recorded, where the height distribution of each gender is normally distributed but with unknown parameters. Of interest is the *difference* between the two unknown population means, $\mu_1 - \mu_2$.
    (a) The logical estimator of $\mu_1 - \mu_2$ is $\overline{X} - \overline{Y}$, the difference of the sample averages of the two samples. By determining the mean and variance of $\overline{X} - \overline{Y}$, show that

$$Z = \frac{(\overline{X} - \overline{Y}) - (\mu_1 - \mu_2)}{\sqrt{\dfrac{\sigma_1^2}{n} + \dfrac{\sigma_2^2}{m}}}$$

has a standard normal distribution.

    (b) Let z^* be the z critical value such that $P(-z^* < Z < z^*) = c$, where Z is as above and $100c\%$ is the prescribed confidence level. Rewrite the inequalities to provide a confidence interval for the difference of population means $\mu_1 - \mu_2$.

(c) If n and m are large (say, ≥ 40), replacing σ_1 and σ_2 with S_1 and S_2 (the two sample standard deviations) under the square root in (a) adds little extra variability; the resulting standardized variable still has approximately a standard normal distribution. Make this substitution to obtain a large-sample z confidence interval for $\mu_1 - \mu_2$ that can be implemented in practice.

(d) The article "Gender Differences in Individuals with Comorbid Alcohol Dependence and Post-Traumatic Stress Disorder" (*Amer. J. Addiction*, 2003: 412-423) reported the accompanying data in total score in the Obsessive-Compulsive Drinking Scale.

Gender	Sample size	Sample mean	Sample SD
Male	44	19.93	7.74
Female	40	16.26	7.58

Calculate and interpret a 95% confidence interval for the difference in the true mean scores for males and females.

5.4   Testing Hypotheses About a Population Mean

We have seen that a parameter can be estimated from sample data, either by a single number (a point estimate) or an entire interval of plausible values (a confidence interval). Frequently, however, the objective of an investigation is not to estimate a parameter but to decide which of two contradictory claims about the parameter is correct. Methods for accomplishing this comprise the part of statistical inference called *hypothesis testing*.

5.4.1   Hypotheses and Test Procedures

A **statistical hypothesis**, or just *hypothesis*, is a claim or assertion either about the value of a single parameter (population characteristic or characteristic of a probability distribution), about the values of several parameters, or about the form of an entire probability distribution. Examples include
- The claim that $\mu = \$350$, where μ is the true average one-term textbook expenditure for students at a university
- The assertion that $p < .50$, where p is the proportion of children who have a food allergy of some sort
- The claim that $\mu_1 - \mu_2 > 3$, where μ_1 is the true average fuel efficiency (mpg) of all current model year Honda Accords equipped with a 4-cylinder engine and μ_2 is the analogous characteristic for Accords equipped with a 6-cylinder engine.
   In any hypothesis-testing problem, there are two contradictory hypotheses under consideration. One hypothesis might be the claim $\mu = \$350$ and the other $\mu \neq \$350$,

or the two contradictory statements might be $p \geq .50$ and $p < .50$. The objective is to decide, based on sample information, which of the two hypotheses is correct. There is a familiar analogy to this in a criminal trial. One claim is the assertion that the accused individual is innocent. In the US judicial system, this is the claim that is initially believed to be true. Only in the face of strong evidence to the contrary should the jury reject this claim in favor of the alternative assertion that the accused is guilty. In this sense, the claim of innocence is the favored or protected hypothesis, and the burden of proof is placed on those who believe in the alternative claim.

Similarly, in testing statistical hypotheses, the problem will be formulated so that one of the claims is initially favored. This initially favored claim will not be rejected in favor of the alternative claim unless sample evidence contradicts it and provides strong support for the alternative assertion.

DEFINITION

The **null hypothesis**, denoted by H_0, is the claim that is initially assumed to be true (the "prior belief" claim). The **alternative hypothesis**, denoted by H_a, is the assertion that is contradictory to H_0.

The null hypothesis will be rejected in favor of the alternative hypothesis only if sample evidence suggests that H_0 is false. If the sample does not strongly contradict H_0, we will continue to believe in the plausibility of the null hypothesis. The two possible conclusions from a hypothesis-testing analysis are then *reject H_0* or *fail to reject H_0*.

A **test of hypotheses** is a method for using sample data to decide whether the null hypothesis should be rejected. Thus we might test H_0: $\mu = 350$ against the alternative H_a: $\mu \neq 350$. Only if sample data strongly suggests that μ is something other than 350 should the null hypothesis be rejected. In the absence of such evidence, H_0 should not be rejected since it is still judged to be plausible.

Sometimes an investigator does not want to accept a particular assertion unless and until data can provide strong support for the assertion; in that situation, this assertion will be the investigator's alternative hypothesis H_a. As an example, suppose a company is considering putting a new additive in the dried fruit that it produces. The true average shelf life with the current additive is known to be 200 days. With μ denoting the true average shelf life with the *new* additive, the company would not want to make a change unless evidence strongly suggested that μ exceeds 200. An appropriate problem formulation would involve testing H_0: $\mu = 200$ against H_a: $\mu > 200$. The conclusion that a change is justified is identified with H_a, and it would take conclusive evidence to justify rejecting H_0 and switching to the new additive.

Scientific research often involves trying to decide whether a current theory should be replaced by a more plausible and satisfactory explanation of the phenomenon under investigation. A conservative approach is to identify the current theory with H_0 and the researcher's alternative explanation with H_a. Rejection of the current theory will then occur only when evidence is much more consistent with the new theory. In many situations, H_a is referred to as the "research hypothesis," since it is the claim that the researcher would really like to validate. The word *null* means "of no value, effect, or consequence," which suggests that H_0 should be identified with the hypothesis of no change (from current opinion), no difference, no improvement, and so on. Suppose, for example, that 10% of all computer circuit boards produced by a manufacturer during a recent period were defective. An engineer has suggested a change in the production process in the belief that it will result in a reduced defective rate. Let p denote the true proportion of defective boards resulting from the changed process. Then the research hypothesis, on which the burden of proof is placed, is the assertion that $p < .10$. Thus the alternative hypothesis is H_a: $p < .10$.

In our treatment of hypothesis testing, H_0 will generally be stated as an equality claim. When the parameter of interest is a population mean μ, the null hypothesis will have the form H_0: $\mu = \mu_0$, where μ_0 is a specified number called the *null value* (value claimed for μ by the null hypothesis). For example, let μ represent the true average breaking strength of nylon string of a certain type. If a particular application requires that μ exceed 100 N and the string will not be used unless there is compelling evidence that this is the case, the natural alternative hypothesis is H_a: $\mu > 100$. It would then make sense to select as the null hypothesis the assertion that $\mu \leq 100$. However, we will instead simplify the null hypothesis to H_0: $\mu = 100$. The rationale for using this simplified null hypothesis is that any reasonable decision procedure for deciding between H_0: $\mu = 100$ and H_a: $\mu > 100$ will also be reasonable for deciding between the claim that $\mu \leq 100$ and H_a, and should lead to exactly the same conclusion for any particular sample. The use of a simplified H_0 is preferred because it has certain technical benefits, which will be apparent shortly.

The alternative to the null hypothesis H_0: $\mu = \mu_0$ will look like one of the following three assertions:

1. H_a: $\mu > \mu_0$ (in which case the implicit null hypothesis is $\mu \leq \mu_0$)
2. H_a: $\mu < \mu_0$ (so the implicit null hypothesis states that $\mu \geq \mu_0$)
3. H_a: $\mu \neq \mu_0$

5.4.2   Test Procedures for Hypotheses About a Population Mean μ

The decision as to whether H_0 should be rejected is based on the analysis of data x_1, $x_2, \ldots, x_n$ resulting from a random sample of the population. A sensible strategy at this point would be to calculate the sample mean $\bar{x}$ and reject the null hypothesis if its value is too far from μ_0 in the appropriate direction. For example, in the scenario involving breaking strength of nylon string, a value of $\bar{x}$ considerably larger than 100 would suggest that H_0 is false and should be rejected. But an $\bar{x}$ value *less* than 100 would not incline us to reject H_0 in favor of H_a, since a sample mean less than 100 would certainly not convince us that the population mean μ is more than 100.

Rather than base a decision on $\bar{x}$ itself, let's standardize $\bar{x}$ assuming that the null hypothesis is true:

$$t = \frac{\bar{x} - \mu_0}{s/\sqrt{n}}$$

If we knew the value of the population standard deviation σ, we'd use it rather than the sample standard deviation s, but in practice this is almost never the case. Continuing with the nylon string scenario, $t = (\bar{x} - 100)/(s/\sqrt{n})$. For $n = 25$ and sample data $\bar{x} = 108.5$, $s = 12.14$, we calculate $t = 8.5/2.428 = 3.50$. The interpretation is that the value of the sample mean, 108.5, is 3.5 estimated standard errors from what we'd expect it to be *if the null hypothesis were true*. In general, t is the distance between the sample mean and what we'd expect it to be if H_0 were true, expressed in standard deviations.

Now let's see if we can identify which values of t are at least as contradictory to H_0 as the value calculated from the available sample data. Again focusing on the nylon string situation, because the alternative hypothesis states that the population mean *exceeds* 100, any value of $\bar{x}$ greater than 108.5 argues even more strongly against H_0 than does the 108.5 resulting from our sample. And any $\bar{x}$ greater than 108.5 corresponds to a value of t that exceeds 3.50. So values of t that are at least 3.50 are at least as contradictory to H_0 as 3.50 itself.

As another example, now suppose that μ represents the mean IQ for a large population of children, and consider the rival hypotheses H_0: $\mu = 100$ and H_a: $\mu \neq 100$. Because 100 is the generally accepted value of mean IQ in the USA, the alternative hypothesis here states that the average for the designated population of children is different from this accepted value. Suppose a sample of 225 children gives a sample mean IQ of 98.6 and a sample standard deviation of 16.15, from which $t = (98.6 - 100)/(16.15/\sqrt{225}) = -1.30$. The average IQ in the sample is 1.3 estimated standard errors smaller than what would be expected were the null hypothesis true. To decide which values of t are at least as contradictory to H_0 as -1.30, first consider which values of $\bar{x}$ are at least as contradictory to H_0 as 98.6. Not only is any value 98.6 or smaller in this category, but also any value that is at least 101.4—that is, any value at least as far from 100 in *either* direction (because $\neq$ appears in the alternative hypothesis). Thus any value of t that is either ≤ -1.30 or ≥ 1.30 is at least as contradictory to H_0 as our calculated $t = -1.30$.

5.4.3   *P*-Values and the One-Sample *t* Test

Before data have been obtained, the sample mean and sample standard deviation are random variables, which we have previously denoted by $\overline{X}$ and S, respectively. Substituting these for $\bar{x}$ and s in the formula for t gives what is called the **test statistic**

$$T = \frac{\overline{X} - \mu_0}{S/\sqrt{n}},$$

which is also a random variable (that is, its value is subject to uncertainty prior to obtaining the sample data).

If the population distribution is normal, Gosset's Theorem from Sect. 5.3 implies that *when the null hypothesis is true, T has a t distribution with $n-1$ degrees of freedom*. In the case of the nylon string scenario, $T = (\overline{X} - 100)/(S/\sqrt{n})$ would have a t_{n-1} distribution when $H_0: \mu = 100$ is true (assuming that the population strength distribution is normal). For the previously given sample information, the calculated value of the test statistic was 3.50. Now consider the probability, calculated assuming that the null hypothesis is true, of obtaining a test statistic value at least as contradictory to the null hypothesis as the value 3.50 resulting from our sample data:

$$\begin{aligned} P_{H_0}(T \geq 3.50) &= P(\text{a } t_{24} \text{ random variable is at least } 3.50) \\ &= \text{the area under the } t_{24} \text{ curve to the right of } 3.50 \\ &= .001 \, (\text{from software}) \end{aligned}$$

That is, if the null hypothesis is true, there is only a .1% chance of obtaining a sample at least as contradictory to the null hypothesis as our sample. So our sample is among the .1% of all samples most contradictory to H_0.

Recall that $t = -1.30$ in the IQ scenario. Again assuming a normal population distribution, the probability of getting a value of T at least as contradictory to H_0 when H_0 is true is

$$\begin{aligned} P_{H_0}(T \leq -1.30 \text{ or } T \geq 1.30) &= P(\text{a } t_{224} \text{ rv is} \leq -1.30 \text{ or} \geq 1.30) \\ &= (\text{area under the } t_{224} \text{ curve to the left of } -1.30) + \\ &\qquad (\text{area under the } t_{224} \text{ curve to the right of } 1.30) \\ &= 2(\text{area under the } t_{224} \text{ curve to the right of } 1.30) \\ &\approx 2(\text{area under the } z \text{ curve to the right of } 1.30) \\ &= .1936 \end{aligned}$$

So when the null hypothesis is true, almost 20% of all samples would result in a test statistic value that is at least as contradictory to H_0 as the one resulting from our sample. This implies that our sample is not very contradictory to H_0.

DEFINITION

The **P-value** is the probability, calculated assuming that the null hypothesis is true, of obtaining a value of the test statistic at least as contradictory to H_0 as the value calculated from the available sample. The smaller the P-value, the

---

(continued)

more the data contradicts the null hypothesis, so H_0 should be rejected in favor of H_a if the P-value is sufficiently small.

More specifically, select a number α reasonably close to 0; then reject the null hypothesis if P-value $\leq \alpha$ and do not reject the null hypothesis if P-value $> \alpha$. The selected α is called the **significance level** of the test.

The most frequently employed values of the significance level are $\alpha = .05$, .01, and .001. We shall say more about the choice of α shortly.

ONE-SAMPLE T TEST

Consider testing the null hypothesis H_0: $\mu = \mu_0$ based on a random sample X_1, X_2, ..., X_n from a normal population distribution (the plausibility of the normality assumption should be checked by examining a normal probability plot). The test statistic is

$$T = \frac{\overline{X} - \mu_0}{S/\sqrt{n}}$$

The calculated value of this test statistic is $t = (\bar{x} - \mu_0)/(s/\sqrt{n})$. The determination of the P-value depends on the choice of H_a as follows:

Alternative Hypothesis	P-value
H_a: $\mu > \mu_0$	Area under the t_{n-1} curve to the right of t
H_a: $\mu < \mu_0$	Area under the t_{n-1} curve to the left of t
H_0: $\mu \neq \mu_0$	2·(Area under the t_{n-1} curve to the right of $|t|$)

The test procedure when the alternative hypothesis is H_a: $\mu > \mu_0$ is referred to as an *upper-tailed test*, because the P-value is the area captured in the upper tail of the relevant t curve (i.e., to the right of t). Analogously, the test procedure for the second case is called a *lower-tailed test*, and the procedure in the third case is a *two-tailed test*. Figure 5.10 illustrates the determination of the P-value in the three different cases.

Appendix Table A.6 provides information about tail areas under various t curves. The calculated value of t (to the accuracy of the tenths digit) appears along the left margin, and there is a different column for each number of df. For example, the entry at the intersection of the $t = 2.4$ row and the 15 df column is .015, the area under the 15 df t curve to the right of 2.4. By symmetry, this is also the area under the 15 df t curve to the left of -2.4. Various software packages will allow for more decimal accuracy in t and the corresponding areas.

Example 5.16 Correct alignment of the tibial and femoral components is an important factor in determining favorable long-term results of total knee

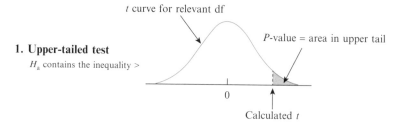

1. Upper-tailed test

H_a contains the inequality >

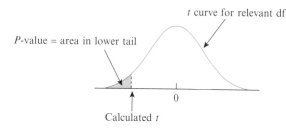

2. Lower-tailed test

H_a contains the inequality <

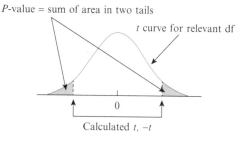

3. Two-tailed test

H_a contains the inequality ≠

Fig. 5.10 *P*-values for *t* tests

arthroplasty (TKA). It is generally accepted that the tibial component should be placed perpendicular to the anatomical axis of the tibia. The article "Simple Method for Confirming Tibial Osteotomy During Total Knee Arthroplasty" (*Sports Medicine, Arthroscopy, Rehabilitation, Therapy, and Technology*, 2012, 4:44) reported that for a sample of 35 TKAs, the sample mean varus angle of the tibial osteotomy was 89.45° and the sample standard deviation was 1.62°. The authors of the cited article carried out a one-sample *t* test to see whether the true average angle differed from 90° (presumably after examining a normal probability plot of the data). The relevant hypotheses are H_0: $\mu = 90$ versus H_a: $\mu \neq 90$.

The calculated value of the test statistic is

$$t = (89.46 - 90)/\left(1.62/\sqrt{35}\right) = -1.97$$

The inequality in H_a implies that the test is two-tailed, so the *P*-value is twice the area under the t_{34} curve to the right of 1.97. The entry in the 2.0 row and 35 df column of Table A.6 is .027, so the *P*-value is approximately 2(.027) = .054 (the article reports .055; notice that we have had to round both the test statistic and the df in order to use the *t* table).

Thus with a significance level of .05, the null hypothesis cannot be rejected because P-value $= .054 > .05 = \alpha$. This is what allowed the investigators to conclude that "there was no significant difference from the target angle of $90°$." ■

Recall from the previous section on confidence intervals that when the sample size n is large, the standardized variable $(\overline{X} - \mu)/(S/\sqrt{n})$ has approximately a standard normal distribution even if the population distribution is not normal. The implication here is that we can relabel our test statistic as $Z = (\overline{X} - \mu_0)/(S/\sqrt{n})$. Then the prescription in the one-sample t box for obtaining the P-value is modified by replacing t_{n-1} and t by z. That is, the P-value for these large-sample tests is an appropriate z curve area.

Example 5.17 The recommended daily intake of calcium for adults ages 18–30 is 1000 mg/day. The article "Dietary and Total Calcium Intakes Are Associated with Lower Percentage Total Body and Truncal Fat in Young, Healthy Adults" (*J. of the Amer. College of Nutr.*, 2011: 484–490) reported the following summary data for a sample of 76 healthy Caucasian males from southwestern Ontario, Canada: $n = 76$, $\bar{x} = 1093$, $s = 477$. Let's carry out a test at significance level .01 to see whether the population mean daily intake exceeds the recommended value. The relevant hypotheses are H_0: $\mu = 1000$ versus H_a: $\mu > 1000$.
    The calculated value of the test statistic is

$$z = (1093 - 1000)/\left(477/\sqrt{76}\right) = 1.70$$

The resulting P-value is the area under the standard normal curve to the right of 1.70 (the inequality in H_a implies that the test is upper-tailed). From Table A.3, this area is $1 - \Phi(1.70) = 1 - .9554 = .0446$. Because this P-value is larger than .01, H_0 cannot be rejected. There is not compelling evidence to conclude at significance level .01 that the population mean daily intake exceeds the recommended value (even though the sample mean does so). Note that the opposite conclusion would result from using a significance level of .05. But the smaller α that we used requires more persuasive evidence from the data before rejecting H_0. ■

5.4.4   Errors in Hypothesis Testing and the Power of a Test

When a jury is called upon to render a verdict in a criminal trial, there are two possible erroneous conclusions to be considered: convicting an innocent person, or letting a guilty person go free. Similarly, in statistical hypothesis testing there are two potential errors whose consequences must be considered when reaching a conclusion.

> **DEFINITION**
> A **Type I error** involves rejecting the null hypothesis H_0 when it is true.
> A **Type II error** involves not rejecting H_0 when it is false.

Since in the US judicial system the null hypothesis (a priori belief) is that the accused is innocent, a Type I error is analogous to convicting an innocent person. It would be nice if test procedures could be developed that offered 100% protection against committing both a Type I error and a Type II error. This is an impossible goal, however, because a conclusion is based on sample data rather than a census of the entire population. There is always some chance that the sample will be unrepresentative of the population and lead to an incorrect conclusion. The best we can hope for is a test procedure for which it is unlikely that either a Type I or a Type II error will be committed.

Let's reconsider the calcium intake scenario of the previous example. We employed a significance level of $\alpha = .01$, and so we could reject H_0 only if P-value $\le .01$. The P-value in the case of this upper-tailed large-sample test is the area under the standard normal curve to the right of the calculated z. Table A.3 shows that the z-value 2.33 captures an upper-tail area of .01 (look inside the table for a cumulative area of .9900). The P-value (captured upper-tail area) will therefore be at most .01 if and only if z is at least 2.33; see Fig. 5.11.

Thus the probability of committing a Type I error—rejecting H_0 when it is true—is the probability that the value of the test statistic Z will be at least 2.33 when H_0 is true. Now the key fact: because we created Z by subtracting the null value μ_0 when standardizing, Z has a standard normal distribution when H_0 is true. So

$$P(\text{Type I error}) = P(\text{rejecting } H_0 \text{ when } H_0 \text{ is true})$$
$$= P(Z \ge 2.33 \text{ when } Z \text{ is a standard normal rv}) = .01 = \alpha$$

This is true not only for the z test of Example 5.17 but also for the t tests described earlier and, in fact, for any test procedure.

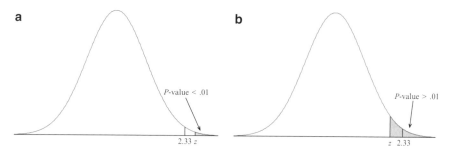

Fig. 5.11 P-values for an upper-tailed large-sample test: (**a**) P-value $<.01$ if $z > 2.33$; (**b**) P-value $>.01$ if $z < 2.33$

> **PROPOSITION**
> The significance level α that is employed when H_0 is rejected iff P-value $\leq \alpha$ is also the probability that the test results in a Type I error.

Thus a test with significance level .01 is one for which there is a 1% chance of committing a Type I error, whereas using a significance level of .05 results in a test with a Type I error probability of .05. The smaller the significance level, the less likely it is that the null hypothesis will be rejected when it is true. A smaller significance level makes it harder for the null hypothesis to be rejected and therefore less likely that a Type I error will be committed.

It is natural to ask at this point why a significance level of .05 should ever be employed when a significance level of .01 can be used. More generally, why use a test with a larger significance level—larger probability of a Type I error—when a smaller level is available? The answer lies in something that we have not yet explicitly considered: the likelihood of committing a Type II error. Let's denote the probability of a Type II error by β. That is,

$$\beta = P(\text{not rejecting } H_0 \text{ when } H_a \text{ is true})$$

This notation is actually somewhat misleading: whereas for any particular test there is a single value of α (a consequence of having H_0 be a statement of equality), there are in fact many different values of β. The alternative hypothesis in the calcium intake situation was H_a: $\mu > 1000$. So this would be true if μ were 1010 or 1050 or 1100 or in fact any value exceeding 1000. Nevertheless, for any particular way of H_0 being false and H_a true, it can be shown that α and β are inversely related: changing the test procedure by decreasing α in order to make the chance of a Type I error smaller has the inevitable consequence of making a Type II error more likely. Conversely, using a larger significance level will make it less likely that the null hypothesis will fail to be rejected when, in fact, it is false.

Let μ' denote some particular value of μ for which H_a is true. For example, for the hypotheses H_0: $\mu = 90$ versus H_a: $\mu \neq 90$ from Example 5.16, we might be interested in determining β when the true angle is $91°$. Then $\mu' = 91$ and we wish $\beta(91)$. The value of β depends on several factors:

- How far the alternative value of interest μ' is from μ_0 [$\beta(\mu')$ decreases as μ' moves further away from μ_0]
- The sample size n [$\beta(\mu')$ decreases as n, and therefore df, increases]
- The value of the population standard deviation σ [the larger the value of σ, the more difficult it is for H_0 to be rejected, and so the larger is $\beta(\mu')$]
- The significance level α [making α smaller increases $\beta(\mu')$]

Calculating β for the one-sample t test by hand is quite difficult. This is because when $\mu = \mu'$ rather than the null value μ_0, the density function that describes the distribution of the test statistic T is exceedingly complicated. Fortunately statistical software comes to our rescue. Rather than work directly with β, the most commonly used software packages involve a quantity called *power*.

> **DEFINITION**
>
> Suppose the null and alternative hypotheses are assertions about the value of some parameter θ, with the null hypothesis having the form H_0: $\theta = \theta_0$ and the alternative hypothesis obtained by replacing $=$ in H_0 by one of the three inequalities $>$, $<$ or $\neq$. Let θ' denote some particular value of θ for which H_a is true. Then the **power** at the value θ' for a test of these hypotheses is the probability of rejecting H_0 when $\theta = \theta'$, which is $1 - \beta(\theta')$. The power of the test when $\theta = \theta_0$ is also the probability that H_0 is rejected, which in this case is the significance level α.

Thus we want the power to be close to 0 when the null hypothesis is true and close to 1 when the null hypothesis is false. A "powerful" test is one that has high power for alternative values of the parameter, and thus good ability to detect departures from the null hypothesis.

Example 5.18 The true average voltage drop from collector to emitter of insulated gate bipolar transistors of a certain type is supposed to be at most 2.5 V. An investigator selects a sample of $n = 10$ such transistors and uses the resulting voltages as a basis for testing H_0: $\mu = 2.5$ versus H_a: $\mu > 2.5$ using a t test with significance level $\alpha = .05$. If the standard deviation of the voltage distribution is $\sigma = .1$ V, how likely is it that H_0 will not be rejected when in fact $\mu = 2.55$ or when $\mu = 2.6$? And what happens to the power and β if the sample size is increased to 20? The sampsizepwr function in Matlab provides the following information:

μ'	n	Power
2.55	10	.4273
2.55	20	.6951
2.6	10	.8975
2.6	20	.9961

So in the case $\mu' = 2.55$, β is roughly .57 when the sample size is 10 and roughly .30 when the sample size is 20. Clearly these Type II error probabilities are rather large. If it is important to detect such a departure from H_0, the test does not have good power to do so. Software can also be used to determine what value of the sample size n is necessary to produce a sufficiently large power and correspondingly small β. For example, when $\mu' = 2.55$, a sample size of $n = 36$ is required to produce a power of .90. ∎

As Example 5.18 illustrates, the power of a test can be disappointingly small for an alternative value of the parameter that represents an important departure from the null hypothesis. Too often investigators are content to specify a comfortingly small value of α without paying attention to power and β. This can easily result in a test which has poor ability to detect when H_0 is false. Given the availability and capabilities of statistical software packages, such a sin is unpardonable!

5.4.5   Software for Hypothesis Test Calculation

The t.test and ttest functions in R and Matlab, respectively, mentioned at the end of Sect. 5.3 can be used to automatically perform the one-sample t test described in this section (in fact, that is the primary purpose of these functions).

Example 5.19 The accompanying data on cube compressive strength (MPa) of concrete specimens appeared in the article "Experimental Study of Recycled Rubber-Filled High-Strength Concrete" (*Magazine of Concrete Res.*, 2009: 549-556):

112.3	97.0	92.7	86.0	102.0
99.2	95.8	103.5	89.0	86.7

Suppose the concrete will be used for a particular application unless there is strong evidence that the true average strength is less than 100 MPa. Should the concrete be used? Test at the .05 significance level.

Let μ denote the true average cube compressive strength of this concrete. We wish to test the hypotheses H_0: $\mu = 100$ versus H_a: $\mu < 100$. A probability plot indicates the data are consistent with a normally distributed population. Figure 5.12 shows the hypothesis test implemented in R and Matlab.

Both R and Matlab give a one-sided P-value of .1315 at 9 df. Since .1315 > .05, at the .05 significance level we fail to reject H_0: there is insufficient evidence to conclude the true mean strength of this concrete is less than 100 MPa. As a consequence, the concrete *should* be used.

In Fig. 5.12a, R gives the computed value of the test statistic, $t = -1.1937$, as well as the sample mean, $\bar{x} = 96.92$ MPa, and a (one-sided) CI for μ of $(-\infty, 101.6497)$. (See Exercises 46–47 from the previous section for information on such bounds.) In Matlab, the significance level of .05 is a required input; the 'left'

a
```
> x<-c(112.3,97.0,97.7,86.0,
102.0,99.2,95.8,103.5
,89.0,86.7)
> t.test(x,mu=100,alternative="less")

        One Sample t-test

data:  x
t = -1.1937, df = 9, p-value = 0.1315
alternative hypothesis: true mean is
less than 100
95 percent confidence interval:
    -Inf 101.6497
sample estimates:
mean of x
    96.92
```

b
```
>> x=[112.3,97.0,97.7,86.0,
102.0,99.2,95.8,103.5,
89.0,86.7];
>> [H,P]=ttest(x,100,.05,'left')
H =
    0
P =
    0.1315
```

Fig. 5.12 Performing the hypothesis test of Example 5.19: (**a**) R; (**b**) Matlab

argument instructs Matlab to perform a lower-tailed test. As seen in Fig. 5.12b, Matlab then returns two items: the P-value, and also a logical bit denoted H indicated whether to reject H_0. (The H $= 0$ output tells the user not to reject H_0 at the specified α level.) ∎

Calculations of power, as well as sample size required to achieve a prescribed power level, are available through the samplesizepwr function in Matlab and the pwr package in R. The former is part of the Matlab Statistics Toolbox; the latter is not part of the R base package and must be downloaded and installed.

5.4.6 Exercises: Section 5.4 (51–76)

51. For each of the following assertions, state whether it is a legitimate statistical hypothesis and why:
    (a) H: $\sigma > 100$
    (b) H: $\hat{P} = .45$
    (c) H: $S \leq .20$
    (d) H: $\sigma_1/\sigma_2 < 1$
    (e) H: $\overline{X} - \overline{Y} = 5$
    (f) H: $\lambda \leq .01$, where λ is the parameter of an exponential distribution used to model component lifetime
52. For the following pairs of assertions, indicate which do not comply with our rules for setting up hypotheses and why (the subscripts 1 and 2 differentiate between quantities for two different populations or samples):
    (a) H_0: $\mu = 100$, H_a: $\mu > 100$
    (b) H_0: $\sigma = 20$, H_a: $\sigma \leq 20$
    (c) H_0: $p \neq .25$, H_a: $p = .25$
    (d) H_0: $\mu_1 - \mu_2 = 25$, H_a: $\mu_1 - \mu_2 > 100$
    (e) H_0: $S_1^2 = S_2^2$, H_a: $S_1^2 \neq S_2^2$
    (f) H_0: $\mu = 120$, H_a: $\mu = 150$
    (g) H_0: $\sigma_1/\sigma_2 = 1$, H_a: $\sigma_1/\sigma_2 \neq 1$
    (h) H_0: $p_1 - p_2 = -.1$, H_a: $p_1 - p_2 < -.1$
53. To determine whether the girder welds in a new performing arts center meet specifications, a random sample of welds is selected, and tests are conducted on each weld in the sample. Weld strength is measured as the force required to break the weld. Suppose the specifications state that mean strength of welds should exceed 100 lb/in^2; the inspection team decides to test H_0: $\mu = 100$ versus H_a: $\mu > 100$. Explain why it might be preferable to use this H_a rather than $\mu < 100$.
54. Let μ denote the true average radioactivity level (picocuries per liter). The value 5 pCi/L is considered the dividing line between safe and unsafe water. Would you recommend testing H_0: $\mu = 5$ versus H_a: $\mu > 5$ or H_0: $\mu = 5$ versus

H_a: $\mu < 5$? Explain your reasoning. [*Hint*: Think about the consequences of a Type I and Type II error for each possibility.]

55. For which of the given P-values would the null hypothesis be rejected when performing a level .05 test?
    (a) .001
    (b) .021
    (c) .078
    (d) .047
    (e) .148

56. Pairs of P-values and significance levels, α, are given. For each pair, state whether the observed P-value would lead to rejection of H_0 at the given significance level.
    (a) P-value $= .084$, $\alpha = .05$
    (b) P-value $= .084$, $\alpha = .10$
    (c) P-value $= .003$, $\alpha = .01$
    (d) P-value $= .039$, $\alpha = .01$

57. Give as much information as you can about the P-value of a t test in each of the following situations:
    (a) Upper-tailed test, df $= 8$, $t = 2.0$
    (b) Lower-tailed test, df $= 11$, $t = -2.4$
    (c) Two-tailed test, df $= 15$, $t = -1.6$
    (d) Upper-tailed test, df $= 19$, $t = -.4$
    (e) Upper-tailed test, df $= 5$, $t = 5.0$
    (f) Two-tailed test, df $= 40$, $t = -4.8$

58. The paint used to make lines on roads must reflect enough light to be clearly visible at night. Let μ denote the true average reflectometer reading for a new type of paint under consideration. A test of H_0: $\mu = 20$ versus H_a: $\mu > 20$ will be based on a random sample of size n from a normal population distribution. What conclusion is appropriate in each of the following situations?
    (a) $n = 15$, test statistic value $= 3.2$, $\alpha = .05$
    (b) $n = 9$, test statistic value $= 1.8$, $\alpha = .01$
    (c) $n = 24$, test statistic value $= -.2$

59. Let μ denote the mean reaction time to a certain stimulus. For a large-sample z test of H_0: $\mu = 5$ versus H_a: $\mu > 5$, find the P-value associated with each of the given values of the z test statistic.
    (a) 1.42
    (b) .90
    (c) 1.96
    (d) 2.48
    (e) −.11

60. Newly purchased tires of a certain type are supposed to be filled to a pressure of 35 lb/in². Let μ denote the true average pressure. Find the P-value associated with each given z statistic value for testing H_0: $\mu = 35$ versus the alternative H_a: $\mu \neq 35$.
    (a) 2.10
    (b) −1.75

(c) $-.55$
(d) 1.41
(e) -5.3

61. A pen has been designed so that true average writing lifetime under controlled conditions (involving the use of a writing machine) is at least 10 h. A random sample of 18 pens is selected, the writing lifetime of each is determined, and a normal probability plot of the resulting data supports the use of a one-sample t test.

(a) What hypotheses should be tested if the investigators believe a priori that the design specification has been satisfied?

(b) What conclusion is appropriate if the hypotheses of part (a) are tested, $t = -2.3$, and $\alpha = .05$?

(c) What conclusion is appropriate if the hypotheses of part (a) are tested, $t = -1.8$, and $\alpha = .01$?

(d) What should be concluded if the hypotheses of part (a) are tested and $t = -3.6$?

62. Lightbulbs of a certain type are advertised as having an average lifetime of 750 h. The price of these bulbs is very favorable, so a potential customer has decided to go ahead with a purchase arrangement unless it can be conclusively demonstrated that the true average lifetime is smaller than what is advertised. A random sample of 50 bulbs was selected and the lifetime of each bulb determined. These 50 light bulbs had a sample mean lifetime of 738.44 h with a sample standard deviation of 38.20 h. What conclusion would be appropriate for a significance level of .05?

63. Automatic identification of the boundaries of significant structures within a medical image is an area of ongoing research. The paper "Automatic Segmentation of Medical Images Using Image Registration: Diagnostic and Simulation Applications" (*J. of Medical Engr. and Tech.*, 2005: 53–63) discussed a new technique for such identification. A measure of the accuracy of the automatic region is the average linear displacement (ALD). The paper gave the following ALD observations for a sample of 49 kidneys (units of pixel dimensions).

1.38	0.44	1.09	0.75	0.66	1.28	0.51
0.39	0.70	0.46	0.54	0.83	0.58	0.64
1.30	0.57	0.43	0.62	1.00	1.05	0.82
1.10	0.65	0.99	0.56	0.56	0.64	0.45
0.82	1.06	0.41	0.58	0.66	0.54	0.83
0.59	0.51	1.04	0.85	0.45	0.52	0.58
1.11	0.34	1.25	0.38	1.44	1.28	0.51

(a) Is it plausible that ALD is at least approximately normally distributed? Must normality be assumed prior to testing hypotheses about true average ALD? Explain.

(b) The authors commented that in most cases the ALD is better than or of the order of 1.0. Does the data in fact provide strong evidence for concluding that true average ALD under these circumstances is less than 1.0? Carry out an appropriate test of hypotheses.

64. A dynamic cone penetrometer (DCP) is used for measuring material resistance to penetration (mm/blow) as a cone is driven into pavement or subgrade. Suppose that for a particular application it is required that the true average DCP value for a certain type of pavement be less than 30. The pavement will not be used unless there is conclusive evidence that the specification has been met. Test the appropriate hypotheses using the following data ("Probabilistic Model for the Analysis of Dynamic Cone Penetrometer Test Values in Pavement Structure Evaluation," *J. of Testing and Evaluation*, 1999: 7–14):

14.1	14.5	15.5	16.0	16.0	16.7	16.9	17.1	17.5	17.8
17.8	18.1	18.2	18.3	18.3	19.0	19.2	19.4	20.0	20.0
20.8	20.8	21.0	21.5	23.5	27.5	27.5	28.0	28.3	30.0
30.0	31.6	31.7	31.7	32.5	33.5	33.9	35.0	35.0	35.0
36.7	40.0	40.0	41.3	41.7	47.5	50.0	51.0	51.8	54.4
55.0	57.0								

65. The article "Uncertainty Estimation in Railway Track Life-Cycle Cost" (*J. of Rail and Rapid Transit*, 2009) presented the following data on time to repair (min) a rail break in the high rail on a curved track of a certain railway line.

159	120	480	149	270	547	340	43	228	202	240	218

A normal probability plot of the data shows a reasonably linear pattern, so it is plausible that the population distribution of repair time is at least approximately normal. The sample mean and standard deviation are 249.7 and 145.1, respectively. Is there compelling evidence for concluding that true average repair time exceeds 200 min? Carry out a test of hypotheses using a significance level of .05.

66. Have you ever been frustrated because you could not get a container of some sort to release the last bit of its contents? The article "Shake, Rattle, and Squeeze: How Much Is Left in That Container?" (*Consumer Reports*, May 2009: 8) reported on an investigation of this issue for various consumer products. Suppose five 6.0 oz tubes of toothpaste of a particular brand are randomly selected and squeezed until no more toothpaste will come out. Then each tube is cut open and the amount remaining is weighed, resulting in the following data (consistent with what the cited article reported): .53, .65, .46, .50, .37. Does it appear that the true average amount left is less than 10% of the advertised net contents?

(a) Check the validity of any assumptions necessary for testing the appropriate hypotheses.

(b) Carry out a test of the appropriate hypotheses using a significance level of .05. Would your conclusion change if a significance level of .01 had been used?

(c) Describe in context Type I and II errors, and say which error might have been made in reaching a conclusion.

67. A random sample of soil specimens was obtained, and the amount of organic matter (%) in the soil was determined for each specimen, resulting in the accompanying data (from "Engineering Properties of Soil," *Soil Science*, 1998: 93–102).

1.10	5.09	0.97	1.59	4.60	0.32	0.55	1.45
0.14	4.47	1.20	3.50	5.02	4.67	5.22	2.69
3.98	3.17	3.03	2.21	0.69	4.47	3.31	1.17
0.76	1.17	1.57	2.62	1.66	2.05		

The values of the sample mean and standard deviation are 2.481 and 1.616, respectively. Does this data suggest that the true average percentage of organic matter in such soil is something other than 3%? Carry out a test of the appropriate hypotheses at significance level .10. Would your conclusion be different if $\alpha = .05$ had been used? [*Note:* A normal probability plot of the data shows an acceptable pattern in light of the reasonably large sample size.]

68. Glycerol is a major by-product of ethanol fermentation in wine production and contributes to the sweetness, body, and fullness of wines. The article "A Rapid and Simple Method for Simultaneous Determination of Glycerol, Fructose, and Glucose in Wine" (*American J. of Enology and Viticulture*, 2007: 279–283) includes the following observations on glycerol concentration (mg/ml) for samples of standard-quality (uncertified) white wines: 2.67, 4.62, 4.14, 3.81, 3.83. Suppose the desired concentration value is 4. Does the sample data suggest that true average concentration is something other than the desired value? Carry out a test of appropriate hypotheses using the one-sample t test with a significance level of .05.

69. Exercise 41 gave $n = 26$ observations on escape time (seconds) for oil workers in a simulated exercise, from which the sample mean and sample standard deviation are 370.69 and 24.36, respectively. Suppose the investigators had believed a priori that true average escape time would be at most 6 min. Does the data contradict this prior belief? Assuming normality, test the appropriate hypotheses using a significance level of .05.

70. Minor surgery on horses under field conditions requires a reliable short-term anesthetic producing good muscle relaxation, minimal cardiovascular and respiratory changes, and a quick, smooth recovery with minimal aftereffects so that horses can be left unattended. The article "A Field Trial of Ketamine Anesthesia in the Horse" (*Equine Vet. J.*, 1984: 176–179) reports that for a sample of $n = 73$ horses to which ketamine was administered under certain conditions, the sample average lateral recumbency (lying-down) time was 18.86 min and the standard deviation was 8.6 min. Does this data suggest that

true average lateral recumbency time under these conditions is less than 20 min? Test the appropriate hypotheses at level of significance .10.

71. The recommended daily dietary allowance for zinc among males older than age 50 years is 15 mg/day. The article "Nutrient Intakes and Dietary Patterns of Older Americans: A National Study" (*J. Gerontol.*, 1992: M145–150) reports the following summary data on intake for a sample of males age 65–74 years: $n = 115, \bar{x} = 11.3$, and $s = 6.43$. Does this data indicate that average daily zinc intake in the population of all males age 65–74 falls below the recommended allowance?

72. The industry standard for the amount of alcohol poured into many types of drinks (e.g., gin for a gin and tonic, whiskey on the rocks) is 1.5 oz. Each individual in a sample of 8 bartenders with at least 5 years of experience was asked to pour rum for a rum and coke into a short, wide (tumbler) glass, resulting in the following data:

| 2.00 | 1.78 | 2.16 | 1.91 | 1.70 | 1.67 | 1.83 | 1.48 |

(Summary quantities agree with those given in the article "Bottoms Up! The Influence of Elongation on Pouring and Consumption Volume," *J. Consumer Res.*, 2003: 455–463.)

(a) Carry out a test of hypotheses to decide whether there is strong evidence for concluding that the true average amount poured differs from the industry standard.

(b) Does the validity of the test you carried out in (a) depend on any assumptions about the population distribution? If so, check the plausibility of such assumptions.

73. Before agreeing to purchase a large order of polyethylene sheaths for a particular type of high-pressure oil-filled submarine power cable, a company wants to see conclusive evidence that the true standard deviation of sheath thickness is less than .05 mm. What hypotheses should be tested, and why? In this context, what are the Type I and Type II errors?

74. Many older homes have electrical systems that use fuses rather than circuit breakers. A manufacturer of 40-amp fuses wants to make sure that the mean amperage at which its fuses burn out is in fact 40. If the mean amperage is lower than 40, customers will complain because the fuses require replacement too often. If the mean amperage is higher than 40, the manufacturer might be liable for damage to an electrical system due to fuse malfunction. To verify the amperage of the fuses, a sample of fuses is to be selected and inspected. If a hypothesis test were to be performed on the resulting data, what null and alternative hypotheses would be of interest to the manufacturer? Describe Type I and Type II errors in the context of this problem situation.

75. Water samples are taken from water used for cooling as it is being discharged from a power plant into a river. It has been determined that as long as the mean temperature of the discharged water is at most 150 °F, there will be no negative effects on the river's ecosystem. To investigate whether the plant

is in compliance with regulations that prohibit a mean discharge-water temperature above 150°, 50 water samples will be taken at randomly selected times, and the temperature of each sample recorded. The resulting data will be used to test the hypotheses $H_0: \mu = 150°$ versus $H_a: \mu > 150°$. In the context of this situation, describe Type I and Type II errors. Which type of error would you consider more serious? Explain.

76. A regular type of laminate is currently being used by a manufacturer of circuit boards. A special laminate has been developed to reduce warpage. The regular laminate will be used on one sample of specimens and the special laminate on another sample, and the amount of warpage will then be determined for each specimen. The manufacturer will then switch to the special laminate only if it can be demonstrated that the true average amount of warpage for that laminate is less than for the regular laminate. State the relevant hypotheses, and describe the Type I and Type II errors in the context of this situation.

5.5   Inferences for a Population Proportion

The previous two sections illustrated the methods of confidence intervals and hypothesis testing for an unknown mean, μ. In this section, we will apply those same ideas to drawing inferences about an unknown probability or population proportion.

Let p denote the proportion of "successes" in a population, where success identifies an individual or object that has a specified property. Equivalently, p is the probability that a randomly selected individual or object is a success. A random sample of n individuals is to be selected, and X denotes the number of successes in the sample. The natural estimator of p is $\hat{P} = X/n$, the sample fraction of successes. As derived in Sect. 2.4 and discussed earlier in this chapter, $E(\hat{P}) = p$ (unbiasedness) and $\mathrm{SD}(\hat{P}) = \sqrt{p(1-p)/n}$; moreover, provided $np \geq 10$ and $n(1-p) \geq 10$, $\hat{P}$ has approximately a normal distribution.

5.5.1   Confidence Intervals for p

Since $\hat{P}$ is approximately normal, standardizing $\hat{P}$ by subtracting p and dividing by $\sigma_{\hat{P}}$ implies that, for example,

$$P\left(-1.96 < \frac{\hat{P} - p}{\sqrt{p(1-p)/n}} < 1.96\right) \approx .95$$

A confidence interval for p results from replacing each $<$ by $=$ and solving the resulting quadratic equation for p. After some tedious algebra, this gives the two roots

$$p = \frac{\left(\hat{P} + 1.96^2/2n\right) \pm 1.96\sqrt{\hat{P}\left(1 - \hat{P}\right)/n + 1.96^2/4n^2}}{1 + 1.96^2/n}$$

These form the endpoints of an approximate 95% CI for p. The more general formula is given in the following proposition.

ONE-PROPORTION Z INTERVAL
Let $\hat{p}$ be the fraction of successes in a random sample of size n. Then a confidence interval for the true/population proportion p has endpoints

$$\tilde{p} \pm z^* \; \frac{\sqrt{\hat{p}(1 - \hat{p})/n + (z^*)^2/4n^2}}{1 + (z^*)^2/n} \tag{5.3}$$

where z^* is the standard normal critical value for the specified confidence level (e.g., $z^* = 1.96$ for 95% confidence) and $\tilde{p}$ is the adjusted sample proportion of successes defined by $\tilde{p} = \left[\hat{p} + (z^*)^2/2n\right]/\left[1 + (z^*)^2/n\right]$.

This is often referred to as the **score confidence interval** for p.

If the sample size n is very large, then all the terms in Expression (5.3) of order $1/n$ are negligible compared to the others. Keeping only the dominant terms, Eq. (5.3) is approximated by

$$\hat{p} \pm z^* \cdot \sqrt{\frac{\hat{p}(1 - \hat{p})}{n}} \tag{5.4}$$

This approximate CI (Eq. 5.4) has the format $\hat{p} \pm z^* \cdot \hat{\sigma}_{\hat{p}}$, similar to the large-sample CI for μ presented in Sect. 5.3, and is the one that for decades has appeared in introductory statistics textbooks. It clearly has a much simpler and more appealing form than Eq. (5.3), so why bother with the score interval at all?

Suppose we use $z = 1.96$ in the traditional formula (5.4). Then our *nominal* confidence level (the one we think we're buying by using that z critical value) is approximately 95%. So before a sample is selected, the probability that the random interval includes the actual value of p (i.e., the *coverage probability*) should be about .95. But it turns out that the actual coverage probability for this interval can differ considerably from the nominal probability .95, particularly when p is not close to .5. This is, generally speaking, a deficiency of the traditional interval—the actual confidence level can be quite different from the nominal level even for reasonably large sample sizes. Recent research has shown that the score interval (Eq. 5.3) rectifies this behavior—for virtually all sample sizes and values of p, its actual confidence level will be quite close to the nominal level specified by the choice of z. This is due largely to the fact that the interval (in particular, the midpoint $\widetilde{p}$) is shifted a bit toward .5 compared to the traditional interval. This is especially important when p is close to 0 or 1.

In addition, the score interval can be used with nearly all sample sizes and parameter values. It is thus not necessary to check the conditions $n\hat{p} \geq 10$ and $n(1 - \hat{p}) \geq 10$ which would be required were the traditional interval employed. So rather than asking when n is large enough for Eq. (5.4) to yield a good approximation to Eq. (5.3), our recommendation is that the score CI should always be used unless the sample size is extremely large (such as in simulations, where $n = 10{,}000$ or more is typical). The slight additional tediousness of the computation is outweighed by the desirable properties of the interval.

Example 5.20 A Gallup poll published June 28, 2013 reported that 41% of US adults surveyed felt that the most important factor in choosing which college or university to attend should be the percentage of graduates who are able to get a good job. (This was the most popular response; cost of tuition was a close second.) The survey was based on a random sample of $n = 1012$ adults; we will assume the *number* who gave the above response is $x = 415$, so that $\hat{p} = 415/1012 = .4101$, matching the survey. Let p denote the proportion of *all* US adults that feel this same way, for which $\hat{p}$ is our point estimate. A confidence interval for p with a confidence level of approximately 95% is

$$\frac{.4101 + 1.96^2/2(1012)}{1 + 1.96^2/1012} \pm 1.96 \frac{\sqrt{(.4101)(.5899)/1012 + 1.96^2/(4 \cdot 1012^2)}}{1 + 1.96^2/1012}$$

$$= .4103 \pm .0302 = (.3801, .4405)$$

Hence, we are 95% confident that between 38 and 44% of all US adults feel that the percentage of graduates that get good jobs is the most important factor when choosing a college or university. The traditional interval is

$$.4101 \pm 1.96\sqrt{\frac{(.410)(.590)}{1012}} = .4101 \pm .0303 = (.3798, .4404)$$

These two intervals are practically identical because $n = 1012$ is so large. ∎

Example 5.21 The article "Repeatability and Reproducibility for Pass/Fail Data" (*J. Testing Eval.*, 1997: 151–153) reported that in $n = 48$ trials in a particular laboratory, 16 resulted in ignition of a particular type of substrate by a lighted cigarette. Let p denote the long-run proportion of all such trials that would result in ignition. A point estimate for p is $\hat{p} = 16/48 = .333$. A 95% confidence interval for p is

$$\frac{.333 + 1.96^2/96}{1 + 1.96^2/48} \pm 1.96 \frac{\sqrt{(.333)(.667)/48 + 1.96^2/(4 \cdot 48^2)}}{1 + 1.96^2/48} = .346 \pm .129$$

$$= (.217, .475)$$

So, the researchers can be 95% confident that between 21.7 and 47.5% of all trials under the same conditions will result in ignition. This interval isn't very precise—its width is nearly 26 percentage points—as a consequence of the relatively small sample size. If the researchers wanted a narrower interval, they would need to use a larger n (which, of course, requires more time and money).

The traditional CI formula (5.4) gives

$$.333 \pm 1.96\sqrt{(.333)(.667)/48} = .333 \pm .133 = (.200, .466)$$

These two intervals are somewhat different because $n = 48$ is not very large. ∎

5.5.2   Hypothesis Testing for p

Analogous to hypothesis testing for a population mean μ, tests for p concern deciding which of two competing hypotheses about the value of p is correct. The null hypothesis will always be written in the form

$$H_0\colon p = p_0$$

where p_0 is the null value for the parameter p (i.e., the value claimed for p by the null hypothesis). The alternative hypothesis has one of three forms, depending on context:

$$H_a\colon p > p_0 \quad H_a\colon p < p_0 \quad H_a\colon p \neq p_0$$

Inferences about p are again based on the value of a sample proportion, $\hat{P}$. When H_0 is true, $E(\hat{P}) = p_0$ and $\mathrm{SD}(\hat{P}) = \sqrt{p_0(1 - p_0)/n}$. Moreover, when n is large and H_0 is true, the test statistic

$$Z = \frac{\hat{P} - p_0}{\sqrt{p_0(1 - p_0)/n}}$$

has approximately a standard normal distribution. The P-value of the hypothesis test is then determined in an analogous manner to those of the one-sample t test in Sect. 5.4, except that calculation is made using the z table rather than a t distribution.

ONE-PROPORTION Z TEST

Consider testing the null hypothesis H_0: $p = p_0$ based on a random sample of size n. Let $\hat{P}$ denote the proportion of "successes" in the sample. The test statistic is

$$Z = \frac{\hat{P} - p_0}{\sqrt{p_0(1 - p_0)/n}}$$

Provided $np_0 \geq 10$ and $n(1 - p_0) \geq 10$, Z has approximately a standard normal distribution when H_0 is true. Let z denote the calculated value of the test statistic. The calculation of the P-value depends on the choice of H_a as follows:

Alternative Hypothesis	P-value
H_a: $p > p_0$	$1 - \Phi(z)$
H_a: $p < p_0$	$\Phi(z)$
H_a: $p \neq p_0$	$2[1 - \Phi(|z|)]$

Illustrations of these P-values are essentially identical to those in Fig. 5.10.

Example 5.22   Obesity is an increasing problem in America among all age groups. The Centers for Disease Control and Prevention (CDCP) reported in 2012 that 35.7% of US adults are obese (a body mass index exceeding 30; this index is a measure of weight relative to height). Physicians at a large hospital in Los Angeles measured the body mass index of 122 randomly selected patients and found that 38 of them should be classified as obese. Do the hospital's data suggest that the true proportion of adults served by this hospital who are obese is less than the national figure of 35.7%? Let's carry out a test of hypotheses using $\alpha = .05$.

The parameter of interest is p = the proportion of all adults served by this hospital who are obese. The competing hypotheses are

H_0: $p = .357$ (the hospital's obesity rate matches the national rate)
H_a: $p < .357$ (the hospital's obesity rate is less than the national rate)

Since $np_0 = 122(.357) = 43.6 \geq 10$ and $n(1 - p_0) = 122(1 - .357) = 78.4 \geq 10$, the one-proportion z test may be applied. With $\hat{p} = 38/122 = .311$, the calculated value of the test statistic is

$$z = \frac{\hat{p} - p_0}{\sqrt{p_0(1 - p_0)/n}} = \frac{.311 - .357}{\sqrt{.357(1 - .357)/122}} = -1.05$$

So, the observed sample proportion is about one standard deviation below what we would expect *if the null hypothesis is true*. The P-value of the test is the probability of obtaining a test statistic value at least that low:

$$P\text{-value} = P(Z \leq -1.05) = \Phi(-1.05) = .1469$$

Since the P-value of .1469 is greater than the significance level .05, we fail to reject H_0. On the basis of the observed data, we cannot conclude that the obesity rate of the population served by this hospital is less than the national rate of 35.7%. ∎

As was the case for inferences on μ, it is desirable to calculate the power of a hypothesis test concerning a population proportion p. The power of our one-sample z test depends on how far the true value of p is from the null value p_0, the sample size, and the selected significance level. The details of such power calculations, which many software packages can perform automatically, are developed in Exercises 96 and 97 of this section.

Inferences about p when n is small can be based directly on the binomial distribution. There are also procedures available for making inferences about a difference $p_1 - p_2$ between two population proportions (e.g., the proportion of all female versus male students that make the honor roll at your school). Please consult the reference by Devore and Berk for more information.

5.5.3 Software for Inferences about p

The `prop.test` function in R will calculate the traditional CI (Eq. 5.4) for a population proportion and perform a one-proportion z test upon request. Figure 5.13

a
```
> prop.test(16,48)

        1-sample proportions test with continuity correction

data:  16 out of 48, null probability 0.5
X-squared = 4.6875, df = 1, p-value = 0.03038
alternative hypothesis: true p is not equal to 0.5
95 percent confidence interval:
 0.2080794 0.4851357
sample estimates:
        p
0.3333333
```

b
```
> prop.test(38,122,p=.357,"less")

        1-sample proportions test with continuity correction

data:  38 out of 122, null probability 0.357
X-squared = 0.9121, df = 1, p-value = 0.1698
alternative hypothesis: true p is less than 0.357
95 percent confidence interval:
 0.0000000 0.3881457
sample estimates:
        p
0.3114754
```

Fig. 5.13 Inferences on p in R: (**a**) Example 5.21; (**b**) Example 5.22

shows output corresponding to Examples 5.21 and 5.22. The 95% CI in Fig. 5.13a for p is roughly (.208, .485); the difference between this interval and the traditional interval provided in Example 5.21 is due to rounding and an adjustment made automatically in R called *Yates' continuity correction*. The inputs to `prop.test` in Fig. 5.13b include not only the raw data x and n, but also the null value of p and the direction of the test. The resulting P-value, .1698, is close to the value of .1469 obtained in Example 5.22. Again, the disparity comes from a combination of rounding and the continuity correction. Unfortunately, to the authors' knowledge, there are no one-proportion z intervals or z tests built into Matlab.

5.5.4 Exercises: Section 5.5 (77–97)

77. In a sample of 1000 randomly selected consumers who had opportunities to send in a rebate claim form after purchasing a product, 250 of these people said they never did so ("Rebates: Get What You Deserve," *Consumer Reports*, May 2009: 7). Reasons cited for their behavior included too many steps in the process, amount too small, missed deadline, fear of being placed on a mailing list, lost receipt, and doubts about receiving the money. Calculate and interpret a 95% confidence level for the true proportion of such consumers who never apply for a rebate.

78. A *Wireless News* article (July 6, 2008) found that 62% of people surveyed would use a Bluetooth device while driving in order to comply with the law. The survey was based upon a random sample of 600 cell phone users. Construct a 95% confidence interval for the proportion of all cell phone users who will use Bluetooth technology while driving.

79. The article "Limited Yield Estimation for Visual Defect Sources" (*IEEE Trans. Semicon. Manuf.*, 1997: 17–23) reported that, in a study of a particular wafer inspection process, 356 dies were examined by an inspection probe and 201 of these passed the probe. Assuming a stable process, calculate a 99% confidence interval for the proportion of all dies that pass the probe.

80. The technology underlying hip replacements has changed as these operations have become more popular (over 250,000 in the USA in 2008). Starting in 2003, highly durable ceramic hips were marketed. Unfortunately, for too many patients the increased durability has been counterbalanced by an increased incidence of squeaking. The May 11, 2008, issue of the *New York Times* reported that in one study of 143 individuals who received ceramic hips between 2003 and 2005, 10 of the hips developed squeaking. Calculate and interpret 95% confidence interval for the true proportion of such hips that develop squeaking.

81. The Pew Forum on Religion and Public Life reported on December 9, 2009, that in a survey of 2003 American adults, 25% said they believed in astrology. Calculate and interpret a confidence interval at the 99% confidence level for the proportion of all adult Americans who believe in astrology.

82. Reconsider the score CI (Eq. 5.3) for p, and focus on a confidence level of 95%. Show that the endpoints agree quite well with those of the traditional interval (Eq. 5.4) once two successes and two failures have been appended to the sample, i.e., Eq. (5.4) based on $(x + 2)$ S's in $(n + 4)$ trials. [*Hint:* $1.96 \approx 2.$]

83. It is often important in planning studies to know in advance what sample size is required to estimate an unknown proportion to within a certain margin of error.
    (a) Suppose we wish to achieve a bound B on the margin of error of a CI for p. By equating the margin of error in the "very large n" CI Eq. (5.4) to B and solving for n, show that the required sample size is

    $$n = \frac{(z^*)^2 \hat{p} (1 - \hat{p})}{B^2}$$

    (b) A state legislator wishes to survey residents of her district to see what proportion of the electorate is aware of her position on using state funds to pay for abortions. If the legislator has strong reason to believe that at least 2/3 of the electorate know of her position, how large a sample size would you recommend in order to estimate the true proportion to within $\pm$ 5 percentage points? Assume 95% confidence.
    (c) What sample size is necessary if the 95% CI for p is to have width of at most .10 irrespective of $\hat{p}$? [*Hint:* What value of $\hat{p}$ makes the expression for n as large as possible?]

84. Write a function in Matlab or R to implement (Eq. 5.3). Your function should have three inputs: the number of successes x, the sample size n, and the desired confidence level. The output of the function should be the endpoints of the CI.

85. Natural cork in wine bottles is subject to deterioration, and as a result wine in such bottles may experience contamination. The article "Effects of Bottle Closure Type on Consumer Perceptions of Wine Quality" (*Amer. J. of Enology and Viticulture*, 2007: 182–191) reported that, in a tasting of commercial chardonnays, 16 of 91 bottles were considered spoiled to some extent by cork-associated characteristics. Does this data provide strong evidence for concluding that more than 15% of all such bottles are contaminated in this way? Carry out a test of hypotheses using a significance level of .10.

86. It is known that roughly 2/3 of all human beings have a dominant right foot or eye. Is there also right-sided dominance in kissing behavior? The article "Human Behavior: Adult Persistence of Head-Turning Asymmetry" (*Nature*, 2003: 771) reported that in a random sample of 124 kissing couples, both people in 80 of the couples tended to lean more to the right than to the left. Does the result of the experiment suggest that the 2/3 figure is implausible for kissing behavior? State and test the appropriate hypotheses.

87. The article referenced in Exercise 85 also reported that in a sample of 106 wine consumers, 22 (20.8%) thought that screw tops were an acceptable substitute for natural corks. Suppose a particular winery decided to use screw

tops for one of its wines unless there was strong evidence to suggest that fewer than 25% of wine consumers found this acceptable.

(a) Using a significance level of .10, what would you recommend to the winery?

(b) For the hypotheses tested in (a), describe in context what the Type I and II errors would be, and say which type of error might have been committed.

88. With domestic sources of building supplies running low several years ago, roughly 60,000 homes were built with imported Chinese drywall. According to the article "Report Links Chinese Drywall to Home Problems" (*New York Times*, November 24, 2009), federal investigators identified a strong association between chemicals in the drywall and electrical problems, and there is also strong evidence of respiratory difficulties due to the emission of hydrogen sulfide gas. An extensive examination of 51 homes found that 41 had such problems. Suppose these 51 were randomly sampled from the population of all homes having Chinese drywall. Does the data provide strong evidence for concluding that more than 50% of all homes with Chinese drywall have electrical/environmental problems? Carry out a test of hypotheses using $\alpha = .01$.

89. A common characterization of obese individuals is that their body mass index is at least 30 [BMI = weight/(height)^2 when height is in meters and weight is in kilograms]. The article "The Impact of Obesity on Illness Absence and Productivity in an Industrial Population of Petrochemical Workers" (*Annals of Epidemiology*, 2008: 8–14) reported that in a sample of female workers, 262 had BMIs of less than 25, 159 had BMIs that were at least 25 but less than 30, and 120 had BMIs exceeding 30. Is there compelling evidence for concluding that more than 20% of the individuals in the sampled population are obese?

(a) State and test appropriate hypotheses using a significance level of .05.

(b) Explain in the context of this scenario what constitutes Type I and II errors.

90. The article "Analysis of Reserve and Regular Bottlings: Why Pay for a Difference Only the Critics Claim to Notice?" (*Chance*, Summer 2005, pp. 9–15) reported on an experiment to investigate whether wine tasters could distinguish between more expensive reserve wines and their regular counterparts. Wine was presented to tasters in four containers labeled A, B, C, and D, with two of these containing the reserve wine and the other two the regular wine. Each taster randomly selected three of the containers, tasted the selected wines, and indicated which of the three he/she believed was different from the other two. Of the $n = 855$ tasting trials, 346 resulted in correct distinctions (either the one reserve that differed from the two regular wines or the one regular wine that differed from the two reserves). Does this provide compelling evidence for concluding that tasters of this type have some ability to distinguish between reserve and regular wines? State and test the relevant hypotheses. Are you particularly impressed with the ability of tasters to distinguish between the two types of wine?

91. The article "Heavy Drinking and Polydrug Use Among College Students" (*J. of Drug Issues*, 2008: 445–466) stated that 51 of the 462 college students in a sample had a lifetime abstinence from alcohol. Does this provide strong evidence for concluding that more than 10% of the population sampled had completely abstained from alcohol use? Test the appropriate hypotheses. [*Note:* The article used more advanced statistical methods to study the use of various drugs among students characterized as light, moderate, and heavy drinkers.]

92. Scientists have recently become concerned about the safety of Teflon cookware and various food containers because perfluorooctanoic acid (PFOA) is used in the manufacturing process. An article in the July 27, 2005, *New York Times* reported that of 600 children tested, 96% had PFOA in their blood. According to the FDA, 90% of all Americans have PFOA in their blood. Does the data on PFOA incidence among children suggest that the percentage of all children who have PFOA in their blood exceeds the FDA percentage for all Americans? Carry out an appropriate test of hypotheses at the $\alpha = .05$ level.

93. A manufacturer of nickel–hydrogen batteries randomly selects 100 nickel plates for test cells, cycles them a specified number of times, and determines that 14 of the plates have blistered. Does this provide compelling evidence for concluding that more than 10% of all plates blister under such circumstances? State and test the appropriate hypotheses using a significance level of .05. In reaching your conclusion, what type of error might you have committed?

94. A random sample of 150 recent donations at a blood bank reveals that 82 were type A blood. Does this suggest that the actual percentage of type A donations differs from 40%, the percentage of the population having type A blood? Carry out a test of the appropriate hypotheses using a significance level of .01. Would your conclusion have been different if a significance level of .05 had been used?

95. The article "Statistical Evidence of Discrimination" (*J. Amer. Statist. Assoc.*, 1982: 773–783) discusses the court case *Swain v. Alabama* (1965), in which it was alleged that there was discrimination against blacks in grand jury selection. Census data suggested that 25% of those eligible for grand jury service were black, yet a random sample of 1050 people called to appear for possible duty yielded only 177 blacks. Using a level .01 test, does this data argue strongly for a conclusion of discrimination?

96. Consider testing hypotheses $H_0: p = p_0$ versus $H_a: p < p_0$. Suppose that, in fact, the true value of the parameter p is p', where $p' < p_0$ (so H_a is true).

(a) Show that the expected value and variance of the test statistic Z in the one-proportion z test are

$$E(Z) = \frac{p' - p_0}{\sqrt{p_0(1-p_0)/n}} \qquad \mathrm{Var}(Z) = \frac{p'(1-p')/n}{p_0(1-p_0)/n}$$

(b) It can be shown that P-value $\leq \alpha$ iff $Z \leq -z_\alpha$, where $-z_\alpha$ denotes the α quantile of the standard normal distribution (i.e., $\Phi(-z_\alpha) = \alpha$). Show that the power of the lower-tailed one-sample z test when $p = p'$ is given by

$$\Phi\left(\frac{p_0 - p' - z_\alpha\sqrt{p_0(1-p_0)/n}}{\sqrt{p'(1-p')/n}}\right)$$

(c) A package-delivery service advertises that at least 90% of all packages brought to its office by 9 a.m. for delivery in the same city are delivered by noon that day. Let p denote the true proportion of such packages that are delivered as advertised and consider the null hypotheses $H_0: p = .9$ versus the alternative $H_a: p < .9$. If only 80% of the packages are delivered as advertised, how likely is it that a level .01 test based on $n = 225$ packages will detect such a departure from H_0?

97. Because of variability in the manufacturing process, the actual yielding point of a sample of mild steel subjected to increasing stress will usually differ from the theoretical yielding point. Let p denote the true proportion of samples that yield before their theoretical yielding point. If on the basis of a sample it can be concluded that more than 20% of all specimens yield before the theoretical point, the production process will have to be modified.

(a) If 15 of 60 specimens yield before the theoretical point, what is the P-value when the appropriate test is used, and what would you advise the company to do?

(b) If the true percentage of "early yields" is actually 50% (so that the theoretical point is the median of the yield distribution) and a level .01 test is used, what is the probability that the company concludes a modification of the process is necessary? [*Hint:* Refer back to the previous exercise. Modify the expression in part (b) to accommodate an upper-tailed test.]

5.6 Bayesian Inference

Throughout this chapter, we have regarded parameters such as μ, σ, p, and λ as having an unknown but single, fixed value. This is often referred to as the classical or *frequentist* approach to statistical inference. However, there is a different paradigm, called *subjective* or *Bayesian inference*, in which an unknown parameter is assigned a distribution of possible values, analogous to a probability distribution. This distribution reflects all available information—past experience, intuition, common sense—about the parameter prior to observing the data. For this reason, it is called the **prior distribution** of the parameter.

DEFINITION
A **prior distribution** for a parameter θ, denoted $\pi(\theta)$, is a probability distribution on the set of possible values for θ. In particular, if the possible values of the parameter θ form an interval I, then $\pi(\theta)$ is a pdf that must satisfy

$$\int_I \pi(\theta)d\theta = 1$$

Similarly, if θ is potentially any value in a discrete set D, then $\pi(\theta)$ is a pmf that must satisfy

$$\sum_{\theta \in D} \pi(\theta) = 1$$

Example 5.23 Consider the parameter μ = the mean GPA of all students at your university. Since GPAs are always between 0.0 and 4.0, μ must also lie in this interval. But common sense tells you that μ is unlikely to be below 2.0, or very few people would graduate, and it would be likewise surprising to find μ much above 3.5. This "prior belief" can be expressed mathematically as a prior distribution for μ on the interval $I = [0, 4]$. If our best guess a priori is that $\mu \approx 2.5$, then our prior distribution $\pi(\mu)$ should be centered around 2.5. The variability of the prior distribution we select should reflect how sure we feel about our initial information.

If we feel very sure that μ is near 2.5, then we should select a prior distribution for μ that has less variation around that value. On the other hand, if we are less certain, this can be reflected by a prior distribution with much greater variability. Figure 5.14 illustrates these two cases.

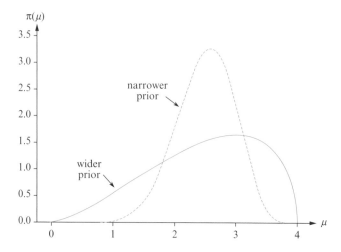

Fig. 5.14 Two prior distributions for a parameter: a more diffuse prior (less certainty) and a more concentrated prior (more certainty)   ∎

5.6.1   The Posterior Distribution of a Parameter

The key to Bayesian inference is having a mathematically rigorous way to incorporate the actual sample data. Suppose we observe values $x_1, \ldots, x_n$ from a distribution depending on the unknown parameter θ for which we have selected some prior distribution. Then a Bayesian statistician wants to "update" her or his belief about the distribution of θ, taking into account both prior belief and the observed x_is. This is achieved using a form of Bayes' Theorem for random variables.

DEFINITION

Suppose $X_1, \ldots, X_n$ have joint pdf $f(x_1, \ldots, x_n; \theta)$ and the unknown parameter θ has been assigned a continuous prior distribution $\pi(\theta)$. Then the **posterior distribution** of θ, given the observations $X_1 = x_1, \ldots, X_n = x_n$, is

$$\pi\left(\theta \mid x_1, \ldots, x_n\right) = \frac{\pi(\theta)f(x_1, \ldots, x_n; \theta)}{\displaystyle\int_{-\infty}^{\infty} \pi(\theta)f(x_1, \ldots, x_n; \theta)\,d\theta} \tag{5.5}$$

The integral in the denominator of Eq. (5.5) insures that the posterior distribution is a valid probability density for θ.

If $X_1, \ldots, X_n$ are discrete, the joint pdf is replaced by their joint pmf and integration by summation.

Notice that constructing the posterior distribution of a parameter requires a specific probability model $f(x_1, \ldots, x_n; \theta)$ for the observed data. In Example 5.23, it would not be enough to simply observe the GPAs of a random sample of n students; one must specify the underlying distribution, with mean μ, from which those GPAs are drawn.

Example 5.24 Emissions of subatomic particles from a radiation source are often modeled as a Poisson process. As we shall see in Chap. 7, this implies that the time between successive emissions follows an exponential distribution. In practice, the parameter λ of this distribution is typically unknown. If researchers believe a priori that the average time between emissions is about half a second, so $\lambda \approx 2$, a prior distribution with a mean around 2 might be selected for λ. One example is the following gamma distribution, which has mean (and variance) of 2:

$$\pi(\lambda) = \lambda e^{-\lambda}, \quad \lambda > 0$$

Notice the gamma distribution lies on the interval $(0, \infty)$, which is also the set of possible values for the unknown parameter λ.

The times $X_1, \ldots, X_5$ between five particle emissions will be recorded; it is these variables that have an exponential distribution with the unknown parameter λ (equivalently, mean $1/\lambda$). Because the X_is are also independent, their joint pdf is

$$f(x_1, \ldots, x_5; \lambda) = f(x_1; \lambda) \cdots f(x_5; \lambda) = \lambda e^{-\lambda x_1} \cdots \lambda e^{-\lambda x_n} = \lambda^5 e^{-\lambda \Sigma x_i}$$

Applying Eq. (5.5) with these two components, the posterior distribution of λ given the observed data is

$$\pi\left(\lambda \mid x_1, \ldots, x_5\right) = \frac{\pi(\lambda)f(x_1, \ldots, x_5; \lambda)}{\displaystyle\int_{-\infty}^{\infty} \pi(\lambda)f(x_1, \ldots, x_5; \lambda)d\lambda} = \frac{\lambda e^{-\lambda} \cdot \lambda^5 e^{-\lambda \Sigma x_i}}{\displaystyle\int_0^{\infty} \lambda e^{-\lambda} \cdot \lambda^5 e^{-\lambda \Sigma x_i}d\lambda}$$

$$= \frac{\lambda^6 e^{-\lambda(1+\Sigma x_i)}}{\displaystyle\int_0^{\infty} \lambda^6 e^{-\lambda(1+\Sigma x_i)}d\lambda}$$

Suppose the five observed inter-emission times are $x_1 = 0.66$, $x_2 = 0.48$, $x_3 = 0.44$, $x_4 = 0.71$, $x_5 = 0.56$. The sum of these five times is $\sum x_i = 2.85$, and so the posterior distribution simplifies to

$$\pi\left(\lambda \mid 0.66, \ldots, 0.56\right) = \frac{\lambda^6 e^{-3.85\lambda}}{\displaystyle\int_0^{\infty} \lambda^6 e^{-3.85\lambda}d\lambda} = \frac{3.85^7}{6!}\lambda^6 e^{-3.85\lambda}$$

The integral in the denominator was evaluated using the gamma integral formula (3.5) from Chap. 3; as noted previously, the purpose of this integral is to guarantee that the posterior distribution of λ is a valid probability density. As a function of λ, we recognize this as a gamma distribution with parameters $\alpha = 7$ and $\beta = 1/3.85$. The prior and posterior density curves of λ appear in Fig. 5.15.

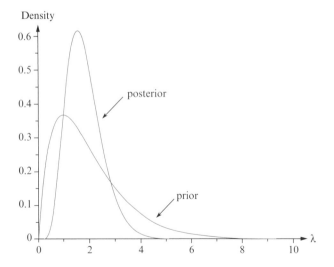

Fig. 5.15 Prior and posterior distribution of λ for Example 5.24   ∎

Example 5.25 A 2010 National Science Foundation study found that 488 out of 939 surveyed adults incorrectly believe that antibiotics kill viruses (they only kill bacteria). Let θ denote the proportion of *all* US adults that hold this mistaken view. Imagine that an NSF researcher, in advance of administering the survey, believed (hoped?) the value of θ was roughly 1 in 3, but he was very uncertain about this belief. Since any proportion must lie between 0 and 1, the beta family of distributions from Sect. 3.5 provides a natural source of priors for θ. One such beta distribution, with an expected value of 1/3, is the Beta(2, 4) model whose pdf is

$$\pi(\theta) = 20\theta(1 - \theta)^3 \quad 0 < \theta < 1$$

The data mentioned at the beginning of the example can be considered either a random sample of size 939 from the Bernoulli distribution or, equivalently, a single observation from the binomial distribution with $n = 939$. Let $X =$ the number of US adults in a random sample of 939 that believe antibiotics kill viruses. Then $X \sim \text{Bin}(939, \theta)$, and the pmf of X is $p(x; \theta) = \binom{939}{x}\theta^x(1 - \theta)^{939-x}$. Substituting the observed value $x = 488$, Eq. (5.5) gives the posterior distribution of θ as

$$\pi(\theta \mid X = 488) = \frac{\pi(\theta)p(488; \theta)}{\int \pi(\theta)p(488; \theta)d\theta} = \frac{20\theta(1 - \theta)^3 \cdot \binom{939}{488}\theta^{488}(1 - \theta)^{451}}{\int_0^1 20\theta(1 - \theta)^3 \cdot \binom{939}{488}\theta^{488}(1 - \theta)^{451}d\theta}$$

$$= \frac{\theta^{489}(1 - \theta)^{454}}{\int_0^1 \theta^{489}(1 - \theta)^{454}d\theta} = c \cdot \theta^{489}(1 - \theta)^{454} \quad 0 < \theta < 1$$

Recall that the constant c, which equals the reciprocal of the integral in the denominator, serves to insure that the posterior distribution $\pi(\theta|X = 488)$ integrates to 1. Rather than evaluate the integral, we can simply recognize the expression $\theta^{489}(1 - \theta)^{454}$ as a standard beta distribution, specifically with parameters $\alpha = 490$ and $\beta = 455$, that's just missing the constant of integration in front. It follows that the posterior distribution of θ given $X = 488$ must be Beta(490, 455); if we require c, it can be copied directly from the beta pdf. (This trick comes in handy quite often in Bayesian statistics: if we can recognize a posterior distribution as being the "kernel" of a particular probability distribution, then it must necessarily be that distribution.)

The prior and posterior density curves for θ are displayed in Fig. 5.16. While the prior distribution is centered around 1/3 and exhibits a great deal of uncertainty (variability), the posterior distribution of θ is centered much closer to the sample proportion of incorrect answers, $488/939 \approx .52$, with considerably less uncertainty.

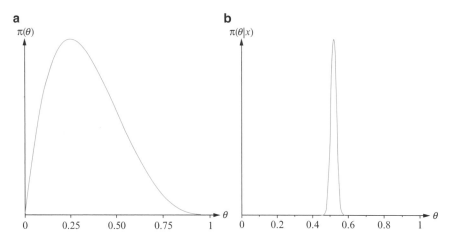

Fig. 5.16 Density curves for the parameter θ in Example 5.25: (**a**) prior Beta(2, 4); (**b**) posterior Beta(490, 455) ∎

5.6.2   Inferences from the Posterior Distribution

Inferences about an unknown parameter can be obtained from its posterior distribution. The most common Bayesian point estimate for a parameter θ is the mean of its posterior distribution:

$$\hat{\theta} = E\big(\theta \mid x_1, \ldots, x_n\big)$$

An interval $[a, b]$ having posterior probability .95 gives a 95% **credibility interval**, the Bayesian analogue of a 95% confidence interval (but the interpretation is different). Typically one selects the middle 95% of the posterior distribution, i.e., the endpoints of a 95% credibility interval are ordinarily the .025 and .975 quantiles of the posterior distribution.

Example 5.26 (Example 5.24 continued) Given the observed values of $X_1, \ldots, X_5$, we previously found that the mean emission rate λ has a Gamma(7, 1/3.85) posterior distribution. Thus, the mean of the posterior distribution of λ is

$$\hat{\lambda} = E\big(\lambda \mid 0.66, \ldots, 0.56\big) = \alpha\beta = 7(1/3.85) = 1.82$$

This isn't too different from the researchers' prior belief that $\lambda \approx 2$. A 95% credibility interval for λ requires determining the .025 and .975 quantiles of the Gamma(7, 1/3.85) model; using statistical software, $\eta_{.025} = 0.7310$ and $\eta_{.975} = 3.3921$. Under the Bayesian interpretation, having observed the five aforementioned inter-emission times, there is a 95% posterior probability that λ is between 0.7310 and 3.3921 emissions per second. ∎

Example 5.27 (Example 5.25 continued) The posterior distribution of the parameter θ = the proportion of all US adults that incorrectly believe antibiotics kill viruses was found to have a Beta(490, 455) distribution. A point estimate of θ is the mean of this distribution:

$$\hat{\theta} = E(\theta \mid X = 488) = \frac{\alpha}{\alpha + \beta} = \frac{490}{490 + 455} = \frac{490}{945} = .5185$$

Notice this is quite close to the traditional estimate $x/n = 488/939 = .5197$; in general, when n is large the mean of the posterior distribution of a parameter will be quite similar to its more traditional, frequentist estimate.

The .025 and .975 quantiles of this beta distribution are $\eta_{.025} = .4866$ and $\eta_{.975} = .5503$. So, after observing the results of the NSF survey, there is a 95% posterior probability that θ is between .4866 and .5503. ∎

5.6.3 Further Comments on Bayesian Inference

In most cases, the role of the observed values in shaping the posterior distribution of a parameter θ increases as the sample size n increases. More precisely, it can be shown that under very general conditions, as $n \to \infty$ the mean of the posterior distribution will converge to the true value of θ while the variance of the posterior distribution of θ converges to zero:

$$E(\theta \mid X_1, \ldots, X_n) \to \theta \qquad \mathrm{Var}(\theta \mid X_1, \ldots, X_n) \to 0$$

The second property manifests itself in our two previous examples: the variability of the posterior distribution of λ based on $n = 5$ observations was still rather substantial, while the posterior distribution of θ based on a sample of size $n = 939$ was quite concentrated.

Since traditional estimators such as $\hat{P}$ and $\overline{X}$ converge to the true values of corresponding parameters (e.g., p or μ) by the Law of Large Numbers, it follows that Bayesian and frequentist estimates will typically be quite close when n is large. This is true both for the point estimates and the interval estimates. But when n is small—a common occurrence in Bayesian methodology—parameter estimates can differ drastically between the two methods. This is especially true if the researcher's prior belief is very far from what's actually true (e.g., believing a proportion is around 1/3 when it's really more than .5).

It should be emphasized that, even if the confidence interval is nearly the same as the credibility interval for a parameter, they have different interpretations. To interpret the Bayesian credibility interval, we say that there is a 95% *probability* that the parameter θ is in the interval. However, for the frequentist confidence interval such a probability statement does not make sense: as we discussed in Sect. 5.3, neither the parameter θ nor the endpoints of the interval are considered random under the traditional, frequentist view.

In the examples of this section, prior distributions were chosen partially by matching the mean of a distribution to someone's a priori "best guess" about the value of the parameter. We also mentioned at the beginning of the section that the variability of the prior distribution often reflects the strength of that belief. In practice, there is a third consideration for choosing a prior distribution: the ability to apply Eq. (5.5) in a simple fashion. Ideally, we would like to choose a prior distribution from a family (gamma, beta, etc.) such that the posterior distribution is from that same family. When this happens we say that the prior distribution is **conjugate** to the data distribution.

In Example 5.24, the prior $\pi(\lambda)$ is the Gamma(2, 1) pdf; we determined, using Eq. (5.5), that the posterior distribution was Gamma(7, 1/3.85). It can be shown in general (Exercise 104) that any gamma distribution is conjugate to an exponential data distribution. Similarly, the prior and posterior distributions of θ in Example 5.25 were Beta(2, 4) and Beta(490, 455), respectively. Exercise 105 shows that any beta distribution is conjugate to a binomial (or Bernoulli) data distribution. If the data is normally distributed with known σ, then a normal prior for μ results in a normal posterior. The case of unknown σ is more complicated; see Section 14.4 of the reference by Devore & Berk.

5.6.4   Exercises: Section 5.6 (98–106)

98. Nationwide, IQs have a normal distribution with mean 100 and standard deviation 15. Let $X_1, \ldots, X_n$ represent the IQs of a random sample of first graders, which we assume also come from a normal distribution having $\sigma = 15$ but possibly a different mean μ. Assign a $N(110, 7.5)$ prior distribution to μ.
    (a) Find the posterior distribution of μ.
    (b) Here are the actual IQ scores of a random sample of $n = 18$ first graders:

113	108	140	113	115	146	136	107	108
119	132	127	118	108	103	103	122	111

Calculate a point estimate of μ using the posterior distribution.
    (c) Calculate and interpret 95% credibility interval for μ.
    (d) Calculate a one-sample z 95% confidence interval for μ using the 18 observations with $\sigma = 15$, and compare with the credibility interval of (b).

99. The number of customers arriving during a 1-h period at an ice cream shop is modeled by a Poisson distribution with unknown parameter μ. Based on past experience, the owner believes that the average number of customers in 1 h is about 15.
    (a) Assign a prior to μ from the gamma family of distributions, such that the mean of the prior is 15 and the standard deviation is 5 (reflecting moderate uncertainty).
    (b) The number of customers in ten randomly selected 1-h intervals is recorded:

| 16 | 9 | 11 | 13 | 17 | 17 | 8 | 15 | 14 | 16 |

Find the posterior distribution of μ.

(c) Find and interpret a 95% credibility interval for μ.

100. In a study of 70 restaurant bills, 40 of the 70 were paid using cash. Let p denote the population proportion paying cash.

(a) Assuming a beta prior distribution for p with $\alpha = 2$ and $\beta = 2$, obtain the posterior distribution of p.

(b) Repeat (a) on with α and β positive and close to 0.

(c) Calculate a 95% credibility interval for p using (b). Is your interval compatible with $p = .5$?

(d) Calculate a 95% confidence interval for p using Eq. (5.3), and compare with the result of (c).

(e) Compare the interpretations of the credibility interval and the confidence interval.

(f) Based on the prior in (b), test the hypothesis $p \le .5$ using the posterior distribution to find $P(p \le .5)$.

101. For the data of Example 5.25 assume a Beta(2, 4) prior distribution and assume that the 939 observations are a random sample from the Bernoulli distribution. Use Eq. (5.5) to derive the posterior distribution, and compare your answer with the result of Example 5.25.

102. *Laplace's rule of succession* says that if there have been n Bernoulli trials and they have all been successes, then the probability of a success on the next trial is $(n + 1)/(n + 2)$. For the derivation, Laplace used a Beta(1, 1) prior for the parameter p.

(a) Show that, if a Beta(1, 1) prior is assigned to p and there are n successes in n trials, then the posterior mean of p is $(n + 1)/(n + 2)$.

(b) Explain (a) in terms of total successes and failures; that is, explain the result in terms of two prior trials plus n later trials.

(c) Laplace applied his rule of succession to compute the probability that the sun will rise tomorrow using 5000 years, or $n = 1,826,214$ days of history in which the sun rose every day. Is Laplace's method equivalent to including two prior days when the sun rose once and failed to rise once? Criticize the answer in terms of total successes and failures.

103. Suppose you have a random sample $X_1, X_2, \ldots, X_n$ from the Poisson distribution with mean μ. If the prior distribution for μ has a gamma distribution with parameters α and β, show that the posterior distribution is also gamma distributed. What are its parameters?

104. Suppose you have a random sample $X_1, X_2, \ldots, X_n$ from the exponential distribution with parameter λ. If the prior distribution for λ has a gamma distribution with parameters α and β, show that the posterior distribution is also gamma distributed. What are its parameters?

105. Suppose $X \sim \text{Bin}(n, p)$, where the probability parameter p is unknown. If the prior distribution for p has a beta distribution with parameters α and β, show that the posterior distribution is also beta distributed. What are its parameters?

106. Consider a random sample $X_1, X_2, \ldots, X_n$ from the normal distribution with mean 0 and variance $\sigma^2 = 1/\tau$. (The parameter $\tau = 1/\sigma^2$ is called the **precision** of the normal distribution.) Assume a gamma-distributed prior for τ and show that the posterior distribution of τ is also gamma. What are its parameters?

5.7   Supplementary Exercises (107–138)

107. At time $t = 0$, there is one individual alive in a certain population. A **pure birth process** then unfolds as follows. The time until the first birth is exponentially distributed with parameter λ. After the first birth, there are two individuals alive. The time until the first gives birth again is exponential with parameter λ, and similarly for the second individual. Therefore, the time until the next birth is the minimum of two exponential (λ) variables, which is exponential with parameter 2λ. Similarly, once the second birth has occurred, there are three individuals alive, so the time until the next birth is an exponential rv with parameter 3λ, and so on (the memoryless property of the exponential distribution is being used here). Suppose the process is observed until the sixth birth has occurred and the successive birth times are 25.2, 41.7, 51.2, 55.5, 59.5, 61.8 (from which you should calculate the times between successive births). Derive the mle of λ. [*Hint*: The likelihood is a product of exponential terms.]

108. When the sample standard deviation S is based on a random sample from a normal population distribution, it can be shown that

$$E(S) = \sqrt{2/(n-1)}\,\Gamma(n/2)\sigma/\Gamma[(n-1)/2]$$

Use this to obtain an unbiased estimator for σ of the form cS. What is c when $n = 20$?

109. Each of n specimens is to be weighed twice on the same scale. Let X_i and Y_i denote the two observed weights for the ith specimen. Suppose X_i and Y_i are independent of each other, each normally distributed with mean value μ_i (the true weight of specimen i) and variance σ^2.

(a) Show that the mle of σ^2 is $\hat{\sigma}^2 = \sum (X_i - Y_i)^2/(4n)$ [*Hint*: If $\bar{z} = (z_1 + z_2)/2$, then $\sum (z_i - \bar{z})^2 = (z_1 - z_2)^2/2$.]

(b) Is the mle $\hat{\sigma}^2$ an unbiased estimator of σ^2? Find an unbiased estimator of σ^2. [*Hint*: For any rv Z, $E(Z^2) = \text{Var}(Z) + [E(Z)]^2$. Apply this to $Z = X_i - Y_i$.]

110. The Principle of Unbiased Estimation has been criticized on the grounds that in some situations the only unbiased estimator is patently ridiculous. Here is one such example. Suppose that the number of major defects X on a randomly selected vehicle has a Poisson distribution with parameter μ. You are going to

purchase two such vehicles and wish to estimate $\theta = P(X_1 = 0, X_2 = 0) = e^{-2\mu}$, the probability that neither of these vehicles has any major defects. Your estimate is based on observing the value of X for a single vehicle. Denote this estimator by $\hat{\theta} = g(X)$. Write the equation implied by the condition of unbiasedness, $E[g(X)] = e^{-2\mu}$, cancel $e^{-\mu}$ from both sides, then expand what remains on the right-hand side in an infinite series, and compare the two sides to determine $g(X)$. If $X = 200$, what is the estimate? Does this seem reasonable? What is the estimate if $X = 199$? Is this reasonable?

111. Let X, the payoff from playing a certain game, have pmf

$$p(x; \theta) = \begin{cases} \theta & x = -1 \\ (1 - \theta)^2 \theta^x & x = 0, 1, 2, \ldots \end{cases}$$

(a) Verify that $p(x; \theta)$ is a legitimate pmf, and determine the expected payoff. [Hint: Look back at the properties of a geometric random variable discussed in Chap. 2.]

(b) Let $X_1, \ldots, X_n$ be the payoffs from n independent games of this type. Determine the mle of θ. [Hint: Let Y denote the number of observations among the n that equal -1, and write the likelihood as a single expression in terms of $\sum x_i$ and y.]

112. The reaction time (RT) to a stimulus is the interval of time commencing with stimulus presentation and ending with the first discernible movement of a certain type. The article "Relationship of Reaction Time and Movement Time in a Gross Motor Skill" (Percept. Motor Skills, 1973: 453–454) reports that the sample average RT for 16 experienced swimmers to a pistol start was .214 s and the sample standard deviation was .036 s. Making any necessary assumptions, derive a 90% CI for true average RT for all experienced swimmers.

113. For each of 18 preserved cores from oil-wet carbonate reservoirs, the amount of residual gas saturation after a solvent injection was measured at water flood-out. Observations, in percentage of pore volume, were

23.5	31.5	34.0	46.7	45.6	32.5
41.4	37.2	42.5	46.9	51.5	36.4
44.5	35.7	33.5	39.3	22.0	51.2

(See "Relative Permeability Studies of Gas-Water Flow Following Solvent Injection in Carbonate Rocks," Soc. Petrol. Eng. J., 1976: 23–30.)

(a) Is it plausible that the sample was selected from a normal population distribution?

(b) Calculate a 98% CI for the true average amount of residual gas saturation.

114. Aphid infestation of fruit trees can be controlled either by spraying with pesticide or by inundation with ladybugs. In a particular area, four different groves of fruit trees are selected for experimentation. The first three groves are

sprayed with pesticides 1, 2, and 3, respectively, and the fourth is treated with ladybugs, with the following results on yield:

Treatment	n_i (number of trees)	$\bar{x}_i$ (bushels/tree)	s_i
1	100	10.5	1.5
2	90	10.0	1.3
3	100	10.1	1.8
4	120	10.7	1.6

Let μ_i = the true average yield (bushels/tree) after receiving the ith treatment. Then

$$\theta = \frac{1}{3}(\mu_1 + \mu_2 + \mu_3) - \mu_4$$

measures the difference in true average yields between treatment with pesticides and treatment with ladybugs. When n_1, n_2, n_3, and n_4 are all large, the estimator $\hat{\theta}$ obtained by replacing each μ_i by $\bar{X}_i$ is approximately normal. Use this to derive a large-sample $100(1 - \alpha)\%$ CI for θ, and compute the 95% interval for the given data.

115. It is important that face masks used by firefighters be able to withstand high temperatures because firefighters commonly work in temperatures of 200–500 °F. In a test of one type of mask, 11 of 55 masks had lenses pop out at 250°. Construct a 90% CI for the true proportion of masks of this type whose lenses would pop out at 250°.

116. A journal article reports that a sample of size 5 was used as a basis for calculating a 95% CI for the true average natural frequency (Hz) of delaminated beams of a certain type. The resulting interval was (229.764, 233.504). You decide that a confidence level of 99% is more appropriate than the 95% level used. What are the limits of the 99% interval? [*Hint*: Use the center of the interval and its width to determine $\bar{x}$ and s.]

117. Chronic exposure to asbestos fiber is a well-known health hazard. The article "The Acute Effects of Chrysotile Asbestos Exposure on Lung Function" (*Envir. Res.*, 1978: 360–372) reports results of a study based on a sample of construction workers who had been exposed to asbestos over a prolonged period. Among the data given in the article were the following (ordered) values of pulmonary compliance (cm³/cm H_2O) for each of 16 subjects 8 months after the exposure period (pulmonary compliance is a measure of lung elasticity, or how effectively the lungs are able to inhale and exhale):

167.9	180.8	184.8	189.8	194.8	200.2
201.9	206.9	207.2	208.4	226.3	227.7
228.5	232.4	239.8	258.6		

(a) Is it plausible that the population distribution is normal?
(b) Compute a 95% CI for the true average pulmonary compliance after such exposure.

118. A triathlon consisting of swimming, cycling, and running is one of the more strenuous amateur sporting events. The article "Cardiovascular and Thermal Response of Triathlon Performance" (*Medicine and Science in Sports and Exercise*, 1988: 385–389) reports on a research study involving nine male triathletes. Maximum heart rate (beats/min) was recorded during performance of each of the three events. For swimming, the sample mean and sample standard deviation were 188.0 and 7.2, respectively. Assuming that the heart-rate distribution is (approximately) normal, construct a 98% CI for true mean heart rate of triathletes while swimming.

119. An April 2009 survey of 2253 American adults conducted by the Pew Research Center's Internet & American Life Project revealed that 1262 of the respondents had at some point used wireless means for online access.
    (a) Calculate and interpret a 95% CI for the proportion of all American adults who at the time of the survey had used wireless means for online access.
    (b) What sample size is required if the desired width of the 95% CI is to be at most .04, irrespective of the sample results? [*Hint:* See Exercise 83.]

120. High concentration of the toxic element arsenic is all too common in groundwater. The article "Evaluation of Treatment Systems for the Removal of Arsenic from Groundwater" (*Practice Periodical of Hazardous, Toxic, and Radioactive Waste Mgmt.*, 2005: 152–157) reported that for a sample of $n = 5$ water specimens selected for treatment by coagulation, the sample mean arsenic concentration was 24.3 mg/L, and the sample standard deviation was 4.1. The authors of the cited article used t-based methods to analyze their data, so hopefully had reason to believe that the distribution of arsenic concentration was normal.
    (a) Calculate and interpret a 95% CI for true average arsenic concentration in all such water specimens.
    (b) Predict the arsenic concentration for a single water specimen in a way that conveys information about precision and reliability. (See Exercise 49.)

121. Let θ_1 and θ_2 denote the mean weights for animals of two different species. An investigator wishes to estimate the ratio θ_1/θ_2. Unfortunately the species are extremely rare, so the estimate will be based on finding a single animal of each species. Let X_i denote the weight of the species i animal ($i = 1, 2$), assumed to be normally distributed with mean θ_i and standard deviation 1.
    (a) What is the distribution of the variable $(\theta_2 X_1 - \theta_1 X_2)/\sqrt{\theta_1^2 + \theta_2^2}$? Show that this variable depends on θ_1 and θ_2 only through θ_1/θ_2 (divide numerator and denominator by θ_2).
    (b) Since the variable in (a) is normally distributed, we have

$$P\left(-1.96 < (\theta_2 X_1 - \theta_1 X_2)/\sqrt{\theta_1^2 + \theta_2^2} < 1.96\right) = .95$$

Now replace $<$ by $=$ and solve for θ_1/θ_2. Then show that a confidence interval results if $x_1^2 + x_2^2 \geq 1.96^2$, whereas if this inequality is not satisfied, the resulting *confidence set* is the complement of an interval.

122. Let $X_1, X_2, \ldots, X_n$ be a random sample from a uniform distribution on the interval $[0, \theta]$.

Then if $Y = \max(X_i)$, the techniques of Sect. 4.9 show that Y has density function

$$f(y) = \begin{cases} \dfrac{n}{\theta^n} y^{n-1} & 0 \leq y \leq \theta \\ 0 & \text{otherwise} \end{cases}$$

(a) Use $f(y)$ to verify that

$$P\left[\theta \cdot (\alpha/2)^{1/n} \leq Y \leq \theta \cdot (1 - \alpha/2)^{1/n}\right] = 1 - \alpha$$

and use this to derive a $100(1 - \alpha)\%$ CI for θ.

(b) Verify that $P(\theta \cdot \alpha^{1/n} \leq Y \leq \theta) = 1 - \alpha$, and derive a $100(1 - \alpha)\%$ CI for θ based on this probability statement.

(c) Which of the two intervals derived in (a) and (b) is shorter? If your waiting time for a morning bus is uniformly distributed and observed waiting times are $x_1 = 4.2$, $x_2 = 3.5$, $x_3 = 1.7$, $x_4 = 1.2$, and $x_5 = 2.4$, obtain a 95% CI for θ by using the shorter of the two intervals.

123. Consider 95% CIs for two different parameters θ_1 and θ_2, and let A_i $(i = 1, 2)$ denote the event that the value of θ_i is included in the random interval that results in the CI. Thus $P(A_i) = .95$.

(a) Suppose that the data on which the CI for θ_1 is based is independent of the data used to obtain the CI for θ_2 (e.g., we might have $\theta_1 = \mu$, the population mean height for American females, and $\theta_2 = p$, the proportion of all Kodak digital cameras that don't need warranty service). What can be said about the *simultaneous* (i.e., *joint*) confidence level for the two intervals? That is, how confident can we be that the first interval contains the value of θ_1 and that the second contains the value of θ_2? [*Hint*: Consider $P(A_1 \cap A_2)$.]

(b) Now suppose the data for the first CI is not independent of that for the second one. What now can be said about the simultaneous confidence level for both intervals? [*Hint*: Consider $P(A_1' \cup A_2')$, the probability that at least one interval fails to include the value of what it is estimating. Now use the fact that $P(A_1' \cup A_2') \leq P(A_1') + P(A_2')$ [why?] to show that the probability that *both* random intervals include what they are estimating is at least .90. The generalization of the bound on $P(A_1' \cup A_2')$ to the probability of a k-fold union is one version of the *Bonferroni* inequality.]

(c) What can be said about the simultaneous confidence level in (b) if the confidence level for each interval separately is $100(1 - \alpha)\%$? What can be said about the simultaneous confidence level if a $100(1 - \alpha)\%$ CI is computed separately for each of k parameters $\theta_1, \ldots, \theta_k$?

124. Let $X_1, \ldots, X_n$ be a random sample from a continuous probability distribution having median η (so that $P(X_i \leq \eta) = P(X_i \geq \eta) = .5$). Let Y_1 and Y_n be the smallest and largest order statistic for the sample (i.e., $Y_1 = \min(X_i)$ and $Y_n = \max(X_i)$).

(a) Show that

$$P(Y_1 \leq \eta \leq Y_n) = 1 - \left(\frac{1}{2}\right)^{n-1}$$

so that (Y_1, Y_n) is a $100(1 - \alpha)\%$ confidence interval for η with $\alpha = (1/2)^{n-1}$. [*Hint*: Use the same arguments employed in Sect. 4.9 to derive the cdfs of Y_1 and Y_n.]

(b) For each of six normal male infants, the amount of the amino acid alanine (mg/100 mL) was determined while the infants were on an isoleucine-free diet, resulting in the following data:

2.84	3.54	2.80	1.44	2.94	2.70

Compute a 97% CI for the true median amount of alanine for infants on such a diet ("The Essential Amino Acid Requirements of Infants," *Amer. J. of Nutrition*, 1964: 322–330).

(c) Let Y_2 and Y_{n-1} denote the second-smallest and second-largest of the X_is, respectively. What is the confidence level of the interval (Y_2, Y_{n-1}) for η?

125. One method for straightening wire before coiling it to make a spring is called "roller straightening." The article "The Effect of Roller and Spinner Wire Straightening on Coiling Performance and Wire Properties" (*Springs*, 1987: 27–28) reports on the tensile properties of wire. Suppose a sample of 16 wires is selected and each is tested to determine tensile strength (N/mm^2). The resulting sample mean and standard deviation are 2160 and 30, respectively.

(a) The mean tensile strength for springs made using spinner straightening is 2150 N/mm^2. What hypotheses should be tested to determine whether the mean tensile strength for the roller method exceeds 2150?

(b) Assuming that the tensile strength distribution is approximately normal, what test statistic would you use to test the hypotheses in part (a)?

(c) What is the value of the test statistic for this data?

(d) What is the P-value for the value of the test statistic computed in part (c)?

(e) For a level .05 test, what conclusion would you reach?

126. A new method for measuring phosphorus levels in soil is described in the article "A Rapid Method to Determine Total Phosphorus in Soils" (*Soil Sci. Amer. J.*, 1988: 1301–1304). Suppose a sample of 11 soil specimens, each with a true phosphorus content of 548 mg/kg, is analyzed using the new

method. The resulting sample mean and standard deviation for phosphorus level are 587 and 10, respectively.

(a) Is there evidence that the mean phosphorus level reported by the new method differs significantly from the true value of 548 mg/kg? Use $\alpha = .05$.

(b) What assumptions must you make for the test in part (a) to be appropriate?

127. The article "Orchard Floor Management Utilizing Soil-Applied Coal Dust for Frost Protection" (*Agric. Forest Meteorol.*, 1988: 71–82) reports the following values for soil heat flux of eight plots covered with coal dust.

| 34.7 | 35.4 | 34.7 | 37.7 | 32.5 | 28.0 | 18.4 | 24.9 |

The mean soil heat flux for plots covered only with grass is 29.0. Assuming that the heat-flux distribution is approximately normal, does the data suggest that the coal dust is effective in increasing the mean heat flux over that for grass? Test the appropriate hypotheses using $\alpha = .05$.

128. The article "Caffeine Knowledge, Attitudes, and Consumption in Adult Women" (*J. Nutrit. Ed.*, 1992: 179–184) reports the following summary data on daily caffeine consumption for a sample of adult women: $n = 47$, $\bar{x} = 215$ mg, $s = 235$ mg, and range $= 5$–1176.

(a) Does it appear plausible that the population distribution of daily caffeine consumption is normal? Is it necessary to assume a normal population distribution to test hypotheses about the value of the population mean consumption? Explain your reasoning.

(b) Suppose it had previously been believed that mean consumption was at most 200 mg. Does the given data contradict this prior belief? Test the appropriate hypotheses at significance level .10.

129. The accompanying observations on residual flame time (seconds) for strips of treated children's nightwear were given in the article "An Introduction to Some Precision and Accuracy of Measurement Problems" (*J. Test. Eval.*, 1982: 132–140). Suppose a true average flame time of at most 9.75 had been mandated. Does the data suggest that this condition has not been met? Carry out an appropriate test after first investigating the plausibility of assumptions that underlie your method of inference.

9.85	9.93	9.75	9.77	9.67	9.87	9.67
9.94	9.85	9.75	9.83	9.92	9.74	9.99
9.88	9.95	9.95	9.93	9.92	9.89	

130. The incidence of a certain type of chromosome defect in the US adult male population is believed to be 1 in 75. A random sample of 800 individuals in US penal institutions reveals 16 who have such defects. Can it be concluded that the incidence rate of this defect among prisoners differs from the presumed rate for the entire adult male population?

(a) State and test the relevant hypotheses using $\alpha = .05$.

(b) What type of error might you have made in reaching a conclusion?

131. In an investigation of the toxin produced by a certain poisonous snake, a researcher prepared 26 different vials, each containing 1 g of the toxin, and then determined the amount of antitoxin needed to neutralize the toxin. The sample average amount of antitoxin necessary was found to be 1.89 mg, and the sample standard deviation was .42. Previous research had indicated that the true average neutralizing amount was 1.75 mg/g of toxin. Does the new data contradict the value suggested by prior research? Test the relevant hypotheses. Does the validity of your analysis depend on any assumptions about the population distribution of neutralizing amount? Explain.

132. The sample average unrestrained compressive strength for 45 specimens of a particular type of brick was computed to be 3107 psi, and the sample standard deviation was 188. The distribution of unrestrained compressive strength may be somewhat skewed. Does the data strongly indicate that the true average unrestrained compressive strength is less than the design value of 3200? Test using $\alpha = .001$.

133. To test the ability of auto mechanics to identify simple engine problems, an automobile with a single such problem was taken in turn to 72 different car repair facilities. Only 42 of the 72 mechanics who worked on the car correctly identified the problem. Does this strongly indicate that the true proportion of mechanics who could identify this problem is less than .75? Test the appropriate hypotheses.

134. The December 30, 2009, the *New York Times* reported that in a survey of 948 American adults who said they were at least somewhat interested in college football, 597 said the Bowl Championship System should be replaced by a playoff similar to that used in college basketball (in fact, a playoff system replaced the BCS starting with the 2014 season). Does this provide compelling evidence for concluding that a majority of all such individuals favored replacing the BCS with a playoff at that time? Test the appropriate hypotheses.

135. An article in the November 11, 2005, issue of the San Luis Obispo *Tribune* reported that researchers making random purchases at California Wal-Mart stores found scanners coming up with the wrong price 8.3% of the time. Suppose this was based on 200 purchases. The National Institute for Standards and Technology says that in the long run at most two out of every 100 items should have incorrectly scanned prices. Carry out an appropriate hypothesis test to decide whether the NIST benchmark is not satisfied. [*Caution*: Are the conditions for a one-proportion z test met? If not, what distribution can be used instead?]

136. Annual holdings turnover for a mutual fund is the percentage of a fund's assets that are sold during a particular year. Generally speaking, a fund with a low value of turnover is more stable and risk averse, whereas a high value of turnover indicates a substantial amount of buying and selling in an attempt to take advantage of short-term market fluctuations. Here are values of turnover for a sample of 20 large-cap blended funds extracted from Morningstar.com:

| 1.03 | 1.23 | 1.10 | 1.64 | 1.30 | 1.27 | 1.25 | 0.78 | 1.05 | 0.64 |
| 0.94 | 2.86 | 1.05 | 0.75 | 0.09 | 0.79 | 1.61 | 1.26 | 0.93 | 0.84 |

(a) Would you use the one-sample t test to decide whether there is compelling evidence for concluding that the population mean turnover is less than 100%? Explain.

(b) A normal probability plot of the 20 ln(turnover) values shows a very pronounced linear pattern, suggesting it is reasonable to assume that the turnover distribution is lognormal. Recall that X has a lognormal distribution if $\ln(X)$ is normally distributed with mean value μ and standard deviation σ. Because μ is also the median of the $\ln(X)$ distribution, e^{μ} is the median of the X distribution. Use this information to decide whether there is compelling evidence for concluding that the median of the turnover population distribution is less than 100%.

137. When $X_1, X_2, \ldots, X_n$ are independent Poisson variables, each with parameter μ, and n is large, the sample mean $\overline{X}$ has approximately a normal distribution with $E(\overline{X}) = \mu$ and $\text{Var}(\overline{X}) = \mu/n$. This implies that

$$Z = \frac{\overline{X} - \mu}{\sqrt{\mu/n}}$$

has approximately a standard normal distribution. For testing $H_0: \mu = \mu_0$, we can replace μ by μ_0 in the equation for Z to obtain a test statistic. This statistic is actually preferred to the large-sample statistic with denominator $S/\sqrt{n}$ (when the X_is are Poisson) because it is tailored explicitly to the Poisson assumption. If the number of requests for consulting received by a certain statistician during a 5-day work week has a Poisson distribution and the total number of consulting requests during a 36-week period is 160, does this suggest that the true average number of weekly requests exceeds 4.0? Test using $\alpha = .02$.

138. When the population distribution is normal and n is large, the sample standard deviation S has approximately a normal distribution with $E(S) \approx \sigma$ and $\text{Var}(S) \approx \sigma^2/(2n)$. We already know that in this case, for any n, $\overline{X}$ is normal with $E(\overline{X}) = \mu$ and $\text{Var}(\overline{X}) = \sigma^2/n$.

(a) Assuming that the underlying distribution is normal, what is an approximately unbiased estimator of the 99th percentile $\theta = \mu + 2.33\sigma$?

(b) When the X_is are normal, it can be shown that $\overline{X}$ and S are independent rvs (one measures location whereas the other measures spread). Use this to compute $\text{Var}(\hat{\theta})$ and $\sigma_{\hat{\theta}}$ for the estimator $\hat{\theta}$ of part (a). What is the estimated standard error $\hat{\sigma}_{\hat{\theta}}$?

(c) Write a test statistic formula for testing $H_0: \theta = \theta_0$ that has approximately a standard normal distribution when H_0 is true. If soil pH is normally distributed in a certain region and 64 soil samples yield $\bar{x} = 6.33$, $s = .16$, does this provide strong evidence for concluding that at most 99% of all possible samples would have a pH of less than 6.75? Test using $\alpha = .01$.

Markov Chains

6

This chapter explores the properties of a broadly applicable probability model called a *Markov chain*, named after Russian mathematician A. A. Markov (1856–1922). Markov observed that many real-world phenomena can be modeled as a sequence of "transitions" from one "state" to another, with each transition having some associated uncertainty. For example, a taxi driver might "transition" between several towns (or zones within a large city); each time he drops off a passenger, he can't be certain where his next fare will want to go. Similarly, a gambler might think of her winnings as transitioning from one "state"—really, a dollar amount—to another; with each round of the game she plays, she cannot be certain whether that dollar amount will go up or down (though, obviously, she hopes it goes up!). The same could be said for modeling the daily closing prices of a stock: each new day, there is uncertainty about whether that stock will "transition" to a higher or lower value, and this uncertainty could be modeled using the tools of probability.

In all of these examples, aside from the probability model for how transitions occur, one extra piece of information is critical: the current "state" (where the taxi driver is, how much money the gambler has). After all, if the gambler is making $5 wagers, how much money she might have after the next game depends on how much she has now—if she currently holds $45 in chips, then at the end of the upcoming round she can only have $40 or $50 on an even bet. The model structure proposed by Markov applies to situations where *only* knowledge of the current state, and the nature of transitions, is necessary—we don't care how our gambler arrived at $45 in chips, only that that's how much she currently possesses.

Section 6.1 introduces basic notation for Markov chains and provides a rigorous definition of the property alluded to in the previous paragraph. In Sects. 6.2 and 6.3 we explain how the use of matrix notation can facilitate Markov chain computations. Section 6.4 focuses on a special class of Markov chains, so-called *regular chains*, which have a rather exceptional property embodied in the Steady-State Theorem. Section 6.5 considers a different class of Markov chains, those with one or more "inescapable" states, such as a gambler going broke. Finally, Sect. 6.6 discusses the simulation of Markov chains using software.

M.A. Carlton and J.L. Devore, *Probability with Applications in Engineering, Science, and Technology*, Springer Texts in Statistics, DOI 10.1007/978-1-4939-0395-5_6,
© Springer Science+Business Media New York 2014

6.1   Terminology and Basic Properties

Markov chains provide a model for sequential information that allows future outcomes to depend on previous ones, albeit in a very specific way (the defining *Markov property*). Researchers in numerous fields employ Markov chains to model the phenomena they study. Recent examples include predicting changes in electricity demand; modeling the motion of sperm whales off the Galapagos Islands; Chinese citizens changing their cell phone service; keeping track of inpatient bed usage at hospitals; monitoring patterns in Web browser histories to deploy better-targeted advertising; the evolution of drought conditions over time; and the dynamics of capital assets.

This first section introduces the basic vocabulary and notation of Markov chains. We begin with the following classic (if slightly artificial) example, which will serve as a thread throughout the chapter.

Example 6.1   A city has three different taxi zones, numbered 1, 2, and 3. A taxi driver operates his cab in all three zones. The probability that his next passenger has a destination in a particular one of these zones depends on where the passenger is picked up. Specifically, whenever the taxi driver is in zone 1, the probability his next passenger is going to zone 1 is .3, to zone 2 is .2, and to zone 3 is .5. Starting in zone 2, the probability his next passenger is going to zone 1 is .1, to zone 2 is .8, and to zone 3 is .1. Finally, whenever he is in zone 3, the probability his next passenger is going to zone 1 is .4, to zone 2 is .4, and to zone 3 is .2. These probabilities are encapsulated in the **state diagram** in Fig. 6.1.

In every such state diagram, the sum of the probabilities on branches exiting any state must equal 1. For example, in Fig. 6.1 the probabilities exiting state 2 (i.e., zone 2) are .1, .8, and .1. We include in this calculation the probability .8 indicated by a "loop" in the state diagram, which simply means that the taxi driver has .8 probability of staying in zone 2 once he has dropped off a fare in zone 2.

Define X_0 to be the zone in which the taxi driver starts and X_n $(n \geq 1)$ to be the zone where he drops off his nth fare. Since $X_0, X_1, X_2, \ldots$ "occur" in sequence, they are often referred to as a *chain*. More precisely, this particular sequence is a *finite-*

Fig. 6.1 State diagram for
Example 6.1

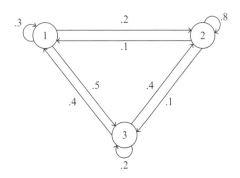

state, discrete-time, time-homogeneous Markov chain. Each of these terms will be explained shortly.   ∎

In Example 6.1, each of the X_n for $n \geq 0$ assumes the value 1, 2, or 3 according to the destination zone. The zones collectively constitute the **states** of our chain, and so the **state space** is {zone 1, zone 2, zone 3}, although we will often drop the state names and just use the integers {1, 2, 3}. States can be identified with physical locations, levels (such as high/medium/low), dollar amounts, or just about anything else. We'll sometimes refer to the X_n as random variables, even though they are not necessarily numerical (which goes against the definition from Chap. 2). The random variable X_0 is called the **initial state** of the chain. A **discrete-space** chain is one for which the number of possible states is finite or countably infinite. If there are finitely many possible states, we have a **finite-state** chain.

Since time was indexed by the discrete listing $n = 0, 1, 2, \ldots$, the sequence of zones the taxi driver visited in Example 6.1 is called a **discrete-time** chain. Section 7.7 gives an overview of **continuous-time** chains, often indexed as {X_t: $t \in [0, \infty)$}, which are useful for modeling behavior continuously in time rather than just at discrete time points (e.g., tracking over time the number of people looking at a particular Web site). The taxi driver chain is also **time-homogeneous**, in that the specified probabilities do not change over time. One could imagine a different, more complicated model where the probabilities specified in Example 6.1 apply during morning hours but not in the evening, so that the probability of taking a fare from zone 1 to zone 3 is .5 for $n = 1$ (beginning of the work day) but is .1, say, for $n = 20$ (end of his shift). See Exercises 78 and 79 for examples of nonhomogeneous Markov chains.

Example 6.2 This is a simple version of the famous *Gambler's Ruin* problem, which we previously considered in Exercise 145 of Chap. 1. Allan and Beth play a succession of independent games for $1 each. Suppose Allan starts with $2 and Beth with $1, and the chance of Allan winning $1 is p on each game. Ties are not allowed, so the chance of Beth winning $1 on any particular game is $1 - p$. They compete until one of the two players goes broke (has $0).

For $n = 0, 1, 2, \ldots$, define $X_n =$ the amount of money Allan has after n games. The initial state has been specified as $X_0 = \$2$; Allan's successive holdings X_0, X_1, X_2, $\ldots$ form our chain. The state space for X_n is {$0, $1, $2, $3} or just {0, 1, 2, 3}, so we again have a finite-state chain. The state space and the specified probabilities are illustrated by the state diagram in Fig. 6.2. Notice we have included two "loops" with probability 1 at $0 and $3—these reflect the constraint that the game stops once Allan reaches one of these dollar amounts. That is, once Allan is "at" $3, he will stay at $3, and the same goes for $0. Also, it will always be understood that if no arrow points from state i to state j in such a diagram, then the probability of moving from state i immediately into state j (i.e., in one time step) is zero.

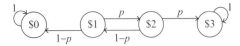

Fig. 6.2 State diagram for Example 6.2   ■

Example 6.3 *A random walk*. Imagine a marker initially placed at 0 on the number line. A fair coin is flipped repeatedly; each head moves the marker one integer to the right, while each tail moves it one integer to the left. Let $X_0 = 0$, the initial state, and $X_n =$ the marker's position after n coin flips for $n \geq 1$. Each member of the chain can only take on a finite set of values: X_1 is either $+1$ or -1, X_2 is one of -2, 0, or 2, and so on. However, the collection of *all* possible states across *all* time indices comprises the entire set of integers: $\{\dots, -3, -2, -1, 0, 1, 2, 3, \dots\}$. Thus, this so-called "random walk" is an infinite-state (though still discrete-state) chain; it is partially illustrated in Fig. 6.3.

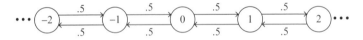

Fig. 6.3 State diagram for Example 6.3   ■

6.1.1   The Markov Property

All of the preceding examples have an important feature known as the *Markov property*. Loosely speaking, it says that in order to know where the chain will go next (say, X_{n+1}), it suffices to know where the chain is now (the value of X_n). In particular, once the current state is specified, the path that brought the chain to that state is irrelevant. Consider, for example, the random walk of Example 6.3: if for any particular n we have $X_n = 4$, then we know $X_{n+1} = 3$ or 5 with probability .5 each. It does not matter whether the chain arrived at 4 quickly ($0 \to 1 \to 2 \to 3 \to 4$) or by a more circuitous route; the probability distribution of the next state in the chain is the same. This notion is formalized in the following definition.

DEFINITION

Let $X_0, X_1, X_2, \dots$ be a sequence of random variables (a chain) on some discrete state space. The sequence is said to have the **Markov property** if, for any time index n and any set of (not necessarily distinct) states $s_0, s_1, \dots, s_n, s_{n+1}$,

$$P\left(X_{n+1} = s_{n+1} \mid X_0 = s_0, X_1 = s_1, \dots, X_n = s_n\right) = P\left(X_{n+1} = s_{n+1} \mid X_n = s_n\right)$$

(6.1)

Such a sequence $\{X_n: n = 0, 1, 2, \dots\}$ is called a **Markov chain**.

The conditional probabilities specified in Expression (6.1) are called the **one-step transition probabilities** of the chain, or sometimes just *transition probabilities*. These are precisely the probabilities specified in Examples 6.1–6.3. It's critical to recognize that these are *conditional* probabilities: they specify the likelihood of the next member of the chain X_{n+1} being in any particular state, *given* the current state of the chain X_n.

Example 6.4 (Example 6.1 continued) The sequence of successive zones visited by our taxi driver is characterized by nine one-step transition probabilities. For example, it is stated that the driver "transitions" from zone 1 to zone 3 with probability .5, which means that for any time index n,

$$P(X_{n+1} = 3 \mid X_n = 1) = .5$$

This probability does not depend on the value of n, because the chain is time-homogeneous. Instead of writing $P(X_{n+1} = 3 | X_n = 1) = .5$, we will sometimes abbreviate with $P(1 \to 3) = .5$ to emphasize the idea of transitioning from one state to another. Thus, the complete set of one-step transition probabilities for the taxi driver is

$$P(1 \to 1) = .3 \qquad P(1 \to 2) = .2 \qquad P(1 \to 3) = .5$$

$$P(2 \to 1) = .1 \qquad P(2 \to 2) = .8 \qquad P(2 \to 3) = .1$$

$$P(3 \to 1) = .4 \qquad P(3 \to 2) = .4 \qquad P(3 \to 3) = .2 \qquad \blacksquare$$

Example 6.5 (Example 6.2 continued) The changing fortunes of Allan are governed by six (non-zero) transition probabilities:

$$P(1 \to 0) = 1 - p \qquad P(1 \to 2) = p \qquad P(2 \to 1) = 1 - p \qquad P(2 \to 3) = p$$
$$P(0 \to 0) = 1 \qquad P(3 \to 3) = 1$$

The last two probabilities above correspond to termination of the sequence of games. From a mathematical (if not practical) perspective, they communicate the idea that the chain marches on even when gameplay has ended (e.g., $2 \to 3 \to 3 \to 3 \ldots$). That is, the conditional probability $P(3 \to 3) = P(X_{n+1} = 3 | X_n = 3) = 1$ indicates that if Allan has all \$3 at stake after n games, he will retain his \$3 while some imaginary future gameplay continues (the $(n+1)$st game, the $(n+2)$nd game, etc.). This convention eliminates the need to "stop" the Markov chain at some particular time point n. We'll elaborate much more on this in Sect. 6.5.

In addition, there are ten one-step transition probabilities that equal zero; for example, according to the rules of Gambler's Ruin, $P(1 \to 3) = 0$, and $P(3 \to x) = 0$ for $x \in \{0, 1, 2\}$. In general, a finite-state Markov chain with s states is specified by s^2 one-step transition probabilities, although it is quite common for many (if not most) of these to be zero. $\blacksquare$

Example 6.6 Markov chains are often used to model changing weather conditions; research literature in both meteorology and climate science is rife with Markov chain applications. The article "To Ski or Not to Ski: Estimating Transition Matrices to Predict Tomorrow's Snowfall Using Real Data" (*J. of Statistics Educ.*, vol. 18, no. 3, 2010) provides data for several US cities on the daily transitions between "snow days," defined by a snow depth of at least 50 mm, and "green days" (snow depth < 50 mm). Let X_n represent the snow status, either S for snow or G for green, on the nth recorded day. For New York City, the following one-step transition probabilities are provided:

$$P(G \to G) = .964 \quad P(G \to S) = .036 \quad P(S \to G) = .224 \quad P(S \to S) = .776$$

If today is a "green day" in New York, then there is a 96.4% chance that tomorrow's snow depth will also be below 50 mm, based on the available weather data (which, incidentally, stretches back to the year 1912 for New York). On the other hand, as the author notes, "the presence of a significant snow depth (accumulation) on the current day in Central Park (New York) has an approximately 1 in 5 chance of melting before the next day." ∎

Not all sequences of random variables possess the Markov property. In econometrics (statistical methodology applied to economic scenarios), for example, most models for the closing price X_{n+1} of a stock on the $(n+1)$st day of trading incorporate not only the previous day's closing price X_n but also information from many previous days (the data X_{n-1}, X_{n-2}, and so on). The likelihood that X_{n+1} will be $5 higher than X_n may depend on the stock's behavior over all of last week, not just where it closed on day n.

That said, in some instances a model that includes more than a one-time-step dependence can be modified by reconfiguring the state space in such a way that it satisfies the Markov property. This expansion of states is illustrated in the next example.

Example 6.7 The weather model presented in Example 6.6 satisfies the Markov property; in particular, it assumes that one can model tomorrow's weather based on today's conditions without incorporating any previous information. A more realistic model might assume that tomorrow's snow depth depends on today's *and* yesterday's weather. Suppose, for example, that tomorrow will be a snow day with probability .8 if both yesterday and today were snow days; with probability .6 if today was a snow day but yesterday was a green day; with probability .3 if it was green today and snowy yesterday; and with probability .1 if both previous days were green.

Once again let $X_n =$ the "state" of the weather on day n: G for green day, S for snow day. Then the sequence $X_0, X_1, X_2, \ldots.$ of weather states does *not* satisfy the Markov property, because the conditional distribution of X_{n+1} given all previous weather information depends on both X_n and X_{n-1} (the previous two days' weather conditions). Let's make the following modification: define Y_n to be the ordered pair $Y_n = $ (day n weather, day $n+1$ weather) $= (X_n, X_{n+1})$

So, for example, if snow depth was ≥ 50 mm on day 4 but < 50 mm on day 5, then $Y_4 = (S, G)$. The weather on day 6 depends on these previous 2 days, but they are now both contained in a single "variable," Y_4. In other words, Y_5 can be modeled entirely by knowing Y_4: Y_5's first entry, X_5, matches the second entry of Y_4, and the probability distribution of the second entry of Y_5 (i.e., X_6) is determined by the rules given at the beginning of this example.

With this modification, the sequence $Y_0, Y_1, Y_2, \ldots$ forms a Markov chain. The state space of this chain is not $\{S, G\}$, but rather $\{(G, G), (G, S), (S, G), (S, S)\}$. The earlier weather rules can be expressed as one-step transition probabilities for this chain:

$$P((S, S) \rightarrow (S, S)) = .8 \qquad P((S, G) \rightarrow (G, S)) = .3$$
$$P((G, S) \rightarrow (S, S)) = .6 \qquad P((G, G) \rightarrow (G, S)) = .1$$

Four other transition probabilities can be found by considering the complements of the given transition events. The final eight transition probabilities (with four states, there are $4^2 = 16$ total one-step transition probabilities) are all 0, e.g., $P((S, G) \rightarrow (S, S)) = 0$, because if $Y_n = (S, G)$ then it was "green" on day $n + 1$ ($X_{n+1} = G$), meaning the first entry of Y_{n+1} must also be G. ∎

The remainder of this chapter will focus almost exclusively on finite-state, discrete-time, time-homogeneous chains; these are the most commonly encountered models in practice. The case of infinite-state chains, including the random walk of Example 6.3, is considered in several more advanced texts; see, for example, the book *Introduction to Probability Models* by Ross listed in the references.

6.1.2   Exercises: Section 6.1 (1–10)

1. The article "Markov Chain Models of Negotiators' Communication" (*Encyclopedia of Peace Psychology* 2012: 608-612) describes the following set-up for the back and forth dialogue between two negotiators. If at any stage a negotiator engages in a cooperative strategy, the other negotiator will respond with a cooperative strategy with probability .6. Otherwise, the response is described as a competitive strategy. Similarly, there is probability .7 that a competitive strategy offered at any stage of the negotiations will be met by another competitive strategy. Let $X_n = $ the strategy employed at the nth stage of the negotiation. Identify the state space for the chain, specify its one-step transition probabilities, and draw the corresponding state diagram.

2. Imagine m balls being exchanged between two adjacent chambers (left and right) according to the following rules. At each time step, one of the m balls is

randomly selected and moved to the opposite chamber, i.e., if the selected ball is currently in the right chamber, it will be moved to the left one, and vice versa. Let $X_n =$ the number of balls in the left chamber after the nth exchange. (This is called an *Ehrenfest chain*, a model often used to describe the movement of gas molecules.)

(a) Identify the state space of this chain.

(b) Suppose $m = 3$. Specify the one-step transition probabilities for this chain. [*Hint*: It might be helpful to draw the two chambers and the possible positions of the three balls.]

(c) Draw the state diagram corresponding to (b).

(d) Generalize the probabilities in (b) to the case of m balls.

3. A certain machine used in a manufacturing process can be in one of three states: fully operational ("full"), partially operational ("part"), or broken. If the machine is fully operational today, there's a .7 probability it will be fully operational again tomorrow, a .2 chance it will be partially operational tomorrow, and otherwise tomorrow it will be broken. If the machine is partially operational today, there is a .6 probability it will continue to be partially operational tomorrow and otherwise it will be broken (because the machine is never repaired in its partially operational state). Finally, if the machine is broken today, there is a .8 probability it will be repaired to fully operational status tomorrow; otherwise, it remains broken. Let $X_n =$ the state of the machine on day n.

(a) Identify the state space of this chain.

(b) Determine the complete set of one-step transition probabilities, and draw the corresponding state diagram.

4. Michelle will flip a coin until she gets heads four times in a row. Define $X_0 = 0$ and, for $n \geq 1$, $X_n =$ the number of heads in the current streak of heads after the nth flip.

(a) If the first seven flips result in the sequence *HTHHHTH*, determine the values of $X_1, X_2, \ldots, X_7$. [*Hint*: Each time Michelle flips tails, the streak is reset to 0.]

(b) Is this an example of a Markov chain? Explain why or why not.

(c) Identify the state space of the chain. Treat reaching four heads in a row in the same manner that the $3 state was treated in the Gambler's Ruin scenario of Example 6.2.

(d) Assume $P(H) = p$ for this particular coin. Determine the one-step transition probabilities of this chain, and draw the corresponding state diagram.

5. A single cell has probability p of dividing into two cells and probability $1 - p$ of dying without dividing. Once two new cells have been created, each has the same probability p of splitting in two, independent of the other. In this fashion, cells continue to divide, either indefinitely or until all cells are dead (extinction of the cell line). Let $X_n =$ the number of cells in the nth generation, with $X_0 = 1$ to reflect the initial, single cell.

(a) What are the possible numerical values of X_1, and what are their probabilities?

(b) What are the possible numerical values of X_2?

(c) Determine the one-step transition probabilities for this chain. That is, given there are x cells in the nth generation ($X_n = x$), determine the conditional probability distribution of X_{n+1}.

[*Note*: This is an example of a *branching process*, commonly known as a *Galton-Watson process*. See Exercise 163 at the end of Chap. 4 for information on determining the probability of eventual extinction.]

6. Imagine a set of stacked files, such as papers on your desk. Occasionally, you will need to retrieve one of these files, which you will find by "sequential search": looking at the first paper in the stack, then the second, and so on until you find the document you require. A sensible *sequential search algorithm* is to place the most recently retrieved file at the top of the stack, the idea being that files accessed more often will "rise to the top" and thus require less searching in the long run. For simplicity's sake, imagine such a scenario with just three files, labeled A, B, C.

(a) Let X_n represent the sequence of the entire stack after the nth search. For example, if the files are initially stacked A on top of B on top of C, then $X_0 = ABC$. Determine the state space for this chain.

(b) If $X_0 = ABC$, list all possible states for X_1. [*Hint*: One of the three files will be selected and rise to the front of the stack. Is every arrangement listed in (a) possible, starting from ABC?]

(c) Suppose that, at any given time, there is probability p_A that file A must be retrieved, p_B that file B must be retrieved, and similarly for p_C ($= 1 - p_A - p_B$). Determine all of the non-zero one-step transition probabilities.

7. Social scientists have used Markov chains to study "social mobility," the movement of people between social classes, for more than a century. In a typical such model, states are defined as social classes, e.g., lower class, middle class, and upper class. The time index n refers to a familial generation, so if X_n represents a man's social class, then X_{n-1} is his father's social class, X_{n-2} his grandfather's, and so on.

(a) In this context, what would it mean for X_n to be a Markov chain? In particular, would that imply that a grandfather's social class has no bearing on his grandson's? Explain.

(b) What would it mean for this chain to be time-homogeneous? Does that seem realistic? Explain why or why not.

8. The article "Markov Chain Models for Delinquency: Transition Matrix Estimation and Forecasting" (*Appl. Stochastic Models Bus. Ind.*, 2011: 267-279) classifies loan status into four categories: current (payments are up-to-date), delinquent (payments are behind but still being made), loss (payments have stopped permanently), and paid (the loan has been paid off). Let X_n = the status of a particular loan in its nth month, and assume (as the authors do) that X_n is a Markov chain.

(a) Suppose that, for one particular loan type, $P(\text{delinquent} \rightarrow \text{current}) = .1$ and $P(\text{current} \rightarrow \text{delinquent}) = .3$. Interpret these probabilities.

(b) According to the definitions of the "loss" and "paid" states, what are $P(\text{loss} \rightarrow \text{loss})$ and $P(\text{paid} \rightarrow \text{paid})$? [*Hint*: Refer back to Example 6.2.]

(c) Draw the state diagram for this Markov chain.

(d) What would it mean for this Markov chain to be time-homogeneous? Does that seem realistic? Explain.

9. The article cited in Exercise 1 also suggests a more complex negotiation model, wherein the strategy employed at the nth stage (cooperative or competitive) is predicted not only by the immediately preceding action but also the one before it. So, negotiator A's next strategy is determined not only by negotiator B's most recent move, but also by A's choice just before that. Again, let $X_n =$ the negotiating strategy used at the nth stage.

(a) Is X_n a Markov chain? Explain.

(b) How could you modify this example to create a Markov chain? What additional information would you need to completely specify this chain? [*Hint*: See Example 6.7.]

10. Let $X_0, X_1, X_2, \ldots$ be a sequence of *independent* discrete rvs taking values in some common state space.

(a) Show that X_n satisfies the Markov property. (That is, all sequences of independent rvs on a common state space are trivially discrete-space Markov chains.)

(b) What additional condition(s), if any, must be satisfied for X_n to be a time-homogeneous Markov chain?

6.2   The Transition Matrix and the Chapman–Kolmogorov Equations

Section 6.1 introduced the notion of a Markov chain and its characteristic one-step transition probabilities. In this section, we will develop a systematic way to determine the probability that a chain moves from one state to another in *two* steps (or three or four ...) by considering all the intermediate paths the chain may have taken. Such calculations are facilitated by aggregating the transition probabilities into a matrix.

6.2.1   The Transition Matrix

The one-step transition probabilities for the taxi driver example were displayed in Example 6.4 as a 3×3 array. It would be more efficient to simply specify the probabilities themselves in that same format, with the understanding that the probability in the ith row and jth column indicates the transition probability $P(i \rightarrow j)$, the chance the taxi driver takes his next fare to zone j given that he picks up the fare in zone i. Such a representation will be critical to understanding how various multistep transition probabilities are calculated.

DEFINITION

Let $X_0, X_1, X_2, \ldots$ be a finite-state, time-homogeneous Markov chain, and index the states of the chain by the positive integers $1, 2, \ldots, s$. The (one-step) **transition matrix** of the Markov chain is the $s \times s$ matrix $\mathbf{P}$ whose (i, j)th entry is given by

$$p_{ij} = P(i \rightarrow j) = P(X_{n+1} = j \mid X_n = i)$$

for $i = 1, \ldots, s$ and $j = 1, \ldots, s$.

Example 6.8 (Example 6.4 continued) The one-step transition matrix for our taxi driver example is

$$\mathbf{P} = \begin{bmatrix} .3 & .2 & .5 \\ .1 & .8 & .1 \\ .4 & .4 & .2 \end{bmatrix}$$

which is identical in format to the display in Example 6.4. The entries are interpreted as the preceding definition suggests, e.g., the upper left entry (first row, first column) of the matrix is

$$p_{11} = P(1 \rightarrow 1) = P(X_{n+1} = 1 \mid X_n = 1) = .3,$$

i.e., the conditional probability that his next fare is dropped off somewhere in zone 1 given that the taxi is currently in zone 1. ∎

Example 6.9 (Example 6.5 continued) For the Gambler's Ruin scenario with a total available fortune of $3, rather than label the four possible states as 1, 2, 3, 4, it's more natural to use state labels 0, 1, 2, and 3 corresponding to Allan's fortune at any particular time. The transition probabilities specified previously may be written as the following 4×4 matrix:

$$\mathbf{P} = \begin{array}{c} 0 \\ 1 \\ 2 \\ 3 \end{array} \begin{bmatrix} 1 & 0 & 0 & 0 \\ 1-p & 0 & p & 0 \\ 0 & 1-p & 0 & p \\ 0 & 0 & 0 & 1 \end{bmatrix}$$

The labels along the left-hand side of the matrix indicate the ordering of the states for the purpose of creating this matrix; they are not, strictly speaking, a part of $\mathbf{P}$. For example, $P(X_{n+1} = 1 \mid X_n = 2) = P(\text{Allan loses the next game}) = 1 - p$, while $P(X_{n+1} = 3 \mid X_n = 0) = 0$. ∎

Example 6.10 (Example 6.6 continued) The snow depth model has only two states, S (snowy day) and G ("green" day). The one-step transition probabilities given for New York City can be summarized by the following 2×2 matrix:

$$\mathbf{P} = \begin{matrix} G \\ S \end{matrix} \begin{bmatrix} .964 & .036 \\ .224 & .776 \end{bmatrix}$$

∎

Notice that the entries of every row in all of the preceding transition matrices sums to 1. This will always be the case: given that the chain is currently in some state i, it has to go somewhere in its next step (even if that entails remaining in state i). That is, for any state i and any time index n, we must have

$$\sum_{j=1}^{s} p_{ij} = \sum_{j=1}^{s} P(i \rightarrow j) = \sum_{j=1}^{s} P(X_{n+1} = j \mid X_n = i) = 1$$

6.2.2   Computation of Multistep Transition Probabilities

We now turn to the determination of multistep transition probabilities. Given that a Markov chain is currently in state i, what is the probability it will be in state j two steps later (i.e., after two transitions)? Three steps later? We begin with the following definition.

DEFINITION
Let $X_0, X_1, X_2, \ldots$ be a time-homogeneous Markov chain. For any positive integer k, the **k-step transition probabilities** are defined by

$$P^{(k)}(i \rightarrow j) = P(X_{n+k} = j \mid X_n = i) \tag{6.2}$$

where i and j range across the states of the chain (typically $1, \ldots, s$). For $k = 1$, we will typically revert to the previous notation: $P^{(1)}(i \rightarrow j) = P(i \rightarrow j)$.

The superscript (k) in Expression (6.2) does not indicate taking the kth power; it is simply notation representing "in k steps." The next example illustrates how these k-step transition probabilities can be calculated, and how they can be represented compactly in terms of powers of the one-step transition matrix.

Example 6.11 (Example 6.8 continued) Suppose our taxi driver just dropped off a fare in zone 3, so that that is his current state. What is the probability that his *second* fare, counting from now, takes him to zone 1? That is, we wish to determine $P(X_{n+2} = 1 | X_n = 3) = P^{(2)}(3 \rightarrow 1)$. The calculation method is suggested by Fig. 6.4. Consider all the possible destinations of the $(n+1)$st fare, i.e., all the intermediate steps the taxi driver could take from zone 3 to zone 1, and then employ the Law of Total Probability (applied here to conditional probabilities).

The partitioning events in the Law of Total Probability are the possible states at time $n+1$:

Fig. 6.4 Transitioning from state 3 to state 1 in two time steps

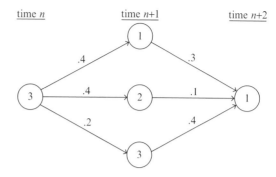

$$P^{(2)}(3 \rightarrow 1) = P(X_{n+2} = 1 \mid X_n = 3)$$
$$= P(X_{n+1} = 1 \mid X_n = 3)P(X_{n+2} = 1 \mid X_n = 3, X_{n+1} = 1)$$
$$+ P(X_{n+1} = 2 \mid X_n = 3)P(X_{n+2} = 1 \mid X_n = 3, X_{n+1} = 2)$$
$$+ P(X_{n+1} = 3 \mid X_n = 3)P(X_{n+2} = 1 \mid X_n = 3, X_{n+1} = 3)$$

By the Markov property, $P(X_{n+2} = 1 | X_n = 3,\ X_{n+1} = 1) = P(X_{n+2} = 1 | X_{n+1} = 1)$, and the other two probabilities involving conditioning on X_n and X_{n+1} simplify analogously. Thus,

$$P^{(2)}(3 \rightarrow 1) = P(X_{n+1} = 1 \mid X_n = 3)P(X_{n+2} = 1 \mid X_{n+1} = 1)$$
$$+ P(X_{n+1} = 2 \mid X_n = 3)P(X_{n+2} = 1 \mid X_{n+1} = 2)$$
$$+ P(X_{n+1} = 3 \mid X_n = 3)P(X_{n+2} = 1 \mid X_{n+1} = 3)$$
$$= P(3 \rightarrow 1)P(1 \rightarrow 1) + P(3 \rightarrow 2)P(2 \rightarrow 1) + P(3 \rightarrow 3)P(3 \rightarrow 1)$$
$$= (.4)(.3) + (.4)(.1) + (.2)(.4) = .24$$

For later reference, the last expression could be written in terms of the elements of the transition matrix $\mathbf{P}$; specifically, it's $p_{31}p_{11} + p_{32}p_{21} + p_{33}p_{31}$.

Similarly, the conditional probability that his second fare wants to be dropped off in zone 2 is computed by

$$P^{(2)}(3 \rightarrow 2) = P(3 \rightarrow 1)P(1 \rightarrow 2) + P(3 \rightarrow 2)P(2 \rightarrow 2) + P(3 \rightarrow 3)P(3 \rightarrow 2)$$
$$= \sum_{j=1}^{3} P(3 \rightarrow j)P(j \rightarrow 2) = \sum_{j=1}^{3} p_{3j}p_{j2}$$
$$= (.4)(.2) + (.4)(.8) + (.2)(.4) = .48$$

Finally, the probability the taxi driver finds himself back in zone 3 after two fares is

$$P^{(2)}(3 \rightarrow 3) = \sum_{j=1}^{3} p_{3j}p_{j3} = (.4)(.5) + (.4)(.1) + (.2)(.2) = .28$$

This makes sense, since the taxi driver must arrive in one of the three zones at time $n + 2$, and $1 - (.24 + .48) = .28$.   ∎

The sums of products of matrix entries that appear repeatedly in the preceding example should look familiar: they are the same manner of computation that arises when one matrix is multiplied by another (or, here, a matrix is multiplied by itself). Indeed, consider what happens if we multiply the one-step transition matrix **P** from Example 6.8 by itself:

$$\mathbf{P}^2 = \mathbf{PP} = \begin{bmatrix} .3 & .2 & .5 \\ .1 & .8 & .1 \\ .4 & .4 & .2 \end{bmatrix} \begin{bmatrix} .3 & .2 & .5 \\ .1 & .8 & .1 \\ .4 & .4 & .2 \end{bmatrix} = \begin{bmatrix} .31 & .42 & .27 \\ .15 & .70 & .15 \\ .24 & .48 & .28 \end{bmatrix}$$

The entries in the bottom row—.24, .48, .28—are precisely the two-step transition probabilities computed in Example 6.11. Specifically, the (3, 1) entry of $\mathbf{P}^2$ is $P^{(2)}(3 \rightarrow 1) = .24$, the (3, 2) entry of $\mathbf{P}^2$ is $P^{(2)}(3 \rightarrow 2) = .48$, and the (3, 3) entry of $\mathbf{P}^2$ is $P^{(2)}(3 \rightarrow 3) = .28$. It should come as no surprise that the other six entries of $\mathbf{P}^2$ follow the same pattern: the (i, j) entry of $\mathbf{P}^2$ is equal to $P^{(2)}(i \rightarrow j)$. Hence, we can obtain all nine two-step transition probabilities with a single matrix computation (which, of course, can be facilitated by Matlab or other matrix-capable software).

The foregoing result can be generalized to an arbitrary fixed number of steps: to find the *three*-step transition probabilities, for example, one only needs to compute the matrix $\mathbf{P}^3$. It is not necessary to consider explicitly the many different paths by which the Markov chain could transition from state i to state j in three steps and add up all the corresponding probabilities (this is, secretly, what the threefold matrix multiplication does). The most general result is often referred to as the set of *Chapman–Kolmogorov Equations*.

CHAPMAN–KOLMOGOROV EQUATIONS
If a Markov chain has one-step transition matrix **P**, then the k-step transition probabilities are the entries of the matrix $\mathbf{P}^k$. Specifically,

$$P^{(k)}(i \rightarrow j) = \text{the } (i, j) \text{ entry of } \mathbf{P}^k$$

Example 6.12 (Example 6.11 continued) Back to our intrepid taxi driver: if he just dropped off a fare in zone 2, what is the probability that he will be in zone 1 two fares later? That is, we wish to determine the two-step transition probability $P(X_{n+2} = 1 | X_n = 2) = P^{(2)}(2 \rightarrow 1)$. According to the Chapman–Kolmogorov Equations, this is simply the (2, 1) entry of the foregoing matrix $\mathbf{P}^2$:

$$P^{(2)}(2 \rightarrow 1) = .15$$

Now consider a longer-term question: If the taxi driver starts the day in zone 3 and transports ten fares before lunch, what is the probability he ends up "back home" (i.e., in zone 3) for lunch? The goal is to find $P(X_{10} = 3 | X_0 = 3) = P^{(10)}(3 \rightarrow 3)$, which could involve summing up a terrifying number of intermediate travel options (19,683 of them, to be precise!). But the Chapman–Kolmogorov Equations, coupled with computer software, makes light work of the problem. With the aid of Matlab, the tenth power of $\mathbf{P}$ is found to be

$$\mathbf{P}^{10} = \begin{bmatrix} .2004 & .5993 & .2004 \\ .1998 & .6004 & .1998 \\ .2002 & .5996 & .2002 \end{bmatrix}$$

The desired probability is just the $(3, 3)$ entry of this 10-step transition matrix: $P^{(10)}(3 \rightarrow 3) = .2002$.   ∎

Example 6.13 The report "Research and Application by Markov Chain Operators in the Mobile Phone Market" (Second International Conference on Artificial Intelligence, Management Science and Electronic Commerce (AIMSEC), 2011) details an analysis of customer loyalty and movement between China's three major cell phone service providers: (1) China Telecom, (2) China Unicom, and (3) China Mobile. A "transition" in this setting refers to an opportunity for a customer to renew his or her contract with a current provider or else switch to one of the other two companies. The report includes the following one-step transition matrix, with the companies numbered as above:

$$\mathbf{P} = \begin{bmatrix} .84 & .06 & .10 \\ .08 & .82 & .10 \\ .10 & .04 & .86 \end{bmatrix}$$

The entries along the main diagonal indicate customer loyalty, e.g., 84% of China Telecom customers stick with that company when their contract expires.

Suppose a customer is currently with China Unicom. What is the probability she will be with the same service provider three contracts from now? In other words, what is $P^{(3)}(2 \rightarrow 2)$? According to the Chapman–Kolmogorov Equations, we need the $(2, 2)$ entry of $\mathbf{P}^3$. That matrix is computed to be

$$\mathbf{P}^3 = \begin{bmatrix} .6310 & .1352 & .2338 \\ .1920 & .5742 & .2338 \\ .2267 & .1006 & .6727 \end{bmatrix}$$

from which we may extract $P^{(3)}(2 \rightarrow 2) = .5742$.

It's important to distinguish this probability from the answer to a more restrictive question: what's the chance she stays with China Unicom for *all* of her next three cell phone contracts? This probability can be represented as $P(X_{n+1} = 2 \cap X_{n+2} = 2 \cap X_{n+3} = 2 | X_n = 2)$ or, less formally, as $P(2 \rightarrow 2 \rightarrow 2 \rightarrow 2)$.

Applying the Markov property gives $[P(2 \rightarrow 2)]^3 = p_{22}^3 = (.82)^3 = .5514$. This probability is slightly lower than $P^{(3)}(2 \rightarrow 2) = .5742$, since the latter accounts for the possibility that the customer switches companies at some intermediate stage(s) but ends up back with China Unicom three contracts later. ∎

Example 6.14 (Example 6.9 continued) Suppose in our earlier Gambler's Ruin example that $p = .55$; that is, Allan has a 55% chance of winning any particular \$1 game. The one- and two-step transition matrices are as follows:

$$
\mathbf{P} = \begin{matrix} 0 \\ 1 \\ 2 \\ 3 \end{matrix}
\begin{bmatrix} 1 & 0 & 0 & 0 \\ .45 & 0 & .55 & 0 \\ 0 & .45 & 0 & .55 \\ 0 & 0 & 0 & 1 \end{bmatrix}
\qquad
\mathbf{P}^2 = \begin{matrix} 0 \\ 1 \\ 2 \\ 3 \end{matrix}
\begin{bmatrix} 1 & 0 & 0 & 0 \\ .45 & .2475 & 0 & .3025 \\ .2025 & 0 & .2475 & .55 \\ 0 & 0 & 0 & 1 \end{bmatrix}
$$

As before, Allan starts with \$2. Looking at the \$2 row (i.e., the third row) of $\mathbf{P}^2$, there is a .2025 probability he has gone broke after two games. This is easy to compute by hand: since he could only lose \$2 in two games by losing twice, the chance is $(.45)^2 = .2025$. The chance that he is back to where he started after two games (i.e., $X_2 = \$2$) is the (\$2, \$2) entry of $\mathbf{P}^2$: .2475. This also could have occurred in just one way: $\$2 \rightarrow \$1 \rightarrow \$2$, for which the two-step transition probability is $(.45)(.55) = .2475$. Notice that the (\$2, \$1) entry of $\mathbf{P}^2$ is 0, i.e., $P^{(2)}(\$2 \rightarrow \$1) = 0$. Since exactly \$1 exchanges hands at the end of each game, it's impossible for Allan to transition from \$2 to \$1 in exactly two steps. Finally, observe that the (\$2, \$3) entry of both matrices is .55, so $P(\$2 \rightarrow \$3) = P^{(2)}(\$2 \rightarrow \$3) = .55$. That's because the game ends when Allan has all \$3 at stake, which he could achieve in one step with probability $p = .55$. Having done so, he will "stay at \$3" in the imaginary second game/step, i.e., from a mathematical perspective, the observed sequence of the Markov chain steps X_0, X_1, and X_2 is $\$2 \rightarrow \$3 \rightarrow \$3$, with the second transition occurring with probability 1.

A natural concern from Allan's perspective is the likelihood that he will *eventually* win. One way to estimate that probability is to look at the chance Allan has arrived at the \$3 "state" after some large number of steps. (This works because once he has \$3, he will always remain at \$3.) Matlab can easily calculate high powers of small matrices; we requested $\mathbf{P}^{75}$:

$$
\mathbf{P}^{75} = \begin{bmatrix} 1 & 0 & 0 & 0 \\ .5980 & 0 & .0000 & .4020 \\ .2691 & .0000 & 0 & .7309 \\ 0 & 0 & 0 & 1 \end{bmatrix}
$$

The two entries that read .0000 indicate that the probability is not strictly 0, but rather is 0 to four decimal places. From this matrix, we have that

$$P\big(\text{Allan eventually has \$3} \mid X_0 = \$2\big) \approx P\big(\text{Allan has \$3 after 75 steps} \mid X_0 = \$2\big)$$
$$= P^{(75)}(\$2 \to \$3) = .7309$$

Had Allan started with just \$1, he would have a roughly .4020 chance of eventually winning all the money.

In Sect. 6.5, we will present an exact method for determining the probability that Allan eventually wins (or loses) his competition with Beth.   ∎

6.2.3   Exercises: Section 6.2 (11–22)

11. The authors of the article "The Fate of Priority Areas for Conservation in Protected Areas: A Fine-Scale Markov Chain Approach" (*Envir. Mgmnt.*, 2011: 263–278) postulated the following model for landscape changes in the forest regions of Italy. Each "pixel" on a map is classified as forested (F) or non-forested (NF). For any specific pixel, X_n represents its status n years after 2000 (so X_1 corresponds to 2001, X_2 to 2002, and so on). Their analysis showed that a pixel has a 90% chance of being forested next year if it is forested this year and an 11% chance of being forested next year if it non-forested this year; moreover, data in the twenty-first century are consistent with the assumptions of a Markov chain.
    (a) Construct the one-step transition matrix for this chain, with states $1 = F$ and $2 = NF$.
    (b) If a map pixel was forested in the year 2000, what is the probability it was still forested in 2002? 2013?
    (c) If a map pixel was non-forested in the year 2000, what is the probability it was still non-forested in 2002? 2013?
    (d) The article's authors use this model to project forested status for several Italian regions in the years 2050 and 2100. Comment on the assumptions required for these projections to be valid.

12. A large automobile insurance company classifies its customers into four risk categories (1 being the lowest risk, aka best/safest drivers, 4 being the worst/highest risk; premiums are assigned accordingly). Each year, upon renewal of a customer's insurance policy, the risk category may change depending on the number of accidents in the previous year. Actuarial data suggest the following: category 1 customers stay in category 1 with probability .9 and move to categories 2, 3, 4 with probabilities .07, .02, and .01, respectively. Category 2 customers shift to category 1 (based on having no accidents last year) with probability .8 and rise to risk categories 3 and 4 with probabilities .15 and .05, respectively. Similarly, category 3 customers transition to 2 and 4 with

probabilities .7 and .3, while category 4 customers stay in that risk category with probability .4 and move to category 3 otherwise.

(a) Let X_n denote a customer's risk category for his/her nth year with the insurance company. Construct the one-step transition matrix for this Markov chain.

(b) If a customer starts in category 1, what is the probability she falls into risk category 2 five years later?

(c) If a customer is currently in risk category 4, determine the probability he will be a category 1 driver in k years, for $k = 1, 2, 3, 4, 5, 6$.

(d) What is the probability that a driver currently in category 1 remains in that category for each of the next 5 years?

13. The article cited in Example 6.6 also gives the following one-step transition matrix, with the same definitions of states, for Willow City, ND:

$$\mathbf{P} = \begin{array}{c} \\ G \\ S \end{array} \begin{array}{cc} G & S \\ \left[\begin{array}{cc} .933 & .067 \\ .012 & .988 \end{array} \right] \end{array}$$

(a) Contrast Willow City with New York City: where is snow more likely to stay on the ground for an extended time period? Explain.

(b) If today is a snowy day in Willow City, what is the probability it will also be a snowy day there 2 days from now? three days from now?

(c) If today is a snowy day in Willow City, what is the probability it will continue to be snowy for the next 4 days in a row?

14. I (author Carlton) have a six-room house whose configuration is depicted in the accompanying diagram. When my sister and her family visit, I often play hide-and-seek with my young nephew, Lucas. Consider the following situation: Lucas counts to ten in Room 1, while I run and hide in Room 6. Lucas' "strategy," as much as he has one, is such that standing in any room of the house, he is equally likely to next visit any of the adjacent rooms, regardless of where he has searched previously. (The exception, of course, is if he enters Room 6, in which case he discovers me and the round of hide-and-seek is over.)

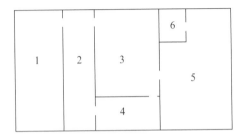

(a) Let $X_n =$ the nth room Lucas visits (with $X_0 = 1$, his starting point). Construct the one-step transition matrix for the corresponding Markov chain.

(b) What is the probability that his third room-to-room transition will take him into Room 2?

(c) What is the fewest number of time steps (i.e., room transitions) required for Lucas to find me?

(d) What is the probability that, after 12 time steps, he still hasn't found me?

15. Refer back to Exercise 1 in the previous section. Consider two negotiators, A and B, who employ strategies according to the Markov chain model described.

(a) Construct the one-step transition matrix for the Markov chain $X_n =$ strategy employed at the nth stage of a negotiation, assuming the states are (1) cooperative and (2) competitive.

(b) If negotiator A employs a cooperative strategy at some stage, what is the probability she uses a competitive strategy the next time? [Don't forget that A's turns are two time steps apart, since B counter-negotiates in between.]

(c) Now introduce a third state, (3) end of the negotiation. Assume that a Markov chain model with the following one-step transition matrix applies:

$$P = \begin{bmatrix} .6 & .2 & .2 \\ .3 & .4 & .3 \\ 0 & 0 & 1 \end{bmatrix}$$

Given that the initial strategy presented is cooperative, what is the probability the negotiations end within three time steps?

(d) Refer back to (c). Given that the initial strategy presented is competitive, what is the probability the negotiations end within three time steps?

16. Sarah, a statistician at a large Midwestern polling agency, owns four umbrellas. Initially, two of them are at her home and two are at her office. Each morning, she takes an umbrella with her to work (assuming she has any at home) if and only if it's currently raining, which happens on 20% of mornings. Each evening, she takes an umbrella from work to home (again, assuming any are available) if and only if it's raining when she leaves work, which happens on 30% of all evenings. Assume weather conditions, including morning and evening on the same day, are independent (in the Midwest, that's not unrealistic). Let $X_n =$ the number of umbrellas Sarah has at home at the end of her nth work day (i.e., once she's back at home).

(a) Identify the state space for this chain.

(b) Assume Sarah has two umbrellas at home tonight. By considering all possible weather conditions tomorrow morning and tomorrow evening, determine the one-step transition probabilities for the number of umbrellas she'll have at home tomorrow night.

(c) Repeat the logic of (b) to determine the complete one-step transition matrix for this chain. Be careful when considering the two extreme cases!

17. Refer back to the previous exercise.

(a) Given that Sarah has two umbrellas at home (and two at work) as of Sunday night, what is the probability she'll have exactly two umbrellas at home the following Friday night? What is the probability she'll have *at least* two umbrellas at home the following Friday night?

(b) Given that Sarah has two umbrellas at home Sunday night, what are the chances she won't have an umbrella to take with her to work the following Thursday morning when a surprise thunderstorm moves through the area?

(c) Assume again that Sarah has two umbrellas at home at the start of the week. Determine the *expected* number of umbrellas she has at home at the end of Monday and at the end of Tuesday. [*Hint*: X_n is a discrete rv; if $X_0 = 2$, then the probability distribution of X_n appears in the corresponding row of $\mathbf{P}^n$.]

18. A box always contains three marbles, each of which is green or yellow. At regular intervals, one marble is selected at random from the box and removed, while another is put in its place according to the following rules: a green marble is replaced by a yellow marble with probability .3 (and otherwise by another green marble), while a yellow marble is equally likely to be replaced by either color. Let X_n = the number of green marbles in the box after the nth swap.

(a) What are the possible values of X_n?

(b) Construct the one-step transition matrix for this Markov chain.

(c) If all three marbles currently in the box are green, what is the probability the same will be true three swaps from now?

(d) If all three marbles currently in the box are green, what is the probability that the fourth marble selected from the box will be green? [*Hint*: Use part (c). Be careful not to confuse the color of the marble *selected* on the fourth swap with the color of the one that replaces it!]

19. A Markov chain model for customer visits to an auto repair shop is described in the article "Customer Lifetime Value Prediction by a Markov Chain Based Data Mining Model: Application to an Auto Repair and Maintenance Company in Taiwan" (*Scientia Iranica*, 2012: 849-855). Customers make between 0 and 4 visits to the repair shop each year; for any customer that made exactly i visits last year, the number of visits s/he will make next year follows a Poisson distribution with parameter μ_i. (The event "4 visits" is really "4 or more visits," so the probability of 4 visits next year is calculated as $1 - \sum_{x=0}^{3} p(x; \mu_i)$ from the appropriate Poisson pmf.) Parameter values cited in the article, which were estimated from real data, appear in the accompanying table.

i	0	1	2	3	4
μ_i	1.938696	1.513721	1.909567	2.437809	3.445738

(a) Construct the one-step transition matrix for the chain X_n = number of repair shop visits by a randomly selected customer in the nth observed year.

(b) If a customer made two visits last year, what is the probability that s/he makes two visits next year and two visits the year after that?

(c) If a customer made no visits last year, what is the probability s/he makes a total of exactly two visits in the next 2 years?

20. The four vans in a university's vanpool are maintained at night by a single mechanic, who can service one van per night (assuming any of them need repairs). Suppose that there is a 10% chance that a van working today will need service by tonight, independent of the status of the other vans. We wish to model $X_n =$ the number of vans *unavailable* for service at the beginning of the nth day.

(a) Suppose all four vans were operational as of this morning. Find the probability that exactly j of them will be unusable tomorrow morning for $j = 0, 1, 2, 3$. [*Hint*: The number of unusable vans for tomorrow will be 1 less than the number that break down today, unless that's 0, because the mechanic can fix only one van per night. What is the probability distribution of $Y =$ the number of vans that break down today, assuming all 4 worked this morning?]

(b) Suppose three vans were operational as of this morning, and one was broken. Find the probabilities $P(1 \rightarrow j)$ for this chain. [*Hint*: Assume the broken van will be fixed tonight. Then the number of unavailable vans tomorrow morning is the number that break down today, out of the three currently functioning.]

(c) Use reasoning similar to that of (a) and (b) to determine the complete one-step transition matrix for this Markov chain.

21. Refer back to the previous exercise.

(a) If all four vans were operational as of Monday morning, what is the probability exactly three vans will be usable Wednesday morning? Thursday morning? Friday morning?

(b) A *backlog* occurs whenever $X_n \geq 1$, indicating that some vans will be temporarily out of commission because the mechanic could not get to them the previous night. Assuming there was no backlog as of Monday morning, what is the probability a backlog exists Tuesday morning? Answer the same question for Wednesday, Thursday, and Friday mornings.

(c) How do the probabilities in (b) change if there was a backlog of 1 van as of Monday morning?

22. Consider a Markov chain with state space $\{1, 2, \ldots, s\}$. Show that, for any positive integers m and n and any states i and j,

$$P_{ij}^{(m+n)} = \sum_{k=1}^{s} P_{ik}^{(m)} P_{kj}^{(n)}$$

This is an alternative version of the Chapman–Kolmogorov Equations. [*Hint*: Write the left-hand side as $P(X_{m+n} = j | X_0 = i)$, and consider all the possible states after m transitions.]

6.3   Specifying an Initial Distribution

Thus far, every probability we have considered in this chapter (i.e., all the one-, two-, and higher-step transition probabilities) has been conditional. For example the entries of any one-step transition matrix indicate $P(X_{n+1} = j | X_n = i)$. In this section, we briefly explore unconditional probabilities, which result from specifying a distribution for the rv X_0, the initial state of the chain. We will consider two cases: modeling the initial state X_0 as a random variable, and treating X_0 as having a fixed/known value.

Example 6.15 (Example 6.11 continued) The never-ending saga of the taxi driver continues! Imagine this poor fellow sleeps in his taxi, so from his perspective each new day starts in a "random" zone. Specifically, suppose for now that he has a 20% chance of waking up in zone 1, a 50% chance of waking up in zone 2, and a 30% chance of waking up in zone 3. That is, we have assigned the following **initial distribution** to the Markov chain:

$$
\begin{array}{c|ccc}
i & 1 & 2 & 3 \\
\hline
P(X_0 = i) & .2 & .5 & .3
\end{array}
\tag{6.3}
$$

Notice that, unlike the conditional probabilities that comprise the transition matrix of the Markov chain, this initial distribution (6.3) specifies the *unconditional* (aka marginal) distribution for the rv X_0. In what follows, we will sometimes refer to the bottom row of (6.3) as the "initial probability vector" of X_0.

Now consider the rv X_1, the destination of the taxi driver's first fare. The probability his first fare wants to go somewhere in zone 3 can be determined via the Law of Total Probability:

$$
\begin{aligned}
P(X_1 = 3) &= P(X_0 = 1)P(X_1 = 3 \mid X_0 = 1) + P(X_0 = 2)P(X_1 = 3 \mid X_0 = 2) \\
&\quad + P(X_0 = 3)P(X_1 = 3 \mid X_0 = 3) \\
&= \sum_{i=1}^{3} [P(X_0 = i)P(i \to 3)] = \sum_{i=1}^{3} [P(X_0 = i)p_{i3}] \\
&= (.2)(.5) + (.5)(.1) + (.3)(.2) = .21
\end{aligned}
$$

As indicated in the intermediate step, this unconditional probability can be computed by taking the product of the initial probability vector [.2 .5 .3], regarded as a 1×3 matrix, with the third column of the one-step transition matrix **P**. Similarly, the (unconditional) probability that his first fare wants to be dropped off in zone 2 is

$$
\begin{aligned}
P(X_1 = 2) &= \sum_{i=1}^{3} [P(X_0 = i)P(i \to 2)] = \sum_{i=1}^{3} [P(X_0 = i)p_{i2}] \\
&= (.2)(.2) + (.5)(.8) + (.3)(.4) = .56
\end{aligned}
$$

The foregoing computation is the product of the initial probability vector with the second column of **P**. Finally, the probability that the first fare is taken to zone 1 equals .23, which can be computed either as a similar product or by observing that $1 - (.21 + .56) = .23$. All together, the unconditional pmf of the rv X_1 is

i	1	2	3
$P(X_1 = i)$	.23	.56	.21

Clearly, the most efficient way to determine the distribution of X_1 is to compute all three products simultaneously through matrix multiplication. If we multiply the transition matrix **P** on the left by a 1×3 row vector containing the initial probabilities for X_0, we obtain

$$[.2 \quad .5 \quad .3]\mathbf{P} = [.2 \quad .5 \quad .3]\begin{bmatrix} .3 & .2 & .5 \\ .1 & .8 & .1 \\ .4 & .4 & .2 \end{bmatrix} = [.23 \quad .56 \quad .21]$$

∎

The method illustrated in the preceding example can be generalized to find the unconditional distribution of the state X_n in the chain after any number of transitions n, starting with a specified initial distribution for X_0.

THEOREM

Let $X_0, X_1, \ldots, X_n, \ldots$ be a Markov chain with state space $\{1, \ldots, s\}$ and one-step transition matrix **P**. Let $\mathbf{v}_0 = [v_{01} \ldots v_{0s}]$ be a $1 \times s$ vector specifying the initial distribution of the chain, i.e., $v_{0k} = P(X_0 = k)$ for $k = 1, \ldots, s$. If $\mathbf{v}_1$ denotes the vector of marginal (i.e., unconditional) probabilities associated with X_1, then

$$\mathbf{v}_1 = \mathbf{v}_0\mathbf{P}$$

More generally, if $\mathbf{v}_n$ denotes the $1 \times s$ vector of marginal probabilities for X_n, then

$$\mathbf{v}_n = \mathbf{v}_0\mathbf{P}^n$$

Proof The formula $\mathbf{v}_1 = \mathbf{v}_0\mathbf{P}$ can be established using the same computational approach displayed in Example 6.15. Now consider $\mathbf{v}_2$, the vector of unconditional probabilities for X_2. By the same reasoning as in Example 6.15, we have

$$\mathbf{v}_2 = \mathbf{v}_1\mathbf{P}$$

The substitution $\mathbf{v}_1 = \mathbf{v}_0\mathbf{P}$ then yields $\mathbf{v}_2 = (\mathbf{v}_0\mathbf{P})\mathbf{P} = \mathbf{v}_0\mathbf{P}^2$. Continuing by induction, we have for general n that $\mathbf{v}_n = \mathbf{v}_{n-1}\mathbf{P} = (\mathbf{v}_0\mathbf{P}^{n-1})\mathbf{P} = \mathbf{v}_0\mathbf{P}^n$, as claimed. ∎

With the aid of software such as Matlab, the unconditional distributions of future states of the Markov chain can be computed very quickly once the initial distribution is specified. For example, as a continuation of Example 6.15, the probability vector for X_2, the destination of the driver's second fare, is given by

$$\mathbf{v}_2 = \mathbf{v}_1\mathbf{P} = \begin{bmatrix} .23 & .56 & .21 \end{bmatrix} \begin{bmatrix} .3 & .2 & .5 \\ .1 & .8 & .1 \\ .4 & .4 & .2 \end{bmatrix} = \begin{bmatrix} .209 & .578 & .213 \end{bmatrix}$$

or, equivalently,

$$\mathbf{v}_2 = \mathbf{v}_0\mathbf{P}^2 = \begin{bmatrix} .2 & .5 & .3 \end{bmatrix} \begin{bmatrix} .3 & .2 & .5 \\ .1 & .8 & .1 \\ .4 & .4 & .2 \end{bmatrix}^2 = \begin{bmatrix} .209 & .578 & .213 \end{bmatrix}$$

That is, assuming that the initial distribution specified in Example 6.15 is correct, the taxi driver has a 20.9% chance of taking his second fare to zone 1, a 57.8% chance of taking him/her to zone 2, and a 21.3% chance of being in zone 3 after two fares.

Example 6.16 As you probably learned in high school biology, Austro-Hungarian scientist Gregor Mendel studied the inheritance of characteristics within plant species, particularly peas. Suppose one particular pea plant can either be green or yellow, which is determined by a single gene with green (G) dominant over yellow (g). That is, the genetic material determining a plant's color (its "genotype") can be one of three pairings—GG, Gg, or gg—depending on which types were passed on by the parent plants. To say that green is "dominant" over yellow means that the plant's visible color—its "phenotype"—will be green unless that gene is completely absent from the plant (so plants with GG or Gg genotype appear green, while only gg plants are yellow).

Consider cross-breeding with a yellow plant, whose genotype is therefore known to be gg. Mendel's laws of genetic recombination can be expressed by the following transition matrix, where X_n is the genotype of an nth-generation plant resulting from cross-breeding with a gg plant:

$$\mathbf{P} = \begin{array}{c} GG \\ Gg \\ gg \end{array} \begin{bmatrix} 0 & 1 & 0 \\ 0 & .5 & .5 \\ 0 & 0 & 1 \end{bmatrix}$$

For example, crossing $GG \times gg$ yields Gg with probability 1, while $Gg \times gg$ results in Gg or gg with probability .5 each.

Suppose our initial population of plants (to be cross-bred with the pure yellow specimens) has the following genotype distribution: 70% GG, 20% Gg and 10% gg. The initial probability vector associated with this "0th generation" is $\mathbf{v}_0 = [.7 \ .2 \ .1]$. The probabilities associated with the first generation of cross-bred plants is $\mathbf{v}_1 = \mathbf{v}_0\mathbf{P} = [0 \ .8 \ .2]$, meaning that 80% of first-generation plants are expected be Gg and the remaining 20% gg. Notice that GG plants cannot exist past the first generation, since cross-breeding with gg plants makes such a recombination impossible.

Similarly, the second-generation probabilities are given by $v_2 = v_1 P = v_0 P^2 = [0\ .4\ .6]$, so that within two generations gg plants should be the majority (60% gg compared to 40% Gg). As cross-breeding with pure gg plants continues, that genotype will increase in relative proportion (80% in generation 3, 90% in generation 4), until eventually the dominant G allele dies out. ∎

6.3.1 A Fixed Initial State

The case is which the initial state X_0 is fixed or known rather than random can be handled by forming a "degenerate" initial probability distribution.

Example 6.17 (Example 6.15 continued) Suppose that our taxi driver lives in zone 3 and always goes home at night, which means that he starts each new day in zone 3. Starting with certainty in zone 3 means that $P(X_0 = 3) = 1$, while $P(X_0 = 1) = P(X_0 = 2) = 0$. Written as a pmf, the distribution of X_0 is

i	1	2	3
$P(X_0 = i)$	0	0	1

Equivalently, the probability vector for X_0 is $v_0 = [0\ 0\ 1]$. From the original description of the Markov chain (Example 6.1), the initial state being zone 3 implies that $X_1 = 1$ with probability .4, $X_1 = 2$ with probability .4, and $X_1 = 3$ with probability .2. This same result can be obtained by applying the theorem of this section:

$$v_1 = v_0 P = \begin{bmatrix} 0 & 0 & 1 \end{bmatrix} \begin{bmatrix} .3 & .2 & .5 \\ .1 & .8 & .1 \\ .4 & .4 & .2 \end{bmatrix} = \begin{bmatrix} .4 & .4 & .2 \end{bmatrix}$$

Notice that left-multiplying P by the vector $[0\ 0\ 1]$ simply extracts the third row of P. Similarly, the pmf of X_5, the destination of the fifth passenger, is given by

$$v_5 = v_0 P^5 = \begin{bmatrix} 0 & 0 & 1 \end{bmatrix} \begin{bmatrix} .3 & .2 & .5 \\ .1 & .8 & .1 \\ .4 & .4 & .2 \end{bmatrix}^5 = \begin{bmatrix} 0 & 0 & 1 \end{bmatrix} \begin{bmatrix} .2115 & .5767 & .2118 \\ .1938 & .6125 & .1938 \\ .2073 & .5858 & .2070 \end{bmatrix}$$

$$= \begin{bmatrix} .2073 & .5858 & .2070 \end{bmatrix}$$

The matrix P^5 was computed by Matlab. The row vector v_5 is simply the third row of P^5, because the chain begins in zone 3 with probability 1. So, starting the day at home in zone 3, the taxi driver finds himself in zone 1, 2, or 3 after five fares with probabilities .2073, .5858, and .2070, respectively. ∎

Example 6.18 (Example 6.14 continued) As before, we can use a high power of the one-step transition matrix, say $\mathbf{P}^{75}$, to approximate the long-term behavior of our Gambler's Ruin Markov chain. Suppose as before that $p = .55$ and Allan's initial stake is \$2. We can express the latter as $\mathbf{v}_0 = [0\ 0\ 1\ 0]$; recall that the states, in order, are \$0, \$1, \$2, \$3. Then the probability distribution of X_{75} is

$$\mathbf{v}_{75} = \mathbf{v}_0 \mathbf{P}^{75} = [0\ 0\ 1\ 0]\mathbf{P}^{75} = \text{the third (i.e., \$2) row of } \mathbf{P}^{75} = [.2691\ .0000\ 0\ .7309]$$

If Allan begins the competition with \$2 (and Beth with \$1), there is a 73.09% chance he will end up with all the money within 75 games, and a 26.91% chance he will end up broke after 75 games. As discussed previously, the competition will almost certainly end long before a 75th game, but for purposes of forecasting long-run behavior we imagine that when either player goes broke, game-play continues but no further money is exchanged.

Suppose instead that Allan's initial stake is just \$1, while Beth starts with \$2. Then Allan's initial "distribution" is specified by $\mathbf{v}_0 = [0\ 1\ 0\ 0]$, meaning $P(X_0 = \$1) = 1$ while $P(X_0 = \$0, \$2, \$3) = 0$. After 75 plays, we now have

$$\mathbf{v}_{75} = \mathbf{v}_0 \mathbf{P}^{75} = [0\ 1\ 0\ 0]\mathbf{P}^{75} = \text{the second (i.e., \$1) row of } \mathbf{P}^{75} = [.5980\ 0\ .0000\ .4020]$$

Starting with \$1, Allan has a 40.2% chance of winning the competition (i.e., ending up with \$3) and a 59.8% chance of being "ruined." ∎

6.3.2 Exercises: Section 6.3 (23–30)

23. Refer back to Exercise 1 of this chapter. Suppose that Negotiator A goes first and that 75% of the time she begins negotiations with a cooperative strategy. (Consider this to be time index 0.)
    (a) Determine the (unconditional) probability that Negotiator B's first strategy will also be cooperative.
    (b) Determine the (unconditional) probability that Negotiator B's second strategy will be cooperative. [*Hint*: Which time index corresponds to his second move?]
24. Refer back to the Ehrenfest chain model described in Exercise 2 with $m = 3$ balls. The possible states of the chain $X_n =$ number of balls in the left chamber after the nth exchange are $\{0, 1, 2, 3\}$.
    (a) Suppose that all four possible initial states are equally likely. Determine the probability distributions of X_1 and X_2.
    (b) Suppose instead that each of the three balls is initially equally likely to be placed in the left or right chamber. In this situation, what is the initial distribution of the chain?
    (c) Using the initial distribution specified in (b), determine the unconditional distributions of X_1 and X_2. What do you notice?

25. Information bits (0s and 1s) in a binary communication system travel through a long series of relays. At each relay, a "bit-switching" error might occur. Suppose that at each relay, there is a 4% chance of a 0 bit being switched to a 1 bit and a 5% chance of a 1 becoming a 0. Let X_0 = a bit's initial parity (0 or 1), and let X_n = the bit's parity after traversing the nth relay.
    (a) Construct the one-step transition matrix for this chain. [*Hint*: There are only two states, 0 and 1.]
    (b) Suppose the input stream to this relay system consists of 80% 0s and 20% 1s. Determine the proportions of 0s and 1s exiting the first relay.
    (c) Under the same conditions as (b), determine the proportions of 0s and 1s exiting the fifth relay.

26. Refer to the genetic recombination scenario of Example 6.16. Suppose that plants will now be cross-bred with known hybrids (i.e., those with genotype Gg). Mendel's laws imply the following transition matrix for such breeding:

$$\mathbf{P} = \begin{array}{c} GG \\ Gg \\ gg \end{array} \begin{bmatrix} .5 & .5 & 0 \\ .25 & .5 & .25 \\ 0 & .5 & .5 \end{bmatrix}$$

Again assume the initial population genotype distribution of plants to be cross-bred with these hybrids is 70% GG, 20% Gg, and 10% gg.
    (a) Determine the genotype distribution of the first generation of plants resulting from this cross-breeding experiment.
    (b) Determine the genotype distributions of the second, third, and fourth generations.

27. Refer to the weather scenario described in Example 6.6 and Example 6.10. Suppose today's weather forecast for New York City gives a 20% chance of experiencing a snowy day.
    (a) Let X_0 denote today's weather condition. Express the information provided as an initial probability vector for X_0.
    (b) Determine the (unconditional) likelihoods of a snowy day and a green day tomorrow, using the one-step transition probabilities specified in Example 6.6.
    (c) Based on today's forecast and the transition probabilities, what is the chance New York City will experience a "green day" 1 week (7 days) from now?

28. The article "Option Valuation Under a Multivariate Markov Chain Model" (Third International Joint Conference on Computational Science and Optimization, 2010) includes information on the dynamic movement of certain assets between three states: (1) up, (2) middle, and (3) down. For a particular class of assets, the following one-step transition probabilities were estimated from available data:

$$P = \begin{bmatrix} .4069 & .3536 & .2395 \\ .3995 & .5588 & .0417 \\ .5642 & .0470 & .3888 \end{bmatrix}$$

Suppose that the initial valuation of this asset class found that 31.4% of such assets were in the "up" dynamic state, 40.5% were "middle," and the remainder were "down."

(a) What is the initial probability vector for this chain?

(b) Determine the unconditional probability distribution of X_1, the asset dynamic state one time step after the initial valuation.

(c) Determine the unconditional probability distribution of X_2, the asset dynamic state two time steps after the initial valuation.

29. Refer back to Exercise 23, and now suppose that Negotiator A always opens talks with a competitive strategy.

(a) What is the probability vector for X_0, Negotiator A's initial strategy?

(b) Without performing any matrix computation, determine the distribution of X_1, Negotiator B's first strategy choice.

(c) What is the probability Negotiator A's second strategy is cooperative? competitive?

30. Transitions between sleep stages are described in the article "Multinomial Logistic Estimation of Markov-Chain Models for Modeling Sleep Architecture in Primary Insomnia Patients" (*J. Pharmacokinet. Pharmacodyn.*, 2010:137–155). The following one-step transition probabilities for the five stages awake (AW), stage 1 sleep (ST1), stage 2 sleep (ST2), slow-wave sleep (SWS), and rapid-eye movement sleep (REM) were obtained from a graph in the article:

$$P = \begin{matrix} AW \\ ST1 \\ ST2 \\ SWS \\ REM \end{matrix} \begin{bmatrix} .90 & .09 & .01 & .00 & .00 \\ .21 & .40 & .34 & .02 & .03 \\ .02 & .02 & .84 & .09 & .03 \\ .02 & .02 & .22 & .72 & .02 \\ .04 & .04 & .05 & .00 & .87 \end{bmatrix}$$

The time index of the Markov chain corresponds to half-hour intervals (i.e., $n=1$ is 30 min after the beginning of the study, $n=2$ is 60 min in, etc.). Initially, all patients in the study were awake.

(a) Let v_0 denote the probability vector for X_0, the initial state of a patient in the sleep study. Determine v_0.

(b) Without performing any matrix computations, determine the distribution of patients' sleep states 30 min (one time interval) into the study.

(c) Determine the distribution of patients' sleep states 4 h into the study. [*Hint*: What time index corresponds to the 4-h mark?]

6.4 Regular Markov Chains and the Steady-State Theorem

In previous sections, we have alluded to the long-term behavior of certain Markov chains. In some cases, such as Gambler's Ruin, we anticipate that the chain will eventually reach, and remain in, one of several "absorbing" states (we'll discuss these in Sect. 6.5). Our taxi driver, in contrast, should continually move around, but perhaps something can be said about how much time he will spend in each of the three zones over the course of many, many fares. It turns out that the taxi driver example belongs to a special class of Markov chains, called *regular chains*, for which the long-run behavior "stabilizes" in some sense and can be determined analytically.

6.4.1 Regular Chains

> **DEFINITION**
>
> A finite-state Markov chain with one-step transition matrix $\mathbf{P}$ is said to be a **regular chain** if there exists a positive integer n such that all of the entries of the matrix $\mathbf{P}^n$ are positive.
>
> In other words, for a regular Markov chain there is some positive integer n such that every state can be reached from every state (including itself) in exactly n steps.

It's straightforward to show that if all the entries of $\mathbf{P}^n$ are positive, then so are all of the entries of $\mathbf{P}^{n+1}$, $\mathbf{P}^{n+2}$, and so on (Exercise 37). Our taxi driver example is a regular chain, since all nine entries of $\mathbf{P}$ itself are positive. The next example shows that a regular chain may have some one-step transition probabilities equal to zero.

Example 6.19 Internet users' browser histories can be modeled as Markov chains, where the "states" are different Web sites (or classes of Web sites) and transitions occur when users move from one Web site to another. The article "Evaluating Variable-Length Markov Chain Models for Analysis of User Web Navigation Sessions" (*IEEE Trans. Knowl. Data Engr.* 2007: 441-452) discusses increasingly complex models of this type. Suppose for simplicity that Web sites are grouped into five categories: (1) social media, (2) e-mail, (3) news and sports, (4) online retailers, and (5) other (use your imagination). Consider a Markov chain model for users' transitions between these five categories whose state diagram is depicted in Fig. 6.5.

Notice that, according to this model, not every state can access all five states in one step, because many one-step transition probabilities are zero. The one-step transition matrix $\mathbf{P}$ of this Markov chain is as follows:

Fig. 6.5 State diagram for
Example 6.19

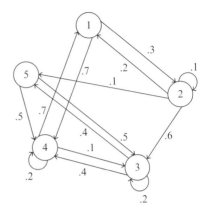

$$
\mathbf{P} = \begin{bmatrix}
0 & .3 & 0 & .7 & 0 \\
.2 & .1 & .6 & 0 & .1 \\
0 & 0 & .2 & .4 & .4 \\
.7 & 0 & .1 & .2 & 0 \\
0 & 0 & .5 & .5 & 0
\end{bmatrix}
$$

Eleven of the twenty-five entries in **P** are zero. However, consider several higher powers of this matrix:

$$
\mathbf{P}^2 = \begin{bmatrix}
.55 & .03 & .25 & .14 & .03 \\
.02 & .07 & .23 & .43 & .25 \\
.28 & 0 & .28 & .36 & .08 \\
.14 & .21 & .04 & .57 & .04 \\
.35 & 0 & .15 & .30 & .20
\end{bmatrix}, \quad
\mathbf{P}^3 = \begin{bmatrix}
.104 & .168 & .097 & .528 & .103 \\
.315 & .013 & .256 & .317 & .099 \\
.252 & .084 & .132 & .420 & .112 \\
.441 & .063 & .211 & .248 & .037 \\
.210 & .105 & .160 & .465 & .060
\end{bmatrix}
$$

Since every entry of $\mathbf{P}^3$ is positive, by definition we have a regular Markov chain. Every state can reach every state (including itself) in exactly three moves. ∎

In contrast, Gambler's Ruin is *not* a regular Markov chain. It is not possible for Allan to go from \$2 to \$1 in an even number of moves, so the (\$2, \$1) entry of $\mathbf{P}^n$ is zero whenever n is even. Similarly, Allan cannot go from \$2 back to \$2 in an odd number of steps, so the (\$2, \$2) entry of $\mathbf{P}^n$ equals zero for every odd exponent n. Thus, there exists no positive integer n for which all sixteen entries of $\mathbf{P}^n$ are positive. (In fact, six other entries of $\mathbf{P}^n$ must always be 0: $P^{(n)}(0 \to j) = 0$ for $j \neq 0$ and $P^{(n)}(3 \to j) = 0$ for $j \neq 3$, since both \$0 and \$3 are "absorbing" states.) Another non-regular Markov chain, one that does not have any absorbing states, is given in the following example.

Example 6.20 Unlike our taxi driver, bus drivers follow a well-defined route. Consider a bus route from campus (state 1), to the nearby student housing complex

Fig. 6.6 State diagram for
Example 6.20

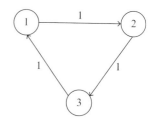

(state 2), to downtown (state 3), and then back to campus. The associated Markov chain cycles endlessly: $1 \rightarrow 2 \rightarrow 3 \rightarrow 1 \rightarrow 2 \rightarrow 3 \rightarrow 1$ .... Figure 6.6 shows the corresponding state diagram.

The one-step transition matrix for this chain is

$$\mathbf{P} = \begin{bmatrix} 0 & 1 & 0 \\ 0 & 0 & 1 \\ 1 & 0 & 0 \end{bmatrix}$$

Direct computation shows that

$$\mathbf{P}^2 = \begin{bmatrix} 0 & 0 & 1 \\ 1 & 0 & 0 \\ 0 & 1 & 0 \end{bmatrix} \quad \text{and} \quad \mathbf{P}^3 = \begin{bmatrix} 1 & 0 & 0 \\ 0 & 1 & 0 \\ 0 & 0 & 1 \end{bmatrix} = \mathbf{I},$$

where $\mathbf{I}$ denotes the 3×3 identity matrix. Hence, $\mathbf{P}^4 = \mathbf{P}^3\mathbf{P} = \mathbf{IP} = \mathbf{P}$; $\mathbf{P}^5 = \mathbf{P}^3\mathbf{P}^2 = \mathbf{IP}^2 = \mathbf{P}^2$; $\mathbf{P}^6 = \mathbf{P}^3\mathbf{P}^3 = \mathbf{II} = \mathbf{I}$; and so on. That is, the n-step transition matrix $\mathbf{P}^n$ equals one of $\mathbf{P}, \mathbf{P}^2$, or $\mathbf{I}$ for every positive integer n, and all three of these contain some zero entries. Therefore, this is *not* a regular Markov chain. ∎

6.4.2 The Steady-State Theorem

What's so special about regular chains? The transition matrices of regular Markov chains exhibit a rather interesting property. Consider a very high power of the transition matrix for our taxi driver, computed with the aid of Matlab:

$$\mathbf{P} = \begin{bmatrix} .3 & .2 & .5 \\ .1 & .8 & .1 \\ .4 & .4 & .2 \end{bmatrix} \Rightarrow \mathbf{P}^{100} = \begin{bmatrix} .2000 & .6000 & .2000 \\ .2000 & .6000 & .2000 \\ .2000 & .6000 & .2000 \end{bmatrix}$$

Notice that every row of $\mathbf{P}^{100}$ is identical: roughly, each one is $[.2\ .6\ .2]$. What's more, raising $\mathbf{P}$ to even higher powers yields the same matrix to several decimal places. That is, $\mathbf{P}^{101}, \mathbf{P}^{102}$, and so on are all roughly equal to $\mathbf{P}^{100}$. Something similar occurs for the regular Markov chain of Example 6.19:

$$P = \begin{bmatrix} 0 & .3 & 0 & .7 & 0 \\ .2 & .1 & .6 & 0 & .1 \\ 0 & 0 & .2 & .4 & .4 \\ .7 & 0 & .1 & .2 & 0 \\ 0 & 0 & .5 & .5 & 0 \end{bmatrix} \Rightarrow P^{100} = \begin{bmatrix} .2844 & .0948 & .1659 & .3791 & .0758 \\ .2844 & .0948 & .1659 & .3791 & .0758 \\ .2844 & .0948 & .1659 & .3791 & .0758 \\ .2844 & .0948 & .1659 & .3791 & .0758 \\ .2844 & .0948 & .1659 & .3791 & .0758 \end{bmatrix}$$

Again, every row of P^{100} is the same, and replacing 100 by an even higher power gives the same result (i.e., to several decimal places $P^{100} = P^{101} = P^{102} = \ldots$). These are two examples of the central result in the theory of Markov chains, the so-called *Steady-State Theorem*.

STEADY-STATE THEOREM
Let P be the one-step transition matrix of a finite-state, *regular* Markov chain. Then the matrix limit

$$\Pi = \lim_{n \to \infty} P^n \qquad (6.4)$$

exists. Moreover, the rows of the limiting matrix Π are identical, with all positive entries.

The proof of the Steady-State Theorem is beyond the scope of this book; interested readers may consult the text by Karlin and Taylor listed in the references.

If we let $\pi = [\pi_1 \cdots \pi_s]$ denote each of the identical rows of the limiting matrix Π in Eq. (6.4), π is called the **steady-state distribution** of the Markov chain. Thus, for the taxi driver example, the steady-state distribution is $\pi = [.2\ .6\ .2]$, while the steady-state distribution for the Web browsing Markov chain in Example 6.19 is $\pi = [.2844\ .0948\ .1659\ .3791\ .0758]$.

A Markov chain does not have to be regular for the limit of P^n to exist as $n \to \infty$. For example, computing progressively larger powers of the one-step transition matrix for the Gambler's Ruin scenario of Example 6.14 shows that, for large n,

$$P^n \approx P^{75} = \begin{bmatrix} 1 & 0 & 0 & 0 \\ .5980 & 0 & .0000 & .4020 \\ .2691 & .0000 & 0 & .7309 \\ 0 & 0 & 0 & 1 \end{bmatrix}$$

That is, the limit of P^n exists and is, at least to four decimal places, equal to the matrix displayed above. However, unlike in the case of a regular Markov chain, the rows of this limiting matrix are not identical and the matrix includes several zeros. We will consider in more detail Markov chains of this type in the next section.

The transition matrix of a "periodic" Markov chain, such as the one in Example 6.20, does not have a limit. This is not surprising, since periodic functions in general do not have long-run limits but rather cycle through their possible values.

6.4.3   Interpreting the Steady-State Distribution

The steady-state distribution $\boldsymbol{\pi}$ of a regular Markov chain can be interpreted in several ways. We present four different interpretations here; verifications of the second and fourth statements can be found in the Karlin and Taylor text.

1. If the "current" state of the Markov chain is observed after a large number of transitions, there is an approximate probability π_j of the chain being in state j. That is, for large n, $P(X_n = j) \approx \pi_j$. Moreover, this holds regardless of the initial distribution of the chain (i.e., the unconditional distribution of the initial state X_0).

The first sentence is essentially the definition of $\boldsymbol{\pi}$ stemming from the Steady-State Theorem.

2. The long-run proportion of time the Markov chain visits the jth state is π_j.

To be more precise, for any state j let $N_j(n)$ denote the number of times the chain visits state j in its first n transitions; that is,

$$N_j(n) = \#\{1 \leq k \leq n : X_k = j\}$$

Then it can be shown that $N_j(n)/n$, the *proportion* of time the Markov chain spends in state j among the first n transitions, converges in probability to π_j.

3. If we assign $\boldsymbol{\pi}$ to be the initial distribution of X_0, then the distribution of X_n is also $\boldsymbol{\pi}$ for any subsequent number of transitions n. For this reason, $\boldsymbol{\pi}$ is customarily referred to as the **stationary distribution** of the Markov chain.

To prove Statement 3, first let $\boldsymbol{\Pi}$ denote the matrix in Eq. (6.4), each of whose rows is $\boldsymbol{\pi}$. Now write $\mathbf{P}^{n+1} = \mathbf{P}^n\mathbf{P}$ and take the limit of both sides as $n \to \infty$:

$$\mathbf{P}^{n+1} = \mathbf{P}^n\mathbf{P} \Rightarrow \lim_{n\to\infty} \mathbf{P}^{n+1} = \lim_{n\to\infty} [\mathbf{P}^n\mathbf{P}] = \left[\lim_{n\to\infty} \mathbf{P}^n\right]\mathbf{P} \Rightarrow \boldsymbol{\Pi} = \boldsymbol{\Pi}\mathbf{P}$$

Each side of the last equation is an $s \times s$ matrix; equating the top rows of these two matrices, we have $\boldsymbol{\pi} = \boldsymbol{\pi}\mathbf{P}$. (You could just as well equate any other row, since all the rows of $\boldsymbol{\Pi}$ are the same.)

Now, assign the steady-state distribution to X_0: $\mathbf{v}_0 = \boldsymbol{\pi}$. Then the (unconditional) distribution of X_1, using the results of Sect. 6.3, is $\mathbf{v}_1 = \boldsymbol{\pi}\mathbf{P}$, which we have established equals $\boldsymbol{\pi}$. Continuing by induction, we have for any n that the unconditional distribution of X_n is $\mathbf{v}_n = \mathbf{v}_{n-1}\mathbf{P} = \boldsymbol{\pi}\mathbf{P} = \boldsymbol{\pi}$, completing the proof.

4. The expected number of transitions required to return to the jth state, beginning in the jth state, is equal to $1/\pi_j$. This is called the **mean recurrence time** for state j.

Compare this result to the mean of a geometric rv from Chap. 2: the expected number of trials (replications) required to first observe an event whose probability is p equals $1/p$. The difference is that the geometric model assumes the trials are independent, while a Markov chain model assumes that successive states of the chain are dependent (as specified by the Markov property). But if we think of "return to the jth state" as our event of interest, then Statement 1 implies that (at least for large n) the probability of this event is roughly π_j, and so it seems reasonable that the average number of tries/steps it will take to achieve this event will be $1/\pi_j$.

Example 6.21 The steady-state distribution of the taxi driver example is the 3×1 vector $\pi = [.2 \ .6 \ .2]$. For now, this relies on the computation of $\mathbf{P}^{100}$ above; shortly, we will present a derivation of this vector that does not require raising $\mathbf{P}$ to a high power. From the preceding descriptions, we conclude all of the following:

1. Regardless of where the taxi driver starts his day, for large n there is about a 20% chance his nth fare will be dropped off in zone 1, a 60% chance that that fare will go to zone 2, and a 20% chance for zone 3.
2. In the long run, the taxi driver drops off about 20% of his fares in zone 1, about 60% in zone 2, and about 20% in zone 3.
3. Suppose the taxi driver sleeps in his cab, thus waking up each day in a "random" zone, and we assign to X_0 (his point of origin tomorrow, say) the initial distribution $\mathbf{v}_0 = \pi = [.2 \ .6 \ .2]$. The unconditional distribution of X_1, the destination of tomorrow's first fare, is

$$\mathbf{v}_1 = \mathbf{v}_0 \mathbf{P} = [.2 \quad .6 \quad .2] \begin{bmatrix} .3 & .2 & .5 \\ .1 & .8 & .1 \\ .4 & .4 & .2 \end{bmatrix}$$

By direct computation, the first entry of $\mathbf{v}_1$ is $(.2)(.3) + (.6)(.1) + (.2)(.4) = .2$; the second entry is $(.2)(.2) + (.6)(.8) + (.2)(.4) = .6$; and the last is .2. That is, $\mathbf{v}_1 = [.2 \ .6 \ .2] = \pi$, and so X_1 has the same distribution as X_0. The same will hold for X_2, X_3, and so on.

4. If the driver starts from his home in zone 3, then on the average the number of fares he handles until he is brought back to zone 3 is given by $1/\pi_3 = 1/(.2) = 5$. That is, the mean recurrence time for state 3 (zone 3) is five transitions. ∎

6.4.4   Efficient Computation of Steady-State Probabilities

The preceding examples of regular Markov chains and the resulting steady-state distributions may suggest that one determines π by computing a high power of the transition matrix $\mathbf{P}$, preferably with software, and then extracting any row of the resulting matrix (all of which will be the same, according to the Steady-State Theorem). Fortunately there is a more direct technique for determining π. The method was hinted at in the proof of Statement 3 above: the steady-state distribution π satisfies the matrix equation $\pi\mathbf{P} = \pi$. In fact, something stronger is true.

THEOREM

Let $\mathbf{P}$ be the one-step transition matrix of a regular Markov chain on the state space $\{1, \ldots, s\}$. The steady-state distribution of the Markov chain is the *unique* solution $\boldsymbol{\pi} = [\pi_1 \cdots \pi_s]$ to the system of equations formed by

$$\boldsymbol{\pi}\mathbf{P} = \boldsymbol{\pi} \quad \text{and} \quad \pi_1 + \cdots + \pi_s = 1 \tag{6.5}$$

Proof Statement 3 above and the fact that $\boldsymbol{\pi}$ is a probability vector (because it's the limit of probability vectors) ensures that $\boldsymbol{\pi}$ itself satisfies Eq. (6.5). We must show that any *other* vector satisfying both equations in Eq. (6.5) is, in fact, $\boldsymbol{\pi}$. To that end, let $\mathbf{w}$ be any $1 \times s$ vector satisfying the two conditions $\mathbf{w}\mathbf{P} = \mathbf{w}$ and $\sum w_i = 1$. Similar to earlier derivations, we have $\mathbf{w}\mathbf{P}^2 = (\mathbf{w}\mathbf{P})\mathbf{P} = \mathbf{w}\mathbf{P} = \mathbf{w}$ and, by induction, $\mathbf{w}\mathbf{P}^n = \mathbf{w}$ for any positive integer n. Taking the limit of both sides as $n \to \infty$, the Steady-State Theorem implies that $\mathbf{w}\boldsymbol{\Pi} = \mathbf{w}$.

Now expand $\mathbf{w}\boldsymbol{\Pi}$:

$$\mathbf{w}\boldsymbol{\Pi} = [w_1 \cdots w_s] \begin{bmatrix} \pi_1 & \cdots & \pi_s \\ \vdots & \vdots & \vdots \\ \pi_1 & \cdots & \pi_s \end{bmatrix} = \left[(w_1\pi_1 + \cdots + w_s\pi_1) \quad \cdots \quad (w_1\pi_s + \cdots + w_s\pi_s) \right]$$

$$= \left[(\Sigma w_i)\pi_1 \quad \cdots \quad (\Sigma w_i)\pi_s \right] = (\Sigma w_i)\left[\pi_1 \cdots \pi_s\right] = (\Sigma w_i)\boldsymbol{\pi}$$

Since $\sum w_i = 1$ by assumption, we have $\mathbf{w}\boldsymbol{\Pi} = \boldsymbol{\pi}$. It was established above that $\mathbf{w}\boldsymbol{\Pi} = \mathbf{w}$, and so we conclude that $\mathbf{w} = \boldsymbol{\pi}$, as originally claimed. ∎

Example 6.22 Consider again the Markov chain model for snowy days (S) and non-snowy or "green" days (G) in New York City, begun in Example 6.6. The one-step transition matrix was given by

$$\mathbf{P} = \begin{matrix} G \\ S \end{matrix} \begin{bmatrix} .964 & .036 \\ .224 & .776 \end{bmatrix}$$

Since all the entries of $\mathbf{P}$ are positive, this is a regular Markov chain. The preceding theorem can be used to determine the steady-state probabilities $\boldsymbol{\pi} = [\pi_1 \ \pi_2]$. The equations in Eq. (6.5), written out long-hand, are

$$.964\pi_1 + .224\pi_2 = \pi_1$$
$$.036\pi_1 + .776\pi_2 = \pi_2$$
$$\pi_1 + \pi_2 = 1$$

Substituting $\pi_2 = 1 - \pi_1$ into the first equation gives $.964\pi_1 + .224(1 - \pi_1) = \pi_1$; solving for π_1 produces $\pi_1 = .224/.260 = .8615$ and then $\pi_2 = 1 - .8615 = .1385$. For the season to which this model applies, in the long run New York City has at least 50 mm of snow on 86.15% of days and less than 50 mm on the other 13.85% of days.

It's important to note that the top two equations alone, i.e., those provided by the relationship $\pi P = \pi$, do not uniquely determine the value of the vector π. The first equation is equivalent to $.224\pi_2 = .036\pi_1$ (subtract $.964\pi_1$ from both sides), but so is the second equation (subtract $.776\pi_2$ from both sides). The final equation, requiring the entries of π to sum to 1, is necessary to obtain a unique solution. ∎

Expression (6.5) may be reexpressed as a single matrix equation. Taking a transpose,

$$\pi P = \pi \Rightarrow P^T \pi^T = \pi^T = I\pi^T \Rightarrow (P^T - I)\pi^T = 0,$$

where $\mathbf{0}$ is an $s \times 1$ vector of zeros. The requirement $\pi_1 + \cdots + \pi_s = 1$ can be rendered in matrix form as $[1 \cdots 1]\pi^T = [1]$, and so the system of Eq. (6.5) can be expressed with the augmented matrix

$$\begin{bmatrix} 1 & \cdots & 1 & \vdots & 1 \\ & & & \vdots & 0 \\ & P^T - I & & \vdots & \vdots \\ & & & \vdots & 0 \end{bmatrix} \tag{6.6}$$

Example 6.23 (Example 6.21 continued) To analytically determine the steady-state distribution of our taxi driver example, first construct the matrix $P^T - I$:

$$P^T - I = \begin{bmatrix} .3 & .1 & .4 \\ .2 & .8 & .4 \\ .5 & .1 & .2 \end{bmatrix} - \begin{bmatrix} 1 & 0 & 0 \\ 0 & 1 & 0 \\ 0 & 0 & 1 \end{bmatrix} = \begin{bmatrix} -.7 & .1 & .4 \\ .2 & -.2 & .4 \\ .5 & .1 & -.8 \end{bmatrix}$$

Second, form the augmented matrix indicated in Expression (6.6), and then finally use Gauss-Jordan elimination to find its reduced row echelon form (e.g., with the `rref` command in Matlab):

$$\begin{bmatrix} 1 & 1 & 1 & \vdots & 1 \\ -.7 & .1 & .4 & \vdots & 0 \\ .2 & -.2 & .4 & \vdots & 0 \\ .5 & .1 & -.8 & \vdots & 0 \end{bmatrix} \xrightarrow{\text{RREF}} \begin{bmatrix} 1 & 0 & 0 & \vdots & .2 \\ 0 & 1 & 0 & \vdots & .6 \\ 0 & 0 & 1 & \vdots & .2 \\ 0 & 0 & 0 & \vdots & 0 \end{bmatrix}$$

From the right-hand matrix, we infer that $\pi_1 = .2$, $\pi_2 = .6$, and $\pi_3 = .2$. This matches our earlier deduction from the matrix P^{100}. ∎

Example 6.24 For the Internet browser scenario of Example 6.19, the steady-state distribution can be determined as follows:

$$\begin{bmatrix} 1 & \cdots & 1 & 1 \\ & & & 0 \\ \mathbf{P}^{\mathsf{T}} - \mathbf{I} & & \vdots \\ & & & 0 \end{bmatrix} = \left[\begin{array}{ccccc|c} 1 & 1 & 1 & 1 & 1 & 1 \\ -1 & .2 & 0 & .7 & 0 & 0 \\ .3 & -.9 & 0 & 0 & 0 & 0 \\ 0 & .6 & -.8 & .1 & .5 & 0 \\ .7 & 0 & .4 & -.8 & .5 & 0 \\ 0 & .1 & .4 & 0 & -1 & 0 \end{array}\right] \xrightarrow{\text{RREF}} \left[\begin{array}{ccccc|c} 1 & 0 & 0 & 0 & 0 & .2840 \\ 0 & 1 & 0 & 0 & 0 & .0948 \\ 0 & 0 & 1 & 0 & 0 & .1659 \\ 0 & 0 & 0 & 1 & 0 & .3791 \\ 0 & 0 & 0 & 0 & 1 & .0758 \\ 0 & 0 & 0 & 0 & 0 & 0 \end{array}\right]$$

That is, $\pi_1 = .2840$, $\pi_2 = .0948$, and so on; these match the results suggested earlier by considering $\mathbf{P}^{100}$. In the long run, about 28.40% percent of Web pages visited by Internet users under consideration are social media sites, 9.48% are for checking e-mail, 16.59% are news and sports Web sites, etc. Also, when a user finishes checking her or his e-mail online, the average number of Web sites visited until s/he checks e-mail again is $1/\pi_2 = 1/.0948 = 10.55$ (including the second login to e-mail).   ∎

6.4.5   Irreducible and Periodic Chains

The existence of a stationary distribution is not unique to regular Markov chains.

> **DEFINITION**
>
> Let i and j be two (not necessarily distinct) states of a Markov chain. State j is **accessible** from state i (or, equivalently, i can **access** j) if $P^{(n)}(i \to j) > 0$ for some integer $n \geq 0$.[1] A Markov chain is **irreducible** if every state is accessible from every other state.

It should be clear that every regular chain is irreducible (do you see why?). However, the reverse is not true: an irreducible Markov chain need not be a regular chain. Consider the cyclic chain of Example 6.20: the bus can access any of the three locations it visits (campus, housing, downtown) from any other location, so the chain is irreducible. However, as discussed earlier in this section, the chain is definitely not regular. The Ehrenfest chain model developed in Exercise 2 is another example of an irreducible but not regular chain; see Exercise 43 at the end of this section.

It can be shown that *any finite-state, irreducible Markov chain has a stationary distribution.* That is, if $\mathbf{P}$ is the transition matrix of an irreducible chain, there exists a row vector π such that $\pi\mathbf{P} = \pi$; moreover, there is a unique such vector satisfying the additional constraint $\sum \pi_i = 1$. For example, the cyclic bus route chain of

---

[1] For $n = 0$, the symbol $P^{(0)}(i \to j)$ is interpreted as the probability of going from i to j in zero steps, and so necessarily $P^{(0)}(i \to i) = 1$ for all i and $P^{(0)}(i \to j) = 0$ for $i \neq j$. In particular, this means every state i is, by definition, accessible from itself.

Example 6.20 has stationary distribution $\pi = [1/3 \ 1/3 \ 1/3]$, as seen by the computation

$$\pi P = [1/3 \ \ 1/3 \ \ 1/3] \begin{bmatrix} 0 & 1 & 0 \\ 0 & 0 & 1 \\ 1 & 0 & 0 \end{bmatrix} = [1/3 \ \ 1/3 \ \ 1/3] = \pi$$

So, if the bus is equally likely to be at any of its three locations right now, it is also equally likely to be at any of those three places after the next transition (the "stationary" interpretation of π). This is true even though the chain is not regular, so the Steady-State Theorem does not apply.

If an s-state Markov chain is irreducible but not regular, then every state can access every other state but there exists no integer n for which all s^2 probabilities $P^{(n)}(i \rightarrow j)$ are positive. The only way this can occur is if the chain exhibits some sort of "periodic" behavior, e.g., when one group of states can access some states only in an even number of steps and others only in an odd number of steps. Formally, the **period** of a state i is defined as the greatest common divisor of all positive integers n such that $P^{(n)}(i \rightarrow i) > 0$; if that gcd equals 1, then state i is called *aperiodic*. All three states in the cyclic chain above have period 3, because for every state the period is gcd(3, 6, 9, ...) = 3. It can be shown that every state in an irreducible chain has the same period; the chain is called aperiodic if that common period is 1 and is called periodic otherwise.

As noted previously, for any regular Markov chain there exists an integer n such that all the entries of $\mathbf{P}^n$, $\mathbf{P}^{n+1}$, $\mathbf{P}^{n+2}$, and so on are positive. Since the gcd of the set $\{n, n+1, n+2, \ldots\}$ is 1, it immediately follows that every regular Markov chain is aperiodic. The following theorem characterizes regularity for finite-state chains.

> **THEOREM**
> A finite-state Markov chain is regular if, and only if, it is both irreducible and aperiodic.

The "only if" direction of the theorem is established in the earlier paragraphs of this sub-section. The converse statement, that all irreducible and aperiodic finite-state chains are regular, can be proved using a result called the *Frobenius coin-exchange theorem* (we will not present the proof here).

6.4.6   Exercises: Section 6.4 (31–43)

31. Refer back to Mendel's plant breeding experiments in Example 6.16 and Exercise 26.
    (a) Do the genotypes formed by successive cross-breeding with pure recessive plants gg, as in Example 6.16, form a regular Markov chain?

(b) Do the genotypes formed by successive cross-breeding with hybrid plans Gg, as in Exercise 26, form a regular Markov chain?

32. Refer back to Exercise 2. Assume $m = 3$ balls are being exchanged between the two chambers. Is the Markov chain $X_n =$ number of balls in the left chamber a regular chain?

33. Refer back to Example 6.13 regarding cell phone contracts in China.
(a) Determine the steady-state probabilities of this chain.
(b) In the long run, what proportion of Chinese cell phone users will have contracts with China Mobile?
(c) A certain cell phone customer currently has a contract with China Telecom. On the average, how many contract changes will s/he make before signing with China Telecom again?

34. The article "Markov Chain Model for Performance Analysis of Transmitter Power Control in Wireless MAC Protocol" (Twenty-first International Conference on Advanced Networking and Applications, 2007) describes a Markov chain model for the state of a communication channel using a particular "slotted non-persistent" (SNP) protocol. The channel's possible states are (1) idle, (2) successful transmission, and (3) collision. For particular values of the authors' proposed four-parameter model, the following transition matrix results:

$$\mathbf{P} = \begin{bmatrix} .50 & .40 & .10 \\ .02 & .98 & 0 \\ .12 & 0 & .88 \end{bmatrix}$$

(a) Verify that $\mathbf{P}$ is the transition matrix of a regular Markov chain.
(b) Determine the steady-state probabilities for this channel.
(c) What proportion of time is this channel idle, in the long run?
(d) What is the average number of time steps between successive collisions?

35. Refer back to Exercise 3.
(a) Construct the one-step transition matrix $\mathbf{P}$ of this chain.
(b) Show that $X_n =$ the machine's state (full, part, broken) on the nth day is a regular Markov chain.
(c) Determine the steady-state probabilities for this chain.
(d) On what proportion of days is the machine fully operational?
(e) What is the average number of days between breakdowns?

36. Refer back to Exercise 6, and assume three files A, B, C are to be repeatedly requested. Suppose that 60% of requests are for file A, 10% for file B, and 30% for C. Let $X_n =$ the stacked order of the files (e.g., ABC) after the nth request.
(a) Construct the transition matrix $\mathbf{P}$ for this chain. (The one-step transition probabilities were established in Exercise 6(c).)
(b) Determine the steady-state probability for the stack ABC.
(c) Show that, in general, the steady-state probability for ABC is given by

$$\pi_{ABC} = \frac{p_A \cdot p_B}{p_B + p_C}$$

where $p_A = P$(file A is requested) and p_B and p_C are defined similarly. (The other five steady-state probabilities can be deduced by changing the subscripts appropriately.)

37. Let **P** be the one-step transition matrix of a Markov chain. Show that if all the entries of $\mathbf{P}^n$ are positive for some positive integer n, then so are all the entries of $\mathbf{P}^{n+1}$, $\mathbf{P}^{n+2}$, and so on. [*Hint*: Write $\mathbf{P}^{n+1} = \mathbf{P} \cdot \mathbf{P}^n$ and consider how the (i, j)th entry of $\mathbf{P}^{n+1}$ is obtained.]

38. Refer back to Exercise 19.
    (a) Consider a new customer. By definition, s/he made no visits to the repair shop last year. What is his/her expected number of visits this year?
    (b) Now suppose a car owner has been a customer of this repair shop for many years. What is the expected number of shop visits s/he will make next year?

39. Consider a Markov chain with just two states, 0 and 1, with one-step transition probabilities $\alpha = P(0 \rightarrow 1)$ and $\beta = P(1 \rightarrow 0)$.
    (a) Assuming $0 < \alpha < 1$ and $0 < \beta < 1$. Determine the steady-state probabilities of states 0 and 1 in terms of α and β.
    (b) What happens if α and/or β equals 0 or 1?

40. *Occupational prestige* describes how particular jobs are regarded by society and is often used by sociologists to study class. The article "Social Mobility in the United States as a Markov Process" (*J. for Economic Educators*, v. 8 no. 1 (2008): 15-37) investigates the occupational prestige of fathers and sons. Data provided in the article can be used to derive the following transition matrix for occupational prestige classified as low (L), medium (M), or high (H):

$$\mathbf{P} = \begin{array}{c} \text{L} \\ \text{M} \\ \text{H} \end{array} \begin{bmatrix} .5288 & .2096 & .2616 \\ .3688 & .2530 & .3782 \\ .2312 & .1738 & .5950 \end{bmatrix}$$

    (a) Which occupational prestige "state" is the most likely to self-replicate (i.e., father and son are in the same category)? Which is the least likely?
    (b) Determine the steady-state distribution of this Markov chain.
    (c) Interpret the distribution in (b), assuming the model specified by the matrix is valid across many generations.
    [*Note*: The authors actually used 11 categories of occupational prestige; we have collapsed these into three categories for simplicity.]

41. The two ends of a wireless communication system can each be inactive (0) or active (1). Suppose the two nodes act independently, each as a Markov chain with the transition probabilities specified in Exercise 39. Let $X_n =$ the "combined" state of the two relays at the nth time step. The state space for this chain is $\{00, 01, 10, 11\}$, e.g., state 01 corresponds to an inactive transmitter with an active receiver. (Performance analysis of such systems is described in "Energy-

Efficient Markov Chain-Based Duty Cycling Schemes for Greener Wireless Sensor Networks," *ACM J. on Emerging Tech. in Computing Systems* (2012):1-32.)

(a) Determine the transition matrix for this chain. [*Hint*: Use independence to uncouple the two states, e.g., $P(00 \to 10) = P(0 \to 1) \cdot P(0 \to 0)$.]

(b) Determine the steady state distribution of this chain.

(c) As the authors note, "a connection is feasible only when both wireless nodes are active." What proportion of time is a connection feasible under this model?

42. A particular gene has three expressions: *AA*, *Aa*, and *aa*. When two individuals mate, one half of each parent's gene is contributed to the offspring (and each half is equally likely to be donated). For example, an *AA* mother can only donate *A* while an *Aa* father is equally likely to donate *A* or *a*, resulting in a child that is either *AA* or *Aa*. Suppose that the population proportions of *AA*, *Aa*, and *aa* individuals are p, q, and r, respectively (so $p + q + r = 1$). Consider the offspring of a randomly selected individual; specifically, let $X_n =$ the gene expression of the oldest child in his or her nth generation of descendants (whom we assume will have at least one offspring).

(a) Assume the nth-generation individual's mate is selected at random from the genetic population described above. Show that $P(X_{n+1} = AA | X_n = AA) = p + q/2$, $P(X_{n+1} = Aa | X_n = AA) = q/2 + r$,   and $P(X_{n+1} = aa | X_n = AA) = 0$.   [*Hint*: Apply the Law of Total Probability.]

(b) Using the same method as in (a), determine the other one-step transition probabilities and construct the transition matrix **P** of this chain.

(c) Verify that X_n is a regular Markov chain.

(d) Suppose there exists some $\alpha \in [0, 1]$ such that $p = \alpha^2$, $q = 2\alpha(1 - \alpha)$, and $r = (1 - \alpha)^2$. (In this context, $\alpha = P(A \text{ allele})$.) Show that $\boldsymbol{\pi} = [p \; q \; r]$ is the stationary distribution of this chain. (This fact is called the *Hardy-Weinberg law*; it establishes that the rules of genetic recombination result in a long-run stable distribution of genotypes.)

43. Refer back to the Ehrenfest chain model of Exercises 2 and 24. Once again assume that $m = 3$ balls are being exchanged between the two chambers.

(a) Explain why this is an irreducible chain, but not a regular chain.

(b) Explain why each state has period equal to 2.

(c) Show that the vector [1/8 3/8 3/8 1/8] is a stationary distribution for this chain. (Thus, even though the chain is not regular and the transition matrix **P** does not have a limit, there still exists a stationary distribution due to irreducibility.)

6.5   Markov Chains with Absorbing States

The Gambler's Ruin scenario, begun in Example 6.2, has the feature that the chain "terminates" when it reaches either of two states ($0 or $3 in our version of the competition). As we've noted previously, it's mathematically advantageous to imagine that the Markov chain actually continues in these cases, just never leaving the state 0 or 3; one such sample path is

$$2 \to 1 \to 2 \to 1 \to 0 \to 0 \to 0 \to 0 \to \dots$$

In this section, we first investigate states from which a Markov chain can never exit and the time it takes to arrive in one of those states.

DEFINITION

A state j of a Markov chain is called an **absorbing state** if

$$P(j \to j) = 1.$$

Equivalently, j is an absorbing state if the (j, j)th entry of the one-step transition matrix of the chain is 1.

The states 0 and 3 are both absorbing states in our Gambler's Ruin example. In contrast, the taxi driver example has no absorbing states. The next example shows that some care must be taken in identifying absorbing states.

Example 6.25 Anyone who has applied for a bank loan knows that the process of eventual approval (or rejection) involves many steps and, occasionally, a lot of complex negotiation. Figure 6.7 illustrates the possible route of a set of loan documents from (1) document initiation to (6) final approval or rejection. The intermediate steps (2)–(5) represent various exchanges between underwriters, loan officers, and the like. In this particular chain, two such individuals (at states 3 and 5) have the authority to make a final decision, though the agent at state 3 may elect to return the documents for further discussion.

The one-step transition matrix of this chain is

$$\mathbf{P} = \begin{bmatrix} 0 & .5 & 0 & .5 & 0 & 0 \\ 0 & 0 & .5 & .5 & 0 & 0 \\ 0 & .5 & 0 & 0 & 0 & .5 \\ 0 & 0 & .5 & 0 & .5 & 0 \\ 0 & 0 & 0 & 0 & 0 & 1 \\ 0 & 0 & 0 & 0 & 0 & 1 \end{bmatrix}$$

Fig. 6.7 State diagram for Example 6.25

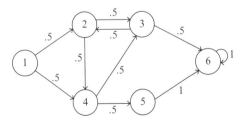

Although the number 1 appears twice in $\mathbf{P}$, only state 6 is an absorbing state of this chain. Indeed, $p_{66} = P(6 \rightarrow 6) = 1$; however, state 5 is *not* an absorbing state because $p_{55} = P(5 \rightarrow 5) = 0$. Rather, the fifth row of $\mathbf{P}$ indicates that if the chain ever enters state 5, it will necessarily pass in the next transition into state 6 (where, as it happens, it will be "absorbed").   ∎

To be clear, a Markov chain may have no absorbing states (the taxi driver), a single absorbing state (Example 6.25), or multiple absorbing states (Gambler's Ruin).

6.5.1   Time to Absorption

When a Markov chain has one or more absorbing states, it is natural to ask how long it will take to reach an absorbing state. Of course, the answer depends on where (i.e., in which state) the Markov chain begins. For any *non-absorbing* state i, define a random variable T_i by

T_i = number of transitions until the Markov chain reaches an absorbing state, starting in state i

This rv T_i is called the **time to absorption from state i**; the possible values of T_i are $1, 2, 3, 4, \ldots$. As we shall now illustrate, the distribution of T_i can be approximated from the k-step transition matrices $\mathbf{P}^k$ for $k = 1, 2, 3, \ldots$. For simplicity's sake, consider first a Markov chain with a single absorbing state, which we will call a. Then the (i, a)th entry of $\mathbf{P}$ is the probability of transitioning directly from state i into the absorbing state a, which is therefore also the probability that T_i equals 1:

$$P(i \rightarrow a) = P(T_i = 1)$$

Since T_i is always a positive integer, this also equals $P(T_i \leq 1)$, a fact which will prove important shortly. Now consider the (i, a)th entry of $\mathbf{P}^2$, which represents $P^{(2)}(i \rightarrow a)$. There are two ways the Markov chain could transition from i to a in two steps:

$$i \rightarrow \text{any non-absorbing state} \rightarrow a \ (T_i = 2), \quad \text{or}$$

$$i \rightarrow a \rightarrow a \ (T_i = 1)$$

Therefore, the two-step probability $P^{(2)}(i \rightarrow a)$ does not represent the chance T_i equals 2, but rather the chance that T_i is *at most* 2. That is,

$$P^{(2)}(i \rightarrow a) = P(T_i \leq 2).$$

Following the same pattern, the k-step transition probability $P^{(k)}(i \rightarrow a)$ is equal to $P(T_i \leq k)$ for any positive integer k.

If the Markov chain has two absorbing states a_1 and a_2, say, then the chance of being absorbed from state i in one step is simply the sum $P(i \rightarrow a_1) + P(i \rightarrow a_2)$, since those two events are mutually exclusive (you can only arrive in one state). Similarly, the probability $P(T_i \leq 2)$ is determined by adding $P^{(2)}(i \to a_1) + P^{(2)}(i \to a_2)$, and so on. The general result is stated in the following theorem.

THEOREM

Consider a finite-state Markov chain, and let A denote the (non-empty) set of absorbing states. For any state $i \notin A$, define $T_i =$ the number of transitions, starting in state i, until the chain arrives in some absorbing state. Then the cdf of T_i is given by

$$F_{T_i}(k) = P(T_i \leq k) = \sum_{a \in A} P^{(k)}(i \to a) \quad k = 1, 2, 3, \ldots$$

In the special case of a single absorbing state, a, this simplifies to

$$F_{T_i}(k) = P(T_i \leq k) = P^{(k)}(i \to a)$$

The probability distribution of T_i (i.e., the pmf of the rv T_i) can then be determined from the cdf.

Example 6.26 (Example 6.25 continued) Let's consider the rv T_1, the absorption time from state 1 (i.e., the number of steps from loan document initiation to the bank's final decision). From the one-step transition matrix **P**, we know that

$$F_{T_1}(1) = P(T_1 \leq 1) = P(T_1 = 1) = P(1 \to 6) = p_{16} = 0.$$

The $(1,6)$ entry of $\mathbf{P}^2$ is also zero, so $F_{T_1}(2) = P(T_1 \leq 2) = P^{(2)}(1 \to 6) = 0$. Software was used to obtain the matrices $\mathbf{P}^3, \ldots, \mathbf{P}^{12}$, resulting in the following values for the $(1,6)$ entry.

k	1	2	3	4	5	6	7	8	9	10	11	12
$F_{T_1}(k)$	0	0	.5	.6875	.75	.8594	.8984	.9336	.9570	.9707	.9810	.9873

The accompanying table is, of course, an incomplete description of the cdf of T_1, since this process could theoretically be continued indefinitely. Next, because the rv T_1 is integer-valued, its pmf is easily determined from the cdf:

$$P(T_1 = 2) = P(T_1 \leq 2) - P(T_1 \leq 1) = F_{T_1}(2) - F_{T_1}(1) = 0 - 0 = 0$$

$$P(T_1 = 3) = P(T_1 \leq 3) - P(T_1 \leq 2) = F_{T_1}(3) - F_{T_1}(2) = .5 - 0 = .5$$

$$P(T_1 = 4) = P(T_1 \leq 4) - P(T_1 \leq 3) = F_{T_1}(4) - F_{T_1}(3) = .6875 - .5 = .1875$$

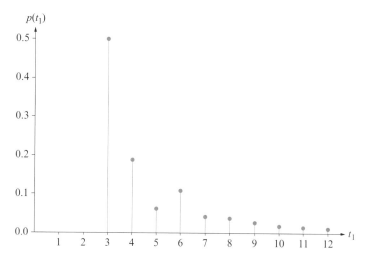

Fig. 6.8 The (incomplete) pmf of T_1 from Example 6.26

The first 12 probabilities in the pmf of T_1 are as follows (their sum is .9873):

k	1	2	3	4	5	6	7	8	9	10	11	12
$p_{T_1}(k)$	0	0	.5	.1875	.0625	.1094	.0390	.0352	.0234	.0137	.0103	.0063

This incomplete pmf is graphed in Fig. 6.8. Notice that T_1 must be at least 3, which is consistent with the state diagram in Fig. 6.7: it takes at least three steps to get from state 1 to state 6 (one of $1 \to 2 \to 3 \to 6$, $1 \to 4 \to 3 \to 6$, or $1 \to 4 \to 5 \to 6$).

A call to the bank determines that the documents are in the hands of the underwriter indicated by state 4. So, let's now consider the rv $T_4 =$ time to absorption (completion of the process) starting from state 4. Based on the state diagram, it seems reasonable to anticipate that it will typically take less time to reach state 6 starting from state 4 than it did when the chain began in state 1. Reading off the (4, 6) entries of $\mathbf{P}$, $\mathbf{P}^2$, …, $\mathbf{P}^{12}$ yields the cdf values in the accompanying table; subtraction as before then gives the corresponding pmf values.

k	1	2	3	4	5	6	7	8	9	10	11	12
$F_{T_4}(k)$	0	.75	.75	.8125	.9063	.9219	.9531	.9688	.9785	.9863	.9907	.9939
$p_{T_4}(k)$	0	.75	0	.0625	.0938	.0156	.0312	.0157	.0097	.0078	.0044	.0032

Notice that, starting in state 4, the chain is quite likely to be absorbed into state 6 in exactly two steps (either $4 \to 5 \to 6$ or $4 \to 3 \to 6$, with probabilities .5 and .25, respectively), and that it is impossible to move from 4 to 6 in exactly three steps. ∎

Example 6.27 In the Gambler's Ruin scenario with $p = .55$, how many games will Allan and Beth play against each other before one player goes broke? Recall that the transition matrix $\mathbf{P}$ is set from Allan's perspective, and that he begins with \$2. Thus, the rv of interest is T_2, the number of transitions (aka games), starting from Allan having \$2, until the competition ends because Allan either has \$0 or \$3. The one- and two-step transition matrices of this chain appear in Example 6.14. Hence

$$P(T_2 \le 1) = P(2 \to 0) + P(2 \to 3) = 0 + .55 = .55$$

$$P(T_2 \le 2) = P^{(2)}(2 \to 0) + P^{(2)}(2 \to 3) = .2025 + .55 = .7525$$

In general, the cumulative probability $P(T_2 \le k)$ can be determined by adding the $(2,0)$ and $(2,3)$ entries of the k-step transition matrix $\mathbf{P}^k$. These values were determined with the aid of software for $k = 1$ through 10 and are summarized in the accompanying table.

k	1	2	3	4	5	6	7	8	9	10
$F_{T_2}(k)$	.55	.7525	.8886	.9387	.9724	.9848	.9932	.9962	.9983	.9991
$p_{T_2}(k)$	.55	.2025	.1361	.0501	.0337	.0124	.0084	.0030	.0021	.0008

It's important to notice that T_2 indicates the number of steps required to enter *some* absorbing state (here, either \$0 or \$3), not the number of steps to enter a particular such state. ∎

6.5.2 Mean Time to Absorption

With $T_i = $ time to absorption starting from state i, the expected value of T_i is called the **mean time to absorption (MTTA) from state i**:

$\mu_i = E(T_i) = $ expected number of transitions until the Markov chain reaches an absorbing state, starting in state i

For each of the preceding examples, the incomplete pmf can be used to approximate the MTTA from state i. Consider the Markov chain in Example 6.26:

$$\begin{aligned}
\mu_1 = E(T_1) &= \sum_{k=1}^{\infty} k \cdot p_{T_1}(k) \approx \sum_{k=1}^{12} k \cdot p_{T_1}(k) \\
&= 1(0) + 2(0) + 3(.5) + 4(.1875) + \cdots + 11(.0103) + 12(.0063) \\
&= 4.31
\end{aligned}$$

To a hopefully reasonable approximation, on average the chain requires 4.31 transitions, starting in state 1, to be absorbed into state 6. Similarly, the mean time to absorption from state 4 is approximately

$$\mu_4 = \sum_{k=1}^{\infty} k \cdot p_{T_4}(k) \approx \sum_{k=1}^{12} k \cdot p_{T_4}(k)$$

$$= 1(0) + 2(.75) + \cdots + 11(.0044) + 12(.0032) = 2.91$$

For the Gambler's Ruin competition with $p = .55$ and Allan's initial stake at \$2, the pmf displayed in Example 6.27 gives

$$\mu_2 \approx 1(.55) + 2(.2025) + \cdots + 10(.0008) = 1.92$$

That is, if Allan starts with \$2 and $p = .55$, the expected length of the Gambler's Ruin competition is approximated to be 1.92 games.

In all such approximations, two things should be clear. First, the estimated means are *smaller than* the correct values, since the sums used are truncated versions of the correct summations and every term is nonnegative. So, in the Gambler's Ruin scenario, $\mu_2 > 1.92$. Second, the more terms we include in the truncated sum, the closer the approximation will be to the correct mean time to absorption from that state. Of course, additional terms require overcoming the practical hurdle of computing successively higher powers of the matrix **P**. With software, one could in practice use this method to get a very good approximation to the MTTA.

Exercise 56 presents a different approximation method that always yields a better approximation to the mean time to absorption; moreover, it relies directly on the cdf values and thus does not require computing differences to form the pmf. But this is still an approximation; what we would really like is an explicit method for determining the exact mean time to absorption from various states in the chain. The following theorem provides such a result.

MTTA THEOREM

Suppose a finite-state Markov chain with one-step transition matrix **P** has r non-absorbing states (and at least one absorbing state). Suppose further that there exists a path from every non-absorbing state into some absorbing state.

Let **Q** be the $r \times r$ sub-matrix of **P** corresponding to the non-absorbing states of the chain. Then the mean times to absorption from these states are given by the matrix formula

$$\boldsymbol{\mu} = (\mathbf{I} - \mathbf{Q})^{-1} \mathbf{1},$$

where $\mu_i = $ MTTA from the ith state in the **Q** sub-matrix, $\boldsymbol{\mu} = (\mu_1, \ldots, \mu_r)^{\mathrm{T}}$, **I** is the $r \times r$ identity matrix, and $\mathbf{1} = (1, \ldots, 1)^{\mathrm{T}}$.

This theorem not only provides the exact mean times to absorption (as opposed to the earlier approximations) but also computes *all* MTTAs simultaneously. A proof of the MTTA Theorem will be presented shortly, but first we illustrate its use with our two ongoing examples.

Example 6.28 For the bank loan Markov chain in Example 6.25, state 6 is the only absorbing state, so there are $r = 5$ non-absorbing states. The sub-matrix corresponding to these non-absorbing states is

$$\mathbf{Q} = \begin{bmatrix} 0 & .5 & 0 & .5 & 0 \\ 0 & 0 & .5 & .5 & 0 \\ 0 & .5 & 0 & 0 & 0 \\ 0 & 0 & .5 & 0 & .5 \\ 0 & 0 & 0 & 0 & 0 \end{bmatrix}$$

This can be obtained by "crossing out" the row and column of $\mathbf{P}$ corresponding to absorbing state 6. Let $\mu_i = E(T_i)$ be the mean time to absorption from state i for $i = 1, 2, 3, 4, 5$. Then, according to the MTTA Theorem,

$$\boldsymbol{\mu} = (\mathbf{I} - \mathbf{Q})^{-1}\mathbf{1}$$

$$= \begin{bmatrix} 1 & -.5 & 0 & -.5 & 0 \\ 0 & 1 & -.5 & -.5 & 0 \\ 0 & -.5 & 1 & 0 & 0 \\ 0 & 0 & -.5 & 1 & -.5 \\ 0 & 0 & 0 & 0 & 1 \end{bmatrix}^{-1} \begin{bmatrix} 1 \\ 1 \\ 1 \\ 1 \\ 1 \end{bmatrix}$$

$$= \begin{bmatrix} 1 & 1 & 1 & 1 & .5 \\ 0 & 1.6 & 1.2 & .8 & .4 \\ 0 & .8 & 1.6 & .4 & .2 \\ 0 & .4 & .8 & 1.2 & .6 \\ 0 & 0 & 0 & 0 & 1 \end{bmatrix} \begin{bmatrix} 1 \\ 1 \\ 1 \\ 1 \\ 1 \end{bmatrix} = \begin{bmatrix} 4.5 \\ 4.0 \\ 3.0 \\ 3.0 \\ 1.0 \end{bmatrix}$$

The inverse of $\mathbf{I} - \mathbf{Q}$ was determined using software.

So, for example, the mean time to absorption from state 1 is $\mu_1 = 4.5$ transitions, slightly larger than our earlier approximation of 4.31. On the average, it takes 4.5 steps to arrive at a loan decision starting from the time the loan documents are initiated. Similarly, the earlier estimate $\mu_4 \approx 2.91$ was a little off from the correct answer of $\mu_4 = 3$. The last entry of the vector $\boldsymbol{\mu}$ is obvious from the design of the chain: since state 5 transitions immediately into state 6 with certainty, T_5 is identically equal to 1, and so its mean is 1. ∎

Example 6.29 Consider once again our Gambler's Ruin scenario, this time with an arbitrary probability p that Allan triumphs over Beth in any one game. The only two non-absorbing states are \$1 and \$2, so the required sub-matrix $\mathbf{Q}$ consists of the "center four" entries of the original 4×4 transition matrix:

$$\mathbf{P} = \begin{matrix} 0 \\ 1 \\ 2 \\ 3 \end{matrix} \begin{bmatrix} 1 & 0 & 0 & 0 \\ 1-p & 0 & p & 0 \\ 0 & 1-p & 0 & p \\ 0 & 0 & 0 & 1 \end{bmatrix} \Rightarrow \mathbf{Q} = \begin{matrix} 1 \\ 2 \end{matrix} \begin{bmatrix} 0 & p \\ 1-p & 0 \end{bmatrix}$$

There is a simple inverse formula for a 2×2 matrix:

$$\begin{bmatrix} a & b \\ c & d \end{bmatrix}^{-1} = \frac{1}{ad - bc} \begin{bmatrix} d & -b \\ -c & a \end{bmatrix} \qquad (6.7)$$

Applying Eq. (6.7) and the MTTA Theorem,

$$\boldsymbol{\mu} = (\mathbf{I} - \mathbf{Q})^{-1} \mathbf{1}$$

$$= \begin{bmatrix} 1 & -p \\ -1+p & 1 \end{bmatrix}^{-1} \begin{bmatrix} 1 \\ 1 \end{bmatrix} = \frac{1}{(1)(1) - (-p)(-1+p)} \begin{bmatrix} 1 & p \\ 1-p & 1 \end{bmatrix} \begin{bmatrix} 1 \\ 1 \end{bmatrix}$$

$$= \frac{1}{1-p+p^2} \begin{bmatrix} 1+p \\ 2-p \end{bmatrix}$$

Hence

$$\mu_1 = \frac{1+p}{1-p+p^2} \qquad \mu_2 = \frac{2-p}{1-p+p^2}$$

Since we have always started Allan with \$2, let's explore μ_2 further. If $p = 1$, so Allan cannot lose, then $\mu_2 = (2-1)/(1-1+1^2) = 1$. This is logical, since Allan would automatically transition from \$2 to \$3 in 1 step/game and the competition would be over. Similarly, substituting $p = 0$ into this expression gives $\mu_2 = 2$, reflecting the fact that if Allan cannot win games then the chain must necessarily proceed along the path $2 \to 1 \to 0$, a total of two transitions. For $p = .55$, the numerical case illustrated earlier, we have

$$\mu_2 = \frac{2 - .55}{1 - .55 + .55^2} = \frac{1.45}{.7525} = \frac{580}{301} = 1.92691$$

which is quite close to our previous approximation of 1.92.

For what value of p is the competition expected to take the longest? Using calculus, one can find the maximum of μ_2 with respect to p, which turns out to occur at $p = 2 - \sqrt{3} \approx .268$. If Allan begins with \$2 and has a .268 probability of winning each game, the expected duration of the competition is maximized, specifically with $\mu_2 = 1 + 2/\sqrt{3} \approx 2.155$ games. ∎

Proof of the MTTA Theorem For notational ease, let 1, 2, ..., r be the non-absorbing states of the chain. Also, let A denote the set of absorbing states (which, if the Markov chain has s total states, could be enumerated as $r + 1, \ldots, s$). Starting in any non-absorbing state i, consider the first transition of the chain. If the chain transitions into any member of A, then it has been "absorbed" in one step and so $T_i = 1$. On the other hand, if the chain transitions into any *non*-absorbing state j (including back into i itself), then the expected number of steps to absorption is $1 + E(T_j)$, where the 1 accounts for the step just taken and T_j represents the time to absorption starting from the *new* state j. Apply the Law of Total Expectation:

$$E(T_i) = 1 \cdot P(i \to A) + \sum_{j=1}^{r} \left(1 + E(T_j)\right) \cdot P(i \to j)$$

$$= P(i \to A) + \sum_{j=1}^{r} P(i \to j) + \sum_{j=1}^{r} E(T_j) \cdot P(i \to j)$$

Since the state space of the Markov chain is $A \cup \{1, 2, \ldots, r\}$, the first two terms in the expression above must sum to 1. Thus, we have $\mu_i = 1 + \sum_{j=1}^{r} \mu_j P(i \to j)$ for $i = 1, 2, \ldots r$. Stacking these equations and rewriting slightly, we have

$$\mu_1 = P(1 \to 1)\mu_1 + \cdots + P(1 \to r)\mu_r + 1$$
$$\mu_2 = P(2 \to 1)\mu_1 + \cdots + P(2 \to r)\mu_r + 1$$
$$\vdots \qquad\qquad\qquad \vdots$$
$$\mu_r = P(r \to 1)\mu_1 + \cdots + P(r \to r)\mu_r + 1$$

This stack can be written more compactly as $\boldsymbol{\mu} = \mathbf{Q}\boldsymbol{\mu} + \mathbf{1}$. Solving for $\boldsymbol{\mu}$ yields the desired result.   ∎

The MTTA Theorem requires that every non-absorbing state can reach (at least) one absorbing state. That is, the set of absorbing states must be *accessible* from every non-absorbing state. What would happen if this were not the case?

Example 6.30 In the Markov chain depicted in Fig. 6.9, 4 is an absorbing state, but it is only accessible from state 3. It is clear that the chain will eventually be absorbed into state 4 if $X_0 = 3$ and will never be absorbed into state 4 if $X_0 = 1$ or 2. So, where does the MTTA Theorem break down?

The one-step transition matrix $\mathbf{P}$ for this chain, the resulting sub-matrix $\mathbf{Q}$ for the non-absorbing states, and the matrix $\mathbf{I} - \mathbf{Q}$ required for calculating mean times to absorption are

$$\mathbf{P} = \begin{array}{c} 1 \\ 2 \\ 3 \\ 4 \end{array}\begin{bmatrix} .5 & .5 & 0 & 0 \\ .4 & .6 & 0 & 0 \\ 0 & 0 & .7 & .3 \\ 0 & 0 & 0 & 1 \end{bmatrix} \qquad \mathbf{Q} = \begin{array}{c} 1 \\ 2 \\ 3 \end{array}\begin{bmatrix} .5 & .5 & 0 \\ .4 & .6 & 0 \\ 0 & 0 & .7 \end{bmatrix} \qquad \mathbf{I} - \mathbf{Q} = \begin{bmatrix} .5 & -.5 & 0 \\ -.4 & .4 & 0 \\ 0 & 0 & .3 \end{bmatrix}$$

The matrix $\mathbf{I} - \mathbf{Q}$ is not invertible; this can be seen by noting that the first and second rows are multiples of each other, or by computing the determinant and discovering that $\det(\mathbf{I} - \mathbf{Q}) = 0$. Because $(\mathbf{I} - \mathbf{Q})^{-1}$ does not exist, the formula from the MTTA Theorem cannot be applied.

Fig. 6.9 State diagram for
Example 6.30

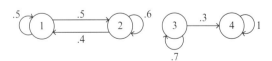

Recall that the cdf of T_1 can be determined from the appropriate entries of the k-step transition matrices; specifically, since the only absorbing state of this chain is state 4,

$$F_{T_1}(k) = P(T_1 \leq k) = P^{(k)}(1 \rightarrow 4) = \text{the } (1,4) \text{ entry of } \mathbf{P}^k$$

The $(1, 4)$ entry of the matrix $\mathbf{P}$ above is 0, so $P(T_1 \leq 1) = 0$. But since state 4 is not accessible from state 1, the $(1, 4)$ entry of every transition matrix $\mathbf{P}^k$ is 0. Thus, $P(T_1 \leq k) = 0$ for all positive integers k and $p_{T_1}(k) = 0 - 0 = 0$ for all k. Since the probabilities associated with T_1 sum to zero and not 1, T_1 is not actually a valid rv (and so, in particular, has no mean).   ∎

In general, when the set of absorbing states is not accessible from one or more non-absorbing states, the matrix $\mathbf{I} - \mathbf{Q}$ will be singular (i.e., not invertible). If a subset of the non-absorbing states can access the absorbing states (that's true for state 3 in Example 6.30), one can apply the MTTA Theorem if one defines $\mathbf{Q}$ to be the sub-matrix of $\mathbf{P}$ corresponding to those states that can access the absorbing states.

6.5.3 Mean First Passage Times

We now briefly turn our attention back to regular Markov chains. In Sect. 6.4, we saw that one interpretation of the probability π_j from the Steady-State Theorem is that $1/\pi_j$ represents the expected number of transitions necessary for the chain to return to state j given that it starts there—the mean recurrence time for state j. With a clever use of the MTTA Theorem, we can also determine the expected number of transitions required for the chain to transition from a state i to a *different* state j—the **mean first passage time** from i to j.

Example 6.31 (Example 6.23 continued) For the ubiquitous taxi driver example, it was found that the steady-state probability for zone 3 is $\pi_3 = .2$ and, thus, the expected number of fares until the driver returns to zone 3 is $1/\pi_3 = 1/.2 = 5$.

But suppose the taxi driver just dropped off a fare in zone 1 (or zone 2). He wonders how long it will take him to get back home to zone 3 for lunch. More precisely, he wishes to know the expected number of fares required to reach zone 3, *starting from some other state* (i.e., different than zone 3).

The trick to answering the taxi driver's question—i.e., to determine the mean first passage time for zone 3 when beginning in zone 1 or zone 2—is to pretend that zone 3 is an absorbing state, and then invoke the MTTA Theorem. Modify the original one-step transition matrix $\mathbf{P}$ of the Markov chain so that zone 3 is absorbing state, and label the new matrix $\tilde{\mathbf{P}}$:

$$\mathbf{P} = \begin{bmatrix} .3 & .2 & .5 \\ .1 & .8 & .1 \\ .4 & .4 & .2 \end{bmatrix} \quad \Rightarrow \quad \tilde{\mathbf{P}} = \begin{bmatrix} .3 & .2 & .5 \\ .1 & .8 & .1 \\ 0 & 0 & 1 \end{bmatrix}$$

Now proceed as before: the sub-matrix for the non-absorbing states, which in $\widetilde{\mathbf{P}}$ are zone 1 and zone 2, is

$$\mathbf{Q} = \begin{bmatrix} .3 & .2 \\ .1 & .8 \end{bmatrix},$$

from which

$$\boldsymbol{\mu} = (\mathbf{I} - \mathbf{Q})^{-1}\mathbf{1} = \begin{bmatrix} .7 & -.2 \\ -.1 & .2 \end{bmatrix}^{-1} \begin{bmatrix} 1 \\ 1 \end{bmatrix} = \cdots = \begin{bmatrix} 3.33 \\ 6.67 \end{bmatrix}$$

Thus, starting in zone 1, the average number of trips required for the taxi driver to get home to zone 3 is 3.33, while it takes twice that long on the average if he's starting from zone 2. ∎

6.5.4 Probabilities of Eventual Absorption

As discussed in Example 6.29 in the context of Gambler's Ruin, when a Markov chain has multiple absorbing states one can only speak of the mean time to absorption into the *set* of absorbing states, not any particular absorbing state (e.g., not time to \$0 separate from time to \$3). We can, however, ask about the *probability* of eventual absorption into state \$0, as opposed to eventual absorption into state \$3.

> **DEFINITION**
> Let a be an absorbing state of a Markov chain and let i be a non-absorbing state. The **probability of eventual absorption into a from state i**, denoted $\pi(i \rightarrow a)$, is defined by
> $$\pi(i \rightarrow a) = \lim_{n \to \infty} P^{(n)}(i \rightarrow a)$$

That is, $\pi(i \rightarrow a)$ is defined to be the limit of the (i, a) entry of $\mathbf{P}^n$ as $n \to \infty$. This is consistent with our previous efforts to determine the probability of eventual absorption by examining $\mathbf{P}^{75}$ or $\mathbf{P}^{100}$. But rather than approximate these probabilities by taking a high power of $\mathbf{P}$, we now present an explicit method for determining them.

Before illustrating the method for determining $\pi(i \rightarrow a)$, a few observations are in order. First, if state a is not accessible from state i, then $P^{(n)}(i \rightarrow a) = 0$ for all n and the limit is also zero, i.e., $\pi(i \rightarrow a) = 0$ when i cannot access a. This occurred in Example 6.30, with state 4 not being accessible from states 1 or 2.

Second, if the Markov chain has a single absorbing state a, then $\pi(i \rightarrow a) = 1$ for every state i that can access a. That is, a chain with an accessible absorbing state will always eventually be absorbed. This would be the case, for instance, in Example 6.25: it is a sure thing that the chain will eventually arrive at (and stay in) state 6, irrespective of where the chain begins. So, the interesting cases of determining $\pi(i \to a)$ are for Markov chains with multiple absorbing states, such as Gambler's Ruin.

Third, suppose we extended the preceding definition to non-absorbing states. That is, what can be said about

$$\lim_{n \to \infty} P^{(n)}(i \to j)$$

when j is *not* an absorbing state? If the Markov chain has any absorbing states (and assuming at least one of these is accessible from i), then the chain will eventually get absorbed and so $P^{(n)}(i \to j) \to 0$. If we have a regular Markov chain—which, in particular, means there are no absorbing states—then the Steady State Theorem tells us $P^{(n)}(i \to j) \to \pi_j$, a steady-state probability that is independent of i. For other cases, such as the cyclic chain of Example 6.20, the limit of $P^{(n)}(i \to j)$ may not exist at all.

On to the calculation: as in the proof of the MTTA Theorem, rearrange the states so that the non-absorbing states of the Markov chain are 1, 2, ..., r and the absorbing states are $r + 1$, ..., s. Then the one-step transition matrix $\mathbf{P}$ can be partitioned as follows:

$$
\mathbf{P} \;=\;
\begin{array}{c}
1 \\ \vdots \\ r \\ r+1 \\ \vdots \\ s
\end{array}
\left[
\begin{array}{cc|cc}
 & \mathbf{Q} & & \mathbf{R} \\[1ex]
\hline
 & \mathbf{O} & & \mathbf{I}
\end{array}
\right]
\tag{6.8}
$$

Expression (6.8) is sometimes called the **canonical form** of a Markov chain. In Eq. (6.8), $\mathbf{Q}$ is the $r \times r$ sub-matrix for the non-absorbing states, as before. The matrix $\mathbf{O}$ in the lower left of Eq. (6.8) consists entirely of zeros, since that quadrant of $\mathbf{P}$ indicates the probabilities of transitioning from an absorbing state ($r + 1, \ldots, s$) to a *non*-absorbing state (1, ..., r). Similarly, $\mathbf{I}$ is the $(s - r) \times (s - r)$ identity matrix, since its diagonal entries correspond to $P(a \to a)$ for the absorbing states and its off-diagonal entries to impossible events (transitions from one absorbing state to another). The "remainder" matrix $\mathbf{R}$ indicates the transition probabilities from the non-absorbing states into the absorbing states and can have (fairly) arbitrary entries.

The probabilities of eventual absorption into every absorbing state from every non-absorbing state are provided by the following theorem.

THEOREM

Consider a Markov chain with non-absorbing states $1, \ldots, r$ and absorbing states $r+1, \ldots, s$. Define sub-matrices $\mathbf{Q}$ and $\mathbf{R}$ of the one-step transition matrix $\mathbf{P}$ as in Eq. (6.8). Suppose further than every absorbing state is accessible from every non-absorbing state. Then the probabilities of eventual absorption are given by

$$\boldsymbol{\Pi} = (\mathbf{I} - \mathbf{Q})^{-1}\mathbf{R},$$

where $\mathbf{I}$ is the $r \times r$ identity matrix and $\boldsymbol{\Pi}$ is an $r \times (s - r)$ matrix whose entries are the probabilities $\pi(i \to a)$ for $i = 1, \ldots, r$ and $a = r+1, \ldots s$.

Some guidance for the proof of this theorem can be found in Exercise 57.

Example 6.32 (Example 6.29 continued) To apply the previous theorem to our Gambler's Ruin example, we need to reorder the states, so that non-absorbing states $1 and $2 come first while absorbing states $0 and $3 come last. The canonical form of $\mathbf{P}$, along with the relevant sub-matrices $\mathbf{Q}$ and $\mathbf{R}$, are

$$\mathbf{P} = \begin{array}{c} 1 \\ 2 \\ 0 \\ 3 \end{array}\left[\begin{array}{cc|cc} 0 & p & 1-p & 0 \\ 1-p & 0 & 0 & p \\ \hline 0 & 0 & 1 & 0 \\ 0 & 0 & 0 & 1 \end{array}\right], \quad \mathbf{Q} = \begin{bmatrix} 0 & p \\ 1-p & 0 \end{bmatrix}, \quad \mathbf{R} = \begin{bmatrix} 1-p & 0 \\ 0 & p \end{bmatrix}$$

Applying the previous theorem, along with the inverse formula (6.7) for a 2×2 matrix,

$$\boldsymbol{\Pi} = (\mathbf{I} - \mathbf{Q})^{-1}\mathbf{R} = \begin{bmatrix} 1 & -p \\ -1+p & 1 \end{bmatrix}^{-1} \begin{bmatrix} 1-p & 0 \\ 0 & p \end{bmatrix}$$

$$= \frac{1}{1-p+p^2}\begin{bmatrix} 1 & p \\ 1-p & 1 \end{bmatrix}\begin{bmatrix} 1-p & 0 \\ 0 & p \end{bmatrix} = \begin{bmatrix} \dfrac{1-p}{1-p+p^2} & \dfrac{p^2}{1-p+p^2} \\[2mm] \dfrac{1-2p+p^2}{1-p+p^2} & \dfrac{p}{1-p+p^2} \end{bmatrix}$$

Reading off the entries of the matrix $\boldsymbol{\Pi}$, we have

$$\pi(\$1 \to \$0) = \frac{1-p}{1-p+p^2} \qquad \pi(\$1 \to \$3) = \frac{p^2}{1-p+p^2}$$

$$\pi(\$2 \to \$0) = \frac{1-2p+p^2}{1-p+p^2} \qquad \pi(\$2 \to \$3) = \frac{p}{1-p+p^2}$$

In particular, if Allan starts with \$2, the probability he will eventually win the competition is $\pi(\$2 \to \$3) = p/(1 - p + p^2)$. As a check, this probability equals zero when $p = 0$ (Allan never wins games) and equals one when $p = 1$ (Allan always wins games). If $p = .55$, as in several of the previous examples in this chapter,

$$\pi(\$2 \to \$3) = \frac{.55}{1 - .55 + .55^2} = \frac{.55}{.7525} = \frac{220}{301} \approx .7309$$

Notice that this is, to four decimal places, the probability we approximated by computing $\mathbf{P}^{75}$ with software and thereby obtaining $P^{(75)}(\$2 \to \$3)$. ∎

The matrices $\mathbf{R}$ and $\mathbf{\Pi}$ in Example 6.32 are square, but this is not necessarily the case in other scenarios. In general, $\mathbf{Q}$ is an $r \times r$ matrix (hence, square), but the dimensions of both $\mathbf{R}$ and $\mathbf{\Pi}$ are $r \times (s - r)$.

6.5.5 Exercises: Section 6.5 (44–58)

44. Explain why a Markov chain with one or more absorbing states cannot be a regular chain.
45. A local community college offers a three-semester athletics training (AT) program. Suppose that at the end of each semester, 75% of students successfully move on to the next semester (or to graduation from the third semester) and 25% are required to repeat the most recent semester.
    (a) Construct a transition matrix to represent this scenario. The four states are (1) first semester, (2) second semester, (3) third semester, (4) graduate.
    (b) What is the probability a student graduates the program within three semesters? Four semesters? Five semesters?
    (c) What is the average number of semesters required to graduate from this AT program?
    (d) According to this model, what is the probability of eventual graduation? Does that seem realistic?
46. Refer back to the previous exercise. Now suppose that at the end of each semester, 75% of students successfully move on to the next semester (or to graduation from the third semester), 15% flunk out of the program, and 10% repeat the most recent semester.
    (a) Construct a transition matrix to represent this updated situation by adding a fifth state, (5) flunk out. [*Hint*: Two of the five states are absorbing.]
    (b) What is the probability a student exits the program, either by graduating or flunking out, within three semesters? Four semesters? Five semesters?
    (c) What is the average number of semesters students spend in this program before exiting (again, either by graduating or flunking out)?
    (d) What proportion of students that enter the program eventually graduate? What proportion eventually flunk out?

(e) Given that a student has passed the first two semesters (and, so, is currently in her third-semester courses), what is the probability she will eventually graduate?

47. The article "Utilization of Two Web-Based Continuing Education Courses Evaluated by Markov Chain Model" (*J. Am. Med. Inform. Assoc.* 2012: 489-494) compared students' flow between pages of an online course for two different Web layouts in two different health professions classes. In the first week of the classes, students could visit (1) the homepage, (2) the syllabus, (3) the introduction, and (4) chapter 1 of the course content. Each student was tracked until s/he either reached chapter 1 or exited without reading chapter 1 (call the latter state 5). For one version of the Web content in one class, the following transition matrix was estimated:

$$\mathbf{P} = \begin{bmatrix} 0 & 1 & 0 & 0 & 0 \\ .21 & 0 & .33 & .05 & .41 \\ .09 & .15 & 0 & .67 & .09 \\ 0 & 0 & 0 & 1 & 0 \\ 0 & 0 & 0 & 0 & 1 \end{bmatrix}$$

When students log into the course, they are always forced to begin on the homepage.

(a) Identify the absorbing state(s) of this chain.

(b) Let T_1 = the number of transitions students take, starting from the homepage, until the either arrive at chapter 1 or exit early. Determine $P(T_1 \leq k)$ for $k = 1, 2, \ldots, 10$.

(c) Use (b) to approximate the pmf of T_1, and then approximate the mean time to absorption starting from the class homepage.

(d) Determine the (true) mean time to absorption starting from the homepage.

(e) What proportion of students eventually got to chapter 1 in the first week? What proportion exited the course without visiting chapter 1?

48. Refer back to the previous exercise. After some content redesign, the same Web-based health professions course was run a second time. The first-week transition probabilities for the revised course were as follows:

$$\mathbf{P} = \begin{bmatrix} 0 & 1 & 0 & 0 & 0 \\ .15 & 0 & .43 & .06 & .36 \\ .09 & .16 & 0 & .66 & .09 \\ 0 & 0 & 0 & 1 & 0 \\ 0 & 0 & 0 & 0 & 1 \end{bmatrix}$$

(a) How did the redesign affect the average amount of time students spent in the course (at least as measured by the number of Web page visits within a session)?

(b) Did the redesign improve the chances that students would get to the chapter 1 content before exiting the system?

49. In Exercise 4, we introduced a game in which Michelle will flip a coin until she gets heads four times in a row. Define $X_0 = 0$ and, for $n \geq 1$, X_n = the number of heads in the current streak of heads after the nth flip.
    (a) Construct the one-step transition matrix **P** for this chain, on the state space $\{0, 1, 2, 3, 4\}$. What is special about state 4?
    (b) Let T_0 denote the total number of coin flips required by Michelle to achieve four heads in a row. Construct the cdf of T_0, $P(T_0 \leq k)$, for $k = 1, 2, \ldots, 15$. [*Hint:* The cdf values for $k = 1, 2, 3$ should be obvious.]
    (c) Michelle will win a prize if she can get four heads in a row within 10 coin flips. What is the probability she wins the prize?
    (d) Use (b) to construct an incomplete pmf of T_0. Then use this incomplete pmf to approximate both the mean and standard deviation of T_0.
    (e) What is the (exact) expected number of coin flips required for Michelle to get four heads in a row?

50. Refer back to Exercise 8. The article referenced in that exercise provides the following transition matrix for the states (1) current, (2) delinquent, (3) loss, and (4) paid, for a certain class of loans:

$$\mathbf{P} = \begin{bmatrix} .95 & .04 & 0 & .01 \\ .15 & .75 & .07 & .03 \\ 0 & 0 & 1 & 0 \\ 0 & 0 & 0 & 1 \end{bmatrix}$$

   (a) Identify the absorbing state(s).
   (b) Determine the mean time to absorption for a loan customer who is current on payments, and for a customer who is delinquent.
   (c) If a loan customer is current on payments, what is the probability s/he will eventually pay off the loan? What is the probability the loan company will suffer a loss on this account?
   (d) Answer (c) for customers who are delinquent on their loans.

51. Refer back to Exercise 15(c). Calculate and interpret the mean times to absorption for this chain. For which opening strategy, cooperative or competitive, is the negotiation process longer on the average?

52. Refer back to Exercise 14. Assuming Lucas begins searching for his uncle in room 1 and his uncle is hiding in room 6, what is the expected number of rooms Lucas will visit in order to "win" this round of hide-and-seek?

53. Modify the Gambler's Ruin example of this section to a $4 total stake. That is, Allan may start with $x_0 = \$1$, $\$2$, or $\$3$, and Beth has $\$(4 - x_0)$ initially. As usual, let p denote the probability Allan wins any single game.
    (a) Construct the one-step transition matrix.
    (b) Determine the mean times to absorption for each of Allan's possible starting values, as functions of p.
    (c) Determine the probability Allan eventually wins, starting with $1 or $2 or $3, as functions of p.

54. Refer back to the Ehrenfest chain model introduced in Exercise 2. Suppose there are $m = 3$ balls being exchanged between the two chambers. If the left chamber is currently empty, what is the expected number of exchanges until it is full (i.e., all 3 balls are on the left side)?

55. Refer back to Exercise 40. If a man has a low-prestige occupation, what is the expected number of generations required for him to have an offspring with a high-prestige occupation?

56. Exercise 48 of Chap. 2 established the following formula for the mean of a rv X whose possible values are positive integers:

$$E(X) = 1 + \sum_{x=1}^{\infty} [1 - F(x)],$$

where $F(x)$ is the cdf of X. Hence, if the values $F(1), F(2), \ldots, F(x^*)$ are known for some integer x^*, the mean of X can be approximated by $1 + \sum_{x=1}^{x^*} [1 - F(x)]$.

(a) Refer back to Example 6.26. Use the given cdf values and the above expression with $x^* = 12$ to approximate $E(T_1)$, the mean time to absorption starting in state 1. How does this compare to the pmf-based approximation in the example? How does it compare to the exact answer, 4.5?

(b) Repeat part (a), starting in state 4 of the bank loan Markov chain.

(c) Will this method always under-approximate the true mean of the rv, or can you tell? Explain.

[*Note:* It can be shown that this "cdf method" of approximating the mean will always produce a higher value than the truncated sum of $x \cdot p(x)$.]

57. This exercise outlines a proof of the formula $\mathbf{\Pi} = (\mathbf{I} - \mathbf{Q})^{-1}\mathbf{R}$ for the probabilities of eventual absorption. You will want to refer back to Eq. (6.8), as well as the proof of the MTTA Theorem.

(a) Starting in a non-absorbing state i, the chain will eventually be absorbed into absorbing state a if either (1) the chain transitions immediately into a, or (2) the chain transitions into any non-absorbing state and then eventually is absorbed into state a. Use this to explain why

$$\pi(i \to a) = P(i \to a) + \sum_{j \in A'} P(i \to j)\pi(j \to a),$$

where A' denotes the set of non-absorbing states of the chain.

(b) The equation in (a) holds for all $i \in A'$ and all $a \in A$. Show that this collection of equations can be rewritten in matrix form as $\mathbf{\Pi} = \mathbf{R} + \mathbf{Q}\mathbf{\Pi}$, and then solve for $\mathbf{\Pi}$. (You may assume the matrix $\mathbf{I} - \mathbf{Q}$ is invertible.)

58. The matrix $(\mathbf{I} - \mathbf{Q})^{-1}$ arises in several contexts in this section. This exercise provides an interpretation of its entries. Consider a Markov chain with at least one absorbing state, and assume that every non-absorbing state can access at least one absorbing state. As before, A and A' will denote the sets of absorbing and non-absorbing states, respectively.

(a) Consider any two non-absorbing states i and j. Let μ_{ij} denote the expected number of visits to state j, starting in state i, before the chain is absorbed. (When $j = i$, the initial state is counted as one visit.) Mimicking the proof of the MTTA Theorem, show that

$$\mu_{ii} = \sum_{a \in A} 1 \cdot P(i \to a) + \sum_{k \in A'} (1 + \mu_{ki}) \cdot P(i \to k)$$
$$= 1 + \sum_{k \in A'} \mu_{ki} \cdot P(i \to k)$$

(b) Using similar reasoning, show that for $i \neq j$,

$$\mu_{ij} = \sum_{k \in A'} \mu_{kj} P(i \to k)$$

(c) Let $\mathbf{M}$ be the square matrix whose (i, j)th entry is μ_{ij}. Combine (a) and (b) to establish the equation $\mathbf{M} = \mathbf{I} + \mathbf{Q}\mathbf{M}$, and solve for $\mathbf{M}$.

6.6   Simulation of Markov chains

A typical Markov chain simulation requires two elements: the one-step transition matrix, $\mathbf{P}$, and an indication of the initial state X_0 (either as a fixed state value or as a rv with a probability distribution). The actual simulation of a single realization of the chain $X_0, X_1, X_2, \ldots$ then amounts to repeated selections from the transitional probability distributions specified by elements of $\mathbf{P}$. Simulation of Markov chains allow us to confirm theoretical results and, more importantly, determine properties of Markov chains that are not covered by the theorems of this chapter or other theoretical results.

The main step in any Markov chain simulation is to simulate a value for the next step, X_{n+1}, based on the transition probabilities coming out of the current step X_n. Let's start with the initial state X_0. Suppose for one particular run of the simulation, X_0 has been assigned the state i, either because that's the fixed initial state or because a single draw from some initial distribution $\mathbf{v}_0$ yielded i. Conditional on $X_0 = i$, the distribution of X_1 is determined by the transition probabilities $P(i \to j)$ for $j = 1, 2, 3, \ldots$, which appear in the ith row of $\mathbf{P}$. Thus, one needs to extract the ith row of $\mathbf{P}$ and use it as the basis for a single discrete simulation. If the result of this simulation is $X_1 = j$, then the jth row of $\mathbf{P}$ can be accessed to simulate X_2, and so on.

Example 6.33 Let's simulate a typical day in the life of our taxi driver. Although a real taxi driver does not have the same number of fares each day, for purposes of this first simulation we'll assume that he takes exactly 25 fares in 1 day.

Suppose first that the driver begins each day in a random zone X_0, as in Example 6.15, specifically with the initial distribution $p(1) = .2$, $p(2) = .5$, $p(3) = .3$. We

begin by simulating a single value from this initial distribution. Once that is
determined, our program should then simulate a single value of X_1 using the row
of **P** corresponding to the value of X_0, then do the same for X_2 based on the
simulated value of X_1, and so on. Figure 6.10 shows Matlab and R code for such
a simulation.

a

```
P=[.3 .2 .5; .1 .8 .1; .4 .4 .2];

states=[1 2 3];
v0=[.2 .5 .3];
X=randsample(states,1,true,v0);
current=X;
for i=1:25
    nextstate=
randsample(states,1,true,P(current,:));
    X=[X nextstate];
    current=nextstate;
end
```

b

```
P <- matrix(c(.3,.2,.5,.1,.8,.1,
.4,.4,.2),nrow=3,ncol=3,byrow=TRUE)
states <- c(1,2,3)
v0 <- c(.2,.5,.3)
X <- sample(states,1,TRUE,v0)
current <- X
for (i in 1:25){
    nextstate <-
sample(states,1,TRUE,P[current,])
    X <- c(X,nextstate)
    current <- nextstate
}
```

Fig. 6.10 Code for Example 6.33: (**a**) Matlab; (**b**) R

In Matlab, P(current,:) calls for the row of **P** specified by the numerical
index current; the code P[current,] performs the same task in R. The
output of both of these programs is a vector, X, containing the sequence of states
for the Markov chain (beginning with X_0). For example, one run of the above
program in R yielded the following output:

> X
[1] 2 1 3 1 2 2 2 2 1 1 1 1 3 1 3 3 1 3 2 2 2 2 3 3 1 3

The randomly selected initial state was $X_0 = 2$, followed by $X_1 = 1, X_2 = 3, \ldots$,
and finally $X_{25} = 3$. (The symbol [1] at the left is not the initial state; this is just R's
way of denoting the beginning of X.) If we weren't interested in the initial state
of the chain, the code could easily be modified not to store X_0, in which case
the indices of the output vector would match the time indices of the Markov chain
(i.e., the subscripts on $X_1, X_2, \ldots, X_{25}$).

To make X_0 a fixed initial state instead of a true random variable, one need
simply replace the two lines of code specifying the initial probability vector and the
first random selection. In the Matlab code, the third and fourth lines could be
replaced by the statement X=3; to fix the taxi driver's initial state as zone 3. A
similar comment applies to the R code. And, again, one could choose whether or not
to store the initial state as part of the output vector.   ∎

It is important not to confuse the *number of transitions of the chain* with the
number of runs of the simulation. In Example 6.33, both programs simulate the
chain through 25 transitions, but only a single run. If it's our desire to keep track of
the chain's behavior across many different runs, analogous to the simulations
described at the ends of Chaps. 1–4, then we must add an additional layer of
code, typically in the form of a surrounding "for" or "while" loop.

Example 6.34 As an illustration of the Steady-State Theorem, consider the model for Web users' browser histories discussed in Example 6.19 (refer back to that example to see the one-step transition matrix). Let's simulate the distribution of X_{75}, the Web site category of a user's 75th visited page. For variety's sake, suppose users are equally likely to start surfing the Web in any one of the five Web site categories; recall that the initial distribution of a regular chain will not affect its long-run behavior. The programs displayed in Fig. 6.11 perform 10,000 runs of simulating this Markov chain up through X_{75}. Purely to save space, the code to create **P** has been suppressed in Fig. 6.11, but it is very similar to what appears in Fig. 6.10.

In the fourth line of code, we have employed a shortcut version of the `randsample` and `sample` functions in Matlab and R, respectively, to randomly and uniformly select a single random integer from 1 to 5 (this is the initial state). Both programs store the state of the Markov chain after 75 transitions in the vector named X75 for each of 10,000 runs. (Notice that the intermediate states $X_1, \ldots, X_{74}$ are not permanently stored.)

The 10,000 simulated values of X_{75} from one execution of the Matlab program are summarized in the accompanying table, along with the steady-state probabilities for this chain determined in Sect. 6.4.

j	1	2	3	4	5
# of times	2822	1004	1599	3816	759
$\hat{P}(X_{75} = j)$	.2822	.1004	.1599	.3816	.0759
π_j	.2840	.0948	.1659	.3791	.0758

The estimated and theoretical steady-state probabilities are quite similar. Remember that these two rows of probabilities should differ slightly for *two* reasons: first, this is only a *simulation* of 10,000 values of the rv X_{75}, so there is natural simulation error; second, the steady-state probabilities indicate the behavior of X_n as $n \to \infty$, and we don't expect the rv X_{75} to have exactly this distribution (although it should be close).

a
```
P=not shown;
X75=zeros(10000,1);
for i=1:10000
    current=randsample(5,1);
    for j=1:75
        nextstate=
randsample(1:5,1,true,P(current,:));
        current=nextstate;
    end
    X75(i)=current;
end
```

b
```
P <- not shown
X75 <- NULL
for (i in 1:10000){
    current <- sample(5,1)
    for (j in 1:75){
        nextstate <-
sample(1:5,1,TRUE,P[current,])
        current <- nextstate
    }
    X75[i] <- current
}
```

Fig. 6.11 Code for Example 6.34: (**a**) Matlab; (**b**) R   ■

Section 6.6 introduced the notions of time to absorption and mean time to absorption for Markov chains with one or more absorbing states. We can also use simulation to explore properties of time-to-absorption variables and first-passage times.

Example 6.35 Consider again the bank loan application process described in Example 6.25, with lone absorbing state 6 (ultimate acceptance or rejection of the application), and the random variable $T_1 =$ time to absorption from state 1 (document initiation). To simulate the distribution of T_1, one begins the chain in state 1 and continues to simulate its transitions until it arrives in state 6. The simulation program now must keep track of how many transitions occur, rather than just where the chain ends up. Figure 6.12 shows Matlab and R code for this purpose; again, to save space, the code for entering the matrix **P** is not included.

a

```
P=not shown;
T1=zeros(10000,1);
  for i=1:10000
      current=1;
      steps=0;
      while curret~=6
          steps=steps+1;
          nextstate=
randsample(1:6,1,true,P(current,:));
          current=nextstate;
      end
      T1(i)=steps;
  end
```

b

```
P <- not shown
T1 <- NULL
for (i in 1:10000){
    current <- 1
    steps <- 0
    while (current!=6){
        steps <- steps+1
        nextstate <-
sample(1:6,1,TRUE,P[current,])
        current <- nextstate
    }
    T1[i] <- steps
}
```

Fig. 6.12 Code for Example 6.35: (**a**) Matlab; (**b**) R

The simulated distribution of T_1 from one execution of the R code in Fig. 6.12b appears in Fig. 6.13. These particular 10,000 simulated values had a sample mean of 4.508 and a sample standard deviation of 2.276. Notice the sample mean is quite close to the theoretical expectation, $E(T_1) = 4.5$, determined in Sect. 6.5.

Clearly, the sample mean of the simulated T_1 values is a better estimate of $E(T_1)$ than the approach utilizing the truncated pmf presented in the previous section. Of course, neither is strictly necessary since the mean of T_1 can be found explicitly using the MTTA Theorem. The new information provided by the simulation is a measure of the *variability* of T_1: we estimate the standard deviation of T_1 to be 2.276, whereas no simple matrix formula exists for its theoretical standard deviation.   ■

The preceding examples employed simulations primarily to confirm theoretical results established in earlier sections. (Or, perhaps better put, our earlier theoretical results validate the simulations!) The final two examples of this section indicate situations where we must rely on simulation methods to approximate values of desired quantities.

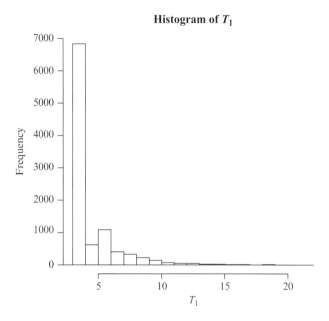

Fig. 6.13 Simulation distribution of T_1 in Example 6.35

Example 6.36 Refer back to Example 6.13, which described Chinese cell phone users' transitions between three major carriers. Suppose users may renew or change contracts annually, and that annual plans for the three carriers cost the following (in $US): 550 for China Telecom, 600 for China Unicom, and 525 for China Mobile. Assume that, because of governmental regulations, these prices will remain the same for the next 10 years. If last year the market shares of the three carriers were .4, .2, and .4, respectively and all contracts are about to come up for renewal, what is the average amount a Chinese cell phone customer will pay over the next decade?

We will employ a Markov chain simulation to model the behavior of customers' carrier choices for 10 consecutive years. Critically, we must keep track of how much money a customer spends each year—that is, our three states now have associated quantitative values. (This is a common instance where simulation proves useful.) Let $Y =$ the total cost of ten 1-year calling plans for a Chinese cell phone customer. Figure 6.14 shows code for simulating Y using the techniques of this section.

An initial state x0 is first determined using the specified initial probability distribution (here, $\mathbf{v}_0 = [.4 \ .2 \ .4]$). Then, ten steps of the Markov chain are simulated; each of these states is temporarily held in nextstate. The vector AnnualCost stores the cost of a 1-year calling plan by calling the appropriate element of the Prices vector. Once a 10-year chain has been simulated, those annual costs are summed and stored as a simulated value of Y.

a
```
P=[.84 .06 .1;.08 .82 .1;.1 .04 .86];

Prices=[550 600 525];
Y=zeros(10000,1);
for i=1:10000
    v0=[.4 .2 .4];
    AnnualCost=zeros(10,1);
    x0=randsample(1:3,1,true,v0);
    current=x0;
    for n=1:10
        nextstate=
randsample(1:3,1,true,P(current,:));
        AnnualCost(n)=
Prices(nextstate);
        current=nextstate;
    end
    Y(i)=sum(AnnualCost);
end
```

b
```
P <- matrix(c(.84,.06,.1,.08,.82,.1,
.1,.04,.86),nrow=3,ncol=3,byrow=TRUE)
Prices <- c(550,600,525)
Y <- NULL
for (i in 1:10000){
    v0 <- c(.4,.2,.4)
    AnnualCost <- NULL
    x0 <- sample(1:3,1,TRUE,v0)
    current <- x0
    for (n in 1:10){
        nextstate <-
sample(1:3,1,TRUE,P[current,])
        AnnualCost[n] <-
Prices[nextstate];
        current <- nextstate
    }
    Y[i]=sum(AnnualCost)
}
```

Fig. 6.14 Code for Example 6.36: (**a**) Matlab; (**b**) R

A histogram of the 10,000 simulated Y values appears in Fig. 6.15. Notice that the distribution of Y has three spikes, at \$5250, \$5500, and \$6000. These correspond to customers who keep the same carrier all 10 years; the large probabilities along the main diagonal of the transition matrix indicate reasonably strong customer loyalty. For this particular run, the simulated values of Y had a sample mean and standard deviation of \$5503.40 and \$199.32, respectively, from which we can be 95% confident, using the methods of Sect. 5.3, that μ_Y is between \$5499.49 and \$5507.31.

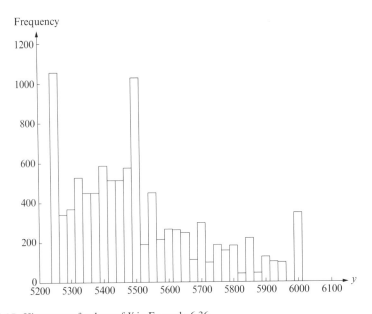

Fig. 6.15 Histogram of values of Y in Example 6.36   ∎

Example 6.37 Our taxi driver now makes one last appearance (hopefully to applause). He starts each morning at home in zone 3. Methods from Sects. 6.4 and 6.5 allow us to determine the expected number of fares required for him to return home, or to reach one of the other two zones. But how long does it take him, in the typical day, to visit *all three zones*? Let

T_{all} = number of transitions required to visit every state at least once (not counting the initial state, X_0)

To simulate T_{all}, our program must keep track of which states have been visited thus far. Once all states/zones have been visited, the numerical value of T_{all} for that simulation run can be recorded. Figure 6.16 shows appropriate code.

a
```
P =[.3 .2 .5;.1 .8 .1;.4,.4,.2];

Tall=zeros(10000,1);
for i=1:10000
    current=3;
    visits=[0 0 0];
    Talltemp=0;
    while (sum(visits)<3)
        nextstate=randsample(1:3,1,
                true,P(current,:));
        visits(nextstate)=1;
        current=nextstate;
        Talltemp=Talltemp+1;
    end
    Tall(i)=Talltemp;
end
```

b
```
P <- matrix(c(.3,.2,.5,.1,.8,.1,
.4,.4,.2),nrow=3,ncol=3,byrow=TRUE)
Tall <- NULL
for (i in 1:10000){
    current <- 3
    visits <- c(0,0,0)
    Talltemp <- 0
    while (sum(visits)<3){
        nextstate <-
        sample(1:3,1,TRUE,P[current,])
        visits[nextstate] <- 1
        current <- nextstate
        Talltemp <- Talltemp+1
    }
    Tall[i] <- Talltemp
}
```

Fig. 6.16 Code for Example 6.37: (**a**) Matlab; (**b**) R

In both programs, a vector called `visits` keeps a record of which states the chain visits within that particular run. When chain j is visited ($j = 1, 2, 3$), the jth entry of `visits` switches from 0 to 1. Once all three entries of `visits` equal 1, as detected by its sum, the while loop terminates and the temporary count of transitions (`Talltemp`) is stored in `Tall`. The result of the program is 10,000 simulated values of the rv T_{all}, stored in the vector `Tall`.

Figure 6.17 displays a histogram of the 10,000 values resulting from running the Matlab program in Fig. 6.16a. The sample mean and standard deviation of these 10,000 values were $\bar{x} = 8.1674$ and $s = 5.8423$. Hence, we estimate the average number of fares required for the taxi driver to visit all three zones to be 8.1674, with an estimated standard error of $s/\sqrt{n} = 5.8423/\sqrt{10,000} = 0.058423$. Using the techniques of Chap. 5, we can say with 95% confidence that μ_{all}, the true mean of T_{all}, lies in the interval

$$\bar{x} \pm 1.96 \frac{s}{\sqrt{n}} = 8.1674 \pm 1.96(0.058423) = (8.053, 8.282)$$

Fig. 6.17 Simulation distribution of the rv T_{all} in Example 6.37

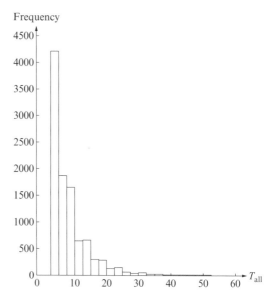

Among the 10,000 simulated values of T_{all}, 4204 were at most 5 (so, 3 or 4 or 5). Hence, the estimated probability that the taxi driver visits all three zones within his first five fares is

$$\hat{p} = \hat{P}\,(T_{all} \leq 5) = \frac{4204}{10,000} = .4204$$

The estimated standard error of this estimate is given by $\sqrt{\hat{p}\,(1 - \hat{p}\,)/n} = .0049$. Hence we are 95% confident that the true probability $P(T_{all} \leq 5)$ lies in the interval $.4204 \pm 1.96(.0049) = (.4108, .4300)$. ■

6.6.1 Exercises: Section 6.6 (59–66)

59. Refer back to Exercise 3. Suppose this machine produces 150 units on days when it is fully operational, 75 units per day when partially operational, and 0 units when broken. Consider a month with 20 work days, and assume the machine ended the previous month fully operational.

   (a) Write a simulation of the rv Y = the number of units produced by this machine in 20 work days.

   (b) Create a histogram of simulated values of Y for at least 10,000 simulation runs.

   (c) Construct a 95% confidence interval for the mean number of units produced by this machine across 20 work days.

   (d) Construct a 95% confidence interval for the probability that the machine produces at least 2000 units in such a month.

60. Four friends A, B, C, and D are notorious for sharing rumors amongst themselves. Being very gossipy but not particularly bright, each friend is equally likely to share a rumor with any of the other three friends, even if that friend has already heard it. (For example, if friend B most recently heard the rumor, each of friends A, C, and D is equally likely to hear it next, regardless of how B came to hear the rumor!) Let X_n = the nth person in this foursome to hear a particular rumor.
   (a) Construct the one-step transition matrix for this Markov chain.
   (b) Friend A has just overheard a particularly nasty rumor about a classmate and is eager to share it with the other three friends. Let T equal the number of times the rumor is repeated within the foursome until all of them have heard the rumor. Write a program to simulate T, and use your program to estimate $E(T)$.

61. A state lottery official has proposed the following system for a new game. In the first week of a new year, a $10 million prize is available. If nobody gets the winning lottery numbers correct and wins the prize that week, the value doubles to $20 million for the second week; otherwise, the prize for the second week is also $10 million. Each week, the prize value doubles if nobody wins it and returns to $10 million otherwise. Suppose that there is a 40% chance that someone in the state wins the lottery prize each week, irrespective of the current value of the prize. Let X_n = the value of the lottery prize in the nth week of the year.
   (a) Determine the one-step transition probabilities for this chain. [*Hint*: Given the value of X_n, X_{n+1} can only be one of two possible dollar amounts.]
   (b) Let M be the maximum value the lottery prize achieves over the course of a 52-week year. Simulate at least 10,000 values of the rv M, and report the sample mean and SD of these simulated values. [*Hint*: Given the large state space of this Markov chain, don't attempt to construct the transition matrix. Instead, code the probabilities in (a) directly.]
   (c) Let Y be the total amount paid out by the lottery in a 52-week year. Simulate at least 10,000 values of the rv Y, and report a 95% confidence interval for $E(Y)$.
   (d) Repeat (c), but now assume the probability of a winner is .7 each week rather than .4. Should the lottery commission make it easier or harder for someone to win each week? Explain.

62. Write a Markov chain simulation program with the following specifications. The inputs should be the transition matrix **P**, an initial state x_0, and the number of steps n. The output should be a single realization of $X_1, X_2, \ldots, X_n$, as either a row vector or a column vector.

63. Refer back to Exercise 12. Suppose that the typical annual premium for a category 1 (safest) customer is $500; for category 2, $600; for category 3, $1000; and for category 4 (riskiest driver), $1500.
   (a) Use a Markov chain simulation to estimate the distribution of the rv Y_1 = total premium paid by a customer over 10 years with the insurance company, assuming s/he starts in category 1. Create a histogram of values for Y_1, and construct a 95% confidence interval for $E(Y_1)$.
   (b) Repeat (a) assuming instead that the customer starts as a category 3 driver.

64. Write a simulation program for Gambler's Ruin. The inputs should be $a =$ Allan's initial stake, $b =$ Beth's initial stake (so $a+b$ is the total stake), $p =$ the probability Allan defeats Beth in any single game, and $N =$ the number of tournaments to be simulated. The program should output two N-by-1 vectors: one recording the number of games played for each of the N runs, and one indicating who won each time. Use your program to determine (a) the average tournament length and (b) the probability Allan eventually wins for the settings $a = b = \$5$ and $p = .4$. Give 95% confidence intervals for both answers.

65. Example 6.3 describes a (one-dimensional) random walk. This is sometimes called a *simple random walk*.

   (a) Write a program to simulate the first 100 steps of a random walk starting at $X_0 = 0$. [*Hint:* If $X_n = s$, then $X_{n+1} = s \pm 1$ with probability 1/2 each.]
   (b) Run your program in (a) 10,000 times, and use the results to estimate the probability that a random walk returns to its origin at any time within the first 100 steps.
   (c) Let $R_0 =$ the number of returns to the origin in the first 100 steps of the random walk, not counting its initial state. Use your simulation to (1) create a histogram of simulated values of R_0 and (2) construct a 95% confidence interval for $E(R_0)$.

66. A *two-dimensional random walk* is a model for movement along the integer lattice in the xy-plane, i.e., points (x, y) where x and y are both integers. The "walk" begins at $X_0 = (0, 0)$. At each time step, a move is made one unit left or right (probability 1/2 each) and, independently, one unit up or down (also equally likely). If $X_n =$ the (x, y)-coordinates of the chain after n steps, then X_n is a Markov chain.

   (a) Write a program to simulate the first 100 steps of a two-dimensional random walk. [*Hint:* The x- and y-coordinates of a two-dimensional random walk are each simple random walks. Since they are independent, the x- and y-movements can be simulated separately.]
   (b) Use your program in (a) to estimate the probability that a two-dimensional random walk returns to its origin within the first 100 steps. Use at least 10,000 runs.
   (c) Use your program in (a) to estimate $E(R_0)$, where $R_0 =$ the number of times the walk returns to $(0, 0)$ in the first 100 steps.

6.7   Supplementary Exercises (67–82)

67. A hamster is placed into the six-chambered circular habitat shown in the accompanying figure. Sitting in any chamber, the hamster is equally likely to next visit either of the two adjacent chambers. Let $X_n =$ the nth chamber visited by the hamster.

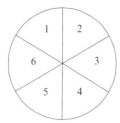

(a) Construct the one-step transition matrix for this Markov chain.
(b) Is this a regular Markov chain?
(c) Intuitively, what should the stationary probabilities of this chain be? Verify these are indeed its stationary probabilities.
(d) Given that the hamster is currently in chamber 3, what is the expected number of transitions it will make until it returns to chamber 3?
(e) Given that the hamster is currently in chamber 3, what is the expected number of transitions it will make until it arrives in chamber 6?

68. Teenager Mike wants to borrow the car. He can ask either parent for permission to take the car. If he asks his mom, there is a 20% chance she will say "yes," a 30% chance she will say "no," and a 50% chance she will say, "ask your father." Similarly, the chances of hearing "yes"/"no"/"ask your mother" from his dad are .1, .2, and .7, respectively. Imagine Mike's efforts can be modeled as a Markov chain with states (1) talk to Mom, (2) talk to Dad, (3) get the car ("yes"), (4) strike out ("no"). Assume that once either parent has said "yes" or "no," Mike's begging is done.
   (a) Construct the one-step transition matrix for this Markov chain.
   (b) Identify the absorbing state(s) of the chain.
   (c) Determine the mean times to absorption.
   (d) Determine the probability that Mike will eventually get the car if (1) he asks Mom first and (2) he asks Dad first. Whom should he ask first?

69. Refer back to Exercise 14. Suppose Lucas starts in room 1 and proceeds as described in that exercise; however, his mean-spirited uncle has snuck out of the house entirely, leaving Lucas to search interminably. So, in particular, if Lucas enters room 6 of the house, his next visit will necessarily be to room 5. (This really happened one summer!)
   (a) Determine the transition matrix for this chain.
   (b) Verify that this Markov chain is regular.
   (c) Determine the steady-state probabilities of this chain.
   (d) What proportion of time in the long run does Lucas spend in room 2?
   (e) What is the average number of room transitions between Lucas' visits to room 1?

70. Refer back to Exercises 20 and 21.
   (a) Suppose all four vans were operational as of Monday morning. What is the expected backlog—that is, the expected number of vans needing repair—as of Friday evening?

(b) Suppose instead that two of the four vans were down for repairs Monday morning. Now what is the expected backlog as of Friday evening?

71. Five Mercedes E550 vehicles are shipped to a local dealership. The dealer sells one E550 in any week with probability .3 and otherwise sells none in that week. When all E550s in stock have been sold, the dealer requests a new shipment of five such cars, and it takes 1 week for that delivery to occur. Let $X_n =$ the number of Mercedes E550s at this dealership n weeks after the initial delivery of five cars.

   (a) Construct the transition matrix for this chain. [*Hint*: The states are 0, 1, 2, 3, 4, 5.]
   (b) Determine the steady-state probabilities for this chain.
   (c) On the average, how many weeks separate successive orders of five E550s?

72. Refer back to the previous exercise. Let $m =$ the number of Mercedes E550s delivered to the dealership at one time (both initially and subsequently), and let $p =$ the probability an E550 is sold in any particular week ($m = 5$ and $p = .3$ in the previous exercise). Determine the steady-state probabilities for this chain and then the average number of weeks between vehicle orders.

73. Sports teams can have long streaks of winning (or losing) seasons, but occasionally a team's fortunes change quickly. Suppose that each team in the population of all college football teams can be classified as (1) weak, (2) medium, or (3) strong, and that the following one-step transition probabilities apply to the Markov chain $X_n =$ a team's strength n seasons from now:

$$\mathbf{P} = \begin{bmatrix} .8 & .2 & 0 \\ .2 & .6 & .2 \\ .1 & .2 & .7 \end{bmatrix}$$

   (a) If a college football team is weak this season, what is the minimum number of seasons required for it to become strong?
   (b) If a team is strong this season, what is the probability it will also be strong four seasons from now?
   (c) What is the average number of seasons that must pass for a weak team to become a strong team?
   (d) What is the average number of seasons that must pass for a strong team to become a weak team?

74. Jay and Carol enjoy playing tennis against each other. Suppose we begin watching them when they are at *deuce*. This means the next person to win a point earns *advantage*. If that same person scores the next point, then s/he wins the game; otherwise, the game returns to deuce.

   (a) Construct a transition matrix to describe the status of the game after n points have been scored (starting at deuce). [*Hint*: There are five states:

(1) Jay wins, (2) advantage Jay, (3) deuce, (4) advantage Carol, (5) Carol wins.]

(b) Suppose Carol is somewhat better than Jay and has a 60% chance of winning any particular point. Determine (1) the probability Carol eventually wins and (2) the expected number of points to be played, starting at deuce. [*Hint*: This should bear surprising similarity to a game played earlier in the chapter by Allan and Beth!]

75. The authors of the article "Pavement Performance Modeling Using Markov Chain" (*Proc. ISEUSAM*, 2012: 619–627) developed a system for classifying pavement segments into five categories: (1) Very good, (2) Good, (3) Fair, (4) Bad, and (5) Very bad. Analysis of pavement samples led to the construction of the following transition matrix for the Markov chain X_n = pavement condition n years from now:

$$\begin{bmatrix} .958 & .042 & 0 & 0 & 0 \\ 0 & .625 & .375 & 0 & 0 \\ 0 & 0 & .797 & .203 & 0 \\ 0 & 0 & 0 & .766 & .234 \\ 0 & 0 & 0 & 0 & 1 \end{bmatrix}$$

Notice that a pavement segment either maintains its condition or goes down by one category each year.

(a) The evaluation of one particular stretch of road led to the following initial probability vector (what the authors call a "condition matrix"): [.3307 .2677 .2205 .1260 .0551]. Use the Markov chain model to determine the condition matrix of this same road section 1, 2, and 3 years from now.

(b) "Very bad" road segments require repairs before they are again usable; the authors' model applies to unrepaired road. What is the average time (number of years) that a very good road can be used before it degrades into very bad condition? Make the same determination for good, fair, and bad roads.

(c) Suppose one road segment is randomly selected from the area to which the condition matrix in (a) applies. What is the expected amount of time until this road segment becomes very bad? [*Hint*: Use the results of part (b).]

76. A *constructive memory agent* (CMA) is an autonomous software unit that uses its interactions not only to change its data ("memory") but also its fundamental indexing systems for that data ("structure"). The article "Understanding Behaviors of a Constructive Memory Agent: A Markov Chain Analysis" (*Knowledge-Based Systems*, 2009: 610–621) describes a study of one such CMA as it moved between nine different stages of learning. (Stage 1 is sensation and perception; later stages add on other behaviors such as hypothesizing, neural network activation, and validation. Consult the article for details.) The accompanying state diagram mirrors the one given in the article for the authors' first experiment.

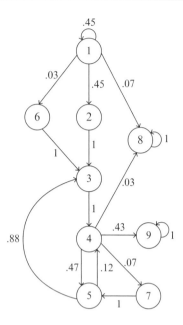

(a) Construct the transition matrix for this chain.
(b) What are the absorbing states of the chain?
(c) All CMA processes begin in stage 1. What is the mean time to absorption for such a process? Here, "time" refers to the number of transitions from one learning stage to another. [*Note*: In this particular experiment, absorbing states correspond to any instance of so-called "inductive" learning.]
(d) Starting in stage 1, what is the probability a CMA will end the experiment in state 8 (constructive learning plus inductive learning)?

77. The authors of the article "Stationarity of the Transition Probabilities in the Markov Chain Formulation of Owner Payments on Projects" (*ANZIAM J.*, v. 53, 2012: C69-C89) studied payment delays in road construction in Australia. States for any payment were defined as follows: k weeks late for $k = 0, 1, 2, 3$; paid (pd), an absorbing state; and "to be resolved" (tbr), meaning the payment was at least 1 month late, which the authors treated as another absorbing state. For one particular project, the following $\mathbf{Q}$ and $\mathbf{R}$ matrices were given for the canonical form of the one-step transition matrix:

$$\mathbf{Q} = \begin{matrix} 0 \\ 1 \\ 2 \\ 3 \end{matrix} \begin{bmatrix} 0 & 1 & 0 & 0 \\ 0 & 0 & .959 & 0 \\ 0 & 0 & 0 & .897 \\ 0 & 0 & 0 & 0 \end{bmatrix} \qquad \mathbf{R} = \begin{matrix} \\ 0 \\ 1 \\ 2 \\ 3 \end{matrix} \begin{matrix} \text{pd} \quad\; \text{tbr} \\ \begin{bmatrix} 0 & 0 \\ .041 & 0 \\ .013 & 0 \\ .804 & .196 \end{bmatrix} \end{matrix}$$

(a) Construct the complete 6×6 transition matrix $\mathbf{P}$ for this Markov chain.

(b) Draw the state diagram of this Markov chain.

(c) Determine the mean time to absorption for payment that is about to come due (i.e., one that is presently 0 weeks late).

(d) What is the probability a payment is eventually made, as opposed to being classified as "to be resolved"?

(e) Consider the two probabilities $P(0 \to 1)$ and $P(3 \to \text{pd})$. What is odd about each of these values? (The authors of the article offer no explanation for the irregularity of these two particular probabilities.)

78. In a *nonhomogeneous* Markov chain, the conditional distribution of X_{n+1} depends on both the previous state X_n and the current time index n. As an example, consider the following method for randomly assigning subjects one at a time to either of two treatment groups, A or B. If n patients have been assigned a group so far, and a of them have been assigned to treatment A, the probability the next patient is assigned to treatment group A is

$$P\big((n+1)\text{st patient assigned to } A \mid a \text{ out of first } n \text{ in } A\big) = \frac{n - a + 1}{n + 2}$$

Hence, the first patient is assigned to A with probability $(0 - 0 + 1)/(0 + 2) = 1/2$; if the first patient was assigned to A, then the second patient is also assigned to A with probability $(1 - 1 + 1)/(1 + 2) = 1/3$. This assignment protocol ensures that each next patient is more likely to be assigned to the smaller group. Let $X_n = $ the number of patients in treatment group A after n total patients have been assigned $(X_0 = 0)$. To simplify matters, assume there are only 4 patients in total to be randomly assigned.

(a) Let $\mathbf{P}_1$ denote the transition matrix from $n = 0$ to $n = 1$. Assume the state space of the chain is $\{0, 1, 2, 3, 4\}$. Construct $\mathbf{P}_1$. [*Hint*: Since $X_0 = 0$, only the first row of $\mathbf{P}_1$ is really relevant. To make this a valid transition matrix, treat the "impossible" states 1, 2, 3, and 4 as absorbing states.]

(b) Construct $\mathbf{P}_2$, the transition matrix from $n = 1$ to $n = 2$. Use the same hint as above for states 2, 3, and 4, which are impossible at time $n = 1$.

(c) Following the pattern of (a) and (b), construct the matrices $\mathbf{P}_3$ and $\mathbf{P}_4$.

(d) For a nonhomogeneous chain, the multistep transition probabilities can be calculated by multiplying the aforementioned matrices from left to right, e.g., the 4-step transition matrix for this chain is $\mathbf{P}_1\mathbf{P}_2\mathbf{P}_3\mathbf{P}_4$. Calculate this matrix, and then use its first row to determine the likelihoods of 0, 1, 2, 3, and 4 patients being randomly assigned to treatment group A using this method.

[*Note*: Random assignment strategies of this type were originally investigated in the article "Forcing a Sequential Experiment to be Balanced," *Biometrika* (1971): 403-417.]

79. A communication channel consists of five relays through which all messages must pass. Suppose that bit switching errors of either kind (0 to 1, or 1 to 0) occur with probability .02 at the first relay. The corresponding probabilities for the other four relays are .03, .02, .01, and .01, respectively. If we define

X_n = the parity of a bit after traversing the nth relay, then $X_0, X_1, \ldots, X_5$ forms a nonhomogeneous Markov chain.

(a) Determine the one-step transition matrices $\mathbf{P}_1$, $\mathbf{P}_2$, $\mathbf{P}_3$, $\mathbf{P}_4$, and $\mathbf{P}_5$.

(b) What is the probability that a 0 bit entering the communication relay system also exits as a 0 bit? [*Hint*: Refer back to the previous exercise for information on nonhomogeneous Markov chains.]

80. Consider the two-state Markov chain described in Exercise 39, whose one-step transition matrix is given by

$$\begin{array}{cc} & \begin{array}{cc} 0 & \quad 1 \end{array} \\ \begin{array}{c} 0 \\ 1 \end{array} & \begin{bmatrix} 1 - \alpha & \alpha \\ \beta & 1 - \beta \end{bmatrix} \end{array}$$

for some $0 < \alpha$, $\beta < 1$. Use mathematical induction to show that the k-step transition probabilities are given by

$$P^{(k)}(0 \to 0) = \delta^k + \left(1 - \pi\right)\left(1 - \delta^k\right) \qquad P^{(k)}(0 \to 1) = \pi\left(1 - \delta^k\right)$$
$$P^{(k)}(1 \to 0) = \left(1 - \pi\right)\left(1 - \delta^k\right) \qquad P^{(k)}(1 \to 1) = \delta^k + \pi\left(1 - \delta^k\right)$$

where $\pi = \alpha/(\alpha + \beta)$ and $\delta = 1 - \alpha - \beta$. [*Note*: Applications of these multistep probabilities are discussed in "Epigenetic Inheritance and the Missing Heritability Problem," *Genetics*, July 2009: 845-850.]

81. A 2012 report ("A Markov Chain Model of Land Use Change in the Twin Cities, 1958-2005," available online) provided a detailed analysis from maps of Minneapolis-St. Paul, MN over the past half-century. The Twin Cities area was divided into 610,988 "cells," and each cell was classified into one of ten categories: (1) airports, (2) commercial, (3) highway, (4) industrial, (5) parks, (6) public, (7) railroads, (8) residential, (9) vacant, (10) water. The report's authors found that X_n = classification of a randomly selected cell was well modeled by a time-homogeneous Markov chain when a time increment of about 8 years is employed. The accompanying matrix shows the one-step transition probabilities from 1997 (n) to 2005 ($n + 1$); rows and columns are in the same order as the sequence of states described above.

$$\begin{bmatrix}
.7388 & .0010 & .0068 & .0010 & .0325 & .0131 & .0000 & .0055 & .1984 & .0029 \\
.0001 & .8186 & .0201 & .0560 & .0045 & .0227 & .0002 & .0413 & .0350 & .0015 \\
.0004 & .0107 & .9544 & .0054 & .0058 & .0031 & .0002 & .0094 & .0105 & .0001 \\
.0004 & .0710 & .0099 & .8371 & .0082 & .0086 & .0011 & .0106 & .0517 & .0014 \\
.0022 & .0036 & .0031 & .0025 & .9128 & .0062 & .0002 & .0116 & .0364 & .0214 \\
.0001 & .0193 & .0100 & .0384 & .0569 & .7364 & .0004 & .0223 & .1091 & .0071 \\
.0000 & .0065 & .0142 & .0201 & .0110 & .0032 & .9139 & .0168 & .0130 & .0013 \\
.0000 & .0024 & .0024 & .0009 & .0041 & .0023 & .0002 & .9634 & .0230 & .0013 \\
.0004 & .0141 & .0099 & .0156 & .0513 & .0057 & .0002 & .0988 & .7920 & .0120 \\
.0001 & .0010 & .0003 & .0014 & .0136 & .0001 & .0000 & .0055 & .0096 & .9684
\end{bmatrix}$$

(a) In 2005, the distribution of cell categories (out of the 610,988 total cells) was as follows:

[4047 20,296 16,635 24,503 74,251 18,820 1505 195,934 200,837 54,160]

The order of the counts matches the category order above, e.g., 4047 cells were part of airports and 54,160 cells were located on water. Use the transition probabilities to predict the land use distribution of the Twin Cities region in 2013.

(b) Determine the predicted land use distribution for the years 2021 and 2029 (remember, each time step of the Markov chain is 8 years). Then determine the percent change from 1995 to 2029 in each of the ten categories (similar computations were made in the cited report).

(c) Though it's unlikely that land use evolution will remain the same forever, imagine that the one-step probabilities can be applied in perpetuity. What is the projected long-run land use distribution in Minneapolis-St. Paul?

82. In the article "Reaching a Consensus" (*J. Amer. Stat. Assoc.*, 1974: 118-121), Morris DeGroot considers the following situation: s statisticians must reach an agreement about an unknown population distribution, F. (The same method, he argues, could be applied to opinions about the numerical value of a parameter, as well as many nonstatistical scenarios.) Let $F_{10}, \ldots, F_{s0}$ represent their initial opinions. Each statistician then revises his belief about F as follows: the ith individual assigns a "weight" p_{ij} to the opinion of the jth statistician ($j = 1, \ldots s$), where $p_{ij} \geq 0$ and $p_{i1} + \cdots + p_{is} = 1$. He then updates his own belief about F to

$$F_{i1} = p_{i1}F_{10} + \cdots + p_{is}F_{s0}$$

This updating is performed simultaneously by all s statisticians (so, $i = 1, 2, \ldots, s$).

(a) Let $\mathbf{F}_0 = (F_{10}, \ldots, F_{s0})^\mathsf{T}$, and let $\mathbf{P}$ be the $s \times s$ matrix with entries p_{ij}. Show that the vector of updated opinions $\mathbf{F}_1 = (F_{11}, \ldots, F_{s1})^\mathsf{T}$ is given by $\mathbf{F}_1 = \mathbf{P}\mathbf{F}_0$.

(b) DeGroot assumes that updates to the statisticians' beliefs continue iteratively, but that the weights do not change over time (so, $\mathbf{P}$ remains the same). Let $\mathbf{F}_n$ denote the vector of opinions after n updates. Show that $\mathbf{F}_n = \mathbf{P}^n \mathbf{F}_0$.

(c) The group is said to *reach a consensus* if the limit of $\mathbf{F}_n$ exists as $n \to \infty$ *and* each entry of that limit vector is the same (so all individuals' opinions converge toward the same belief). What would be a sufficient condition on the weights in $\mathbf{P}$ for the group to reach a consensus?

(d) DeGroot specifically considers four possible weight matrices:

$$\mathbf{P}_A = \begin{bmatrix} \frac{1}{2} & \frac{1}{2} \\ \frac{1}{4} & \frac{3}{4} \end{bmatrix} \qquad \mathbf{P}_B = \begin{bmatrix} \frac{1}{2} & \frac{1}{2} & 0 \\ \frac{1}{4} & \frac{3}{4} & 0 \\ \frac{1}{3} & \frac{1}{3} & \frac{1}{3} \end{bmatrix} \qquad \mathbf{P}_C = \begin{bmatrix} 1 & 0 \\ 0 & 1 \end{bmatrix} \qquad \mathbf{P}_D = \begin{bmatrix} \frac{1}{2} & \frac{1}{2} & 0 & 0 \\ \frac{1}{2} & \frac{1}{2} & 0 & 0 \\ 0 & 0 & \frac{1}{2} & \frac{1}{2} \\ 0 & 0 & \frac{1}{2} & \frac{1}{2} \end{bmatrix}$$

Discuss what each one indicates about the statisticians' views on each other, and determine for which matrices the group ultimately reaches a consensus. If a consensus is reached, write out the consensus "answer" as a linear combination of $F_{10}, \ldots, F_{s0}$.

Random Processes

7

In Chap. 1, we introduced the concept of a random event: a collection of one or more outcomes resulting from a random experiment (e.g., a randomly selected device works for 1000 h, or a randomly selected person has brown hair). In Chaps. 2–4, we studied random variables: numerical values resulting from random experiments (the number of flaws on a randomly selected wafer, the number of wins in 5 games of chance, the mass of a randomly selected object). In this chapter, we look at *random processes*, also called *stochastic processes* ("stochastic" is a synonym for "random"): time-dependent functions resulting from random phenomena.

For example, consider modeling the number of people logged into a particular server over the course of the day. Since the exact times at which individuals log in are generally unpredictable, we might reasonably apply a model which treats logins as "random." In particular, at any specific, fixed point in time—say, noon—we can model the number of people logged in by an appropriate (discrete) random variable. The new concept in Chap. 7 is to model the evolution of that random count over time. This gives us two dimensions of interest: the random variable itself (here, the count of logins) and time.

Among the most common applications of random processes in engineering is that of *random noise*, a term for the disparity between what a received signal should "ideally" look like and what actually arrives at the receiver. Our ability to accurately model this distortion or noise will enable us to filter out some (hopefully large) proportion of that noise, thereby recovering a cleaner signal.

In Sect. 7.1, we look at classifications of random processes according to whether the variable dimension and/or the time dimension are modeled as discrete or continuous. In Sect. 7.2, we connect previous ideas of mean, standard deviation, and so on to this new world of random processes. Section 7.3 introduces the concept of a *stationary* random process and the special class of *wide-sense stationary processes*; these will be the backbone of signal processing in Chap. 8. Sections 7.4–7.7 consider several specific classes of random processes: discrete-time, Poisson, Gaussian, and continuous-time Markov.

M.A. Carlton and J.L. Devore, *Probability with Applications in Engineering, Science, and Technology*, Springer Texts in Statistics, DOI 10.1007/978-1-4939-0395-5_7, © Springer Science+Business Media New York 2014

7.1   Types of Random Processes

In Chap. 2, we defined a random variable as a rule that associates a number with each outcome in the sample space of some experiment. For example, we may associate with each outcome of the experiment of rolling two dice an integer X between 2 and 12 indicating the sum of the two up-facing sides. Any single realization of this experiment results in a specific number, a sample value of X. We define random processes analogously.

> **DEFINITION**
> For a given sample space $\mathscr{S}$ of some experiment, a **random process** is any rule that associates a time-dependent function with each outcome in $\mathscr{S}$. Any such function that may result is a **sample function** of the random process. The collection of all possible sample functions is called the **ensemble** of the random process.

Figure 7.1 illustrates this definition. Analogous to our notation for random variables, we will denote a (continuous-time) random process by $X(t)$, while the lower-case $x(t)$ indicates a particular sample function.

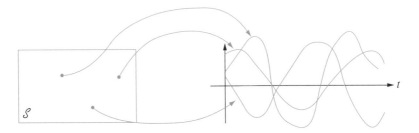

Fig. 7.1 A random process

Example 7.1 Some communication systems use *phase-shift keying* to transmit information. A quaternary phase-shift keying (QPSK) system can transmit four distinct symbols (often used to encode two bits at a time: 00, 01, 10, 11). The four symbols are distinguished by varying the phase at which they are transmitted; specifically, for $k = 1, 2, 3, 4$, the kth symbol is transmitted for T seconds with the wave

$$x_k(t) = \cos(2\pi f_0 t + \pi/4 + k\pi/2), \qquad 0 \le t \le T \qquad (7.1)$$

for some predetermined frequency f_0. If we consider the transmission of a single randomly selected symbol, we may let $X(t)$ denote the corresponding transmitted wave. Each function $x_k(t)$ in Expression (7.1) is a sample function; the set of these four functions comprises the ensemble of $X(t)$ and is displayed in Fig. 7.2 for $0 \le t \le 4$.

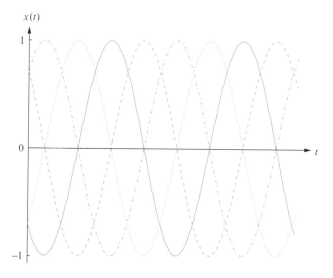

Fig. 7.2 Ensemble of the QPSK process in Example 7.1   ■

Example 7.2 Imagine the fluctuation in the value of Apple Inc. stock (symbol: AAPL) during the next 8-h trading day, measuring time from the opening bell on Wall Street. Since that fluctuation cannot be predicted precisely, we may reasonably model the stock's value by an appropriate random process $X(t)$; the ensemble of $X(t)$ would be subject to the constraint $X(0) = $ yesterday's closing value. Two examples of possible sample functions appear in Fig. 7.3, where we have assumed a previous day's closing value of $580. The ensemble of $X(t)$ consists of all possible paths that the price of Apple stock could hypothetically take tomorrow, starting at $580 per share. Economists and statisticians use a variety of *time series* models to forecast the behaviors of such random processes.

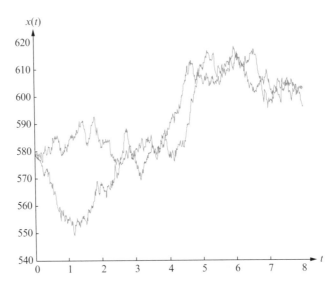

Fig. 7.3 Two sample functions for a stock price's fluctuation   ∎

Example 7.3 Consider modeling the number of people $N(t)$ logged in to a specific server at time t (perhaps measured from midnight). Since logins and logouts are unpredictable, we might reasonably apply a random process model to $N(t)$. Figure 7.4 shows one possible sample function; notice that, since our variable is integer-valued, the function "jumps" rather than varying continuously. In this context, the ensemble of $N(t)$ consists of all nonnegative integer-valued functions $n(t)$ that might hypothetically arise from successive logins and logouts.

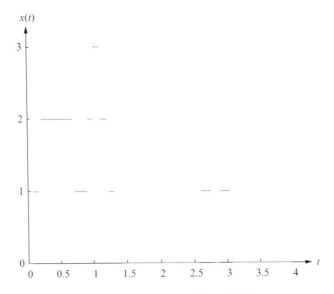

Fig. 7.4 A sample function for the random process of Example 7.3   ∎

Example 7.4 A dust particle lands on the surface of the water in a glass. For simplicity's sake, consider observing the motion of the particle only in the vertical direction (relative to our orientation) as time progresses. If we define the particle's initial position as 0, we have a random process $Y(t) =$ the vertical position of the particle t seconds after landing on the water.

Figure 7.5a shows one possible sample function for this particle motion, while Fig. 7.5b shows 100 different sample functions and thus approximates the ensemble of $Y(t)$. Notice that, as t increases, the particle has greater potential to be farther away from its origin, the line $y = 0$, since the particle has had more time to move. However, the particle will naturally "wiggle," and so a typical sample function will return to its origin multiple times, rather than "flying off" away from 0.

This is an example of *Brownian motion*, a model physicists regularly use for the seemingly random motion of electrons and various microscopic particles. Brownian motion, in turn, is an example of a *Gaussian process*; we will study Gaussian processes (in particular, Brownian motion) in Sect. 7.6.

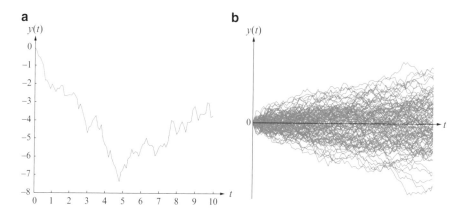

Fig. 7.5 Brownian motion: (**a**) a single sample function, (**b**) 100 sample functions   ■

7.1.1   Classification of Processes

As mentioned in the introduction to this chapter, we can classify random processes according to whether the variable and time dimensions are modeled as discrete or continuous. We call $X(t)$ a **discrete-space process** if its set of possible values at any time t is finite or countably infinite. Otherwise, $X(t)$ is a **continuous-space process.**

Example 7.1 and Example 7.2 illustrate continuous-space processes, since the variables (height of the sinusoid, value of the stock) may take on a continuum of values. In contrast, we have a discrete-space process in Example 7.3, since the only possible values of $N(t)$ are the countable set $\{0, 1, 2, \ldots\}$. These classifications are consistent with our usage of the terms *discrete* and *continuous* in Chaps. 2 and 3.

The difference between discrete- and continuous-space processes is less important than distinguishing how we model time. All of our above examples are **continuous-time processes**, because time is measured on a continuous scale, typically $[0, \infty)$ or $[0, T]$ for some fixed T. In contrast, imagine recording the value of Apple stock at the end of each day (or the number of people logged into a server at the end of each hour). Treating the variable as random, we would have a sequence $X_1, X_2, X_3, \ldots$, where X denotes the value of the variable and the index n corresponds to the nth instance of measuring the process. The listing $X_1, X_2, \ldots$, or more simply X_n, is a **discrete-time random process**, also called a **random sequence.** We already saw a special type of random sequence, Markov chains, in Chap. 6; we will consider general discrete-time processes more carefully in Sect. 7.4. Throughout the rest of this chapter as well as Chap. 8, the term "random process" will always refer to a continuous-time process unless indicated otherwise.

7.1.2 Random Processes Regarded as Random Variables

In most of the figures in this section, you will notice we have displayed time, t, on the horizontal axis, while the "random" behavior is illustrated in the vertical direction. You may find it helpful to think of these as the "time direction" and "random direction," respectively. To model a random process, we must truly understand its behavior in the "random direction." Toward that understanding, consider Fig. 7.5b, which shows the ensemble of a Brownian motion process. Fix a time point—say, $t = 1$. Looking in the vertical direction, we have a collection of "heights" corresponding to the numerical values of the many sample functions $y(t)$ displayed in the figure evaluated at $t = 1$. These many values of $y(1)$ form a probability distribution in the vertical direction: they show possible values of $Y(1)$, and the underlying random experiment that generated these sample functions determines the relative likelihoods of those values. It is in that sense that the vertical axis of our graphs is the "random direction."

More simply (and perhaps more usefully) put, we make the following observation: At any *fixed* time point t_0, the ensemble of a random process $X(t)$ forms a probability distribution; that is, $X(t_0)$ is a *random variable*.

Example 7.5 An intended signal may have the form $v_0 + a\cos(\omega_0 t + \theta_0)$, but amplitude variation may occur (due to natural current or voltage variation). We can define a random process by

$$X(t) = v_0 + A \cos(\omega_0 t + \theta_0)$$

where A is a random variable whose distribution describes the amplitude variation. Figure 7.6 illustrates part of the ensemble of $X(t)$ when the model for amplitude variation is a uniform distribution on $[-1, 1]$, for the specifications $v_0 = 0$, $\omega_0 = 2\pi$, and $\theta_0 = 0$. That is, $X(t) = A\cos(2\pi t)$ with $A \sim \text{Unif}[-1, 1]$.

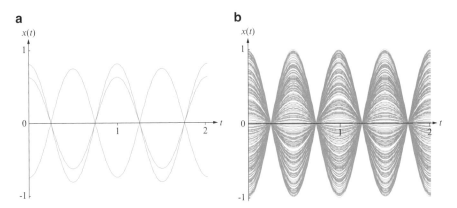

Fig. 7.6 The ensemble of $X(t) = A\cos(2\pi t)$: (**a**) three sample functions; (**b**) hundreds of sample functions

At time $t = 0$ (the far left edge of the graph), we may write $X(0) = A\cos(2\pi 0) = A\cos(0) = A$. Since $A \sim \text{Unif}[-1, 1]$ and $X(0) = A$, clearly $X(0) \sim \text{Unif}[-1, 1]$. That matches what our eyes see in the graph at $t = 0$: the values in the vertical direction seem "evenly" distributed on $[-1, 1]$. This same distribution can be seen at $t = 0.5$, 1, 1.5, 2, ….

In contrast, $X(1/3) = A\cos(2\pi/3) = -.5A \sim \text{Unif}[-.5,.5]$. This, too, is visible in the graph: at $t = 1/3 \approx .33$, the vertical expanse of the graph is not from -1 to 1 but rather from $-.5$ to $.5$.

Finally, at $t = 1.75$ we have $X(1.75) = A\cos(7\pi/2) = A(0) \equiv 0$, i.e., $X(1.75)$ equals 0 with probability 1 (i.e., for every member of the ensemble). We see in the graph that all functions of the form $x(t) = a\cos(2\pi t)$ indeed equal 0 at $t = 1.75$ (as well as at $t = 0.25, 0.75$, and 1.25). ∎

Example 7.6 (Example 7.4 continued) A Brownian motion process $Y(t)$ is partially characterized by the fact that, at any time t, $Y(t)$ has a Gaussian (i.e., normal) distribution with mean 0 and variance αt, for some constant α. In Fig. 7.5, we used the parameter $\alpha = 1$ to generate the graph. Thus, in Fig. 7.5b the probability distribution displayed in the vertical direction at time $t = 1$ is Gaussian with a mean of 0 and a variance of $(1)(1) = 1$, i.e., a standard normal distribution. In contrast, looking at time $t = 9$, $Y(9)$ is also Gaussian with mean zero but with standard deviation $\sqrt{\alpha t} = \sqrt{(1)(9)} = 3$, i.e., $Y(9) \sim N(0, 3)$. The increase in the variability of the ensemble as t increases is apparent in Fig. 7.5b. The Gaussian nature of the model is reflected by the fact that we see a greater concentration of values nearer the $y = 0$ line and a sparser set of values far from that midline. ∎

In the previous two examples, we have focused on the probability distribution of $X(t)$ at a single fixed time point, t. In fact, a random process is characterized by its simultaneous behavior at *all* time points. To be precise, a random process $X(t)$ is characterized only if we know the joint distribution of $X(t_1), \ldots, X(t_r)$ for all sets of time points $t_1 < \ldots < t_r$ and $r = 1, 2, 3, 4, \ldots$. The "joint behavior" of a random process, particularly at two points in time, will be explored in depth in the next section.

7.1.3 Exercises: Section 7.1 (1–10)

1. Classify each of the following processes as discrete-time or continuous-time, and discrete-space or continuous-space.
   (a) The temperature in downtown Chicago throughout a day
   (b) The number of customers in line at a certain store throughout the day
   (c) The high temperature in downtown Chicago for each day in a year
   (d) The total number of customers served each day at a certain store
2. Classify each of the following processes as discrete-time or continuous-time, and discrete-space or continuous-space.
   (a) The baud rate of a modem, recorded every 60 s
   (b) The number of people logged into Facebook throughout the day
   (c) The operational state, denoted 1 or 0, of a certain machine recorded at the end of each hour
   (d) The noise (in dB) in an audio signal measured throughout transmission
3. For each of the processes in Exercise 1, sketch two possible sample functions.
4. For each of the processes in Exercise 2, sketch two possible sample functions.
5. Consider the server login scenario of Example 7.3. Assuming $N(0) = 0$, sketch sample functions for $N(t)$ in each of the following cases:
   (a) The login rate exceeds the logout rate.
   (b) The logout rate exceeds the login rate.
   (c) The login and logout rates are equal.
6. *Correlated bit noise.* Let X_n be a sequence of random bits (0s and 1s) constructed as follows: $X_0 = 0$ or 1 with probability .5 each. For $n \geq 1$, $X_n = X_{n-1}$ with probability .9 and $X_n = 1 - X_{n-1}$ with probability .1.
   (a) Write out and sketch two examples of possible sample functions of X_n for $n = 0, \ldots, 10$.
   (b) Which sample function is more likely to be observed: 01100101010, or 00011110000? Explain.
   (c) Find the distribution of X_n at time $n = 1$.
7. *Binary phase-shift keying* (BPSK) is a simplified version of the QPSK system described in Example 7.1. One version of the system transmits the bit b, 0 or 1, with the waveform

$$x_b(t) = \cos(2\pi f_0 t + \pi + b\pi) \qquad 0 \leq t \leq T$$

   for suitable choices of frequency f_0 and time duration T. For purposes of this example, assume $f_0 = 1$ and $T = 1$.
   (a) Sketch the ensemble of this process.
   (b) Can the two bits be distinguished at time $t = 0.25$ s? Why or why not?

(c) Suppose random bit noise with $P(0)=.8$ and $P(1)=.2$ is transmitted via BPSK, and call the resulting random process $X(t)$. Find the probability distributions of $X(0)$ and of $X(.5)$.

8. Consider a random process $X(t)$ defined by $X(t)=A\cos(\pi t)+B\sin(\pi t)$, where A and B are iid $N(5, 2)$ rvs.
   (a) Graph a sample function of $X(t)$.
   (b) Find the probability distributions of $X(1/4)$ and of $X(1/2)$.
   (c) Find the *joint* pdf of $X(1/4)$ and $X(1/2)$. [*Hint:* Refer back to Sect. 4.7.]

9. A gambler plays roulette conservatively: she bets on black every time, which gives her probability 18/38 of winning on each spin. Define a random sequence $X_n =$ the number of wins she has after the nth spin for $n = 1, 2, 3, \ldots$.
   (a) Is X_n a discrete-space or continuous-space sequence?
   (b) Sketch two possible sample functions (sequences) for $n = 1, \ldots, 10$.
   (c) What is the probability distribution of X_n for fixed n?

10. Refer to Example 7.2. Suppose Apple stock has value \$580 at time $t = 0$, that the stock's value increases an average of 25 cents per day, and that the variation around that increasing trend can be described by a Brownian motion process with parameter $\alpha = 20$ (see Example 7.6).
   (a) Write an expression for $X(t)$, starting at time $t = 0$, in terms of t and the process $Y(t)$ from Example 7.6.
   (b) Sketch two sample functions of $X(t)$.
   (c) Find the probability distribution of Apple's stock at the end of 1 week of trading $(t = 5)$ and at the end of 2 weeks' trading $(t = 10)$.

7.2   Properties of the Ensemble: Mean and Autocorrelation Functions

In the previous section, we introduced the notion of a random process $X(t)$. We emphasized that, for a fixed time value t, $X(t)$ is a random variable possessing some probability distribution. Moreover, if we look at two fixed time points t and s, the two random variables $X(t)$ and $X(s)$ are usually not independent, and we can attempt to describe their joint probability distribution. In this section, we explore these ideas further.

7.2.1   Mean and Variance Functions

At any particular time t, the random variable $X(t)$ has a probability distribution and thus has both a mean value and variance. Since $X(t)$ for fixed t is a random variable, we should be able to calculate its mean using the techniques of Chaps. 2 and 3. Such a mean value exists for every time t, and the mean might not be the same at every time t, i.e., the mean of $X(t)$ may vary with t. Then considering all values of t gives a mean *function*. Similar comments apply to the variance and standard deviation of t.

DEFINITION

The **mean function** of a random process $X(t)$ is given by

$$\mu_X(t) = E[X(t)],$$

where $E[X(t)]$ is the expected value of the random variable $X(t)$ for the fixed time point t.

Similarly, we define the **variance function** of $X(t)$ by

$$\sigma_X^2(t) = \text{Var}(X(t)) = E\left[(X(t) - \mu_X(t))^2\right] = E\left[X^2(t)\right] - [\mu_X(t)]^2$$

and the **standard deviation function** of $X(t)$ by $\sigma_X(t) = \sqrt{\text{Var}(X(t))}$.

Notice that the mean, variance, and standard deviation functions are *nonrandom* functions of the time variable t, just as the mean, variance, and standard deviation of a random variable are numbers and not random quantities. It's vital to keep in mind that the mean function of a random process is taking an average *with respect to the ensemble* (i.e., in the "random direction") and not with respect to time.

Example 7.7 Reconsider Example 7.5 from Sect. 7.1, where the random process $X(t)$ was defined by the equation $X(t) = v_0 + A\cos(\omega_0 t + \theta_0)$. To find the mean and variance functions of $X(t)$, we apply the properties of expected value and variance established in earlier chapters. Remembering that time t is fixed, the entire term $\cos(\omega_0 t + \theta_0)$ may be treated as a constant, from which we obtain

$$\mu_X(t) = E[X(t)] = E[v_0 + A\cos(\omega_0 t + \theta_0)] = v_0 + E(A) \cdot \cos(\omega_0 t + \theta_0)$$

$$\sigma_X^2(t) = \text{Var}(X(t)) = \text{Var}(v_0 + A\cos(\omega_0 t + \theta_0)) = \text{Var}(A) \cdot \cos^2(\omega_0 t + \theta_0)$$

Remember that we must square a multiplicative constant for variance.

In the case where $A \sim \text{Unif}[-1, 1]$ illustrated in Fig. 7.6, we have $E(A) = (-1 + 1)/2 = 0$. Thus the mean function of $X(t)$ is $\mu_X(t) = v_0 + 0\cos(\omega_0 t + \theta_0) = v_0$. We can see this in Fig. 7.6: at any fixed time point t, the average of the values in the vertical ("random") direction is clearly zero, the value of v_0 for that graph. If we imagine vertically averaging these functions, we would arrive at the constant function $f(t) = 0$, as claimed.

In this same case, the variance of A is given by $\text{Var}(A) = (1 - (-1))^2/12 = 1/3$, whence

$$\sigma_X^2(t) = \frac{1}{3}\cos^2(\omega_0 t + \theta_0)$$

Thus, the variability of $X(t)$ in the vertical direction increases and decreases in a periodic manner as we vary t. This, too, can be seen in Fig. 7.6: the vertical spread varies with t, and this variability is the largest at $t=0, 0.5, 1$, and so on, when the cosine function is maximal. The standard deviation function of $X(t)$ is

$$\sigma_X(t) = \sqrt{\frac{1}{3}\cos^2(\omega_0 t + \theta_0)} = \frac{1}{\sqrt{3}}|\cos(\omega_0 t + \theta_0)|$$

The absolute value ensures that our standard deviation is always nonnegative. ∎

Example 7.8 In the previous example, we modeled amplitude variation with a uniform distribution. However, this is not a realistic model for most observed amplitude variation. Engineers frequently model amplitude variation A from a signal with a *Rayleigh distribution*. One example of a Rayleigh pdf is given by

$$f_A(a) = \begin{cases} ae^{-a^2/2} & a > 0 \\ 0 & \text{otherwise} \end{cases} \tag{7.2}$$

The graph of Eq. (7.2) appears in Fig. 7.7a, illustrating that a small amplitude is more likely than a large one. Notice that this model only provides positive values for the amplitude. Figure 7.7 shows the ensemble of $X(t) = A\cos(2\pi t)$ when A has the pdf specified in Eq. (7.2).

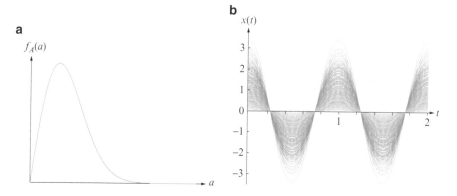

Fig. 7.7 (a) a Rayleigh pdf; (b) the resulting ensemble of $X(t) = A\cos(2\pi t)$

It's clear from the graph that the mean function is not zero; rather, it appears to be itself a sinusoid. (See if you can estimate the amplitude of the mean function by looking at $t=0$ on the graph.) Borrowing from Example 7.7, it's still true that $X(t)$ has mean and variance functions given by $\mu_X(t) = v_0 + E(A)\cos(\omega_0 t + \theta_0)$ and $\sigma_X^2(t) = \text{Var}(A)\cos^2(\omega_0 t + \theta_0)$, respectively. Using calculus, it can be shown the pdf in Eq. (7.2) has expected value $\sqrt{\pi/2} \approx 1.253$; hence, the mean function of the random process displayed in Fig. 7.7 is $\mu_X(t) \approx 1.253\cos(t)$, which is indeed a sinusoid. ∎

Example 7.9 *Signal plus noise.* A deterministic (i.e., nonrandom) signal $s(t)$ incurs noise during transmission, in which case the received message may have the form $Y(t) = s(t) + N(t)$. The term $N(t)$ is called the "noise component" of the received signal, and a variety of models can be used to describe such noise. Figure 7.8 below illustrates (part of) the ensemble of

$$Y(t) = 3\cos(2\pi t + \pi/2) + N(t),$$

where $N(t)$ is *Gaussian noise* with mean 0 and standard deviation 1 (that is, at each fixed time point t, $N(t)$ is standard normal).

Let's first determine the probability distribution of $Y(t)$ at both $t = 0.25$ and $t = 2$. With $s(t) = 3\cos(2\pi t + \pi/2)$, $Y(0.25) = s(0.25) + N(0.25) = -3 + N(0.25)$.   Since $N(0.25)$ has a Gaussian distribution with mean 0 and variance 1, it follows that $Y(0.25)$ is also Gaussian, but with a mean of -3 and standard deviation 1. Similarly, $Y(2) = s(2) + N(2) = 0 + N(2) = N(2)$, so $Y(2)$ is standard normal. We can visualize both of these distributions by looking vertically in Fig. 7.8.

Fig. 7.8 The ensemble for Example 7.9

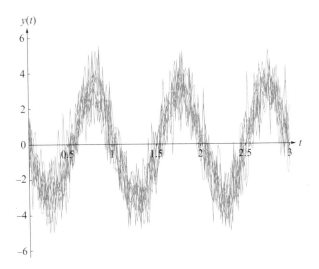

To find the mean and variance functions of $Y(t)$, note that $s(t)$ is to be treated as a constant with respect to the ensemble. We find that

$$\mu_Y(t) = E[Y(t)] = E[s(t) + N(t)] = s(t) + E[N(t)] = s(t) + 0 = s(t)$$

That is, the mean function of this random process is just the original signal, $s(t)$; this is the sinusoid that "carves down the middle" of Fig. 7.8. Finally, since $s(t)$ is an additive constant in the expression for $Y(t)$, $\sigma_Y^2(t) = \text{Var}(Y(t)) = \text{Var}(s(t) + N(t)) = \text{Var}(N(t)) = 1$. The amount of variability around the signal is the same at every point t in the process.

Notice that the distribution of the noise component, $N(t)$, is $N(0, 1)$ at every point t. But be careful: saying a process has *the same distribution* at every point t is very different from saying $N(t)$ is a constant! ∎

Example 7.10 *Signal plus noise, round two.* Let's modify the previous example by specifying that the spread of the noise component $N(t)$ varies with time; specifically, suppose that $N(t)$ is Gaussian with mean 0 and variance t. The ensemble of the resulting random process $Y(t)$ appears in Fig. 7.9. The mean function of $Y(t)$ is still $s(t)$; however, following the derivation in Example 7.9, we find $\text{Var}(Y(t)) = \text{Var}(N(t)) = t$, so $\sigma_Y(t) = \sqrt{t}$.

Fig. 7.9 The ensemble for Example 7.10

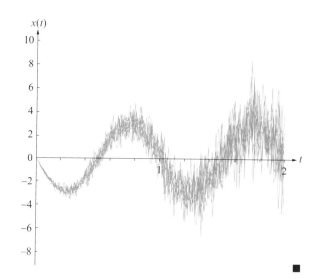

∎

7.2.2 Autocovariance and Autocorrelation Functions

The mean and variance describe the distribution of a single random variable. In the context of a random process, the mean and variance functions contain information about the behavior of the ensemble at each single point in time. But it should be clear that for two different times t and s, the random variables $X(t)$ and $X(s)$ will typically be related. A complete statistical analysis of a random process should include an exploration of that relationship. To that end, we now extend the notion of covariance from Chap. 4 to a random process.

DEFINITION
The **autocovariance function** of a random process $X(t)$ is defined by

$$C_{XX}(t, s) = \text{Cov}(X(t), X(s)) = E[(X(t) - \mu_X(t))(X(s) - \mu_X(s))]$$

Notice that the autocovariance function is a nonrandom function of *two* time points, t and s.

We can interpret the autocovariance function of $X(t)$ much as $\text{Cov}(X, Y)$ was interpreted back in Chap. 4. When $C_{XX}(t, s) > 0$, above-average values of $X(t)$ tend to be associated with above-average values of $X(s)$. That is, when $X(t)$ is above its mean function at time t, it also tends to be above its mean function at time s (and vice versa). If $C_{XX}(t, s) < 0$, then above-average values of the random process at time t are associated with *below*-average values at time s (and vice versa).

Properties of the autocovariance function follow directly from the properties previously derived for covariance. We provide a partial listing here.

PROPOSITION
Let $C_{XX}(t, s)$ denote the autocovariance function of a random process $X(t)$.
1. $C_{XX}(t, s) = C_{XX}(s, t)$
2. $C_{XX}(t, s) = E[X(t)X(s)] - \mu_X(t)\mu_X(s)$
3. $\sigma_X^2(t) = \text{Var}(X(t)) = \text{Cov}(X(t), X(t)) = C_{XX}(t, t) = E[X^2(t)] - \mu_X^2(t)$

In the engineering literature, $E[X(t)X(s)]$ in property 2 is called the **autocorrelation function** of $X(t)$ and is denoted $R_{XX}(t, s)$. Although it will be vital to our study of signal processing in Chap. 8, don't confuse this with the *correlation coefficient* from Chap. 4; in particular, the sign of $R_{XX}(t, s)$ does not indicate the direction of the association between $X(t)$ and $X(s)$, and the magnitude of $R_{XX}(t, s)$ is not bounded by 1.

Example 7.11 Let's find the autocovariance function of the random process $X(t) = A\cos(2\pi t)$ from Examples 7.5 and 7.7. We will illustrate two methods here. Since we already have the mean function of $X(t)$, we can calculate the autocorrelation function and then apply property 2 from the preceding proposition:

$$R_{XX}(t, s) = E[X(t)X(s)] = E[A\cos(2\pi t)A\cos(2\pi s)] = E[A^2\cos(2\pi t)\cos(2\pi s)]$$
$$= E(A^2)\cos(2\pi t)\cos(2\pi s) \Rightarrow$$
$$C_{XX}(t, s) = E[X(t)X(s)] - \mu_X(t)\mu_X(s)$$
$$= E(A^2)\cos(2\pi t)\cos(2\pi s) - E(A)\cos(2\pi t) \cdot E(A)\cos(2\pi s)$$
$$= [E(A^2) - (E(A))^2]\cos(2\pi t)\cos(2\pi s)$$
$$= \text{Var}(A)\cos(2\pi t)\cos(2\pi s)$$

Alternatively, we can manipulate the covariance expression directly by applying its distributive properties from Chap. 4:

$$
\begin{aligned}
C_{XX}(t, s) &= \mathrm{Cov}\big(X(t), X(s)\big) = \mathrm{Cov}\big(A\cos(2\pi t), A\cos(2\pi s)\big) \\
&= \mathrm{Cov}(A, A)\cos(2\pi t)\cos(2\pi s) \\
&= \mathrm{Var}(A)\cos(2\pi t)\cos(2\pi s)
\end{aligned}
$$

As a check, substituting $s = t$ gives $C_{XX}(t, t) = \mathrm{Var}(A)\cos^2(2\pi t)$, which matches the expression for $\sigma_X^2(t)$ we found in Example 7.7 (with $\omega_0 = 2\pi$ and $\theta_0 = 0$).   ∎

Example 7.12 Let's now consider a sinusoid with *phase variation*, rather than amplitude variation. Define a random process $X(t)$ by

$$
X(t) = A_0 \cos(\omega_0 t + \Theta) \tag{7.3}
$$

where the phase shift Θ is a rv, uniformly distributed on the interval $(-\pi, \pi]$. The amplitude A_0 and fundamental frequency $\omega_0 \neq 0$ are constants. Figure 7.10 shows several sample functions for this random process with $A_0 = 1$ and $\omega_0 = 2\pi$.

Until now, we have managed to compute means and variances without any calculus; however, because the random process $X(t)$ defined by Eq. (7.3) is a *nonlinear* function (cosine) of a random variable, we must rely on calculus here. Specifically, we apply the formula for the expectation of a function of a continuous random variable, presented in Chap. 3:

$$
\begin{aligned}
\mu_X(t) &= E\big[X(t)\big] = E\big[A_0\cos(\omega_0 t + \Theta)\big] \\
&= \int_{-\infty}^{\infty} A_0\cos(\omega_0 t + \theta) f_\Theta(\theta)\, d\theta = \int_{-\pi}^{\pi} A_0\cos(\omega_0 t + \theta)\frac{1}{\pi - (-\pi)}\, d\theta \\
&= \frac{A_0}{2\pi}\int_{-\pi}^{\pi} \cos(\omega_0 t + \theta)\, d\theta = \frac{A_0}{2\pi}(0) = 0
\end{aligned}
$$

The last integral equals zero because, as a function of θ, it represents the integration of a cosine through one period. A mean function identically equal to zero coincides with what we see in Fig. 7.10.

Since the mean function is zero, the autocovariance and autocorrelation functions will be identical. Calculation of these functions requires a trig identity:

Fig. 7.10 Sample functions for the phase-variation process in Example 7.12

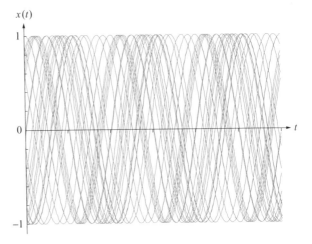

$$
\begin{aligned}
C_{XX}(t,s) &= E\big[X(t)X(s)\big] - (0)\,(0) = E\big[X(t)X(s)\big] \\
&= E\big(A_0 \cos(\omega_0 t + \Theta) \cdot A_0 \cos(\omega_0 s + \Theta)\big) \\
&= A_0^2 E\big[\cos(\omega_0 t + \Theta)\cos(\omega_0 s + \Theta)\big] \\
&= A_0^2 E\left[\frac{1}{2}\{\cos(\omega_0 t + \Theta + \omega_0 s + \Theta) + \cos(\omega_0 t + \Theta - [\omega_0 s + \Theta])\}\right] \\
&= \frac{A_0^2}{2} E\big[\cos(\omega_0 t + \omega_0 s + 2\Theta) + \cos(\omega_0 t - \omega_0 s)\big] \\
&= \frac{A_0^2}{2} E\big[\cos(\omega_0 t + \omega_0 s + 2\Theta)\big] + \frac{A_0^2}{2}\cos(\omega_0 t - \omega_0 s) \\
&= \frac{A_0^2}{2}\cdot 0 + \frac{A_0^2}{2}\cos(\omega_0 t - \omega_0 s) = \frac{A_0^2}{2}\cos(\omega_0 t - \omega_0 s)
\end{aligned}
$$

The last expected value equals zero because it represents an integral of a cosine through two periods. Finally, the variance function is given by

$$
\sigma_X^2(t) = C_{XX}(t,t) = \frac{A_0^2}{2}\cos(\omega_0 t - \omega_0 t) = \frac{A_0^2}{2}\cos(0) = \frac{A_0^2}{2}
$$

Notice that the variance function of $X(t)$ is a constant (same spread for all t), which agrees with Fig. 7.10.   ■

7.2.3   The Joint Distribution of Two Random Processes

Some applications involve the consideration of two random processes $X(t)$ and $Y(t)$. We may then be concerned not only with their individual distributions but also their *joint* behavior. This is especially true when $Y(t)$ is the result of some action taken on $X(t)$, such as passing the random signal $X(t)$ through an appropriate filter. To quantify their relationship, we define the **cross-covariance function of $X(t)$ with $Y(t)$** by $C_{XY}(t, s) = \text{Cov}(X(t),Y(s))$ and the **cross-correlation function of $X(t)$ with $Y(t)$** by $R_{XY}(t, s) = E[X(t)Y(s)]$. These two functions are, not surprisingly, connected by the formula $C_{XY}(t, s) = R_{XY}(t, s) - \mu_X(t)\mu_Y(s)$.

> **DEFINITION**
>
> Two random processes $X(t)$ and $Y(t)$ are **independent** if, for all fixed t and s, the random variables $X(t)$ and $Y(s)$ are independent rvs as defined in Chap. 4. $X(t)$ and $Y(t)$ are **uncorrelated** if, for all t and s, $C_{XY}(t, s) = 0$. Finally, $X(t)$ and $Y(t)$ are **orthogonal** if $R_{XY}(t, s) = 0$ for all t and s.

Notice in these definitions that properties must hold for all times t and s. For example, the independence of $X(t)$ and $Y(t)$ requires that the random variables $X(2)$ and $Y(10)$ be independent, as must $X(2)$ and $Y(2)$ be, and so on. A similar comment applies to being uncorrelated or orthogonal.

As in Chap. 4, independence is a stronger condition than zero correlation:

$$X(t) \text{ and } Y(t) \text{ independent} \Rightarrow X(t) \text{ and } Y(t) \text{ uncorrelated,}$$

but the converse is false. If $X(t)$ and $Y(t)$ are uncorrelated, it follows from the definition of covariance that $E[X(t)Y(s)] = E[X(t)]E[Y(s)]$ for all t and s. Thus, being uncorrelated does not imply being orthogonal (nor vice versa); however, if either random process has mean identically equal to zero, then the properties of being uncorrelated and orthogonal are equivalent.

7.2.4   Exercises: Section 7.2 (11–24)

11. Consider the QPSK system described in Example 7.1 as a model for random noise. Suppose the four possible symbols to be transmitted are equally likely to occur, i.e., we have a random process

$$X(t) = \cos(2\pi f_0 t + \pi/4 + K\pi/2)$$

where K is 0, 1, 2, or 3 with probability .25 each.
    (a) Find the mean function of $X(t)$. Simplify as much as possible.
    (b) Find the variance function of $X(t)$. Are your answers consistent with Fig. 7.2?

12. Show that $C_{XX}(t, s) = E[X(t)X(s)] - \mu_X(t)\mu_X(s)$.
13. Consider the random process $X(t) = v_0 + A\cos(\omega_0 t + \theta_0)$ from Example 7.7. Find the autocovariance function and autocorrelation function of $X(t)$.
14. Let $X(t) = At + B$, where A and B are independent random variables with $A \sim \text{Unif}[0, 6]$ and $B \sim \text{Unif}[-10, 10]$.
    (a) Describe the ensemble of $X(t)$.
    (b) Determine the mean function of $X(t)$.
    (c) Determine the autocovariance function of $X(t)$.
    (d) Determine the autocorrelation function of $X(t)$.
    (e) Determine the variance function of $X(t)$.
15. Let $N(t)$ be a Gaussian noise process as in Example 7.9, with mean 0 for all t and autocovariance function $C_{NN}(t, s) = e^{-|s - t|}$.
    (a) Verify that $N(t)$ has variance 1 for all t.
    (b) If $N(t) > 0$, would you predict that $N(s) > 0$ or $N(s) < 0$? Explain.
    (c) Determine the correlation coefficient ρ of $N(10)$ and $N(12)$.
    (d) Determine the probability distribution of $N(12) - N(10)$.
16. Let $N(t)$ be the Gaussian noise process of Example 7.10, with mean function 0 and variance function t.
    (a) Calculate $P(N(1) > .5)$ and $P(N(4) > .5)$.
    (b) Could the autocovariance function of $N(t)$ be $e^{-|s - t|}$? Why or why not?
    (c) Suppose the autocovariance function of $N(t)$ is $\min(t, s)$, i.e., $C_{NN}(t, s) = t$ for $t \le s$ and s for $t > s$. Find the correlation coefficient between $N(t)$ and $N(s)$. [*Hint*: consider the two cases $t \le s$ and $t > s$.]
    (d) Determine the probability distribution of $N(s) - N(t)$.
17. Consider the phase-variation random process (7.3) with $A_0 = 1$ and $\omega_0 = 2\pi$.
    (a) Use the results of Example 7.12 to show that, for fixed t, $X(t)$ does *not* have a uniform distribution. [*Hint*: What is the interval of possible values for $X(t)$? If $X(t)$ were uniform, what would its variance be?]
    (b) Use the transformation method of Sect. 3.7 to show that the rv $Y = X(0)$ has an *arcsine distribution*:

$$f_Y(y) = \frac{1}{\pi}\frac{1}{\sqrt{1 - y^2}} \qquad -1 < y < 1$$

    [*Note*: It can be shown that $X(t)$ has this same distribution for all t.]
18. Let $A(t)$ be a random process, and define an "amplitude modulated" version of $A(t)$ by $X(t) = A(t)\cos(\omega_0 t + \Theta)$, where $\Theta \sim \text{Unif}(-\pi, \pi]$ and is independent of $A(t)$, and ω_0 is a constant.
    (a) Determine the mean function of $X(t)$.
    (b) Determine the autocorrelation function of $X(t)$.
    (c) Determine the cross-correlation of $A(t)$ and $X(t)$.
    [*Hint*: Use the results of Example 7.12.]
19. Consider a "signal plus noise" process where both components are random: $X(t) = S(t) + N(t)$. Assume $S(t)$ and $N(t)$ are uncorrelated random processes.

Determine each of the following functions in terms of the mean, autocorrelation, etc. of $S(t)$ and $N(t)$.

(a) The mean function of $X(t)$.

(b) The autocorrelation function of $X(t)$.

(c) The autocovariance function of $X(t)$.

(d) The variance function of $X(t)$.

20. Consider the random process $X(t) = S(t) + N(t)$ from the previous exercise. Find the cross-correlation between the signal component $S(t)$ and the overall process $X(t)$.

21. Consider two random processes $X(t)$ and $Y(t)$.

(a) Show that if $X(t)$ and $Y(t)$ are uncorrelated random processes, then $E[X(t)Y(s)] = \mu_X(t)\mu_Y(s)$.

(b) Show that if $X(t)$ and $Y(t)$ are uncorrelated random processes and $X(t)$ has mean function equal to zero, then $X(t)$ and $Y(t)$ are orthogonal.

22. Let $A(t)$ and $B(t)$ be iid processes, i.e., $A(t)$ and $B(t)$ are independent processes with the same mean function $\mu(t)$, autocovariance function $C(t, s)$, etc. Define a pair of new random processes by

$$X(t) = A(t) + B(t)$$
$$Y(t) = A(t) - B(t)$$

(a) Find the mean functions of $X(t)$ and $Y(t)$.

(b) Find the autocovariance functions of $X(t)$ and $Y(t)$.

(c) Find the cross-covariance function $C_{XY}(t, s)$.

23. Let Θ be a uniformly distributed rv on $(-\pi, \pi]$. Define two random processes $X(t) = \cos(\omega_0 t + \Theta)$ and $Y(t) = \sin(\omega_0 t + \Theta)$.

(a) Find the cross-correlation and cross-covariance of $X(t)$ and $Y(t)$.

(b) Are $X(t)$ and $Y(t)$ orthogonal random processes? Uncorrelated random processes? Independent random processes?

24. Let $R_{XY}(t, s)$ be the cross-correlation function of $X(t)$ with $Y(t)$, and define the cross-correlation of $Y(t)$ with $X(t)$ by $R_{YX}(t, s)$. Show that $R_{YX}(t, s) = R_{XY}(s, t)$. Show that a similar relationship holds for cross-covariance.

7.3   Stationary and Wide-Sense Stationary Processes

When modeling certain random processes, particularly those representing noise, it facilitates the analysis if the statistical properties of the process remain the same across time. This turns out to be true for some, though certainly not all, models. We will make this notion more precise shortly, but first let's revisit three of the examples from the previous section. The relevant graphs are presented in Fig. 7.11 below. Figure 7.11a shows the ensemble of the phase-variation random process from Example 7.12. Notice that the probability distribution of $X(t)$—remember, that's the distribution in the vertical direction—appears to be the same at each time point t. Figure 7.11b shows just the noise component, $N(t)$, from

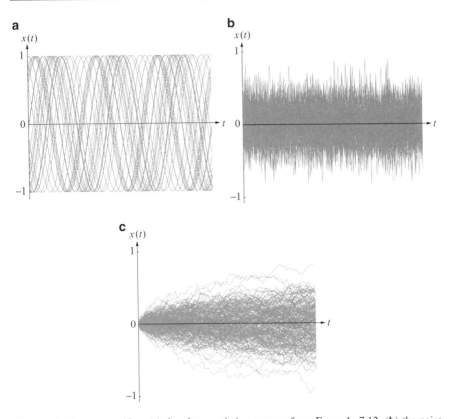

Fig. 7.11 Three ensembles: (**a**) the phase-variation process from Example 7.12; (**b**) the noise component from Example 7.9; (**c**) the noise component from Example 7.10

Example 7.9. Again, we see roughly the same ensemble behavior at every time t, suggesting the process' statistical properties do not change over time. In contrast, consider the noise component $N(t)$ from Example 7.10, displayed here in Fig. 7.11c. While the mean of $N(t)$ is constant, its variance clearly increases with t; this model does *not* possess the property of interest.

We now formalize the notion of stable behavior over time.

> **DEFINITION**
> A random process $X(t)$ is **(strict-sense) stationary** if all of its statistical properties are invariant with respect to time. More precisely, $X(t)$ is stationary if, for any time points $t_1, \ldots, t_r$ and any value τ, the joint distribution of $X(t_1)$, $\ldots, X(t_r)$ is the same as the joint distribution of $X(t_1 + \tau), \ldots, X(t_r + \tau)$.

This definition requires that the statistical behavior of $X(t)$ remain the same if we "translate" the random process τ time units. In particular, it requires that $X(t_1)$ and

$X(t_1 + \tau)$ have the same distribution for all t_1 and τ; it follows that $X(t)$ must have the same mean, standard deviation, etc. at all times t. This corresponds to what we see in Fig. 7.11a, b, but not c. Notice, however, that the definition requires more: since the *joint* distribution of $X(t_1)$ and $X(t_2)$ must be translation-invariant, it follows that the autocovariance function of $X(t)$ must be translation-invariant as well. We certainly cannot determine this from a visual inspection of the ensemble.

In fact, it is rarely practical to determine whether a particular random process model is strict-sense stationary, since it requires an unlimited number of comparisons (joint distributions of r variables at all time-points and for all possible r). Fortunately, a weaker version of stationarity suffices for the purposes of many analyses.

DEFINITION
A random process $X(t)$ is **wide-sense stationary (WSS)** if the following two conditions hold:
1. The mean function of $X(t)$, $\mu_X(t)$, is a constant.
2. The autocovariance function of $X(t)$, $C_{XX}(t, s)$, depends only on $s - t$.

We interpret condition 2 as follows: the degree of association between $X(t)$ and $X(s)$, as measured by covariance, depends on how far apart the two times s and t are, but not where those times are located on an absolute scale. So, for example, the covariance between $X(3)$ and $X(10)$ is the same as the covariance between $X(23)$ and $X(30)$ when condition 2 is satisfied (since, in both cases, $s - t = 7$).

Condition 2 of this definition can be stated more cleanly if we re-parameterize the second time variable. Let's write $s = t + \tau$, so that τ represents the difference between the two times s and t. Then wide-sense stationarity requires that the autocovariance function $C_{XX}(t, t + \tau)$ depend only on τ (and not on t). In fact, with this notation, we can define a wide-sense stationary process to be one such that both $\mu_X(t)$ and $C_{XX}(t, t + \tau)$ are independent of t.

Before looking at some examples, we note that the defining conditions can be restated in terms of the autocorrelation function, R_{XX}: a random process $X(t)$ is WSS iff (1) $\mu_X(t)$ is a constant and (2) $R_{XX}(t, t + \tau)$ depends only on τ.

Example 7.13 Is the amplitude-variation random process, $X(t) = A\cos(2\pi t)$, wide-sense stationary? The graphs in Figs. 7.6 and 7.7 clearly indicate not. Indeed, in Example 7.11 we found the autocovariance of this random process to be $C_{XX}(t, s) = \mathrm{Var}(A)\cos(2\pi t)\cos(2\pi s)$, which depends separately on t and s, not just their difference. Therefore, the amplitude-variation random process is not WSS. ∎

Example 7.14 Is the phase-variation random process $X(t) = A_0\cos(\omega_0 t + \Theta)$ from Example 7.12 wide-sense stationary? Using the results of Example 7.12, we can check the two required conditions:

1. $\mu_X(t) = 0$, a constant. Thus the first condition is satisfied.

2. $C_{XX}(t, s) = \dfrac{A_0^2}{2} \cos(\omega_0 t - \omega_0 s)$, so

$$C_{XX}(t, t + \tau) = \frac{A_0^2}{2} \cos(\omega_0 t - \omega_0(t + \tau)) = \frac{A_0^2}{2} \cos(-\omega_0 \tau) = \frac{A_0^2}{2} \cos(\omega_0 \tau).$$

Since $C_{XX}(t, t+\tau)$ depends only on τ and not on t, the second condition is met. Therefore, $X(t)$ is indeed wide-sense stationary.   ∎

Example 7.15 Let A and B be iid mean-zero random variables, and define a random process by

$$X(t) = A \cos(\omega_0 t) + B \sin(\omega_0 t) \tag{7.4}$$

for some frequency ω_0. Is $X(t)$ wide-sense stationary? The mean function is

$$\mu_X(t) = E[A \cos(\omega_0 t) + B \sin(\omega_0 t)] = E[A] \cos(\omega_0 t) + E[B] \sin(\omega_0 t)$$
$$= 0 \cos(\omega_0 t) + 0 \sin(\omega_0 t) = 0$$

Since the mean of $X(t)$ is a constant, the first condition is met. Next, let's consider the autocovariance function. Using the distributive properties of covariance,

$$
\begin{aligned}
C_{XX}(t, s) =\ & \mathrm{Cov}\big(A \cos(\omega_0 t) + B \sin(\omega_0 t), A \cos(\omega_0 s) + B \sin(\omega_0 s)\big) \\
=\ & \mathrm{Cov}\big(A \cos(\omega_0 t), A \cos(\omega_0 s)\big) + \mathrm{Cov}\big(A \cos(\omega_0 t), B \sin(\omega_0 s)\big) \\
& + \mathrm{Cov}\big(B \sin(\omega_0 t), A \cos(\omega_0 s)\big) + \mathrm{Cov}\big(B \sin(\omega_0 t), B \sin(\omega_0 s)\big) \\
=\ & \mathrm{Cov}(A, A) \cos(\omega_0 t) \cos(\omega_0 s) + \mathrm{Cov}(A, B) \cos(\omega_0 t) \sin(\omega_0 s) \\
& + \mathrm{Cov}(B, A) \sin(\omega_0 t) \cos(\omega_0 s) + \mathrm{Cov}(B, B) \sin(\omega_0 t) \sin(\omega_0 s)
\end{aligned}
$$

Since A and B are independent, $\mathrm{Cov}(A, B) = \mathrm{Cov}(B, A) = 0$; since they're identically distributed, $\mathrm{Cov}(A, A) = \mathrm{Cov}(B, B) = \sigma^2$, the common variance of A and B. Using a trig identity, we arrive at

$$C_{XX}(t, s) = \sigma^2 \cos(\omega_0 t) \cos(\omega_0 s) + \sigma^2 \sin(\omega_0 t) \sin(\omega_0 s) = \sigma^2 \cos(\omega_0 t - \omega_0 s)$$
$$= \sigma^2 \cos(\omega_0[t - s])$$

Since this depends only on the difference in the two times t and s, the second condition is met. (In fact, we may simplify the autocovariance expression further, to $\sigma^2 \cos(\omega_0[t - (t + \tau)]) = \sigma^2 \cos(-\omega_0 \tau) = \sigma^2 \cos(\omega_0 \tau)$.) Therefore, *yes*, $X(t)$ in Expression (7.4) is wide-sense stationary.   ∎

Example 7.16 (Example 7.15 continued) In the previous example, we proved that any random process of the form (7.4) is WSS, provided A and B are iid with mean zero.

As a curious example, suppose A and B are independent and each is equally likely to be $+1$ or -1; this particular distribution has mean 0 and variance 1. Since

A and B are iid with mean 0, the random process $X(t)$ in Eq. (7.4) is WSS, with mean 0, autocovariance function $C_{XX}(t, t+\tau) = \sigma^2\cos(\omega_0\tau) = \cos(\omega_0\tau)$, and thus variance $\sigma_X^2(t) = C_{XX}(t, t) = 1$. However, since A and B each can only take on the two values ± 1, the entire ensemble of $X(t)$ consists of just four functions:

$$X(t) = \pm\cos(\omega_0 t) \pm \sin(\omega_0 t)$$

This ensemble appears in Fig. 7.12; its appearance does not match earlier pictures of WSS processes. But it is nonetheless true that the mean of the vertical coordinates at any time-point t is 0, the variance is 1, and that the covariance between any two time points τ units apart is $\cos(\omega_0\tau)$.

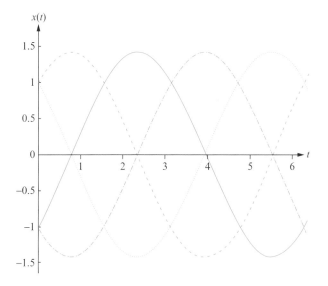

Fig. 7.12 The ensemble of $X(t)$ in Example 7.16

In part, the lesson here is that we cannot rely on a visualization of a random process to determine whether it's wide-sense stationary—despite appearances, this really is a WSS process. That said, it's clear that the probability *distribution* of $X(t)$ is not the same for all t, e.g., the possible values of the process are $\{-1, 0, 1\}$ at some t-coordinates and $\{-1/\sqrt{2}, 1/\sqrt{2}\}$ at others. So, while this random process is *wide*-sense stationary, it is certainly not *strict*-sense stationary.   ∎

In Chap. 8, we will study relationships between the behavior of the input $X(t)$ to a filter and the resulting output $Y(t)$; we will often require that $X(t)$ be WSS. Two random processes $X(t)$ and $Y(t)$ are called **jointly wide-sense stationary** if (1) $X(t)$ is WSS, (2) $Y(t)$ is WSS, and (3) the cross-covariance function $C_{XY}(t, t+\tau)$ does not depend on t. (Equivalently, $X(t)$ and $Y(t)$ are jointly WSS if they are both WSS processes and $R_{XY}(t, t+\tau)$ is independent of t.)

7.3.1 Properties of Wide-Sense Stationary Processes

By definition, a WSS random process has a constant mean (function). In fact, such a process also has constant variance, because $\sigma_X^2(t) = C_{XX}(t,t)$ from a proposition in the previous section and wide-sense stationarity requires that this covariance not depend on t. With that and our previous discussions in mind, we adopt the following notational conventions for WSS processes.

NOTATION

Suppose $X(t)$ is a wide-sense stationary process. Then we denote its statistical functions as follows:

$$\mu_X = E\big[X(t)\big]$$
$$\sigma_X^2 = \text{Var}\big(X(t)\big)$$
$$C_{XX}(\tau) = \text{Cov}\big(X(t), X(t+\tau)\big) = R_{XX}(\tau) - \mu_X^2$$
$$R_{XX}(\tau) = E\big[X(t)X(t+\tau)\big] = C_{XX}(\tau) + \mu_X^2$$

Next we present some important properties of these functions.

PROPOSITION

Let $X(t)$ be a wide-sense stationary process with autocovariance function $C_{XX}(\tau)$ and autocorrelation function $R_{XX}(\tau)$.

Properties of $C_{XX}(\tau)$:
1. $C_{XX}(0) = E\big[X^2(t)\big] - \mu_X^2 = \sigma_X^2 = \text{Var}(X(t))$.
2. $C_{XX}(-\tau) = C_{XX}(\tau)$; that is, the autocovariance function is symmetric in τ.
3. $|C_{XX}(\tau)| \le C_{XX}(0)$ for every τ; that is, the autocovariance function achieves its largest value at $\tau = 0$.
4. If $X(t)$ is periodic, so is $C_{XX}(\tau)$, and with the same period.
5. If $X(t)$ is ergodic[1] and has no periodic component, then $C_{XX}(\tau) \to 0$ as $|\tau| \to \infty$.

Properties of $R_{XX}(\tau)$:
1. $R_{XX}(0) = E[X^2(t)]$, called the **mean square value** of $X(t)$.
2. $R_{XX}(-\tau) = R_{XX}(\tau)$; that is, the autocorrelation function is symmetric in τ.

(continued)

---

[1] Loosely speaking, a random process is *ergodic* if its time and ensemble properties "match." We will define ergodicity more carefully later in this section; for now, you may assume the processes referenced in this section are ergodic unless noted otherwise.

3. $|R_{XX}(\tau)| \leq R_{XX}(0)$ for every τ; that is, the autocorrelation function achieves its largest value at $\tau = 0$.
4. If $X(t)$ is periodic, so is $R_{XX}(\tau)$, and with the same period.
5. If $X(t)$ is ergodic and has no periodic component, then $R_{XX}(\tau) \to \mu_X^2$ as $|\tau| \to \infty$.

Proof We begin with the properties of $C_{XX}(\tau)$. Property 1 follows from the covariance shortcut formula. To prove Property 2, recall from Chap. 4 that covariance is symmetric in its arguments:

$$C_{XX}(\tau) = \text{Cov}(X(t), X(t+\tau)) = \text{Cov}(X(t+\tau), X(t))$$

Because $X(t)$ is WSS, the right-most expression depends only on the difference in the two times; specifically, $\text{Cov}(X(t+\tau), X(t)) = C_{XX}(t - [t+\tau]) = C_{XX}(-\tau)$. This establishes the result.

Property 3 is left as an exercise (see Exercise 38). We note that Property 3 makes intuitive sense: since covariance measures the association between two variables, and τ represents the time distance between these two variables, covariance should be largest when that time difference is as small as possible. That is, the behaviors of $X(t)$ and $X(t+\tau)$ should be more closely related when τ is small than when τ is large.

Properties 1–3 for the autocorrelation function follow automatically, since $R_{XX}(\tau)$ and $C_{XX}(\tau)$ only differ by the constant μ_X^2.

Toward proving property 4, suppose $X(t)$ is periodic with period d, so $X(t) = X(t+d)$ for all t. Then, for any τ,

$$\begin{aligned}
R_{XX}(\tau + d) &= E\big[X(t)X(t + \tau + d)\big] \\
&= E\big[X(t)X(t + \tau)\big] \qquad \text{because } X(t + \tau + d) = X(t + \tau) \\
&= R_{XX}(\tau)
\end{aligned}$$

which shows that $R_{XX}(\tau)$ is also periodic with period d. The analogous property holds for autocovariance, because subtracting μ_X^2 to get C_{XX} from R_{XX} does not affect periodicity.

A formal proof of Property 5 is beyond the scope of this book; however, the paragraph above regarding Property 3 should give some intuition for why covariance should vanish as $|\tau| \to \infty$. Some further information about "ergodicity" appears at the end of this section. ∎

It's important to note that while every autocovariance and autocorrelation function for WSS processes satisfy the properties listed in this proposition, these properties do not completely characterize such functions. That is to say, there exist functions that satisfy properties 1–5 but are *not* valid autocovariance/autocorrelation functions. We'll explore this further in Chap. 8, when we connect autocorrelation and autocovariance functions to the power spectrum of a random signal. (For a preview, see Exercise 40.)

Example 7.17 Suppose $X(t)$ is a wide-sense stationary random process with autocorrelation function

$$R_{XX}(\tau) = 100 + \frac{16}{1 + \tau^2}$$

Let's determine as much as we can about the other statistical properties of $X(t)$. First, the mean square value of $X(t)$ is $R_{XX}(0) = 100 + 16 = 116$ (Property 1). Next, $X(t)$ clearly has no periodic component; otherwise, $R_{XX}(\tau)$ would also (Property 4). Thus we may apply Property 5:

$$\mu_X^2 = \lim_{|\tau| \to \infty} R_{XX}(\tau) = \lim_{|\tau| \to \infty} \left[100 + \frac{16}{1 + \tau^2} \right] = 100 + 0 = 100$$

from which μ_X either equals $+10$ or -10; notice that we cannot determine which is correct from $R_{XX}(\tau)$. We can, however, determine the autocovariance function:

$$C_{XX}(\tau) = R_{XX}(\tau) - \mu_X^2 = 100 + \frac{16}{1 + \tau^2} - 100 = \frac{16}{1 + \tau^2}$$

Notice this autocovariance function goes to 0 as $|\tau| \to \infty$, as guaranteed by Property 5. Finally, the variance of this random process is given by $\sigma_X^2 = C_{XX}(0) = 16$, and the standard deviation is $\sigma_X = 4$. ∎

Example 7.18 *Partitioning a random process.* Suppose $X_1(t)$ and $X_2(t)$ are independent, zero-mean, WSS random processes with autocorrelation functions $R_{11}(\tau) = 2000\text{tri}(10,000\tau)$ and $R_{22}(\tau) = 650\cos(40,000\pi\tau)$, respectively.[2]

Define a new random process by $X(t) = X_1(t) + X_2(t) + 40$. The mean function of $X(t)$ is

$$\mu_X(t) = E[X(t)] = E[X_1(t) + X_2(t) + 40] = E[X_1(t)] + E[X_2(t)] + 40$$
$$= 0 + 0 + 40 = 40$$

Determining the autocorrelation function requires some significant algebraic work:

$$R_{XX}(t,s) = E[X(t)X(s)] = E[(X_1(t) + X_2(t) + 40)(X_1(s) + X_2(s) + 40)]$$
$$= E[X_1(t)X_1(s)] + E[X_1(t)X_2(s)] + E[40X_1(t)] + E[X_2(t)X_1(s)]$$
$$+ E[X_2(t)X_2(s)] + E[40X_2(t)] + E[40X_1(s)] + E[40X_2(s)] + E[1600]$$
$$(7.5)$$

Four of the terms in Expression (7.5) may be simplified by removing the constant 40, e.g., $E[40X_1(t)] = 40E[X_1(t)] = 40(0) = 0$, since $X_1(t)$ is a mean-zero process. The other three similar terms are also 0. Using the independence assumption, we can

---

[2] Readers not familiar with the triangular or "tri" function should consult Appendix B.

rewrite the second term in Eq. (7.5) as $E[X_1(t)X_2(s)]=E[X_1(t)]E[X_2(s)]=(0)(0)=0$. The last term in the middle line of Eq. (7.5) is 0 for the same reason. In fact, only three terms do not vanish:

$$
\begin{aligned}
R_{XX}(t,s) &= E\big[X_1(t)X_1(s)\big] + E\big[X_2(t)X_2(s)\big] + E\big[1600\big] \\
&= R_{11}(t,s) + R_{22}(t,s) + 1600 \\
&= R_{11}(\tau) + R_{22}(\tau) + 1600 \qquad \text{because } X_1(t) \text{ and } X_2(t) \text{ are WSS} \\
&= 2000\mathrm{tri}(10{,}000\tau) + 650\cos(40{,}000\pi\tau) + 1600 \qquad (\tau = s - t)
\end{aligned}
$$

That is, the autocorrelation function of $X(t)$ equals the sum of the autocorrelations of $X_1(t)$ and $X_2(t)$, plus the square of the constant term ($40^2 = 1600$). Since the mean of $X(t)$ is a constant, 40, and the autocorrelation function of $X(t)$ depends only on τ, the random process $X(t)$ is indeed wide-sense stationary.

Finally, we can easily find the autocovariance function (which, since $X(t)$ is WSS, only depends on τ):

$$
C_{XX}(\tau) = R_{XX}(\tau) - \mu_X^2 = R_{XX}(\tau) - 40^2 = 2000\mathrm{tri}(10{,}000\tau) + 650\cos(40{,}000\pi\tau)
$$

Notice that, since $X_1(t)$ and $X_2(t)$ have mean zero, the two terms in $C_{XX}(\tau)$ are, in fact, their respective autocovariance functions.

Let's examine this example further. The random process $X(t)$ consists of three components: $X_1(t)$, which is not periodic (since $R_{11}(\tau)$ isn't); $X_2(t)$, which is periodic; and a constant. It's often the case that we can *partition* a random process in this manner:

$$
X(t) = \{\text{aperiodic components}\} + \{\text{periodic components}\} + \{\text{constant}\}
$$

Any or all of these three elements may be present, and the first two components may themselves be sums of other parts, e.g., the sum of several sinusoids with different periods may comprise the "periodic components" piece. In engineering language, the constant term is called the **dc offset**: if $X(t)$ represents a current waveform, the constant term is the direct current in $X(t)$, while the other two components comprise the alternating current (ac) of $X(t)$.

Now look at $R_{XX}(\tau)$, which also consists of three parts: an aperiodic part (also called the *dissipative* component, since this is the term that goes to 0 as $|\tau| \to \infty$), a periodic part, and the square of the dc offset. In general, the autocorrelation function of a WSS process can be decomposed into

$$
R_{XX}(\tau) = \{\text{dissipative components}\} + \{\text{periodic components}\} + \{\text{constant}\}
$$

$$(7.6)$$

The constant term in Eq. (7.6) is called the **dc power offset**, since it reflects the power that results if the dc offset were to pass through a 1-Ω resistance (viz., $P = I^2 R = 40^2(1) = 1600$). Finally, the autocovariance function of $X(t)$ includes only the first two parts of Eq. (7.6), dissipative and periodic components, and not the dc power offset. In a sense, $C_{XX}(\tau)$ tells us something about the ac power in our random current waveform. We will explore these ideas much further in Chap. 8. ∎

7.3.2   Ergodic Processes

As we'll see in this chapter and the next, it is desirable to understand the statistical properties of a random process—mean, variance, and so on. But because these are properties of the ensemble, a problem arises: in practice, we generally only observe a single realization of the process, and so we must somehow reconstruct the ensemble properties from this one signal. Thankfully, many stationary random processes have the feature that their time and ensemble properties match (e.g., the time average of a single realization equals the ensemble mean). A process with this feature is called *ergodic*.

To give some intuition for this concept, imagine you have a large collection of identical dice. If you did not know how dice behave, you would have two options: roll all of the dice once, or roll one die many times. These should give us roughly the same information, but the first method illustrates *ensemble* properties (many realizations at one point in time) while the second gives *time* properties (one sampled die observed again and again over time). This process is ergodic. The benefit of ergodicity here is that we can learn about the behavior of all dice by purchasing (and repeatedly rolling) just a single die.

To make the property of ergodicity more precise, we need to introduce the time-dimension analogues of mean, autocorrelation, etc. For a fixed value $T > 0$, the average of a function $x(t)$ over the interval $[-T, T]$ is defined in calculus by

$$\langle x(t) \rangle_T = \frac{1}{2T} \int_{-T}^{T} x(t)dt$$

By allowing T to approach infinity, we may define the **time average** of a function $x(t)$ by

$$\langle x(t) \rangle = \lim_{T \to \infty} \langle x(t) \rangle_T = \lim_{T \to \infty} \frac{1}{2T} \int_{-T}^{T} x(t)dt$$

If the function $x(t)$ is periodic, the time average defined above is equal to the average of $x(t)$ across one period.

The time average of a *random* process $X(t)$ is defined by replacing $x(t)$ in the above expression with $X(t)$. The result is a quantity that clearly does not depend on time (since we have integrated dt) but which may vary across different members of the ensemble (i.e., the time average $\langle X(t) \rangle$ is still a random quantity).

DEFINITION

A random process $X(t)$ is **mean ergodic** if its time average and ensemble average are the same, i.e., if

$$\langle X(t) \rangle = E[X(t)]$$

in the mean square sense.[3]

Since the time average of $X(t)$ does not depend on t, a necessary condition for mean ergodicity is for $E[X(t)]$ to be a constant. Generally speaking, a random process must be stationary in order to be ergodic, but this is not always sufficient.

Example 7.19 Let $X(t) = A_0\cos(\omega_0 t + \Theta)$, where $\Theta \sim \text{Unif}(-\pi, \pi]$ and $\omega_0 \neq 0$. We found in Example 7.12 that $E[X(t)] = 0$. Now consider the time average:

$$\langle X(t) \rangle_T = \langle A_0 \cos(\omega_0 t + \Theta) \rangle_T$$

$$= \frac{1}{2T} \int_{-T}^{T} A_0 \cos(\omega_0 t + \Theta)dt = \frac{A_0}{2\omega_0 T}[\sin(\omega_0 T + \Theta) - \sin(-\omega_0 T + \Theta)]$$

Since the term in brackets is bounded no matter the value of Θ, the factor of T in the denominator implies that $\langle X(t) \rangle_T \to 0$ as $T \to \infty$, whence $\langle X(t) \rangle = 0$. Therefore, by definition, $X(t)$ is mean ergodic. ∎

Example 7.20 Consider a random dc signal, $X(t) = X$. That is, any particular sample function of the random process is some constant x, but that constant varies from realization to realization. This is trivially a stationary process; in particular, $E[X(t)] = E[X] = \mu_X$, a constant. However, the time average of X on $[-T, T]$ is just X, so $\langle X(t) \rangle = X$, which is not a constant. Therefore, $\langle X(t) \rangle \neq E[X(t)]$, and $X(t)$ is not mean ergodic. (To be precise, the time and ensemble averages would be equal in the mean-square sense if we had $E[(X - \mu_X)^2] = 0$, i.e., if X had zero variance.)

This matches with intuition: if the level of the dc signal X varies across different realizations of the process, then a single realization, $X(t) = x$, cannot tell us anything about the statistical behavior of the signal. ∎

The preceding example indicates the most common situation wherein a random process is stationary but not ergodic: some element of the process is random but not time-varying. See Example 7.21 below, as well as Exercise 36, for other such examples.

---

[3] A rv Y equals a constant c in the **mean square sense** if $E[(Y - c)^2] = 0$. This is equivalent to requiring that $E(Y) = c$ and $\text{Var}(Y) = 0$. In the definition above, $Y = \langle X(t) \rangle$ is the rv and $c = E[X(t)]$ is the constant.

Similar to the definition of time average, we may define the **time autocorrelation** of a random process by

$$\langle X(t)X(t+\tau)\rangle = \lim_{T\to\infty} \langle X(t)X(t+\tau)\rangle_T = \lim_{T\to\infty} \frac{1}{2T}\int_{-T}^{T} X(t)X(t+\tau)dt$$

$X(t)$ is said to be **autocorrelation ergodic** if $\langle X(t)X(t+\tau)\rangle = E[X(t)X(t+\tau)]$ in the mean square sense. Since the time autocorrelation does not depend on t, a random process can only be autocorrelation ergodic if its autocorrelation function $R_{XX}(t, t+\tau)$ is also free of t (the second condition for wide-sense stationarity).

Example 7.21 (Example 7.19 continued) We showed in Example 7.14 that the random process $X(t) = A_0\cos(\omega_0 t + \Theta)$ has autocorrelation function $R_{XX}(\tau) = (A_0^2/2)\cos(\omega_0 \tau)$. Applying an appropriate trig identity, we can also find its time autocorrelation:

$$\langle X(t)X(t+\tau)\rangle_T = \frac{1}{2T}\int_{-T}^{T} A_0\cos(\omega_0 t)\cdot A_0\cos(\omega_0(t+\tau))dt$$

$$= \frac{A_0^2}{2T}\int_{-T}^{T}\frac{1}{2}[\cos(\omega_0\tau) + \cos(\omega_0(2t+\tau))]dt$$

$$= \frac{A_0^2}{2T}\cdot\frac{1}{2}\cos(\omega_0\tau)\left[T-(-T)\right] + \frac{A_0^2}{2T}\int_{-T}^{T}\cos(\omega_0(2t+\tau))dt$$

$$= \frac{A_0^2}{2}\cos(\omega_0\tau) + \frac{A_0^2}{8\omega_0 T}\left[\cos\left(\omega_0(2T+\tau)\right) - \cos\left(\omega_0(-2T+\tau)\right)\right]$$

Taking the limit as $T\to\infty$, the second term above goes to zero, and we have $\langle X(t)X(t+\tau)\rangle = (A_0^2/2)\cos(\omega_0\tau)$, the same as the autocorrelation function of $X(t)$. Hence, $X(t)$ is autocorrelation ergodic.

However, suppose we replace the constant amplitude A_0 with a random amplitude A, i.e., a random variable not varying with time. Then calculations similar to those in Example 7.12 and above show that

$$R_{XX}(\tau) = \frac{E[A^2]}{2}\cos(\omega_0\tau) \quad \text{while} \quad \langle X(t)X(t+\tau)\rangle = \frac{A^2}{2}\cos(\omega_0\tau)$$

These are *not* equal—that is, now $X(t)$ is *not* autocorrelation ergodic—unless it happens that $E[A^2] = A^2$ in the mean square sense, which is true iff $\text{Var}(A) = 0$. ∎

7.3.3   Exercises: Section 7.3 (25–40)

25. Define a random process $X(t) = V + A_0\cos(\omega_0 t + \Theta)$, where V and Θ are independent random variables; $\Theta \sim \text{Unif}(-\pi, \pi]$; and V has mean μ_V and variance σ_V^2. (That is, $X(t)$ models a signal with both phase and dc variation.)
    (a) Find the mean function of $X(t)$.
    (b) Find the autocorrelation function of $X(t)$.
    (c) Is $X(t)$ wide-sense stationary?

26. Define a random process $X(t) = A\cos(\omega_0 t + \Theta)$, where A and Θ are independent random variables; $\Theta \sim \text{Unif}(-\pi, \pi]$; and A has mean μ_A and variance σ_A^2. (That is, $X(t)$ models a signal with both phase and amplitude variation.)
    (a) Find the mean function of $X(t)$.
    (b) Find the autocorrelation function of $X(t)$.
    (c) Is $X(t)$ wide-sense stationary?

27. Determine whether each of the following functions could potentially be the autocovariance function of a WSS random process.
    (a) $\cos(\tau)$
    (b) $\sin(\tau)$
    (c) $1/(1 + \tau^2)$
    (d) $e^{-|\tau|}$

28. Determine whether each of the following functions could potentially be the autocovariance function of a WSS random process.
    (a) $e^{-|\tau+1|}$
    (b) τ^2
    (c) $\text{tri}(\tau)$, defined by $\text{tri}(\tau) = 1 - |\tau|$ for $|\tau| \le 1$ and 0 otherwise
    (d) $\text{sinc}(\tau)$, defined by $\text{sinc}(0) = 1$ and $\text{sinc}(\tau) = \sin(\pi\tau)/(\pi\tau)$ for $\tau \ne 0$

29. Let A and B be iid random variables, and define a random process by $X(t) = A\cos(\omega_0 t) + B\sin(\omega_0 t)$. Is $X(t)$ necessarily WSS? Why or why not?

30. Define $X(t) = At + B$, where A and B are independent, $A \sim \text{Unif}[-3, 3]$, and $B \sim \text{Unif}[-10, 10]$.
    (a) Find the mean function of $X(t)$.
    (b) On the basis of (a), can you determine whether $X(t)$ is WSS? If so, what is your determination?
    (c) Find the variance function of $X(t)$.
    (d) On the basis of (c), can you determine whether $X(t)$ is WSS? If so, what is your determination?

31. Let $A(t)$ and $B(t)$ be jointly wide-sense stationary random processes, and define a new process by $X(t) = A(t) + B(t)$. Find the mean and autocovariance functions of $X(t)$. Is $X(t)$ WSS?

32. Let $A(t)$ and $B(t)$ be jointly wide-sense stationary random processes, and define a pair of new processes by

$$X(t) = A(t) + B(t)$$
$$Y(t) = A(t) - B(t)$$

Are $X(t)$ and $Y(t)$ jointly wide-sense stationary?

33. A wide-sense stationary process $Y(t)$ has mean -7 and autocovariance function $C_{YY}(\tau) = 50\cos(100\pi\tau) + 8\cos(600\pi\tau)$.
    (a) Does $Y(t)$ have any periodic components? How can you tell?
    (b) Find $\text{Cov}(Y(0), Y(0.01))$.
    (c) Find the autocorrelation function of $Y(t)$.
    (d) Find the mean square value of $Y(t)$.
    (e) Find the variance of $Y(t)$.

34. A wide-sense stationary process $X(t)$ has autocorrelation function $R_{XX}(\tau) = 60 + 125e^{-|\tau|/100}$.
    (a) Does $X(t)$ have any periodic components? How can you tell?
    (b) Find the mean square value of $X(t)$.
    (c) Find the mean of $X(t)$, if possible.
    (d) Find the autocovariance function of $X(t)$.
    (e) Find $\text{Cov}(X(10), X(15))$.
    (f) Find the standard deviation of $X(t)$.

35. Consider the random process $X(t) = A\cos(\omega_0 t) + B\sin(\omega_0 t)$, where A and B are iid random variables with mean zero. In Example 7.15, we showed that $X(t)$ is wide-sense stationary. Is $X(t)$ mean ergodic?

36. Let $X(t) = A \cdot Y(t)$, where A is a random variable and $Y(t)$ is an ergodic, WSS random process independent of A.
    (a) Find the mean and autocorrelation of $X(t)$ in terms of the properties of A and $Y(t)$. Is $X(t)$ WSS?
    (b) Show that the autocovariance function of $X(t)$ is given by $C_{XX}(\tau) = E(A^2)C_{YY}(\tau) + \sigma_A^2\,\mu_Y^2$.
    (c) Find the time average of $X(t)$. Is $X(t)$ mean ergodic?
    (d) Assume $Y(t)$ has no periodic component, so its autocovariance function goes to 0 as $|\tau| \to \infty$. Does the same hold true for the autocovariance function of $X(t)$? Why is this not a violation of property 5 of WSS processes?

37. Recall from Chap. 4 that the *correlation coefficient* of two rvs X and Y is given by $\rho(X, Y) = \text{Cov}(X, Y)/\sigma_X\sigma_Y$. For a WSS random process $X(t)$, find the correlation coefficient of $X(t)$ and $X(t+\tau)$ in terms of the autocovariance function $C_{XX}(\tau)$.

38. Let $X(t)$ be a WSS random process.
    (a) Prove that the autocovariance function $C_{XX}(\tau)$ satisfies $|C_{XX}(\tau)| \leq C_{XX}(0)$. [*Hint*: Use the previous exercise.]
    (b) Prove that the autocovariance function $R_{XX}(\tau)$ satisfies $|R_{XX}(\tau)| \leq R_{XX}(0)$.

39. Let $X(t)$ and $Y(t)$ be jointly wide-sense stationary. Show that $R_{XY}(\tau) = R_{YX}(-\tau)$ and $C_{XY}(\tau) = C_{YX}(-\tau)$.

40. Let $X(t)$ be a WSS random process.
    (a) Show that for any constants $a_1, \ldots, a_n$ and any time points $t_1, \ldots, t_n$,

$$\text{Var}(a_1 X(t_1) + \cdots + a_n X(t_n)) = \sum_{j=1}^{n} \sum_{k=1}^{n} a_j a_k C_{XX}(t_j - t_k)$$

(b) Explain why a valid autocovariance function must be **positive semi-definite**, i.e., $C_{XX}(\tau)$ must satisfy $\sum_{j=1}^{n} \sum_{k=1}^{n} a_j a_k C_{XX}(t_j - t_k) \geq 0$ for all constants $a_1, \ldots, a_n$ and times $t_1, \ldots, t_n$.

(c) Now consider the "rectangular function" defined by

$$\text{rect}(\tau) = \begin{cases} 1 & |\tau| \leq 1/2 \\ 0 & \text{otherwise} \end{cases}$$

Show that $\text{rect}(\tau)$ satisfies the properties of an autocovariance function expressed in the main proposition of this section.

(d) By considering $n = 3$, $a_1 = a_3 = 1$, $a_2 = -1$, $t_1 = -.3$, $t_2 = 0$, $t_3 = .3$, show that $\text{rect}(\tau)$ is not positive semi-definite and, hence, cannot be the autocovariance function of any WSS random process.

[*Note*: It can be shown that positive semi-definiteness completely characterizes the collection of all valid autocovariance functions. That is, any valid autocovariance function is automatically positive semi-definite, as shown in part (b), and for any positive semi-definite function C there exists a WSS random process whose autocovariance function is $C(\tau)$.]

7.4   Discrete-Time Random Processes

Much of what we have discussed in the previous two sections applies equally to discrete-time random processes (aka random sequences). After reviewing definitions and properties for the discrete-time case, we introduce a few specific examples of important discrete-time models.

Recall from Sect. 7.1 that a random sequence is simply a list of random variables X_1, X_2, and so on; we write X_n for the general term. (We may also define a sequence indexed by the entire set of integers: $\ldots, X_{-2}, X_{-1}, X_0, X_1, X_2, \ldots$.) The subscript takes the place of the time index t from earlier. The sequence can also be written as $X[1], X[2], \ldots$ or $X[n]$ to mirror the continuous-time notation; the square brackets will remind us that we're working on a discrete-time scale.

Example 7.22 Let's return to the value of Apple Inc. stock, but now consider only recording the closing price at the end of each trading day (starting, say, January 2 of next year). If we define X_n to be the closing price of Apple stock on the nth recorded day, then we can model $X_1, X_2, \ldots$ as a random sequence. Figure 7.13 shows one possible sample function of this random sequence, assuming the closing price just prior to our day 1 was \$580. In effect, we have converted a continuous-time process (analog) into a discrete-time process (digital) by *sampling* the process from Example 7.2 at designated times.

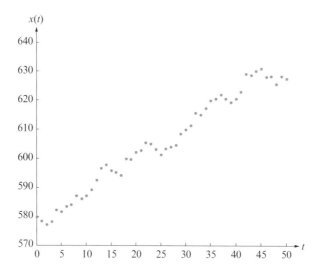

Fig. 7.13   A sample function of the random sequence of Example 7.22   ■

For any random sequence, we can define several statistical functions for times $n = 1, 2, 3, \ldots$ as follows:

- Mean function: $\mu_X[n] = E[X_n]$
- Variance function: $\sigma_X^2[n] = \text{Var}(X_n)$
- Standard deviation function: $\sigma_X[n] = \sqrt{\text{Var}(X_n)}$
- Autocovariance function: $C_{XX}[n, m] = \text{Cov}(X_n, X_m)$
- Autocorrelation function $R_{XX}[n, m] = E[X_n X_m]$

The relationships between these functions established in the continuous-time case still hold, e.g., $\sigma_X^2[n] = C_{XX}[n, n]$, and $C_{XX}[n, m] = R_{XX}[n, m] - \mu_X[n]\mu_X[m]$.

A random sequence is **(strict-sense) stationary** if the joint distribution of $X[n_1], \ldots, X[n_r]$ equals the joint distribution of $X[n_1 + k], \ldots, X[n_r + k]$ for any indices $n_1, \ldots, n_r$ and any integer k. A random sequence is **wide-sense stationary** if (1) $\mu_X[n]$ is a constant, μ_X, and (2) $C_{XX}[n, m]$ depends only on the difference $m - n$; if we call this difference k, we may then denote the autocovariance function as $C_{XX}[k]$. As was true for continuous-time processes, we may make an equivalent definition regarding the mean and autocorrelation functions.

Example 7.23   Any Markov chain from Chap. 6 is an example of a random sequence, provided the states for the chain are truly quantitative (e.g., counts or dollar amounts and not indicators for locations). Figure 7.14 shows two sample

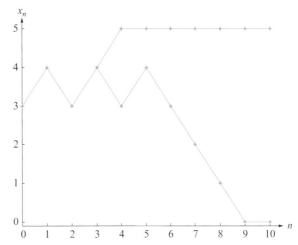

Fig. 7.14 Two sample functions of a Gambler's Ruin random sequence

functions for the Gambler's Ruin chain used repeatedly in Chap. 6, assuming Allan's initial stake is \$3, Beth's is \$2, and $p = .5$; the connecting line segments are just to help distinguish the two iterations. For each nonnegative index n, X_n equals the amount of money held by Allan after n games have been played.

The mean function $\mu_X[n]$ is the mean value of Allan's fortune after n games have been played. For example, with an initial stake of \$3 ($X_0 = 3$), X_1 equals \$2 or \$4 with probability .5 each, so $\mu_X[1] = \$2(.5) + \$4(.5) = \$3$. By considering the outcomes of the first two games, we find X_2 to be \$1, \$3, or \$5 with probabilities .25, .5, and .25, respectively; this gives $\mu_X[2] = \$3$ as well. In fact, it turns out that $\mu_X[n] = \$3$ for all n under the specified conditions, even though the probability distribution of X_n changes with n. (In particular, the distribution of X_n converges to $p(0) = .4$ and $p(5) = .6$, i.e., Allan's long run chance of winning all \$5 at stake is 60%.)

Similarly, we can compute the variance function of X_n; using the distributions described in the previous paragraph, it's straightforward to show that $\sigma_X^2[0] = 0$, $\sigma_X^2[1] = 1$, $\sigma_X^2[2] = 2$, and $\sigma_X^2[n] \to 6$ as $n \to \infty$. That the variance function is not constant is sufficient to show that the Gambler's Ruin Markov chain is *not* a WSS random sequence. ∎

7.4.1 Special Discrete Sequences

Perhaps the simplest type of random sequence is the **Bernoulli random sequence**: let $X_1, X_2, \ldots$ be iid, with each X_n following a Bernoulli(p) distribution. A sample function of a Bernoulli random sequence with $p = .6$ appears in Fig. 7.15. By grace of the variables being iid, a Bernoulli random sequence is trivially (strict-sense) stationary; in particular, we have

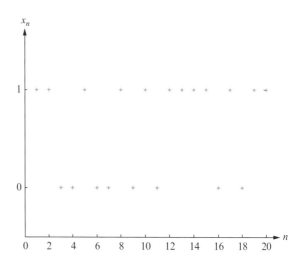

Fig. 7.15 A sample function of a Bernoulli random sequence

$$\mu_X[n] = E[X_n] = p, \sigma_X^2[n] = \text{Var}(X_n) = p(1-p), \text{and}$$

$$C_{XX}[n, m] = \text{Cov}(X_n, X_m) = \begin{cases} p(1-p) & m = n \\ 0 & m \neq n \end{cases}$$

A more general iid sequence will also be strict-sense stationary, but the formulas for the mean and variance will, of course, depend on the underlying common distribution of the X_n.

From an iid sequence, we can construct a much more interesting random sequence as follows: define $S_1 = X_1$, $S_2 = X_1 + X_2$, and so on, so that $S_n = S_{n-1} + X_n = \sum_{i=1}^{n} X_i$ for all $n \geq 2$. The resulting sequence of partial sums $S_1, S_2, S_3, \ldots$ is called a **random walk**. For example, from Chap. 2 we know the sum of iid Bernoulli rvs is binomial; hence, if X_n represents a Bernoulli random sequence, then the corresponding random walk S_n has a Bin(n, p) distribution for each n.

We can use the properties of iid sums to derive some general properties of random walks.

PROPOSITION

Let $X_1, X_2, \ldots$ be an iid sequence with common mean μ_X and common variance σ_X^2. Let S_n be the associated random walk, i.e., $S_n = X_1 + \cdots + X_n$ for every n. Then

1. $\mu_S[n] = E[S_n] = n\mu_X$
2. $\sigma_S^2[n] = \text{Var}(S_n) = n\sigma_X^2$
3. $C_{SS}[n, m] = \min(n, m)\sigma_X^2$

Proof   The proofs of properties 1 and 2 were given in advance of the Central Limit Theorem discussion in Sect. 4.5. To prove property 3, assume $m > n$ and proceed as follows:

$$
\begin{aligned}
C_{SS}[n,m] &= \text{Cov}(S_n, S_m) = \text{Cov}(X_1 + \cdots + X_n, X_1 + \cdots + X_m) \\
&= \text{Cov}(X_1 + \cdots + X_n, X_1 + \cdots + X_n + X_{n+1} + \cdots + X_m) \\
&= \text{Cov}(X_1 + \cdots + X_n, X_1 + \cdots + X_n) + \text{Cov}(X_1 + \cdots + X_n, X_{n+1} + \cdots + X_m) \\
&= \text{Cov}(S_n, S_n) + \text{Cov}(X_1 + \cdots + X_n, X_{n+1} + \cdots + X_m)
\end{aligned}
$$

In the third line, we have used the distributive property of covariance. The first term, $\text{Cov}(S_n, S_n)$, is simply $\text{Var}(S_n)$ (the covariance of S_n with itself). In the second term, the X_is ($i = 1$ to n) in the first argument are independent of the X_js ($j = n+1$ to m) in the second argument; therefore, that covariance equals zero. Thus, we have $C_{SS}[n,m] = \text{Var}(S_n) = n\sigma_X^2$.

This holds for $m > n$; if $m < n$, the same argument yields $C_{SS}[n,m] = m\sigma_X^2$. Therefore, for general n and m we may write $C_{SS}[n,m] = \min(n,m)\sigma_X^2$.   ∎

Example 7.24   Let X_1, X_2, ... be iid, with each X_i being $+1$ or -1 with equal probability. The resulting random walk S_n is called the *simple symmetric random walk in one dimension*. Since each "step" X_i has mean 0 and variance 1, it follows that $\mu_S[n] = n(0) = 0$ and $\sigma_S^2[n] = n(1) = n$. Several sample functions of S_n are shown in Fig. 7.16; the connecting line segments are just to help distinguish different iterations. Notice in Fig. 7.16b that the variability in S_n increases with n, but not linearly; this corresponds to the fact that $\text{SD}(S_n) = \sqrt{n}$. We also see, especially for larger n, that values of the ensemble of S_n are more concentrated near 0 and sparser at the edges. This is a consequence of the Central Limit Theorem: since the X_is are iid, their sum S_n becomes increasingly normal as n increases. (In fact, this is true for any random walk.)

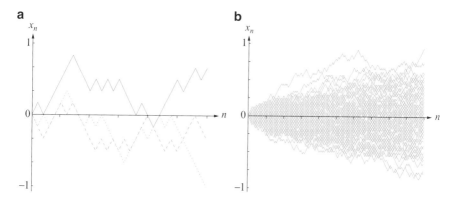

Fig. 7.16   Simple symmetric random walks: (**a**) the first 30 steps for three sample functions; (**b**) the first 200 steps for 100 sample functions   ∎

7.4.2   Exercises: Section 7.4 (41–52)

41. Let T_n denote the high temperature (°F) in Sacramento, CA on the nth day of the year. Consider the following model:

$$T_n = 75 + 25 \sin\left(\frac{2\pi}{365}(n - 150)\right) + 4\varepsilon_n,$$

where the ε_ns are a sequence of iid $N(0, 1)$ rvs.
    (a) Determine the probability that the high temperature on February 28 exceeds 60 °F.
    (b) Find $E[T_n]$.
    (c) Find $C_{TT}[n, m]$.
    (d) Is T_n a WSS random sequence? Should it be?
42. The output of a certain amplifier, sampled every second, has the form

$$X[n] = A_0 \sin(\omega_0 n) + Z[n],$$

where the noise component $Z[n]$ is a sequence of iid $N(0, \sigma)$ rvs for some $\sigma > 0$.
    (a) Find the mean function of $X[n]$.
    (b) Find the autocovariance function of $X[n]$.
    (c) Is $X[n]$ wide-sense stationary?
43. A gambler plays roulette, betting \$5 on black every time (so, she has probability 18/38 of winning on any particular spin). The gambler receives \$5 for each win and gives up \$5 for each loss.
    (a) Define a random sequence S_n = the number of games this gambler has won after n spins. Find the mean, variance, autocovariance, and autocorrelation function of S_n.
    (b) Define a random sequence Y_n = the amount of money this gambler has won after n spins. Find the mean, variance, autocovariance, and autocorrelation function of Y_n.
    (c) What is the probability the gambler is "ahead" after 10 spins (i.e., $Y_{10} > 0$)?
44. Gravel is being loaded onto rail cars by a dump truck for long-distance transport. Let X_n equal the amount of gravel (in tons) emptied onto the rail car by the nth dump truck run, and assume the X_n are iid Unif[15, 17].
    (a) Define $S_n = X_1 + \cdots + X_n$. Interpret S_n in this context.
    (b) Find the mean, variance, autocorrelation, and autocovariance functions of S_n.
    (c) Use the Central Limit Theorem to approximate both the distribution of S_6 and $P(S_6 \geq 100)$, the chance the dump truck will be able to fill a 100-ton rail car in 6 runs.
45. Let $X(t)$ be a WSS random process. For some fixed $T_s > 0$ define $X[n] = X(nT_s)$, so that $X[n]$ is a "sampled" version of $X(t)$. Show that the random sequence $X[n]$ is also WSS.

46. A *subsample* of a random sequence $X[n]$ is obtained by observing every kth element of the sequence, for some integer $k > 1$. The resulting random sequence, $Y[n]$, is given by $Y[n] = X[kn]$.
    (a) Find the mean and autocorrelation functions of $Y[n]$ in terms of those of $X[n]$.
    (b) If $X[n]$ is WSS, is the subsample $Y[n]$ also WSS?

47. Let X_n be a WSS random sequence, and define a *simple moving average* sequence Y_n by

$$Y_n = \frac{X_n + X_{n-1}}{2}$$

   (a) Find the mean function of Y_n.
   (b) Find the autocovariance function of Y_n.
   (c) Is Y_n wide-sense stationary?
   (d) Find the variance function of Y_n.

48. Let X_n be a sequence of iid random variables, and consider a new random sequence Y_n given by

$$Y_n = \frac{1}{2}X_n + \frac{1}{4}X_{n-1} + \frac{1}{8}X_{n-2}$$

   Is Y_n WSS?

49. Let the random sequence $\ldots, X_{-2}, X_{-1}, X_0, X_1, X_2, \ldots$ be iid, with mean μ and variance σ^2. Define a *first-order autoregressive* sequence Y_n by

$$Y_n = \alpha Y_{n-1} + X_n,$$

   where $|\alpha| < 1$.
   (a) Show that, for $N > 0$, $Y_n = \alpha^N Y_{n-N} + \sum_{i=0}^{N-1} \alpha^i X_{n-i}$.
   (b) Let $N \to \infty$ in (a) to conclude that $Y_n = \sum_{i=0}^{\infty} \alpha^i X_{n-i}$.
   (c) Find the mean function of Y_n.
   (d) Find the autocovariance function of Y_n.
   (e) Is Y_n wide-sense stationary?
   (f) Find the correlation coefficient $\rho(Y_n, Y_{n+k})$.

50. *Correlated bit noise.* Let X_n be a sequence of random bits (0s and 1s) constructed as follows: $X_0 = 0$ or 1 with probability .5 each. For $n \geq 1$, $X_n = X_{n-1}$ with probability .9 and $X_n = 1 - X_{n-1}$ with probability .1. (In the language of Chap. 6, this is a Markov chain with a symmetric transition matrix.)
   (a) Find the pmf of X_1, and argue that this is also the pmf of X_n for all $n \geq 1$.
   (b) Is X_n a WSS sequence?
   (c) Find the mean and variance functions of X_n.
   (d) It can be shown, using techniques from Chap. 6, that for $k \geq 0$

$$P(X_{n+k} = 1 | X_n = 1) = \frac{1 + .8^k}{2}$$

Use this to find $R_{XX}[k]$ and $C_{XX}[k]$.

51. Let $X[n]$ be a random sequence whose time index n ranges across all integers $\ldots, -2, -1, 0, 1, 2, \ldots$. Similar to the continuous-time case, the *time average* of $X[n]$ over $\{-N, \ldots, 0, \ldots, N\}$ is defined by

$$\langle X[n] \rangle_N = \frac{1}{2N + 1} \sum_{n=-N}^{N} X[n],$$

which is just the arithmetic mean of $X[-N], \ldots, X[N]$. The (overall) time average of $X[n]$ is then defined as a limit: $\langle X[n] \rangle = \lim_{N \to \infty} \langle X[n] \rangle_N$. Assume $X[n]$ is a WSS random sequence with mean μ_X and autocovariance function $C_{XX}[k]$.

(a) Show that, for all integers $N \geq 0$, $E[\langle X[n] \rangle_N] = \mu_X$.

(b) Show that $\mathrm{Var}\left(\langle X[n] \rangle_N\right) = \dfrac{1}{2N + 1} \displaystyle\sum_{k=-2N}^{2N} C_{XX}[k] \left(1 - \dfrac{|k|}{2N + 1}\right)$. [*Hint*: Use the relationship $\mathrm{Var}(Y) = \mathrm{Cov}(Y, Y)$ and the distributive property of covariance to create a double sum (with indices m and n, say). Then make the change of variable $k = m - n$ and rearrange the terms to create a single sum.]

52. Refer back to the previous exercise. A WSS random sequence $X[n]$ is called *mean ergodic* if its time average $\langle X[n] \rangle_N$ converges to μ_X as $N \to \infty$, in the sense that

$$\lim_{n \to \infty} E\left[\left(\langle X[n] \rangle_N - \mu_X\right)^2\right] = 0$$

(a) Use the previous exercise to show that $X[n]$ is mean ergodic if

$$\frac{1}{2N + 1} \sum_{k=-2N}^{2N} C_{XX}[k] \to 0 \text{ as } N \to \infty.$$

(b) Let $X[n]$ be WSS with autocovariance function $C_{XX}[k] = \alpha \cdot \rho^{|k|}$ for some $\alpha > 0$ and $|\rho| < 1$. Show that $X[n]$ is mean ergodic.

7.5   Poisson Processes

In Sect. 2.5 we indicated that, under rather general conditions, the Poisson distribution furnishes a probability model for the number of events of some sort (logins to a server, arrivals of radioactive pulses, flaws on the surface of a wafer, etc.) that occur in some fixed interval of time or region of space. We now present a more formal development of conditions that lead to the Poisson distribution in such contexts, and then explore properties of this event process.

> **DEFINITION**
> Consider the experiment of observing randomly occurring events of some type continuously over time. Define $X(0) = 0$, and define $X(t)$ for $t > 0$ to be the number of events that occur in the time interval $(0, t]$. $X(t)$ is called a **Poisson (counting) process** if it satisfies the following two conditions:
> 1. The numbers of events occurring in nonoverlapping time intervals are independent.
> 2. There exists a parameter $\lambda > 0$, called the **rate** of the process, such that the number of events occurring in any interval of length τ has a Poisson distribution with mean $\lambda\tau$.

Later in this section, we present an alternative definition of a Poisson counting process which does not explicitly assume that the event count follows a Poisson distribution.

Condition 1 states that a Poisson counting process $X(t)$ has **independent increments**: if $(s_1, t_1]$ and $(s_2, t_2]$ are two intervals of time with $t_1 \leq s_2$, so that the intervals do not overlap, then the number of events that occur in the first interval is independent of the number of events that occur in the second interval. That is, the "increment" $X(t_1) - X(s_1)$ is independent of $X(t_2) - X(s_2)$.

Condition 2 states that for any $t > 0$ and $\tau > 0$, the number of events in the interval $(t, t + \tau]$ has a Poisson distribution with mean $\lambda\tau$. Since this count is represented by $X(t + \tau) - X(t)$, we may write $X(t + \tau) - X(t) \sim \text{Poisson}(\lambda\tau)$. Since the distribution of this "increment" depends only on τ and not t, we say that a Poisson counting process has **stationary increments**. By substituting $t = 0$ and $\tau = t$ into this expression, we have that $X(t) \sim \text{Poisson}(\lambda t)$, so that at each time t the process itself has a Poisson distribution. It follows that

$$\mu_X(t) = \lambda t \quad \text{and} \quad \sigma_X^2(t) = \lambda t \quad \text{(since mean and variance are equal for Poisson)}$$

A graph of a sample function of $X(t)$ appears in Fig. 7.17. It is clear both from Fig. 7.17 and the formulas above that a Poisson counting process is *not* stationary.

Let us now derive the autocovariance function of $X(t)$. We will rely on a clever trick; namely, for $t < s$ we will split the interval $(0, s]$ into the smaller intervals $(0, t]$ and $(t, s]$. A similar "trick" was used for a random walk in the previous section. Begin as follows:

$$C_{XX}(t, s) = \text{Cov}(X(t), X(s)) = \text{Cov}(X(t), X(t) + [X(s) - X(t)])$$
$$= \text{Cov}(X(t), X(t)) + \text{Cov}(X(t), X(s) - X(t))$$

In the second argument of the covariance function, we have separated $X(s)$, the count of the number of events in $(0, s]$, into two pieces: $X(t)$, the number of events in $(0, t]$, and $X(s) - X(t)$, which represents the number of events in the interval $(t, s]$. Now, we simplify: the first term, $\text{Cov}(X(t), X(t))$, is simply $\text{Var}(X(t))$; the second term, thanks to Condition 1, represents the covariance of two independent

Fig. 7.17 A sample function of a Poisson (counting) process

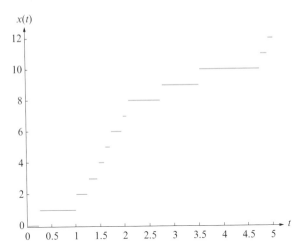

counts (since the intervals $(0, t]$ and $(t, s]$ don't overlap). Thus, the second term is zero, and we have $C_{XX}(t, s) = \text{Var}(X(t)) + 0 = \lambda t$ for $t < s$.

If $s < t$, we can use the same argument to find the covariance equals λs; therefore, the general expression for the autocovariance function of $X(t)$ is

$$C_{XX}(t, s) = \lambda \min(t, s)$$

Example 7.25 Database queries to a certain data warehouse occur randomly throughout the day. On average, 0.8 queries arrive per second during regular business hours. Assume a Poisson process model is applicable here.

First, consider the number of queries in the first five seconds, $X(5)$. The rv $X(5)$ is Poisson with mean $\lambda t = 0.8(5) = 4$. Thus, for example, the chance of exactly three queries in the first five seconds is

$$P(X(5) = 3) = \frac{e^{-4}4^3}{3!} = .195$$

Next, let's find the probability of exactly 1 query in the first second and exactly 2 queries in the four seconds thereafter, which requires the independent increments property: the number of queries in the first second and the number of queries in the four seconds thereafter are independent. More formally, $X(1)$ and $X(5) - X(1)$ are independent Poisson random variables with means $0.8(1) = 0.8$ and $0.8(4) = 3.2$, respectively. Hence,

$$P(X(1) = 1 \cap X(5) - X(1) = 2) = P(X(1) = 1) \cdot P(X(5) - X(1) = 2)$$

$$= \frac{e^{-0.8}0.8^1}{1!} \cdot \frac{e^{-3.2}3.2^2}{2!} = .075$$

Finally, consider the random variables $X(10)$ and $X(30)$. These two rvs are *not* independent: the time intervals $(0, 10]$ and $(0, 30]$ overlap. In fact, it should be obvious that $X(30)$ depends upon $X(10)$, since $X(30)$ counts the number of queries in the first 10 s, $X(10)$, plus the number of additional queries in the 20 s thereafter. Intuition suggests that the two random variables are positively correlated, and we verify this now.

The mean of $X(10)$ is $0.8(10) = 8$; since $X(10)$ is Poisson, its standard deviation is then $\sqrt{8}$. Similarly, $E[X(30)] = 0.8(30) = 24$ and $SD(X(30)) = \sqrt{24}$. We can find the covariance of $X(10)$ and $X(30)$ through the autocovariance function above:

$$\text{Cov}(X(10), X(30)) = C_{XX}(10, 30) = \lambda \min(10, 30) = 0.8(10) = 8$$

Finally, the correlation coefficient of $X(10)$ and $X(30)$ is

$$\text{Corr}(X(10), X(30)) = \frac{\text{Cov}(X(10), X(30))}{\text{SD}(X(10))\text{SD}(X(30))} = \frac{8}{\sqrt{8}\sqrt{24}} = \frac{1}{\sqrt{3}} \approx .577$$

We're not surprised to find a moderate, positive relationship between these two variables. As you might guess, the correlation coefficient will be largest when the two time intervals (here, $(0, 10]$ and $(0, 30]$) overlap the most; if the time intervals of two increments only overlap to a very small degree, the resulting correlation coefficient will likewise be small (but still positive). ∎

7.5.1 Relation to Exponential and Gamma Distributions

Because the events in a Poisson process occur "at random," there is a second type of random variable we may wish to model: the *time between events*. Consider Fig. 7.18, which illustrates a Poisson process: the symbols along the time axis indicate the occurrences of events. Along the time axis, we have indicated several random variables: $T_1 =$ the time until the first event occurs, measured from $t = 0$; $T_2 =$ the time between the first and second events; $T_3 =$ the time between the second and third events; and so on. These random variables $T_1, T_2, \ldots$ are called the **interarrival times** of the process. Unlike the Poisson count of events, which is discrete, each of these random time lengths is a *continuous* random variable. Thanks to the following theorem, their probability distribution is known.

> **THEOREM**
>
> Suppose events occur in accordance with the conditions of a Poisson counting process. Define $T_1 =$ the time until the first event occurs and, for $n \geq 2$, $T_n =$ the time between the occurrence of the $(n-1)$th and nth events. Also, define $Y_n =$ the time until the nth event occurs, starting at $t = 0$. Then

(continued)

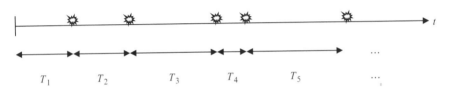

Fig. 7.18 Interarrival times in a Poisson process

1. $T_1, T_2, \ldots$ are independent exponential random variables with parameter λ (mean $1/\lambda$) and
2. Y_n is a gamma random variable with parameters $\alpha = n$ and $\beta = 1/\lambda$ (aka the *Erlang* distribution).

Proof Since the time intervals spanned by T_1, T_2, etc. do not overlap, the T_n are independent by condition 1 of a Poisson process (independent increments). To find the distribution of T_1, start with its cdf: for $t > 0$,

$$F_{T_1}(t) = P(T_1 \le t) = 1 - P(T_1 > t)$$
$$= 1 - P(\text{no events occur in the time interval } (0, t])$$
$$= 1 - P(X(t) = 0) \qquad \text{where } X(t) \sim \text{Poisson}(\lambda t)$$
$$= 1 - \frac{e^{-\lambda t}(\lambda t)^0}{0!} = 1 - e^{-\lambda t} \Rightarrow$$

$$f_{T_1}(t) = \frac{d}{dt} F_{T_1}(t) = \frac{d}{dt}\left[1 - e^{-\lambda t}\right] = \lambda e^{-\lambda t}$$

This is the exponential(λ) pdf, as claimed. The distribution of T_2 is also exponential(λ) because, thanks to Condition 1 of the definition, we may "restart the clock" when the first event occurs and derive the pdf of T_2 in the exact same manner as above. Propagating this idea forward, we have that $T_n \sim$ exponential(λ) for all n.

As for Y_n, notice we may write $Y_n = T_1 + \cdots + T_n$, which implies Y_n is the sum of n iid exponential(λ) rvs. Exercise 65 in Sect. 4.3 showed, using moment-generating functions, that the sum of n iid exponential(λ) rvs has a gamma(n, $1/\lambda$) distribution. ∎

Exercise 72 offers a direct proof of Statement 2 of the preceding theorem, without relying on moment-generating functions or Statement 1.

Example 7.26 Consider again the database queries described in Example 7.25. Rather than investigate the number of queries in preset time intervals, let's look at the random arrival times themselves. The average time between successive queries can actually be deduced without the preceding theorem: if queries arrive at 0.8

queries/second on average, then the mean time between queries is clearly just the reciprocal: $1/0.8 = 1.25$ s. If we let $T = $ the time between queries, we now know that $T \sim \text{exponential}(0.8)$, whence $E(T) = 1/\lambda = 1/0.8 = 1.25$ s *and* $\text{SD}(T) = 1/\lambda = 1.25$ s as well (remember that an exponential random variable has identical mean and standard deviation).

Next, let $Y_{50} = $ time to the 50th query, starting at the beginning of regular business hours. The preceding theorem tells us $Y_{50} \sim \text{gamma}(50,\ 1/0.8)$, so $E(Y_{50}) = 50(1/0.8) = 62.5$. We expect the 50th query to arrive 62.5 s into regular business hours. The arrival time of the 50th query has a standard deviation of

$$\text{SD}(Y_{50}) = \sqrt{\alpha\beta^2} = \sqrt{50(1/0.8)^2} = \sqrt{78.125} = 8.84\,\text{s}.$$

If 50 or more queries arrive within the first minute, system users will experience a significant backlog in subsequent minutes because of processing time. What is the probability this happens? A backlog occurs iff $Y_{50} \le 1\ \text{min} = 60$ s. The probability of this event, evaluated using software, is

$$P(Y_{50} \le 60) = \int_0^{60} \frac{1}{(50-1)!(1/0.8)^{50}} x^{50-1} e^{-0.8x}\, dx = \cdots = .4054$$

Alternatively, return to the original Poisson process: a backlog occurs iff $X(60)$, the number of queries in the first 60 s, is 50 or more. Since $X(60)$ has a Poisson distribution with mean $60(0.8) = 48$,

$$P(X(60) \ge 50) = 1 - P(X(60) \le 49) = 1 - \sum_{x=0}^{49} \frac{e^{-48}48^x}{x!} = 1 - .5946 = .4054$$

■

We have described a Poisson process as modeling the count of events that occur "at random" across time. This notion can actually be made more precise: *in a Poisson process, given that an event has occurred by time t_0, it is equally likely to have occurred anywhere in the interval $(0, t_0]$.* To see this, suppose we know that exactly one event has occurred by time t_0, so $X(t_0) = 1$. Conditional on that knowledge, let's find the distribution of the random variable $T_1 = $ arrival time of this event. Begin with the (conditional) cdf: for $t \le t_0$,

$$P\left(T_1 \leq t \mid X(t_0) = 1\right) = \frac{P(T_1 \leq t \cap X(t_0) = 1)}{P(X(t_0) = 1)}$$

$$= \frac{P\left(1 \text{ event occurred in } (0, t] \text{ and none in } (t, t_0]\right)}{P(X(t_0) = 1)}$$

$$= \frac{P(X(t) = 1 \cap X(t_0) - X(t) = 0)}{P(X(t_0) = 1)}$$

$$= \frac{\dfrac{e^{-\lambda t}(\lambda t)^1}{1!} \dfrac{e^{-\lambda(t_0 - t)}(\lambda(t_0 - t))^0}{0!}}{\dfrac{e^{-\lambda t_0}(\lambda t_0)^1}{1!}} = \cdots = \frac{t}{t_0}$$

Differentiating with respect to t, we find the conditional distribution of T, given $X(t_0) = 1$, is $1/t_0$, the uniform pdf on $(0, t_0]$.

Generalizing this argument, conditional on $X(t_0) = n$ (i.e., on exactly n events occurring in $(0, t_0]$) each of the n event occurrence times is uniformly distributed on $(0, t_0]$. Moreover, the n times are independent of one another. In light of this uniform distribution property, it is fair to say that for a Poisson process, events really do occur "at random."

7.5.2   Combining and Decomposing Poisson Processes

In Sect. 4.3, we showed that the Poisson distribution is *additive*, i.e., the sum of two independent Poisson rvs is again Poisson distributed. This result immediately generalizes to Poisson counting processes.

> **PROPOSITION**
> Let $X_1(t)$ and $X_2(t)$ be independent Poisson processes with rate parameters λ_1 and λ_2, respectively. Define a new random process by $X(t) = X_1(t) + X_2(t)$. Then $X(t)$ is also a Poisson process, with rate parameter $\lambda_1 + \lambda_2$. (This theorem can be extended to the sum of k independent Poisson processes for $k > 2$ as well.)

Example 7.27 Two roads feed into the northbound lanes on the Anderson Street Bridge. During rush hour, the number of vehicles arriving from the first road can be modeled by a Poisson process with a rate parameter of 10 per minute, while arrivals from the second road form an independent Poisson process with rate 8 cars per minute. If we let $X(t)$ denote the total number of cars entering the northbound lanes, then $X(t)$ is also a Poisson process, with rate parameter $10 + 8 = 18$ vehicles per minute. Hence, the probability that a total of more than 100 vehicles will arrive via the two feeder roads in the first 5 min of rush hour is given by

$$P(X(5) > 100) = 1 - P(X(5) \leq 100) = 1 - \sum_{x=0}^{100} \frac{e^{-18(5)}[18(5)]^x}{x!} = 1 - .865 = .135$$

This calculation is much simpler than considering all the possible ways the two individual Poisson processes could total more than 100 (e.g., 55 vehicles on the first road and 48 on the second road, and so on).   ■

The foregoing proposition and example show that we can combine separate Poisson processes into a single Poisson process. Interestingly, we can also do the reverse: if we can categorize the events of a Poisson process (e.g., arrivals of people separated into women's arrivals and men's arrivals), then we can *decompose* the overall process into two "smaller" processes. We make this more precise in the next proposition.

PROPOSITION

Suppose events occur according to the conditions of a Poisson process, and that each event can be classified as either Type 1 or Type 2. Suppose that each event is Type 1 with probability p, independent of the types of all other events and independent of the number of events that have occurred. Define two random processes: $X_1(t) =$ number of Type 1 events up to time t, and $X_2(t) =$ number of Type 2 events up to time t. Then
1. $X_1(t)$ is a Poisson process with rate parameter $p\lambda$;
2. $X_2(t)$ is a Poisson process with rate parameter $(1-p)\lambda$; and
3. $X_1(t)$ and $X_2(t)$ are independent.

Proof We will derive the joint distribution of $X_1(t)$ and $X_2(t)$, i.e., $P(X_1(t) = x$ and $X_2(t) = y)$, for arbitrary nonnegative integers x and y. The event $\{X_1(t) = x$ and $X_2(t) = y\}$ is equivalent to the event

$$\{X(t) = x + y \text{ and exactly } x \text{ of these } x + y \text{ events are Type 1}\} = A \cap B$$

where $X(t)$ denotes the overall Poisson process. The second part of this event follows a binomial model: we have a fixed number of trials $(x + y)$, each with two basic outcomes (Type 1 or Type 2), plus independent trials and constant probability by assumption. Combining that with the known Poisson distribution of $X(t)$ and the Multiplication Rule $P(A \cap B) = P(A)P(B|A)$ gives

$P(X(t)=x+y$ and exactly x of these $x+y$ events are Type 1)

$$= \frac{e^{-\lambda t}(\lambda t)^{x+y}}{(x+y)!} \cdot \binom{x+y}{x} p^x(1-p)^y = \frac{e^{-\lambda t}(\lambda t)^{x+y}p^x(1-p)^y}{x!y!}$$

$$= \frac{e^{-[p+(1-p)]\lambda t}(\lambda t)^{x+y}p^x(1-p)^y}{x!y!} = \frac{e^{-p\lambda t}(p\lambda t)^x}{x!} \frac{e^{-(1-p)\lambda t}((1-p)\lambda t)^y}{y!}$$

We recognize these two functions as the pmfs of a Poisson($p\lambda t$) distribution and a Poisson($(1-p)\lambda t$) distribution, respectively. Moreover, since the joint pmf of $X_1(t)$ and $X_2(t)$ separates into the product of individual pmfs, $X_1(t)$ and $X_2(t)$ are independent. ∎

Example 7.28   At a certain large hospital, patients enter the emergency room at a mean rate of 15 per hour. Suppose 20% of patients arrive in critical condition, i.e., they require immediate treatment. Assume patient arrivals meet the conditions of a Poisson process.

Let's first find the probability that more than 50 patients arrive in the next 4 h. Let $X(t)$ denote the Poisson process of patient arrivals (regardless of condition). Then $X(4)$ has a Poisson distribution with mean $\mu = \lambda t = 15(4) = 60$, so

$$P(X(4) > 50) = 1 - P(X(4) \le 50) = 1 - \sum_{x=0}^{50} \frac{e^{-60}60^x}{x!} = 1 - .108 = .892$$

Next, we find the probability that more than 10 critical patients arrive in the next 4 h. Let $X_1(t)$ denote the number of critical ("Type 1") patients that arrive within t hours. By the previous proposition, $X_1(t)$ is a Poisson process with rate parameter $p\lambda = .20(15) = 3$, so $X_1(4)$ is Poisson with mean $3(4) = 12$. Thus,

$$P(X_1(4) > 10) = 1 - P(X_1(4) \le 10) = 1 - \sum_{x=0}^{10} \frac{e^{-12}12^x}{x!} = 1 - .347 = .653$$

Finally, to find the probability that more than 10 critical patients and more than 40 noncritical patients arrive in the next 4 h, let $X_2(t)$ denote the number of noncritical ("Type 2") patients that arrive within t hours. Then $X_2(t)$ is also a Poisson process, but with rate parameter $(1-p)\lambda = (1-.20)(15) = 12$. Thus $X_2(4) \sim$ Poisson(48); moreover, $X_2(4)$ is independent of $X_1(4)$. Therefore,

$$P(X_1(4) > 10 \cap X_2(4) > 40) = P(X_1(4) > 10) \cdot P(X_2(4) > 40) = (.653)(.862) = .563$$

The calculation of $P(X_2(4) > 40)$ is similar to those displayed above.   ∎

7.5.3   Alternative Definition of a Poisson Process

The definition of a Poisson process at the beginning of this section almost seems a tautology, since we said $X(t)$ is a Poisson process if it has a Poisson distribution. The following theorem provides an alternative way to define a Poisson counting process.

THEOREM
Consider the experiment of observing randomly occurring events of some type along continuous time. Define $X(0) = 0$, and define $X(t)$ for $t > 0$ to be the number of events that occur in the time interval $(0, t]$. Suppose $X(t)$ has the following properties:
1. $X(t)$ has independent and stationary increments.
2. There exists $\lambda > 0$ such that in any time interval of length h, the probability that exactly one event occurs is $\lambda h + o(h)$.[4]
3. The probability of more than one event occurring in an interval of length h is $o(h)$.
Then $X(t)$ is a Poisson counting process with rate parameter λ.

Proof   Because of the stationarity assumption, it suffices to consider a time interval beginning at time 0. Let $P_k(t)$ denote the probability that exactly k events occur in the interval $[0, t]$. First consider $P_0(t + h)$, the probability that no events occur during the first $t + h$ units of time. In order for this to happen, no events must occur in $[0, t]$ and also no events must occur during the next h units of time. Since these two time intervals are nonoverlapping, the number of events that occur in the first interval is independent of the number that occur in the second interval. Thus

$$P_0(t + h) = P_0(t) \cdot P(\text{no events in an interval of length } h)$$
$$= P_0(t) \cdot \left[1 - P(\text{exactly one event}) - P(\text{at least two events})\right]$$
$$= P_0(t) \cdot \left[1 - \left(\lambda h + o(h)\right) - o(h)\right]$$
$$= P_0(t) \cdot \left[1 - \lambda h - o(h) - o(h)\right] = P_0(t) \cdot \left[1 - \lambda h + o(h)\right]$$

Rearranging this expression gives

$$\frac{P_0(t + h) - P_0(t)}{h} = -\lambda P_0(t) + \frac{o(h)}{h}$$

Now taking the limit as $h \to 0$ gives the derivative of $P_0(t)$:

$$P_0'(t) = -\lambda P_0(t)$$

This differential equation has the unique solution $P_0(t) = ce^{-\lambda t}$, where the constant c is determined by the initial condition $P_0(0) = 1$. This implies that $c = 1$ and thus that $P_0(t) = e^{-\lambda t}$, which sure enough is the probability of no events when the distribution is Poisson with parameter λt.

---

[4] Readers not familiar with $o(h)$ notation should consult Appendix B.

Now consider $P_k(t)$ for general k. In order to have k events occur in the interval $[0, t+h]$, it must be the case that either (1) k events occur in $[0, t]$ and none in the next h time units, or (2) $k-1$ events occur in $[0, t]$ and one occurs in the next h time units, or (3) For $l \geq 2$, $k-l$ occur in $[0, t]$ and l occur in the next h time units. By condition 3 in the theorem, the probability of the event in (3) is $o(h)$. Writing $P_k(t+h)$ as a sum of probabilities corresponding to cases (1), (2), and (3), rearranging, dividing by h, and taking the limit as $h \to 0$ gives the following system of differential equations:

$$P_k'(t) = -\lambda P_k(t) + \lambda P_{k-1}(t) \qquad k = 1, 2, 3, \ldots$$

Letting $Q_k(t) = P_k(t)e^{\lambda t}$, the above differential equation becomes $Q_k'(t) = \lambda Q_{k-1}(t)$. This system can be solved recursively starting with $Q_0(t) = 1$ (because $P_0(t) = e^{-\lambda t}$) to give $Q_k(t) = \lambda^k t^k / k!$, whence

$$P_k(t) = \frac{(\lambda t)^k e^{-\lambda t}}{k!} \qquad k = 0, 1, 2, \ldots \qquad \blacksquare$$

In this chapter, we are discussing *temporal* stochastic processes—that is, random processes that are functions of time. For some of these processes, in particular the Poisson counting process, there exist *spatial* analogues. A **spatial Poisson process** models the occurrence of "random" events in space, rather than in time (e.g., the location of flaws on an integrated circuit, or of trees in a forest). Analogous to the preceding theorem, suppose these random events meet the following conditions: (1) the numbers of events in nonoverlapping regions of space are independent; (2) the probability of exactly one event in a region of area h is $\lambda h + o(h)$ for some $\lambda > 0$; and (3) the probability of more than one event is a region of area h is $o(h)$. Then a similar proof to the one above shows that the random variable $X(R) =$ number of events in region R has a Poisson distribution with mean $\lambda \cdot (\text{area of } R)$.

7.5.4 Nonhomogeneous Poisson Processes

The Poisson process considered thus far is characterized by a constant rate λ at which events occur per unit time. A generalization of this is to suppose that the probability of exactly one event occurring in the interval $(t, t+h]$ is $\lambda(t) \cdot h + o(h)$ for some function $\lambda(t)$. That is, we replace λ in condition 2 of the preceding theorem with a nonnegative function $\lambda(t)$. It can then be shown that the number of events occurring during an interval $(t_1, t_2]$ has a Poisson distribution with mean

$$E[X(t_2) - X(t_1)] = \int_{t_1}^{t_2} \lambda(t) dt \qquad (7.7)$$

The occurrence of events over time in this situation is called a **nonhomogeneous Poisson process**, and $\lambda(t)$ is called the **intensity function** of the process. Notice that the special case $\lambda(t) = \lambda$ (a constant) returns us to the usual, "homogeneous" case; in particular, from Eq. (7.7) we immediately have $\mu_X(t) = \lambda t$.

Example 7.29 The article "Inference Based on Retrospective Ascertainment" (*J. Amer. Statist. Assoc.*, 1989: 360-372) considers the intensity function

$$\lambda(t) = e^{a+bt}$$

as appropriate for events involving transmission of HIV (the AIDS virus) via blood transfusions. Suppose that $a = 2$ and $b = 0.6$ (close to values suggested in the paper), with time in years. What is the expected number of events in the first 4 years? In the time interval (2, 6]? What is the probability that at most 20 events occur in first 18 months?

To determine the expectation in any interval, we apply Eq. (7.7). The expected number of events in the interval (0, 4] is

$$E[X(4) - X(0)] = \int_0^4 e^{2+0.6t}\, dt = 123.44,$$

while the expected number of events in (2, 6] equals $\int_2^6 e^{2+0.6t}\, dt = 409.82$. Notice that the expected numbers of events for these two time intervals are quite different, even though each interval has length 4 years; this illustrates that a nonhomogeneous Poisson process does *not* have stationary increments.

Finally, the number of events in the first 18 months (1.5 years), $X(1.5)$, has a Poisson distribution with parameter

$$\mu = E[X(1.5)] = E[X(1.5) - X(0)] = \int_0^{1.5} e^{2+0.6t}\, dt = 17.975$$

Therefore, $P(X(1.5) \le 20) = \sum_{x=0}^{20} e^{-17.975} 17.975^x / x! = .733.$ ∎

7.5.5   The Poisson Telegraphic Process

We end this section with a brief discussion of the **Poisson telegraphic process** (or **Poisson telegraph**), a popular model in engineering for noise in a binary channel. Suppose we have events occurring according to the conditions of a Poisson process. Define a new random process, $N(t)$, as follows: $N(0) = -1$ with probability .5 and $+1$ with probability .5; when a random event occurs, $N(t)$ switches parity (i.e., from -1 to $+1$ or vice versa). A sample function appears in Fig. 7.19; the x's through the middle indicate the time occurrences of the random events (notice these are precisely where the process switches parity).

The statistical properties of the Poisson telegraph are catalogued in the following proposition.

Fig. 7.19 One sample
function of a Poisson
telegraph

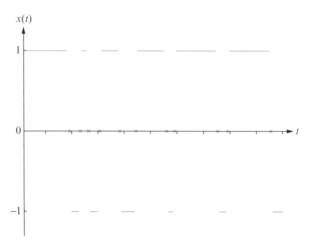

> **PROPOSITION**
> Let $N(t)$ be a Poisson telegraphic process with rate parameter λ.
> 1. For all $t \geq 0$, $N(t)$ is $+1$ or -1 with probability .5 each. (Thus, a Poisson telegraph has the same distribution at all time-points.)
> 2. $\mu_N(t) = 0$ and $\sigma_N(t) = 1$ for all $t \geq 0$.
> 3. $N(t)$ is WSS, and $R_{NN}(\tau) = C_{NN}(\tau) = e^{-2\lambda|\tau|}$.

The proofs of these statements are left as exercises (see Exercises 69–70 at the end of this section).

We more commonly think of the symbols 0 and 1 in binary communication, rather than -1 and $+1$. The Poisson telegraph described above can be easily modified: let $N^*(t) = .5[N(t) + 1]$, so that $N^*(t)$ takes on the values 0 and 1. We call $N^*(t)$ a **Poisson 0-1 telegraph**. In Exercise 71 of this chapter, you are asked to derive the properties of the Poisson 0-1 telegraph.

7.5.6   Exercises: Section 7.5 (53–72)

53. The number of requests for assistance received by a towing service is a Poisson process with rate $\lambda = 4$ per hour.
    (a) Compute the probability that exactly ten requests are received during a particular 2-h period.
    (b) If the operators of the towing service take a 30-min break for lunch, what is the probability that they do not miss any calls for assistance?
    (c) How many calls would you expect during their break?
54. During the daily lunch rush, arrivals at the drive-thru at a nearby fast food restaurant follow a Poisson process with a rate of 0.8 customers per minute.

(a) What is the expected number of customers in 1 h, and what is the corresponding standard deviation?

(b) The drive-thru's workers can't handle more than 10 customers in any 5-min span. Determine the probability that too many customers arrive for the workers to handle between 12:15 p.m. and 12:20 p.m.

(c) A customer has just arrived. What is the probability another customer will arrive within the next 30 s?

(d) The 100th lunch customer, starting at 12:00 p.m., gets a free meal. What is the expected arrival time of that lucky customer, and what is the standard deviation of that time?

55. Packets arrive at a certain node on the university's intranet at 10 packets per minute, on average. Assume packet arrivals meet the assumptions of a Poisson process.

(a) Calculate the probability that exactly 15 packets arrive in the next 2 min.

(b) Find an expression for the probability that more than 75 packets arrive in the next 5 min.

(c) Calculate the probability that the next packet will arrive in less than 15 s.

(d) What is the average time between successive packet arrivals?

(e) Calculate the probability that the fifth packet arrives in less than 45 s.

56. The article "Reliability-Based Service-Life Assessment of Aging Concrete Structures" (*J. Struct. Engrg.*, 1993: 1600–1621) suggests that a Poisson process can be used to represent the occurrence of structural loads over time. Suppose the mean time between occurrences of loads is .5 year.

(a) How many loads can be expected to occur during a 2-year period?

(b) What is the probability that more than five loads occur during a 2-year period?

(c) How long must a time period be so that the probability of no loads occurring during that period is at most .1?

57. Travelers arrive at an airport shuttle station according to a Poisson process with rate λ. The shuttle vehicle will depart only once k travelers have arrived. Assuming that there are no travelers waiting at time 0, what is the expected duration of time until the next shuttle vehicle departs?

58. The parking lot for a local ballpark has two entrances (east and west). In the hour before a game, cars entering the lot from east and west form two independent Poisson processes with rates 10 per minute and 15 per minute, respectively.

(a) What is the expected number of cars entering the parking lot in any 10-min span, and what is the corresponding standard deviation?

(b) In any particular minute, what is the probability exactly 12 cars enter from each side?

(c) What is the probability that exactly 24 cars enter the lot in any particular minute?

(d) Write an expression for the probability that, in any particular minute, the same number of cars enter through the east side and the west side.

59. Orders are submitted to a certain online business according to a Poisson process with rate 3 orders per hour.
    (a) Given that 4 orders are submitted during the time interval [0, 2], what is the probability that 10 orders are submitted in the interval [0, 5]?
    (b) More generally, consider two fixed times $s < t$ and two nonnegative integers $m < n$. Given that m orders are submitted by time s, what is the probability that n orders are submitted by time t?
60. Automobiles arrive at a vehicle equipment inspection station according to a Poisson process with rate $\lambda = 10$ per hour. Suppose that with probability .5 an arriving vehicle will have no equipment violations.
    (a) What is the probability that exactly ten vehicles arrive during the hour and all ten have no violations?
    (b) For any fixed $y \geq 10$, what is the probability that exactly y vehicles arrive during the hour, of which ten have no violations?
    (c) What is the probability that ten "no-violation" cars arrive during the next hour? [*Hint*: Sum the probabilities in (b) from $y = 10$ to ∞.]
61. A certain component is subject to electrical surges over time. Suppose that these surges occur according to a Poisson process with rate λ. Suppose also that with probability p, any particular surge will disable the component. What is the probability that the component survives (is not disabled) throughout the period $[0, t]$? [*Hint*: Make appropriate independence assumptions.]
62. Suppose events occur according to a Poisson process with rate λ.
    (a) Given that n events have occurred in the interval [0, n], what is probability that x of these events occurred in [0, 1]? [*Hint:* Let $X(t)$ be the Poisson process, and write the conditional probability of interest in terms of $X(t)$. Then apply the definition of conditional probability.]
    (b) Given that n events have occurred in the interval [0, n], what is the *limiting* conditional distribution of the number of events in [0, 1] as $n \to \infty$?
63. There is one hospital at the northern end of a particular county and another hospital at the southern end of the county. Suppose that arrivals to each hospital's emergency room occur according to a Poisson process with the same rate λ and that the two arrival processes are independent of one another. Starting at time $t = 0$, let Y be the elapsed time until at least one arrival has occurred at each of the two emergency rooms. Determine the probability distribution of Y.
64. Suppose that flaws occur along a cable according to a Poisson process with parameter λ. A segment of this cable of length Y is removed, where Y has an exponential distribution with parameter θ. Determine the distribution of the number of flaws that occur in this random-length segment. [*Hint*: Let X be the number of flaws on this segment. Condition on $Y = y$ to obtain $P(X = x | Y = y)$. Then "uncondition" using the Law of Total Probability (multiply by the pdf of Y and integrate). The gamma integral (3.5) will prove useful.]
65. Starting at time $t = 0$, certain events occur at random with inter-arrival times T_1, T_2, and so on as in Fig. 7.18. Define $X(t) =$ the number of arrivals in (0, t]; if we

assume the T_n are iid (but not necessarily exponentially distributed), then $X(t)$ is called a *renewal process*.

(a) Show that a renewal process whose inter-arrival times are iid exponential rvs is a Poisson process. That is, show that if the T_n are iid exponential(λ) rvs then $X(t)$ has a Poisson(λt) distribution.

(b) The *elementary renewal theorem* states that, for any renewal process,

$$\lim_{t \to \infty} \frac{E[X(t)]}{t} = \frac{1}{E[T_n]}$$

Show that this is trivially true for a Poisson process.

[*Note*: A stronger version of the renewal theorem actually shows $X(t)/t$ converges in probability to the constant $1/E[T_n]$.]

66. Let $X(t)$ count the number of events of a certain type in the time interval $(0, t]$, and suppose $X(t)$ can be modeled by a nonhomogeneous Poisson process with intensity function $\lambda(t)$.

(a) Does $X(t)$ have stationary increments? Why or why not?

(b) Find the mean and variance functions of $X(t)$.

(c) What is the probability that no events occur in the time interval $(0, t]$?

67. A certain repair facility is open for 8 h on a particular day. Customers arrive according to a nonhomogeneous Poisson process with rate (per hour) function $\lambda(t) = t$ for $0 \leq t < 1$, $= 1$ for $1 \leq t < 7$, and $= 8 - t$ for $7 \leq t \leq 8$.

(a) What is the probability that no customers arrive in both the first and last hours and that 4 customers arrive in the middle 6 h?

(b) What is the probability that the same number of customers arrive in the first hour, middle 6 h, and last hour?

68. During the first round of enrollment, students begin registering for classes at the beginning of each hour. There's a mad rush at the beginning of the hour, and then logins taper off. Let $X(t) =$ the number of logins t minutes into the hour, and suppose $X(t)$ can be modeled by a nonhomogeneous Poisson process with intensity function $\lambda(t) = 500/(t + 1)^2$ for $0 < t < 60$.

(a) What is the expected number of students that will log into the registration system in the first 5 min of the hour? In the last 5 min of the hour?

(b) What is the probability that no students log in during the last 5 min of an hour?

(c) The registration system will crash if more than 450 students log in during any 5-min period. What is the probability that this occurs in the first 5 min of an hour? (You will need to use software or a Central Limit Theorem approximation to determine this probability.)

69. Consider a Poisson telegraphic process $N(t)$ with rate parameter λ.

(a) Let $p = P(\text{an even number of events occur in } (0, t])$. Explain why, for $t > 0$,

$$P\big(N(t) = +1 \big| N(0) = +1\big) = p \quad \text{and} \quad P\big(N(t) = +1 \big| N(0) = -1\big) = 1 - p.$$

(b) Use (a) and the Law of Total Probability to show that $P(N(t) = +1) = .5$ for all $t \geq 0$. (Since the only other possible value of $N(t)$ is -1, this establishes property 1 of the last proposition of this section.)

(c) Establish property 2 of the Poisson telegraph, i.e., that $\mu_N(t) = 0$ and $\sigma_N(t) = 1$ for all $t \geq 0$.

70. (a) Consider a Poisson process with parameter λ. Show that the probability that an even number of events $(0, 2, 4, \ldots)$ occurs in any interval $(t, t + \tau]$ is equal to $(1 + e^{-2\lambda\tau})/2$.

(b) Let $N(t)$ be a Poisson telegraphic process with parameter λ. By considering the possible values of the product $N(t)N(t + \tau)$, show that the autocorrelation function of $N(t)$ is $e^{-2\lambda|\tau|}$. [*Hint:* Use (a).]

71. A *Poisson 0-1 telegraph* $N^*(t)$ is constructed as follows: $N^*(0)$ equals 0 or 1 with probability .5 each, and then $N^*(t)$ switches parity upon the occurrence of an event in a Poisson process. Find the pmf, mean, variance, autocovariance function, and autocorrelation function of $N^*(t)$. [*Hint:* Use the relationship $N^*(t) = \frac{1}{2}[N(t) + 1]$, where $N(t)$ is an ordinary Poisson telegraph.]

72. This exercise outlines a proof that the time to the nth event of a Poisson process has an Erlang distribution.

(a) Let Y_n denote the time to the nth event in a Poisson process. Argue that, for any time $y > 0$, $P(Y_n > y) = P(\text{fewer than } n \text{ events occur in the time interval } (0, y])$.

(b) Suppose the Poisson process has rate parameter λ. Use (a) to write an expression for the cdf of Y_n. [*Hint:* the right-hand side of (a) can be written as a finite sum using the definition of a Poisson process.]

(c) Differentiate your answer to part (b) to obtain the pdf of Y_n, and verify that it is an Erlang pdf with parameters n and λ (aka the gamma distribution with $\alpha = n$ and $\beta = 1/\lambda$).

7.6   Gaussian Processes

We introduced the normal or Gaussian distribution in Chap. 3 and then extended it to a multivariate distribution in Chap. 4. Here, we consider the extension of the normal distribution to random processes. Engineers commonly use such models for noise in audio signals and the (seemingly) random motion of small particles. We'll explore both of these applications shortly.

> **DEFINITION**
>
> A random process $X(t)$ is a **Gaussian process** if for all time points, $t_1, \ldots, t_n$ the random variables $X(t_1), \ldots, X(t_n)$ have a multivariate normal distribution (as defined in Sect. 4.7). In particular, the distribution of $X(t)$ at any time point t is normal.

As discussed in Sect. 4.7, we can also characterize a joint Gaussian distribution by requiring that all linear combinations of the random variables be Gaussian. Applying that characterization here, we have an alternate definition of a Gaussian process: $X(t)$ is a Gaussian process iff all linear combinations of $X(t_1)$, ..., $X(t_n)$ have a normal distribution for $n = 1, 2, 3, ...$, and all time-points $t_1, ..., t_n$.

In Sect. 7.3, we distinguished *strict-sense* stationary processes from *wide-sense* stationary processes. We noted that a strict-sense stationary process is automatically WSS, but not vice versa. However, suppose that a Gaussian process $X(t)$ is WSS: this implies the mean and covariance structure of $X(t)$ are time-invariant. But we know from Sect. 4.7 that mean and covariance completely characterize a joint Gaussian distribution; all other statistical properties can be derived from these two. Thus, if a Gaussian process is WSS, *all* of its statistical properties must be time-invariant.

> **PROPOSITION**
> Suppose $X(t)$ is a Gaussian process. Then $X(t)$ is wide-sense stationary if, and *only* if, $X(t)$ is strict-sense stationary.

Example 7.30 The noise $X(t)$ (measured in decibels) in an audio signal is modeled as a wide-sense stationary Gaussian process, with mean zero and autocovariance function

$$C_{XX}(\tau) = 0.04e^{-|\tau|/10}$$

Let's first investigate the distributions of $X(3)$ and $X(8)$, the noise three and eight seconds into the audio signal, respectively. Since $X(t)$ is a Gaussian process, by definition $X(3)$ is a Gaussian random variable; we merely have to specify its mean and variance. We are given that $X(t)$ is a mean-zero process, so in particular $E[X(3)] = 0$. We can extract the variance of $X(3)$ from the autocovariance using a property of WSS processes:

$$\text{Var}(X(3)) = \sigma_X^2 = C_{XX}(0) = 0.04e^{-|0|/10} = 0.04$$

Therefore, $X(3) \sim N(0, 0.2)$. Moreover, since $X(t)$ is stationary, this is also the distribution of $X(8)$.

Next, notice that $X(8) - X(3)$ is a linear combination of $X(3)$ and $X(8)$; therefore, since $X(t)$ is a Gaussian process, the random variable $X(8) - X(3)$ is also Gaussian. Its mean is simply $E[X(8) - X(3)] = 0 - 0 = 0$. Computing the variance takes a bit more effort:

$$\text{Var}(X(8) - X(3)) = \text{Var}(X(8)) + (-1)^2 \text{Var}(X(3)) + 2(1)(-1)\text{Cov}(X(8), X(3))$$
$$= \text{Var}(X(8)) + \text{Var}(X(3)) - 2\text{Cov}(X(8), X(3))$$
$$= C_{XX}(0) + C_{XX}(0) - 2C_{XX}(-5) \quad \text{since } \tau = 3 - 8 = -5$$
$$= 0.04 + 0.04 - 2 \cdot 0.04 e^{-|-5|/10}$$
$$= 0.08 \left(1 - e^{-1/2}\right) = .0315$$

Therefore, $X(8) - X(3) \sim N(0, \sqrt{.0315})$. Finally, we can use this distribution to find the probability the noise at $t=8$ is more than 0.3 dB above the noise at $t=3$:

$$P(X(8) > 0.3 + X(3)) = P(X(8) - X(3) > 0.3) = 1 - P(X(8) - X(3) \le 0.3)$$
$$= 1 - \Phi\left(\frac{0.3 - 0}{\sqrt{.0315}}\right) = 1 - \Phi(1.69) = .0455$$

$\blacksquare$

7.6.1   Brownian Motion

In Example 7.4, we introduced the idea of Brownian motion, a model for the seemingly random behavior of a dust particle on a liquid surface. Physicists also use the Brownian motion model to describe a variety of physical processes, including the motion of some celestial bodies in response to gravitational forces. A precise mathematical construction of Brownian motion is beyond the scope of this book; however, we characterize Brownian motion in the following definition.

> **DEFINITION**
> A (one-dimensional) **Brownian motion process** (also called a **Wiener process**) with parameter $\alpha > 0$, denoted $B(t)$, is a Gaussian process with the following properties:
> 1. $\mu_B(t) \equiv 0$; that is, Brownian motion is a mean-zero process.
> 2. $\sigma_B^2(t) = \alpha t$, so $\sigma_B(t) = \sqrt{\alpha t}$.
> 3. $B(t)$ has stationary and independent increments.
> If $\alpha = 1$, $B(t)$ is called **standard Brownian motion.**[5]

It can be shown that $B(t + \tau) - B(t) \sim N(0, \sqrt{\alpha \tau})$ for any $\tau > 0$, reflecting the stationary increments property. Figure 7.20 shows several sample functions of

---

[5] Albert Einstein showed in 1905 from physical considerations that the conditional pdf of $B(t_0 + t)$ given $B(t_0) = x$ must satisfy the partial differential equation $\partial f / \partial t = \frac{1}{2}\alpha \cdot \partial^2 f / \partial x^2$, where the "diffusion constant" α involves a gas constant, temperature, a coefficient of friction, and Avogadro's number. He also showed that the unique solution to this PDE is the normal pdf.

Fig. 7.20 Brownian motion

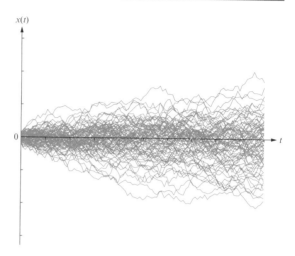

Brownian motion. It is important to note that while $B(t)$ has stationary *increments*, it is not itself a stationary process (the variance of Brownian motion depends on t).

Because Brownian motion is *not* a stationary process, we expect the autocovariance function will depend on "absolute" time (t and s) rather than "relative" time τ. It can be shown that

$$C_{BB}(t,s) = R_{BB}(t,s) = \alpha \cdot \min(t,s)$$

The derivation is similar to that of the Poisson process autocovariance function in the previous section.

Brownian motion actually shares several features with the Poisson process of the previous section: both have initial value 0 (with probability 1), stationary and independent increments, and variance proportional to time. In fact, it can be shown (see Exercise 85 at the end of this section) that any random process with constant initial value along with stationary and independent increments must necessarily have a variance function that's linear in t.

Since Brownian motion is a one-dimensional random process, clearly it can only describe particle motion in a single direction. The random motion of particles on a surface or through space can be described by 2- or 3-dimensional Brownian motion processes, for which it's assumed the motion along each dimensional axis is an independent, one-dimensional Brownian motion process.

Example 7.31 Consider the movement of a particle along a single axis, governed by Brownian motion with parameter $\alpha = 4$. Let's begin by identifying the probability distribution of the particle's displacement from time $t = 1$ s to time $t = 4$ s. If we write $B(t)$ for the process, then we wish to know the distribution of $B(4) - B(1)$. Applying the comment below the definition of Brownian motion with $\tau = 4 - 1 = 3$, we have $B(4) - B(1) \sim N(0, \sqrt{12})$. (An alternative derivation uses a similar approach to Example 7.30 and the autocovariance function mentioned above.)

The particle's displacement from time $t=2$ s to time $t=5$ s has this same distribution, since both increments span a time length of $\tau=3$ and, by Property 3 of the definition, $B(t)$ has stationary increments. However, the increments $B(5) - B(2)$ and $B(4) - B(1)$ are *not* independent: while Property 3 states that Brownian motion has independent increments, the two time intervals (2, 5] and (1, 4] overlap.

Finally, what is the probability that the particle is displaced more than 10 units in the time interval (1, 4]? Since the question does not indicate whether the displacement is positive or negative (relative to the axis), we're really interested in determining $P(|B(4) - B(1)| > 10)$. Because the distribution of $B(4) - B(1)$ is symmetric about 0, we may proceed as follows:

$$
\begin{aligned}
P(|B(4) - B(1)| > 10) &= 2P(B(4) - B(1) > 10) \qquad \text{by symmetry} \\
&= 2[1 - P(B(4) - B(1) \le 10)] \\
&= 2\left[1 - \Phi\left(\frac{10-0}{\sqrt{12}}\right)\right] = 2[1 - \Phi(2.89)] = .0038
\end{aligned}
$$

∎

7.6.2   Brownian Motion as a Limit

The Brownian motion process described above can actually be constructed as the limit of a discrete-time random process—specifically, the simple symmetric random walk S_n of Example 7.24. We will shrink both the time increment and the size of a jump in this random walk as follows: for some $h>0$ and $\Delta x>0$, suppose at times h, $2h$, $3h$, etc. the walk moves $+\Delta x$ or $-\Delta x$ with probability .5 each. Then, with [ ] denoting the greatest integer function, the random process $B(t)$ defined by

$$
B(t) = (\Delta x)X_1 + \cdots + (\Delta x)X_{[t/h]} = (\Delta x)S_{[t/h]}
$$

indicates the location of the random walk at time t. The coefficient Δx changes the motion increment from ± 1 to $\pm \Delta x$; the time index $n = [t/h]$ equals the number of moves (equivalently, the number of h-second time intervals) in the interval $[0, t]$.

From the properties of the random walk,

$$
\begin{aligned}
\mu_B(t) = E[\Delta x S_{[t/h]}] = \Delta x(0) = 0 \quad \text{and} \quad \sigma_B^2(t) &= (\Delta x)^2 \mathrm{Var}(S_{[t/h]}) = (\Delta x)^2 \cdot [t/h] \\
&= (\Delta x)^2[t/h]
\end{aligned}
$$

Moreover, the Central Limit Theorem tells us that, for large values of $[t/h]$, the distribution of $B(t)$ is approximately normal.

Up to now, the choices of h and Δx have been arbitrary. But suppose we choose $\Delta x = \sqrt{\alpha h}$ for some $\alpha>0$. Then, if we shrink h to 0 (effectively moving from discrete time to continuous time), $B(t)$ will be normally distributed with mean 0 and variance

$$\lim_{h \to 0} \alpha h[t/h] = \alpha t$$

The properties of independent and stationary increments clearly follow as consequences of the iid steps in the random walk. Thus, $B(t)$ becomes a Brownian motion process as $h \to 0$.

7.6.3 Further Properties of Brownian Motion

Consider some fixed value $x_0 > 0$ and a fixed time t_0. The maximum value of $B(t)$ during the time interval $0 \le t \le t_0$ is a random variable M. What is the probability that M exceeds the threshold x_0? Figure 7.21 shows two sample paths, one for which the level x_0 is exceeded prior to t_0 and one for which this does not occur.

Let's focus on a path $b(t)$ that does reach x_0 during the specified time interval. Corresponding to this path, we now create a new path by reflecting about the line $y = x_0$ the part of $b(t)$ that lies to the right of the *first* time it reaches x_0. Denote the first time at which the original path reaches x_0 by T. Then for $t \le T$ the new path— call it $b^*(t)$—is identical to the original path. But for any time $t > T$, it's easy to show that the reflected path is given by $b^*(t) = 2x_0 - b(t)$. The original path and its reflected path are illustrated in Fig. 7.22.

Notice that the maximum level for each of these paths exceeds x_0, and exactly one of these two paths has level exceeding x_0 at time t_0. That is, for every sample path that exceeds level x_0 at time t_0, there are two sample paths whose maxima on $[0, t_0]$ exceed x_0, the original path and the reflected path. Put another way, for each

Fig. 7.21 A sample path for which $M = \max_{0 \le t \le t_0} B(t) > x_0$ and a path for which $M < x_0$

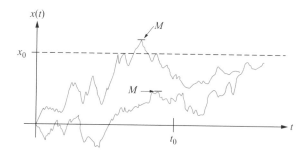

Fig. 7.22 A sample path crossing x_0 before time t_0 and its paired reflected path

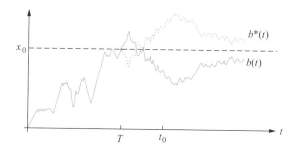

pair of sample paths whose level exceeds x_0 some time before t_0, one being the reflection of the other about x_0 subsequent to its first "hitting time," there is exactly one sample path satisfying $B(t_0) > x_0$.

Now given that $B(T) = x_0$, consider the pdf of the level $B(t)$ at some time subsequent to T. Because Brownian motion has independent increments, the process "begins anew" at time T, except that its Gaussian behavior starts at x_0 rather than at 0. The symmetry of the normal distribution implies that the pdf at a level above x_0 at the future time is the same as the pdf at a level that is below x_0 by the same amount. That is, the original path and the reflected path are equally likely. Melding this equally likely property with the result of the previous paragraph, establishing a one to one correspondence between pairs of reflected paths crossing x_0 and paths whose level exceeds x_0 at time t_0, gives the following result.

PROPOSITION
Let $B(t)$ be Brownian motion and, for $t_0 > 0$, let $M = \max_{0 \le t \le t_0} B(t)$. Then

$$P(M > x_0) = 2P(B(t_0) > x_0) = 2\left[1 - \Phi\left(\frac{x_0}{\sqrt{\alpha t_0}}\right)\right] \qquad (7.8)$$

The second equality in the proposition comes from the fact that $B(t_0) \sim N(0, \sqrt{\alpha t_0})$. Replacing x_0 with m on both sides of Eq. (7.8) and differentiating with respect to M, we can determine the pdf of this rv:

$$f_M(m) = \frac{2}{\sqrt{2\pi \alpha t_0}} e^{-m^2/(2\alpha t_0)} \qquad m > 0$$

The foregoing proposition also allows us to obtain the distribution of the random variable T = the first time at which the process hits level x_0. To see this, note that a sample path will first hit level x_0 before or at time t_0 iff the maximum level of the path during the time interval from 0 to t_0 is at least x_0. In symbols, $T \le t_0$ iff $M \ge x_0$. Since the probability of the latter event is what appears in the last proposition box, we immediately have the following result.

PROPOSITION
Let T be the first time that a Brownian motion process reaches level x_0. Then

$$F_T(t) = P(T \le t) = 2\left[1 - \Phi\left(\frac{x_0}{\sqrt{\alpha t}}\right)\right],$$

from which it follows that

$$f_T(t) = \frac{x_0}{\sqrt{2\pi\alpha}} t^{-3/2} e^{-x_0^2/2\alpha t} \qquad t > 0$$

Fig. 7.23 pdf of the hitting time, T, for Brownian motion

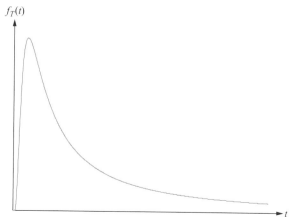

$f_T(t)$

Figure 7.23 shows the probability distribution of the "hitting time" T. Exercise 84 asks you to show this is a valid probability distribution and to derive the pdf from the cdf.

7.6.4 Variations on Brownian Motion

Let $B(t)$ denote a standard Brownian motion process (i.e., with $\alpha = 1$), and let μ and $\sigma > 0$ be constants. **Brownian motion with drift** is the process $X(t) = \mu t + \sigma B(t)$. $X(t)$ also has stationary an independent increments; an increment $X(t+\tau) - X(t)$ is normally distributed with mean $\mu\tau$ and variance $\sigma^2\tau$. Brownian motion with drift has many interesting applications and properties. For example, suppose the drift parameter μ is negative. Then over time $X(t)$ will tend toward ever lower values. It can be shown that M, the maximum level attained over all time $t \geq 0$, has an exponential distribution with parameter $2|\mu|/\sigma^2$.

Standard Brownian motion and Brownian motion with drift both allow for positive and negative values of the process. Thus they are not typically acceptable models for the behavior of the price of some asset over time. A stochastic process $Z(t)$ is called **geometric Brownian motion with drift parameter α** if $X(t) = \ln[Z(t)]$ is a Brownian motion process with drift having mean parameter $\mu = \alpha - \sigma^2/2$ and standard deviation parameter σ. Since $Z(t) = \exp(X(t))$, a geometric Brownian motion process will be nonnegative. Any particular sample path will show random fluctuations about a long-term exponential decay or growth curve.

Geometric Brownian motion is a popular model for the pricing of assets. Let $X(t)$ be the price of an asset at time t. The ratio $X(t)/X(0)$ is the proportion by which the price has increased or decreased between time 0 and time t. In the same way that we obtained Brownian motion as a limit of a simple symmetric random walk, geometric Brownian motion can be obtained as a limit in which the price at each time point either increases by a multiplicative factor u or goes down by another particular multiplicative factor d. The limit is taken as the number of price changes during

(0, t] gets arbitrarily large, while the factors u and d get closer and closer to 1 and the two probabilities associated with u and d approach .5. Geometric Brownian motion is the basis for the famous Black-Scholes option pricing formula that is used extensively in quantitative finance. This formula specifies a fair price for a contract allowing an investor to purchase an asset at some future time point for a particular price (e.g., a contract permitting an investor to purchase 100 shares of Facebook stock at a price of $20 per share 3 months from now).

7.6.5   Exercises: Section 7.6 (73–85)

73. Let $X(t)$ be a wide-sense stationary Gaussian process with mean $\mu_X = 13$ and autocovariance function $C_{XX}(\tau) = 9\cos(\tau/5)$.
    (a) Calculate $P(X(10) < 5)$.
    (b) Calculate $P(X(10) < X(8) + 2)$.
74. Let $Y(t)$ be a WSS Gaussian process with mean $\mu_Y = -5$ and autocorrelation function $R_{YY}(\tau) = (25\tau^2 + 34)/(1 + \tau^2)$. Determine each of the following:
    (a) $\text{Var}(Y(t))$
    (b) $P(Y(2) > 5)$
    (c) $P(|Y(2)| > 5)$
    (d) $P(Y(6) - Y(2) > 5)$
75. The voltage noise $N(t)$ in a certain analog signal is modeled by a Gaussian process with mean 0 V and autocorrelation function $R_{NN}(t, s) = 1 - |t - s|/10$ for $|t - s| \le 10$ (and zero otherwise).
    (a) Is $N(t)$ stationary? How can you tell?
    (b) Determine $P(|N(t)| > 1)$.
    (c) Determine $P(|N(t+5) - N(t)| > 1)$.
    (d) Determine $P(|N(t+15) - N(t)| > 1)$.
    [*Hint* for (b)–(d): Does your answer depend on t?]
76. The text *Gaussian Processes for Machine Learning* (2nd ed., 2006) discusses applications of the "regression-style" model $Y(t) = \beta_0 + \beta_1 t + X(t) + \varepsilon$, where β_0 and β_1 are constants, the "error term" ε is a $N(0, \sigma)$ rv, and $X(t)$ is a Gaussian random process with mean 0 and covariance function

$$C_{XX}(t, s) = \kappa^2 e^{-\lambda(t-s)^2}$$

for suitable choices of the parameters $\kappa > 0$ and $\lambda > 0$. $X(t)$ and ε are assumed to be independent.
    (a) Is $X(t)$ wide-sense stationary?
    (b) Is $Y(t)$ a Gaussian process?
    (c) Determine the mean, variance, and autocovariance functions of $Y(t)$. Is $Y(t)$ WSS?
    (d) What effect does the parameter κ have on $Y(t)$? That is, how would the behavior of $Y(t)$ be different for large κ versus small κ?
    (e) What effect does the parameter λ have on $Y(t)$?

77. Consider the following model for the temperature $X(t)$, in °F, measured t hours after midnight on August 1, in Bakersfield, CA:

$$X(t) = 80 + 20\cos\left(\frac{\pi}{12}(t - 15)\right) + B(t),$$

where $B(t)$ is a Brownian motion process with parameter $\alpha = .2$.
(a) Determine the mean and variance functions of $X(t)$. Interpret these functions in the context of the example.
(b) According to this model, what is the probability that the temperature at 3 pm on August 1 will exceed 102 °F?
(c) Repeat part (b) for 3 p.m. on August 5.
(d) What is the probability that the temperatures at 3 p.m. on August 1 and August 5 will be within 1 °F of each other?

78. Brownian motion is sometimes used in finance to model short-term asset price fluctuation. Suppose the price (in dollars) of a barrel of crude oil varies according to a Brownian motion process; specifically, suppose the *change* in a barrel's price t days from now is modeled by Brownian motion $B(t)$ with $\alpha = .15$.
(a) Find the probability that the price of a barrel of crude oil has changed by more than $1, in either direction, after 5 days.
(b) Repeat (a) for a time interval of 10 days.
(c) Given that the price has increased by $1 in seven days, what is the probability the price will be another dollar higher after an additional seven days?

79. Refer to the weather model in Exercise 77. Suppose a meteorologist uses the mean function of $X(t)$ as her weather forecast over the next week.
(a) Over the next five days, what is the probability that the actual temperature in Bakersfield will exceed the meteorologist's prediction by more than 5 °F? [*Hint:* What part of $X(t)$ represents her prediction error?]
(b) What is the probability that the actual temperature will exceed her prediction by 5 °F for the first time by midnight on August 3 (i.e., two days after $t = 0$)?

80. Refer to Exercise 78, and suppose the initial price of crude oil is $110 per barrel.
(a) Over the next 30 days, what is the probability the maximum price of a barrel of crude oil will exceed $115?
(b) Determine the probability that the price of crude oil will hit $120 for the first time within the next 60 days.

81. The motion of a particle in two dimensions (e.g., a dust particle on a liquid surface) can be modeled by using Brownian motion in each direction, horizontal and vertical. That is, if $(X(t), Y(t))$ denotes the position of a particle at time t, starting at $(0, 0)$, we assume $X(t)$ and $Y(t)$ are independent Brownian motion processes with common parameter α. This is sometimes called *two-dimensional Brownian motion*.

(a) Suppose a certain particle moves in accordance with two-dimensional Brownian motion with parameter $\alpha = 5$. Find the probability the particle is more than 3 units away from $(0, 0)$ in each dimension at time $t = 2$.

(b) For the particle in (a), find the probability the particle is more than 3 units away from $(0, 0)$ radially (i.e., by Euclidean distance) at time $t = 2$. [*Hint:* It can be shown that the sum of squares of two independent $N(0, \sigma)$ rvs has an exponential distribution with parameter $\lambda = 1/(2\sigma^2)$.]

(c) *Three-dimensional Brownian motion*, a model for particulate movement in space, assumes that each location coordinate $(X(t), Y(t), Z(t))$ is an independent Brownian motion process with common parameter α. Suppose a certain particle's motion follows three-dimensional Brownian motion with parameter $\alpha = 0.2$. Find the probability that (1) the particle is more than 1 unit away from $(0, 0, 0)$ in each dimension at time $t = 4$, and (2) the particle is more than 1 unit away from $(0, 0, 0)$ radially at time $t = 4$. [*Hint:* The sum of squares of three independent $N(0, \sigma)$ rvs has a gamma distribution with $\alpha = 3/2$ and $\beta = 2\sigma^2$.]

82. Some forms of thermal voltage noise can be modeled by an *Ornstein-Uhlenbeck process* $X(t)$, which is the solution to the "stochastic differential equation" $X'(t) + \kappa X(t) = \sigma B'(t)$, where $B(t)$ is standard Brownian motion and κ, $\sigma > 0$ are constants. With the initial condition $X(0) = 0$, It can be shown that $X(t)$ is a Gaussian process with mean 0 and autocovariance function

$$C_{XX}(t, s) = \frac{\sigma^2}{2\kappa}\left[e^{-\kappa|s-t|} - e^{-\kappa(s+t)}\right]$$

(a) Is the Ornstein-Uhlenbeck process wide-sense stationary? Why or why not?
(b) Find the variance of $X(t)$. What happens to the variance of $X(t)$ as $t \to \infty$?
(c) Let $s = t + \tau$. What happens to $C_{XX}(t, t + \tau)$ as $t \to \infty$?
(d) For $s > t$, determine the conditional distribution of $X(s)$ given $X(t)$.

83. A *Gaussian white noise process* is a Gaussian process $N(t)$ with mean $\mu_N(t) = 0$ and autocorrelation function $R_{NN}(\tau) = (N_0/2)\delta(\tau)$, where $N_0 > 0$ is a constant and $\delta(\tau)$ is the Dirac delta function (see Appendix B).

(a) Is Gaussian white noise a stationary process?
(b) Define a new random process, $X(t)$, as the integrated version of $N(t)$:

$$X(t) = \int_0^t N(s)\,ds$$

Find the mean and autocorrelation functions of $X(t)$. Is $X(t)$ stationary?

84. Consider the "hitting-time" distribution $F_T(t) = 2[1 - \Phi(x_0/\sqrt{\alpha t})]$, $t > 0$, for Brownian motion presented in the last proposition of this section.

(a) Show that $F_T(t)$ is a valid cdf for a nonnegative rv by proving that (1) $F_T(t) \to 0$ as $t \to 0^+$, (2) $F_T(t) \to 1$ as $t \to \infty$, and (3) $F_T(t)$ is an increasing function of t.

(b) Find the median of this hitting-time distribution.

(c) Derive the pdf of T from the cdf.

(d) Does the mean of this hitting-time distribution exist?

85. Let $X(t)$ be a random process with stationary and independent increments and $X(0)$ a constant.

(a) Take the variance of both sides of the expression

$$X(t + \tau) - X(0) = [X(t + \tau) - X(t)] + [X(t) - X(0)]$$

and use the properties of $X(t)$ to show that $\mathrm{Var}(X(t+\tau)) = \mathrm{Var}(X(t)) + \mathrm{Var}(X(\tau))$.

(b) The only solution to the functional relation $g(t + \tau) = g(t) + g(\tau)$ is a linear function: $g(t) = at$ for some constant a. Apply this fact to part (a) to conclude that any random process with a constant initial value and stationary and independent increments must have linear variance. (This includes both Brownian motion as well as the Poisson counting process of the previous section.)

7.7   Continuous-Time Markov Processes

Recall from Chap. 6 that a discrete-time Markov chain is a sequence of random variables X_0, X_1, X_2, ... satisfying the *Markov property* on some state space (typically a set of integers). In this section, we continue to assume that the state space consists of either a finite or infinite set of integers, but now the index set of possible subscripts consists of all t for which $t \geq 0$. Thus we have not only X_5 = the state of the process at time $t = 5$ and X_{12} = the state of the process at time 12, but also $X_{6.5}$, the state of the process at time $t = 6.5$, $X_{27.249}$, and so on. For example, X_t might be the number of customers in a service facility of some sort, where $t = 0$ is the time at which the facility opened; we then track the number of customers in continuous time rather than just at times 0, 1, 2, and so on. To be consistent with the notation of Chap. 6, we will write the random process as X_t, with time as a subscript, but we could just as well use the $X(t)$ notation from earlier sections of this chapter.

As before, the Markov property says that once we know the state of the process at some time t, the probability distribution of future states does not depend on the state of the process at any time prior to t. That is, given that $X_t = i$, the values of X_u for $u > t$ do not depend on the values of X_u for $u < t$. In particular, whenever we have times $t_1 < t_2 < \ldots < t_n < t < u$,

$$P(X_u = j | X_{t_1} = i_1, \ldots, X_{t_n} = i_n, X_t = i) = P(X_u = j | X_t = i)$$

In general, the **transition probabilities** $P(X_u = j | X_t = i)$ might depend not only on the time increment $u - t$ but also on t itself. For example, it might be the case that

$P(X_{10} = j | X_{7.5} = i)$ is different from $P(X_{15.5} = j | X_{13} = i)$ even though the increment is 2.5 time units in both expressions. As in the discrete-time case, we will assume throughout this section that our Markov processes are **time homogeneous**, i.e., for any time increment $h > 0$, the transition probability $P(X_{t+h} = j | X_t = i)$ depends on h but not on t, so that we may write

$$P_{ij}(h) = P(X_{t+h} = j | X_t = i)$$

Thus $P_{ij}(h)$ is the conditional probability that the state of the process h time units into the future will be j given that the process is presently in state i.

Paralleling the discrete-time case, here we also have the Chapman–Kolmogorov equations; these describe how $P_{ij}(t + h) = P(X_{t+h} = j | X_0 = i)$ is obtained by conditioning on the state of the process after t time units have elapsed:

$$\begin{aligned}
P_{ij}(t + h) &= P(\text{process is in state } j \text{ after } t + h \text{ time units} | \text{now in state } i) \\
&= \sum_k P(\text{process is in state } k \text{ after } t \text{ time units and in} \\
&\qquad\quad \text{state } j \text{ after } h \text{ additional units} | \text{now in state } i) \\
&= \sum_k P(X_{t+h} = j | X_t = k, X_0 = i) \cdot P(X_t = k | X_0 = i) \\
&= \sum_k P(X_{t+h} = j | X_t = k) \cdot P(X_t = k | X_0 = i) \\
&= \sum_k P_{ik}(t) \cdot P_{kj}(h)
\end{aligned}$$

The second-to-last equality is by virtue of the Markov property: conditional on the process being in state k at time t, the state at any previous time is irrelevant to the chance of being in state j at the future time.

7.7.1 Transition Rates and Sojourn Times

It is reasonable to assume that each transition probability $P_{ij}(t)$ is a continuous function of t, so that such probabilities change smoothly as t does. Since $P_{ii}(0) = 1$ and $P_{ij}(0) = 0$ for $i \neq j$, this implies that as h approaches 0, $P_{ii}(h)$ approaches 1 and $P_{ij}(h)$ approaches 0 when $i \neq j$. Rather amazingly, it turns out that all $P_{ij}(t)$ are differentiable, and in particular are differentiable at 0:

$$P_{ii}'(0) = \lim_{h \to 0} \frac{P_{ii}(h) - P_{ii}(0)}{h} = \lim_{h \to 0} \frac{P_{ii}(h) - 1}{h} = -q_i$$

(it is convenient to denote this derivative by $-q_i$ because the numerator $P_{ii}(h) - 1$ is negative and so the limit itself will be negative; q_i is then positive), and

$$P'_{ij}(0) = \lim_{h \to 0} \frac{P_{ij}(h) - P_{ij}(0)}{h} = \lim_{h \to 0} \frac{P_{ij}(h)}{h} = q_{ij} \quad \text{for } i \neq j$$

In our development of the Poisson process Sect. 7.5, we employed $o(h)$ notation to represent a quantity that for small h is negligible compared to h (see also Appendix B). Using this notation in combination with the preceding derivative expressions, we have that for h close to 0,

$$P(X_{t+h} = i \mid X_t = i) = P_{ii}(h) = 1 - q_i h + o(h)$$

$$P(X_{t+h} = j \mid X_t = i) = P_{ij}(h) = q_{ij}h + o(h) \quad i \neq j$$

A continuous-time Markov process is characterized by these various transition probability derivatives at $t = 0$, which are collectively called the **infinitesimal parameters** of the process.

Because a process that's in state i at time t must be somewhere at time $t + h$, $\sum_j P_{ij}(h) = 1$ for any i. Taking the derivative of both sides and evaluating at 0 gives a simple relationship between q_i and the various q_{ij}s for $i \neq j$[6]:

$$
\begin{aligned}
1 &= \sum_j P_{ij}(h) = P_{ii}(h) + \sum_{j \neq i} P_{ij}(h) \Rightarrow \\
0 &= P'_{ii}(h) + \sum_{j \neq i} P'_{ij}(h) \Rightarrow \\
0 &= P'_{ii}(0) + \sum_{j \neq i} P'_{ij}(0) = -q_i + \sum_{j \neq i} q_{ij} \Rightarrow \\
q_i &= \sum_{j \neq i} q_{ij}
\end{aligned}
\tag{7.9}
$$

Example 7.32 Our old friend the Poisson process with rate parameter λ is an especially simple case of a continuous-time Markov process. After remaining in state i for a time that is exponentially distributed with parameter λ, the only possible transition is to state $i + 1$ when the next event occurs. So, $P_{ii}(h) = P(\text{no events in an interval of length } h) = e^{-\lambda h}$, whence $P'_{ii}(h) = -\lambda e^{-\lambda h}$ and $q_i = -P'_{ii}(0) = \lambda$. Similarly, $P_{i,i+1}(h) = 1 - P(\text{no events in interval}) = 1 - e^{-\lambda h}$ implies $q_{i,i+1} = P'_{i,i+1}(0) = \lambda$, and $q_{ij} = 0$ for $j \neq i + 1$. Notice that these values indeed satisfy Eq. (7.9). ∎

In the preceding example, it was determined that $q_{i,i+1} = P'_{i,i+1}(0) = \lambda$. In light of the fact that λ represents the rate at which events occur (i.e., the count increases by 1), this seems quite reasonable—after all, we know from calculus that a derivative also represents a rate of change. In general, we can interpret the q_{ij}s in this fashion: q_{ij} represents the rate at which a Markov process transitions from state i to state j over some very short time interval. Hence, the q_{ij}s are called the **instantaneous transition rates** of the Markov process.

---

[6] When the state space is infinite in extent, it can sometimes happen that $q_i = \infty$. This will not occur for the situations considered herein.

Example 7.33 A nurse checks in on patients in three hospital rooms and also spends time at the nurses' station. Identify these "states" as $0 =$ nurses' station and 1, 2, $3 =$ the three patient rooms. Let X_t denote the nurse's location t hours into her shift. Figure 7.24 shows an example of the nurse's transitions between these four states across continuous time. In this figure, she begins her shift at the nurse's station $(X_0 = 0)$, spends some time there, and then moves periodically from room to room.

Fig. 7.24 One realization of 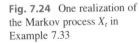 the Markov process X_t in Example 7.33

Suppose the nurse walks from the nurses' station to room 3 an average of twice per hour; this is the *rate* at which she transitions from state 0 to state 3. But the foregoing discussion also indicated that the derivative q_{ij} represents the rate of change from state i to state j. Therefore, we have that $q_{03} = 2$ (two such transitions per hour). A complete description of the nurse's movements requires specifying all of the other instantaneous transition rates as well; let's say those are

$$
\begin{array}{llll}
& q_{01} = 4 & q_{02} = .5 & q_{03} = 2 \\
q_{10} = 3 & & q_{12} = 4 & q_{13} = 1 \\
q_{20} = 3 & q_{21} = .5 & & q_{23} = 4 \\
q_{30} = 3 & q_{31} = 0 & q_{32} = 1 &
\end{array}
$$

The remaining four infinitesimal parameters of this Markov process model can be determined using Eq. (7.9). For example,

$$q_0 = \sum_{j \neq 0} q_{0j} = q_{01} + q_{02} + q_{03} = 4 + .5 + 2 = 6.5$$

Similarly, $q_1 = 8$, $q_2 = 7.5$, and $q_3 = 4$. ∎

The parameters $q_0, q_1, \ldots$ are not transition rates like the q_{ij}s, since they are associated with time intervals in which the process stays in the same state. (Unlike in the case of discrete-time Markov chains from Chap. 6, here we do not speak of "transitions" from a state into itself.) Rather, an interpretation of the q_is will be provided by the main theorem of this section, coming up shortly.

7.7.2   Sojourn Times and Transitions

The time durations spent in various states are important features of a Markov process. A continuous time interval spent in one state is called a **sojourn time** of the process. In Fig. 7.24, five sojourn times are visible: the nurse spends a while at her station (state 0), then some time in patient room 1, then back at her station, over into room 3, and finally into room 2. These sojourn times are clearly continuous random variables, but what are their distributions?

Think back to Example 7.32. From the results of Sect. 7.5, the distribution of time that a Poisson process spends in state i before moving to state $i + 1$—that is, the sojourn time between the occurrence of the ith event and the $(i + 1)$st event—is exponential with parameter λ. Our next theorem says that sojourn times in *any* continuous-time Markov process are also exponentially distributed, and specifies the probabilities of moving to various other states once a state transition occurs.

THEOREM

1. A sojourn time for state i of a continuous-time Markov chain has an exponential distribution, with parameter $\lambda = q_i$.
2. Once a sojourn in state i has ended, the process next moves to a particular state $j \neq i$ with probability q_{ij}/q_i.

Proof   Let's first consider the distribution of $T =$ sojourn time in state i, and in particular the probability that this sojourn time is at least $t + h$, where h is very small. In order for this event to occur, the process must remain in state i continuously throughout the time period of length t and then continue in this state for an additional h time units. That is, with F denoting the cumulative distribution function of T and $G(x) = 1 - F(x)$,

$$G(t+h) = 1 - F(t+h) = P(T \geq t+h) = P(X_u = i \text{ for } 0 \leq u \leq t \text{ and for } t \leq u \leq t+h)$$
$$= P(X_u = i \text{ for } t \leq u \leq t+h | X_u = i \text{ for } 0 \leq u \leq t) \cdot P(X_u = i \text{ for } 0 \leq u \leq t)$$
$$= P(X_u = i \text{ for } t \leq u \leq t+h | X_t = i) \cdot P(X_u = i \text{ for } 0 \leq u \leq t)$$

$$\text{by the Markov property}$$

$$= P(X_u = i \text{ for } t \leq u \leq t+h | X_t = i) \cdot P(T \geq t)$$
$$= P(X_u = i \text{ for } t \leq u \leq t+h | X_t = i) \cdot G(t)$$

Now, the probability $P(X_u = i \text{ for } t \leq u \leq t+h | X_t = i)$ is not quite $P_{ii}(h)$, because the latter includes both the chance of remaining in state i throughout $[t, t+h]$ and also of making multiple transitions that bring the process back to state i by the end of this time interval. But because h is small, the probability of two or more transitions is negligible compared to the likelihood of either making a single transition (to some other state j) or remaining in state i. That is, these two probabilities differ by a term that is $o(h)$. Therefore we have

$$G(t+h) = [P_{ii}(h) + o(h)] \cdot G(t) = [1 - q_i h] \cdot G(t) + o(h) \Rightarrow$$

$$\frac{G(t+h) - G(t)}{h} = -q_i + \frac{o(h)}{h}$$

Taking the limit as $h \to 0$ results in the differential equation $G'(t) = -q_i$, whose solution is $G(t) = e^{-q_i t}$. Therefore, the cdf of T is $F(t) = 1 - e^{-q_i t}$, an exponential cdf with parameter q_i. This proves the first part of the theorem.

For the second part, we consider the probability that the process is in state j after a short interval of time, given that it is in state i at the beginning of that interval and is *not* in state i at the end of the interval:

$$P(X_{t+h} = j | X_t = i, X_{t+h} \neq i) = \frac{P(X_{t+h} = j, X_t = i, X_{t+h} \neq i)}{P(X_t = i, X_{t+h} \neq i)}$$

$$= \frac{P(X_t = i, X_{t+h} = j)}{P(X_t = i, X_{t+h} \neq i)} \quad \text{because } \{X_{t+h} = j, X_{t+h} \neq i\} = \{X_{t+h} = j\}$$

$$= \frac{P(X_{t+h} = j | X_t = i) P(X_t = i)}{P(X_{t+h} \neq i | X_t = i) P(X_t = i)}$$

$$= \frac{P(X_{t+h} = j | X_t = i)}{P(X_{t+h} \neq i | X_t = i)} = \frac{P_{ij}(h)}{1 - P_{ii}(h)}$$

If we now divide both numerator and denominator by h and take the limit as h approaches 0, the result is $P(\text{next in } j | \text{currently in } i) = q_{ij}/q_i$, as asserted. ∎

Example 7.34 (Example 7.33 continued) According to the preceding theorem, the time intervals spent by the nurse at the nurses' station are exponentially distributed with parameter $\lambda = q_0 = 6.5$. Hence, the average length of time she spends there is $1/\lambda = 1/6.5 \text{ h} \approx 9.23$ min. Similarly, the average sojourn time in patient room 3 is $1/q_3 = 1/4 \text{ h} = 15$ min. When the nurse leaves her station, the likelihoods that she next visits rooms 1, 2, and 3 are, respectively,

$$\frac{q_{01}}{q_0} = \frac{4}{6.5} = \frac{8}{13}, \qquad \frac{q_{02}}{q_0} = \frac{.5}{6.5} = \frac{1}{13}, \qquad \frac{q_{03}}{q_0} = \frac{2}{6.5} = \frac{4}{13}$$

Similarly, when the nurse leaves patient room 1, there is a 3/8 chance she'll return to the nurses' station, a 4/8 probability of moving on to room 2, and a 1/8 chance of checking the patients in room 3.

Notice that we could also obtain these probabilities by rescaling the appropriate row of the array in Example 7.33 so that the entries sum to 1. In general, the transition probabilities exiting a sojourn spent in state i are proportional to the instantaneous transition rates out of state i. ∎

Example 7.35 Consider a machine that goes back and forth between working condition (state 0) and needing repair (state 1). Suppose that the duration of working condition time is exponential with parameter α and the duration of repair time is exponential with parameter β. From the preceding theorem, $q_0 = \alpha$ and $q_1 = \beta$. Equation (7.9) then implies that $q_{01} = \alpha$ and $q_{10} = \beta$, from which we infer that

$$P_{01}(h) = \alpha h + o(h) \quad \text{and} \quad P_{10}(h) = \beta h + o(h)$$

That is, for very small values of h, the chance of transitioning from working condition to needing repair in the next h time units is roughly αh, while P(working at time $t+h$ | being repaired at time t) $\approx \beta h$ for h small.

Notice also that once a sojourn in the working state ($i = 0$) has ended, the process moves to the repair state ($j = 1$) with probability $q_{01}/q_0 = \alpha/\alpha = 1$. This makes sense, since a machine leaving the working condition has nowhere else to go except into repair. The same is true if the roles of i and j are reversed. ∎

Example 7.36 A commercial printer has four machines of a certain type. Because there are only 3 employees trained to operate this kind of machine, at most three of the four can be in operation at any given time. Once a machine starts to operate, the time until it fails is exponentially distributed with parameter α (so the mean time until failure is $1/\alpha$). There are unfortunately only two employees who can repair these machines, each of whom works on just one machine at a time. So if three machines need repair at any given time, only two of these will be undergoing repair, and if all four machines need repair, two will be waiting to start the repair process. Time necessary to repair a machine is exponentially distributed with parameter β (thus mean time to repair is $1/\beta$).

Let X_t be the number of machines that are in operation at time t. Possible values of X_t (i.e., the states of the process) are 0, 1, 2, 3, and 4. If the system is currently in state 1, 2, 3, or 4, one possible state transition results from one of the working machines suddenly breaking down. Alternatively, if the system is currently in state 0, 1, 2, or 3, the next transition might result from one of the machines in repair finishing the repair process. These possible transitions are depicted in the state diagram in Fig. 7.25.

The eight non-zero instantaneous transition probabilities must be determined. Two of them follow the derivation from the Poisson process in Example 7.32:

$q_{10} = \alpha$ [i.e., $P_{10}(h) = \alpha h + o(h)$, corresponding to the one working machine going down]

$q_{34} = \beta$ [in state 3, only one machine is currently being repaired]

Next, consider the transition from 2 working machines to just 1. For a time interval of length h,

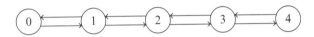

Fig. 7.25 State diagram for Example 7.36

$$P_{21}(h) = P\left(X_{t+h} = 1 \,\big|\, X_t = 2\right) = P\left(1^{\text{st}} \text{ working machine breaks down} \cup\right.$$
$$\left.2^{\text{nd}} \text{ breaks down}\right)$$
$$= P(1^{\text{st}} \text{ breaks}) + P(2^{\text{nd}} \text{ breaks}) - P(\text{both break})$$
$$= \left(1 - e^{-\alpha h}\right) + \left(1 - e^{-\alpha h}\right) - \left(1 - e^{-\alpha h}\right)^2$$

The term $(1 - e^{-\alpha h})$ comes from the fact that for an exponentially distributed rv T, $P(T \le t + h | T \ge t) = 1 - e^{-\alpha h}$. Differentiating and substituting $h = 0$ gives

$$q_{21} = P'_{21}(0) = \cdots = 2\alpha$$

When exactly two machines are working, the instantaneous failure rate is twice that of a single machine (because, in effect, twice as many things can go wrong).

By similar reasoning, $q_{32} = 3\alpha$, and also $q_{43} = 3\alpha$, because although four machines are in operating condition, only three are actually operating. Finally, $q_{01} = 2\beta$ (none of the machines are working, but only two are undergoing repair), and likewise $q_{12} = q_{23} = 2\beta$.

From Eq. (7.9), the parameters of the exponential sojourn distributions are

$$q_0 = 2\beta \quad q_1 = \alpha + 2\beta \quad q_2 = 2\alpha + 2\beta \quad q_3 = 3\alpha + \beta \quad q_4 = 3\alpha$$

So, for example, the length of a time interval in which exactly three machines are operating has an exponential distribution with $\lambda = 3\alpha + \beta$, and the expected duration of such an interval is $1/(3\alpha + \beta)$.   ■

A continuous-time Markov chain for which the only possible transitions from state i are either to state $i - 1$ or state $i + 1$ is called a **birth and death process**. The Poisson process is an example of a pure birth process—no deaths are allowed. In Example 7.36, a birth occurs when a machine finishes repair, and a death occurs when a machine breaks down. Thus, starting from state 0 only a birth is possible, starting from state 4 only a death is possible, and either a birth or a death is possible when starting from state 1, 2, or 3.

7.7.3   Long-Run Behavior of Markov Processes

Consider first a continuous-time Markov chain for which the state space is finite and consists of the states 0, 1, 2, 3, …, N. Then we already know that

$$\lim_{h \to 0} \frac{P_{ii}(h) - 1}{h} = -q_i, \quad \lim_{h \to 0} \frac{P_{ij}(h)}{h} = q_{ij} \quad \text{for} \quad i \ne j$$

Let's now create a matrix of these parameters—the exponential sojourn parameters and the instantaneous transition rates—in which the diagonal elements are the $-q_i$s and the off-diagonal elements are the q_{ij}s. Here is the matrix in the case $N = 4$:

$$
\mathbf{Q} =
\begin{bmatrix}
-q_0 & q_{01} & q_{02} & q_{03} & q_{04} \\
q_{10} & -q_1 & q_{12} & q_{13} & q_{14} \\
q_{20} & q_{21} & -q_2 & q_{23} & q_{24} \\
q_{30} & q_{31} & q_{32} & -q_3 & q_{34} \\
q_{40} & q_{41} & q_{42} & q_{43} & -q_4
\end{bmatrix}
$$

Equation (7.9) implies that the sum of every row in this matrix of parameters is zero (since each q_i is the sum of the other q_{ij}s in its row). This matrix $\mathbf{Q}$ is sometimes called the **generator matrix** of the Markov process.

Next, define a transition matrix $\mathbf{P}(t)$ whose (i, j)th entry is the transition probability $P_{ij}(t) = P(X_t = j | X_0 = i)$. Analogous to the discrete case, the Chapman–Kolmogorov equations can be rendered in terms of the transition matrix: $\mathbf{P}(t + h) = \mathbf{P}(t)\mathbf{P}(h)$. We now consider the derivative of the transition matrix at time t:

$$
\mathbf{P}'(t) = \lim_{h \to 0} \frac{\mathbf{P}(t + h) - \mathbf{P}(t)}{h} = \lim_{h \to 0} \frac{\mathbf{P}(t)\mathbf{P}(h) - \mathbf{P}(t)}{h} = \mathbf{P}(t)\left\{ \lim_{h \to 0} \frac{\mathbf{P}(h) - \mathbf{I}}{h} \right\}
$$

From the earlier derivatives, the limit of the matrix inside braces is precisely the generator matrix $\mathbf{Q}$. Thus we obtain the following system of so-called "forward" differential equations in the transition probabilities:

$$
\mathbf{P}'(t) = \mathbf{P}(t)\mathbf{Q} \tag{7.10}
$$

where $\mathbf{P}'(t)$ is the matrix of derivatives of the transition probabilities.

As in the discrete case, if the chain is irreducible, i.e., all states communicate with one another, then $P_{ij}(t) > 0$ for every pair of states and $\lim_{t \to \infty} P_{ij}(t)$ exists and equals a value π_j independent of the initial state. Thus as $t \to \infty$, $\mathbf{P}'(t)$ approaches a matrix consisting entirely of 0s (because the probabilities themselves are approaching constants independent of t) and $\mathbf{P}(t)$ itself approaches a matrix each of whose rows is $\boldsymbol{\pi} = [\pi_0, \pi_1, \ldots, \pi_N]$. Applying these statements to Eq. (7.10) and taking the top row (or any row) of each side, the vector of stationary probabilities must then satisfy $\mathbf{0} = \boldsymbol{\pi}\mathbf{Q}$, as well as $\sum \pi_j = 1$. Slight rearrangement of the equations gives

$$
\begin{aligned}
\pi_0 q_0 &= \pi_1 q_{10} + \pi_2 q_{20} + \ldots + \pi_N q_{N0} \\
\pi_1 q_1 &= \pi_0 q_{01} + \pi_2 q_{21} + \ldots + \pi_N q_{N1} \\
&\quad\cdots \\
\pi_N q_N &= \pi_0 q_{0N} + \pi_1 q_{1N} + \ldots + \pi_{N-1} q_{N-1,N} \\
1 &= \pi_0 + \pi_1 + \pi_2 + \ldots + \pi_N
\end{aligned}
$$

Consider the first of these equations. The left hand side gives the long-run rate at which the process leaves state 0, and the right hand side is the sum of rates at which the process goes from some other state to state 0. So, the equation says that the long-run rate out of that state equals the long-run rate into the state. The other equations have analogous interpretations.

Example 7.37 (Example 7.36 continued) The generator matrix for the printing machines scenario is

$$
\mathbf{Q} = \begin{bmatrix}
-2\beta & 2\beta & 0 & 0 & 0 \\
\alpha & -(\alpha + 2\beta) & 2\beta & 0 & 0 \\
0 & 2\alpha & -(2\alpha + 2\beta) & 2\beta & 0 \\
0 & 0 & 3\alpha & -(3\alpha + \beta) & \beta \\
0 & 0 & 0 & 3\alpha & -3\alpha
\end{bmatrix}
$$

The stationary distribution of the chain satisfies $\mathbf{0} = \boldsymbol{\pi}\mathbf{Q}$. Expanding these matrices, the resulting system of equations is

$$
-2\beta\pi_0 + \alpha\pi_1 = 0,\ 2\beta\pi_0 - (\alpha + 2\beta)\pi_1 + 2\alpha\pi_2 = 0,\ 2\beta\pi_1 - (2\alpha + 2\beta)\pi_2 + 3\alpha\pi_3 = 0
$$

$$
2\beta\pi_2 - (3\alpha + \beta)\pi_3 + 3\alpha\pi_4 = 0,\ \ \beta\pi_3 - 3\alpha\pi_4 = 0
$$

The first equation immediately gives $\pi_1 = (2\beta/\alpha)\pi_0$. Then substituting this expression for π_1 into the second equation and doing a bit of algebra results in $\pi_2 = (2\beta^2/\alpha^2)\pi_0$. Now substitute this expression into the third equation and solve for π_3 in terms of π_0, and then obtain an expression for π_4 in terms of π_0. The stationary probabilities are now

$$
\pi_0,\quad \pi_1 = \frac{2\beta}{\alpha}\pi_0,\quad \pi_2 = \frac{2\beta^2}{\alpha^2}\pi_0,\quad \pi_3 = \frac{4\beta^3}{3\alpha^3}\pi_0,\quad \pi_4 = \frac{4\beta^4}{9\alpha^4}\pi_0
$$

Finally, the fact that the sum of all five πs equals 1 gives an expression for π_0:

$$
\pi_0 = \frac{1}{1 + \frac{2\beta}{\alpha} + \frac{2\beta^2}{\alpha^2} + \frac{4\beta^3}{3\alpha^3} + \frac{4\beta^4}{9\alpha^4}}
$$

Consider two different specific cases: (1) $\alpha = 1$, $\beta = 2$, (2) $\alpha = 2$, $\beta = 1$. In the first case, machines get repaired more quickly than they fail, and in the second case the opposite is true. Here are the stationary probabilities:
(1) $\pi_0 = .0325$, $\pi_1 = .1300$, $\pi_2 = .2600$, $\pi_3 = .3466$, $\pi_4 = .2310$
(2) $\pi_0 = .3711$, $\pi_1 = .3711$, $\pi_2 = .1856$, $\pi_3 = .0619$, $\pi_4 = .0103$
In the first case, the mean number of machines in operating condition is $\sum i\pi_i = 2.614$, and in the second case it is only .969.   ∎

Under quite general conditions, the forward system of differential equations (7.10) is valid for a birth and death process even when the state space is infinite (i.e., when there is no upper bound on the population size). Furthermore, the stationary distribution exists and has a rather simple form. Let

$$\theta_0 = 1, \quad \theta_1 = \frac{q_{01}}{q_{10}}, \quad \theta_2 = \frac{q_{01}q_{12}}{q_{10}q_{21}}, \quad \theta_3 = \frac{q_{01}q_{12}q_{23}}{q_{10}q_{21}q_{32}}, \quad \ldots \qquad (7.11)$$

Then $\pi_1 = \theta_1\pi_0, \pi_2 = \theta_2\pi_0, \pi_3 = \theta_3\pi_0, \ldots$, and $\pi_0 = 1/\Sigma\theta_i$ provided that the sum in the denominator is finite.

Example 7.38 Customers arrive at a service facility according to a Poisson process with rate parameter λ (so the times between successive arrivals are independent and exponentially distributed, each with parameter λ). The facility has only one server, and the service time for any particular customer is exponentially distributed with parameter μ. This is often referred to as an **M/M/1 queue**, where M stands for Markovian. Let X_t represent the number of customers in the system at time t.

The mean time between successive arrivals is $1/\lambda$, and the mean time for a service to be completed is $1/\mu$. Intuitively if $1/\mu > 1/\lambda$ (i.e., if $\mu < \lambda$), then customers will begin to pile up in the system and there won't be a limiting distribution because the number of customers in the system will grow arbitrarily large over time. We therefore restrict consideration to the case $\lambda < \mu$ (the case in which $\lambda = \mu$ is a bit tricky).

The infinitesimal birth parameters are $q_{i,i+1} = \lambda$ for $i = 0, 1, 2, \ldots$, since a birth occurs when a new customer enters the facility, and the infinitesimal death parameters are $q_{i+1,i} = \mu$ for $i = 1, 2, 3, \ldots$, since a death occurs when a customer finishes service. Substituting into Eq. (7.11),

$$\theta_k = \frac{\lambda^k}{\mu^k} \quad k = 0, 1, 2, 3, \ldots, \qquad \pi_0 = \frac{1}{\displaystyle\sum_{k=0}^{\infty} \left(\frac{\lambda}{\mu}\right)^k} = 1 - \frac{\lambda}{\mu}$$

$$\pi_k = \pi_0\theta_k = \left(1 - \frac{\lambda}{\mu}\right)\left(\frac{\lambda}{\mu}\right)^k \quad k = 0, 1, 2, 3, \ldots$$

This is similar to a geometric distribution with $p = 1 - \lambda/\mu$, except that the terms start at $k = 0$ rather than $k = 1$. Nevertheless, we can quickly determine that the mean number of customers in the system is $\Sigma k\pi_k = (\lambda/\mu)/(1 - \lambda/\mu) = \lambda/(\mu - \lambda)$. ∎

7.7.4 Explicit Form of the Transition Matrix

The forward system of differential equations was obtained by decomposing the time interval from 0 to $t+h$ into the interval from 0 to t and the interval from t to $t+h$. A "backward" system of equations results from considering the two intervals $[0, h]$ and $(h, t+h]$ and again using the Chapman–Kolmogorov equations: $\mathbf{P}(t+h) = \mathbf{P}(h)\mathbf{P}(t)$. The derivative of the transition matrix at time t is then

$$\lim_{h \to 0} \frac{\mathbf{P}(t+h) - \mathbf{P}(t)}{h} = \lim_{h \to 0} \frac{\mathbf{P}(h)\mathbf{P}(t) - \mathbf{P}(t)}{h} = \left\{ \lim_{h \to 0} \frac{\mathbf{P}(h) - \mathbf{I}}{h} \right\} \mathbf{P}(t)$$

The matrix limit is again $\mathbf{Q}$, giving the following system of equations:

$$\mathbf{P}'(t) = \mathbf{Q}\mathbf{P}(t) \tag{7.12}$$

Contrast the "backward" equation (7.12) with the "forward" equation (7.10): the two matrices on the right-hand side are simply reversed. Of course, in general matrices do not commute, so one equation does not follow from the other; that both $\mathbf{Q}\mathbf{P}(t)$ and $\mathbf{P}(t)\mathbf{Q}$ equal $\mathbf{P}'(t)$ is a special property of Markov processes.

Now recall from calculus that a solution to the equation $f'(t) = cf(t)$ is $f(t) = e^{ct}$, and also that the infinite series expansion for e^{ct} is $1 + \sum_{k=1}^{\infty} c^k t^k / k!$. By analogy, the solution to our system of backward equations (7.12) is

$$\mathbf{P}(t) = e^{\mathbf{Q}t} = \mathbf{I} + \sum_{k=1}^{\infty} \frac{\mathbf{Q}^k t^k}{k!}$$

Example 7.39 (Example 7.35 continued) Let's return to the scenario involving a single machine which is either working or undergoing repair, where time until failure has an exponential distribution with parameter α and repair time is exponentially distributed with parameter β. The matrix of infinitesimal parameters is

$$\mathbf{Q} = \begin{bmatrix} -\alpha & \alpha \\ \beta & -\beta \end{bmatrix}$$

It is easily verified that $\mathbf{Q}^k = [-(\alpha + \beta)]^{k-1}\mathbf{Q}$, from which

$$\mathbf{P}(t) = \mathbf{I} - \frac{1}{\alpha + \beta} \sum_{k=1}^{\infty} \frac{[-(\alpha + \beta)]^k t^k}{k!} \mathbf{Q}$$

$$= \mathbf{I} - \frac{1}{\alpha + \beta} \left[e^{-(\alpha+\beta)t} - 1 \right] \mathbf{Q}$$

$$= \begin{bmatrix} 1 - \dfrac{\alpha}{\alpha + \beta}\left(1 - e^{-(\alpha+\beta)t}\right) & \dfrac{\alpha}{\alpha + \beta}\left(1 - e^{-(\alpha+\beta)t}\right) \\ \dfrac{\beta}{\alpha + \beta}\left(1 - e^{-(\alpha+\beta)t}\right) & 1 - \dfrac{\beta}{\alpha + \beta}\left(1 - e^{-(\alpha+\beta)t}\right) \end{bmatrix}$$

We now have a completely explicit formula for the transition probabilities of the Markov process for any time duration t. Notice that the sum of each row in the transition matrix is 1, as required.

This explicit form of $\mathbf{P}(t)$ also allows us to investigate the chain's long-run behavior. Specifically, as $t \to \infty$,

$$\mathbf{P}(t) \rightarrow \mathbf{I} + \frac{1}{\alpha + \beta}\mathbf{Q} = \begin{bmatrix} \beta/(\alpha+\beta) & \alpha/(\alpha+\beta) \\ \beta/(\alpha+\beta) & \alpha/(\alpha+\beta) \end{bmatrix}$$

Thus the stationary distribution is given by $\pi_0 = \beta/(\alpha+\beta)$ and $\pi_1 = \alpha/(\alpha+\beta)$, which could also have been obtained by solving $\boldsymbol{\pi}\mathbf{Q} = \mathbf{0}$, $\pi_0 + \pi_1 = 1$. ∎

7.7.5 Exercises: Section 7.7 (86–97)

86. The authors of the article "A Multi-State Markov Model for a Short-Term Reliability Analysis of a Power Generating Unit" (*Reliab. Eng. and Sys. Safety,* 2012: 1–6) modeled the transitions of a particular coal-fired generating unit through four states, characterized by the unit's capacity: $0 =$ complete failure, (0 MW of power), $1 = 247$ MW, $2 = 482$ MW, and $3 = 575$ MW (full power). Observation of the unit over an extended period of time yielded the following instantaneous transition rates:

$$q_{01} = .0800 \qquad q_{02} = .0133 \qquad q_{03} = 0$$
$$q_{10} = .0294 \qquad\qquad\qquad q_{12} = .3235 \qquad q_{13} = .0294$$
$$q_{20} = 0 \qquad q_{21} = .0288 \qquad\qquad\qquad q_{23} = .3558$$
$$q_{30} = .0002 \qquad q_{31} = .0001 \qquad q_{32} = .0007$$

   (a) Determine the complete generator matrix $\mathbf{Q}$ of this Markov process.
   (b) Determine the stationary probabilities of this process.
   (c) What is the long-run expected output of this particular unit, in megawatts?

87. Potential customers arrive at a service facility according to a Poisson process with rate λ. However, an arrival will enter the facility only if there is no one already being served, and otherwise will disappear (there is no waiting room!). Once a customer enters the facility, service is carried out in two stages. The time to complete the first stage of service is exponentially distributed with parameter λ_1. A customer completing the first stage of service immediately enters the second stage, where the distribution of time to complete service is exponential with parameter λ_2.

   (a) Define appropriate states, and then identify the q_is and q_{ij}s.
   (b) Determine the stationary probabilities when $\lambda = 1$, $\lambda_1 = 3$, $\lambda_2 = 2$.
   (c) Determine the stationary probabilities when $\lambda = 1$, $\lambda_1 = 2$, $\lambda_2 = 3$.
   (d) Determine the stationary probabilities when $\lambda = 4$, $\lambda_1 = 2$, $\lambda_2 = 1$.

88. Return to the scenario of the previous exercise, and now suppose that the facility has a waiting area that will accommodate one customer. A customer in the waiting area cannot begin the first stage of service until the previous customer has completed both stages.

   (a) Define appropriate states, and then identify the q_is and q_{ij}s. [*Hint:* The chain now has five possible states.]
   (b) Determine the stationary probabilities when $\lambda = 1$, $\lambda_1 = 3$, $\lambda_2 = 2$.
   (c) Determine the stationary probabilities when $\lambda = 1$, $\lambda_1 = 2$, $\lambda_2 = 3$.
   (d) Determine the stationary probabilities when $\lambda = 4$, $\lambda_1 = 2$, $\lambda_2 = 1$.

89. Reconsider the scenario of Exercise 87. Now suppose that a customer who finishes stage 2 service leaves the facility with probability .8, but with probability .2 returns to stage 1 for rework because of deficient service and then proceeds again to stage 2.
    (a) Define appropriate states, and then identify the q_is and q_{ij}s.
    (b) Determine the stationary probabilities when $\lambda = 1$, $\lambda_1 = 3$, $\lambda_2 = 2$.
    (c) Determine the stationary probabilities when $\lambda = 1$, $\lambda_1 = 2$, $\lambda_2 = 3$.
    (d) Determine the stationary probabilities when $\lambda = 4$, $\lambda_1 = 2$, $\lambda_2 = 1$.
    (e) What is the expected total time that a customer remains in the facility once he/she has entered?

90. The *Yule Process* is a special case of a birth and death process in which only births occur; each member of the population at time t has probability $\beta h + o(h)$ of giving birth to an additional member during a short time interval of length h independently of what happens to any other member of the population at that time (so there is no interaction among population members). Let X_t denote the population size at time t.
    (a) Show that if the population size is currently n, then the probability of a birth in the next interval of length h is $n\beta h + o(h)$, and that the probability of no births in the next interval of length h is $1 - n\beta h + o(h)$.
    (b) Relate $P_{ij}(t+h)$ to the transition probabilities at time t, and take an appropriate limit to establish a differential equation for $P_{ij}(t)$. [*Hint:* If $X_{t+h} = j$ and h is small, there are only two possible values for X_t. Your answer should relate $P'_{ij}(t)$, $P_{ij}(t)$, and $P_{i,j-1}(t)$.]
    (c) Assuming that there is one individual alive at time 0, show that a solution to the differential equation in (b) is $P_{1n}(t) = e^{-\beta t}(1 - e^{-\beta t})^{n-1}$. (In fact, this is the only solution satisfying the initial condition.)
    (d) Determine the expected population size at the t, assuming $X_0 = 1$. [*Hint:* What type of probability distribution is $P_{1n}(t)$?]

91. Another special case of a birth and death process involves a population consisting of N individuals. At time $t = 0$ exactly one of these individuals is infected with a particular disease, and the other $N - 1$ are candidates for acquiring the disease (susceptibles). Once infected, an individual remains so forever. In any short interval of time h, the probability that any particular infected individual will transmit the disease to any particular non-diseased individual is $\beta h + o(h)$. Let X_t represent the number of infected individuals at time t. Specify the birth parameters for this process. [*Hint:* Use the differential equation from the last exercise.]

92. At time $t = 0$ there are N individuals in a population. Let X_t represent the number of individuals alive at time t. A *linear pure death process* is one in which the probability that any particular individual alive at time t dies in a short interval of length h is $\beta h + o(h)$; no births can occur, deaths occur independently, and there is no immigration into the population.
    (a) Obtain a differential equation for the transition probabilities of this process, and then show that the solution is $P_{Nn}(t) = \binom{N}{n} e^{-n\beta t}(1 - e^{-\beta t})^{N-n}$.

(b) What is the expected population size at time t? [*Hint:* According to (a), what type of probability distribution is $P_{Nn}(t)$?]

93. A radioactive substance emits particles over time according to a Poisson process with parameter λ. Each emitted particle has an exponentially distributed lifetime with parameter β, and the lifetime of any particular particle is independent of that of any other particle. Let X_t be the number of particles that exist at time t. Assuming that $X_0 = 0$, specify the parameters of this birth and death process.

94. Consider a machine shop that has three machines of a particular type. The time until any one of these machines fails is exponentially distributed with mean lifetime 10 h, and machines fail independently of one another. The shop has a single individual capable of repairing these machines. Once a machine fails, it will immediately begin service provided that the other two machines are still working; otherwise it will wait in a repair queue until the repair person has finished work on any other machines that need service. Time to repair is exponentially distributed with expected repair time 2 h. Obtain the stationary probabilities and determine the expected number of machines operating under stationary conditions.

95. A system consists of two independent components connected in parallel, so the system will function as long as at least one of the components functions. Component A has an exponentially distributed lifetime with parameter α_0. Once it fails, it immediately goes into repair, and its repair time is exponentially distributed with parameter α_1. Similarly, component B has an exponentially distributed lifetime with parameter β_0 and an exponentially distributed repair time with parameter β_1. Determine the stationary probabilities for the corresponding continuous time Markov chain, and then the probability that the system is operating.

96. The article "Optimal Preventive Maintenance Rate for Best Availability with Hypo-Exponential Failure Distribution" (*IEEE Trans. on Reliability*, 2013: 351-361) describes the following model for maintenance of a particular machine. The machine naturally has three states: $0 =$ "up" (i.e., fully operational), $1 =$ first stage degraded, and $2 =$ second stage degraded. A machine in state 2 requires corrective maintenance, which restores the machine to the "up" state. But the machine's operators can voluntarily put a machine currently in states 0 or 1 into one other state, $3 =$ preventive maintenance. The cited article gives the following instantaneous transition rates:

$$
\begin{array}{llll}
 & q_{01} = \lambda_1 & q_{02} = 0 & q_{03} = \delta \\
q_{10} = 0 & & q_{12} = \lambda_2 & q_{13} = \delta \\
q_{20} = \mu & q_{21} = 0 & & q_{23} = 0 \\
q_{30} = m & q_{31} = 0 & q_{32} = 0 &
\end{array}
$$

The parameter $\delta \geq 0$ is called the *trigger rate* for preventive maintenance and is controlled by the machine's operator.

(a) Draw a state diagram for this Markov process.

(b) Interpret the parameters λ_1, λ_2, μ, and m.

(c) Determine the stationary distribution of this chain.

> The machine can be operated in both states 0 and 1, and so the *availability* of the machine, $A(\delta)$, is defined to be the sum of the stationary probabilities for those two states.

(d) Show that

$$A(\delta) = \left[1 + \frac{\delta}{m} + \frac{\lambda_1 \lambda_2}{\mu(\lambda_1 + \lambda_2 + \delta)} \right]^{-1}$$

(e) Determine the value of δ that maximizes the long-run proportion of time the machine is available for use. [*Hint:* You'll have to consider two separate cases, depending on whether a certain quadratic equation has any positive solutions.]

97. A discrete-time Markov chain, i.e., the type investigated in Chap. 6, can be obtained from a continuous-time chain by sampling it every h time units. That is, for $n = 0, 1, 2, \ldots$ we define

$$Y_n = X_{nh},$$

where X_t is a Markov process. For example, the nurse's movements in Example 7.33 could be observed every 6 min ($h = 1/10$ h), and a discrete-time Markov chain could be defined by $Y_n = X_{n/10} =$ the nurse's location at the nth observed time.

(a) Let **P** be the one-step transition matrix for Y_n, so the (i, j)th entry of **P** is $p_{ij} = P(Y_{n+1} = j | Y_n = i)$. Show that $p_{ij} \approx q_{ij} h$ for $i \neq j$ and $p_{ii} \approx (1 - q_i)h$, where the q_{ij}s and q_is are the infinitesimal parameters of X_t and the approximations are on the order $o(h)$.

(b) Suppose Y_n is a regular chain, and that the one-step transition probabilities in part (a) are exact (rather than just $o(h)$-approximate). Show that the stationary distribution of Y_n is identical to that of its continuous-time version X_t. [*Hint:* Use part (a) to show that the equations $\pi \mathbf{P} = \pi$ from Chap. 6 and $\pi \mathbf{Q} = \mathbf{0}$ from this section are one and the same.]

7.8 Supplementary Exercises (98–114)

98. Let $X(t)$ be a WSS random process.
    (a) Show that

$$\mathrm{Var}[X(t+\tau)-X(t)] = E\Big[(X(t+\tau)-X(t))^2\Big] = 2[R_{XX}(0)-R_{XX}(\tau)]$$
$$= 2[C_{XX}(0)-C_{XX}(\tau)].$$

(b) Show that if $C_{XX}(d)=C_{XX}(0)$ for any $d\neq 0$, then $X(t)$ is *mean square periodic*, i.e., $E[(X(t+d)-X(t))^2]=0$.
(c) Show that if $C_{XX}(d)=C_{XX}(0)$ for any $d\neq 0$, then $C_{XX}(\tau)$ is periodic. (A similar property holds for R_{XX}.) [*Hint:* Consider the covariance of $X(0)$ and $X(\tau+d)-X(\tau)$, and use the fact that $|\mathrm{Cov}(U, V)|\leq \mathrm{SD}(U)\cdot \mathrm{SD}(V)$ for any two rvs U and V.]

99. Consider the following model for binary voltage noise: let V_1, V_2, ... be independent rvs with $P(V_n=+1)=P(V_n=-1)=.5$. Then define $X(t)=V_n$ for $n-1\leq t<n$, i.e., V_1 is transmitted for $0\leq t<1$, V_2 is transmitted for $1\leq t<2$, and so on.
    (a) Find the mean function of $X(t)$.
    (b) Find the autocovariance function of $X(t)$. [*Hint:* Consider separate cases depending on whether or not t and s lie in the same unit interval, e.g., $[1, 2)$.]

100. Modify the previous exercise as follows: let $T_0 \sim \mathrm{Unif}[0, 1]$ be independent of the V_ns. Then the random process $X(t)$ equals V_1 for $T_0\leq t<T_0+1$, V_2 is transmitted for $T_0+1\leq t<T_0+2$, and so on.
    (a) Find the mean function of $X(t)$.
    (b) Find the autocorrelation function of $X(t)$. [*Hint:* First find the conditional distribution of $X(t)$ and $X(s)$ given $T_0=t_0$.]

101. Define a collection of random processes by $X_k(t)=A_k\cos(\omega_k t)+B_k\sin(\omega_k t)$ for $k=1, 2, \ldots, n$, where the coefficients $A_1, \ldots, A_n, B_1, \ldots, B_n$ are iid $\mathrm{Unif}[-1, 1]$ rvs, and the frequencies $\omega_1, \ldots, \omega_n$ are constants. Let $Y(t)=X_1(t)+\cdots+X_n(t)$.
    (a) Find the mean and autocovariance functions of $X_k(t)$ for $k=1, 2, \ldots, n$.
    (b) Find the mean and autocovariance functions of $Y(t)$. Is $Y(t)$ WSS?

102. Let $\Theta_1, \ldots, \Theta_n$ be iid $\mathrm{Unif}(-\pi, \pi]$ rvs and define a random process $X(t)$ by

$$X(t)=\sum_{k=1}^{n} a_k\sin(\omega_k t+\Theta_k)$$

Is $X(t)$ wide-sense stationary?

103. Let $X(t)=\cos(\Omega t+\Theta)$, where Ω and Θ are independent rvs, $\Theta\sim\mathrm{Unif}(-\pi, \pi]$, and Ω equals ω_k with probability p_k for $k=1, 2, \ldots, n$ (i.e., Ω is a discrete rv).
    (a) Find the mean function of $X(t)$.
    (b) Find the autocovariance function of $X(t)$.
    (c) Is $X(t)$ WSS?

104. Let $X(t)$ be a WSS random process, and let $Y(t) = X(t - d)$, a d-second delayed
     version of $X(t)$.
     (a) Find the mean and autocorrelation functions of $Y(t)$ in terms of those
         of $X(t)$.
     (b) Is $Y(t)$ WSS?
     (c) Find the cross-correlation $R_{XY}(t, t + \tau)$. Are $X(t)$ and $Y(t)$ jointly WSS?

105. A rotor within a certain manufacturing machine must be replaced every 125 h
     of use, on average. Let X_n denote the lifetime of the nth rotor ($n = 1, 2, 3, \ldots$),
     and suppose the X_ns are iid exponential rvs with mean 125 h. (In this context,
     the "time" index n actually counts rotors, not hours or some other time unit.)
     (a) Define $S_n = X_1 + \cdots + X_n$. Interpret S_n in this context.
     (b) Find the mean, variance, autocorrelation, and autocovariance functions of
         S_n.
     (c) Use the Central Limit Theorem to determine the approximate distribution
         of S_{50} and to approximate $P(S_{50} \geq 6240)$, the chance that 50 rotors will be
         sufficient to operate the machine for 3 years (40 h per week, 52 weeks
         a year).

106. Let $X(t)$ be a WSS random process with mean μ_X and autocovariance function
     $C_{XX}(\tau)$.
     (a) Show that $E[\langle X(t) \rangle_T] = \mu_X$ for all T. [*Note:* Since this is the ensemble
         average of $X(t)$, it follows that $X(t)$ is mean ergodic iff $\mathrm{Var}(\langle X(t) \rangle_T) \to 0$ as
         $T \to \infty$.]
     (b) It is straightforward to show that

     $$\mathrm{Var}\left(\langle X(t)_T \rangle\right) = \frac{1}{4T^2} \int_{-T}^{T} \int_{-T}^{T} C_{XX}(s - t) dt ds$$

     Make the substitution $\tau = s - t$ to prove

     $$\mathrm{Var}\left(\langle X(t) \rangle_T\right) = \frac{1}{2T} \int_{-2T}^{2T} C_{XX}(\tau) \left(1 - \frac{|\tau|}{2T}\right) d\tau,$$

     so that $X(t)$ is mean ergodic iff this integral converges to 0 as $T \to \infty$.
     [This can be a useful test for ergodicity when a model is specified in terms
     of its covariance function and no explicit form of $X(t)$ is available.]
     (c) Show that $X(t)$ is mean ergodic if $\dfrac{1}{2T} \int_{-2T}^{2T} C_{XX}(\tau) d\tau \to 0$ as $T \to \infty$.

107. Let X_n be a WSS random sequence, and define $Y_n = X_n - X_{n-1}$. Is Y_n also
     WSS?

108. Let X_n be iid, with mean 0 and variance σ^2. Define a *kth-order moving average* sequence Y_n by

$$Y_n = a_1 X_n + \cdots + a_k X_{n-k+1}$$

where the nonnegative constants a_i are such that $a_1 + \cdots + a_k = 1$.
(a) Find the mean function of Y_n.
(b) Find the variance function of Y_n.
(c) Find the autocovariance function of Y_n.
(d) Is Y_n wide-sense stationary?
(e) Find the correlation coefficient $\rho(Y_n, Y_{n+k})$.

109. Suppose that noise impulses occur on a telephone line at random, with a mean rate λ per second. Assume the occurrence of noise impulses meet the conditions of a Poisson process.
(a) Find the probability that no noise impulses occur during the transmission of a t-second message.
(b) Suppose that the message is encoded so that errors caused by a single noise impulse can be corrected. What is the probability that a t-second message is either error-free or correctable?
(c) Suppose the error correction protocols can reset themselves so long as successive noise impulses are more than ε seconds apart. What is the probability the next noise impulse will be corrected?

110. A bus has just departed from a certain New York City bus stop. Passengers for the next bus arrive according to a Poisson process with rate 3 per minute. Suppose the arrival time Y of the next bus has a uniform distribution on the interval $[0, 5]$.
(a) Given that $Y = y$, what is the expected number of passengers at the stop for this next bus?
(b) Use the result of (a) along with the Law of Total Expectation to determine the expected number of passengers at this stop when the next bus arrives.
(c) Given that $Y = y$, determine the (conditional) variance of the number of passengers at the stop for this next bus. Then use the Law of Total Variance to determine the standard deviation of the number of passengers at this stop for the next bus.

111. Starting at time $t = 0$, commuters arrive at a subway station according to a Poisson process with rate λ per minute. The subway fare is \$2. Suppose this fare is "exponentially discounted" back to time 0; that is, if a commuter arrives at time t, the resulting discounted fare is $2e^{-\alpha t}$, where α is the "discount rate."
(a) If five commuters arrive in the first t_0 minutes, what is the expected value of the total discounted fare collected from these five individuals? [*Hint:* Recall that for a Poisson process, conditional on any particular number of events occurring in some time interval, each event occurrence time is uniformly distributed over that interval.]

(b) What is the expected value of the total discounted fare collected from customers who arrive in the first t_0 minutes? [*Hint*: Conditioning on the number of commuters who arrive, use an expected value argument like that employed in (a), and then apply the Law of Total Probability.]

112. Individuals enter a museum exhibit according to a Poisson process with rate λ. The amount of time any particular individual spends in this exhibit is a random variable having an exponential distribution with parameter θ, and these exhibit-viewing times are independent of one another. Let $Y(t)$ denote the number of individuals who have entered the exhibit prior to time t and are still viewing the exhibit, and let $Z(t)$ denote the number of individuals who have entered the exhibit and departed by time t.
    (a) Obtain an expression for $P(Y(t) = 6$ and $Z(t) = 4)$.
    (b) Generalize the argument leading to the expression of (a) to obtain the joint pmf of the two random variables $Y(t)$ and $Z(t)$.

113. According to the article "Reliability Evaluation of Hard Disk Drive Failures Based on Counting Processes" (*Reliability Engr. and System Safety*, 2013: 110–118), particles accumulating on a disk drive come from two sources, one external and the other internal. The article proposed a model in which the internal source contains a number of loose particles M having a Poisson distribution with mean value μ; when a loose particle releases, it immediately enters the drive, and the release times are iid with cumulative distribution function $G(t)$. Let $X(t)$ denote the number of loose particles not yet released at a particular time t. Show that $X(t)$ has a Poisson distribution with parameter $\mu[1 - G(t)]$. [*Hint*: Let $Y(t)$ denote the number of particles accumulated on the drive from the internal source by time t, so that $X(t) + Y(t) = M$. Obtain an expression for $P(X(t) = x, Y(t) = y)$, and then sum over y.]

114. Suppose the strength of a system is a nonnegative rv Y with pdf $g(y)$. The system experiences shocks over time according to a Poisson process with rate λ. Let X_i denote the magnitude of the ith shock, and suppose the X_is are iid with cdf $F(x)$. If when the ith shock occurs, $X_i > Y$, then the system immediately fails; otherwise it continues to operate as though nothing happened. Let $S(t)$ denote the number of shocks in $[0, t]$, and let T denote the system lifetime.
    (a) Determine the probability $P(T > t | Y = y$ and $S(t) = n)$. [*Hint:* Your answer should involve y, n, and the cdf F.]
    (b) Apply the Law of Total Probability along with (a), to determine $P(T > t | Y = y)$.
    (c) Obtain an integral expression for the probability that the system lifetime exceeds t. [*Hint:* Write $P(T > t)$ as a double integral involving the joint pdf of T and Y. Then simplify to a single integral using (b).]
    (Based on the article "On Some Comparisons of Lifetimes for Reliability Analysis," *Reliability Engr. and Safety Analysis*, 2013: 300–304.)

Introduction to Signal Processing

8

The previous chapter introduced the concept of a random process and explored in depth the temporal (i.e., time-related) properties of such processes. Many of the specific random processes introduced in Chap. 7 are used in modern engineering to model noise or other unpredictable phenomena in signal communications. In this chapter, we investigate the *frequency*-related properties of random processes, with a particular emphasis on power and filtering.

Section 8.1 introduces the *power spectral density*, which describes how the power in a random signal is distributed across all possible frequencies. This first section also discusses so-called *white noise* processes, which are best described in terms of a frequency distribution. In Sect. 8.2, we look at filters; or, more precisely, linear, time-invariant (LTI) systems. We explore some techniques for filtering random signals, including the use of so-called "ideal" filters. Finally, Sect. 8.3 reexamines these topics in the context of discrete-time signals.

We assume throughout this chapter that readers have some familiarity with (nonrandom) signals and frequency representations. In particular, knowledge of Fourier transforms and LTI systems will be critical to understanding our exposition. Appendix B includes a brief summary of the properties of Fourier transforms; Sect. 8.2 includes a short discussion of LTI systems.

8.1 Power Spectral Density

In Chap. 7, we considered numerous models for random processes $X(t)$ as well as several ways to quantify the statistical properties of such processes (the mean, variance, autocovariance, and autocorrelation functions). All of these statistical functions describe the behavior of $X(t)$ in the time domain. Now we turn our attention to properties of a random process that can be described in the *frequency domain*.

At the outset, some basic notation and conventions must be established. First, the letter j will denote $\sqrt{-1}$ in order to be consistent with engineering practice (some

M.A. Carlton and J.L. Devore, *Probability with Applications in Engineering, Science, and Technology*, Springer Texts in Statistics, DOI 10.1007/978-1-4939-0395-5_8, © Springer Science+Business Media New York 2014

readers may be more familiar with the symbol i). Second, we will denote frequency by f, whose units are Hertz (1/s). For those more familiar with radian frequency ω, the two are of course related by $\omega = 2\pi f$. Third, throughout this chapter $X(t)$ will represent a random current waveform through a 1-Ω impedance. This is a standard convention in signal processing; it has the advantage that we can talk about current and voltage interchangeably (since $V = IR$). Finally, we will assume that all random processes are wide-sense stationary (WSS) unless otherwise noted, because this is a key assumption for the main theorem of this section.

Our ultimate goal is to describe how the *power* in a random process is distributed across the frequency spectrum. From basic electrical engineering, we know $P = I^2 R$, where $P = $ power, $I = $ current $= X(t)$, and $R = $ resistance $= 1\ \Omega$. Hence, we may think of $I^2 R = X^2(t)$ as the "instantaneous power" in the random process at time t.

DEFINITION
Let $X(t)$ be a WSS random process. The **(ensemble) average power** (also called the **expected power**) of $X(t)$, denoted by P_X, is

$$P_X = E\left[X^2(t)\right]$$

The average power of $X(t)$ is related to its autocorrelation function by

$$P_X = R_{XX}(0)$$

Notice we may write P_X rather than $P_X(t)$, i.e., the ensemble average power in $X(t)$ does not vary with time. This is due to the assumption of wide-sense stationarity. To see why P_X equals $R_{XX}(0)$, recall that for WSS processes we have $R_{XX}(\tau) = E[X(t)X(t+\tau)]$, which does not depend on t. Setting $\tau = 0$ immediately gives $R_{XX}(0) = E[X^2(t)] = P_X$.

Example 8.1 In Chap. 7, we introduced the "phase variation" random process $X(t) = A_0\cos(\omega_0 t + \Theta)$, where the phase shift Θ is uniformly distributed on the interval $(-\pi, \pi]$. We showed that $X(t)$ is WSS, with mean $\mu_X = 0$ and autocovariance function $C_{XX}(\tau) = (A_0^2/2)\cos(\omega_0\tau)$, from which $R_{XX}(\tau) = (A_0^2/2)\cos(\omega_0\tau)$ as well. Thus, the ensemble average power in the phase variation process is

$$P_X = R_{XX}(0) = \frac{A_0^2}{2}\cos\left(\omega_0 \cdot 0\right) = \frac{A_0^2}{2}$$

This formula for the average power of a sinusoid is well known to electrical engineers. ∎

Now we turn to describing how the expected power P_X in a random process is distributed across the frequency domain. For example, is this power concentrated at just a few frequencies, or across a very large frequency band? Typically in engineering, we move from the time domain t to the frequency domain f by taking the *Fourier transform* of our time-dependent function. Because of some technical issues related to the existence of certain integrals that arise in connection with random processes, we must proceed carefully here. To begin, define a truncated version of a random process $X(t)$ by

$$X_T(t) = \begin{cases} X(t) & |t| \leq T \\ 0 & \text{otherwise} \end{cases}$$

This function is square-integrable with respect to t, and so its Fourier transform exists. Define

$$F_T(f) = \mathscr{F}\{X_T(t)\} = \int_{-\infty}^{\infty} X_T(t)e^{-j2\pi ft}\,dt = \int_{-T}^{T} X(t)e^{-j2\pi ft}\,dt$$

Parseval's Theorem then connects the integrals of $X_T(t)$ and $F_T(f)$: $\int_{-\infty}^{\infty} |F_T(f)|^2\,df = \int_{-\infty}^{\infty} |X_T(t)|^2\,dt = \int_{-T}^{T} X^2(t)\,dt$, where the absolute value bars denote the magnitude of a possibly complex number. (Since $X^2(t)$ is real-valued and nonnegative, those bars may be dropped.) Divide both sides by $2T$:

$$\int_{-\infty}^{\infty} \frac{|F_T(f)|^2}{2T}\,df = \int_{-T}^{T} \frac{X^2(t)}{2T}\,dt = \frac{1}{2T}\int_{-T}^{T} X^2(t)\,dt \tag{8.1}$$

Since the right-most expression in Eq. (8.1) gives the average power in $X(t)$ across the interval $[-T, T]$, so does the far left term, and it follows that the integrand $|F_T(f)|^2/2T$ describes how that average power is distributed in the frequency domain. In fact, $|F_T(f)|^2$ has units of energy, and so the units on $|F_T(f)|^2/2T$ are energy/time = power. We still need to remove the truncation of the original $X(t)$, and it is desirable to take the ensemble average of this power representation.

DEFINITION
The **power spectral density (psd)**, or **power spectrum**, of a random process $X(t)$ is defined by

$$S_{XX}(f) = \lim_{T \to \infty} E\left[\frac{|F_T(f)|^2}{2T}\right]$$

As may be evident by the preceding development, applying this definition is typically extremely difficult in practice. Thankfully, for wide-sense stationary processes, there is a simpler method for calculating the power spectral density $S_{XX}(f)$. The formula, presented in the following theorem, is hinted at by the fact that the average power itself can be found through the autocorrelation function, $P_X = R_{XX}(0)$, as noted before. It was first discovered by Albert Einstein but is more commonly attributed to Norbert Wiener and Aleksandr Khinchin.

> **WIENER–KHINCHIN THEOREM**
>
> If $X(t)$ is a wide-sense stationary random process, then
>
> $$S_{XX}(f) = \mathscr{F}\{R_{XX}(\tau)\}$$

A proof of this theorem appears at the end of this section.

Example 8.2 Let $X(t) = 220\cos(2000\pi t + \Theta)$, an example of the phase variation process from Example 8.1 (with $A_0 = 220$ and $\omega_0 = 2000\pi$). The expected power of this signal is $A_0^2/2 = 24{,}200 = 24.2$ kW. It's clear from the formula for $X(t)$ that it broadcasts this 24.2 kW signal at a single frequency of 2000π radians, or 1 kHz. Thus, we anticipate that all the power in $X(t)$ is concentrated at 1 kHz. To verify this, use the autocorrelation function from Example 8.1 and apply the Wiener–Khinchin Theorem:

$$S_{XX}(f) = \mathscr{F}\{R_{XX}(\tau)\} = \mathscr{F}\left\{\frac{A_0^2}{2}\cos(\omega_0\tau)\right\} = \mathscr{F}\{24{,}200\cos(2000\pi\tau)\}$$

Use the linear property of Fourier transforms, and then apply the known transform of the cosine function:

$$S_{XX}(f) = 24{,}200\,\mathscr{F}\{\cos(2000\pi\tau)\}$$

$$= 24{,}200 \cdot \frac{1}{2}\left[\delta(f - 1000) + \delta(f + 1000)\right]$$

$$= 12{,}100\left[\delta(f - 1000) + \delta(f + 1000)\right]$$

where $\delta()$ denotes an impulse function (see Appendix B for more information). Figure 8.1 shows this power spectral density, which consists of two impulses located at ± 1000 Hz, each with intensity 12,100. Of course, in practice, the frequency -1000 Hz is really the same as $+1000$ Hz, and so the power spectrum of this random process in the *positive* frequency domain is located solely at 1000 Hz; the impulse at this lone frequency carries intensity 12,100 $+ 12{,}100 = 24{,}200$, the ensemble average power of the signal.

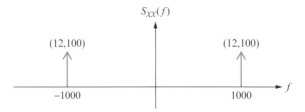

Fig. 8.1 The power spectral density of Example 8.2   ∎

Because the Fourier transform results in part of the power spectral density being represented at negative frequencies, the psd is sometimes called a *two-sided power spectrum*. Next we will explore this property and others more thoroughly.

8.1.1 Properties of the Power Spectral Density

The following proposition describes several basic properties of $S_{XX}(f)$ and indicates how the psd is related to average power.

> **PROPOSITION**
> Let $S_{XX}(f)$ be the power spectral density of a WSS random process $X(t)$.
> 1. $S_{XX}(f)$ is real-valued and nonnegative.
> 2. $S_{XX}(f)$ is an even function, i.e., $S_{XX}(-f)=S_{XX}(f)$. (This is the "two-sided" nature of the psd.)
> 3. $\displaystyle\int_{-\infty}^{\infty} S_{XX}(f)\,df = P_X$, the ensemble average power in $X(t)$.

Proof Property 1 follows from the definition of the psd: even though the Fourier transform $F_T(f)$ may be complex-valued, $|F_T(f)|^2/2T$ must be real and nonnegative. Since $S_{XX}(f)$ is the limit of the expected value of $|F_T(f)|^2/2T$, it must also be real and nonnegative.

To prove property 2, we invoke the Wiener–Khinchin Theorem. Since $R_{XX}(\tau)$ is even, we can simplify its Fourier transform:

$$S_{XX}(f) = \mathscr{F}\{R_{XX}(\tau)\} = \int_{-\infty}^{\infty} R_{XX}(\tau)e^{-j2\pi f\tau}d\tau = \int_{-\infty}^{\infty} R_{XX}(\tau)\cos(2\pi f\tau)d\tau \quad (8.2)$$

The sine component of the complex exponential drops out because $R_{XX}(\tau)$ is even. From Eq. (8.2) it is clear that $S_{XX}(f)$ is both real-valued and an even function of f, since cosine is even.

Property 3 also follows from the Wiener–Khinchin Theorem: writing the autocorrelation function as the inverse Fourier transform of the pdf, we have

$$R_{XX}(\tau) = \mathscr{F}^{-1}\{S_{XX}(f)\} = \int_{-\infty}^{\infty} S_{XX}(f)e^{+j2\pi f\tau}\,df \Rightarrow$$
$$P_X = R_{XX}(0) = \int_{-\infty}^{\infty} S_{XX}(f)e^{+j2\pi f(0)}\,df = \int_{-\infty}^{\infty} S_{XX}(f)\,df$$

∎

The foregoing proposition gives some insight into the interpretation of the power spectral density. As stated previously, $S_{XX}(f)$ describes how the ensemble average power in $X(t)$ is distributed across the frequency domain. Since power must be real and nonnegative, so must the power spectrum. Property 3 shows why $S_{XX}(f)$ is called a "density": if we integrate this function across its entire domain, we recover the total (expected) power in the signal, P_X, much in the same way that integrating a pdf from $-\infty$ to ∞ returns the total probability of 1. Property 3 also indicates the appropriate units for the psd: since the integral is performed with respect to f (Hertz) and the result is power (watts), *the correct units for the power spectral density are watts per Hertz (W/Hz).*

Property 2 makes precise the two-sided nature of a psd. The power spectrum $S_{XX}(f)$ will always be symmetric about $f = 0$; this is a by-product of how Fourier transforms are computed and the fact that autocorrelation functions are always symmetric in τ. But we must make sense of it in terms of "true" (i.e., nonnegative) frequencies. Look back at Example 8.2: that particular phase variation random process had a power spectral density consisting of two impulses, each of intensity 12,100 W/Hz, at ± 1 kHz. We can interpret the impulse at -1000 by mentally "folding" the power spectrum along the vertical axis, left to right, so that the two impulses line up at $+1$ kHz with a total intensity of 24,200 W/Hz. Integrating that impulse df recovers the ensemble average power of 24.2 kW.

Our next example illustrates a more general psd, including components other than impulses.

Example 8.3: *Partitioning a power spectrum.* Suppose $X_1(t)$ and $X_2(t)$ are independent, zero-mean, WSS random processes with autocorrelation functions

$$R_{11}(\tau) = 2000\mathrm{tri}(10{,}000\tau), R_{22}(\tau) = 650\cos(40{,}000\pi\tau)$$

Define a new random process by $X(t) = X_1(t) + X_2(t) + 40$. We encountered this random process in Example 7.18, from which we know that $X(t)$ is WSS with a mean of $\mu_X = 40$ and an autocorrelation function of

$$R_{XX}(\tau) = R_{11}(\tau) + R_{22}(\tau) + 40^2$$
$$= 2000\text{tri}(10{,}000\tau) + 650\cos(40{,}000\pi\tau) + 1600$$

First, let's find the ensemble average power in $X(t)$:

$$P_X = R_{XX}(0) = 2000 + 650 + 1600 = 4250\,\text{W}$$

Recall from Example 7.18 that $X(t)$ consists of three pieces: the aperiodic component $X_1(t)$, the periodic component $X_2(t)$, and the dc offset of 40. These deliver a total of 4.25 kW of power: 2000 W from $X_1(t)$, 650 W from $X_2(t)$, and 1600 W from the dc power offset (recall that we always assume $R = 1\ \Omega$, so $P = I^2R = 40^2(1) = 1600$ for that term).

Next, let's see how this 4250 W of power is distributed in the frequency domain by determining the power spectral density of $X(t)$. Apply the Wiener–Khinchin Theorem:

$$S_{XX}(f) = \mathscr{F}\{R_{XX}(\tau)\} = \mathscr{F}\{2000\text{tri}(10{,}000\tau) + 650\cos(40{,}000\pi\tau) + 1600\}$$
$$= 2000\,\mathscr{F}\{\text{tri}(10{,}000\tau)\} + 650\,\mathscr{F}\{\cos(40{,}000\pi\tau)\} + \mathscr{F}\{1600\}$$

To evaluate each of these three Fourier transforms, we use the table of Fourier pairs in Appendix B. The last two are straightforward, while the transform of $\text{tri}(10{,}000\tau)$ requires the rescaling property with $a = 10{,}000$. Since the Fourier transform pair of $\text{tri}(t)$ is $\text{sinc}^2(f)$, the ultimate result is

$$S_{XX}(f) = 2000 \cdot \frac{1}{|10{,}000|}\text{sinc}^2\left(\frac{f}{10{,}000}\right)$$
$$+ 650 \cdot \frac{1}{2}\left[\delta(f - 20{,}000) + \delta(f + 20{,}000)\right] + 1600\delta(f)$$
$$= 0.2\text{sinc}^2\left(\frac{f}{10{,}000}\right) + 325\left[\delta(f - 20{,}000) + \delta(f + 20{,}000)\right] + 1600\delta(f)$$

A graph of this power spectrum appears in Fig. 8.2. Notice the graph is symmetric about the vertical axis $f = 0$, as guaranteed by property 2 of the previous proposition. The psd consists of three elements, corresponding to the three components of the original signal. The power spectrum of the aperiodic component appears as a continuous function (a true "density") that vanishes as $|f| \to \infty$. This is sometimes referred to as the *dissipative component* of the psd. The periodic component of the signal has psd equal to a pair of impulses (sometimes called split impulses) at its fundamental frequency—here, $40{,}000\pi$ radians, or 20 kHz. Finally, the direct current corresponds to a "frequency" of $f = 0$; thus, the dc power offset of 1600 W is represented by $1600\delta(f)$, an impulse at $f = 0$.

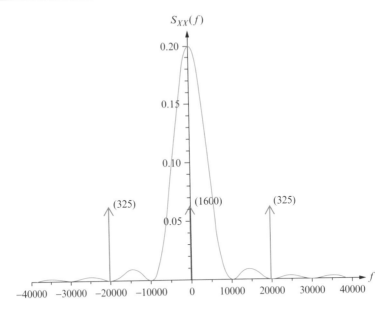

Fig. 8.2 Power spectral density of Example 8.3 ∎

As illustrated in the foregoing example, a power spectral density generally consists of at most three pieces, {dissipative components} + {periodic components} + {dc power offset}, and the last two will be comprised of impulses.

8.1.2   Power in a Frequency Band

Suppose we wish to determine how much of the power in a random signal lies within a particular frequency band; this is, as it turns out, a primary purpose of the psd. For frequencies f_1 and f_2 with $0 < f_1 < f_2$, let $P_X[f_1, f_2]$ denote the expected power in $X(t)$ in the band $[f_1, f_2]$. Then, to account for the two sides of the power spectrum, we calculate as follows:

$$P_X[f_1, f_2] = \int_{f_1}^{f_2} S_{XX}(f)\, df + \int_{-f_2}^{-f_1} S_{XX}(f)\, df = 2 \int_{f_1}^{f_2} S_{XX}(f)\, df \qquad (8.3)$$

The last two expressions in Eq. (8.3) are equal because $S_{XX}(f)$ is an even function. Figure 8.3a shows a generic power spectrum. Figure 8.3b shows the calculation of power in a band, accounting for the two sides of the psd; it's clear that we could simply double the right-hand area and get the same result.

Extra care must be taken to find the power in $X(t)$ below some frequency f_2, i.e., between 0 and f_2 including the possible dc power offset at $f = 0$. When we "fold" the negative frequencies over to the positive side, any power represented by an impulse

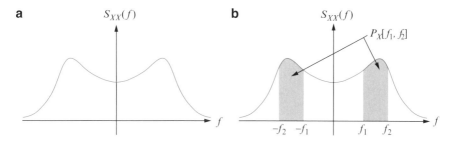

Fig. 8.3 (**a**) A generic power spectral density; (**b**) the ensemble average power in a specified frequency band

at $f = 0$ is *not* duplicated. Therefore, we cannot simply double the entire integral of $S_{XX}(f)$ from 0 to f_2; we must count the dc power offset a single time, and then integrate the rest of the psd. Written mathematically,

$$P_X[0,f_2] = \int_{-f_2}^{f_2} S_{XX}(f)\,df = (\text{dc power offset}) + 2\int_{0^+}^{f_2} S_{XX}(f)\,df \qquad (8.4)$$

The lower limit 0^+ in Eq. (8.4) indicates that the integral term does not include an impulse at zero, should one exist.

Example 8.4 For the random process $X(t)$ in Example 8.3, let's first find the ensemble average power in the band from 10 to 30 kHz. With $f_1 = 10{,}000$ and $f_2 = 30{,}000$, we proceed as follows:

$$P_X[10{,}000, 30{,}000] = 2\int_{10{,}000}^{30{,}000} S_{XX}(f)\,df$$

$$= 2\int_{10{,}000}^{30{,}000}\left[0.2\mathrm{sinc}^2\left(\frac{f}{10{,}000}\right) + 325[\delta(f-20{,}000) + \delta(f+20{,}000)] + 1600\delta(f)\right]df$$

$$= 2\int_{10{,}000}^{30{,}000} 0.2\mathrm{sinc}^2\left(\frac{f}{10{,}000}\right)df + 2\int_{10{,}000}^{30{,}000} 325\delta(f-20{,}000)df$$

$$+ 2\int_{10{,}000}^{30{,}000} 325\delta(f+20{,}000)df + 2\int_{10{,}000}^{30{,}000} 1600\delta(f)df$$

To evaluate the integrals of the three impulses, we use the sifting property (see Appendix B); since the specified frequencies of the last two impulses lie outside the band [10,000, 30,000], those two integrals are zero. The calculation continues

$$P_X[10{,}000, 30{,}000] = 2 \int\limits_{10{,}000}^{30{,}000} 0.2\mathrm{sinc}^2\left(\frac{f}{10{,}000}\right)df + 2(325) + 2(0) + 0$$

$$= 0.4 \int\limits_{10{,}000}^{30{,}000} \frac{\sin^2(\pi f/10{,}000)}{(\pi f/10{,}000)^2} df + 650 = 127.17 + 650 = 777.17\mathrm{W}$$

This last integration of the sinc^2 function requires software (or an advanced calculator). Next, let's find the average power in $X(t)$ concentrated below 10 kHz. We must remember to include the impulse representing the dc power offset at $f = 0$, but only once. Also, we can ignore the impulses at ± 20 kHz, since they lie outside our desired range. Applying Eq. (8.4),

$$P_X[0, 10{,}000] = \int\limits_{-10{,}000}^{10{,}000} S_{XX}(f)df = 1600 + 2 \int\limits_{0^+}^{10{,}000} S_{XX}(f)df$$

$$= 1600 + 2 \int\limits_{0}^{10{,}000} 0.2\,\mathrm{sinc}^2\left(\frac{f}{10{,}000}\right)df$$

$$= 1600 + 1805.65 = 3405.65\,\mathrm{W}$$

Again, a numerical integration tool is required.   ∎

8.1.3 White Noise Processes

As mentioned previously, engineers frequently use random process models in an attempt to describe the noise acquired by an intended signal during transmission. One of the simplest models, called *white noise*, can most easily be described by its frequency representation (as opposed to the time-domain models of Chap. 7).

DEFINITION
A random process $N(t)$ is **(pure) white noise** if there exists a constant $N_0 > 0$, called the *intensity parameter*, such that the psd of $N(t)$ is

$$S_{NN}(f) = \frac{N_0}{2} \qquad -\infty < f < \infty$$

As a special case, $N(t)$ is called **Gaussian white noise** if $N(t)$ is a Gaussian process as defined in Sect. 7.6 and its psd is as above.

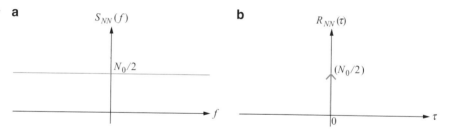

Fig. 8.4 Pure white noise: (**a**) power spectral density; (**b**) autocorrelation function

The power spectral density of pure white noise appears in Fig. 8.4a. A white noise model assumes that all frequencies appear at equal power intensity throughout the entire spectrum. In that sense, it is analogous to white light (all frequencies at equal intensity), which gives white noise its name.

White noise processes can also be partially described in the time domain through the autocorrelation function:

$$R_{NN}(\tau) = \mathscr{F}^{-1}\{S_{XX}(f)\} = \mathscr{F}^{-1}\left\{\frac{N_0}{2}\right\} = \frac{N_0}{2}\delta(\tau)$$

Figure 8.4b shows this autocorrelation function. From property 5 of the main proposition in Sect. 7.3, it follows that the mean of a white noise process is $\mu_N = 0$. (That's also evident from the psd itself, since it lacks an impulse at $f = 0$ that would correspond to a dc power offset.) Thus, the autocovariance function of pure white noise is also $C_{NN}(\tau) = R_{NN}(\tau) = (N_0/2)\delta(\tau)$.

This has a rather curious consequence: since $\delta(\tau) = 0$ for $\tau \neq 0$, $C_{NN}(\tau) = (N_0/2)\delta(\tau)$ implies that the random variables $N(t)$ and $N(t + \tau)$ are uncorrelated except when $\tau = 0$. If $N(t)$ is Gaussian white noise, then $N(t)$ and $N(t + \tau)$ are independent for $\tau \neq 0$ (since uncorrelated implies independent for normal rvs), even if the two times are very close together. That is, a pure white noise process has the property that its location at any given time is completely uncorrelated with, say, its location the nanosecond before!

Although the pure white noise model is commonly used in engineering practice, no such process can exist in physical reality. In order for the description in the preceding paragraph to be true, the process would have to "move" infinitely quickly, thus requiring infinite power. This can be seen directly from the definition:

$$P_X = \int_{-\infty}^{\infty} S_{XX}(f)df = \int_{-\infty}^{\infty} \frac{N_0}{2}df = \infty$$

That is, the area under the curve in Fig. 8.4a is infinite. So, why use a model for a process that cannot exist? As we'll see in Sect. 8.2, when a white noise process is passed through certain filters, the resulting output will have finite power. Thus, if

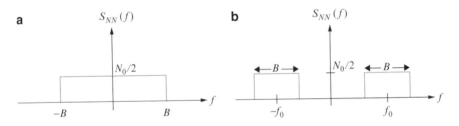

Fig. 8.5 Examples of band-limited white noise: (**a**) lowpass; (**b**) bandpass

we are interested in analyzing the filtered version of our communication noise, using the very simple model of pure white noise for the input is not unreasonable.

Though pure white noise cannot exist, various types of **band-limited white noise** are physically realizable. Two types of band-limited white noise, called *lowpass white noise* and *bandpass white noise*, are depicted in Fig. 8.5.[1] Notice that the area under both of these power spectral densities is finite, and thus the corresponding random processes both have finite power.

8.1.4   Power Spectral Density for Two Processes

For two jointly WSS random processes $X(t)$ and $Y(t)$, the **cross-power spectral density of $X(t)$ with $Y(t)$** is defined by

$$S_{XY}(f) = \mathscr{F}\{R_{XY}(\tau)\},$$

where $R_{XY}(\tau)$ is the cross-correlation function defined in Sect. 7.2. A similar definition can be made for $S_{YX}(f)$. Since $R_{XY}(\tau)$ is generally not an even function of τ, the cross-power spectral density need not be real-valued. When $X(t)$ and $Y(t)$ are orthogonal random processes, $R_{XY}(\tau) = 0$ by definition and so $S_{XY}(f) = 0$. The cross-power spectral density gives information about the distribution of the power generated by combining $X(t)$ and $Y(t)$, above and beyond their individual power spectra, when $X(t)$ and $Y(t)$ are not orthogonal. See Exercise 16.

Proof of the Wiener–Khinchin Theorem   The definition of $S_{XX}(f)$ involves the squared magnitude of a complex function; from the theory of complex numbers, we know that $|z|^2 = z \cdot z^*$, where $*$ denotes the complex conjugate. The proof then proceeds as follows:

---

[1] Readers already familiar with filters will recognize the terms "lowpass" and "bandpass." We will see these terms again in the next section.

$$S_{XX}(f) = \lim_{T \to \infty} E\left[\frac{|F_T(f)|^2}{2T}\right] = \lim_{T \to \infty} \frac{1}{2T} E\left[F_T(f)F_T^*(f)\right]$$

$$= \lim_{T \to \infty} \frac{1}{2T} E\left[\int_{-T}^{T} X(s)e^{-j2\pi fs}\,ds \left(\int_{-T}^{T} X(t)e^{-j2\pi ft}\,dt\right)^*\right]$$

$$= \lim_{T \to \infty} \frac{1}{2T} E\left[\int_{-T}^{T} X(t)e^{-j2\pi fs}\,ds \int_{-T}^{T} X(t)e^{j2\pi ft}\,dt\right]$$

$$= \lim_{T \to \infty} \frac{1}{2T} E\left[\int_{-T}^{T}\int_{-T}^{T} X(s)X(t)e^{-j2\pi f(s-t)}\,dt\,ds\right]$$

Next, pass the expected value into the integrand (which is permissible because the integral converges), and use the fact that wide-sense stationarity implies $E[X(s)X(t)] = R_{XX}(s-t)$:

$$\int_{-T}^{T}\int_{-T}^{T} E\left[X(s)X(t)e^{-j2\pi f(s-t)}\right]dt\,ds = \int_{-T}^{T}\int_{-T}^{T} E[X(s)X(t)]e^{-j2\pi f(s-t)}\,dt\,ds$$

$$= \int_{-T}^{T}\int_{-T}^{T} R_{XX}(s-t)e^{-j2\pi f(s-t)}\,dt\,ds$$

Now make the change of variables $\tau = s - t$ (i.e., $s = t + \tau$), under which the region of integration becomes the parallelogram pictured in Fig. 8.6.

Integrating in the order $dt\,d\tau$ yields the sum of two integrals:

$$S_{XX}(f) = \lim_{T \to \infty} \frac{1}{2T}\left[\int_{-2T}^{0}\int_{-T}^{\tau+T} R_{XX}(\tau)e^{-j2\pi f\tau}\,dt\,d\tau + \int_{0}^{2T}\int_{\tau-T}^{T} R_{XX}(\tau)e^{-j2\pi f\tau}\,dt\,d\tau\right]$$

$$= \lim_{T \to \infty} \frac{1}{2T}\left[\int_{-2T}^{0} R_{XX}(\tau)e^{-j2\pi f\tau}(2T+\tau)\,d\tau + \int_{0}^{2T} R_{XX}(\tau)e^{-j2\pi f\tau}(2T-\tau)\,d\tau\right]$$

$$= \lim_{T \to \infty} \frac{1}{2T}\int_{-2T}^{2T} R_{XX}(\tau)e^{-j2\pi f\tau}(2T-|\tau|)\,d\tau$$

$$= \lim_{T \to \infty}\int_{-2T}^{2T} R_{XX}(\tau)e^{-j2\pi f\tau}\left(1 - \frac{|\tau|}{2T}\right)d\tau$$

$$= \lim_{T \to \infty}\int_{-\infty}^{\infty} R_{XX}(\tau)e^{-j2\pi f\tau}q_T(\tau)\,d\tau$$

Fig. 8.6 Region of
integration for the proof of the
Wiener–Khinchin Theorem

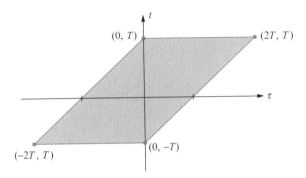

where $q_T(\tau) = 1 - |\tau|/2T$ for $|\tau| \le 2T$ and 0 otherwise. Since $q_T(\tau) \to 1$ as $T \to \infty$ for
all τ, we conclude that

$$S_{XX}(f) = \int_{-\infty}^{\infty} R_{XX}(\tau)e^{-j2\pi f\tau}d\tau = \mathscr{F}\{R_{XX}(\tau)\},$$

as claimed.   ∎

8.1.5 Exercises: Section 8.1 (1–21)

1. The function $\text{rect}(\tau)$ satisfies all the properties of an autocorrelation function for
   a WSS process that were specified in the main proposition of Sect. 7.3: $\text{rect}(\tau)$ is
   even, has its maximum value at 0, vanishes as $\tau \to \infty$. However, $\text{rect}(\tau)$ cannot
   be the autocorrelation of a WSS random process. Why not? [*Hint*: Consider the
   resulting psd.] This demonstrates that the properties listed in that proposition
   do *not* fully characterize the types of functions that can be autocorrelations.
2. Let $A(t)$ be a WSS random process with autocorrelation function $R_{AA}(\tau)$ and
   power spectral density $S_{AA}(f)$. Define an "amplitude modulated" version of
   $A(t)$ by

   $$X(t) = A(t)\cos(2\pi f_0 t + \Theta),$$

   where $\Theta \sim \text{Unif}(-\pi, \pi]$ and is independent of $A(t)$.
   (a) Find the mean and autocorrelation functions of $X(t)$.
   (b) Find the power spectral density of $X(t)$.
   (c) Find an expression for the expected power in $X(t)$.
3. Suppose $X(t)$ is a wide-sense stationary process with the following autocorrelation function:

   $$R_{XX}(\tau) = 250 + 1000\exp\left(-4 \times 10^6 \tau^2\right)$$

   (a) Find and graph the power spectral density of $X(t)$.
   (b) Find the ensemble average power in $X(t)$ between 500 Hz and 1 kHz.
   (c) Find the ensemble average power in $X(t)$ below 200 Hz.

4. Let $A(t)$ be a wide-sense stationary waveform with autocorrelation function $R_{AA}(\tau) = 2400\text{sinc}(2000\tau)$. Define a new random process $X(t)$ by

$$X(t) = 20 + A(t)\cos(5000\pi t + \Theta)$$

where Θ is uniform on $(-\pi, \pi]$ and independent of $A(t)$.
   (a) Find the mean function of $X(t)$.
   (b) Find the autocorrelation function of $X(t)$. Is $X(t)$ WSS?
   (c) Find the expected power in $X(t)$.
   (d) Find and sketch the power spectral density of $X(t)$.
   (e) Find the expected power in $X(t)$ in the frequency band from 2 to 3 kHz.
5. Suppose $X(t)$ is a WSS random process with power spectral density $S_{XX}(f) = 0.2 \exp(-\pi^2 f^2/10^{12})$.
   (a) Sketch the psd, and find the expected power in $X(t)$.
   (b) Find the expected power in $X(t)$ above 10 kHz.
   (c) Find the autocorrelation function of $X(t)$ and verify your answer to (a).
6. Let $X(t)$ be a WSS random process with mean $\mu_X = 32.6$ and autocovariance function $C_{XX}(\tau) = 12{,}160\text{sinc}^2(40{,}000\tau)$.
   (a) Find and sketch the power spectral density of $X(t)$.
   (b) Find the expected power in $X(t)$ below 20 kHz.
   (c) Find the expected power in $X(t)$ between 10 and 30 kHz.
   (d) Find the total expected power in $X(t)$.
7. Let $N(t)$ be *lowpass white noise*, i.e., $N(t)$ is WSS with power spectral density given by $S_{NN}(f) = N_0/2$ for $|f| \leq B$ and 0 otherwise (see Fig. 8.5a).
   (a) Find the expected power in $N(t)$.
   (b) Find the autocorrelation function of $N(t)$.
8. Let $N(t)$ be *bandpass white noise*, i.e., $N(t)$ is WSS with power spectral density given by $S_{NN}(f) = N_0/2$ for $f_0 - B/2 \leq |f| \leq f_0 + B/2$ and 0 otherwise (see Fig. 8.5b).
   (a) Find the expected power in $N(t)$.
   (b) Find the autocorrelation function of $N(t)$.
9. Let $N(t)$ be a Poisson telegraphic process with parameter λ as defined in Sect. 7.5, and consider $Y(t) = A_0 N(t)$ for some constant $A_0 > 0$.
   (a) Find the autocorrelation function of $Y(t)$.
   (b) Find and sketch the power spectral density of $Y(t)$.
   (c) Find the expected power in $Y(t)$.
   (d) What proportion of the expected power in $Y(t)$ lies below the frequency λ Hz?
10. Let $X(t)$ have power spectral density $S_{XX}(f) = N_0 - |f|/A$ for $|f| \leq B$ (and zero otherwise), where $B < N_0 A$.
   (a) Find the expected power in $X(t)$.
   (b) Find the autocorrelation function of $X(t)$.
   [*Hint*: It may be helpful to sketch $S_{XX}(f)$ first.]

11. Suppose a random process $X(t)$ has autocorrelation function $R_{XX}(\tau) = 100e^{-|\tau|} + 50e^{-|\tau-1|} + 50e^{-|\tau+1|}$.
    (a) Find the expected power in $X(t)$.
    (b) Find and sketch the power spectral density of $X(t)$.
    (c) Find the expected power in $X(t)$ below 1 Hz.

12. Let $X(t)$ be a WSS random process, and define a d-second delay of $X(t)$ by $Y(t) = X(t-d)$. Find the mean, autocorrelation, and power spectrum of $Y(t)$ in terms of those of $X(t)$.

13. Let $X(t)$ be a WSS random process, and define a d-second "moving window" process by $W(t) = X(t) - X(t-d)$. Find the mean, autocorrelation, and power spectrum of $W(t)$ in terms of those of $X(t)$.

14. Let $X(t)$ and $Y(t)$ be jointly WSS random processes. Show that $S_{XY}(f) = S_{YX}^{*}(f)$.

15. Let $X(t)$ and $Y(t)$ be orthogonal and WSS random processes, and define $Z(t) = X(t) + Y(t)$.
    (a) Are $X(t)$ and $Y(t)$ jointly WSS? Why or why not?
    (b) Is $Z(t)$ WSS?
    (c) Find the psd of $Z(t)$.

16. Let $X(t)$ and $Y(t)$ be *non*-orthogonal, jointly WSS random processes, and define $Z(t) = X(t) + Y(t)$.
    (a) Find the autocorrelation function of $Z(t)$. Is $Z(t)$ WSS?
    (b) Find the power spectral density of $Z(t)$, and explain why this expression is real-valued.

17. Let $X(t)$ and $Y(t)$ be independent WSS random processes, and define $Z(t) = X(t)Y(t)$.
    (a) Show that $Z(t)$ is also WSS.
    (b) Find the psd of $Z(t)$.

18. *Pink noise*, also called $1/f$ noise, is characterized by the power spectrum $S_{NN}(f) = 1/|f|$ for $f \neq 0$.
    (a) Explain why such a process is not physically realizable.
    (b) Consider a band-limited pink noise process with psd $S_{NN}(f) = 1/|f|$ for $f_0 \leq |f| \leq f_1$. Find the expected power of such a random process.
    (c) A "generalized pink noise" process has the psd $S_{NN}(f) = N_0/(2|f|^{1+\beta})$ for $|f| > f_0$ and 0 otherwise, where $0 < \beta < 1$. Find the expected power of such a random process.

19. *Highpass white noise* is characterized by the power spectrum $S_{NN}(f) = N_0/2$ for $|f| > B$ and 0 otherwise. Is highpass white noise a physically realizable process? Why or why not?

20. The *ac power spectral density* (ac-psd) of a WSS random process is defined as the Fourier transform of its auto*covariance* function:

$$S_{XX}^{ac}(f) = \mathscr{F}\{C_{XX}(\tau)\}$$

(a) By using the relationship between $C_{XX}(\tau)$ and $R_{XX}(\tau)$, develop an equation relating the psd of a random process to its ac-psd.

(b) Find the ac-psd for the random process of Example 8.3.

(c) Explain why the term "ac power spectral density" is appropriate.

21. Exercise 36 of Chap. 7 presented a random process of the form $X(t) = A \cdot Y(t)$, where A is a random variable and $Y(t)$ is an ergodic, WSS random process independent of A. It was shown that $X(t)$ is WSS but *not* ergodic.

(a) Find the psd of $X(t)$.

(b) Find the ac-psd of $X(t)$. (See the previous exercise.)

(c) Does the ac-psd of $X(t)$ include an impulse at zero? What does this say about our interpretation of "dc power offset" for non-ergodic processes?

8.2 Random Processes and LTI Systems

For any communication system to be effective, one must be able to successfully distinguish the intended signal from the noise it encounters during transmission. If we understand enough about the statistical properties of that noise, then in theory a filter can be constructed to minimize noise effects, thereby making the signal easier to "hear." This section gives a very brief overview of filters[2] and then investigates aspects of applying a filter to a random, continuous-time signal.

In communication theory, a *system* refers to anything that operates on a signal. We will denote a generic system by the letter L. If we let $x(t)$ and $y(t)$ denote the input and output of this system, respectively, then we may write

$$y(t) = L[x(t)]$$

where $L[]$ denotes the application of the system to a signal. One particular class of systems is of the greatest interest, since they form the backbone of filtering.

DEFINITION

A **linear, time-invariant (LTI) system** L satisfies the following two properties:

1. (Linearity) For all functions $x_1(t)$ and $x_2(t)$ and all constants a_1 and a_2,
$$L[a_1 x_1(t) + a_2 x_2(t)] = a_1 L[x_1(t)] + a_2 L[x_2(t)]$$

2. (Time invariance) For all $d > 0$, if $y(t) = L[x(t)]$, then $y(t-d) = L[x(t-d)]$.

---

[2] Readers interested in a thorough treatment of filters and other systems should consult the reference by Ambardar.

Part 2 of this definition says, in essence, that it does not matter on an absolute time scale when we apply the LTI system to $x(t)$; the response will be the same, other than the time delay. As it turns out, an LTI system can be completely characterized by its effect on an impulse, essentially because a signal can generally be decomposed into a weighted sum of impulses, and then we may apply linearity. With this in mind, an LTI system is described in the time domain by its **impulse response (function)**, denoted $h(t)$:

$$h(t) = L[\delta(t)]$$

It can be shown (see Chap. 6 of the reference by Ambardar) that if L is an LTI system with impulse response $h(t)$, then the input and output signals of L are related by a convolution operation:

$$y(t) = x(t) \star h(t) = \int_{-\infty}^{\infty} x(s)h(t-s)ds = \int_{-\infty}^{\infty} x(t-s)h(s)ds \qquad (8.5)$$

The same relationship holds for random signals, i.e., if $X(t)$ is the random input to an LTI system and $Y(t)$ the output, then $Y(t) = X(t) \star h(t)$.

The appearance of a convolution operator suggests it would be desirable to apply a transform to Eq. (8.5). The Fourier transform of the impulse response, denoted $H(f)$, is called the **transfer function** of the LTI system:

$$H(f) = \mathscr{F}\{h(t)\}$$

For *deterministic* signals, we may then write $Y(f) = X(f)H(f)$, where $X(f)$ and $Y(f)$ denote the Fourier transforms of $x(t)$ and $y(t)$, respectively. However, Fourier transforms of *random* signals do not exist (due to convergence issues), so the transfer function $H(f)$ cannot be defined as the ratio of the output and input in the frequency domain as one commonly does in other engineering situations. Still, the transfer function will prove critical in determining how the power in a random signal $X(t)$ is "transferred" by an LTI system, as we will see shortly.

8.2.1   Statistical Properties of the LTI System Output

The following proposition summarizes the relationships between the statistical properties of the random input signal $X(t)$ of an LTI system and the corresponding output signal $Y(t)$. Here $X(t)$ is again assumed to be wide-sense stationary.

> **PROPOSITION**
>
> Let L be an LTI system with impulse response $h(t)$ and transfer function $H(f)$. Suppose $X(t)$ is a wide-sense stationary process and let $Y(t)=L[X(t)]$, the output of the LTI system applied to $X(t)$. Then $X(t)$ and $Y(t)$ are *jointly* WSS, with the following properties.
>
> | Time domain | Frequency domain |
> |---|---|
> | 1. $\mu_Y = \mu_X \cdot \displaystyle\int_{-\infty}^{\infty} h(s)\,ds$ | 1. $\mu_y = \mu_x \cdot H(0)$ |
> | 2. $R_{YY}(\tau) = R_{XX}(\tau) \star h(\tau) \star h(-\tau)$ | 2. $S_{YY}(f) = S_{XX}(f) \cdot |H(f)|^2$ |
> | 3. $P_Y = R_{YY}(0)$ | 3. $P_Y = \displaystyle\int_{-\infty}^{\infty} S_{YY}(f)\,df$ |
> | 4. $R_{XY}(\tau) = R_{XX}(\tau) \star h(\tau)$ | 4. $S_{XY}(f) = S_{XX}(f) \cdot H(f)$ |

The quantity $|H(f)|^2$ in property 2 is called the **power transfer function** of the LTI system.

Proof   Using the convolution relationship between $X(t)$ and $Y(t)$,

$$Y(t) = X(t) \star h(t) = \int_{-\infty}^{\infty} X(t-s)h(s)\,ds \Rightarrow$$

$$E[Y(t)] = E\left[\int_{-\infty}^{\infty} X(t-s)h(s)\,ds\right]$$

$$= \int_{-\infty}^{\infty} E[X(t-s)]h(s)\,ds$$

Since $X(t)$ is WSS, the expression $E[X(t-s)]$ is just a constant, μ_X, from which $E[Y(t)] = \mu_X \int_{-\infty}^{\infty} h(s)\,ds$, as desired. Since this expression does not depend on t, we deduce that the mean of $Y(t)$ is constant (and we may denote it μ_Y). This establishes property 1 in the time domain. For the parallel result in the frequency domain, simply note that since $H(f) = \mathscr{F}\{h(t)\}$, it follows from the definition of the Fourier transform that $\int_{-\infty}^{\infty} h(s)\,ds = H(0)$.

A similar (but vastly more tedious) derivation yields property 2 in the time domain (see Exercise 31). The right-hand side establishes that the autocorrelation of $Y(t)$ depends only on τ and not t, and therefore $Y(t)$ is indeed WSS. Hence, the Wiener–Khinchin Theorem applies to $Y(t)$, and taking the Fourier transform of both sides gives

$$\mathscr{F}\{R_{YY}(\tau)\} = \mathscr{F}\{R_{XX}(\tau) \star h(\tau) \star h(-\tau)\} \Rightarrow$$
$$S_{YY}(f) = \mathscr{F}\{R_{XX}(\tau)\}\mathscr{F}\{h(\tau)\}\mathscr{F}\{h(-\tau)\}$$
$$= S_{XX}(f)H(f)H^*(f),$$

where $H^*(f)$ denotes the complex conjugate of $H(f)$. Now, recall that for any complex number z, $z \cdot z^* = |z|^2$. We immediately have $H(f)H^*(f) = |H(f)|^2$, completing property 2 in the frequency domain.

Both the time and frequency versions of property 3 follow immediately from Sect. 8.1 and the fact that $Y(t)$ is WSS. The proofs of property 4 in the time and frequency domain are parallel to those of property 2.   ∎

The frequency domain properties of the previous theorem are the most illuminating. Property 1 says the dc offset of $X(t)$, μ_X, is "transferred" to the dc offset of $Y(t)$ by evaluating the transfer function $H(f)$ at 0. This makes sense, since the dc offset corresponds to the frequency $f = 0$. Notice in particular that if $\mu_X = 0$, necessarily $\mu_Y = 0$; an LTI system cannot introduce a dc offset if none exists in the input signal.

Property 2 states that the power spectrum of the output of an LTI system is obtained from the input psd through multiplication by the quantity $|H(f)|^2$, hence the name "power transfer function." Similar to the preceding discussion about dc offset, observe that if $X(t)$ carries no power at some particular frequency f (so that $S_{XX}(f) = 0$), then $S_{YY}(f)$ will be zero there as well. An LTI system cannot introduce power to *any* frequency that did not appear in the input signal.

Example 8.5 One of the simplest filters is an *RC circuit*, an LTI system whose impulse response is given by

$$h(t) = \frac{1}{RC}e^{-t/RC}u(t)$$

where $u(t)$ is the unit step function, equal to 1 for $t \geq 0$ and zero otherwise. (The product RC of the resistance and the capacitance is called the *time constant* of the circuit, since its units are seconds. The unit step function makes $h(t)$ equal 0 for $t < 0$; engineers call this a *causal filter*.) Suppose we have such a circuit with time constant RC and that we model the input to our system as a pure white noise process with power spectral density $S_{XX}(f) = N_0/2$ W/Hz. Let's investigate the properties of the output, $Y(t)$.

First, since white noise has mean zero, it follows that $\mu_Y = 0$ as well (property 1). Now we need the transfer function of the system:

$$H(f) = \mathscr{F}\{h(t)\} = \frac{1}{RC}\mathscr{F}\left\{e^{-t/RC}u(t)\right\} = \frac{1}{RC}\frac{0!}{(1/RC + j2\pi f)^{0+1}} = \frac{1}{1 + j2\pi fRC}$$

Fig. 8.7 Power spectral
density of $Y(t)$ in Example 8.5

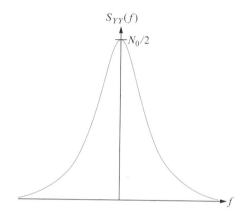

Next, we find the psd of $Y(t)$ using property 2:

$$S_{YY}(f) = S_{XX}(f) \cdot |H(f)|^2 = \frac{N_0}{2} \left| \frac{1}{1 + j2\pi fRC} \right|^2 = \frac{N_0}{2} \cdot \frac{1^2}{1^2 + (2\pi fRC)^2}$$

$$= \frac{N_0/2}{1 + (2\pi fRC)^2}$$

Figure 8.7 displays this power spectral density. Finally, the ensemble average
power of $Y(t)$ is given by

$$P_Y = \int_{-\infty}^{\infty} S_{YY}(f) df = \int_{-\infty}^{\infty} \frac{N_0/2}{1 + (2\pi fRC)^2} df = \frac{N_0}{2} \int_{-\infty}^{\infty} \frac{df}{1 + (2\pi fRC)^2} = \frac{N_0}{4RC}$$

where the integral is evaluated by the substitution $x = 2\pi fRC$ and the fact that the
antiderivative of $1/(1 + x^2)$ is $\arctan(x)$.

We find that, even though the input signal had (theoretically) infinite power, the
output $Y(t)$ has finite power, directly proportional to the intensity of the input and
inversely proportional to the time constant of the circuit. (As an exercise, see if you
can verify that the units on the final expression for power are indeed watts.) ∎

Example 8.6 An LTI system has an impulse response of $h(t) = t^2 e^{-t} u(t)$. The input
to this system is the random process

$$X(t) = V + 500 \cos \left(2 \times 10^6 \pi t + \Theta \right),$$

where V and Θ are independent random variables, Θ is uniformly distributed on
$(-\pi, \pi]$, and V has mean 60 and variance 12. It was shown in Exercise 25 of Chap. 7
that $X(t)$ is WSS, with mean $\mu_X = \mu_V = 60$ and autocorrelation function
$R_{XX}(\tau) = 3612 + 125{,}000\cos(2 \times 10^6 \pi \tau)$. (Depending on whether we choose to
interpret $X(t)$ as a voltage or current waveform, the units on the mean are either
volts or amperes.) Applying the Wiener–Khinchin Theorem, the psd of $X(t)$ is

$$S_{XX}(f) = \mathscr{F}\{R_{XX}(\tau)\}$$
$$= \mathscr{F}\{3612 + 125{,}000\cos\left(2\times 10^6\pi\tau\right)\}$$
$$= 3612\delta(f) + 62{,}500\delta(f - 10^6) + 62{,}500\delta(f + 10^6)$$

Since $X(t)$ consists of a (random) dc offset and a periodic component, the power spectrum of $X(t)$ is comprised entirely of impulses. Now let $Y(t)$ denote the output of the LTI system. To deduce the properties of $Y(t)$ requires the transfer function, $H(f)$, of the LTI system. Using the table of Fourier transforms in Appendix B,

$$H(f) = \mathscr{F}\{h(t)\} = \mathscr{F}\{t^2 e^{-t} u(t)\} = \frac{2!}{(1 + j2\pi f)^{2+1}} = \frac{2}{(1 + j2\pi f)^3}$$

According to property 1 of the earlier proposition, the mean of the output signal $Y(t)$ is given by

$$\mu_Y = \mu_X \cdot H(0) = 60 \cdot \frac{2}{(1 + j2\pi \cdot 0)^3} = 120$$

To find the psd of $Y(t)$, we must first calculate the power transfer function of the LTI system:

$$|H(f)|^2 = \left|\frac{2}{(1 + j2\pi f)^3}\right|^2 = \frac{|2|^2}{\left(|1 + j2\pi f|^2\right)^3} = \frac{4}{\left(1 + (2\pi f)^2\right)^3}$$

Since the input power spectrum consists of impulses, so does the output power spectrum; the coefficients on the impulses are found by evaluating the power transfer function at the appropriate frequencies:

$$S_{YY}(f) = S_{XX}(f)|H(f)|^2$$
$$= 3612\delta(f)|H(f)|^2 + 62{,}500\delta(f - 10^6)|H(f)|^2 + 62{,}500\delta(f + 10^6)|H(f)|^2$$
$$= 3612\delta(f)|H(0)|^2 + 62{,}500\delta(f - 10^6)|H(10^6)|^2 + 62{,}500\delta(f + 10^6)|H(-10^6)|^2$$
$$= 3612\delta(f) \cdot \frac{4}{(1 + 0^2)^3} + 62{,}500\delta(f - 10^6) \cdot \frac{4}{\left(1 + (2\times 10^6\pi)^2\right)^3}$$
$$+ 62{,}500\delta(f + 10^6) \cdot \frac{4}{\left(1 + (-2\times 10^6\pi)^2\right)^3}$$
$$= 14{,}448\delta(f) + 4\times 10^{-36}\delta(f - 10^6) + 4\times 10^{-36}\delta(f + 10^6)$$

The effect of the LTI system is to "ramp up" the dc power and to effectively eliminate the power at 1 MHz. In particular, the expected power in the output signal $Y(t)$ is

$$P_Y = \int\limits_{-\infty}^{\infty} S_{YY}(f)df = 14,448 + 2\left(4 \times 10^{-36}\right) \approx 14.448\text{kW},$$

with essentially all of the power coming from the dc component.   ∎

8.2.2   Ideal Filters

The goal of a filter is, of course, to eliminate ("filter out") whatever noise has accumulated during the transmission of a signal. At the same time, we do not want our filter to affect the intended signal, lest information be lost. Ideally, we would know at what frequencies the noise in our transmission exists, and then a filter would be designed that completely eliminates those frequencies while preserving all others. (If the frequency band of the noise overlaps that of the signal, one can *modulate* the signal so that the two frequency bands are disjoint.)

DEFINITION
An LTI system is an **ideal filter** if there exists some set of frequencies, F_{pass}, such that the system's power transfer function is given by

$$|H(f)|^2 = \begin{cases} 1 & \text{for } f \in F_{\text{pass}} \\ 0 & \text{otherwise} \end{cases}$$

If we let $X(t)$ denote the input to the system (which may consist of both signal and noise) and $Y(t)$ the output, then for an ideal filter we have

$$S_{YY}(f) = S_{XX}(f)|H(f)|^2 = \begin{cases} S_{XX}(f) & \text{for } f \in F_{\text{pass}} \\ 0 & \text{otherwise} \end{cases}$$

In other words, the power spectrum of $X(t)$ within the band F_{pass} is unchanged by the filter, while everything in $X(t)$ lying outside that band is completely eliminated. Thus, the obvious goal is to select F_{pass} to include all frequencies in the signal and exclude all frequencies in the accumulated noise.

Figure 8.8 displays $|H(f)|$ for four different types of ideal filters. To be consistent with the two-sided nature of power spectral densities, we present the graphs for $-\infty < f < \infty$, even though plots starting at $f = 0$ are more common in engineering practice. Figure 8.8a shows a **lowpass filter**, which preserves the signal up to some threshold B. Under our notation, $F_{\text{pass}} = [0, B]$ for an ideal lowpass filter. The ideal **highpass filter** of Fig. 8.8b does essentially the opposite, preserving frequencies *above* B. Figure 8.8c, d illustrate a **bandpass filter** and a **bandstop filter** (also called a **notch filter**), respectively.

The previous section briefly mentioned band-limited white noise processes, wherein we also used the terms "lowpass" and "bandpass." These models inherit

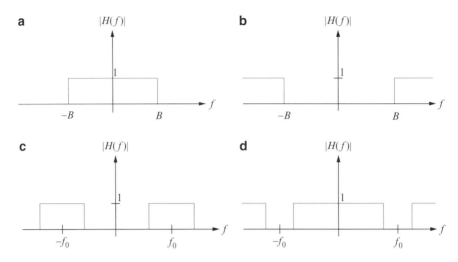

Fig. 8.8 Ideal filters: (**a**) lowpass; (**b**) highpass; (**c**) bandpass; (**d**) bandstop

their names from the aforementioned filters, e.g., if pure white noise passes through an ideal bandpass filter, the result is called bandpass white noise.

Example 8.7 A WSS random signal $X(t)$ with autocorrelation function $R_{XX}(\tau) = 250 + 1500\exp(-1.6 \times 10^9 \tau^2)$ is passed through an ideal lowpass filter with $B = 10$ kHz (i.e., 10^4 Hz). Before considering the effect of the filter, let's investigate the properties of the input signal $X(t)$. The ensemble average power of the input is $P_X = R_{XX}(0) = 250 + 1500 = 1750$ W; moreover, we recognize that 250 W represents the dc power offset while the other 1500 W comes from an aperiodic component. Applying the Wiener–Khinchin Theorem, the input power spectral density is given by

$$S_{XX}(f) = \mathscr{F}\{R_{XX}(\tau)\} = \mathscr{F}\{250 + 1500\exp(1.6 \times 10^9 \tau^2)\}$$
$$= 250\delta(f) + 1500\mathscr{F}\{\exp(-1.6 \times 10^9 \tau^2)\}$$

The second Fourier transform requires the rescaling property; however, we must be careful in identifying the rescaling constant. If we rewrite $1.6 \times 10^9 \tau^2$ as $(4 \times 10^4 \tau)^2$, we see that the appropriate rescaling constant is actually $a = 4 \times 10^4$. Continuing,

$$S_{XX}(f) = 250\delta(f) + 1500\mathscr{F}\{\exp(-(4 \times 10^4 \tau)^2)\}$$
$$= 250\delta(f) + 1500 \cdot \frac{1}{4 \times 10^4}\sqrt{\pi}\exp(-\pi^2(f/4 \times 10^4)^2)$$
$$= 250\delta(f) + \frac{3\sqrt{\pi}}{80}\exp(-\pi^2 f^2/1.6 \times 10^9)$$

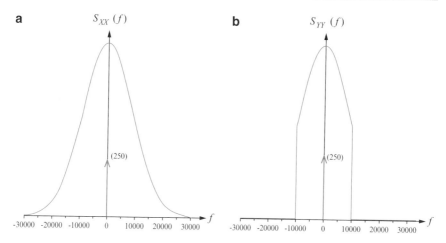

Fig. 8.9 Power spectral densities for Example 8.7: (**a**) input signal; (**b**) output signal

This psd appears in Fig. 8.9a. Now let's apply the filter, and as usual let $Y(t)$ denote the output. Then, based on the preceding discussion, the psd of $Y(t)$ is given by

$$S_{YY}(f) = \begin{cases} S_{XX}(f) & f \in F_{\text{pass}} \\ 0 & \text{otherwise} \end{cases}$$

$$= \begin{cases} 250\delta(f) + \dfrac{3\sqrt{\pi}}{80} e^{-\pi^2 f^2/1.6 \times 10^9} & |f| \leq 10^4 \text{Hz} \\ 0 & \text{otherwise} \end{cases}$$

Figure 8.9b shows the output power spectrum, which is identical to $S_{XX}(f)$ in the preserved band $[0, 10^4]$ and zero everywhere else.

The ensemble average power of the output signal $Y(t)$ is calculated by taking the integral of $S_{YY}(f)$, which in this case requires numerical integration by a calculator or computer:

$$P_Y = \int_{-\infty}^{\infty} S_{YY}(f)df = \int_{-10^4}^{10^4} \left[250\delta(f) + \frac{3\sqrt{\pi}}{80} e^{-\pi^2 f^2/1.6 \times 10^9} \right] df$$

$$= 250 + 2\int_{0}^{10^4} \frac{3\sqrt{\pi}}{80} e^{-\pi^2 f^2/1.6 \times 10^9} df \approx 250 + 1100 = 1350\text{W}$$

∎

In the preceding example, the output power from the ideal filter was less than the input power (1350 W < 1750 W). It should be clear that this will always be the case: it is impossible to achieve a power gain with an ideal filter of any type. At best, if the entire input lies within the preserved band F_{pass}, then the input and output power will be equal.

Of course, in practice one cannot actually construct an "ideal" filter—there is no engineering system that will perfectly cut off a signal at a prescribed frequency. But

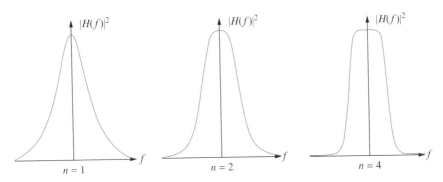

Fig. 8.10 Power transfer functions for Butterworth filters (approximations to ideal filters)

many simple systems can approximate our ideal. For instance, consider Example 8.5: the power transfer function of that RC filter is identical to Fig. 8.7 (except that the height at $f=0$ is 1 rather than $N_0/2$). This bears some weak resemblance to the picture for an ideal lowpass filter in Fig. 8.8a. In fact, a more general class of LTI systems called *Butterworth filters* can achieve an even more "squared off" appearance; the nth-order Butterworth filter has a power transfer function of the form

$$|H(f)|^2 = \frac{\alpha}{1 + (\beta 2\pi f)^{2n}},$$

where the constants α and β can be derived from the underlying circuit. The RC filter of Example 8.5 is a "first-order" (i.e., $n=1$) Butterworth filter. The books by Peebles and Ambardar listed in the references provide more information. Examples of these power transfer functions are displayed in Fig. 8.10.

8.2.3 Signal Plus Noise

For a variety of physical reasons, it is common in engineering practice to assume that communication noise is *additive*, i.e., if our intended signal $X(t)$ experiences noise $N(t)$ during transmission, then the received transmission (prior to any filtering) has the form $X(t)+N(t)$. We assume throughout this subsection that $X(t)$ and $N(t)$ are independent, WSS random processes and that $E[N(t)]=0$ (i.e., the noise component does not contain a dc offset, a standard engineering assumption).[3]

The mean of the input process is given by $E[X(t)+N(t)] = E[X(t)] + E[N(t)] = \mu_X + 0 = \mu_X$, the dc offset of the input signal. Computing the autocorrelation of the input process relies on the assumed independence:

---

[3] Please note: The case of a deterministic signal $x(t)$ must be handled somewhat differently. Consult the reference by Ambardar for details.

$$\begin{aligned}
R_{\text{in}}(\tau) &= E\big[(X(t)+N(t))\cdot(X(t+\tau)+N(t+\tau))\big] \\
&= E\big[X(t)X(t+\tau)\big] + E\big[X(t)N(t+\tau)\big] + E\big[N(t)X(t+\tau)\big] + E\big[N(t)N(t+\tau)\big] \\
&= R_{XX}(\tau) + E\big[X(t)\big]E\big[N(t+\tau)\big] + E\big[N(t)\big]E\big[X(t+\tau)\big] + R_{NN}(\tau) \\
&= R_{XX}(\tau) + E\big[X(t)\big]0 + 0E\big[X(t+\tau)\big] + R_{NN}(\tau) \quad \text{since } \mu_N = 0 \\
&= R_{XX}(\tau) + R_{NN}(\tau)
\end{aligned}$$

Then, by the Wiener–Khinchin Theorem, the input power spectrum is

$$S_{\text{in}}(f) = \mathscr{F}\{R_{XX}(\tau) + R_{NN}(\tau)\} = S_{XX}(f) + S_{NN}(f)$$

Now we imagine passing the random process $X(t)+N(t)$ through some LTI system L (presumably a filter intended to reduce the noise). The foregoing assumptions make the analysis of the output process quite straightforward. To start, the linearity property allows us to regard the system output as the sum of two parts:

$$L[X(t) + N(t)] = L[X(t)] + L[N(t)]$$

That is, we may identify $L[X(t)]$ and $L[N(t)]$ as the output signal and output noise, respectively. These two output processes are also independent and WSS. Letting $H(f)$ denote the transfer function of the LTI system, the mean of the output signal and output noise are, respectively,

$$\mu_{L[X]} = E(L[X(t)]) = \mu_X H(0), \quad \mu_{L[N]} = E(L[N(t)]) = \mu_N H(0) = 0$$

The mean of the overall output process is, by linearity, $\mu_X H(0) + 0 = \mu_X H(0)$. Similarly, the power spectral density of the output process is

$$S_{\text{out}}(f) = S_{\text{in}}(f)|H(f)|^2 = S_{XX}(f)|H(f)|^2 + S_{NN}(f)|H(f)|^2;$$

the two halves of this expression are the psds of the output signal and output noise.

One measure of the quality of the filter (the LTI system) involves comparing the **power signal-to-noise ratio** of the input and output:

$$\text{SNR}_{\text{in}} = \frac{P_X}{P_N} \quad \text{versus} \quad \text{SNR}_{\text{out}} = \frac{P_{L[X]}}{P_{L[N]}}$$

A good filter should achieve a higher SNR_{out} than SNR_{in} by reducing the amount of noise without losing any of the intended signal.

Example 8.8 Suppose a random signal $X(t)$ incurs additive noise $N(t)$ in transmission. Assume the signal and noise components are independent and wide-sense stationary, $X(t)$ has autocorrelation function $R_{XX}(\tau) = 2400 + 45{,}000\,\text{sinc}^2(1800\tau)$, and $N(t)$ has autocorrelation function $R_{NN}(\tau) = 1500e^{-10{,}000|\tau|}$. To filter out the noise, we pass the input $X(t)+N(t)$ through an ideal lowpass filter with band limit 1800 Hz.

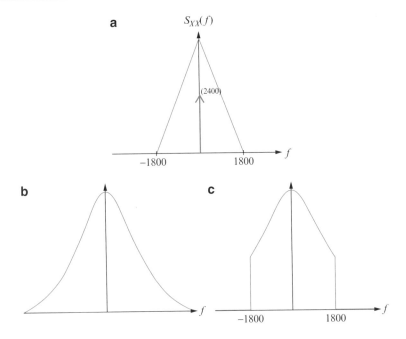

Fig. 8.11 Power spectra for Example 8.8: (**a**) input signal; (**b**) input noise; (**c**) output noise

Our input power signal-to-noise ratio is

$$\text{SNR}_{\text{in}} = \frac{P_X}{P_N} = \frac{R_{XX}(0)}{R_{NN}(0)} = \frac{2400 + 45{,}000}{1500} = 31.6$$

The power spectral density of $X(t)$ is

$$S_{XX}(f) = \mathscr{F}\{R_{XX}(\tau)\} = \mathscr{F}\{2400 + 45{,}000\text{sinc}^2(1800\tau)\}$$
$$= 2400\delta(f) + (45{,}000)\frac{1}{1800}\text{tri}\left(\frac{f}{1800}\right) = 2400\delta(f) + 25\text{tri}\left(\frac{f}{1800}\right)$$

This psd is displayed in Fig. 8.11a. Notice that the entire power spectrum of the input signal lies within the band [0 Hz, 1800 Hz], which is precisely the preserved band of the filter. Therefore, the filter will have no effect on the input signal; in particular, the input and output signal components will have the same power spectral density and the same ensemble average power (47.4 kW).

On the other hand, part of the input noise will be removed by the filter. Begin by finding the psd of the input noise:

$$S_{NN}(f) = \mathscr{F}\{R_{NN}(\tau)\} = \mathscr{F}\left\{1500e^{-10,000|\tau|}\right\} = 1500 \cdot \frac{2(10,000)}{(10,000)^2 + (2\pi f)^2}$$

$$= \frac{3 \times 10^7}{10^8 + (2\pi f)^2}$$

Figure 8.11b shows the psd of the input noise, while in Fig. 8.11c we see the psd of the output noise $L[N(t)]$ resulting from passing $N(t)$ through the ideal filter. The average power in the output noise is

$$P_{L[N]} = 2 \int_0^{1800} \frac{3 \times 10^7}{10^8 + (2\pi f)^2} \, df = \cdots = 808.6\text{W},$$

slightly more than half the original (i.e., input) noise power. As a result, the output power signal-to-noise ratio equals

$$\text{SNR}_{\text{out}} = \frac{P_{L[X]}}{P_{L[N]}} = \frac{47{,}400}{808.6} = 58.6$$

Because the signal and noise power spectra were so similar, it was not possible to filter out very much noise. Assuming our model for the input noise is correct, one solution would be to modulate the signal before transmission to a center frequency in the "tail" of the $S_{NN}(f)$ distribution and then employ a bandpass filter around that center frequency (see Exercise 30). ∎

8.2.4 Exercises: Section 8.2 (22–38)

22. Let $Y(t)$ be the output process from Example 8.5. Find the autocorrelation function of $Y(t)$.
23. A WSS current waveform $X(t)$ with power spectral density $S_{XX}(f) = 0.02$ W/Hz for $|f| \leq 60$ kHz is the input to a filter with impulse response $h(t) = 40e^{-40t}u(t)$. Let $Y(t)$ denote the output current waveform.
    (a) Find the autocorrelation function of the input process $X(t)$. [*Hint:* Draw $S_{XX}(f)$ first.]
    (b) Calculate the ensemble average power in the input process $X(t)$.
    (c) Find the transfer function of this filter.
    (d) Find and graph the power spectral density of the output process $Y(t)$.
    (e) Determine the ensemble average power in the output process $Y(t)$.
24. A Poisson telegraphic process $N(t)$ with parameter $\lambda = 2$ (see Sect. 7.5) is the input to an LTI system with impulse response $h(t) = 2e^{-t}u(t)$.
    (a) Find the power spectral density of $N(t)$.
    (b) Find the transfer function of the LTI system.
    (c) Find the power spectral density of the output process $Y(t) = L[N(t)]$.

25. A white noise process $X(t)$ with power spectral density $S_{XX}(f) = N_0/2$ is the input to an LTI system with impulse response $h(t) = 1$ for $0 \leq t < 1$ (and 0 otherwise). Let $Y(t)$ denote the output.
    (a) Find the mean of $Y(t)$.
    (b) Find the transfer function of the LTI system.
    (c) Find the power spectral density of $Y(t)$.
    (d) Find the expected power of $Y(t)$.

26. The random process $X(t) = A_0\cos(\omega_0 t + \Theta)$, where $\Theta \sim \text{Unif}(-\pi, \pi]$, is the input to an LTI system with impulse response $h(t) = Be^{-Bt}u(t)$. Let $Y(t)$ denote the output.
    (a) Determine the transfer function and power transfer function of this system.
    (b) Find the power spectral density of $Y(t)$.
    (c) Determine the expected power in $Y(t)$. How does that compare to $X(t)$?

27. A WSS random process $X(t)$ with autocorrelation function $R_{XX}(\tau) = 100 + 25e^{-|\tau|}$ is passed through an LTI system having impulse response $h(t) = te^{-4t}u(t)$. Let $Y(t)$ denote the output.
    (a) Find the power spectral density of $X(t)$.
    (b) What is the expected power of $X(t)$?
    (c) Determine the transfer function and power transfer function of this system.
    (d) Find and sketch the power spectral density of $Y(t)$.
    (e) What is the expected power of $Y(t)$?

28. A white noise process $X(t)$ with power spectral density $S_{XX}(f) = N_0/2$ is the input to an LTI system with impulse response $h(t) = e^{-Bt}\sin(\omega_0 t)u(t)$. Let $Y(t)$ denote the output.
    (a) Determine the transfer function of the LTI system.
    (b) Find and sketch the power spectral density of $Y(t)$.

29. Suppose $X(t)$ is a white noise process with power spectral density $S_{XX}(f) = N_0/2$. A filter with transfer function $H(f) = e^{-\alpha|f|}$ is applied to this process; let $Y(t)$ denote the output.
    (a) Find the power spectral density of $Y(t)$.
    (b) Find the autocorrelation function of $Y(t)$.
    (c) Find the expected power of $Y(t)$.

30. Let $X(t)$ be a WSS random process with autocorrelation function $R_{XX}(\tau) = 45{,}000\text{sinc}^2(1800\tau)$; this is the signal from Example 8.8 without the dc offset. Suppose $X(t)$ encounters the noise $N(t)$ described in Example 8.8. Since both $X(t)$ and $N(t)$ are concentrated at low frequencies, it is desirable to *modulate* $X(t)$ and then use an appropriate filter. Consider the following modulation, performed prior to transmission: $X_{\text{mod}}(t) = X(t)\cos(4000\pi t + \Theta)$, where $\Theta \sim \text{Unif}(-\pi, \pi]$. The received signal will be $X_{\text{mod}}(t) + N(t)$, to which an ideal bandpass filter on the spectrum of $X_{\text{mod}}(t)$ will be applied.
    (a) Find the autocorrelation function of $X_{\text{mod}}(t)$.
    (b) Find the power spectral density of $X_{\text{mod}}(t)$.
    (c) Based on (b), what would be the optimal frequency band to "pass" through a filter?

(d) Use the results of Example 8.8 to determine the expected power in $L[N(t)]$, the filtered noise process.

(e) Compare the input and output power signal-to-noise ratios. How do these compare to the SNRs in Example 8.8?

31. Let $X(t)$ be the WSS input to an LTI system with impulse response $h(t)$, and let $Y(t)$ denote the output.

(a) Show that the cross-correlation function $R_{XY}(\tau)$ equals $R_{XX}(\tau)\star h(\tau)$ as stated in the main proposition of this section. [*Hint:* In the definition of $R_{XY}(\tau)$, write $Y(t+\tau)$ as a convolution integral. Rearrange, and then make an appropriate substitution to show that the integrand is equal to $R_{XX}(\tau - s)\cdot h(s)$.]

(b) Show that the autocorrelation function of $Y(t)$ is given by

$$R_{YY}(\tau) = R_{XY}(\tau)\star h(-\tau) = R_{XX}(\tau)\star h(\tau)\star h(-\tau)$$

[*Hint:* Write $Y(t) = X(t)\star h(t)$ in the definition of $R_{YY}(\tau)$. Rearrange, and then make an appropriate substitution to show that the integrand is equal to $R_{XY}(\tau - s)\cdot h(-s)$. Then invoke (a).]

32. A *T-second moving-average filter* has impulse response $h(t) = 1/T$ for $0 \le t \le T$ (and zero otherwise).

(a) Find the transfer function of this filter.

(b) Find the power transfer function of this filter.

(c) Suppose $X(t)$ is a white noise process with power spectral density $S_{XX}(f) = N_0/2$. If $X(t)$ is passed through this moving-average filter and $Y(t)$ is the resulting output, find the power spectral density, expected power, and autocorrelation function of $Y(t)$.

33. Suppose we pass band-limited white noise $X(t)$ with arbitrary parameters N_0 and B through a *differentiator*:

$$Y(t) = L[X(t)] = \frac{d}{dt}X(t)$$

The transfer function of the differentiator is known to be $H(f) = j2\pi f$.

(a) Find the power spectral density of $Y(t)$.

(b) Find the autocorrelation function of $Y(t)$.

(c) What is the ensemble average power of the output?

34. A *short-term integrator* is defined by the input-output relationship

$$Y(t) = L[X(t)] = \frac{1}{T}\int_{t-T}^{t} X(s)\ ds$$

(a) Find the impulse response of this system.

(b) Find the power spectrum of $Y(t)$ in terms of the power spectrum of $X(t)$. [*Hint:* Write the answer to (a) in terms of the rectangular function first.]

35. Let $X(t)$ be WSS, and let $Y(t)$ be the output resulting from the application to $X(t)$ of an LTI system with impulse response $h(t)$ and transfer function $H(f)$.

Define a new random process as the difference between input and output: $D(t) = X(t) - Y(t)$.

(a) Find an expression for the autocorrelation function of $D(t)$ in terms of R_{XX} and h.

(b) Determine the power spectral density of $D(t)$, and verify that your answer is real, symmetric, and nonnegative.

36. An amplitude-modulated waveform can be modeled by the expression $A(t)\cos(100\pi t + \Theta) + N(t)$, where $A(t)$ is WSS and has autocorrelation function $R_{AA}(\tau) = 80\text{sinc}^2(10\tau)$; $\Theta \sim \text{Unif}(-\pi, \pi]$ and is independent of $A(t)$; and $N(t)$ is band-limited white noise, independent of $A(t)$ and Θ, with $S_{NN}(f) = 0.05$ W/Hz for $|f| < 100$ Hz. To filter out the noise, we pass the waveform through an ideal bandpass filter with transfer function $H(f) = 1$ for $40 < |f| < 60$.

Let $X(t) = A(t)\cos(100\pi t + \Theta)$, the signal part of the input.

(a) Find the autocorrelation of $X(t)$.

(b) Find the ensemble average power in $X(t)$.

(c) Find and graph the power spectral density of $X(t)$.

(d) Find the ensemble average power in the signal part of the output.

(e) Find the ensemble average power in $N(t)$.

(f) Find the ensemble average power in the noise part of the output.

(g) Find the power signal-to-noise ratio of the input and the power signal-to-noise ratio of the output. Discuss what you find.

37. A random signal $X(t)$ incurs additive noise $N(t)$ in transmission. The signal and noise components are independent and WSS, $X(t)$ has autocorrelation function $R_{XX}(\tau) = 250{,}000 + 120{,}000\cos(70{,}000\pi\tau) + 800{,}000\text{sinc}(100{,}000\tau)$, and $N(t)$ has power spectral density $S_{NN}(f) = 2.5\times10^{-2}$ W/Hz for $|f| \le 100$ kHz. To filter out the noise, we pass the input $X(t) + N(t)$ through an ideal lowpass filter with transfer function $H(f) = 1$ for $|f| \le 60$ kHz.

(a) Find the ensemble average power in $X(t)$.

(b) Find and sketch the power spectral density of $X(t)$.

(c) Find the power spectral density of $L[X(t)]$.

(d) Find the ensemble average power in $L[X(t)]$.

(e) Find the ensemble average power in $N(t)$.

(f) Find the ensemble average power in $L[N(t)]$. (Think about what the power spectral density of $L[N(t)]$ will look like.)

(g) Find the power signal-to-noise ratio of the input and the power signal-to-noise ratio of the output. Discuss what you find.

38. Let $X(t)$ be a pure white noise process with psd $N_0/2$. Consider an LTI system with impulse response $h(t)$, and let $Y(t)$ denote the output resulting from passing $X(t)$ through this LTI system.

(a) Show that $R_{XY}(\tau) = \frac{N_0}{2}h(\tau)$.

(b) Show that $P_Y = \frac{N_0}{2}E_h$, where E_h is the *energy* in the impulse response function, defined by $E_h = \int\limits_{-\infty}^{\infty} h^2(t)dt$.

8.3 Discrete-Time Signal Processing

Recall from Sect. 7.4 that a random sequence (i.e., a discrete-time random process) X_n is said to be wide-sense stationary if (1) its mean, $\mu_X[n]$, is a constant μ_X and (2) its autocorrelation function, $R_{XX}[n, n+k]$, depends only on the integer-valued time difference k (in which case we may denote the autocorrelation $R_{XX}[k]$). Analogous to the Wiener–Khinchin Theorem, the power spectral density of a WSS random sequence is given by the **discrete-time Fourier transform** of its autocorrelation function:

$$S_{XX}(F) = \sum_{k=-\infty}^{\infty} R_{XX}[k]e^{-j2\pi Fk} \tag{8.6}$$

We use parentheses around the argument F in Eq. (8.6) because $S_{XX}(F)$ is a function on a continuum, even though the random sequence is on a discrete index set (the integers). Similar to the continuous case, it can be shown that $S_{XX}(F)$ is a real-valued, nonnegative, symmetric function of F. (The choice of capital F will be explained toward the end of this section.)

Power spectral densities for random sequences differ from their continuous-time counterparts in one key respect: *the psd of a WSS random sequence is always a periodic function, with period 1*. To see this, recall that $e^{j2\pi k} = 1$ for any integer k, and write

$$S_{XX}(F+1) = \sum_{k=-\infty}^{+\infty} R_{XX}[k]e^{-j2\pi(F+1)k} = \sum_{k=-\infty}^{+\infty} R_{XX}[k]e^{-j2\pi Fk}e^{-j2\pi k}$$

$$= \sum_{k=-\infty}^{+\infty} R_{XX}[k]e^{-j2\pi Fk} = S_{XX}(F)$$

As a consequence, we may recover the autocorrelation function of a WSS random sequence from its power spectrum by taking the inverse Fourier transform of $S_{XX}(F)$ over an interval of length 1:

$$R_{XX}[k] = \int_{-1/2}^{1/2} S_{XX}(F)e^{j2\pi Fk}dF \tag{8.7}$$

This affects how we calculate the power in a random sequence from its power spectral density. Analogous to the continuous-time case, we define the (ensemble) average power of a WSS random sequence X_n by

$$P_X = E\left(X^2[n]\right) = R_{XX}[0] = \int_{-1/2}^{1/2} S_{XX}(F)e^{j2\pi F(0)}dF = \int_{-1/2}^{1/2} S_{XX}(F)dF$$

That is, the expected power in a random sequence is determined by integrating its psd *over one period*, not the entire frequency spectrum.

Example 8.9 Consider the Bernoulli sequence of Sect. 7.4: the X_n are iid Bernoulli rvs, a stationary sequence with $\mu_X = p$, $C_{XX}[0] = \text{Var}(X_n) = p(1-p)$, and $C_{XX}[k] = 0$ for $k \neq 0$. From these, the autocorrelation function is

$$R_{XX}[k] = C_{XX}[k] + \mu_X^2 = \begin{cases} p & k = 0 \\ p^2 & k \neq 0 \end{cases}$$

In particular, $P_X = R_{XX}[0] = p$. To determine the power spectral density, apply Eq. (8.6):

$$S_{XX}(F) = \sum_{k=-\infty}^{+\infty} R_{XX}[k] e^{-j2\pi Fk} = R_{XX}[0] e^{-j2\pi F(0)} + \sum_{k \neq 0} R_{XX}[k] e^{-j2\pi Fk}$$

$$= p + p^2 \sum_{k \neq 0} e^{-j2\pi Fk} = p + p^2 \sum_{k=-\infty}^{+\infty} e^{-j2\pi Fk} - p^2 e^{-j2\pi F(0)}$$

$$= p(1-p) + p^2 \sum_{k=-\infty}^{+\infty} e^{-j2\pi Fk}$$

Engineers will recognize this last summation as an *impulse train* (sometimes called a *sampling function* or *Dirac comb*), from which we have

$$S_{XX}(F) = p(1-p) + p^2 \sum_{n=-\infty}^{+\infty} \delta(F - n)$$

A graph of this periodic function appears in Fig. 8.12; notice it is indeed a nonnegative, symmetric, periodic function with period 1. Since it's sufficient to define the psd of a WSS random sequence on the interval $(-1/2, 1/2)$, we could drop all but one of the impulses and write $S_{XX}(F) = p(1-p) + p^2 \delta(F)$ for $-1/2 < F < 1/2$.

For a more general iid sequence with $E[X_n] = \mu_X$ and $\text{Var}(X_n) = \sigma_X^2$, a similar derivation shows that $S_{XX}(F) = \sigma_X^2 + \mu_X^2 \sum_{n=-\infty}^{+\infty} \delta(F - n)$, or $\sigma_X^2 + \mu_X^2 \delta(F)$ for $-1/2 < F < 1/2$. In particular, if X_n is a mean-zero iid sequence, the psd of X_n is just the constant σ_X^2.

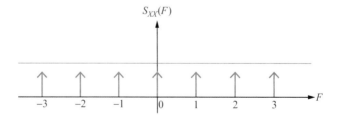

Fig. 8.12 Power spectral density of a Bernoulli sequence (Example 8.9)

Example 8.10 Suppose X_n is a WSS random sequence with power spectral density $S_{XX}(F) = \text{tri}(2F)$ for $-1/2 < F < 1/2$. Let's determine the autocorrelation function of X_n.

The psd may be rewritten as $S_{XX}(F) = (1 - 2|F|)$ for $-1/2 < F < 1/2$, which is shown in Fig. 8.13a. Apply Eq. (8.7):

$$
\begin{aligned}
R_{XX}[k] &= \int_{-1/2}^{1/2} S_{XX}(F)e^{j2\pi Fk}dF = \int_{-1/2}^{1/2} (1 - 2|F|)e^{j2\pi Fk}dF \\
&= \int_{-1/2}^{1/2} (1 - 2|F|)\cos(2\pi Fk)dF \quad \text{(since } 1 - 2|F| \text{ is even)} \\
&= 2\int_{0}^{1/2} (1 - 2F)\cos(2\pi Fk)dF \quad \text{(since the intergrand is even)}
\end{aligned}
$$

For $k = 0$, this is a simple polynomial integral resulting in $R_{XX}[0] = 1/2$, which equals the area under $S_{XX}(F)$, as required. For $k \neq 0$, integration by parts yields

$$
R_{XX}[k] = \frac{1 - \cos(\pi k)}{\pi^2 k^2} = \begin{cases} 2/(\pi^2 k^2) & k \text{ odd} \\ 0 & k \text{ even} \end{cases}
$$

The graph of this autocorrelation function appears in Fig. 8.13b.

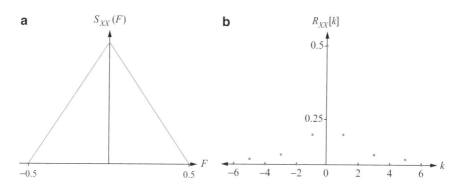

Fig. 8.13 Graphs for Example 8.10: (**a**) Power spectral density; (**b**) autocorrelation function ∎

8.3.1 Random Sequences and LTI Systems

A discrete-time LTI system L has similar properties to those described in the previous section for continuous time. If we let $\delta[n]$ denote the *Kronecker delta function*—i.e., $\delta[0] = 1$ and $\delta[n] = 0$ for $n \neq 0$—then a discrete-time LTI system is characterized by an impulse response[4] function $h[n]$ defined by $h[n] = L[\delta[n]]$. If we let X_n denote the input to the LTI system and Y_n the output, so that $Y_n = L[X_n]$, then Y_n may be computed through discrete-time convolution:

$$Y_n = X_n \star h[n] = \sum_{k=-\infty}^{\infty} X_k h[n-k] = \sum_{k=-\infty}^{\infty} X_{n-k} h[k]$$

Discrete-time LTI systems can be characterized in the frequency domain by a transfer function $H(F)$, defined as the discrete-time Fourier transform of the impulse response:

$$H(F) = \sum_{n=-\infty}^{\infty} h[n] e^{-j2\pi Fn}$$

This transfer function, like the power spectral density, is periodic in F with period 1. The properties of the output sequence Y_n are similar to those for $Y(t)$ in the continuous-time case.

PROPOSITION

Let L be an LTI system with impulse response $h[n]$ and transfer function $H(F)$. Suppose X_n is a wide-sense stationary sequence and let $Y_n = L[X_n]$, the output of the LTI system applied to X_n. Then Y_n is also WSS, with the following properties.

Time domain	Frequency domain
1. $\mu_Y = \mu_X \sum_{n=-\infty}^{\infty} h[n]$	1. $\mu_Y = \mu_X \cdot H(0)$
2. $R_{YY}[k] = R_{XX}[k] \star h[k] \star h[-k]$	2. $S_{YY}(F) = S_{XX}(F) \cdot \lvert H(F) \rvert^2$
3. $P_Y = R_{YY}[0]$	3. $P_Y = \int_{-1/2}^{1/2} S_{YY}(F)\, dF$

Example 8.11 A *moving average* operator can be used to "smooth out" a noisy sequence. The simplest moving average takes the mean of two successive terms: $Y_n = (X_{n-1} + X_n)/2$. This formula is equivalent to passing the sequence X_n through an LTI system with an impulse response given by $h[0] = h[1] = 1/2$ and $h[n] = 0$ otherwise. The transfer function of this LTI system is

---

[4] In this context, the Kronecker delta function is also commonly called the **unit sample response**, since it is strictly speaking not an impulse (its value is well defined at zero). It does, however, share the two key properties of a traditional Dirac delta function (i.e., an impulse): it equals zero for all non-zero inputs, and the sum across its entire domain equals 1.

$$H(F) = \sum_{n=-\infty}^{\infty} h[n]e^{-j2\pi Fn} = \frac{1}{2}e^{-j2\pi F(0)} + \frac{1}{2}e^{-j2\pi F(1)} = \frac{1 + e^{-j2\pi F}}{2},$$

from which the power transfer function is

$$|H(F)|^2 = \left|\frac{1 + e^{-j2\pi F}}{2}\right|^2 = \left|\frac{1 + \cos(2\pi F) - j\sin(2\pi F)}{2}\right|^2$$

$$= \frac{(1 + \cos(2\pi F))^2 + \sin^2(2\pi F)}{2^2} = \frac{1 + \cos(2\pi F)}{2}$$

Notice that the function $(1 + \cos(2\pi F))/2$ is periodic with period 1, as required. Suppose X_n is a WSS random sequence with power spectral density $S_{XX}(F) = N_0$ for $|F| < 1/2$, as depicted in Fig. 8.14a. Then the moving average Y_n has psd equal to

$$S_{YY}(F) = S_{XX}(F) \cdot |H(F)|^2 = N_0 \cdot \frac{1 + \cos(2\pi F)}{2}$$

The graph of this power spectral density appears in Fig. 8.14b. The ensemble average power in Y_n can be determined by integrating this function from $-1/2$ to $1/2$:

$$P_Y = \int_{-1/2}^{1/2} S_{YY}(F)dF = \frac{N_0}{2}\int_{-1/2}^{1/2} [1 + \cos(2\pi F)]dF = \frac{N_0}{2}(1) = \frac{N_0}{2}.$$

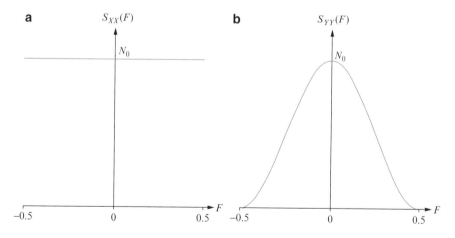

Fig. 8.14 Power spectral density of the moving average in Example 8.11   ∎

8.3.2   Random Sequences and Sampling

Modern electronic systems often work with *digitized* signals: analog signals that have been "sampled" at regular intervals to create a digital (i.e., discrete-time) signal. Suppose we have a continuous-time (analog) signal $X(t)$, which we sample every T_s seconds; T_s is called the **sampling interval**. That is, we only observe $x(t)$ at times $0, \pm T_s, \pm 2T_s$, and so on. Then we can regard our observed (digital) signal as a random sequence $X[n]$ defined by

$$X[n] = X(nT_s) \quad \text{for } n = \ldots, -2, -1, 0, 1, 2, \ldots$$

This is illustrated for a sample function in Fig. 8.15.

The following proposition ensures that the sampled version of a WSS random process is also WSS—and, hence, that the spectral density theory presented in this chapter applies.

PROPOSITION

Let $X(t)$ be a WSS random process, and for some fixed $T_s > 0$ define $X[n] = X(nT_s)$. Then the random sequence $X[n]$ is a WSS random sequence.

The proof was requested in Exercise 45 of Chap. 7.

If the sampling interval is selected judiciously, then we may (in some sense) recover the original signal from the digitized version. This relies on a key result from communication theory called the **Nyquist sampling theorem** for deterministic signals: If a signal $x(t)$ has no frequencies above B Hz, then $x(t)$ is completely determined by its sample values $x[n] = x(nT_s)$ so long as

$$f_s = \frac{1}{T_s} \geq 2B$$

The quantity f_s is called the **sampling rate**. The Nyquist sampling theorem says that a *band-limited* signal (with band limit B) can be completely recovered from its digital version, provided the sampling rate is at least $2B$. For example, a signal with band limit $B = 1$ kHz $= 1000$ Hz must be sampled at least 2,000 times per second; equivalently, the sampling interval T_s can be *at most* $1/(2B) = .0005$ s. The minimum sampling rate, $2B$, is sometimes called the **Nyquist rate** of that signal.

When $T_s \leq 1/(2B)$, as required by the Nyquist sampling theorem, the original deterministic signal $x(t)$ may be reconstructed by the interpolation formula

$$x(t) = \sum_{n=-\infty}^{\infty} x[n]\,\text{sinc}\left(\frac{t - nT_s}{T_s}\right) \tag{8.8}$$

The heart of the Nyquist sampling theorem is the statement that the two sides of Eq. (8.8) are equal.

Fig. 8.15 A smooth signal $x(t)$ and its sampled version $x[n]$ (indicated by asterisks)

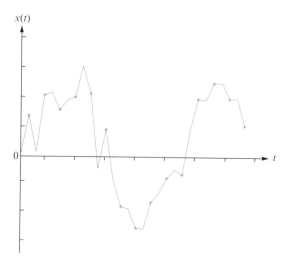

For a band-limited *random* process $X(t)$ with corresponding digital sequence $X[n] = X(nT_s)$, we may define a Nyquist interpolation of $X(t)$ by

$$X_{\text{Nyq}}(t) = \sum_{n=-\infty}^{\infty} X[n]\,\text{sinc}\left(\frac{t - nT_s}{T_s}\right)$$

It can be shown that $X_{\text{Nyq}}(t)$ equals the original $X(t)$ in the "mean square sense," i.e., that

$$E\left[\left(X_{\text{Nyq}}(t) - X(t)\right)^2\right] = 0$$

(This is slightly weaker than saying $X_{\text{Nyq}}(t) = X(t)$; in particular, there may exist a negligible set of sample functions for which the two differ.)

There is a direct connection between the Nyquist sampling rate and the argument F of the discrete-time Fourier transform. Suppose a random process $X(t)$ has band limit B, i.e., the set of frequencies f represented in the spectrum of $X(t)$ satisfies $-B \leq f \leq B$. Provided we use a sampling rate, f_s, at least as great as the Nyquist rate $2B$, we have:

$$-B \leq f \leq B, f_s \geq 2B \quad \Rightarrow \quad -\frac{1}{2} \leq \frac{f}{f_s} \leq \frac{1}{2}$$

If we define $F = f/f_s$, we have a unitless variable whose set of possible values exactly corresponds to that of F in the discrete-time Fourier transform. Said differently, F in the discrete-time Fourier transform represents a *normalized* frequency; we can recover the spectrum of $X(t)$ across its original frequency band by writing $f = F \cdot f_s$. (In some textbooks, you will see the argument of the discrete-time Fourier transform denoted Ω, to indicate radian measure. The variables F and Ω are, of course, related by $\Omega = 2\pi F$.)

8.3.3   Exercises: Section 8.3 (39–50)

39. Let $X[n]$ be a WSS random sequence. Show that the power spectral density of $X[n]$ may be rewritten as

$$S_{XX}(F) = R_{XX}[0] + 2\sum_{k=1}^{\infty} R_{XX}[k]\cos(2\pi kF)$$

40. Let $X(t)$ be a WSS random process, and let $X[n] = X(nT_s)$, the sampled version of $X(t)$. Find the power spectral density of $X[n]$ in terms of the psd of $X(t)$.

41. Suppose $X[n]$ is a WSS random sequence with autocorrelation function $R_{XX}[k] = \alpha^{|k|}$ for some constant $|\alpha| < 1$. Find the power spectral density of $X[n]$. Sketch this psd for $\alpha = -.5$, 0, and $.5$.

42. Consider the correlated bit noise sequence described in Exercise 50 of Chap. 7: X_0 is 0 or 1 with probability $.5$ each and, for $n \geq 1$, $X_n = X_{n-1}$ with probability $.9$ and $1 - X_{n-1}$ with probability $.1$. It was shown in that exercise that X_n is a WSS random sequence with mean $\mu_X = .5$ and autocorrelation function

$$R_{XX}[k] = \frac{1 + .8^{|k|}}{4}$$

(This particular random sequence can be "time reversed" so that X_n is defined for negative indices as well.) Find the power spectral density of this correlated bit noise sequence.

43. A Poisson telegraphic process $N(t)$ with parameter $\lambda = 1$ (see Sect. 7.5) is sampled every 5 s, resulting in the random sequence $X[n] = N(5n)$. Find the power spectral density of $X[n]$.

44. **Discrete-time white noise** is a WSS, mean-zero process such that X_n and X_m are uncorrelated for all $n \neq m$.

    (a) Show that the autocorrelation function of discrete-time white noise is $R_{XX}[k] = \sigma^2\delta[k]$ for some constant $\sigma > 0$, where $\delta[k]$ is the Kronecker delta function.

    (b) Find the power spectral density of discrete-time white noise. Is it what you'd expect?

45. Suppose X_n is a WSS random sequence with the following autocorrelation function:

$$R_{XX}[k] = \begin{cases} 1 & k = 0 \\ \dfrac{1}{2k^2} & k \text{ odd} \\ 0 & \text{otherwise} \end{cases}$$

    Determine the power spectral density of X_n. [*Hint:* Use Example 8.10.]

46. Let X_n be the WSS input to a discrete-time LTI system with impulse response $h[n]$, and let Y_n be the output. Define the *cross-correlation* of X_n and Y_n by $R_{XY}[n, n+k] = E[X_n Y_{n+k}]$.

(a) Show that R_{XY} does not depend on n, and that $R_{XY} = R_{XX} \star h$, where $\star$ denotes discrete-time convolution. (This is the discrete-time version of a result from the previous section.)

(b) The cross-power spectral density $S_{XY}(F)$ of two jointly WSS random sequences X_n and Y_n is defined as the discrete-time Fourier transform of $R_{XY}[k]$. In the present context, show that $S_{XY}(F) = S_{XX}(F)H(F)$, where H denotes the transfer function of the LTI system.

47. The WSS random sequence X_n has power spectral density $S_{XX}(F) = 2P$ for $|F| \leq 1/4$ and 0 for $1/4 < |F| < 1/2$.

(a) Verify that the ensemble average power in X_n is P.

(b) Find the autocorrelation function of X_n.

48. Let X_n have power spectral density $S_{XX}(F)$, and suppose X_n is passed through a discrete-time LTI system with impulse response $h[n] = \alpha^n$ for $n = 0, 1, 2, \ldots$ for some constant $|\alpha| < 1$ (and $h[n] = 0$ otherwise). Let Y_n denote the output sequence.

(a) Find the mean of Y_n in terms of the mean of X_n.

(b) Find the power spectral density of Y_n in terms of the psd of X_n.

49. The system in Example 8.11 can be extended to an M-term simple moving average filter, with impulse response

$$h[n] = \begin{cases} 1/M & n = 0, 1, \ldots, M-1 \\ 0 & \text{otherwise} \end{cases}$$

Let X_n be the WSS input to such a filter, and let Y_n be the output.

(a) Write an expression for Y_n in terms of the X_n.

(b) Determine the transfer function of this filter.

(c) Assuming X_n is a discrete-time white noise process (see Exercise 44), determine the autocorrelation function of Y_n.

50. A more general moving average process has the form

$$Y[n] = \theta_0 X[n] + \theta_1 X[n-1] + \cdots + \theta_M X[n-M]$$

for some integer M and constants $\theta_0, \ldots, \theta_M$. Let the input sequence $X[n]$ be iid, with mean 0 and variance σ^2.

(a) Find the impulse response $h[n]$ of the LTI system that produces $Y[n]$ from $X[n]$.

(b) Find the transfer function of this system.

(c) Find the mean of $Y[n]$.

(d) Find the variance of $Y[n]$.

(e) Find the autocovariance function of $Y[n]$.

Appendix A: Statistical Tables

M.A. Carlton and J.L. Devore, *Probability with Applications in Engineering, Science, and*   725
Technology, Springer Texts in Statistics, DOI 10.1007/978-1-4939-0395-5,
© Springer Science+Business Media New York 2014

A.1 Binomial cdf

Table A.1 Cumulative binomial probabilities $B(x; n, p) = \sum_{y=0}^{x} b(y; n, p)$

(a) $n = 5$

							p						
	0.05	0.10	0.20	0.25	0.30	0.40	0.50	0.60	0.70	0.75	0.80	0.90	0.95
0	.774	.590	.328	.237	.168	.078	.031	.010	.002	.001	.000	.000	.000
1	.977	.919	.737	.633	.528	.337	.188	.087	.031	.016	.007	.000	.000
x **2**	.999	.991	.942	.896	.837	.683	.500	.317	.163	.104	.058	.009	.001
3	1.000	1.000	.993	.984	.969	.913	.812	.663	.472	.367	.263	.081	.023
4	1.000	1.000	1.000	.999	.998	.990	.969	.922	.832	.763	.672	.410	.226

(b) $n = 10$

							p						
	0.05	0.10	0.20	0.25	0.30	0.40	0.50	0.60	0.70	0.75	0.80	0.90	0.95
0	.599	.349	.107	.056	.028	.006	.001	.000	.000	.000	.000	.000	.000
1	.914	.736	.376	.244	.149	.046	.011	.002	.000	.000	.000	.000	.000
2	.988	.930	.678	.526	.383	.167	.055	.012	.002	.000	.000	.000	.000
3	.999	.987	.879	.776	.650	.382	.172	.055	.011	.004	.001	.000	.000
4	1.000	.998	.967	.922	.850	.633	.377	.166	.047	.020	.006	.000	.000
x **5**	1.000	1.000	.994	.980	.953	.834	.623	.367	.150	.078	.033	.002	.000
6	1.000	1.000	.999	.996	.989	.945	.828	.618	.350	.224	.121	.013	.001
7	1.000	1.000	1.000	1.000	.998	.988	.945	.833	.617	.474	.322	.070	.012
8	1.000	1.000	1.000	1.000	1.000	.998	.989	.954	.851	.756	.624	.264	.086
9	1.000	1.000	1.000	1.000	1.000	1.000	.999	.994	.972	.944	.893	.651	.401

(c) $n = 15$

							p						
	0.05	0.10	0.20	0.25	0.30	0.40	0.50	0.60	0.70	0.75	0.80	0.90	0.95
0	.463	.206	.035	.013	.005	.000	.000	.000	.000	.000	.000	.000	.000
1	.829	.549	.167	.080	.035	.005	.000	.000	.000	.000	.000	.000	.000
2	.964	.816	.398	.236	.127	.027	.004	.000	.000	.000	.000	.000	.000
3	.995	.944	.648	.461	.297	.091	.018	.002	.000	.000	.000	.000	.000
4	.999	.987	.836	.686	.515	.217	.059	.009	.001	.000	.000	.000	.000
5	1.000	.998	.939	.852	.722	.402	.151	.034	.004	.001	.000	.000	.000
6	1.000	1.000	.982	.943	.869	.610	.304	.095	.015	.004	.001	.000	.000
x **7**	1.000	1.000	.996	.983	.950	.787	.500	.213	.050	.017	.004	.000	.000
8	1.000	1.000	.999	.996	.985	.905	.696	.390	.131	.057	.018	.000	.000
9	1.000	1.000	1.000	.999	.996	.966	.849	.597	.278	.148	.061	.002	.000
10	1.000	1.000	1.000	1.000	.999	.991	.941	.783	.485	.314	.164	.013	.001
11	1.000	1.000	1.000	1.000	1.000	.998	.982	.909	.703	.539	.352	.056	.005
12	1.000	1.000	1.000	1.000	1.000	1.000	.996	.973	.873	.764	.602	.184	.036
13	1.000	1.000	1.000	1.000	1.000	1.000	1.000	.995	.965	.920	.833	.451	.171
14	1.000	1.000	1.000	1.000	1.000	1.000	1.000	1.000	.995	.987	.965	.794	.537

(continued)

Table A.1 (continued)

(d) $n = 20$

x \ p	0.05	0.10	0.20	0.25	0.30	0.40	0.50	0.60	0.70	0.75	0.80	0.90	0.95
0	.358	.122	.012	.003	.001	.000	.000	.000	.000	.000	.000	.000	.000
1	.736	.392	.069	.024	.008	.001	.000	.000	.000	.000	.000	.000	.000
2	.925	.677	.206	.091	.035	.004	.000	.000	.000	.000	.000	.000	.000
3	.984	.867	.411	.225	.107	.016	.001	.000	.000	.000	.000	.000	.000
4	.997	.957	.630	.415	.238	.051	.006	.000	.000	.000	.000	.000	.000
5	1.000	.989	.804	.617	.416	.126	.021	.002	.000	.000	.000	.000	.000
6	1.000	.998	.913	.786	.608	.250	.058	.006	.000	.000	.000	.000	.000
7	1.000	1.000	.968	.898	.772	.416	.132	.021	.001	.000	.000	.000	.000
8	1.000	1.000	.990	.959	.887	.596	.252	.057	.005	.001	.000	.000	.000
9	1.000	1.000	.997	.986	.952	.755	.412	.128	.017	.004	.001	.000	.000
10	1.000	1.000	.999	.996	.983	.872	.588	.245	.048	.014	.003	.000	.000
11	1.000	1.000	1.000	.999	.995	.943	.748	.404	.113	.041	.010	.000	.000
12	1.000	1.000	1.000	1.000	.999	.979	.868	.584	.228	.102	.032	.000	.000
13	1.000	1.000	1.000	1.000	1.000	.994	.942	.750	.392	.214	.087	.002	.000
14	1.000	1.000	1.000	1.000	1.000	.998	.979	.874	.584	.383	.196	.011	.000
15	1.000	1.000	1.000	1.000	1.000	1.000	.994	.949	.762	.585	.370	.043	.003
16	1.000	1.000	1.000	1.000	1.000	1.000	.999	.984	.893	.775	.589	.133	.016
17	1.000	1.000	1.000	1.000	1.000	1.000	1.000	.996	.965	.909	.794	.323	.075
18	1.000	1.000	1.000	1.000	1.000	1.000	1.000	.999	.992	.976	.931	.608	.264
19	1.000	1.000	1.000	1.000	1.000	1.000	1.000	1.000	.999	.997	.988	.878	.642

(e) $n = 25$

x \ p	0.05	0.10	0.20	0.25	0.30	0.40	0.50	0.60	0.70	0.75	0.80	0.90	0.95
0	.277	.072	.004	.001	.000	.000	.000	.000	.000	.000	.000	.000	.000
1	.642	.271	.027	.007	.002	.000	.000	.000	.000	.000	.000	.000	.000
2	.873	.537	.098	.032	.009	.000	.000	.000	.000	.000	.000	.000	.000
3	.966	.764	.234	.096	.033	.002	.000	.000	.000	.000	.000	.000	.000
4	.993	.902	.421	.214	.090	.009	.000	.000	.000	.000	.000	.000	.000
5	.999	.967	.617	.378	.193	.029	.002	.000	.000	.000	.000	.000	.000
6	1.000	.991	.780	.561	.341	.074	.007	.000	.000	.000	.000	.000	.000
7	1.000	.998	.891	.727	.512	.154	.022	.001	.000	.000	.000	.000	.000
8	1.000	1.000	.953	.851	.677	.274	.054	.004	.000	.000	.000	.000	.000
9	1.000	1.000	.983	.929	.811	.425	.115	.013	.000	.000	.000	.000	.000
10	1.000	1.000	.994	.970	.902	.586	.212	.034	.002	.000	.000	.000	.000
11	1.000	1.000	.998	.980	.956	.732	.345	.078	.006	.001	.000	.000	.000
12	1.000	1.000	1.000	.997	.983	.846	.500	.154	.017	.003	.000	.000	.000
13	1.000	1.000	1.000	.999	.994	.922	.655	.268	.044	.020	.002	.000	.000
14	1.000	1.000	1.000	1.000	.998	.966	.788	.414	.098	.030	.006	.000	.000
15	1.000	1.000	1.000	1.000	1.000	.987	.885	.575	.189	.071	.017	.000	.000
16	1.000	1.000	1.000	1.000	1.000	.996	.946	.726	.323	.149	.047	.000	.000
17	1.000	1.000	1.000	1.000	1.000	.999	.978	.846	.488	.273	.109	.002	.000
18	1.000	1.000	1.000	1.000	1.000	1.000	.993	.926	.659	.439	.220	.009	.000
19	1.000	1.000	1.000	1.000	1.000	1.000	.998	.971	.807	.622	.383	.033	.001
20	1.000	1.000	1.000	1.000	1.000	1.000	1.000	.991	.910	.786	.579	.098	.007
21	1.000	1.000	1.000	1.000	1.000	1.000	1.000	.998	.967	.904	.766	.236	.034
22	1.000	1.000	1.000	1.000	1.000	1.000	1.000	1.000	.991	.968	.902	.463	.127
23	1.000	1.000	1.000	1.000	1.000	1.000	1.000	1.000	.998	.993	.973	.729	.358
24	1.000	1.000	1.000	1.000	1.000	1.000	1.000	1.000	1.000	.999	.996	.928	.723

A.2 Poisson cdf

Table A.2   Cumulative Poisson probabilities $P(x; \mu) = \sum_{y=0}^{x} \frac{e^{-\mu}\mu^y}{y!}$

		.1	.2	.3	.4	.5	μ .6	.7	.8	.9	1.0
	0	.905	.819	.741	.670	.607	.549	.497	.449	.407	.368
	1	.995	.982	.963	.938	.910	.878	.844	.809	.772	.736
	2	1.000	.999	.996	.992	.986	.977	.966	.953	.937	.920
x	3		1.000	1.000	.999	.998	.997	.994	.991	.987	.981
	4				1.000	1.000	1.000	.999	.999	.998	.996
	5							1.000	1.000	1.000	.999
	6										1.000

		2.0	3.0	4.0	5.0	6.0	μ 7.0	8.0	9.0	10.0	15.0	20.0
	0	.135	.050	.018	.007	.002	.001	.000	.000	.000	.000	.000
	1	.406	.199	.092	.040	.017	.007	.003	.001	.000	.000	.000
	2	.677	.423	.238	.125	.062	.030	.014	.006	.003	.000	.000
	3	.857	.647	.433	.265	.151	.082	.042	.021	.010	.000	.000
	4	.947	.815	.629	.440	.285	.173	.100	.055	.029	.001	.000
	5	.983	.916	.785	.616	.446	.301	.191	.116	.067	.003	.000
	6	.995	.966	.889	.762	.606	.450	.313	.207	.130	.008	.000
	7	.999	.988	.949	.867	.744	.599	.453	.324	.220	.018	.001
	8	1.000	.996	.979	.932	.847	.729	.593	.456	.333	.037	.002
	9		.999	.992	.968	.916	.830	.717	.587	.458	.070	.005
	10		1.000	.997	.986	.957	.901	.816	.706	.583	.118	.011
	11			.999	.995	.980	.947	.888	.803	.697	.185	.021
	12			1.000	.998	.991	.973	.936	.876	.792	.268	.039
	13				.999	.996	.987	.966	.926	.864	.363	.066
	14				1.000	.999	.994	.983	.959	.917	.466	.105
	15					.999	.998	.992	.978	.951	.568	.157
x	16					1.000	.999	.996	.989	.973	.664	.221
	17						1.000	.998	.995	.986	.749	.297
	18							.999	.998	.993	.819	.381
	19							1.000	.999	.997	.875	.470
	20								1.000	.998	.917	.559
	21									.999	.947	.644
	22									1.000	.967	.721
	23										.981	.787
	24										.989	.843
	25										.994	.888
	26										.997	.922
	27										.998	.948
	28										.999	.966
	29										1.000	.978
	30											.987

A.3 Standard Normal cdf

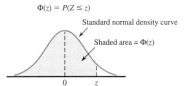

$\Phi(z) = P(Z \leq z)$

Standard normal density curve

Shaded area = $\Phi(z)$

Table A.3 Standard normal curve areas

z	.00	.01	.02	.03	.04	.05	.06	.07	.08	.09
−3.4	.0003	.0003	.0003	.0003	.0003	.0003	.0003	.0003	.0003	.0002
−3.3	.0005	.0005	.0005	.0004	.0004	.0004	.0004	.0004	.0004	.0003
−3.2	.0007	.0007	.0006	.0006	.0006	.0006	.0006	.0005	.0005	.0005
−3.1	.0010	.0009	.0009	.0009	.0008	.0008	.0008	.0008	.0007	.0007
−3.0	.0013	.0013	.0013	.0012	.0012	.0011	.0011	.0011	.0010	.0010
−2.9	.0019	.0018	.0017	.0017	.0016	.0016	.0015	.0015	.0014	.0014
−2.8	.0026	.0025	.0024	.0023	.0023	.0022	.0021	.0021	.0020	.0019
−2.7	.0035	.0034	.0033	.0032	.0031	.0030	.0029	.0028	.0027	.0026
−2.6	.0047	.0045	.0044	.0043	.0041	.0040	.0039	.0038	.0037	.0036
−2.5	.0062	.0060	.0059	.0057	.0055	.0054	.0052	.0051	.0049	.0048
−2.4	.0082	.0080	.0078	.0075	.0073	.0071	.0069	.0068	.0066	.0064
−2.3	.0107	.0104	.0102	.0099	.0096	.0094	.0091	.0089	.0087	.0084
−2.2	.0139	.0136	.0132	.0129	.0125	.0122	.0119	.0116	.0113	.0110
−2.1	.0179	.0174	.0170	.0166	.0162	.0158	.0154	.0150	.0146	.0143
−2.0	.0228	.0222	.0217	.0212	.0207	.0202	.0197	.0192	.0188	.0183
−1.9	.0287	.0281	.0274	.0268	.0262	.0256	.0250	.0244	.0239	.0233
−1.8	.0359	.0352	.0344	.0336	.0329	.0322	.0314	.0307	.0301	.0294
−1.7	.0446	.0436	.0427	.0418	.0409	.0401	.0392	.0384	.0375	.0367
−1.6	.0548	.0537	.0526	.0516	.0505	.0495	.0485	.0475	.0465	.0455
−1.5	.0668	.0655	.0643	.0630	.0618	.0606	.0594	.0582	.0571	.0559
−1.4	.0808	.0793	.0778	.0764	.0749	.0735	.0722	.0708	.0694	.0681
−1.3	.0968	.0951	.0934	.0918	.0901	.0885	.0869	.0853	.0838	.0823
−1.2	.1151	.1131	.1112	.1093	.1075	.1056	.1038	.1020	.1003	.0985
−1.1	.1357	.1335	.1314	.1292	.1271	.1251	.1230	.1210	.1190	.1170
−1.0	.1587	.1562	.1539	.1515	.1492	.1469	.1446	.1423	.1401	.1379
−0.9	.1841	.1814	.1788	.1762	.1736	.1711	.1685	.1660	.1635	.1611
−0.8	.2119	.2090	.2061	.2033	.2005	.1977	.1949	.1922	.1894	.1867
−0.7	.2420	.2389	.2358	.2327	.2296	.2266	.2236	.2206	.2177	.2148
−0.6	.2743	.2709	.2676	.2643	.2611	.2578	.2546	.2514	.2483	.2451
−0.5	.3085	.3050	.3015	.2981	.2946	.2912	.2877	.2843	.2810	.2776
−0.4	.3446	.3409	.3372	.3336	.3300	.3264	.3228	.3192	.3156	.3121
−0.3	.3821	.3783	.3745	.3707	.3669	.3632	.3594	.3557	.3520	.3482
−0.2	.4207	.4168	.4129	.4090	.4052	.4013	.3974	.3936	.3897	.3859
−0.1	.4602	.4562	.4522	.4483	.4443	.4404	.4364	.4325	.4286	.4247
−0.0	.5000	.4960	.4920	.4880	.4840	.4801	.4761	.4721	.4681	.4641

(continued)

Table A.3 (continued)

z	.00	.01	.02	.03	.04	.05	.06	.07	.08	.09
0.0	.5000	.5040	.5080	.5120	.5160	.5199	.5239	.5279	.5319	.5359
0.1	.5398	.5438	.5478	.5517	.5557	.5596	.5636	.5675	.5714	.5753
0.2	.5793	.5832	.5871	.5910	.5948	.5987	.6026	.6064	.6103	.6141
0.3	.6179	.6217	.6255	.6293	.6331	.6368	.6406	.6443	.6480	.6517
0.4	.6554	.6591	.6628	.6664	.6700	.6736	.6772	.6808	.6844	.6879
0.5	.6915	.6950	.6985	.7019	.7054	.7088	.7123	.7157	.7190	.7224
0.6	.7257	.7291	.7324	.7357	.7389	.7422	.7454	.7486	.7517	.7549
0.7	.7580	.7611	.7642	.7673	.7704	.7734	.7764	.7794	.7823	.7852
0.8	.7881	.7910	.7939	.7967	.7995	.8023	.8051	.8078	.8106	.8133
0.9	.8159	.8186	.8212	.8238	.8264	.8289	.8315	.8340	.8365	.8389
1.0	.8413	.8438	.8461	.8485	.8508	.8531	.8554	.8577	.8599	.8621
1.1	.8643	.8665	.8686	.8708	.8729	.8749	.8770	.8790	.8810	.8830
1.2	.8849	.8869	.8888	.8907	.8925	.8944	.8962	.8980	.8997	.9015
1.3	.9032	.9049	.9066	.9082	.9099	.9115	.9131	.9147	.9162	.9177
1.4	.9192	.9207	.9222	.9236	.9251	.9265	.9278	.9292	.9306	.9319
1.5	.9332	.9345	.9357	.9370	.9382	.9394	.9406	.9418	.9429	.9441
1.6	.9452	.9463	.9474	.9484	.9495	.9505	.9515	.9525	.9535	.9545
1.7	.9554	.9564	.9573	.9582	.9591	.9599	.9608	.9616	.9625	.9633
1.8	.9641	.9649	.9656	.9664	.9671	.9678	.9686	.9693	.9699	.9706
1.9	.9713	.9719	.9726	.9732	.9738	.9744	.9750	.9756	.9761	.9767
2.0	.9772	.9778	.9783	.9788	.9793	.9798	.9803	.9808	.9812	.9817
2.1	.9821	.9826	.9830	.9834	.9838	.9842	.9846	.9850	.9854	.9857
2.2	.9861	.9864	.9868	.9871	.9875	.9878	.9881	.9884	.9887	.9890
2.3	.9893	.9896	.9898	.9901	.9904	.9906	.9909	.9911	.9913	.9916
2.4	.9918	.9920	.9922	.9925	.9927	.9929	.9931	.9932	.9934	.9936
2.5	.9938	.9940	.9941	.9943	.9945	.9946	.9948	.9949	.9951	.9952
2.6	.9953	.9955	.9956	.9957	.9959	.9960	.9961	.9962	.9963	.9964
2.7	.9965	.9966	.9967	.9968	.9969	.9970	.9971	.9972	.9973	.9974
2.8	.9974	.9975	.9976	.9977	.9977	.9978	.9979	.9979	.9980	.9981
2.9	.9981	.9982	.9982	.9983	.9984	.9984	.9985	.9985	.9986	.9986
3.0	.9987	.9987	.9987	.9988	.9988	.9989	.9989	.9989	.9990	.9990
3.1	.9990	.9991	.9991	.9991	.9992	.9992	.9992	.9992	.9993	.9993
3.2	.9993	.9993	.9994	.9994	.9994	.9994	.9994	.9995	.9995	.9995
3.3	.9995	.9995	.9995	.9996	.9996	.9996	.9996	.9996	.9996	.9997
3.4	.9997	.9997	.9997	.9997	.9997	.9997	.9997	.9997	.9997	.9998

A.4 Incomplete Gamma Function

Table A.4 The incomplete gamma function $G(x; \alpha) = \int_0^x \frac{1}{\Gamma(\alpha)} y^{\alpha-1} e^{-y} dy$

	α									
	1	2	3	4	5	6	7	8	9	10
1	.632	.264	.080	.019	.004	.001	.000	.000	.000	.000
2	.865	.594	.323	.143	.053	.017	.005	.001	.000	.000
3	.950	.801	.577	.353	.185	.084	.034	.012	.004	.001
4	.982	.908	.762	.567	.371	.215	.111	.051	.021	.008
5	.993	.960	.875	.735	.560	.384	.238	.133	.068	.032
6	.998	.983	.938	.849	.715	.554	.394	.256	.153	.084
7	.999	.993	.970	.918	.827	.699	.550	.401	.271	.170
8	1.000	.997	.986	.958	.900	.809	.687	.547	.407	.283
9		.999	.994	.979	.945	.884	.793	.676	.544	.413
10		1.000	.997	.990	.971	.933	.870	.780	.667	.542
11			.999	.995	.985	.962	.921	.857	.768	.659
12			1.000	.998	.992	.980	.954	.911	.845	.758
13				.999	.996	.989	.974	.946	.900	.834
14				1.000	.998	.994	.986	.968	.938	.891
15					.999	.997	.992	.982	.963	.930

x

A.5 Critical Values for *t* Distributions

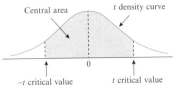

Table A.5 Critical values for *t* distributions

	Central area						
ν	80%	90%	95%	98%	99%	99.8%	99.9%
1	3.078	6.314	12.706	31.821	63.657	318.31	636.62
2	1.886	2.920	4.303	6.965	9.925	22.326	31.598
3	1.638	2.353	3.182	4.541	5.841	10.213	12.924
4	1.533	2.132	2.776	3.747	4.604	7.173	8.610
5	1.476	2.015	2.571	3.365	4.032	5.893	6.869
6	1.440	1.943	2.447	3.143	3.707	5.208	5.959
7	1.415	1.895	2.365	2.998	3.499	4.785	5.408
8	1.397	1.860	2.306	2.896	3.355	4.501	5.041
9	1.383	1.833	2.262	2.821	3.250	4.297	4.781
10	1.372	1.812	2.228	2.764	3.169	4.144	4.587
11	1.363	1.796	2.201	2.718	3.106	4.025	4.437
12	1.356	1.782	2.179	2.681	3.055	3.930	4.318
13	1.350	1.771	2.160	2.650	3.012	3.852	4.221
14	1.345	1.761	2.145	2.624	2.977	3.787	4.140
15	1.341	1.753	2.131	2.602	2.947	3.733	4.073
16	1.337	1.746	2.120	2.583	2.921	3.686	4.015
17	1.333	1.740	2.110	2.567	2.898	3.646	3.965
18	1.330	1.734	2.101	2.552	2.878	3.610	3.922
19	1.328	1.729	2.093	2.539	2.861	3.579	3.883
20	1.325	1.725	2.086	2.528	2.845	3.552	3.850
21	1.323	1.721	2.080	2.518	2.831	3.527	3.819
22	1.321	1.717	2.074	2.508	2.819	3.505	3.792
23	1.319	1.714	2.069	2.500	2.807	3.485	3.767
24	1.318	1.711	2.064	2.492	2.797	3.467	3.745
25	1.316	1.708	2.060	2.485	2.787	3.450	3.725
26	1.315	1.706	2.056	2.479	2.779	3.435	3.707
27	1.314	1.703	2.052	2.473	2.771	3.421	3.690
28	1.313	1.701	2.048	2.467	2.763	3.408	3.674
29	1.311	1.699	2.045	2.462	2.756	3.396	3.659
30	1.310	1.697	2.042	2.457	2.750	3.385	3.646
32	1.309	1.694	2.037	2.449	2.738	3.365	3.622
34	1.307	1.691	2.032	2.441	2.728	3.348	3.601
36	1.306	1.688	2.028	2.434	2.719	3.333	3.582
38	1.304	1.686	2.024	2.429	2.712	3.319	3.566
40	1.303	1.684	2.021	2.423	2.704	3.307	3.551
50	1.299	1.676	2.009	2.403	2.678	3.262	3.496
60	1.296	1.671	2.000	2.390	2.660	3.232	3.460
120	1.289	1.658	1.980	2.358	2.617	3.160	3.373
∞	1.282	1.645	1.960	2.326	2.576	3.090	3.291

A.6 Tail Areas of t Distributions

Table A.6 t curve tail areas

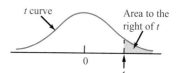

					Degrees of Freedom (ν)							
t	1	2	3	4	5	6	7	8	9	10	11	12
0.0	.500	.500	.500	.500	.500	.500	.500	.500	.500	.500	.500	.500
0.1	.468	.465	.463	.463	.462.	.462	.462	.461	.461	.461	.461	.461
0.2	.437	.430	.427	.426	.425	.424	.424	.423	.423	.423	.423	.422
0.3	.407	.396	.392	.390	.388	.387	.386	.386	.386	.385	.385	.385
0.4	.379	.364	.358	.355	.353	.352	.351	.350	.349	.349	.348	.348
0.5	.352	.333	.326	.322	.319	.317	.316	.315	.315	.314	.313	.313
0.6	.328	.305	.295	.290	.287	.285	.284	.283	.282	.281	.280	.280
0.7	.306	.278	.267	.261	.258	.255	.253	.252	.251	.250	.249	.249
0.8	.285	.254	.241	.234	.230	.227	.225	.223	.222	.221	.220	.220
0.9	.267	.232	.217	.210	.205	.201	.199	.197	.196	.195	.194	.193
1.0	.250	.211	.196	.187	.182	.178	.175	.173	.172	.170	.169	.169
1.1	.235	.193	.176	.167	.162	.157	.154	.152	.150	.149	.147	.146
1.2	.221	.177	.158	.148	.142	.138	.135	.132	.130	.129	.128	.127
1.3	.209	.162	.142	.132	.125	.121	.117	.115	.113	.111	.110	.109
1.4	.197	.148	.128	.117	.110	.106	.102	.100	.098	.096	.095	.093
1.5	.187	.136	.115	.104	.097	.092	.089	.086	.084	.082	.081	.080
1.6	.178	.125	.104	.092	.085	.080	.077	.074	.072	.070	.069	.068
1.7	.169	.116	.094	.082	.075	.070	.065	.064	.062	.060	.059	.057
1.8	.161	.107	.085	.073	.066	.061	.057	.055	.053	.051	.050	.049
1.9	.154	.099	.077	.065	.058	.053	.050	.047	.045	.043	.042	.041
2.0	.148	.092	.070	.058	.051	.046	.043	.040	.038	.037	.035	.034
2.1	.141	.085	.063	.052	.045	.040	.037	.034	.033	.031	.030	.029
2.2	.136	.079	.058	.046	.040	.035	.032	.029	.028	.026	.025	.024
2.3	.131	.074	.052	.041	.035	.031	.027	.025	.023	.022	.021	.020
2.4	.126	.069	.048	.037	.031	.027	.024	.022	.020	.019	.018	.017
2.5	.121	.065	.044	.033	.027	.023	.020	.018	.017	.016	.015	.014
2.6	.117	.061	.040	.030	.024	.020	.018	.016	.014	.013	.012	.012
2.7	.113	.057	.037	.027	.021	.018	.015	.014	.012	.011	.010	.010
2.8	.109	.054	.034	.024	.019	.016	.013	.012	.010	.009	.009	.008
2.9	.106	.051	.031	.022	.017	.014	.011	.010	.009	.008	.007	.007
3.0	.102	.048	.029	.020	.015	.012	.010	.009	.007	.007	.006	.006
3.1	.099	.045	.027	.018	.013	.011	.009	.007	.006	.006	.005	.005
3.2	.096	.043	.025	.016	.012	.009	.008	.006	.005	.005	.004	.004
3.3	.094	.040	.023	.015	.011	.008	.007	.005	.005	.004	.004	.003
3.4	.091	.038	.021	.014	.010	.007	.006	.005	.004	.003	.003	.003
3.5	.089	.036	.020	.012	.009	.006	.005	.004	.003	.003	.002	.002
3.6	.086	.035	.018	.011	.008	.006	.004	.004	.003	.002	.002	.002
3.7	.084	.033	.017	.010	.007	.005	.004	.003	.002	.002	.002	.002
3.8	.082	.031	.016	.010	.006	.004	.003	.003	.002	.002	.001	.001
3.9	.080	.030	.015	.009	.006	.004	.003	.002	.002	.001	.001	.001
4.0	.078	.029	.014	.008	.005	.004	.003	.002	.002	.001	.001	.001

(continued)

Table A.6 (continued)

					Degrees of Freedom (ν)							
t	13	14	15	16	17	18	19	20	21	22	23	24
0.0	.500	.500	.500	.500	.500	.500	.500	.500	.500	.500	.500	.500
0.1	.461	.461	.461	.461	.461	.461	.461	.461	.461	.461	.461	.461
0.2	.422	.422	.422	.422	.422	.422	.422	.422	.422	.422	.422	.422
0.3	.384	.384	.384	.384	.384	.384	.384	.384	.384	.383	.383	.383
0.4	.348	.347	.347	.347	.347	.347	.347	.347	.347	.347	.346	.346
0.5	.313	.312	.312	.312	.312	.312	.311	.311	.311	.311	.311	.311
0.6	.279	.279	.279	.278	.278	.278	.278	.278	.278	.277	.277	.277
0.7	.248	.247	.247	.247	.247	.246	.246	.246	.246	.246	.245	.245
0.8	.219	.218	.218	.218	.217	.217	.217	.217	.216	.216	.216	.216
0.9	.192	.191	.191	.191	.190	.190	.190	.189	.189	.189	.189	.189
1.0	.168	.167	.167	.166	.166	.165	.165	.165	.164	.164	.164	.164
1.1	.146	.144	.144	.144	.143	.143	.143	.142	.142	.142	.141	.141
1.2	.126	.124	.124	.124	.123	.123	.122	.122	.122	.121	.121	.121
1.3	.108	.107	.107	.106	.105	.105	.105	.104	.104	.104	.103	.103
1.4	.092	.091	.091	.090	.090	.089	.089	.089	.088	.088	.087	.087
1.5	.079	.077	.077	.077	.076	.075	.075	.075	.074	.074	.074	.073
1.6	.067	.065	.065	.065	.064	.064	.063	.063	.062	.062	.062	.061
1.7	.056	.055	.055	.054	.054	.053	.053	.052	.052	.052	.051	.051
1.8	.048	.046	.046	.045	.045	.044	.044	.043	.043	.043	.042	.042
1.9	.040	.038	.038	.038	.037	.037	.036	.036	.036	.035	.035	.035
2.0	.033	.032	.032	.031	.031	.030	.030	.030	.029	.029	.029	.028
2.1	.028	.027	.027	.026	.025	.025	.025	.024	.024	.024	.023	.023
2.2	.023	.022	.022	.021	.021	.021	.020	.020	.020	.019	.019	.019
2.3	.019	.018	.018	.018	.017	.017	.016	.016	.016	.016	.015	.015
2.4	.016	.015	.015	.014	.014	.014	.013	.013	.013	.013	.012	.012
2.5	.013	.012	.012	.012	.011	.011	.011	.011	.010	.010	.010	.010
2.6	.011	.010	.010	.010	.009	.009	.009	.009	.008	.008	.008	.008
2.7	.009	.008	.008	.008	.008	.007	.007	.007	.007	.007	.006	.006
2.8	.008	.007	.007	.006	.006	.006	.006	.006	.005	.005	.005	.005
2.9	.006	.005	.005	.005	.005	.005	.005	.004	.004	.004	.004	.004
3.0	.005	.004	.004	.004	.004	.004	.004	.004	.003	.003	.003	.003
3.1	.004	.004	.004	.003	.003	.003	.003	.003	.003	.003	.003	.002
3.2	.003	.003	.003	.003	.003	.002	.002	.002	.002	.002	.002	.002
3.3	.003	.002	.002	.002	.002	.002	.002	.002	.002	.002	.002	.001
3.4	.002	.002	.002	.002	.002	.002	.002	.001	.001	.001	.001	.001
3.5	.002	.002	.002	.001	.001	.001	.001	.001	.001	.001	.001	.001
3.6	.002	.001	.001	.001	.001	.001	.001	.001	.001	.001	.001	.001
3.7	.001	.001	.001	.001	.001	.001	.001	.001	.001	.001	.001	.001
3.8	.001	.001	.001	.001	.001	.001	.001	.001	.001	.000	.000	.000
3.9	.001	.001	.001	.001	.001	.001	.000	.000	.000	.000	.000	.000
4.0	.001	.001	.001	.001	.000	.000	.000	.000	.000	.000	.000	.000

(continued)

Table A.6 (continued)

Degrees of Freedom (ν)

t	25	26	27	28	29	30	35	40	60	120	$\infty(=z)$
0.0	.500	.500	.500	.500	.500	.500	.500	.500	.500	.500	.500
0.1	.461	.461	.461	.461	.461	.461	.460	.460	.460	.460	.460
0.2	.422	.422	.421	.421	.421	.421	.421	.421	.421	.421	.421
0.3	.383	.383	.383	.383	.383	.383	.383	.383	.383	.382	.382
0.4	.346	.346	.346	.346	.346	.346	.346	.346	.345	.345	.345
0.5	.311	.311	.311	.310	.310	.310	.310	.310	.309	.309	.309
0.6	.277	.277	.277	.277	.277	.277	.276	.276	.275	.275	.274
0.7	.245	.245	.245	.245	.245	.245	.244	.244	.243	.243	.242
0.8	.216	.215	.215	.215	.215	.215	.215	.214	.213	.213	.212
0.9	.188	.188	.188	.188	.188	.188	.187	.187	.186	.185	.184
1.0	.163	.163	.163	.163	.163	.163	.162	.162	.161	.160	.159
1.1	.141	.141	.141	.140	.140	.140	.139	.139	.138	.137	.136
1.2	.121	.120	.120	.120	.120	.120	.119	.119	.117	.116	.115
1.3	.103	.103	.102	.102	.102	.102	.101	.101	.099	.098	.097
1.4	.087	.087	.086	.086	.086	.086	.085	.085	.083	.082	.081
1.5	.073	.073	.073	.072	.072	.072	.071	.071	.069	.068	.067
1.6	.061	.061	.061	.060	.060	.060	.059	.059	.057	.056	.055
1.7	.051	.051	.050	.050	.050	.050	.049	.048	.047	.046	.045
1.8	.042	.042	.042	.041	.041	.041	.040	.040	.038	.037	.036
1.9	.035	.034	.034	.034	.034	.034	.033	.032	.031	.030	.029
2.0	.028	.028	.028	.028	.027	.027	.027	.026	.025	.024	.023
2.1	.023	.023	.023	.022	.022	.022	.022	.021	.020	.019	.018
2.2	.019	.018	.018	.018	.018	.018	.017	.017	.016	.015	.014
2.3	.015	.015	.015	.015	.014	.014	.014	.013	.012	.012	.011
2.4	.012	.012	.012	.012	.012	.011	.011	.011	.010	.009	.008
2.5	.010	.010	.009	.009	.009	.009	.009	.008	.008	.007	.006
2.6	.008	.008	.007	.007	.007	.007	.007	.007	.006	.005	.005
2.7	.006	.006	.006	.006	.006	.006	.005	.005	.004	.004	.003
2.8	.005	.005	.005	.005	.005	.004	.004	.004	.003	.003	.003
2.9	.004	.004	.004	.004	.004	.003	.003	.003	.003	.002	.002
3.0	.003	.003	.003	.003	.003	.003	.002	.002	.002	.002	.001
3.1	.002	.002	.002	.002	.002	.002	.002	.002	.001	.001	.001
3.2	.002	.002	.002	.002	.002	.002	.001	.001	.001	.001	.001
3.3	.001	.001	.001	.001	.001	.001	.001	.001	.001	.001	.000
3.4	.001	.001	.001	.001	.001	.001	.001	.001	.001	.000	.000
3.5	.001	.001	.001	.001	.001	.001	.001	.001	.000	.000	.000
3.6	.001	.001	.001	.001	.001	.001	.000	.000	.000	.000	.000
3.7	.001	.001	.000	.000	.000	.000	.000	.000	.000	.000	.000
3.8	.000	.000	.000	.000	.000	.000	.000	.000	.000	.000	.000
3.9	.000	.000	.000	.000	.000	.000	.000	.000	.000	.000	.000
4.0	.000	.000	.000	.000	.000	.000	.000	.000	.000	.000	.000

Appendix B: Background Mathematics

B.1 Trigonometric Identities

$$\cos(a + b) = \cos(a)\cos(b) - \sin(a)\sin(b)$$
$$\cos(a - b) = \cos(a)\cos(b) + \sin(a)\sin(b)$$
$$\sin(a + b) = \sin(a)\cos(b) + \cos(a)\sin(b)$$
$$\sin(a - b) = \sin(a)\cos(b) - \cos(a)\sin(b)$$
$$\cos(a)\cos(b) = \tfrac{1}{2}[\cos(a + b) + \cos(a - b)]$$
$$\sin(a)\sin(b) = \tfrac{1}{2}[\cos(a - b) - \cos(a + b)]$$

B.2 Special Engineering Functions

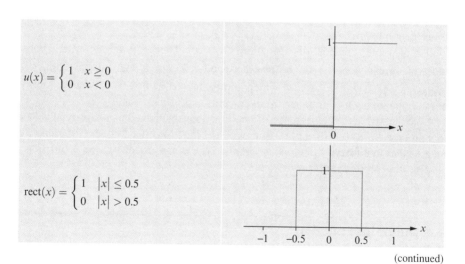

$$u(x) = \begin{cases} 1 & x \geq 0 \\ 0 & x < 0 \end{cases}$$

$$\mathrm{rect}(x) = \begin{cases} 1 & |x| \leq 0.5 \\ 0 & |x| > 0.5 \end{cases}$$

(continued)

M.A. Carlton and J.L. Devore, *Probability with Applications in Engineering, Science, and Technology*, Springer Texts in Statistics, DOI 10.1007/978-1-4939-0395-5, © Springer Science+Business Media New York 2014

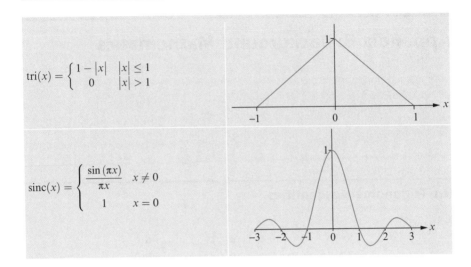

$$\text{tri}(x) = \begin{cases} 1 - |x| & |x| \leq 1 \\ 0 & |x| > 1 \end{cases}$$

$$\text{sinc}(x) = \begin{cases} \dfrac{\sin(\pi x)}{\pi x} & x \neq 0 \\ 1 & x = 0 \end{cases}$$

B.3 $o(h)$ Notation

The symbol $o(h)$ denotes any function of h which has the property that

$$\lim_{h \to 0} \frac{o(h)}{h} = 0$$

Informally, this property says that the value of the function approaches 0 even faster than h approaches 0.

For example, consider the function $f(h) = h^3$. Then $f(h)/h = h^2$, which does indeed approach 0 as $h \to 0$. On the other hand, $f(h) = \sqrt{h}$ does not have the $o(h)$ property, since $f(h)/h = 1/\sqrt{h}$, which approaches ∞ as $h \to 0^+$. Likewise, $\sin(h)$ does not have the $o(h)$ property: from calculus, $\sin(h)/h \to 1$ as $h \to 0$.

Note that the sum or difference of two functions that have this property also has this property: $o(h) \pm o(h) = o(h)$. The two $o(h)$ functions need not be the same as long as they both have the property. Similarly, the product of two such functions also has this property: $o(h) \cdot o(h) = o(h)$.

B.4 The Delta Function

The **Dirac delta function**, $\delta(x)$, also called an **impulse** or **impulse function**, is such that $\delta(x) = 0$ for $x \neq 0$ and

$$\int_{-\infty}^{\infty} \delta(x)dx = 1$$

More generally, an impulse at location x_0 with intensity a is $a \cdot \delta(x - x_0)$. An impulse is often graphed as an arrow, with the intensity listed in parentheses, as in the accompanying figure. The height of the arrow is meaningless; in fact, the "height" of an impulse is $+\infty$.

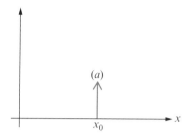

(a)

Properties of the delta function:

Basic integral:	$\int_{-\infty}^{\infty} \delta(x)dx = 1$, so $\int_{-\infty}^{\infty} a\delta(x - x_0)dx = a$
Antiderivative:	$\int_{-\infty}^{x} \delta(t)dt = u(x)$
Rescaling:	$\delta(cx) = \dfrac{\delta(x)}{|c|}$ for $c \neq 0$
Sifting:	$\int_{-\infty}^{\infty} g(x)\delta(x - x_0)dx = g(x_0)$
Convolution:	$g(x) \star \delta(x - x_0) = g(x - x_0)$

B.5 Fourier Transforms

The **Fourier transform** of a function $g(t)$, denoted $\mathscr{F}\{g(t)\}$ or $G(f)$, is defined by

$$G(f) = \mathscr{F}\{g(t)\} = \int_{-\infty}^{\infty} g(t)e^{-j2\pi ft}dt$$

where $j = \sqrt{-1}$. The Fourier transform of $g(t)$ exists provided that the integral of $g(t)$ is absolutely convergent; i.e., $\int_{-\infty}^{\infty} |g(t)|dt < \infty$.

The **inverse Fourier transform** of a function $G(f)$, denoted $\mathscr{F}^{-1}\{G(f)\}$ or $g(t)$, is defined by

$$g(t) = \mathscr{F}^{-1}\{G(f)\} = \int_{-\infty}^{\infty} G(f)e^{+j2\pi ft}\,df$$

Properties of Fourier transforms:

Linearity:	$\mathscr{F}\{a_1g_1(t) + a_2g_2(t)\} = a_1G_1(f) + a_2G_2(f)$
Rescaling:	$\mathscr{F}\{g(at)\} = \dfrac{1}{|a|}G\left(\dfrac{f}{a}\right)$
Duality:	$\mathscr{F}\{g(t)\} = G(f) \Rightarrow \mathscr{F}\{G(t)\} = g(-f)$
Time shift:	$\mathscr{F}\{g(t - t_0)\} = G(f)e^{-j2\pi ft_0}$
Frequency shift:	$\mathscr{F}\{g(t)e^{j2\pi f_0 t}\} = G(f - f_0)$
Time convolution:	$\mathscr{F}\{g_1(t)\star g_2(t)\} = G_1(f)G_2(f)$
Frequency convolution:	$\mathscr{F}\{g_1(t)g_2(t)\} = G_1(f)\star G_2(f)$

Fourier transform pairs:

$g(t)$	$G(f)$
1	$\delta(f)$
$u(t)$	$\dfrac{1}{2}\delta(f) + \dfrac{1}{j2\pi f}$
$\cos(2\pi f_0 t)$	$\dfrac{1}{2}[\delta(f - f_0) + \delta(f + f_0)]$
$\sin(2\pi f_0 t)$	$\dfrac{1}{2j}[\delta(f - f_0) - \delta(f + f_0)]$
$t^k e^{-at}u(t),\ a > 0, k = 0, 1, 2, \ldots$	$\dfrac{k!}{(a + j2\pi f)^{k+1}}$
$e^{-a|t|},\ a > 0$	$\dfrac{2a}{a^2 + (2\pi f)^2}$
e^{-t^2}	$\sqrt{\pi}e^{-\pi^2 f^2}$
$\text{rect}(t)$	$\text{sinc}(f)$
$\text{tri}(t)$	$\text{sinc}^2(f)$

B.6 Discrete-Time Fourier Transforms

The **discrete-time Fourier transform** (DTFT) of a function $g[n]$ is defined by

$$G(F) = \sum_{n=-\infty}^{\infty} g[n]e^{-j2\pi Fn}$$

The DTFT of $g[n]$ exists provided that $g[n]$ is absolutely summable; i.e., $\sum_{n=-\infty}^{\infty} |g[n]| < \infty$.

The **inverse DTFT** of a function $G(F)$ is defined by

$$g[n] = \int_{-1/2}^{1/2} G(F)e^{+j2\pi Fn} dF$$

Properties of DTFTs: (an arrow indicates application of the DTFT)

Periodicity:	$G(F+m) = G(F)$ for all integers m; i.e., $G(F)$ has period 1
Linearity:	$a_1 g_1[n] + a_2 g_2[n] \rightarrow a_1 G_1(F) + a_2 G_2(F)$
Time shift:	$g[n - n_0] \rightarrow G(F)e^{-j2\pi F n_0}$
Frequency shift:	$g[n]e^{j2\pi F_0 n} \rightarrow G(F - F_0)$
Time convolution:	$g_1[n] \star g_2[n] \rightarrow G_1(F)G_2(F)$
Frequency convolution:	$g_1[n]g_2[n] \rightarrow \int_{-1/2}^{1/2} G_1(\phi)G_2(F - \phi)d\phi$ (periodic convolution of G_1 and G_2)

DTFT pairs:

$g[n]$	$G(F)$
1	$\delta(F)$
$\delta[n]$	1
$u[n]$	$\dfrac{1}{2}\delta(F) + \dfrac{1}{1 - e^{-j2\pi F}}$
$\cos(2\pi F_0 n)$	$\dfrac{1}{2}[\delta(F - F_0) + \delta(F + F_0)]$
$\sin(2\pi F_0 n)$	$\dfrac{1}{2j}[\delta(F - F_0) - \delta(F + F_0)]$
$\alpha^{|n|}$, $|\alpha| < 1$	$\dfrac{1 - \alpha^2}{1 + \alpha^2 - 2\alpha\cos(2\pi F)}$
$\alpha^n u[n]$, $|\alpha| < 1$	$\dfrac{1}{1 - \alpha e^{-j2\pi F}}$

Appendix C: Important Probability Distributions

C.1 Discrete Distributions

For discrete distributions, the specified pmf and cdf are valid on the range of the random variable. The cdf and mgf are only provided when simple expressions exist for those functions.

Binomial (n, p) $\qquad\qquad\qquad\qquad\qquad\qquad\qquad\qquad\qquad\qquad$ $X \sim \text{Bin}(n, p)$

range:	$\{0, 1, \ldots, n\}$
parameters:	$n, n = 0, 1, 2, \ldots$ (number of trials)
	$p, 0 < p < 1$ (success probability)
pmf:	$b(x; n, p) = \binom{n}{x} p^x (1 - p)^{n-x}$
cdf:	$B(x; n, p)$ (see Table A.1)
mean:	np
variance:	$np(1 - p)$
mgf:	$(1 - p + pe^t)^n$

Note: The $n = 1$ case is called a Bernoulli distribution.

Geometric (p)

range:	$\{1, 2, 3, \ldots\}$
parameter:	$p, 0 < p < 1$ (success probability)
pmf:	$p(1 - p)^{x-1}$
cdf:	$1 - (1 - p)^x$
mean:	$\dfrac{1}{p}$
variance:	$\dfrac{1 - p}{p^2}$
mgf:	$\dfrac{pe^t}{1 - (1 - p)e^t}$

Note: Other sources defined a geometric rv to be the number of failures preceding the first success in independent and identical trials. See Sect. 2.6 for details.

M.A. Carlton and J.L. Devore, *Probability with Applications in Engineering, Science, and Technology*, Springer Texts in Statistics, DOI 10.1007/978-1-4939-0395-5, © Springer Science+Business Media New York 2014

Hypergeometric (n, M, N) $X \sim \text{Hyp}(n, M, N)$

range: $\{\max(0, n - N + M), \ldots, \min(n, M)\}$

parameters: $n, n = 0, 1, \ldots, N$ (number of trials)

                           $M, M = 0, 1, \ldots, N$ (population number of successes)

                           $N, N = 1, 2, 3, \ldots$ (population size)

pmf: $h(x; n, M, N) = \dfrac{\dbinom{M}{x} \dbinom{N - M}{n - x}}{\dbinom{N}{n}}$

cdf: $H(x; n, M, N)$

mean: $n \cdot \dfrac{M}{N}$

variance: $n \cdot \dfrac{M}{N} \cdot \left(1 - \dfrac{M}{N}\right) \cdot \dfrac{N - n}{N - 1}$

Note: With the understanding that $\dbinom{a}{b} = 0$ for $a < b$, the range of the hypergeometric distribution can be simplified to $\{0, \ldots, n\}$.

Negative Binomial (r, p) $X \sim \text{NB}(r, p)$

range: $\{r, r + 1, r + 2, \ldots\}$

parameters: $r, r = 1, 2, \ldots$ (desired number of successes)

                           $p, 0 < p < 1$ (success probability)

pmf: $nb(x; n, p) = \dbinom{x - 1}{r - 1} p^r (1 - p)^{x - r}$

mean: $\dfrac{r}{p}$

variance: $\dfrac{r(1 - p)}{p^2}$

mgf: $\left[\dfrac{p e^t}{1 - (1 - p)e^t}\right]^r$

Notes: The $r = 1$ case corresponds to the geometric distribution.

Other sources defined a negative binomial rv to be the number of failures preceding the rth success in independent and identical trials. See Sect. 2.6 for details.

Poisson (μ)

range: $\{0, 1, 2, \ldots\}$

parameter: $\mu, \mu > 0$ (expected number of events)

pmf: $p(x; \mu) = \dfrac{e^{-\mu} \mu^x}{x!}$

cdf: $P(x; \mu)$ (see Table A.2)

mean: μ

variance: μ

mgf: $e^{\mu(e^t - 1)}$

C.2 Continuous Distributions

For continuous distributions, the specified pdf and cdf are valid on the range of the random variable. The cdf and mgf are only provided when simple expressions exist for those functions.

Beta (α, β, A, B)

range:	$[A, B]$
parameters:	$\alpha, \alpha > 0$ (first shape parameter)
	$\beta, \beta > 0$ (second shape parameter)
	$A, -\infty < A < B$ (lower bound)
	$B, A < B < \infty$ (upper bound)

pdf:
$$\frac{1}{B-A} \cdot \frac{\Gamma(\alpha+\beta)}{\Gamma(\alpha) \cdot \Gamma(\beta)} \left(\frac{x-A}{B-A}\right)^{\alpha-1} \left(\frac{B-x}{B-A}\right)^{\beta-1}$$

mean:
$$A + (B-A) \cdot \frac{\alpha}{\alpha+\beta}$$

variance:
$$\frac{(B-A)^2 \alpha\beta}{(\alpha+\beta)^2 (\alpha+\beta+1)}$$

Notes: The $A = 0$, $B = 1$ case is called the standard beta distribution. The $\alpha = 1$, $\beta = 1$ case in the uniform distribution.

Exponential (λ)

range:	$(0, \infty)$
parameter:	$\lambda, \lambda > 0$ (rate parameter)
pdf:	$\lambda e^{-\lambda x}$
cdf:	$1 - e^{-\lambda x}$
mean:	$\dfrac{1}{\lambda}$
variance:	$\dfrac{1}{\lambda^2}$
mgf:	$\dfrac{\lambda}{\lambda - t} \qquad t < \lambda$

Note: A second parameter γ, called a *threshold parameter*, can be introduced to shift the density curve away from $x = 0$. In that case, $X - \gamma$ has an exponential distribution.

Gamma (α, β)

range:	$(0, \infty)$
parameters:	$\alpha, \alpha > 0$ (shape parameter)
	$\beta, \beta > 0$ (scale parameter)

pdf:
$$\frac{1}{\Gamma(\alpha)\beta^\alpha} x^{\alpha-1} e^{-x/\beta}$$

cdf:
$$G\left(\frac{x}{\beta}; \alpha\right) \text{ (see Table A.4)}$$

(continued)

mean: $\alpha\beta$

variance: $\alpha\beta^2$

mgf: $\left(\dfrac{1}{1-\beta t}\right)^{\alpha}$ $t < 1/\beta$

Notes: The $\alpha = 1$, $\beta = 1/\lambda$ case corresponds to the exponential distribution.
The $\beta = 1$ case is called the standard gamma distribution.
The $\alpha = n$ (an integer), $\beta = 1/\lambda$ case is called the Erlang distribution.
A third parameter γ, called a *threshold parameter*, can be introduced to shift the density curve away from $x = 0$. In that case, $X - \gamma$ has the two-parameter gamma distribution described above.

Lognormal (μ, σ)

range: $(0, \infty)$

parameters: μ, $-\infty < \mu < \infty$ (first shape parameter)

                            σ, $\sigma > 0$ (second shape parameter)

pdf: $\dfrac{1}{\sqrt{2\pi}\sigma x} e^{-[\ln(x)-\mu]^2/(2\sigma^2)}$

cdf: $\Phi\left[\dfrac{\ln(x) - \mu}{\sigma}\right]$

mean: $e^{\mu + \sigma^2/2}$

variance: $e^{2\mu + \sigma^2} \cdot \left(e^{\sigma^2} - 1\right)$

Note: A third parameter γ, called a *threshold parameter*, can be introduced to shift the density curve away from $x = 0$. In that case, $X - \gamma$ has the two-parameter lognormal distribution described above.

Normal (μ, σ) [or Gaussian (μ, σ)] $X \sim N(\mu, \sigma)$

range: $(-\infty, \infty)$

parameters: μ, $-\infty < \mu < \infty$ (mean)

                            σ, $\sigma > 0$ (standard deviation)

pdf: $\dfrac{1}{\sigma\sqrt{2\pi}} e^{-(x-\mu)^2/(2\sigma^2)}$

cdf: $\Phi\left(\dfrac{x - \mu}{\sigma}\right)$ (see Table A.3)

mean: μ

variance: σ^2

mgf: $e^{\mu t + \sigma^2 t^2/2}$

Note: The $\mu = 0$, $\sigma = 1$ case is called the standard normal or z distribution.

Uniform (A, B) $X \sim \text{Unif}[A, B]$

range:	$[A, B]$
parameters:	$A, -\infty < A < B$ (lower bound)
	$B, A < B < \infty$ (upper bound)
pdf:	$\dfrac{1}{B - A}$
cdf:	$\dfrac{x - A}{B - A}$
mean:	$\dfrac{A + B}{2}$
variance:	$\dfrac{(B - A)^2}{12}$
mgf:	$\dfrac{e^{Bt} - e^{At}}{(B - A)t} \qquad t \neq 0$

Note: The $A = 0, B = 1$ case is called the standard uniform distribution.

Weibull (α, β)

range:	$(0, \infty)$
parameters:	$\alpha, \alpha > 0$ (shape parameter)
	$\beta, \beta > 0$ (scale parameter)
pdf:	$\dfrac{\alpha}{\beta^\alpha} x^{\alpha-1} e^{-(x/\beta)^\alpha}$
cdf:	$1 - e^{-(x/\beta)^\alpha}$
mean:	$\beta \cdot \Gamma\left(1 + \dfrac{1}{\alpha}\right)$
variance:	$\beta^2 \left\{ \Gamma\left(1 + \dfrac{2}{\alpha}\right) - \left[\Gamma\left(1 + \dfrac{1}{\alpha}\right)\right]^2 \right\}$

Note: A third parameter γ, called a *threshold parameter*, can be introduced to shift the density curve away from $x = 0$. In that case, $X - \gamma$ has the two-parameter Weibull distribution described above.

C.3 Matlab and R Commands

Table C.1 indicates the template for Matlab and R commands related to the "named" probability distributions. In Table C.1,

x = input to the pmf, pdf, or cdf
p = left-tail probability (e.g., p = .5 for the median, or .9 for the 90th percentile)
N = simulation size; i.e., the length of the vector of random numbers
pars = the set of parameters, in the order prescribed
name = a text string specifying the particular distribution

Table C.1 Matlab and R syntax for probability distribution commands

	Matlab	R
pmf/pdf	`pdf('name',x,pars)`	`dname(x,pars)`
cdf	`cdf('name',x,pars)`	`pname(x,pars)`
Quantile	`icdf('name',p,pars)`	`qname(p,pars)`
Random #s	`random('name',pars,[N,1])`	`rname(N,pars)`

Table C.2 catalogs the names and parameters for a variety of distributions. Notice in Table C.1 that Matlab takes the name as a text string between single quotes, while R incorporates it into the command name.

Table C.2 Names and parameter sets for major distributions in Matlab and R

Distribution	Matlab		R	
	name	pars	name	pars
Binomial	`bin`	n, p	`binom`	n, p
Geometric[a]	`geo`	p	`geom`	p
Hypergeometric	`hyge`	N, M, n	`hyper`	$M, N - M, n$
Negative binomial[a]	`nbin`	r, p	`nbinom`	r, p
Poisson	`pois`	μ	`pois`	μ
Beta[b]	`beta`	α, β	`beta`	α, β
Exponential	`exp`	$1/\lambda$	`exp`	λ
Gamma	`gamma`	α, β	`gamma`	$\alpha, 1/\beta$
Lognormal	`logn`	μ, σ	`lnorm`	μ, σ
Normal	`norm`	μ, σ	`norm`	μ, σ
Uniform	`unif`	A, B	`unif`	A, B
Weibull	`wbl`	β, α	`weibull`	α, β

[a]The geometric and negative binomial commands in Matlab and R assume that the random variable counts only failures, and not the total number of trials. See Sect. 2.6 or the software documentation for details.

[b]The beta distribution commands in Matlab and R assume a *standard* beta distribution; i.e., with $A = 0$ and $B = 1$.

Answers to Odd-Numbered Exercises

Chapter 1

1. (a) $A \cap B'$ (b) $A \cup B$ (c) $(A \cap B') \cup (B \cap A')$

3. (a) $\mathscr{S} = \{1324, 1342, 1423, 2314, 2341, 2413, 2431, 3124, 3142, 4123, 4132, 3214, 3241, 4213, 4231\}$
   (b) $A = \{1324, 1342, 1423, 1432\}$
   (c) $B = \{2314, 2341, 2413, 2431, 3214, 3241, 4213, 4231\}$
   (d) $A \cup B = \{1324, 1342, 1423, 1432, 2314, 2341, 2413, 2431, 3214, 3241, 4213, 4231\}$
   $A \cap B = \varnothing$
   $A' = \{2314, 2341, 2413, 2431, 3124, 3142, 4123, 4132, 3214, 3241, 4213, 4231\}$

5. (a) $A = \{SSF, SFS, FSS\}$
   (b) $B = \{SSS, SSF, SFS, SSS\}$
   (c) $C = \{SSS, SSF, SFS\}$
   (d) $C' = \{SFF, FSS, FSF, FFS, FFF\}$
   $A \cup C = \{SSS, SSF, SFS, FSS\}$
   $A \cap C = \{SSF, SFS\}$
   $B \cup C = \{SSS, SSF, SFS, FSS\}$
   $B \cap C = \{SSS, SSF, SFS\}$

7. (a) $\{111, 112, 113, 121, 122, 123, 131, 132, 133, 211, 212, 213, 221, 222, 223, 231,$
   $232, 233, 311, 312, 313, 321, 322, 323, 331, 332, 333\}$
   (b) $\{111, 222, 333\}$
   (c) $\{123, 132, 213, 231, 312, 321\}$
   (d) $\{111, 113, 131, 133, 311, 313, 331, 333\}$

9. (a) $\mathscr{S} = \{BBBAAAA, BBABAAA, BBAABAA, BBAAABA, BBAAAAB, BABBAAA,$
   $BABABAA, BABAABA, BABAAAB, BAABBAA, BAABABA, BAABAAB, BAAABBA,$
   $BAAABAB, BAAAABB, ABBBAAA, ABBABAA, ABBAABA, ABBAAAB, ABABBAA,$
   $ABABABA, ABABAAB, ABAABBA, ABAABAB, ABAAABB, AABBBAA, AABBABA,$
   $AABBAAB, AABABBA, AABABAB, AABAABB, AAABBBA, AAABBAB, AAABABB,$
   $AAAABBB\}$
   (b) $\{AAAABBB, AAABABB, AAABBAB, AABAABB, AABABAB\}$

13. (a) .07 (b) .30 (c) .57

15. (a) They are awarded at least one of the first two projects, .36
    (b) They are awarded neither of the first two projects, .64
    (c) They are awarded at least one of the projects, .53
    (d) They are awarded none of the projects, .47
    (e) They are awarded only the third project, .17
    (f) Either they fail to get the first two or they are awarded the third, .75

(continued)

M.A. Carlton and J.L. Devore, *Probability with Applications in Engineering, Science, and Technology*, Springer Texts in Statistics, DOI 10.1007/978-1-4939-0395-5, 749
© Springer Science+Business Media New York 2014

17. (a) .572 (b) .879
19. (a) SAS and SPSS are not the only packages
     (b) .7 (c) .8 (d) .2
21. (a) .8841 (b) .0435
23. (a) .10 (b) .18, .19 (c) .41 (d) .59 (e) .31 (f) .69
25. (a) 1/15 (b) 6/15 (c) 14/15 (d) 8/15
27. (a) .85 (b) .15 (c) .22 (d) .35
29. (a) 1/9 (b) 8/9 (c) 2/9
31. (a) 10,000 (b) .9876 (c) .03 (d) .0337
33. (a) 336 (b) 593,775 (c) 83,160 (d) .140 (e) .002
35. (a) 240 (b) 12 (c) 108 (d) 132 (e) .55, .413
37. (a) .0775 (b) .0082
39. (a) 8008 (b) 3300 (c) 5236 (d) .4121, .6538
41. .2
43. (a) .2967 (b) .0747 (c) .2637 (d) .042
45. (a) 369,600 (b) .00006494
47. (a) 1/15 (b) 1/3 (c) 2/3
51. $P(A|B) > P(B|A)$
53. (a) .50 (b) .0833 (c) .3571 (d) .8333
55. (a) .05 (b) .12 (c) .56, .44 (d) .49, .25 (e) .533 (f) .444, .556
57. .04
59. (a) .50 (b) .0455 (c) .682 (d) .0189
65. (a) 3/4 (b) 2/3
67. (a) .067 (b) .509
71. (a) .765 (b) .235
73. .087, .652, .261
75. .00329
77. .4657 for airline #1, .2877 for airline #2, .2466 for airline #3
81. A_2 and A_3 are independent
83. .1936, .3816
85. .1052
87. .99999969, .226
89. .9981
91. (a) Yes (b) No
93. (a) .343 (b) .657 (c) .189 (d) .216 (e) .3525
95. (a) $P(A) = P(B) = .02$, $P(A \cap B) = .039984$, A and B are not independent
     (b) .04, very little difference
     (c) $P(A \cap B) = .0222$, not close; $P(A \cap B)$ is close to $P(A)P(B)$ when the sample size is
     very small relative to the population size
97. (a) Route #1 (b) .216
99. (a) $1 - (1 - 1/N)^n$
     (b) $n = 3$: .4212, 1/2; $n = 6$: .6651, 1; $n = 10$: .8385, 10/6; the answers are not close
     (c) .1052, $1/9 = .1111$; much closer
101. (a) Exact answer = .46 (b) se $\approx$ .005
103. .8186 (answers will vary)

(continued)

105. $\approx .39, \approx .88$ (answers will vary)

107. $\approx .91$ (answers will vary)

109. $\approx .02$ (answers will vary)

111. (b) $\approx .37$ (answers will vary) (c) $\approx 176,000,000$ (answers will vary; exact $= 176,214,841$)

113. (a) $\approx .20$ (b) $\approx .56$ (answers will vary)

115. (a) $\approx .5177$ (b) $\approx .4914$ (answers will vary)

117. $\approx .2$ (answers will vary)

119. (b) $\pi \approx 4 \cdot \hat{P}(A)$ (numerical answers will vary)

121. (a) 1140 (b) 969 (c) 1020 (d) .85

123. (a) .0762 (b) .143

125. (a) .512 (b) .608 (c) .7835

127. .1074

129. (a) 10^{14} (b) 7.3719×10^{-9}

131. (a) .974 (b) .9754

133. .926

135. (a) .018 (b) .601

137. .156

139. (a) .0625 (b) .15625 (c) .34375 (d) .014

141. (a) .12, .88 (b) .18, .38

143. $1/4 = P(A_1 \cap A_2 \cap A_3) \neq P(A_1) \cdot P(A_2) \cdot P(A_3) = 1/8$

145. (a) $a_0 = 0, a_5 = 1$ (b) $a_2 = (1/2)a_1 + (1/2)a_3$ (c) $a_i = i/5$ for $i = 0, 1, 2, 3, 4, 5$

149. (a) .6923 (b) .52

Chapter 2

1. $x = 0$ for FFF; $x = 1$ for $SFF, FSF,$ and FFS; $x = 2$ for $SSF, SFS,$ and FSS; $x = 3$ for SSS

3. $Z =$ average of the two numbers, with possible values $2/2, 3/2, \ldots, 12/2$; $W =$ absolute value of the difference, with possible values 0, 1, 2, 3, 4, 5

5. No. In Example 2.4, let $Y = 1$ if at most three batteries are examined and let $Y = 0$ otherwise. Then Y has only two values

7. (a) $\{0, 1, 2\ldots, 12\}$; discrete (c) $\{1, 2, 3, \ldots\}$; discrete (e) $\{0, c, 2c, \ldots, 10000c\}$ where c is the royalty per book; discrete g $\{x: m \leq x \leq M\}$ where m and M are the minimum and maximum possible tension; continuous

9. (a) $\{2, 4, 6, 8, \ldots\}$, that is, $\{2(1), 2(2), 2(3), 2(4), \ldots\}$, an infinite sequence; discrete

11. (a) .10 (c) .45, .25

13. (a) .70 (b) .45 (c) .55 (d) .71 (e) .65 (f) .45

15. (a) (1,2), (1,3), (1,4), (1,5), (2,3), (2,4), (2,5), (3,4), (3,5), (4,5) (b) $p(0) = .3, p(1) = .6,$ $p(2) = .1$ (c) $F(x) = 0$ for $x < 0, = .3$ for $0 \leq x < 1, = .9$ for $1 \leq x < 2,$ and $= 1$ for $x \geq 2$

17. (a) .81 (b) .162 (c) it is A; $AUUUA, UAUUA, UUAUA, UUUAA$; .00324

19. $p(0) = .09, p(1) = .40, p(2) = .32, p(3) = .19$

21. (b) $p(x) = .301, .176, .125, .097, .079, .067, .058, .051, .046$ for $x = 1, 2, \ldots, 9$
(c) $F(x) = 0$ for $x < 1, = .301$ for $1 \leq x < 2, = .477$ for $2 \leq x < 3, \ldots, = .954$ for $8 \leq x < 9,$ and $= 1$ for $x \geq 9$
(d) .602, .301

23. (a) .20 (b) .33 (c) .78 (d) .53

25. (a) $p(y) = (1 - p)^y \cdot p$ for $y = 0, 1, 2, 3, \ldots$

(continued)

27. (a) 1234, 1243, 1324, ..., 4321
     (b) $p(0) = 9/24$, $p(1) = 8/24$, $p(2) = 6/24$, $p(3) = 0$, $p(4) = 1/24$

29. (a) 6.45 GB (b) 15.6475 (c) 3.96 GB (d) 15.6475

31. 4.49, 2.12, .68

33. (a) p (b) $p(1 - p)$ (c) p

35. $E[h_3(X)] = \$4.93$, $E[h_4(X)] = \$5.33$, so 4 copies is better

37. $E(X) = (n + 1)/2$, $E(X^2) = (n + 1)(2n + 1)/6$, $\text{Var}(X) = (n^2 - 1)/12$

39. (b) .61 (c) .47 (d) \$2598 (e) \$4064

41. (a) $\mu = -\$2/38$ for both methods (c) single number: $\sigma = \$5.76$; square: $\sigma = \$2.76$

45. $E(X - c) = E(X) - c$, $E(X - \mu) = 0$

47. (a) .25, .11, .06, .04, .01 (b) $\mu = 2.64$, $\sigma = 1.54$; for $k = 2$, the probability is .04, and the bound of .25 is much too conservative; for $k = 3, 4, 5, 10$, the probability is 0, and the bounds are again conservative (c) $\mu = \$0$, $\sigma = \$d$, 0 (d) 1/9, same as the Chebyshev bound (e) there are many, e.g., $p(1) = p(-1) = .02$ and $p(0) = .96$

49. (a) Yes, $n = 10$, $p = 1/6$ (b) Yes, $n = 40$, $p = 1/4$ (c) No (d) No (e) No (f) Yes, assuming the population is very large; $n = 15$, $p = P$(a randomly selected apple weighs > 150 g)

51. (a) .515 (b) .218 (c) .011 (d) .480 (e) .965 (f) .000 (g) .595

53. (a) .354 (b) .115 (c) .918

55. (a) 5 (b) 1.94 (c) .017

57. (a) .403 (b) .787 (c) .774

59. .1478

61. .407, independence

63. (a) .010368 (c) the probability decreases, to .001970 (d) 1500, 259.2

65. (a) .017 (b) .811, .425 (c) .006, .902, .586

67. When $p = .9$, the probability is .99 for A and .9963 for B. If $p = .5$, the probabilities are .75 and .6875, respectively

69. (a) 20, 16 (b) 70, 21

71. (a) $p = 0$ or 1 (b) $p = .5$

73. $P(|X - \mu| \geq 2\sigma) = .042$ when $p = .5$ and $= .065$ when $p = .75$, compared to the upper bound of .25. Using $k = 3$ in place of $k = 2$, these probabilities are .002 and .004, respectively, whereas the upper bound is .11

75. (a) .932 (b) .065 (c) .068 (d) .492 (e) .251

77. (a) .011 (b) .441 (c) .554, .459 (d) .945

79. Poisson(5) (a) .492 (b) .133

81. .271, .857

83. (a) 2.9565, .948 (b) .726

85. (a) .122, .809, .283 (b) 12, 3.464 (c) .530, .011

87. (a) .221 (b) 6,800,000 (c) $p(x; 20.106)$

89. (a) $1/(1 - e^{-\theta})$ (b) $\theta = 2$; .981 (c) 1.26

91. (a) .114 (b) .879 (c) .121 (d) Use the binomial distribution with $n = 15$, $p = .10$

93. (a) $h(x; 15, 10, 20)$ for $x = 5, ..., 10$ (b) .0325 (c) .697

95. (a) $h(x; 10, 10, 20)$ (b) .033 (c) $h(x; n, n, 2n)$

97. (a) .2817 (b) .7513 (c) .4912, .9123

99. (a) $nb(x; 2, .5)$ (b) .188 (c) .688 (d) 2, 4

101. $nb(x; 6, .5)$, 6

103. $nb(x; 5, 6/36)$, 30, 12.2

(continued)

105. (a) 160, 21.9 (b) .6756
107. (a) $.01e^{9t}+.05e^{10t}+.16e^{11t}+.78e^{12t}$ (b) $E(X)=11.71$, $SD(X)=0.605$
109. $M_X(t)=e^t/(2-e^t)$, $E(X)=2$, $SD(X)=\sqrt{2}$
111. Skewness $=-2.20$ (Ex. 107), $+0.54$ (Ex. 108), $+2.12$ (Ex. 109), 0 (Ex. 110)
113. $E(X)=0$, $Var(X)=2$
115. $p(y)=(.25)^{y-1}(.75)$ for $y=1, 2, 3, \ldots$
117. $M_Y(t)=e^{t^2/2}$, $E(Y)=0$, $Var(Y)=1$
121. $E(X)=5$, $Var(X)=4$
123. $M_{n-X}(t)=(p+(1-p)e^t)^n$
125. $M_Y(t)=p^r[1-(1-p)e^t]^{-r}$, $E(Y)=r(1-p)/p$; $Var(Y)=r(1-p)/p^2$
129. mean ≈ 0.5968, sd ≈ 0.8548 (answers will vary)
131. $\approx .9090$ (answers will vary)
133. (a) $\mu \approx 13.5888$, $\sigma \approx 2.9381$ (b) $\approx .1562$ (answers will vary)
135. mean ≈ 3.4152, variance ≈ 5.97 (answers will vary)
137. (b) 142 tickets
139. (a) $\approx .2291$ (b) $\approx \$8696$ (c) $\approx \$7811$ (d) $\approx .2342$, $\approx \$7,767$, $\approx \$7,571$ (answers will vary)
141. (b) probability $\approx .9196$, confidence interval $=(.9143, .9249)$ (answers will vary)
143. (b) 3.114, .405, .636
145. (a) $b(x; 15, .75)$ (b) .686 (c) .313 (d) 11.25, 2.81 (e) .310
147. (a) .013 (b) 19 (c) .266 (d) Poisson with $\mu=500$
149. (a) $p(x; 2.5)$ (b) .067 (c) .109
151. 1.813, 3.05
153. $p(2)=p^2$, $p(3)=(1-p)p^2$, $p(4)=(1-p)p^2$, $p(x)=[1-p(2)-\ldots-p(x-3)](1-p)p^2$
   for $x=5, 6, 7, \ldots$; .99950841
155. (a) .0029 (b) .0767, .9702
157. (a) .135 (b) .00144 (c) $\sum_{x=0}^{\infty} [p(x; 2)]^5$
159. 3.590
161. (a) No (b) .0273
163. (b) $.5\mu_1+.5\mu_2$ (c) $.25(\mu_1-\mu_2)^2+.5(\mu_1+\mu_2)$ (d) .6 and .4 replace .5 and .5, respectively
165. $\mu=.5$
167. $500p+750$, $100\sqrt{p(1-p)}$
169. (a) 2.50 (b) 3.1

Chapter 3

1. (b) .4625; the same (c) .5, .278125
3. (b) .5 (c) .6875 (d) .6328
5. (a) $k=3/8$ (b) .125 (c) .296875 (d) .578125
7. (a) $f(x)=1/4.05$ for $.20 \leq x \leq 4.25$ (b) .3086 (c) .4938 (d) 1/4.05
9. (a) .562 (b) .438, .438 (c) .071
11. (a) .25 (b) .1875 (c) .4375 (d) 1.414 h (e) $f(x)=x/2$ for $0 \leq x < 2$
13. (a) $k=3$ (b) $F(x)=1-1/x^3$ for $x \geq 1$ and $=0$ otherwise (c) .125, .088
15. (a) $F(x)=x^3/8$ for $0 \leq x \leq 2$, $=0$ for $x<0$, $=1$ for $x>2$ (b) .015625 (c) .0137, .0137
   (d) 1.817 min

(continued)

17. (a) .597 (b) .369 (c) $f(x) = [\ln(4) - \ln(x)]/4$ for $0 < x < 4$
19. (a) 1.333 h (b) .471 h (c) $2
21. (a) .8182 ft^3 (b) .3137
23. (a) $A + (B - A)p$ (b) $(A + B)/2$ (c) $(B^{n+1} - A^{n+1})/[(n + 1)(B - A)]$
25. 314.79 m^2
27. 248 °F, 3.6 °F
29. 1/4 min, 1/4 min
31. (c) $\mu_R \approx v/20$, $\sigma_R \approx v/800$ (d) ~100π (e) ~$80\pi^2$
33. $g(x) = 10x - 5$, $M_Y(t) = (e^{5t} - e^{-5t})/10t$, $Y \sim \text{Unif}[-5, 5]$
35. (a) $M_X(t) = .15e^{.5t}/(.15 - t)$, $\mu = 7.167$, variance $= 44.444$ (b) $.15/(.15 - t)$, $\mu = 6.67$, variance $= 44.444$ (c) $M_Y(t) = .15/(.15 - t)$
39. (a) .4850 (b) .3413 (c) .4938 (d) .9876 (e) .9147 (f) .9599 (g) .9104 (h) .0791 (i) .0668 (j) .9876
41. (a) 1.34 (b) -1.34 (c) .675 (d) $-.675$ (e) -1.555
43. (a) .9664 (b) .2451 (c) .8664
45. (a) .4584 (b) 135.8 kph (c) .9265 (d) .3173 (e) .6844
47. (a) .9236 (b) .0021 (c) .1336
49. $.6826 < .9987 \Rightarrow$ the second machine
51. (a) .2514, ~0 (b) ~39.985 ksi
53. $\sigma = .0510$
55. (a) .8664 (b) .0124 (c) .2718
57. (a) .7938 (b) 5.88 (c) 7.938 (d) .2651
59. (a) $\Phi(1.72) - \Phi(.55)$ (b) $\Phi(.55) - [1 - \Phi(1.72)]$
61. (a) .7286 (b) .8643, .8159
63. (a) .9932 (b) .9875 (c) .8064
65. (a) .0392 (b) ~1
69. (a) .1587 (b) .0013 (c) .999937 (d) .00000029
71. (a) 1 (b) 1 (c) .982 (d) .129
73. (a) .1481 (b) .0183
75. (a) 120 (b) $(3/4)\sqrt{\pi}$ (c) .371 (d) .735 (e) 0
77. (a) .424 (b) .567; median < 24 (c) 60 weeks (d) 66 weeks
79. $\eta_p = -\ln(1 - p)/\lambda$, $\eta = .693/\lambda$
81. (a) .5488 (b) .3119 (c) 7.667 s (d) 6.667 s
85. (a) .8257, .8257, .0636 (b) .6637 (c) 172.727 h
89. (a) .9295 (b) .2974 (c) 98.184 ksi
91. (a) $\mu = 9.164$, $\sigma = .38525$ (b) .8790 (c) .4247 (d) no
93. $\eta = e^\mu = 9547$ kg/day/km
95. (a) 3.96, 1.99 months (b) .0375 (c) .7016 (d) 7.77 months (e) 13.75 months (f) 4.522
97. $\alpha = \beta$
99. (b) $\Gamma(\alpha + \beta)\Gamma(m + \beta)/[\Gamma(\alpha + m + \beta)\Gamma(\beta)]$, $\beta/(\alpha + \beta)$
101. Yes, since the pattern in the plot is quite linear
103. Yes
105. Yes
107. Plot $\ln(x)$ versus z percentile. The pattern is somewhat straight, so a lognormal distribution is plausible

(continued)

109. It is plausible that strength is normally distributed, because the pattern is reasonably linear

111. There is substantial curvature in the plot. λ is a scale parameter (as is σ for the normal family)

113. $f_Y(y) = 2/y^3$ for $y > 1$

115. $f_Y(y) = ye^{-y^2/2}$ for $y > 0$

117. $f_Y(y) = 1/16$ for $0 < y < 16$

119. $f_Y(y) = 1/[\pi(1 + y^2)]$ for $-\infty < y < \infty$

121. $Y = g(X) = X^2/16$

123. $f_Y(y) = 1/\left[2\sqrt{y}\right]$ for $0 < y \leq 1$

125. $$f_Y(y) = \begin{cases} 1/(4\sqrt{y}) & 0 < y \leq 1 \\ 1/(8\sqrt{y}) & 1 < y \leq 9 \\ 0 & \text{otherwise} \end{cases}$$

129. (a) $F(x) = x^2/4$, $F^{-1}(u) = 2\sqrt{u}$ (c) $\mu = 1.333$, $\sigma = 0.4714$, $\bar{x}$ and s will vary

131. The inverse cdf is $F^{-1}(u) = \left[\sqrt{1 + 48u} - 1\right]/3$

133. (a) The inverse cdf is $F^{-1}(u) = \tau \cdot [1 - (1 - u)^{1/\theta}]$ (b) $E(X) = 16$, $\bar{x}$ will vary

135. (a) $c = 1.5$ (c) 15,000 (d) $\mu = 3/8$, $\bar{x}$ will vary (e) $\hat{P}(M < .1) = .8760$ (answers will vary)

137. (a) $x = G^{-1}(u) = -\ln(1 - u)$ (b) $\sqrt{2e/\pi} \approx 1.3155$ (c) ~13,155

141. (a) .4 (b) .6 (c) $F(x) = x/25$ for $0 \leq x \leq 25$, $= 0$ for $x < 0$, $= 1$ for $x > 25$ (d) 12.5, 7.22

143. (b) $F(x) = 1 - 16/(x + 4)^2$ for $x > 0$, $= 0$ for $x \leq 0$ (c) .247 (d) 4 years (e) 16.67

145. (a) .6568 (b) 41.56 V (c) .3197

147. (a) .0003 (exact: .00086) (b) .0888 (exact: .0963)

149. (a) 68.03 dB, 122.09 dB (b) .3204 (c) .7642, because the lognormal distribution is not symmetric

151. (a) $F(x) = 1.5(1 - 1/x)$ for $1 \leq x \leq 3$, $= 0$ for $x < 1$, $= 1$ for $x > 3$ (b) .9, .4 (c) 1.648 s (d) .553 s (e) .267 s

153. (a) 1.075, 1.075 (b) .0614, .333 (c) 2.476 mm

155. (b) $95,600, .3300

157. (b) $F(x) = .5e^{2x}$ for $x < 0$, $= 1 - .5e^{-2x}$ for $x \geq 0$ (c) .5, .665, .256, .670

159. (a) $k = (\alpha - 1)5^{\alpha - 1}$, $\alpha > 1$ (b) $F(x) = 1 - (5/x)^{\alpha - 1}$ for $x \geq 5$ (c) $5(\alpha - 1)/(\alpha - 2)$, $\alpha > 2$

161. (b) .4602, .3636 (c) .5950 (d) 140.178 MPa

163. (a) Weibull, with $\alpha = 2$ and $\beta = \sqrt{2}\sigma$ (b) .542

165. .5062

171. (a) 710, 84.423, .684 (b) .376

Chapter 4

1. (a) .20 (b) .42 (c) .70 (d) $p_X(x) = .16, .34, .50$ for $x = 0, 1, 2$; $p_Y(y) = .24, .38, .38$ for $y = 0, 1, 2$; .50 (e) no

3. (a) .15 (b) .40 (c) .22 (d) .17, .46 (e) $p_1(x_1) = .19, .30, .25, .14, .12$ for $x_1 = 0, 1, 2, 3, 4$ (f) $p_2(x_2) = .19, .30, .28, .23$ for $x_2 = 0, 1, 2, 3$ (g) no

5. (a) .0305 (b) .1829 (c) .1073

7. (a) .054 (b) .00018

9. (a) .030 (b) .120 (c) .300 (d) .380 (e) no

11. (a) $k = 3/380,000$ (b) .3024 (c) .3593 (d) $f_X(x) = 10kx^2 + .05$ for $20 \leq x \leq 30$ (e) no

(continued)

13. (a) $p(x,y) = \dfrac{e^{-\mu_1-\mu_2}\mu_1^x\mu_2^y}{x!y!}$ (b) $e^{-\mu_1-\mu_2}[1 + \mu_1 + \mu_2]$ (c) $\dfrac{e^{-\mu_1-\mu_2}}{m!}(\mu_1 + \mu_2)^m$

15. (a) $f(x, y) = e^{-x-y}$ for $x, y \geq 0$ (b) .400 (c) .594 (d) .330

17. (a) $F(y) = (1 - e^{-\lambda y}) + (1 - e^{-\lambda y})^2 - (1 - e^{-\lambda y})^3$ for $y > 0$, $f(y) = 4\lambda e^{-2\lambda y} - 3\lambda e^{-3\lambda y}$ for $y > 0$ (b) $2/3\lambda$

19. (a) .25 (b) $1/\pi$ (c) $2/\pi$ (d) $f_X(x) = \dfrac{2\sqrt{r^2 - x^2}}{\pi r^2}$ for $-r \leq x \leq r$, $f_Y(y) = f_X(y)$, no

21. 1/3

23. (a) .11 (b) $p_X(x) = .78, .12, .07, .03$ for $x = 0, 1, 2, 3$; $p_Y(y) = .77, .14, .09$ for $y = 0, 1, 2$ (c) no (d) 0.35, 0.32 (e) 95.72

25. .15

27. L^2

29. .25 h, or 15 min

31. $-2/3$

33. (a) -3.20 (b) $-.207$

35. (a) .238 (b) .51

37. (a) $\text{Var}(h(X,Y)) = \iint [h(x,y)]^2 \cdot f(x,y)dA - [\iint h(x,y) \cdot f(x,y)dA]^2$ (b) 13.34

43. (a) 87,850, 19,100,116 (b) mean yes, variance no (c) .0027

45. .2877, .3686

47. .0314

49. (a) 45 min (b) 68.33 (c) -1 min, 13.67 (d) -5 min, 68.33

51. (a) 50, 10.308 (b) .0075 (c) 50 (d) 111.5625 (e) 131.25

53. (a) .9616 (b) .0623

55. (a) $E(Y_i) = 1/2$, $E(W) = n(n+1)/4$ (b) $\text{Var}(Y_i) = 1/4$, $\text{Var}(W) = n(n+1)(2n+1)/24$

57. 10:52.76 a.m.

59. (a) mean $= 0$, sd $= \sqrt{2}$

61. (a) $X \sim \text{Bin}(10, 18/38)$ (b) $Y \sim \text{Bin}(15, 18/38)$ (c) $X + Y \sim \text{Bin}(25, 18/38)$ (f) no

65. (a) $\alpha = 2$, $\beta = 1/\lambda$ (c) gamma, $\alpha = n$, $\beta = 1/\lambda$

67. (a) .5102 (b) .000000117

69. (a) $x^2/2$, $x^4/12$ (b) $f(x, y) = 1/x^2$ for $0 < y < x^2 < 1$ (c) $f_Y(y) = 1/\sqrt{y}$ for $0 < y < 1$

71. (a) $p_X(x) = 1/10$ for $x = 0, 1, \ldots, 9$; $p(y|x) = 1/9$ for $y = 0, \ldots, 9$ and $y \neq x$; $p(x, y) = 1/90$ for $x, y = 0, 1, \ldots, 9$ and $y \neq x$ (b) $5 - x/9$

73. (a) $f_X(x) = 2x$, $0 < x < 1$ (b) $f(y|x) = 1/x$, $0 < y < x$ (c) .6 (d) no (e) $x/2$ (f) $x^2/12$

75. (a) $p(x,y) = \dfrac{2!}{x!y!(2 - x - y)!}(.3)^x(.2)^y(.5)^{2-x-y}$ (b) $X \sim \text{Bin}(2, .3)$, $Y \sim \text{Bin}(2, .2)$ (c) $Y/X = x \sim \text{Bin}(2 - x, .2/.7)$ (d) no (e) $(4 - 2x)/7$ (f) $10(2 - x)/49$

77. (a) $x/2$, $x^2/12$ (b) $f(x, y) = 1/x$ for $0 < y < x < 1$ (c) $f_Y(y) = -\ln(y)$ for $0 < y < 1$

79. (a) $.6x$, $.24x$ (b) 60 (c) 60

81. 176 lbs, 12.68 lbs

83. (a) $1 + 4p$, $4p(1 - p)$ (b) $2598, 16,158,196$ (c) $2598(1 + 4p)$, $\sqrt{16518196 + 93071200p - 26998416p^2}$ (d) $2598 and $4064 for $p = 0$; $7794 and $7504 for $p = .5$; $12,990 and $9088 for $p = 1$

85. (a) 12 cm, .01 cm (b) 12 cm, .005 cm (c) the larger sample

87. (a) .9772, .4772 (b) 10

89. 43.29 h

(continued)

91. .9332

93. (a) .8357 (b) no

95. (a) .1894 (b) .1894 (c) 621.5 gallons

97. (a) .0968 (b) .8882

99. .9616

101. $1/\overline{X}$

103. (a) $f(y_1, y_2) = \dfrac{1}{4\pi} e^{-(y_1^2+y_2^2)/4}$ (b) $f_{Y_1}(y_1) = \dfrac{1}{\sqrt{4\pi}} e^{-y_1^2/4}$ (c) yes

105. (a) $y(2 - y)$ for $0 \le y \le 1$ (b) $2(1 - w)$ for $0 \le w \le 1$

107. $4y_3[\ln(y_3)]^2$ for $0 < y_3 < 1$

111. (a) $N(984, 197.45)$ (b) .1379 (c) 1237

113. (a) $N(158, 8.72)$ (b) $N(170, 8.72)$ (c) .4090

115. (a) $.8875x + 5.2125$ (b) 111.5775 (c) 10.563 (d) .0951

117. (a) $2x - 10$ (b) 9 (c) 3 (d) .0228

119. (a) .1410 (b) .1165

121. (a) $R(t) = e^{-t^2}$ (b) .1054 (c) $2t$ (d) 0.886 thousand hours

123. (a) $R(t) = 1 - .125t^3$ for $0 \le t \le 2$, $= 0$ for $t > 2$ (b) $3t^2/(8 - t^3)$ (c) undefined

125. (a) parallel (b) $R(t) = 1 - \left(1 - e^{-t^2}\right)^4$ (c) $h(t) = \dfrac{8te^{-t^2}\left(1 - e^{-t^2}\right)^3}{1 - \left(1 - e^{-t^2}\right)^4}$

127. (a) $[1 - (1 - R_1(t))(1 - R_2(t))][1 - (1 - R_3(t))(1 - R_4(t))][1 - (1 - R_5(t))(1 - R_6(t))]$
(b) 70 h

129. (a) $R(t) = e^{-\alpha\left(t - t^2/[2\beta]\right)}$ for $t \le \beta$, $= e^{-\alpha\beta/2}$ for $t > \beta$ (b) $f(t) = \alpha\left(1 - \tfrac{t}{\beta}\right)e^{-\alpha\left(t - t^2/[2\beta]\right)}$

133. (a) $5y^4/10^5$ for $0 < y < 10$, 8.33 min (b) 6.67 min (c) 5 min (d) 1.409 min

135. (a) .0238 (b) $2,025

137. $\dfrac{n!\Gamma(i + 1/\theta)}{(i - 1)!\Gamma(n + 1/\theta + 1)}$, $\dfrac{n!\Gamma(i + 2/\theta)}{(i - 1)!\Gamma(n + 2/\theta + 1)} - \left[\dfrac{n!\Gamma(i + 1/\theta)}{(i - 1)!\Gamma(n + 1/\theta + 1)}\right]^2$

139. $E(Y_{k+1}) = \eta$

143. (b) $\hat{P}(X \le 1, Y \le 1) = .4154$ (answers will vary), exact $= .42$ (c) mean ≈ 0.4866, sd ≈ 0.6438 (answers will vary)

145. (b) 60,000 (c) 7.0873, 1.0180 (answers will vary) (d) .2080 (answers will vary)

147. (a) $f_X(x) = 12x(1 - x^2)$ for $0 \le x \le 1$, $f(y|x) = 2y/(1 - x)^2$ for $0 \le y \le 1 - x$ (c) we expect 16/9 candidates per accepted value, rather than 6

149. (a) $p_X(100) = .5$ and $p_X(250) = .5$ (b) $p(y|100) = .4, .2, .4$ for $y = 0, 100, 200$; $p(y|250) = .1, .3, .6$ for $y = 0, 100, 200$

151. (a) $N(\mu_1, \sigma_1)$, $N(\mu_2 + \rho\sigma_2/\sigma_1[(x - \mu_1)], \sigma_2\sqrt{1 - \rho^2})$

153. (b) $\hat{\mu} = 196.6193$ h, standard error $= 1.045$ h (answers will vary) (c) .9554, .0021 (answers will vary)

155. $f_T(t) = e^{-t/2} - e^{-t}$ for $t > 0$

157. (a) $k = 3/81,250$ (b) $f_X(x) = k(250x - 10x^2)$ for $0 \le x \le 20$, $= k(450x - 30x^2 + .5x^3)$ for $20 \le x \le 30$; $fY(y) = fX(y)$; not independent (c) .355 (d) 25.969 lb (e) -32.19, $-.894$ (f) 7.66

159. $t = E(X + Y) = 1.167$

163. (c) $p = 1$, because $\mu < 1$; $p = 2/3 < 1$, because $\mu > 1$

(continued)

165. (a) $F(b, d) - F(a, d) - F(b, c) + F(a, c)$
(b) $F(10,6) - F(4,6) - F(10,1) + F(4,1)$; $F(b, d) - F(a-1, d) - F(b, c-1) + F(a-1, c-1)$
(c) At each (x^*, y^*), $F(x^*, y^*)$ is the sum of the probabilities at points (x, y) such that $x \le x^*$ and $y \le y^*$. The table of $F(x, y)$ values is

		x	
		100	**250**
	200	.50	1
y	100	.30	.50
	0	.20	.25

(d) $F(x, y) = .6x^2y + .4xy^3, 0 \le x \le 1; 0 \le y \le 1$; $F(x, y) = 0, x \le 0$;
   $F(x, y) = 0, y \le 0$;
   $F(x, y) = .6x^2 + .4x, 0 \le x \le 1, y > 1$;
   $F(x, y) = .6y + .4y^3, x > 1, 0 \le y \le 1$;
   $F(x, y) = 1, x > 1, y > 1$
   $P(.25 \le x \le .75, .25 \le y \le .75) = .23125$
(e) $F(x, y) = 6x^2y^2, x + y \le 1, 0 \le x \le 1; 0 \le y \le 1, x \ge 0, y \ge 0$
   $F(x, y) = 3x^4 - 8x^3 + 6x^2 + 3y^4 - 8y^3 + 6y^2 - 1, x + y > 1, x \le 1, y \le 1$
   $F(x, y) = 0, x \le 0$; $F(x, y) = 0, y \le 0$;
   $F(x, y) = 3x^4 - 8x^3 + 6x^2, 0 \le x \le 1, y > 1$
   $F(x, y) = 3y^4 - 8y^3 + 6y^2, 0 \le y \le 1, x > 1$
   $F(x, y) = 1, x > 1, y > 1$

167. (a) $2x, x$ (b) 40 (c) 100

169. Undefined, ≈ 0

171. $\dfrac{2}{(1 - 1000t)(2 - 1000t)}$, 1500 h

173. Not valid for 75th percentile, but valid for 50th percentile; sum of percentiles $= (\mu_1 + z\sigma_1)$
   $+ (\mu_2 + z\sigma_2) = \mu_1 + \mu_2 + z(\sigma_1 + \sigma_2)$, percentile of sums $= (\mu_1 + \mu_2) + z\sqrt{\sigma_1^2 + \sigma_2^2}$

175. (a) 2360, 73.7 (b) .9713

177. .9686

179. .9099

181. .8340

183. (a) $\dfrac{\sigma_W^2}{\sigma_W^2 + \sigma_E^2}$ (b) .9999

185. 26, 1.64

187. (a) $g(y_1, y_n) = n(n-1)[F(y_n) - F(y_1)]^{n-2}f(y_1)f(y_n)$ for $y_1 < y_n$
(b) $f(w_1, w_2) = n(n-1)[F(w_1 + w_2) - F(w_1)]^{n-2}f(w_1)f(w_1 + w_2)$,
   $f_{W_2}(w_2) = n(n-1)\displaystyle\int_{-\infty}^{\infty}[F(w_1 + w_2) - F(w_1)]^{n-2}f(w_1)f(w_1 + w_2)\,dw_1$
(c) $n(n-1)w_2^{n-2}(1 - w_2)$ for $0 \le w_2 \le 1$

191. (a) 10/9 (b) 10/8 (c) $1 + Y_2 + \ldots + Y_{10}$, 29.29 boxes (d) 11.2 boxes

Chapter 5

1. (a) $\bar{x} = 113.73$ (b) $\tilde{x} = 113$ (c) $s = 12.74$ (d) .9091 (e) $s/\bar{x} = 11.2$

3. (a) $\bar{x} = 1.3481$ (b) $\bar{x} = 1.3481$ (c) $\bar{x} + 1.28s = 1.7814$ (d) .6736

5. $\hat{\theta}_1 = N\bar{X} = 1{,}703{,}000$, $\hat{\theta}_2 = \tau - N\bar{D} = 1{,}591{,}300$, $\hat{\theta}_3 = \tau \cdot \dfrac{\bar{X}}{\bar{Y}} = 1{,}601{,}438.281$

7. (a) 120.6 (b) 1,206,000 (c) .80 (d) 120

9. (a) $\bar{X}$; $\bar{x} = 2.11$ (b) $\sqrt{\bar{\mu}}/\sqrt{n}$, .119

11. (b) $n\lambda/(n-1)$ (c) $n^2\lambda^2/(n-1)^2(n-2)$

13. (b) $\sqrt{p_1 q_1/n_1 + p_2 q_2/n_2}$ (c) with $\hat{p}_1 = x_1/n_1$ and $\hat{p}_2 = x_2/n_2$, $\sqrt{\hat{p}_1\hat{q}_1/n_1 + \hat{p}_2\hat{q}_2/n_2}$
    (d) $-.245$ (e) .041

17. (a) $\sum X_i^2/2n$ (b) 74.505

19. (b) .444

21. (a) $\hat{p} = 2Y/n - .3$; .2 (c) $(10/7)Y/n - 9/70$

23. (a) $\dfrac{\sqrt{np}+1/2}{\sqrt{n}+1}$, $\dfrac{p(1-p)}{(\sqrt{n}+1)^2}$, $\dfrac{1}{4(\sqrt{n}+1)^2}$; the MSE does not depend on p (b) when p is
    near .5, the MSE from part (a) is smaller; when p is near 0 or 1, the usual estimator
    has lower MSE

25. (a) $\hat{p} = x/n = .15$ (b) yes (c) .4437

27. $\bar{x}, \bar{y}, \bar{x} - \bar{y}$

29. $\hat{p} = r/x = .15$, yes

31. (a) $\hat{\theta} = \displaystyle\sum X_i^2/2n = 74.505$, yes (b) $\hat{\eta} = \sqrt{-2\ln(.5)\hat{\theta}} = 10.163$

33. (a) $\hat{\theta} = \dfrac{n}{-\Sigma \ln(1 - x_i/\tau)}$ (b) $(\theta - 1)\displaystyle\sum_{i=1}^{n} \dfrac{x_i}{\tau - x_i} = n$, subject to $\tau > \max(x_i)$

35. $\hat{\lambda} = \dfrac{n}{\displaystyle\sum_{i=1}^{n}(Y_i/t_i)}$

37. (a) 2.228 (b) 2.131 (c) 2.947 (d) 4.604 (e) 2.492 (f) ~2.715

39. (a) A normal probability plot of these 20 values is quite linear. (b) (23.79, 26.31) (c) yes

41. (a) (357.38, 384.01) (b) narrower

43. (a) Based on a normal probability plot, it is reasonable to assume the sample observations
    came from a normal distribution. (b) (430.51, 446.08); 440 is plausible, 450 is not

45. Interval (c)

47. 26.14

49. (c) (12.10, 31.70)

51. (a) yes (b) no (c) no (d) yes (e) no (f) yes

53. Using H_a: $\mu < 100$ results in the welds being believed in conformance unless proved
    otherwise, so the burden of proof is on the nonconformance claim

55. (a) reject H_0 (b) reject H_0 (c) don't reject H_0 (d) reject H_0 (e) don't reject H_0

57. (a) .040 (b) .018 (c) .130 (d) .653 (e) $<.005$ (f) $\sim.000$

59. (a) .0778 (b) .1841 (c) .0250 (d) .0066 (e) .5438

61. (a) H_0: $\mu = 10$ versus H_a: $\mu < 10$ (b) reject H_0 (c) don't reject H_0 (d) reject H_0

63. (a) no; no, because $n = 49$ (b) H_0: $\mu = 1.0$ versus H_a: $\mu < 1.0$, $z = -5.79$, reject H_0, yes

65. H_0: $\mu = 200$ versus H_a: $\mu > 200$, $t = 1.19$ at ll df, P-value $= .128$, do not reject H_0

67. H_0: $\mu = 3$ versus H_a: $\mu \neq 3$, $t = -1.759$, P-value $= .082$, reject H_0 at $\alpha = .10$ but not at
    $\alpha = .05$

69. H_0: $\mu = 360$ versus H_a: $\mu > 360$, $t = 2.24$ at 25 df, P-value $= .018$, reject H_0, yes

(continued)

71. $H_0: \mu = 15$ versus $H_a: \mu < 15$, $z = -6.17$, P-value ≈ 0, reject H_0, yes

73. $H_0: \sigma = .05$ versus $H_a: \sigma < .05$. Type I error: Conclude that the standard deviation is $<.05$ mm when it is really equal to .05 mm. Type II error: Conclude that the standard deviation is .05 mm when it is really $<.05$

75. Type I: saying that the plant is not in compliance when in fact it is. Type II: conclude that the plant is in compliance when in fact it isn't

77. (.224, .278)

79. (.496, .631)

81. (.225, .275)

83. (b) 342 (c) 385

85. $H_0: p = .15$ versus $H_a: p > .15$, $z = 0.69$, P-value $= .2451$, fail to reject H_0

87. (a) $H_0: p = .25$ versus $H_a: p < .25$, $z = -1.01$, P-value $= .1562$, fail to reject H_0: the winery *should* switch to screw tops (b) Type I: conclude that less than 25% of all customers find screw tops acceptable, when the true percentage is 25%. Type II: fail to recognize that less than 25% of all customers find screw tops acceptable when that's actually true. Type II

89. (a) $H_0: p = .2$ versus $H_a: p > .2$, $z = 1.27$, P-value $= .1020$, fail to reject H_0 (b) Type I: conclude that more than 20% of the population of female workers is obese, when the true percentage is 20%. Type II: fail to recognize that more than 20% of the population of female workers is obese when that's actually true

91. $H_0: p = .1$, $H_a: p > .1$, $z = 0.74$, P-value $\approx .23$, fail to reject H_0

93. $H_0: p = .1$ versus $H_a: p > .1$, $z = 1.33$, P-value $= .0918$, fail to reject H_0; Type II

95. $H_0: p = .25$ versus $H_a: p < .25$, $z = -6.09$, P-value ≈ 0, reject H_0

97. (a) $H_0: p = .2$ versus $H_a: p > .2$, $z = 0.97$, P-value $= .166$, fail to reject H_0, so no modification appears necessary (b) .9974

99. (a) Gamma(9, 5/3) (b) Gamma(145, 5/53) (c) (11.54, 15.99)

101. B(490, 455), the same posterior distribution found in the example

103. Gamma$(\alpha + \Sigma\, x_i, 1/(n + 1/\beta))$

105. Beta$(\alpha + x, \beta + n - x)$

107. $n / \sum kx_k = .0436$

109. No: $E(\hat{\sigma}^2) = \sigma^2/2$

111. (a) expected payoff $= 0$ (b) $\hat{\theta} = \dfrac{\Sigma x_i + 2y}{\Sigma x_i + 2n}$

113. (a) The pattern of points in a normal probability plot (not shown) is reasonably linear, so, yes, normality is plausible. (b) (33.53, 43.79)

115. (.1295, .2986)

117. (a) A normal probability plot lends support to the assumption that pulmonary compliance is normally distributed. (b) (196.88, 222.62)

119. (a) (.539, .581) (b) 2401

121. (a) $N(0, 1)$ (b) $\dfrac{x_1 x_2 \pm 1.96\sqrt{x_1^2 + x_2^2 - (1.96)^2}}{x_2^2 - (1.96)^2}$ provided $x_1{}^2 + x_2{}^2 \geq (1.96)^2$

123. (a) 90.25% (b) at least 90% (c) at least $100(1 - k\alpha)\%$

125. (a) $H_0: \mu = 2150$ versus $H_a: \mu > 2150$ (b) $t = (\bar{x} - 2150)/(s/\sqrt{n})$ (c) 1.33 (d) .107 (e) fail to reject H_0

127. $H_0: \mu = 29.0$ versus $H_a: \mu > 29.0$, $t = .7742$, P-value $= .232$, fail to reject H_0

129. $H_0: \mu = 9.75$ versus $H_a: \mu > 9.75$, $t = 4.75$, P-value ≈ 0. The condition is not met.

(continued)

131. H_0: $\mu = 1.75$ versus H_a: $\mu \neq 1.75$, $t = 1.70$, P-value $= .102$, do not reject H_0; the data
       does not contradict prior research
133. H_0: $p = .75$ versus H_a: $p < .75$, $z = -3.28$, P-value $= .0005$, reject H_0
135.   (a) H_0: $p \leq .02$ versus H_a: $p > .02$; with $X \sim \text{Bin}(200, .02)$, P-value $= P(X \geq 17) =$
       7.5×10^{-7}; reject H_0 here and conclude that the NIST benchmark is *not* satisfied (b) .2133
137. H_0: $\mu = 4$ versus H_a: $\mu > 4$, $z = 1.33$, P-value $= .0918 > .02$, fail to reject H_0

Chapter 6

1.   {cooperative, competitive}; with $1 =$ cooperative and $2 =$ competitive, $p_{11} = .6$, $p_{12} = .4$,
       $p_{22} = .7$, $p_{21} = .3$
3.   (a) {full, part, broken} (b) with $1 =$ fill, $2 =$ part, $3 =$ broken, $p_{11} = .7$, $p_{12} = .2$, $p_{13} = .1$,
       $p_{21} = 0$, $p_{22} = .6$, $p_{23} = .4$, $p_{31} = .8$, $p_{32} = 0$, $p_{33} = .2$
5.   (a) $X_1 = 2$ with prob. p and $= 0$ with prob. $1 - p$ (b) 0, 2, 4
       (c) $P\left(X_{n+1} = 2y \mid X_n = x\right) = \binom{x}{y} p^y (1 - p)^{x-y}$ for $y = 0, 1, \ldots, x$
7.   (a) A son's social status, given his father's social status, has the same probability
       distribution as his social status conditional on all family history; no
       (b) The probabilities of social status changes (e.g., poor to middle class) are the same in
       every generation; no
9.   (a) no (b) define a state space by pairs; probabilities from each pair into the next state
11.   (a) $\begin{bmatrix} .90 & .10 \\ .11 & .89 \end{bmatrix}$ (b) .8210, .5460 (c) .8031, .5006
13.   (a) Willow City: $P(S \to S) = .988 > .776$ (b) .9776, .9685 (c) .9529
15.   (a) $\begin{bmatrix} .6 & .4 \\ .3 & .7 \end{bmatrix}$ (b) .52 (c) .524 (d) .606
17.   (a) .2740, .7747 (b) .0380 (c) 2.1, 2.2
19.   (a) $\begin{bmatrix} .1439 & .2790 & .2704 & .1747 & .1320 \\ .2201 & .3332 & .2522 & .1272 & .0674 \\ .1481 & .2829 & .2701 & .1719 & .1269 \\ .0874 & .2129 & .2596 & .2109 & .2292 \\ .0319 & .1099 & .1893 & .2174 & .4516 \end{bmatrix}$ (b) .0730 (c) .1719
21.   (a) .0608, .0646, .0658 (b) .0523, .0664, .0709, .0725 (c) they increase to .2710, .1320,
       .0926, .0798
23.   (a) .525 (b) .4372
25.   (a) $\begin{bmatrix} .96 & .04 \\ .05 & .95 \end{bmatrix}$ (b) .778 0's, .222 1's (c) .7081 0's, .2919 1's
27.   (a) $\boldsymbol{\pi} = [.80\ .20]$ (b) $P(X_1 = G) = .816$, $P(X_1 = S) = .184$ (c) .8541
29.   (a) $\boldsymbol{\pi} = [0\ 1]$ (b) $P(\text{cooperative}) = .3$, $P(\text{competitive}) = .7$ (c) .39, .61
31.   (a) no (b) yes
33.   (a) (.3681, .2153, .4167) (b) .4167 (c) 2.72
35.   (a) $\begin{bmatrix} .7 & .2 & .1 \\ 0 & .6 & .4 \\ .8 & 0 & .2 \end{bmatrix}$ (b) $\mathbf{P}^2$ has all nonzero entries (c) (8/15, 4/15, 1/5)
       (d) 8/15 (e) 5

(continued)

39. (a) $\pi_0 = \beta/(\alpha + \beta)$, $\pi_1 = \alpha/(\alpha + \beta)$ (b) $\alpha = \beta = 0 \Rightarrow$ the chain is constant; $\alpha = \beta = 1 \Rightarrow$ the chain alternates perfectly; $\alpha = 0$, $\beta = 1 \Rightarrow$ the chain is always 0; $\alpha = 1$, $\beta = 0 \Rightarrow$ the chain is always 1; $\alpha = 0$, $0 < \beta < 1 \Rightarrow$ the chain eventually gets stuck at 0; $0 < \alpha < 1$, $\beta = 0 \Rightarrow$ the chain eventually gets stuck at 1; $0 < \alpha < 1$ and $\beta = 1$ or $\alpha = 1$ and $0 < \beta < 1 \Rightarrow$ the chain is regular, and the answers to (a) still hold

41.

(a) $\begin{array}{c} 00 \\ 01 \\ 10 \\ 11 \end{array} \begin{bmatrix} (1-\alpha)^2 & \alpha(1-\alpha) & \alpha(1-\alpha) & \alpha^2 \\ \beta(1-\alpha) & (1-\alpha)(1-\beta) & \alpha\beta & \alpha(1-\beta) \\ \beta(1-\alpha) & \alpha\beta & (1-\alpha)(1-\beta) & \alpha(1-\beta) \\ \beta^2 & \beta(1-\beta) & \beta(1-\beta) & (1-\beta)^2 \end{bmatrix}$

(b) $\dfrac{\beta^2}{(\alpha+\beta)^2}, \dfrac{\alpha\beta}{(\alpha+\beta)^2}, \dfrac{\alpha\beta}{(\alpha+\beta)^2}, \dfrac{\alpha^2}{(\alpha+\beta)^2}$ (c) $\dfrac{\alpha^2}{(\alpha+\beta)^2}$

45. (a) $\begin{bmatrix} .25 & .75 & 0 & 0 \\ 0 & .25 & .75 & 0 \\ 0 & 0 & .25 & .75 \\ 0 & 0 & 0 & 1 \end{bmatrix}$ (b) .4219, .7383, .8965 (c) 4 (d) 1; no

47.

(a) states 4 and 5

(b)
k	1	2	3	4	5	6	7	8	9	10
$P(T_1 \le k)$	0	.46	.7108	.8302	.9089	.9474	.9713	.9837	.9910	.9949

(c)
k	1	2	3	4	5	6	7	8	9	10
$P(T_1 = k)$	0	.46	.2508	.1194	.0787	.0385	.0239	.0124	.0073	.0039

$\mu \approx 3.1457$

(d) 3.2084 (e) .3814, .6186

49.

(a) $\begin{bmatrix} .5 & .5 & 0 & 0 & 0 \\ .5 & 0 & .5 & 0 & 0 \\ .5 & 0 & 0 & .5 & 0 \\ .5 & 0 & 0 & 0 & .5 \\ 0 & 0 & 0 & 0 & 1 \end{bmatrix}$, 4 is an absorbing state

(b) $P(T_0 \le k) = 0$ for $k = 1, 2, 3$; the probabilities for $k = 4, \ldots, 15$ are .0625, .0938, .1250, .1563, .1875, .2168, .2451, .2725, .2988, .3242, .3487, .3723

(c) .2451

(d) $P(T_0 = k) = 0$ for $k = 1, 2, 3$; the probabilities for $k = 4, \ldots, 15$ are .0625, .03125, .03125, .03125, .03125, .0293, .0283, .0273, .0264, .0254, .0245, .0236; $\mu \approx 3.2531$, $\sigma \approx 3.9897$ (e) 30

51. $\mu_{\text{coop}} = 4.44$, $\mu_{\text{comp}} = 3.89$; cooperative

53.

(a) $\begin{array}{c} 0 \\ 1 \\ 2 \\ 3 \\ 4 \end{array} \begin{bmatrix} 1 & 0 & 0 & 0 & 0 \\ 1-p & 0 & p & 0 & 0 \\ 0 & 1-p & 0 & p & 0 \\ 0 & 0 & 1-p & 0 & p \\ 0 & 0 & 0 & 0 & 1 \end{bmatrix}$

(b) for $x_0 = \$1, \$2, \$3$: $\dfrac{2p^2 + 1}{2p^2 - 2p + 1}, \dfrac{2}{2p^2 - 2p + 1}, \dfrac{2p^2 - 4p + 3}{2p^2 - 2p + 1}$

(c) for $x_0 = \$1, \$2, \$3$: $\dfrac{p^3}{2p^2 - 2p + 1}, \dfrac{p^2}{2p^2 - 2p + 1}, \dfrac{p^3 - p^2 + p}{2p^2 - 2p + 1}$

55. 3.4825 generations

59. (c) (2069,0, 2079.8) (d) (.5993, .6185) (answers will vary)

61. (a) $P(X_{n+1} = 10 \mid X_n = x) = .4$, $P(X_{n+1} = 2x \mid X_n = x) = .6$ (b) mean $\approx \$47.2$ billion, sd $\approx \$2.07$ trillion (c) (\$6.53 billion, \$87.7 billion) (d) (\$618.32 million, \$627.90 million); easier

63. (a) (\$5586.60, \$5632.3) (b) (\$6695.50, \$6773.80) (answers will vary)

(continued)

65. (b) .9224 (answers will vary) (c) (6.89, 7.11) (answers will vary)

67. (a) $\begin{bmatrix} 0 & .5 & 0 & 0 & 0 & .5 \\ .5 & 0 & .5 & 0 & 0 & 0 \\ 0 & .5 & 0 & .5 & 0 & 0 \\ 0 & 0 & .5 & 0 & .5 & 0 \\ 0 & 0 & 0 & .5 & 0 & .5 \\ .5 & 0 & 0 & 0 & .5 & 0 \end{bmatrix}$ (b) no (c) $\boldsymbol{\pi} = \begin{bmatrix} \frac{1}{6} & \frac{1}{6} & \frac{1}{6} & \frac{1}{6} & \frac{1}{6} & \frac{1}{6} \end{bmatrix}$ (d) 6 (e) 9

69. (a) $\begin{bmatrix} 0 & 1 & 0 & 0 & 0 & 0 \\ \frac{1}{3} & 0 & \frac{1}{3} & \frac{1}{3} & 0 & 0 \\ 0 & \frac{1}{3} & 0 & \frac{1}{3} & \frac{1}{3} & 0 \\ 0 & \frac{1}{2} & \frac{1}{2} & 0 & 0 & 0 \\ 0 & 0 & \frac{1}{2} & 0 & 0 & \frac{1}{2} \\ 0 & 0 & 0 & 0 & 1 & 0 \end{bmatrix}$ (b) all entries of $\mathbf{P}^6$ are positive

(c) 1/12, 1/4, 1/4, 1/6, 1/6, 1/12 (d) 1/4 (e) 12

71. (a) $\begin{array}{c} 0 \\ 1 \\ 2 \\ 3 \\ 4 \\ 5 \end{array} \begin{bmatrix} 0 & 0 & 0 & 0 & 0 & 1 \\ .3 & .7 & 0 & 0 & 0 & 0 \\ 0 & .3 & .7 & 0 & 0 & 0 \\ 0 & 0 & .3 & .7 & 0 & 0 \\ 0 & 0 & 0 & .3 & .7 & 0 \\ 0 & 0 & 0 & 0 & .3 & .7 \end{bmatrix}$ (b) .0566, .1887, .1887, .1887, .1887, .1887

(c) 17.67 weeks (including the one week of shipping)

73. (a) 2 seasons (b) .3613 (c) 15 seasons (d) 6.25 seasons

75. (a) $\mathbf{p}_1 = [0.3168 \quad 0.1812 \quad 0.2761 \quad 0.1413 \quad 0.0846]$;
$\mathbf{p}_2 = [0.3035 \quad 0.1266 \quad 0.2880 \quad 0.1643 \quad 0.1176]$;
$\mathbf{p}_3 = [0.2908 \quad 0.0918 \quad 0.2770 \quad 0.1843 \quad 0.1561]$
(b) 35.7 years, 11.9 years, 9.2 years, 4.3 years
(c) 16.6 years

77. (a) $\begin{array}{c} 0 \\ 1 \\ 2 \\ 3 \\ pd \\ tbr \end{array} \begin{bmatrix} 0 & 1 & 0 & 0 & 0 & 0 \\ 0 & 0 & .959 & 0 & .041 & 0 \\ 0 & 0 & 0 & .987 & .013 & 0 \\ 0 & 0 & 0 & 0 & .804 & .196 \\ 0 & 0 & 0 & 0 & 1 & 0 \\ 0 & 0 & 0 & 0 & 0 & 1 \end{bmatrix}$ (c) 3.9055 weeks (d) .8145

(e) payments are always at least 1 week late; most payments are made at the end of 3 weeks

79. (a) $\mathbf{P}_1 = \mathbf{P}_3 = \begin{bmatrix} .98 & .02 \\ .02 & .98 \end{bmatrix}$, $\mathbf{P}_2 = \begin{bmatrix} .97 & .03 \\ .03 & .97 \end{bmatrix}$, $\mathbf{P}_4 = \mathbf{P}_5 = \begin{bmatrix} .99 & .01 \\ .01 & .99 \end{bmatrix}$ (b) .916

81. (a) [3259 22,533 19,469 26,066 81,227 16,701 1511 211,486 171,820 56,916]
(b) [2683 24,119 21,980 27,015 86,100 15,117 1518 223,783 149,277 59,395];
[2261 25,213 24,221 27,526 89,397 13,926 1524 233,533 131,752 61,636];
−44%, +24%, +46%, +12%, +20%, −26%, +1.3%, +19%, +34%, +13.4%
(c) [920 23,202 51,593 21,697 78,402 8988 1445 266,505 65,073 93,160]

Chapter 7

1. (a) Continuous-time, continuous-space (b) continuous-time, discrete-space (c) discrete-time, continuous-space (d) discrete-time, discrete-space

7. (b) No: at time $t = .25$, $x_0(.25) = -\cos(\pi/2) = 0$ and $x_1(.25) = \cos(\pi/2) = 0$
    (c) $X(0) = -1$ with probability .8 and $+1$ with probability .2; $X(.5) = +1$ with probability .8 and -1 with probability .2

9. (a) discrete-space (c) $X_n \sim \text{Bin}(n, 18/38)$

11. (a) 0 (b) 1/2

13. $C_{XX}(t, s) = \text{Var}(A)\cos(\omega_0 t + \theta_0)\cos(\omega_0 s + \theta_0)$, $R_{XX}(t, s) = v_0^2 + v_0 E[A][\cos(\omega_0 t + \theta_0) + \cos(\omega_0 s + \theta_0)] + E[A^2]\cos(\omega_0 t + \theta_0)\cos(\omega_0 s + \theta_0)$

15. (b) $N(s) > 0$, because covariance > 0 (c) $\rho = e^{-2}$ (d) Gaussian, mean 0, variance 1.73

19. (a) $\mu_S(t) + \mu_N(t)$ (b) $R_{SS}(t, s) + \mu_S(t)\mu_N(s) + \mu_N(t)\mu_S(s) + R_{NN}(t, s)$
    (c) $C_{SS}(t, s) + C_{NN}(t, s)$ (d) $\sigma_S^2(t) + \sigma_N^2(t)$

23. (a) $(1/2)\sin(\omega_0(s - t))$ (b) not orthogonal, not uncorrelated, not independent

25. (a) μ_V (b) $E[V^2] + (A_0^2/2)\cos(\omega_0\tau)$ (c) yes

27. (a) yes (b) no (c) yes (d) yes

29. no

31. $\mu_A + \mu_B$, $C_{AA}(\tau) + C_{AB}(\tau) + C_{BA}(\tau) + C_{BB}(\tau)$, yes

33. (a) yes, because its autocovariance has periodic components (b) -42 (c) $500\cos(100\pi\tau) + 8\cos(600\pi\tau) + 49$ (d) 557 (e) 508

35. yes: both the time average and ensemble average are 0

37. $C_{XX}(\tau)/C_{XX}(0)$

41. (a) .0062 (b) $75 + 25\sin\left(\dfrac{2\pi}{365}(59 - 150)\right)$ (c) $16\delta[n - m]$ (d) no, and it shouldn't be

43. (a) $18n/38$, $360n/1444$, $360\min(m, n)/1444$, $(360\min(m, n) + 324mn)/1444$
    (b) $-10n/38$, $36{,}000n/1444$, $36{,}000\min(m, n)/1444$, $(36{,}000\min(n, m) + 100mn)/1444$
    (c) .3141

47. (a) μ_X (b) $\dfrac{1}{4}(2C_{XX}[m - n] + C_{XX}[m - n + 1] + C_{XX}[m - n - 1])$ (c) yes
    (d) $(C_{XX}[0] + C_{XX}[1])/2$

49. (c) $\dfrac{\mu}{1 - \alpha}$ (d) $\dfrac{\alpha^{m-n}\sigma^2}{1 - \alpha^2}$ (e) yes (f) α^k

53. (a) .0993 (b) .1353 (c) 2

55. (a) .0516 (b) $1 - \sum_{x=0}^{75} e^{-50}50^x/x!$ (c) .9179 (d) 6 s (e) .8679

57. k/λ

59. (a) .0911 (b) $\dfrac{e^{3(t-s)}[3(t - s)]^{n-m}}{(n - m)!}$

61. $e^{-p\lambda t}$

63. $f_Y(y) = 2\lambda e^{-\lambda y}(1 - e^{-\lambda y})$ for $y > 0$

67. (a) .0492 (b) .00255

71. pmf: $N(t) = 0$ or 1 with probability 1/2 each for all t; mean $= .5$, variance $= .25$, $C_{NN}(\tau) = .25e^{-2\lambda|\tau|}$, $R_{NN}(\tau) = .25 + .25e^{-2\lambda|\tau|}$

73. (a) .0038 (b) .9535

75. (a) yes (b) .3174 (c) .3174 (d) .4778

77. (a) $E[X(t)] = 80 + 20\cos\left(\dfrac{\pi}{12}(t - 15)\right)$, $\text{Var}(X(t)) = .2t$ (b) .1251 (c) .3372 (d) .1818

(continued)

79. (a) .3078 (b) .1074

81. (a) .1171 (b) .6376 (c) .0181, .7410

83. (a) yes (b) $E[X(t)] = 0$, $R_{XX}(t, s) = (N_0/2)\min(t, s)$, no

87. (a) $0 =$ empty, $1 =$ a person in stage 1, and $2 =$ a person in stage 2; $q_0 = \lambda$, $q_1 = \lambda_1$, $q_2 = \lambda_2$;
$q_{02} = q_{21} = q_{10} = 0$; $q_{01} = \lambda$, $q_{12} = \lambda_1$, $q_{20} = \lambda_2$ (b) $\pi = (6/11, 2/11, 3/11)$
(c) $\pi = (6/11, 3/11, 2/11)$ (d) $\pi = (1/7, 2/7, 4/7)$

89. (a) $q_0 = \lambda$, $q_1 = \lambda_1$, $q_2 = \lambda_2$; $q_{02} = q_{10} = 0$; $q_{01} = \lambda$, $q_{12} = \lambda_1$, $q_{20} = .8\lambda_2$, $q_{21} = .2\lambda_2$
(b) $\pi = (24/49, 10/49, 15/49)$ (c) $\pi = (24/49, 15/49, 10/49)$ (d) $\pi = (2/17, 5/17, 10/17)$
(e) $1.25(1/\lambda_1 + 1/\lambda_2)$

91. $q_i = i\beta$, $q_{i,i+1} = i\beta$ for $i = 1, \ldots, N-1$

93. $q_{i,i+1} = \lambda$ for $i \geq 0$, $q_{i,i-1} = i\beta$ for $i \geq 1$, $q_i = \lambda + i\beta$ for $i \geq 1$

95. $\pi_{00} = \dfrac{\alpha_1\beta_1}{\Sigma}, \pi_{01} = \dfrac{\alpha_1\beta_0}{\Sigma}, \pi_{10} = \dfrac{\alpha_0\beta_1}{\Sigma}, \pi_{11} = \dfrac{\alpha_0\beta_0}{\Sigma}$, where $\Sigma = \alpha_1\beta_1 + \alpha_1\beta_0 + \alpha_0\beta_1 + \alpha_0\beta_0$;
$1 - \pi_{11} = 1 - \dfrac{\alpha_0\beta_0}{\alpha_1\beta_1 + \alpha_1\beta_0 + \alpha_0\beta_1 + \alpha_0\beta_0}$

99. (a) 0 (b) $C_{XX}(t, s) = 1$ if $\text{floor}(t) = \text{floor}(s)$, $= 0$ otherwise

101. (a) 0, $(1/3)\cos(\omega_k\tau)$ (b) 0, $\dfrac{1}{3}\displaystyle\sum_{k=1}^{n} \cos(\omega_k\tau)$, yes

103. (a) 0 (b) $\dfrac{1}{2}\displaystyle\sum_{k=1}^{n} \cos(\omega_k\tau) \cdot p_k$ (c) yes

105. (a) S_n denotes the total lifetime of the machine through its use of the first n rotors.
(b) $\mu_S[n] = 125n$; $\sigma_S^2[n] = 15{,}625n$; $C_{SS}[n, m] = 15{,}625\min(n, m)$; $R_{SS}[n, m] = 15{,}625[\min(n, m) + nm]$ (c) .5040

107. Yes

109. (a) $e^{-\lambda t}$ (b) $e^{-\lambda t}(1 + \lambda t)$ (c) $e^{-\lambda\varepsilon}$

111. (a) $\dfrac{10(1 - e^{-\alpha t_0})}{\alpha t_0}$ (b) $\dfrac{2\lambda}{\alpha}(1 - e^{-\alpha t_0})$

Chapter 8

1. $\mathscr{F}\{R_{XX}(\tau)\} = \text{sinc}(f)$, which is not ≥ 0 for all f

3. (a) $250\delta(f) + \dfrac{\sqrt{\pi}}{2}\exp\left(-\dfrac{\pi^2 f^2}{4 \times 10^6}\right)$ (b) 240.37 W (c) 593.17 W

5. (a) 112,838 W (b) 108,839 W (c) $R_{XX}(\tau) = \dfrac{200{,}000}{\sqrt{\pi}}\exp(-10^{12}\tau^2)$

7. (a) $N_0 B$ (b) $N_0 B\,\text{sinc}(2B\tau)$

9. (a) $A_0^2 e^{-2\lambda|\tau|}$ (b) $\dfrac{2\lambda A_0^2}{\lambda^2 + (2\pi f)^2}$ (c) A_0^2 (d) $\dfrac{2A_0^2}{\pi}\arctan(2\pi)$

11. (a) $100(1 + e^{-1}) \approx 136.8$ W (b) $\dfrac{200}{1 + (2\pi f)^2}[1 + \cos(2\pi f)]$ (c) 126.34 W

13. $\mu_W(t) = 0$, $R_{WW}(\tau) = 2R_{XX}(\tau) - R_{XX}(\tau - d) - R_{XX}(\tau + d)$, $S_{WW}(f) = 2S_{XX}(f)[1 - \cos(2\pi f d)]$

15. (a) Yes (b) Yes (c) $S_{ZZ}(f) = S_{XX}(f) + S_{YY}(f)$

17. (b) $S_{ZZ}(f) = S_{XX}(f) \bigstar S_{YY}(f)$

19. No, because $P_N = \infty$

21. (a) $S_{XX}(f) = E[A^2]S_{YY}(f)$ (b) $S_{XX}^{cc}(f) = E[A^2]S_{YY}(f) - (E[A]\mu_Y)^2\delta(f)$ (c) Yes; our "engineering interpretation" of the elements of a psd are not valid for non-ergodic processes

(continued)

23. (a) $2400\sin(120,000\tau)$ (b) 2400 W (c) $40/(40+j2\pi f)$ (d) $32/(1600+(2\pi f)^2)$ for $|f| \le 60$ kHz
(e) 0.399997 W

25. (a) 0 (b) $(1 - e^{-j2\pi f})/(j2\pi f)$ (c) $(N_0/2)\text{sinc}^2(f)$ (d) $N_0/2$

27. (a) $100\delta(f) + \dfrac{50}{1+(2\pi f)^2}$ (b) 125 W (c) $\dfrac{1}{(4+j2\pi f)^2}, \dfrac{1}{\left(16+(2\pi f)^2\right)^2}$

(d) $\left[100\delta(f) + \dfrac{50}{1+(2\pi f)^2}\right] \cdot \dfrac{1}{\left(16+(2\pi f)^2\right)^2}$ (e) 0.461 W

29. (a) $(N_0/2)e^{-2\alpha|f|}$ (b) $\dfrac{2N_0\alpha}{4\alpha^2+4\pi^2\tau^2}$ (c) $N_0/2\alpha$

33. (a) $2N_0\pi^2 f^2 \text{rect}(f/2B)$ (b) $\dfrac{N_0}{\pi\tau^3}\left[2\pi^2 B^2\tau^2\sin(2\pi B\tau) + 2\pi B\tau\cos(2\pi B\tau) - \sin(2\pi B\tau)\right]$
(c) $4N_0\pi^2 B^3/3$

35. (a) $R_{XX}(\tau) - R_{XX}(\tau) \star h(\tau) - R_{XX}(\tau) \star h(-\tau) + R_{XX}(\tau) \star h(\tau) \star h(-\tau)$
(b) $S_{XX}(f)|1 - H(f)|^2$

37. (a) 1.17 MW (b) $250,000\delta(f) + 60,000[\delta(f-35,000) + \delta(f+35,000)] + 8\text{rect}\left(\dfrac{f}{100,000}\right)$
(c) same as part (b) (d) 1.17 MW (e) 5000 W (f) 3000 W (g) $\text{SNR}_{\text{in}} = 234, \text{SNR}_{\text{out}} = 390$

41. $\dfrac{1-\alpha^2}{1+\alpha^2 - 2\alpha\cos(2\pi F)}$

43. $\dfrac{1-e^{-20\lambda}}{1+e^{-20\lambda} - 2e^{-10\lambda}\cos(2\pi F)}$

45. $1 - \dfrac{\pi^2}{8} + \dfrac{\pi^2}{4}\text{tri}(2F)$

47. (b) $P\text{sinc}(k/2)$

49. (a) $Y_n = (X_{n-M+1} + \ldots + X_n)/M$ (b) $\dfrac{1-e^{-j2\pi FM}}{M(1-e^{-j2\pi F})}$ (c) $\sigma^2\dfrac{M-|k|}{M^2}$ for $|k| = 0, 1, \ldots, M-1$
and zero otherwise

References

Ambardar, Ashok, *Analog and Digital Signal Processing* (2nd ed.), Brooks/Cole Publishing, Pacific Grove, CA, 1999. A thorough treatment of the mathematics of signals and systems, including both discrete- and continuous-time structures.

Bury, Karl, *Statistical Distributions in Engineering*, Cambridge University Press, Cambridge, England, 1999. A readable and informative survey of distributions and their properties.

Crawley, Michael, *The R Book* (2nd ed.), Wiley, Hoboken, NJ, 2012. At more than 1000 pages, carrying it may give you lower back pain, but it obviously contains a great deal of information about the R software.

Gorroochurn, Prakash, *Classic Problems of Probability*, Wiley, Hoboken, NJ, 2012. An entertaining excursion through 33 famous probability problems.

Davis, Timothy A., *Matlab Primer* (8th ed.), CRC Press, Boca Raton, FL, 2010. A good reference for basic Matlab syntax, along with extensive catalogs of Matlab commands.

DeGroot, Morris and Mark Schervish, *Probability and Statistics* (4th ed.), Addison-Wesley, Upper Saddle River, NJ, 2012. Contains a nice exposition of subjective probability and an introduction to Bayesian methods of inference.

Devore, Jay and Ken Berk, *Modern Mathematical Statistics with Applications* (2nd ed.), Springer, New York, 2011. A comprehensive text on statistical methodology designed for upper-level students.

Durrett, Richard, *Elementary Probability for Applications*, Cambridge Univ. Press, London, England, 2009. A very brief (254 pp.) introduction that still finds room for some interesting examples.

Johnson, Norman, Samuel Kotz, and Adrienne Kemp, *Univariate Discrete Distributions* (3rd ed.), Wiley-Interscience, New York, 2005. An encyclopedia of information on discrete distributions.

Johnson, Norman, Samuel Kotz, and N. Balakrishnan, *Continuous Univariate Distributions*, vols. 1–2, Wiley, New York, 1993. These two volumes together present an exhaustive survey of various continuous distributions.

Law, Averill, *Simulation Modeling and Analysis* (4th ed.), McGraw-Hill, New York, 2006. An accessible and comprehensive guide to many aspects of simulation.

Meys, Joris, and Andrie de Vries, *R for Dummies*, For Dummies (Wiley), New York, 2012. Need we say more?

Mosteller, Frederick, Robert Rourke, and George Thomas, *Probability with Statistical Applications* (2nd ed.), Addison-Wesley, Reading, MA, 1970. A very good precalculus introduction to probability, with many entertaining examples; especially good on counting rules and their application.

Nelson, Wayne, *Applied Life Data Analysis*, Wiley, New York, 1982. Gives a comprehensive discussion of distributions and methods that are used in the analysis of lifetime data.

Olofsson, Peter, *Probabilities: The Little Numbers That Rule Our Lives*, Wiley, Hoboken, NJ, 2007. A very non-technical and thoroughly charming introduction to the quantitative assessment of uncertainty.

M.A. Carlton and J.L. Devore, *Probability with Applications in Engineering, Science, and Technology*, Springer Texts in Statistics, DOI 10.1007/978-1-4939-0395-5, © Springer Science+Business Media New York 2014

Peebles, Peyton, *Probability, Random Variables, and Random Signal Principles* (4th ed.), McGraw-Hill, New York, 2001. Provides a short introduction to probability and distributions, then moves quickly into signal processing with an emphasis on practical considerations. Includes some Matlab code.

Ross, Sheldon, *Introduction to Probability Models* (9th ed.), Academic Press, San Diego, CA, 2006. A good source of material on the Poisson process and generalizations, Markov chains, and other topics in applied probability.

Ross, Sheldon, *Simulation* (5th ed.), Academic Press, San Diego, CA, 2012. A tight presentation of modern simulation techniques and applications.

Taylor, Howard M. and Samuel Karlin, *An Introduction to Stochastic Modeling* (3rd ed.), Academic Press, San Diego, CA, 1999. More sophisticated than our book, but with a wealth of information about discrete and continuous time Markov chains, Poisson processes, Brownian motion, and queueing systems.

Winkler, Robert, *Introduction to Bayesian Inference and Decision* (2nd ed.), Probabilistic Publishing, Sugar Land, Texas, 2003. A very good introduction to subjective probability.

Index

M.A. Carlton and J.L. Devore, *Probability with Applications in Engineering, Science, and*   769
Technology, Springer Texts in Statistics, DOI 10.1007/978-1-4939-0395-5,
© Springer Science+Business Media New York 2014

Printed in the United States of America